Selected Titles in This Series

288 **Marisa Fernández and Joseph A. Wolf, Editors,** Global differential geometry: The mathematical legacy of Alfred Gray, 2001

287 **Marlos A. G. Viana and Donald St. P. Richards, Editors,** Algebraic methods in statistics and probability, 2001

286 **Edward L. Green, Serkan Hoşten, Reinhard C. Laubenbacher, and Victoria Ann Powers, Editors,** Symbolic computation: Solving equations in algebra, geometry, and engineering, 2001

285 **Joshua A. Leslie and Thierry P. Robart, Editors,** The geometrical study of differential equations, 2001

284 **Gaston M. N'Guérékata and Asamoah Nkwanta, Editors,** Council for African American researchers in the mathematical sciences: Volume IV, 2001

283 **Paul A. Milewski, Leslie M. Smith, Fabian Waleffe, and Esteban G. Tabak, Editors,** Advances in wave interaction and turbulence, 2001

282 **Arlan Ramsay and Jean Renault, Editors,** Groupoids in analysis, geometry, and physics, 2001

281 **Vadim Olshevsky, Editor,** Structured matrices in mathematics, computer science, and engineering II, 2001

280 **Vadim Olshevsky, Editor,** Structured matrices in mathematics, computer science, and engineering I, 2001

279 **Alejandro Adem, Gunnar Carlsson, and Ralph Cohen, Editors,** Topology, geometry, and algebra: Interactions and new directions, 2001

278 **Eric Todd Quinto, Leon Ehrenpreis, Adel Faridani, Fulton Gonzalez, and Eric Grinberg, Editors,** Radon transforms and tomography, 2001

277 **Luca Capogna and Loredana Lanzani, Editors,** Harmonic analysis and boundary value problems, 2001

276 **Emma Previato, Editor,** Advances in algebraic geometry motivated by physics, 2001

275 **Alfred G. Noël, Earl Barnes, and Sonya A. F. Stephens, Editors,** Council for African American researchers in the mathematical sciences: Volume III, 2001

274 **Ken-ichi Maruyama and John W. Rutter, Editors,** Groups of homotopy self-equivalences and related topics, 2001

273 **A. V. Kelarev, R. Göbel, K. M. Rangaswamy, P. Schultz, and C. Vinsonhaler, Editors,** Abelian groups, rings and modules, 2001

272 **Eva Bayer-Fluckiger, David Lewis, and Andrew Ranicki, Editors,** Quadratic forms and their applications, 2000

271 **J. P. C. Greenlees, Robert R. Bruner, and Nicholas Kuhn, Editors,** Homotopy methods in algebraic topology, 2001

270 **Jan Denef, Leonard Lipschitz, Thanases Pheidas, and Jan Van Geel, Editors,** Hilbert's tenth problem: Relations with arithmetic and algebraic geometry, 2000

269 **Mikhail Lyubich, John W. Milnor, and Yair N. Minsky, Editors,** Laminations and foliations in dynamics, geometry and topology, 2001

268 **Robert Gulliver, Walter Littman, and Roberto Triggiani, Editors,** Differential geometric methods in the control of partial differential equations, 2000

267 **Nicolás Andruskiewitsch, Walter Ricardo Ferrer Santos, and Hans-Jürgen Schneider, Editors,** New trends in Hopf algebra theory, 2000

266 **Caroline Grant Melles and Ruth I. Michler, Editors,** Singularities in algebraic and analytic geometry, 2000

265 **Dominique Arlettaz and Kathryn Hess, Editors,** Une dégustation topologique: Homotopy theory in the Swiss Alps, 2000

264 **Kai Yuen Chan, Alexander A. Mikhalev, Man-Keung Siu, Jie-Tai Yu, and Efim I. Zelmanov, Editors,** Combinatorial and computational algebra, 2000

(Continued in the back of this publication)

Global Differential Geometry: The Mathematical Legacy of Alfred Gray

Alfred Gray, Turin, 24 May 1994

Contents

Foreword xi

Preface xiii

Photograph of Alfred Gray Memorial Bronze xiv

Acknowledgements xv

List of publications of Alfred Gray xvii

List of participants xxvii

Invited contributions

On the formality problem for symplectic manifolds
IVAN K. BABENKO AND ISKANDER A. TAIMANOV 1

Osculating tubes and self-linking for curves on the three-sphere
THOMAS F. BANCHOFF 10

Isoparametric functions and harmonic and minimal unit vector fields
E. BOECKX AND L. VANHECKE 20

The Seiberg-Witten equations and almost-Hermitian geometry
S. K. DONALDSON 32

Sculpture inspired by work with Alfred Gray: Kepler elliptic curves and minimal surface sculptures of the planets
HELAMAN FERGUSON 39

Mesoscopic curvature and hyperbolicity
MISHA GROMOV 58

Stable forms and special metrics
NIGEL HITCHIN 70

A CR-Momentum Ansatz for nilpotent groups
A. HUCKLEBERRY AND M. VÖLLER 90

Can tangent sphere bundles over Riemannian manifolds have strictly positive sectional curvature?
O. KOWALSKI, M. SEKIZAWA, AND Z. VLÁŠEK 110

Volumes of geodesic balls and spheres associated to a metric connection with torsion
VICENTE MIQUEL 119

Special isothermic surfaces and solitons
EMILIO MUSSO AND LORENZO NICOLODI 129

The isoperimetric and Willmore problems
ANTONIO ROS 149

Almost parallel structures
SIMON SALAMON 162

Technical computing with Mathematica
PAUL WELLIN 182

Complex geometry and representations of Lie groups
JOSEPH A. WOLF 202

Communications

On the volume of compact spacelike hypersurfaces with hyperbolic boundary in Minkowski space
JUAN A. ALEDO AND JOSÉ A. PASTOR 239

Stability of zero mean curvature surfaces in flat Lorentzian 4-manifolds
LUIS J. ALÍAS AND BENNETT PALMER 244

Hitchin-Kobayashi correspondence for equivariant bundles on $X \times \mathbb{P}^1$
LUIS ÁLVAREZ-CÓNSUL AND OSCAR GARCÍA-PRADA 249

Hopf vesicles in $\mathbf{S}^3(\mathbf{1})$
J. ARROYO AND O. J. GARAY 258

Invariant nearly Kähler f-structures on homogeneous spaces
VITALY V. BALASHCHENKO 263

An extension of A. Gray's investigations on small geodesic balls
N. BOKAN, M. DJORIĆ, AND U. SIMON 268

An energy minimizing family of unit vector fields on odd-dimensional spheres
VINCENT BORRELLI, FABIANO BRITO, AND OLGA GIL-MEDRANO 273

Is it admissible to linearize the Einstein equation in the presence of matter?
LLUÍS BRUNA AND JOAN GIRBAU 277

Solutions of some conformally invariant equations of the first order
JAROLÍM BUREŠ 281

On biharmonic maps
RENZO CADDEO, STEFANO MONTALDO, AND PAOLA PIU 286

Characterizations of Pfaffian systems. Consequences in dimension five
MARÍA A. CAÑADAS-PINEDO AND CEFERINO RUIZ 291

On a new class of contact Riemannian manifolds
JONG TAEK CHO 296

Description of $(1, 2)$-symplectic metrics on flag manifolds
NIR COHEN, CAIO J. C. NEGREIROS, AND LUIZ A. B. SAN MARTIN 300

Complex surfaces with Betti numbers $b_1=1$, $b_2 > 0$ and finite quotients
GEORGES DLOUSSKY 305

Hypercomplex nilpotent Lie groups
ISABEL G. DOTTI AND ANNA FINO 310

$\mathfrak{C}$-spaces of Iwasawa type and Damek-Ricci spaces
MARIA J. DRUETTA 315

Deformations of constant mean curvature surfaces in half space models
M. ELISA E. L. GALVÃO AND CÉLIA C. GÓES 320

The geometry of the skew-symmetric curvature operator in the complex setting
PETER GILKEY AND RAINA IVANOVA 325

Representing minimal curves in $\mathbb{C}^3$ by differential operators
HUBERT GOLLEK 334

Isometric immersions without positive Ricci curvature
LUIS GUIJARRO 339

Natural Poisson and Jacobi structures on the tangent bundle of a pseudo-Riemannian manifold
JOSEF JANYŠKA 343

On Alfred Gray's elliptical catenoid
HUGO JIMÉNEZ AND SANTIAGO LÓPEZ DE MEDRANO 348

Sub-Riemannian geometry and non-holonomic mechanics
JAIR KOILLER, PAULO R. RODRIGUES, AND PAULO PITANGA 353

Weil algebras of generalized higher order velocities bundles
MIROSLAV KUREŠ 358

The variety of Lie algebras from certain Riemannian viewpoint
JORGE LAURET 363

How to a use MATHEMATICA^{TM} to find cyclic surfaces of constant curvature in Lorentz-Minkowski space
RAFAEL LÓPEZ 371

The Fuglede-Ishihara and Baird-Eells theorems for $p > 1$
E. LOUBEAU 376

Cohomology of nilmanifolds and Gontcharova's theorem
DMITRI V. MILLIONSCHIKOV 381

Asymptotically holomorphic embeddings of contact manifolds in projective spaces
VICENTE MUÑOZ, FRAN PRESAS, AND IGNACIO SOLS 386

Convergent star products on Fréchet linear Poisson algebras of Heisenberg type
HIDEKI OMORI, YOSHIAKI MAEDA, NAOYA MIYAZAKI, AND AKIRA YOSHIOKA 391

Resolutions of 2-polyhedra by fake surfaces and embeddings into $\mathbb{R}^4$
ALEXANDER ONISCHENKO, DUŠAN REPOVŠ, AND ARKADY SKOPENKOV 396

Cayley 4-frames and a quaternion Kähler reduction related to Spin(7)
LIVIU ORNEA AND PAOLO PICCINNI 401

Old and new structures on products of spheres
MAURIZIO PARTON 406

1-Homogeneous sprays in Finsler manifolds
L. DEL RIEGO 411

The Gysin sequence for riemannian flows
JOSÉ IGNACIO ROYO PRIETO 415

On contact groupoids and Legendre bisections
TOMASZ RYBICKI 420

Affine maximal tori intersecting a fixed one
MARCOS SALVAI 425

On the asymptotic series of the heat content
ALESSANDRO SAVO 428

Visualization of flat slant surfaces in $\mathbb{C}^2$
YOSHIHIKO TAZAWA 433

On solvable Lie groups without lattices
ALEKSY TRALLE 437

Round Table Discussion

Where does geometry go? A research and education perspective
JEAN-PIERRE BOURGUIGNON, EUGENIO CALABI, JAMES EELLS, OSCAR GARCIA-PRADA, AND MISHA GROMOV 442

FOREWORD

In September 2000 a Congress on Differential Geometry was held, in Bilbao, in memory of Professor Alfred Gray whose sudden death two years before had come as a shock and sadness to the international community.

Alfred Gray's work covered a wide swathe of differential geometry; both standing in the centre of the classical tradition of the subject and playing a seminal role in many modern developments. We mention only a few of the research topics to which he contributed. For a more detailed account of parts of Gray's work see the article of Salamon in this volume, and also the list of publications that follows.

In Riemannian geometry, Alfred Gray studied the volumes of tubes and comparison theorems, with special reference to the case of Kähler manifolds and the relations with characteristic classes. Much of this work is brought together in his outstanding monograph "Tubes" (1990). He also worked on extensions of the O'Neill theory of Riemannian submersions. Gray was a central figure in the development of almost-complex geometry and made important contributions to the study of symplectic manifolds, starting before the explosion of activity in this area in the 1980's and 1990's. The work of Gray and his collaborators on the de Rham theory of complex and symplectic manifolds initiates a striking interaction between rational homotopy theory, differential topology, differential geometry and complex manifolds. (Complex variables are a theme running through much of Gray's work, and it is interesting to note that his earliest research papers deal with classical complex function theory.) Much of Gray's work was informed by his deep knowledge of Lie groups and homogeneous spaces, and he made important contributions to the study of manifolds with exceptional holonomy groups, again presaging later developments.

Alfred Gray pursued many other mathematical interests alongside these pioneering contributions to modern differential geometry. On the one hand he had a profound knowledge of classical work on the geometry of surfaces in 3-space and on ordinary differential equations. On the other hand he was a leading figure in the use of computers in Pure Mathematics, both as an educational tool and to enable the visualisation of complicated phenomena. These two strands came together in his books "Modern Differential Geometry of Curves and Surfaces" (first published in 1993), and "Introduction to Ordinary Differential Equations with Mathematica" (with Mezzino and Pinsky, 1997), which will surely be models for the way the subjects are taught in the future. A very important feature of Gray's mathematical activity was his commitment to the development of the subject internationally, notably in Belgium, Italy and Spain, where he built up networks of collaborators, colleagues and friends.

The papers in this volume of Proceedings, covering the whole field of differential geometry and in many cases having special connections with Gray's own work and interests, attest to the breadth of his influence and the respect and affection in which he was held. At the same time, these papers may help to give an overall view of the current position of the subject. All the mathematicians who have been involved in the meeting or contributed to this volume will wish to express hearty thanks to the Organisers of this splendid International Congress in Bilbao, and to the Proceedings Editors, for their hard and efficient work.

London, July 2001

Simon Donaldson

Preface

The International Congress on Differential Geometry in Memory of Alfred Gray (1939-1998) took place in Bilbao from September 18 to September 23, 2000. The Congress was organized by the Department of Mathematics of the University of the Basque Country, and sought to display gratitude, respect and affection for Alfred Gray.

This Congress was a very special mathematical event. It reflected a number of active subjects in modern differential geometry, many of them connected with the work and interests of Alfred Gray, such as special Riemannian manifolds, homogeneous spaces, complex structures, symplectic manifolds, geometry of geodesic spheres and tubes and related problems, geometry of surfaces, and computer graphics in differential geometry. A round table was also held with a discussion on geometry from the viewpoints of education and research.

In all there were one hundred and seventy-six participants from twenty-four countries in Europe, Asia, the Americas and indeed all over the world. The programme featured fourteen plenary conferences, a video conference, a round table discussion, sixty oral communications, and a section of 20 poster communications. The accompanying social events provided an inspiring atmosphere for scientific contacts and fruitful discussions.

In the preparation for the Congress, and during the event itself, L.C. de Andrés, R. Ibáñez, M. Macho and L. Ugarte were constantly ready to help.

The present volume contains a foreword by S. Donaldson describing several topics of Alfred Gray's work, many of the lectures given at the Congress and a record of the round table discussions. A list of Alfred Gray's publications and a list of participants are also included.

It is our pleasant duty to thank all the participants who contributed to the success of the Congress, especially those who offered us their manuscripts for publication. We also thank the referees for their careful work. Thanks as well to E. Abbena, L.A. Cordero, M. Djorić, O. Kowalski, A.R. Marinosci, M. Sekizawa, D. Tanré, Gr. Tsagas and L. Vanhecke for their work in collecting complete information on Alfred Gray's publications, to the American Mathematical Society (AMS) for giving us the opportunity to publish these proceedings in the Contemporary Mathematics, and to Ch. Thivierge of the AMS for helpful editorial advice in preparing this volume.

July, 2001

M. Fernández and J.A. Wolf
The Editors

The Alfred Gray Memorial Bronze D_4, K<0, September 2000. Sculpture by Helaman Ferguson, bronze casting by Alfa Arte of Eibar, Spain. Photograph reprinted with permission from Helaman Ferguson.

Acknowledgements

It is my pleasant duty to thank the professors on the Scientific Committee of the International Congress on Differential Geometry in Memory of Alfred Gray (1939-1998): Th. Banchoff, J.P. Bourguignon, S. Donaldson, J. Eells, S. Gindikin, M. Gromov, O. Kowalski, M. Mezzino, S. Novikov, M. Pinsky, A. Ros, S. Salamon, L. Vanhecke and J.A. Wolf. Their continual advice was a key factor in ensuring the international interest of the Congress.

I would also like to thank the members of the Organizing Committee: L.C. de Andrés, L.A. Cordero, A. Ferrández, R. Ibáñez, M. de León, M. Macho, A.M. Naveira and L. Ugarte whose hard work contributed so much to the success of the Conference.

Thanks are also due to all participants and in particular to the professors on the round table: J.P. Bourguignon, E. Calabi, J. Eells, O. García–Prada and M. Gromov, to the main speakers: Th. Banchoff, R. Bryant, H. Ferguson, Th. Friedrich, S. Gindikin, K. Grove, V. Miquel, E. Musso, M. Pinsky, A. Ros, S. Salamon, I. Taimanov, P. Wellin and J.A. Wolf, and to S. Wolfram for his video conference.

Thanks again to H. Ferguson for the special sculpture that he created for the Congress, a Costa type surface which readers can see for themselves on the facing page.

The Congress would not have been possible without the generous funding and ongoing support provided by the Basque Government, the University of the Basque Country, the Spanish Ministry of Science and Technology, Consejo Superior de Investigaciones Científicas, European Mathematical Society, Real Sociedad Matemática Española, Wolfram Research Inc. and many private companies.

I would like to say a very special thank you to the guest of honor, Professor Mary Gray.

During the week of the Congress, we devoted time to recalling the personality and scientific work of Alfred Gray. Although he passed away three years ago, his work and human qualities remain and his memory will be enhanced by the scholarship instituted by Wolfram Research Inc.

Bilbao, July 2001

Marisa Fernández
Chairperson of the Organizing Committee
and co-Editor

List of Publications of Alfred Gray

Books

1. Tubes, Addison-Wesley, Redwood City, CA, 1990. Second Edition, Addison-Wesley, Redwood City, CA, to appear. Russian translation, 1993.

2. Appendix: Using *Mathematica* (Appendix to M. Pinsky's book Partial Differential Equations and Boundary Value Problems with Applications), McGraw-Hill, Inc., New York, NY, 1991.

3. Modern Differential Geometry of Curves and Surfaces, CRC Press, Boca Raton, FL, 1993. Second Edition, CRC Press, Boca Raton, FL, 1998. Spanish and German Editions, 1995. Japanese translation, 1996. Italian Edition, 2000.

4. (with M. Mezzino and M. Pinsky) Introduction to Ordinary Differential Equations with *Mathematica*, TELOS, Springer-Verlag New York, Inc., NY, 1997. CD-ROM included.

Articles in Journals

1. (with S.M. Shah) A note on entire functions and a conjecture of Erdős I, *Bulletin of the American Mathematical Society* **69** (1963), 573–577.

2. (with S.M. Shah) A note on entire functions and a conjecture of Erdős II, *Journal d'Analyse Mathématique* **12** (1964), 83–104.

3. (with S.M. Shah) Asymptotic values of holomorphic functions of irregular growth, *Bulletin of the American Mathematical Society* **71** (1965), 747–749.

4. (with S.M. Shah) Holomorphic functions with gap power series I, *Mathematische Zeitschrift* **86** (1965), 375–394.

5. Minimal varieties and almost Hermitian submanifolds, *Michigan Mathematical Journal* **12** (1965), 273–287.

6. Some examples of almost Hermitian manifolds, *Illinois Journal of Mathematics* **10** (1966), 353–366.

7. (with S.M. Shah) Holomorphic functions with gap power series II, *Journal of Mathematical Analysis and Applications* **13** (1966), 370–382.

8. (with S.M. Shah) Holomorphic functions with gap power series III, *Journal of Mathematics and Mechanics* **16** (1966), 297–309.

9. (with S.M. Shah) Asymptotic values of a holomorphic function with respect to its maximum terms, *Pacific Journal of Mathematics* **18** (1966), 111–120.

10. Spaces of constancy of curvature operators, *Proceedings of the American Mathematical Society* **17** (1966), 897–902.

11. (with R.B. Brown) Vector cross products, *Commentarii Mathematici Helvetici* **42** (1967), 222–236.

12. Pseudo-Riemannian almost product manifolds and submersions, *Journal of Mathematics and Mechanics* **16** (1967), 715–737.

13. (with J.A. Wolf) Homogeneous spaces defined by Lie group automorphisms I, *Journal of Differential Geometry* **2** (1968), 77–114.

14. (with J.A. Wolf) Homogeneous spaces defined by Lie group automorphisms II, *Journal of Differential Geometry* **2** (1968), 115–159.

15. (with S.M. Shah) Borel exceptional values of entire functions, *Mathematische Annalen* **175** (1968), 185–188.

16. Isometric immersions of almost Hermitian manifolds, *Canadian Journal of Mathematics* **21** (1969), 456–459.

17. Principal curvature forms, *Duke Mathematical Journal* **36** (1969), 33–42.

18. Isometric immersions in symmetric spaces, *Journal of Differential Geometry* **3** (1969), 237–244.

19. A generalization of F. Schur's theorem, *Journal of the Mathematical Society of Japan* **21** (1969), 454–457.

20. A note on manifolds whose holonomy group is a subgroup of $Sp(n) \cdot Sp(1)$, *Michigan Mathematical Journal* **16** (1969), 125–128. Correction, **17** (1970), 409.

21. Almost complex submanifolds of the six sphere, *Proceedings of the American Mathematical Society* **20** (1969), 277–279.

22. Kähler submanifolds of homogeneous almost Hermitian manifolds, *Tôhoku Mathematical Journal* **21** (1969), 190–194.

23. Six-dimensional almost complex manifolds defined by means of three-fold vector cross products, *Tôhoku Mathematical Journal* **21** (1969), 614–620.

24. Vector cross products on manifolds, *Transactions of the American Mathematical Society* **141** (1969), 465–504. Correction, **148** (1970), 625.

25. Nearly Kähler manifolds, *Journal of Differential Geometry* **4** (1970), 283–309.

26. Some relations between curvature and characteristic classes, *Mathematische Annalen* **184** (1969/1970), 257–267.

27. (with P.S. Green) Sphere transitive structures and the triality automorphisms, *Pacific Journal of Mathematics* **34** (1970), 83–96.

28. Weak holonomy groups, *Mathematische Zeitschrift* **123** (1971), 290–300.

29. Riemannian manifolds with geodesic symmetries of order 3, *Journal of Differential Geometry* **7** (1972), 343–369.

30. Calculus of variations in complex vector bundles, *Rocky Mountain Journal of Mathematics* **2** (1972), no. 1, 119–130.

31. The volume of a small geodesic ball in a Riemannian manifold, *Michigan Mathematical Journal* **20** (1973), 329–344.

32. The six-dimensional Gauss Bonnet integrand, *Proceedings of the American Mathematical Society* **38** (1973), 374–380.

33. Classification des variétés approximativement kählériennes de courbure sectionnelle holomorphe constante, *Comptes Rendus de l'Académie des Sciences Paris. Série I Mathématique* **279** (1974), 797–800.

34. The structure of nearly Kähler manifolds, *Mathematische Annalen* **223** (1976), no. 3, 233–248.

35. Curvature identities for Hermitian and almost Hermitian manifolds, *Tôhoku Mathematical Journal* **28** (1976), no. 4, 601–612.

36. Compact Kähler manifolds with nonnegative sectional curvature, *Inventiones Mathematicae* **41** (1977), no. 1, 33–43.

37. Vector cross products, *Rendiconti del Seminario Matematico Università e Politecnico di Torino* **35** (1976/77), 69–75.

38. Chern numbers and curvature, *American Journal of Mathematics* **100** (1978), no. 3, 463–476.

39. Einstein-like manifolds which are not Einstein, *Geometriae Dedicata* **7** (1978), 259–280.

40. (with L. Vanhecke) Riemannian geometry as determined by the volumes of small geodesic balls, *Acta Mathematica* **142** (1979), no. 3-4, 157–198.

41. (with L. Vanhecke) Almost Hermitian manifolds with constant holomorphic sectional curvature, *Československá Akademie Věd. Časopis Pro Pěstovaní Matematiky* **104** (1979), no. 2, 170–179.

42. (with L.M. Hervella) The sixteen classes of almost Hermitian manifolds and their linear invariants, *Annali di Matematica pura ed applicata* (IV) **123** (1980), 35–58.

43. (with M. Barros, A.M. Naveira and L. Vanhecke) The Chern numbers of holomorphic vector bundles and formally holomorphic connections of complex vector bundles over almost complex manifolds, *Journal für die Reine und Angewandte Mathematik* **314** (1980), 84–98.

44. (with L. Vanhecke) Oppervlakten van geodetische cirkels op oppervlakken, *Mededelingen van de Koninklijke Academic voor Wetenschappen, Letteren*

en Schone Kunsten van Belgie Klasse der Wetenschappen **42** (1980), no. 1, 1–17.

45. (with E. Abbena and L. Vanhecke) Steiner's formula for the volume of a parallel hypersurface in a Riemannian manifold, *Annali della Scuola Normale Superiore di Pisa* **8** (1981), 473–493.

46. (with L. Vanhecke) The volumes of tubes in a Riemannian manifold, *Rendiconti del Seminario Matematico Università e Politecnico di Torino* **39** (1981), 1–50.

47. (with M. Fernández) Riemannian manifolds with structure group G_2, *Annali di Matematica pura ed applicata* (IV) **132** (1982), 19–45.

48. (with L. Vanhecke) The volumes of tubes about curves in a Riemannian manifold, *Proceedings of the London Mathematical Society* **44** (1982), 215–243.

49. (with T.J. Willmore) Mean-value theorems for Riemannian manifolds, *Proceedings of the Royal Society of Edinburgh* **92A** (1982), 343–360.

50. (with P. Carpenter and T.J. Willmore) The curvature of Einstein symmetric spaces, *Quarterly Journal of Mathematics, Oxford* **33** (1982), 45–64.

51. An estimate for the volume of a tube about a complex hypersurface, *Tensor* **39** (1982), 303–305.

52. Comparison theorems for the volumes of tubes as generalizations of the Weyl's tube formula, *Topology* **21** (1982), no. 2, 201–228.

53. (with M. Pinsky) The mean exit time from a small geodesic ball in a Riemannian manifold, *Bulletin des Sciences Mathématiques* **107** (1983), no. 4, 345–370.

54. Homogeneous almost Hermitian manifolds, Proceedings of the Conference on Differential Geometry on Homogeneous Spaces, Turin, Italy, 1983, *Rendiconti del Seminario Matematico Università e Politecnico di Torino* (1983), 17–58. (Special Issue.)

55. Volumes of tubes about Kähler submanifolds expressed in terms of Chern classes, *Journal of the Mathematical Society of Japan* **36** (1984), no. 1, 23–35.

56. Brownian motion and Riemannian geometry, *Mededelingen van de Koninklijke Academie voor Wetenschappen, Letteren en Schone Kunsten van Belgie Klasse der Wetenschappen* **46** (1984), no. 1, 24–30.

57. A product formula for the volumes of tubes about Kähler submanifolds of complex Euclidean space, Papers from the international meeting in honor of Enzo Martinelli, Rome, Italy, 1983, ed. by G.B. Rizza and F. Succi, *Rivista di Matematica della Università di Parma* (IV) **10** (1984), 109–121.

58. (with L.A. Cordero and M. Fernández) Variétés symplectiques sans structures Kählériennes, *Comptes Rendus de l'Académie des Sciences Paris. Série I Mathématique* **301** (1985), no. 5, 217–218.

59. Volumes of tubes about complex submanifolds of complex projective space, *Transactions of the American Mathematical Society* **291** (1985), no. 2, 437–449.

60. (with L.A. Cordero and M. Fernández) Symplectic manifolds with no Kähler structure, *Topology* **25** (1986), no. 3, 375–380.

61. (with L.A. Cordero, M. Fernández and M. de León) Productos de Massey e Variedades Simplécticas non Kählerianas, *Boletín da Academia Galega de Ciencias* **6** (1987), 137–164.

62. (with L.A. Cordero and M. Fernández) La suite spectrale de Frölicher et les nilvariétés complexes compactes, *Comptes Rendus de l'Académie des Sciences Paris. Série I Mathématique* **305** (1987), no. 17, 753–756.

63. (with M. Fernández and M. Gotay) Compact parallelizable four-dimensional symplectic and complex manifolds, *Proceedings of the American Mathematical Society* **103** (1988), no. 4, 1209–1212.

64. The minimal focal distance of a complex hypersurface in complex projective space, Proceedings of the Conference on Differential Geometry and Topology, Lecce, Italy, 1989, *Note di Matematica* **9** (1989), 119–122. (Supplemento.)

65. (with M. Fernández) Compact symplectic solvmanifolds not admitting complex structures, *Geometriae Dedicata* **34** (1990), no. 3, 295–299.

66. (with L.A. Cordero and M. Fernández) Lie groups with no left invariant complex structures, *Portugaliae Mathematica* **47** (1990), no. 2, 184–190.

67. (with L.C. de Andrés, M. Fernández and J.J. Mencía) Moduli spaces of complex structures on compact four-dimensional nilmanifolds, *Bolletino Unione Matematica Italiana* A (7) **5** (1991), no. 3, 381–389.

68. (with L.A. Cordero and M. Fernández) The Frölicher spectral sequence for compact nilmanifolds, *Illinois Journal of Mathematics* **35** (1991), no. 1, 56–67.

69. (with M. Fernández and J.W. Morgan) Compact symplectic manifolds with free circle actions and Massey products, *Michigan Mathematical Journal* **38** (1991), no. 2, 271–283.

70. (with M. Pinsky) Computer graphics and a new Gibbs's phenomenon for Fourier-Bessel series, *Experimental Mathematics* **1** (1992), 313–316.

71. (with R.J. Knill) Using *Mathematica* to find closed form expressions for approximations to $\sqrt{x}$, *Mathematica in Education and Research* **1** (1992), no. 4, 12–13.

72. (with M. Pinsky) Gibbs's phenomenon for Fourier-Bessel series, *Expositiones Mathematicae* **11** (1993), no. 2, 123–135.

73. (with H. Ferguson and S. Markvorsen) Costa's minimal surface via *Mathematica*, *Mathematica in Education and Research* **5** (1996), no. 1, 5–10.

74. A property of a hypothetical complex structure on the six sphere, *Bolletino Unione Matematica Italiana* B (7) **11** (1997), no. 2 Supplemento, 251–255. (Memorial volume for Franco Tricerri.)

75. (with L.A. Cordero, M. Fernández and L. Ugarte) A general description of the terms in the Frölicher spectral sequence, *Differential Geometry and its Applications* **7** (1997), no. 1, 75–84.

76. (with L.A. Cordero, M. Fernández and L. Ugarte) Nilpotent complex structures on compact nilmanifolds, Proceedings Workshop on Differential Geometry and Topology dedicated to the memory of Franco Tricerri, Palermo, Italy, 1996, ed. by R. Grimaldi and S. Marchiafava, Supplemento ai *Rendiconti del Circolo Matematico di Palermo* serie II (1997), no. 49, 83–100.

77. (with V. Miquel) On Pappus-type theorems on the volume in space forms, Special issue in memory of Alfred Gray (1939–1998), ed. by K. Grove and J. Rosenberg, *Annals of Global Analysis and Geometry* **18** (2000), no. 3-4, 241–254.

78. (with L.A. Cordero, M. Fernández and L. Ugarte) Compact nilmanifolds with nilpotent complex structures: Dolbeault cohomology, *Transactions of the American Mathematical Society* **352** (2000), no. 12, 5405–5433.

79. (with H. Ferguson and S. Ferguson) Costa's minimal surface: sculpture by virtual image projection, to appear.

Articles in Collections

1. (with S.M. Shah) The behavior of an entire function as determined by its maximum terms, Entire Functions and Related Parts of Analysis, *Proceedings of Symposia in Pure Mathematics* **11**, La Jolla, California, 1966, ed. by *S.S. Chern, L. Ehrenpreis, J. Korevaar, W.H.J. Fuchs and L.A. Rubel*, American Mathematical Society, Providence, RI, 1968, pp. 202–209.

2. Integral distributions determined by an immersion, Global Analysis, *Proceedings of Symposia in Pure Mathematics* **15**, Berkeley, California, 1968, ed. by *S.S. Chern and S. Smale*, American Mathematical Society, Providence, RI, 1970, pp. 239–249.

3. Structures on Riemannian manifolds determined by Lie groups which act transitively on spheres, Differentialgeometrie im Grossen, *Tagung, Mathematischen Forschungsinstitut*, Oberwolfach, Germany, 1969, ed. by *W. Klingenberg*, Berichte aus dem Mathematischen Forschungsinstitut Oberwolfach **4**, Bibliographisches Institut, Mannheim-Vienna-Zürich, 1971, pp. 145–153.

4. (with R.B. Brown) Riemannian manifolds with holonomy group *Spin*(9), Differential Geometry (in honor of Kentaro Yano), ed. by *S. Kobayashi, M. Obata and T. Takahashi*, Kinokuniya Book-Store, Tokyo, 1972, pp. 41–59.

5. Geodesic balls in Riemannian product manifolds, Differential Geometry and Relativity, *A volume in honour of André Lichnerowicz on his 60th birthday*, ed. by *M. Cahen and M. Flato*, Mathematical Physics and Applied Mathematics **3**, D. Reidel Publishing Co., Dordrecht-Boston, 1976, pp. 63–66.

6. (with L. Vanhecke) Power series expansions, differential geometry of geodesic spheres and tubes, and mean-value theorems, Geometry and Differential Geometry, *Proceedings of the Conference on Geometry and Differential Geometry*, University Haifa, Haifa, Israel, 1979, ed. by *R. Artzy and I. Vaisman*, Lecture Notes in Mathematics **792**, Springer, Berlin, 1980, pp. 252–259.

7. A generalization of Weyl's tube formula, Global differential Geometry and Global Analysis, *Proceedings Colloquium*, Technical University of Berlin, Berlin, Germany, 1979, ed. by *D. Ferus, W. Kühnel, U. Simon and B. Wegner*, Lecture Notes in Mathematics **838**, Springer, Berlin, 1981, pp. 140–144.

8. Comparison theorems for the volumes of half-tubes as generalizations of the formulas of Steiner, Symposia Mathematica **26**, Rome, Italy, 1980, Academic Press, London, 1982, pp. 181–195.

9. Vector cross products, harmonic maps and the Cauchy Riemann equations, Harmonic Maps, *Proceedings of the N.S.F.-C.B.M.S. Regional Conference*, Tulane University, New Orleans, 1980, ed. by *R.J. Knill, M. Kalka and H.C.J. Sealey*, Lecture Notes in Mathematics **949**, Springer, Berlin, 1982, pp. 57–74.

10. Comparison theorems for volumes in surfaces, Geometry and Topology, *Proceedings of the special year*, University of Maryland, College Park, Maryland, 1983/84, ed. by *J.C. Alexander and J.L. Harer*, Lecture Notes in Mathematics **1167**, Springer, Berlin, 1985, pp. 118–124.

11. Ersatz Chern polynomial and Weyl's tube formula, Differential Geometry, *Proceedings of the Fifth International Colloquium on Differential Geometry*, Santiago de Compostela, Spain, 1984, ed. by *L.A. Cordero*, Research Notes in Mathematics **131**, Pitman, Boston, 1985, pp. 66–75.

12. (with L. Karp and M. Pinsky) The mean exit time from a tube in a Riemannian manifold, Probability Theory and Harmonic Analysis, *Proceedings of the Conference on Probability and Harmonic Analysis*, Cleveland, Ohio, 1983, ed. by *J.A. Chao and W.A. Woyczynski*, Marcel Dekker, New York, 1986, pp. 113–137.

13. (with M. Fernández) The Iwasawa manifold, Differential Geometry, *Proceedings of the First International Symposium on Differential Geometry*, Peñíscola, Spain, 1985, ed. by *A.M. Naveira, A. Ferrández and F. Mascaró*, Lecture Notes in Mathematics **1209**, Springer, Berlin, 1986, pp. 157–159.

14. Weyl's tube formula as the integral of a Bessel function, Topics in Differential Geometry, *Colloquia Mathematica Societatis János Bolyai* **46**, Debrecen, Hungary, 1984, ed. by *J. Szenthe and L. Tamassy*, North-Holland, Amsterdam, 1988, pp. 511–517.

15. (with L.A. Cordero and M. Fernández) The failure of complex and symplectic manifolds to be Kählerian, Differential geometry: geometry in mathematical physics and related topics, *Proceedings of Symposia in Pure Mathematics* **54**, Los Angeles, California, 1990, ed. by *R.E. Greene and S.T. Yau*, American Mathematical Society, Providence, RI, 1993, pp. 107–123.

16. (with L.A. Cordero and M. Fernández) Some differential forms are more exact than others, *Differential Geometry Workshop on Spaces of Geometry*, Guanajuato, Mexico, 1992, ed. by *L. del Riego and C.T.J. Dodson*, Aportaciones Matemáticas, Serie: Notas de Investigación **8**, Sociedad Matemática Mexicana, Mexico City, 1993, pp. 55–65.

17. (with L. A. Cordero, M. Fernández and L. Ugarte) Frölicher spectral sequence of compact nilmanifolds with nilpotent complex structure, New Development in Differential Geometry, *Proceedings of the Conference on Differential Geometry*, Budapest, Hungary, 1996, ed. by *J. Szenthe*, Kluwer Academic Publishers, Dordrecht, 1999, pp. 77–102.

18. (with L.A. Cordero, M. Fernández and L. Ugarte) Dolbeault homotopy theory and compact nilmanifolds, Homotopy and Geometry, *Proceedings Workshop Homotopy and Geometry*, Warsaw, Poland, 1997, ed. by *J. Oprea and A. Tralle*, Banach Center Publications **45**, Polish Academy of Sciences, Institute of Mathematics, Warsaw, 1998, pp. 137–154.

19. (with L.A. Cordero, M. Fernández and L. Ugarte) Nilpotent complex structures, VI Workshop de Otoño, *Proceedings VI Encuentro de Otoño de Geometría y Física*, University of Salamanca, Salamanca, Spain, 1997, ed. by *P.L. García, A. Fernández and J. Mateos*, Memorias de la Real Academia de Ciencias Exactas, Físicas y Naturales, Madrid, to appear.

Articles in Proceedings

1. Invariants of curvature operators of four-dimensional Riemannian manifolds, *Proceedings of the Thirteenth Biennial Seminar of the Canadian Mathematical Congress* **2**, Dalhousie University, Halifax, Canada, 1971, ed. by *J.R. Vanstone*, Canadian Mathematical Congress, Montreal, 1972, pp. 42–65.

2. The Gauss Bonnet integrand of a compact six-dimensional Einstein Kähler manifold, *Proceedings of the C. Caratheodory International Symposium*, Athens, Greece, 1973, ed. by *A. Panagiotopoulos*, Greek Mathematical Society, Athens, 1974, pp. 165–172.

3. Les nombres de Chern d'un fibré vectoriel holomorphe, *Journées de Géométrie Différentielle*, Université de Liège, Belgium, 1976, Fonds National de la Recherche Scientifique, Belgium, 1976, pp. 1–4.

4. Mean values in Riemannian manifolds, *Proceedings of the Fourth International Colloquium on Differential Geometry*, Santiago de Compostela, Spain, 1978, ed. by *L.A. Cordero and L. Hervella*, Cursos y Congresos de la Universidad de Santiago de Compostela **15**, Santiago de Compostela, 1979, pp. 140–146.

5. Weyl's tube formula in the complex case, *Proceedings of the CSSR-GDR-Polish Conference on Differential Geometry and its Applications*, Nove Mesto na Morave, Czech Republic, 1980, ed. by *O. Kowalski*, Charles University, Prague, 1981, pp. 225–227.

6. Volumes of geodesic balls in Riemannian manifolds, *Proceedings of the Summer School on Applied Mathematics and Informatics for Socio-Economic Development*, ed. by *Gr. Tsagas*, Thessaloniki, Greece, 1980, pp. 94–106.

7. (with S.Y. Lee) Product formulas for the volume of tubes, *Proceedings of the Conference on Differential Geometry and its Applications*, **1**, Nove Mesto na Morave, Czech Republic, 1983, ed. by *O. Kowalski*, Charles University, Prague, 1984, pp. 77–85.

8. (with L.A. Cordero, M. Fernández and M. de León) Nilvariedades simplécticas compactas que no admiten estructuras Kählerianas, *Actas VII Reunión Grupo Matemática Expresión Latina* **1**, Coimbra, Portugal, 1985, Departamento de Matemática, Universidade de Coimbra, Coimbra, 1986, pp. 384–387.

9. (with L.C. de Andrés, M. Fernández and J.J. Mencía) Compact manifolds with indefinite Kähler metrics, *Proceedings of the Sixth International Colloquium on Differential Geometry*, Santiago de Compostela, Spain, 1988, ed. by *L.A. Cordero*, Cursos y Congresos de la Universidad de Santiago de Compostela **61**, Santiago de Compostela, 1989, pp. 25–50.

10. (with L.A. Cordero and M. Fernández) Modelos minimales en geometría diferencial, *Proceedings of the Workshop Recent Topics in Differential Geometry*, Puerto de la Cruz, Tenerife, Spain, 1990, ed. by *D. Chinea and J.M. Sierra*, Secretariado Publicaciones Universidad de La Laguna, Tenerife, 1991, pp. 31–41.

11. (with L.A. Cordero and M. Fernández) La estructura fibrada de una nilvariedad compacta, *Proceedings of the Workshop Recent Topics in Differential Geometry*, Puerto de la Cruz, Tenerife, Spain, 1990, ed. by *D. Chinea and J.M. Sierra*, Secretariado Publicaciones Universidad de La Laguna, Tenerife, 1991, pp. 43–64.

12. (with L.A. Cordero, M. Fernández and L. Ugarte) Complex homotopy theory for compact nilmanifolds, *Proceedings of the Workshop on Recent Topics in Differential Geometry*, Santiago de Compostela, Spain, 1997, ed. by *L.A. Cordero and E. García-Rio*, Publicaciones Departamento Geometría y Topología **89**, Universidad de Santiago de Compostela, Santiago de Compostela, 1998, pp. 71–84.

13. (with L.A. Cordero, M. Fernández and L. Ugarte) The holomorphic fiber bundle structure of some compact complex nilmanifolds, *Proceedings of the First International Meeting on Geometry and Topolgy*, Braga, Portugal, 1997, ed. by *A. Pereira do Vale and M.R. Pinto*, Public. Centro de Matemática da Universidade do Minho, Braga, 1998, pp. 207–221.

Software

1. Curves and Surfaces (*Mathematica* and *Maple* programs for the geometry of Curves and Surfaces).
   Available via anonymous ftp from `math.cl.uh.edu/pub/gray`.
   The web site is located at `http://math.cl.uh.edu/~gray`.

2. ODE (with Mike Mezzino) (*Mathematica* programs for solving and plotting ordinary differential equations).
   Available via anonymous ftp from `math.cl.uh.edu/pub/gray`.
   The web site is located at `http://math.cl.uh.edu/ode/ode.html`.

Videos

1. *Strange surfaces*, 10 minute video tape, made at the Geometry Center of the University of Minnesota, 1991.

2. *Classical surfaces*, 20 minute video tape, made at the Geometry Center of the University of Minnesota, 1993.

Editor

1. (with R. Brooks and B. Reinhart) Differential Geometry, Birkhäuser 1983.

2. (with J. Rosenberg) Special Issue on Foliations in Honor of Bruce Reinhart, *Annals of Global Analysis and Geometry* **9** (1991), no. 1.

List of Participants

Ilka AGRICOLA	Humboldt U. Berlin	Germany
Juan A. ALEDO	U. de Castilla-La Mancha	Spain
Luis J. ALIAS	U. de Murcia	Spain
James ANDERSON	American U.	USA
Luis C. de ANDRES	U. del País Vasco	Spain
S. ARDANZA-TREVIJANO	SUNY Binghamton	USA
Josu ARROYO	U. del País Vasco	Spain
A. ARVANITOYEORGOS	The British Council-Athens	Greece
Vitaly BALASHCHENKO	Belarusian State U.	Belarus
Thomas F. BANCHOFF	Brown U.	USA
Helga BAUM	Humboldt U. Berlin	Germany
Christina BENEKI	U. of Patras	Greece
Jurgen BERNDT	U. of Hull	U.K.
Tran Quoc BINH	Debrecen U.	Hungary
Eric BOECKX	Katholieke U. Leuven	Belgium
Gil BOR	CIMAT	Mexico
Alexander BORISENKO	Kharkov National U.	Ukraine
J. Pierre BOURGUIGNON	CNRS–IHÈS	France
Robert L. BRYANT	Duke U.	USA
Volker BUCHHOLZ	Humboldt U. Berlin	Germany
Jarolim BURES	Charles U. Prague	Czech Rep.
Renzo CADDEO	U. degli Studi di Cagliari	Italy
Eugenio CALABI	U. of Pennsylvania	USA
Giovanni CALVARUSO	U. di Lecce	Italy
María A. CAÑADAS-PINEDO	U. de Málaga	Spain
Alfonso CARRIAZO	U. de Sevilla	Spain
Domenico A. CATALANO	U. de Aveiro	Portugal
Pablo M. CHACON MARTIN	U. de Valencia	Spain
Simon G. CHIOSSI	U. di Genova	Italy
Jong Taek CHO	Chonnam National U.	Korea
Sergio CONSOLE	U. di Torino	Italy
Luis A. CORDERO	U. de Santiago de Compostela	Spain
Celso J. da COSTA	U. Federal Fluminense	Brazil
Maciej CZARNECKI	U. Lodzki	Poland
Giusi D'AMBRA	U. degli Studi di Cagliari	Italy
Mirjana DJORIC	U. of Belgrade	Yugoslavia
Georges DLOUSSKY	U. d'Aix-Marseille I	France
Miroslav DOUPOVEC	Tech. U. of Brno	Czech Rep.

M. Josefina DRUETTA	U. Nacional de Córdoba	Argentina
James EELLS	Cambridge U.	U.K.
Claire FERGUSON	Laurel, Maryland	USA
Helaman FERGUSON	Laurel, Maryland	USA
Marisa FERNANDEZ	U. del País Vasco	Spain
M. FERNANDEZ ANDRES	U. de Sevilla	Spain
Angel FERRANDEZ	U. de Murcia	Spain
Gabriele FICI	U. di Palermo	Italy
Anna FINO	U. di Torino	Italy
Mario FIORAVANTI	U. de Cantabria	Spain
Thomas FRIEDRICH	Humboldt U. Berlin	Germany
Atsushi FUJIOKA	Kanazawa U.	Japan
Eduardo GALLEGO	U. Aut. de Barcelona	Spain
M. Elisa L. GALVAO	U. de Mogi das Cruzes	Brazil
Oscar J. GARAY	U. del País Vasco	Spain
Oscar GARCIA-PRADA	U. Aut. de Madrid	Spain
Begoña GARCIA-RAMIRO	U. del País Vasco	Spain
Eduardo GARCIA-RIO	U. de Santiago de Compostela	Spain
Nima GEFFEN	Tel-Aviv U.	Israel
Graziano GENTILI	U. di Firenze	Italy
Olga GIL-MEDRANO	U. de Valencia	Spain
Peter B. GILKEY	U. of Oregon	USA
Simon GINDIKIN	Rutgers U.	USA
Joan GIRBAU	U. Aut. de Barcelona	Spain
Hubert GOLLEK	Humboldt U. Berlin	Germany
Claudio GORODSKI	U. of São Paulo	Brazil
Mary GRAY (Guest of Honor)	American U.	USA
Marina GREBENYUK	U. Kiev	Ukraine
Misha GROMOV	IHÈS	France
Karsten GROVE	U. of Maryland	USA
Luis GUIJARRO	U. Complutense de Madrid	Spain
Hideya HASIMOTO	Nippon Inst. of Tech.	Japan
L. HERNANDEZ LAMONEDA	CIMAT	Mexico
Luis HERVELLA	U. de Santiago de Compostela	Spain
Raúl IBAÑEZ	U. del País Vasco	Spain
Jin-ichi ITOH	Kumamoto U.	Japan
Josef JANYSKA	Masaryk U. Brno	Czech Rep.
Wlodzimierz JELONEK	Cracow U. of Tech.	Poland
Hugo JIMENEZ-PEREZ	U. Nacional Aut. Mexico	Mexico
S. M. B. KASHANI	Tarbiat Modarres U.	Iran
Ines KATH	Humboldt U. Berlin	Germany
Haruo KITAHARA	Kanazawa U.	Japan
Ronald J. KNILL	Tulade U.	USA
Jair KOILLER	Lab. Nac. de Comp. Cientif.	Brazil
Ivan KOLAR	Masaryk U. Brno	Czech Rep.
Oldrich KOWALSKI	Charles U. Prague	Czech Rep.
Laszlo KOZMA	U. of Debrecen	Hungary
Lukas KRUMP	Charles U. Prague	Czech Rep.

Jan KUREK	MC Sklodovska U.	Poland
Miroslav KURES	Brno U. of Tech.	Czech Rep.
Natalie M. KUZUB	Irkutsk State U.	Russia
Jorge LAURET	U. Nacional de Córdoba	Argentina
Thomas LEISTNER	Humboldt U. Berlin	Germany
Felipe LEITNER	Humboldt U. Berlin	Germany
Luc LEMAIRE	U. Libre de Bruxelles	Belgium
Manuel de LEON	C.S.I.C.	Spain
Ana LLUCH	U. Jaume I	Spain
Rafael LOPEZ	U. de Granada	Spain
S. LOPEZ de MEDRANO	UNAM	Mexico
Eric LOUBEAU	U. of Brest	France
Pascual LUCAS	U. de Murcia	Spain
Tsasa LUSALA	Technische U. Berlin	Germany
Armando MACHADO	U. de Lisboa	Portugal
Marta MACHO-STADLER	U. del País Vasco	Spain
Yoshiaki MAEDA	Keio U.	Japan
Domenico MAIOLO	U. di Lecce	Italy
Stefano MARCHIAFAVA	U. di Roma "La Sapienza"	Italy
Anna R. MARINOSCI	U. di Lecce	Italy
José M. MASEDA	U. del País Vasco	Spain
José J. MENCIA	U. del País Vasco	Spain
Francesco MERCURI	U. Estadual de Campinas	Brazil
Josef MIKESH	Palacky U. Olomouc	Czech Rep.
Dmitri MILLIONSCHIKOV	Moscow State U.	Russia
Vicente MIQUEL	U. de Valencia	Spain
Stefano MONTALDO	U. degli Studi di Cagliari	Italy
Simone M. de MORAES	U. de Valencia	Spain
Vicente MUÑOZ	U. Aut. de Madrid	Spain
Emilio MUSSO	U. di L'Aquila	Italy
Antonio M. NAVEIRA	U. de Valencia	Spain
Caio J. C. NEGREIROS	U. Estadual de Campinas	Brazil
Alexander I. NESTEROV	U. de Guadalajara	Mexico
Stana NIKCEVIC	Mathematical Inst. SANU	Yugoslavia
Igor NIKITIN	German Nat. Research Center	Germany
Vicente PALMER	U. Jaume I	Spain
Maurizio PARTON	U. di Pisa	Italy
Ana PEREIRA DO VALE	U. do Minho	Portugal
Domenico PERRONE	U. di Lecce	Italy
Paolo PICCINNI	U. di Roma "La Sapienza"	Italy
Mark J. PINSKY	Northwestern U.	USA
Paola PIU	U. degli Studi di Cagliari	Italy
Olga POKORNA	Czech U. of Agriculture	Czech Rep.
Massimiliano PONTECORVO	U. di Roma Tre	Italy
Ebrahim POURREZA	U. of Tabriz	Iran
Agostino PRASTARO	U. di Roma "La Sapienza"	Italy
Ivan'shin PYOTR	Kazan State U.	Russia
Zoran RAKIC	Mathematics SANU	Yugoslavia
Pablo RAMACHER	Humboldt U. Berlin	Germany

Agustí REVENTOS	U. Aut. de Barcelona	Spain
Ramón REYES	CIMAT	Mexico
Lilia del RIEGO	U. Aut. de San Luis Potosí	Mexico
Paulo R. RODRIGUES	U. Federal Fluminense	Brazil
Antonio ROS	U. de Granada	Spain
José I. ROYO	U. del País Vasco	Spain
Tomasz RYBICKI	Akad. Gorniczo-Hutnicza	Poland
Lev SABININ	Quintana Roo U.	Mexico
Elena SAFIULINA	Tartu U.	Estonia
Simon SALAMON	Politecnico di Torino	Italy
Beatriz SALVADOR	U. Complutense de Madrid	Spain
Marcos SALVAI	U. Nacional de Córdoba	Argentina
Esther SANABRIA	U. de Valencia	Spain
A. SANCHEZ VALENZUELA	C.I.M.A.T.	Mexico
Aitor SANTAMARIA	Oxford U.	U.K.
Joseba SANTISTEBAN	U. del País Vasco	Spain
Alessandro SAVO	U. di Roma "La Sapienza"	Italy
Larissa SBITNEVA	Quintana Roo U.	Mexico
Kouei SEKIGAWA	Niigata U.	Japan
Masami SEKIZAWA	Tokyo Gakugei U.	Japan
Uwe SEMMELMANN	U. of Munich	Germany
Uladzimir SHTUKAR	Mogilev State Tech. U.	Belarus
Arkadi SKOPENKOV	Kolmogorov College	Russia
Gil SOLANES	U. Aut. de Barcelona	Spain
Juan SOUTO	U. Bonn	Germany
Rolf SULANKE	Humboldt U. Berlin	Germany
Tim SWIFT	U. of the West of England	U.K.
Iskander TAIMANOV	Russ. Acad. Sci., Novosibirsk	Russia
Hiroshi TAMARU	Sophia U.	Japan
Corrado TANASI	U. di Palermo	Italy
Yoshihiko TAZAWA	Tokyo Denki U.	Japan
Jiri TOMAS	Technical U. Brno	Czech Rep.
Gabor TOTH	Rutgers U.	USA
Aleksy TRALLE	U. of Warmia and Mazuria	Poland
Kazumi TSUKADA	Ochanomizu U. Tokyo	Japan
Luis UGARTE	U. de Zaragoza	Spain
Lieven VANHECKE	Katholieke U. Leuven	Belgium
Elena VAZQUEZ	U. de Santiago de Compostela	Spain
Pawel WALCZAK	U. Lodzki	Poland
Paul WELLIN	Wolfram Research Inc.	USA
Joseph A. WOLF	U. of California-Berkeley	USA

Contemporary Mathematics
Volume **288**, 2001

On the Formality Problem for Symplectic Manifolds

Ivan K. Babenko and Iskander A. Taimanov

Abstract. We discuss the formality problem for symplectic manifolds. In particular, we discuss the first examples of nonformal simply connected symplectic manifolds which we found in 1998.

1. Examples and constructions of symplectic manifolds.

A smooth manifold M endowed with a closed 2-form ω is called symplectic if this form is nondegenerate. This form is called symplectic. By the Darboux theorem, in canonical coordinates it is written as $\omega = \sum_{i=1}^{n} dx^i \wedge dx^{n+i}$. Therefore any symplectic manifold is even-dimensional and admits an almost complex structure J compatible with the symplectic structure. This means that the inner product $\langle \xi, \eta \rangle = \omega(\xi, J\eta)$ defines a Riemannian metric and J is skew-symmetric with respect to this metric.

The simplest examples of symplectic manifolds are Kähler manifolds with Kähler forms. For example the complex projective spaces $\mathbb{C}P^n$ with the forms $\omega_{\mathrm{FS}} = g_{ij} dz^i \wedge d\bar{z}^j$ where $g_{ij} dz^i d\bar{z}^j$ is the Fubini–Study metric.

Gromov proved that any open almost complex manifold admits a compatible symplectic structure [**14**]. For closed manifolds this is not true since ω^n is the volume form and therefore $[\omega]^k \neq 0 \in H^{2k}(M;\mathbb{R})$ for $k = 1, \dots, n$.

The Gromov–Tischler theorem [**14, 24**] reads that if M is a compact symplectic manifold with an integer symplectic form, $[\omega] \in H^2(M;\mathbb{Z})$, then there is a symplectic embedding $f : (M, \omega) \to (\mathbb{C}P^{2n+1}, \omega_{\mathrm{FS}})$, with $\dim M = n$, such that $f^*(\omega_{\mathrm{FS}}) = \omega$. This implies that any compact symplectic manifold is diffeomorphic to a symplectic submanifold of a complex projective space.

The problem of how symplectic manifolds differ from Kähler manifolds led to the introduction of several geometric methods for constructing symplectic manifolds. These include symplectic fibrations (Thurston [**23**]), symplectic blow-ups (McDuff [**19**]), and fiber connected sums (Gompf [**13**]). The last two constructions were first outlined by Gromov.

There are other constructions by Donaldson [**10**] and Guan [**15**] which involve hard global analysis. We recall that a complex manifold M is called holomorphically symplectic if there is a closed holomorphic $(2,0)$-form ω with maximal rank. Notice that $\omega + \bar{\omega}$ is a symplectic form. Generalizing the construction of $K3$ surfaces Guan

1991 *Mathematics Subject Classification.* 53D05, 55S30.
Key words and phrases. symplectic manifolds, rational homotopy type, formality, Massey products, nilmanifolds.

had found the first examples of closed simply connected holomorphically symplectic manifolds which are not Kählerian. [1]

2. Symplectic manifolds with no Kähler structure.

The first example of a compact symplectic manifold with no Kähler structure were found by Thurston [**23**]. This manifold already appeared in the Kodaira classification of complex surfaces. So it is called the Kodaira–Thurston manifold and its construction is as follows.

Denote by $\mathcal{H}$ the three-dimensional Heisenberg group formed by the upper-triangular matrices

$$\begin{pmatrix} 1 & x & z \\ 0 & 1 & y \\ 0 & 0 & 1 \end{pmatrix}$$

with $x, y, z \in \mathbb{R}$ and the multiplication operation. The matrices with x, y, z integers form a uniform lattice $\mathcal{H}_{\mathbb{Z}}$ in $\mathcal{H}$. On the circle S^1 take a coordinate u defined modulo 1 and put

$$M_{\mathrm{KT}} = (\mathcal{H}/\mathcal{H}_{\mathbb{Z}}) \times S^1, \quad \omega = dx \wedge du + dy \wedge dz.$$

We may also construct M_{KT} as follows. Let T^2 be a two-torus $\mathbb{R}^2/\mathbb{Z}^2$ with linear coordinates z and y defined modulo integers. Take a product of T^2 with an interval $[0,1]$ with a coordinate x. Glue two components of the boundary by the automorphism $\begin{pmatrix} z \\ y \end{pmatrix} \to \begin{pmatrix} 1 & 1 \\ 0 & 1 \end{pmatrix} \begin{pmatrix} z \\ y \end{pmatrix}$. The resulting nilmanifold is $\mathcal{H}/\mathcal{H}_{\mathbb{Z}}$. Its product with the circle S^1 is M_{KT}. It is fibered over the two torus with the coordinates x and u defined modulo integers:

$$T^2_{(y,z)} \to M_{\mathrm{KT}} \to T^2_{(x,u)}.$$

The symplectic form ω splits into the sum of the pullback of the symplectic form $dx \wedge du$ on the base and the form $dy \wedge dz$ whose restriction to the fibers defines symplectic structures. This is a simplest example of a symplectic fibration.

The first Betti number of M_{KT} is odd (it is actually 3). Therefore it has no Kähler structure.

THEOREM 1 (Thurston). *The Kodaira–Thurston manifold is a non-simply connected symplectic manifold with no Kähler structure.*

The Kodaira–Thurston manifold admits a complex structure and moreover an indefinite Kähler metric.

Fernandez, Gotay and Gray studied such symplectic fibrations in detail. They had found examples of symplectic manifolds with no complex structure [**11**] (see also [**12**]).

THEOREM 2 (Fernandez–Gotay–Gray). *Let M^4 be a principal S^1 bundle over a three-manifold N^3 which in turn is a principal S^1-bundle over a two-torus T^2. Let the first Betti number satisfy the inequalities $2 \leq b_1(M^4) \leq 4$. Then*

1) if $b_1 = 2$ then M^4 has symplectic but no complex structures;

2) if $b_1 = 3$ then M^4 has both symplectic and complex structures but no Kähler structure; however it has an indefinite Kähler metric;

3) if $b_1 = 4$ then M^4 is a torus T^4.

[1] We are grateful to M. Gromov who draw our attention to the papers by Guan after our talk.

Some high-dimensional generalizations of the Kodaira–Thurston manifold were found by Cordero, Fernandez and Gray [**8**].

Recently we have found other high-dimensional generalizations of these examples. They appear to be interesting for other reasons [**4**].

Consider the infinite-dimensional algebra W of formal vector fields on the line. It has the basis

$$e_k = x^{k+1}\frac{d}{dx}, \quad k \geq -1,$$

and the commutation relations are $[e_i, e_j] = (j-i)e_{i+j}$. There is a natural filtration $\dots \subset \mathcal{L}_1 \subset \mathcal{L}_0 \subset \mathcal{L}_{-1} \subset W$ where $\mathcal{L}_k$ is a subalgebra spanned by $e_k, e_{k+1}, \dots$. Consider the series of finite-dimensional nilpotent Lie algebras

$$\mathcal{V}_n = \mathcal{L}_1/\mathcal{L}_{n+1}, \ n = 3, 4, \dots,$$

and denote by V_n the corresponding Lie groups. The group V_n is isomorphic to $(\mathcal{V}_n, \times)$ with the polynomial multiplication law $\times$ given by the Campbell–Hausdorff formula. The basis $\{e_1, \dots, e_n\}$ generates a uniform lattice Γ_n and we have a series of nilmanifolds

$$M_n = V_n/\Gamma_n, \ n = 3, 4, \dots, \qquad \dim M_n = n.$$

THEOREM 3 (Babenko–Taimanov). *The nilmanifolds M_{2n} have integer symplectic forms $\omega_{2n} \in H^2(M_{2n}; \mathbb{Z})$.*

We have $V_3 = \mathcal{H}$, $M_3 = \mathcal{H}/\mathcal{H}_Z$, and $M_3 \times S^1 = M_{\mathrm{KT}}$. The manifold M_4 admits no complex structure.

Interesting properties of these manifolds were observed by Buchstaber [**7**] who described them by a sequence $M_{n+1} \to M_n$ of principal S^1-bundles and showed that $H^*(\varprojlim M_n) = H^*(\mathcal{S})$ where $\mathcal{S}$ is the Landweber–Novikov algebra from the complex cobordism theory. These ideas were recently developed by Alaniya [**1**], Artel'nykh [**2**], and Millionshchikov [**20**].

Before speaking about simply connected symplectic manifolds we would like to mention the following theorem by Gompf [**13**] which solved the Kotschick problem.

THEOREM 4 (Gompf). *For any even $n \geq 4$ and any finitely-presented group G there is a compact symplectic n-manifold M such that $\pi_1(M) = G$ and this manifold does not satisfy the Hard Lefschetz Condition. Therefore it is not homotopically equivalent to any Kähler manifold.*

The proof of Gompf uses the fiber connected sums construction. The Hard Lefschetz Condition is valid for Kähler manifolds and says that for any k the multiplication by the k-th power of the cohomology class $[\omega]$ of the Kähler form defines an isomorphism $H^{n-k}(M; \mathbb{C}) \xrightarrow{\times[\omega]^k} H^{n+k}(M; \mathbb{C})$ where $\dim M = 2n$. Gompf had shown that for his manifolds this condition is not satisfied for any 2-dimensional cohomology class.

3. Simply connected symplectic manifolds with no Kähler structure. A symplectic blow-up was used by McDuff for constructing a simply connected compact symplectic manifold with no Kähler structure.

This method is as follows. Let X be a symplectic manifold of dimension $2n$ and Y be a $2k$-dimensional symplectic submanifold of X. There is a symplectic normal bundle ν to Y in X which is a vector bundle over Y with $2(n-k)$ dimensional fibers. On this bundle there is a fiberwise almost complex structure induced by a symplectic structure and thus we identify the fibers with $\mathbb{C}^{n-k}$. The blow up of

$\mathbb{C}^{n-k}$ at the origin results in replacing any nonzero vector ξ by a pair (ξ, l) where l is a line passing through the origin in the direction of ξ. Such a replacement is smoothly extended up to replacing $\mathbb{C}^{n-k}$ by the canonical line bundle E over $\mathbb{C}P^{n-k-1}$. There is a mapping

$$\pi : E \to \mathbb{C}$$

which maps the zero section of E to the origin $0 \in \mathbb{C}$ and maps diffeomorphically the complement of the zero section of E onto $\mathbb{C} \setminus \{0\}$.

Take a small normal neighborhood of Y in X, identify it with a neighborhood of the zero section of ν, and blow up the fibers of ν at the origins. The resulting manifold $\widetilde{X}$ is called the blow-up of X along Y. There is a natural mapping

$$\pi : \widetilde{X} \to X$$

which is a diffeomorphism outside $\pi^{-1}(Y)$. This construction was proposed by Gromov and explained in detail by McDuff [**19**].

THEOREM 5 (Gromov–McDuff). *Given a symplectic form ω on X, there is a symplectic form $\widetilde{\omega}$ on $\widetilde{X}$ such that outside a small neighborhood of $\pi^{-1}(Y)$ we have $\widetilde{\omega} = \pi^*\omega$.*

There is a symplectic embedding of M_{KT} into $(\mathbb{C}P^5, \omega_{\mathrm{FS}})$. It was used by McDuff for constructing the first example of a simply connected compact symplectic manifold with no Kähler structure [**19**]:

THEOREM 6 (McDuff). *The third Betti number of the symplectic blow-up $\widetilde{\mathbb{C}P}^5$ of $\mathbb{C}P^5$ along M_{KT} equals $b_1(M_{KT})$ (which is 3). Therefore $\widetilde{\mathbb{C}P}^5$ has no Kähler structure.*

Many exciting examples coming from the Seiberg–Witten theory are now known in the four-dimensional case however we do not discuss them here.

4. Minimal models and formality.

Minimal models of simply connected and nilpotent spaces were invented by Sullivan [**22**].

Let $\mathcal{M}$ be a free graded commutative algebra over $\mathbb{Q}$ with homogeneous generators $x_1, \dots$ such that $1 \leq \deg x_i \leq \deg x_j$ for $i \leq j$ and for any dimension $l \geq 1$ there are finitely many generators x_j of dimension l. Such an algebra with a differential d is called minimal if $dx_i \in \bigwedge(x_1, \dots, x_{i-1})$ for $i \geq 2$.

In the sequel we consider only connected differential graded (d.g.) algebras $\mathcal{A}$ which means that $\mathcal{A}^0 = \mathbb{Q}$ and the multiplication by the elements from $\mathcal{A}^0$ is identified with the multiplication by rationals.

It is said that a minimal algebra $\mathcal{M}$ is the minimal model for $\mathcal{A}$ if there is a homomorphism of d. g. algebras $f : \mathcal{M} \to \mathcal{A}$ inducing an isomorphism of cohomology: $f^* : H^*(\mathcal{M}) \xrightarrow{=} H^*(\mathcal{A})$. For the algebra of $\mathbb{Q}$-polynomial forms on X its minimal model $\mathcal{M}_X$ is called the minimal model for X.

THEOREM 7 (Sullivan). *Any compact simply connected polyhedron or nilmanifold has a minimal model which is unique up to isomorphism. The algebra $\mathcal{M}_X$ describes up to isomorphism the rational homotopy type of X and in particular we have*

$$\mathrm{Hom}\,(\pi_*(X), \mathbb{Q}) = \mathcal{M}_X / \mathcal{M}_X \wedge \mathcal{M}_X.$$

A minimal algebra $\mathcal{M}$ is called formal if there is a homomorphism of d.g. algebras $f : (\mathcal{M}, d) \to (H^*(\mathcal{M}), 0)$ inducing an isomorphism of cohomology. The space X is called formal if its minimal model is formal. This implies that $\mathcal{M}$ is the minimal model for the cohomology ring $H^*(\mathcal{M})$ with the zero differential.

A d.g. algebra $\mathcal{A}$ is called formal if its minimal model is formal. Notice that if $\mathcal{A}$ is one-connected d.g. algebra, then its minimal model is uniquely reconstructed from $\mathcal{A}$. Therefore the formality of a simply connected space means that its rational homotopy type is reconstructed from its rational cohomology ring. Examples of formal spaces are; global symmetric spaces, classifying spaces of Lie groups, and closed simply connected manifolds of dimension ≤ 6 (Neisendorfer–Miller).

A nilmanifold is formal if and only if it is a torus.

THEOREM 8 (Deligne–Griffiths–Morgan–Sullivan [**9**]). *Kähler manifolds are formal.*

5. On simply connected nonformal symplectic manifolds.

We see that symplectic nilmanifolds different from tori are nonformal. However Lupton and Oprea conjectured that on simply connected compact manifolds the existence of a symplectic structure implies formality [**17**] (see also [**25**]). They also found some additional conditions on the minimal model which together with a symplectic structure imply formality.

The Lupton–Oprea conjecture was disproved in [**3, 4**] by the following result.

THEOREM 9 (Babenko–Taimanov). *For $N \geq 5$ the symplectic blow up $\widetilde{\mathbb{C}P}^N$ of the complex projective space along an embedded Kodaira–Thurston manifold is a simply connected nonformal symplectic manifold.*

The proof is as follows. Nontrivial Massey products obstruct formality and it appeared that if the codimension of a symplectic submanifold $Y \subset X$ is rather large then nontrivial Massey products on Y induce nontrival Massey products on the symplectic blow-up of X along Y. Generically for such an inheritance of a triple product it needs to have $\operatorname{codim} Y \geq 8$. However in [**4**], the extended version of [**3**], we showed that for the special pairs $(\mathbb{C}P^N, M_{\mathrm{KT}})$ this estimate is improved to $\operatorname{codim} M_{\mathrm{KT}} \geq 6$ (that is $N \geq 5$).

6. The Massey products.

Let $\mathcal{A}$ be a differential graded module. That is a graded module $\mathcal{A} = \oplus_{k \geq 0} \mathcal{A}^k$ with the associative multiplication $\wedge$ and with a differential d meeting the conditions $d^2 = 0$ and

$$d(x \wedge y) = dx \wedge y + \bar{x} \wedge dy.$$

The conjugation on $\mathcal{A}$ is defined on homogeneous elements as $x \to \bar{x} = (-1)^p x$ for $x \in \mathcal{A}^p$ and is linearly extended onto the whole ring.

Denote by $M_n(\mathcal{A})$ the algebra of $(n \times n)$-matrices with coefficients from $\mathcal{A}$. For a matrix $A = (a_{ij}) \in M_n(\mathcal{A})$ we mean by $\bar{A}$ the matrix whose elements are conjugates of elements of A: $\bar{A} = (\bar{a}_{ij})$. We have the Leibniz rule:

$$d(A \wedge B) = dA \wedge B + \bar{A} \wedge dB.$$

For any matrix $A \in M_n(\mathcal{A})$ denote by $\operatorname{Ker} A$ the $\mathcal{A}$-submodule of $M_n(\mathcal{A})$ generated by all matrices $(1)_{ij}$ such that $(1)_{ij} A = A(1)_{ij} = 0$. Here $(1)_{ij}$ stands for the matrix such that all its entries but the entry on the intersection of the i-th line and the j-th column vanish and this nonzero entry is 1. In particular, we have $AB = BA = 0$ for any $B \in \operatorname{Ker} A$.

We say that a matrix $A \in M_n(\mathcal{A})$ satisfies the (generalized) Maurer–Cartan equation if

$$dA - \bar{A} \wedge A = 0 \mod \operatorname{Ker} A.$$

In this case the matrix $\mu(A) = dA - \bar{A} \wedge A$ is called the curvature matrix.

The following Bianchi identity holds for any matrix $A \in M_n(A)$:

$$d\mu(A) = \overline{\mu(A)} \wedge A - A \wedge \mu(A).$$

If $\mu(A) \in \operatorname{Ker} A$, then $d\mu(A) = 0$ and the elements of $\mu(A)$ define a (matrix) cohomology class with coefficients from $H^*(\mathcal{A})$ which we call a generalized Massey product $[\mu(A)]$ [**5**].

We introduce the Massey products (but not in the most general situation). We may assume that the elements of $\mathcal{A}$ are also matrices such that the associative multiplication $A \wedge B$ is correctly defined and the Leibniz rule holds.

Example. The ordinary Massey products. Let $\alpha_1, \dots, \alpha_n$ be homogeneous elements of the cohomology ring $H^*(\mathcal{A})$. Let these classes be represented by elements $a_1, \dots, a_n$. For $n \geq 3$ the n-tuple Massey product $\langle \alpha_1, \dots, \alpha_n \rangle$ is correctly defined if there is a matrix $A \in M_{n+1}(\mathcal{A})$ of the form

$$A = \begin{pmatrix} 0 & a_1 & * & \dots & * & * \\ 0 & 0 & a_2 & \dots & * & * \\ & & & \dots & & \\ 0 & 0 & 0 & \dots & 0 & a_n \\ 0 & 0 & 0 & \dots & 0 & 0 \end{pmatrix},$$

satisfying the Maurer–Cartan equation. This means that

$$dA - \bar{A} \wedge A = \begin{pmatrix} 0 & \dots & 0 & \tau \\ 0 & \dots & 0 & 0 \\ & \dots & & \\ 0 & \dots & 0 & 0 \end{pmatrix}.$$

To define the classical product $\langle \alpha_1, \dots, \alpha_n \rangle$ we have to consider all solutions $[\tau]$ to these equations for all initial data $a_1, \dots, a_n$. Therefore it is actually the set of cohomology classes $[\tau]$. For instance for $n = 3$ the triple product of $\alpha_1 \in H^p(\mathcal{A}), \alpha_2 \in H^q(\mathcal{A}), \alpha_3 \in H^r(\mathcal{A})$ is defined modulo $\alpha_1 \wedge H^{q+r-1}(\mathcal{A}) + \alpha_3 \wedge H^{p+q-1}(\mathcal{A})$. Moreover it is not defined for all tuples of cohomology classes $\alpha_1, \dots, \alpha_n$ because for some tuples the Maurer–Cartan equation may have no solutions. Hence

the classical Massey product $\langle \alpha_1, \dots, \alpha_n \rangle$ is the set of all generalized Massey products $[\mu(A)]$ for all matrices A with $[a_{i,i+1}] = \alpha_i$ and $a_{ij} = 0$ for $i \geq j$, $i, j = 1, \dots, (n+1)$.

Such a definition of the ordinary Massey products can be extracted from the paper by Kraines [**16**] and that was actually done by May [**18**] who used it for defining matrix Massey products. The matrix products are defined as ordinary products with replacing elements from $\mathcal{A}$ by multipliable matrices with coefficients from $\mathcal{A}$. Notice, that the given definition of Massey products is rather general. The classical (ordinary and matrix) Massey products correspond to nilpotent upper triangular matrices only.

A generalized Massey product $[\mu(A)]$ is called completely reducible if all coefficients of $[\mu(A)]$ belong to $H^+(\mathcal{A}) \wedge H^+(\mathcal{A})$. A classical product $\langle \alpha_1, \dots, \alpha_n \rangle$ is called trivial if it contains the zero.

THEOREM 10. *If a differential graded algebra $\mathcal{A}$ is formal then*
1) all generalized Massey products $[\mu(A)]$ are completely reducible;
2) all classical (ordinary and matrix) Massey products are trivial.

7. On Massey products in symplectic manifolds.

In [**5**] we developed the observation from [**3, 4**] and described some new classes of nonformal blow-ups of symplectic manifolds.

THEOREM 11 (Babenko–Taimanov). *a) Let a simply connected symplectic manifold X have an irreducible generalized Massey product of dimension k. Then for any symplectic submanifold $Y \subset X$ with* $\operatorname{codim} Y > k$ *the corresponding symplectic blow-up $\widetilde{X}$ also has an irreducible generalized Massey product of dimension k.*

b) Let a symplectic manifold Y have a nontrivial ordinary or matrix triple Massey product. Then for any symplectic embedding $Y \subset X$ of codimension greater thatn or equal 8 *the corresponding symplectic blow-up $\widetilde{X}$ also has a nontrivial triple matrix Massey product.*

c) Let a symplectic manifold Y have a strictly irreducible quadruple matrix Massey product $\langle S_1, S_2, S_3, S_4 \rangle$. Then for any symplectic embedding $Y \subset X$ such that $\operatorname{codim} Y > 2 \operatorname{sdeg} \langle S_1, S_2, S_3, S_4 \rangle$ *the corresponding symplectic blow-up $\widetilde{X}$ has a nontrivial quadruple matrix Massey product.*

Inheritance of triple ordinary Massey products observed in [**3**] was later studied by Rudyak and Tralle in the general situation. They proved statement b) of the theorem for ordinary triple Massey products [**21**].

To explain part c) we recall that a classical product $\langle \alpha_1, \dots, \alpha_n \rangle$ is called strictly irreducible if it does not contain a matrix with coefficients from $H^+(\mathcal{A}) \wedge H^+(\mathcal{A})$ and $\operatorname{sdeg} \alpha$ equal to the maximal degree of elements of a matrix cohomology class α.

Parts b) and c) do not generalized for n-tuple products with $n \geq 5$. Due to a high indeterminacy of higher products they contain many more elements than lower-order products. Typically the indeterminacy substantially grows with the order and reasonable conditions do not guarantee that the product would not contain the zero.

8. Some problems and conjectures.

Now it is clear that a symplectic structure does not strongly restrict topology of symplectic manifolds. The only known obstruction is still the existence of a two-class $[\omega] \in H^*(X)$ such that its powers $[\omega]^k$ are nontrivial for $2k \leq \dim X$.

In [**5**] we propose the following

CONJECTURE 1. *For any finite polyhedron P and any $N \geq 1$ there is a symplectic manifold X and an embedding $f : P \to X$ such that*

$$f_* : \pi_k(P) \to \pi_k(X) \quad \textit{is a monomorphism for } k \leq N.$$

For $N = 1$ and $\dim P = 2$ the conjecture follows from Gompf's results [**13**].

There is another interesting

PROBLEM 1. *Does there exist a symplectic simply connected closed manifold satisfying the Hard Lefschetz Condition and with no Kähler structure ? Or, more strongly, nonformal ?*

Another conjecture proposed by Benson and Gordon [**6**] is very close to the subject of our talk.

CONJECTURE 2. *If a compact quotient of a simply connected completely solvable Lie group by a lattice has a Kähler structure, then it is a complex torus.*

Recall that a solvable group G is called completely solvable if all eigenvalues of the operator ad : $\mathcal{G} \to \mathcal{G}$ on its Lie algebra are real.

During their research the authors were partially supported by the Russian Foundation for Basic Research (grants 99-01-01202a (I.K.B.), 00-01-00915 and 00-15-99252 (I.A.T.)) and the Switzerland National Foundation for Basic Research (I.K.B.)

References

[1] Alaniya, L. A. Cohomologies with local coefficients of some nilmanifolds. Russian Math. Surveys **54**:5 (1999), 1019–1020.
[2] Artel'nykh, I. V. Massey products and the Buchstaber spectral sequence. Russian Math. Surveys **55**:3 (2000).
[3] Babenko, I. K., and Taimanov, I. A. On existence of nonformal simply connected symplectic manifolds, Russian Math. Surveys **53**:5 (1998), 1082–1083; also: math.SG/9810065.
[4] Babenko, I. K., and Taimanov, I. A. On nonformal simply connected symplectic manifolds. Siberian Math. Journal **41**:2 (2000), 204–217; also: math.SG/9811167.
[5] Babenko, I. K., and Taimanov, I. A. Massey products in symplectic manifolds. Math. Sbornik **191**:8 (2000), 3–44 (in Russian); also: math.SG/9911132.
[6] Benson, C., and Gordon, C. Kähler and symplectic structures on nilmanifolds, Topology **27** (1988), 513–518.
[7] Buchstaber, V. M. Groups of polynomial transformations of a line, nonformal symplectic manifolds, and the Landweber-Novikov algebra. Russian Math. Surveys **54**:4 (1999), 837–838.
[8] Cordero, L. A., Fernandez, M., and Gray, A. Symplectic manifolds with no Kähler structure, Topology **25** (1986), 375–380.
[9] Deligne, P., Griffiths, P., Morgan, J., and Sullivan, D. Real homotopy theory of Kähler manifolds, Invent. Math. **19** (1975), 245–274.
[10] Donaldson, S., Symplectic submanifolds and almost-complex geometry, J. Diff. Geom. **44** (1996), 666–705.
[11] Fernandez, M., Gotay, M., and Gray, A. Compact parallelizable four dimensional symplectic and complex manifolds, Proc. Amer. Math. Soc. **103** (1988), 1209–1212.
[12] Fernandez, M., Gray A., and Morgan J.W. Compact symplectic manifolds with free circle actions, and Massey products, Michigan Math. J. **38** (1991), 271–283.
[13] Gompf, R. E. A new construction of symplectic manifolds, Ann. of Math. **142** (1995), 527–595.
[14] Gromov, M. L. A topological technique for the construction of solutions of differential equations and inequalities, Actes Congrés Intern. Math. (Nice, 1970), Gauthier-Villars, Paris, No. 2, 1971, 221–225.
[15] Guan, D., Examples of compact holomorphic symplectic manifolds which are not Kählerian. II, Invent. Math. **121** (1995), 135–145.
[16] Kraines, D., Massey higher products, Trans. of Amer. Math. Soc. **124** (1966), 431–449.
[17] Lupton, G., and Oprea, J. Symplectic manifolds and formality, J. Pure Appl. Algebra **91** (1994), 193–207.
[18] May, J. P., Matric Massey products, J. Algebra. **12** (1969), 533–568.
[19] McDuff, D. Examples of symplectic simply connected manifolds with no Kähler structure, J. Diff. Geom. **20** (1984), 267–277.
[20] Millionshchikov, D. V., Cohomology of nilmanifolds and Gontcharova's theorem. Russian Math. Surveys, to appear.
[21] Rudyak, Yu., and Tralle, A. On Thom spaces, Massey products, and non-formal symplectic manifolds. Internat. Math. Res. Notices **10** (2000), 495–513; also: math.DG/9907035.
[22] Sullivan, J. Infinitesimal computations in topology, Publ. IHES **47** (1978), 269–331.

[23] Thurston, W. Some simple examples of compact symplectic manifolds, Proc. Amer. Math. Soc. **55** (1976), 467–468.
[24] Tischler, D. Closed 2-forms and an embedding theorem for symplectic manifolds, J. Diff. Geom. **12** (1977), 229–235.
[25] Tralle, A., and Oprea, J. Symplectic Manifolds with no Kähler Structure, Lect. Notes in Math. **1661**, Springer, Berlin Heidelberg, 1997.

Département de Mathématiques,, Université Montpellier II,, Place Eugène Bataillon,, 34095 Montpellier Cedex 5, France. On leave of absence from, Department of Mathematics and Mechanics, of the Moscow State University
E-mail address: babenko@math.univ-montp2.fr

Institute of Mathematics,, 630090 Novosibirsk, Russia
E-mail address: taimanov@math.nsc.ru

Contemporary Mathematics
Volume **288**, 2001

Osculating Tubes and Self-Linking for Curves on the Three-Sphere

Thomas F. Banchoff

ABSTRACT. For a smooth closed curve, the osculating tube consists of the union of the osculating circles at points of the curve. For a closed curve on the three-sphere, the self-linking number of the image of the curve under stereographic projection from a point to the opposite hyperplane depends on the point, and the difference between the numbers for two points is related to the algebraic number of times that a curve from one point to the other meets the osculating tube of the curve transversely. The self-linking number of an (n, m)-torus knot on the three-sphere is shown to be nm, answering a question of N. Kuiper.

Introduction

Alfred Gray made many contributions to the geometry of tubes, quite often using computer graphics both for research and for illustrations of his books and papers. In this note, we will deal with a tube that he did not study extensively, connected with the osculating circles of a parametrized curve. For a curve in the plane, the geometry of the osculating tube gives an alternate proof of the well-known fact that the osculating circles of a spiral, with monotonic curvature, are nested, with no two intersecting (This result for plane curves is developed by Alfred Gray in [**6**], pages 111–112).

For a space curve, the osculating tube is connected with the behavior of self-linking numbers under conformal transformations, as studied earlier by the author and James White in [**1**]. In this article, we develop the geometry of this osculating tube for plane curves, then curves on the two-sphere, then curves in three-space, finally we concentrate on the osculating tubes of curves on the three-sphere in four-space. The main result relates the self-linking numbers of stereographic projections of a curve on the three-sphere from two points to the positions of these points relative to the osculating tube of the curve.

2000 *Mathematics Subject Classification.* 53A04, 53A30.

Key words and phrases. Self-Linking Numbers, Stereographic Projection, Osculating Circles, Osculating Tube.

This research was partially supported by the National Science Foundation grants NSF967077 and NSF9814593.

The author would like to thank Davide Cervone and his graphics program `CenterStage` for assistance in producing the illustrations for this article.

The problem of determining the self-linking numbers of torus knots on the 3-sphere was first raised by Nicolaas Kuiper in his paper on hyperbolic 4-manifolds [**8**], page 75, following the work on the same subject by M. Gromov, H. B. Lawson, and W. Thurston [**7**]. In 1991, Kuiper discovered that there was an error in his argument, and communicated that information to the author. The proof of the self-linking formula presented here is basically the one constructed by the author at that time but not published previously. The relationship with osculating tubes is new.

Definition of the Osculating Tube

Let $X(t)$ be a twice-differentiable curve in $\mathbf{R}^n$ defined over the interval $a \leq t \leq b$. Assume that the velocity vector $X'(t)$ is never zero, with length $s'(t)$. We define the *unit tangent vector* $T(t)$ to be the normalized velocity vector, so $X'(t) = s'(t)T(t)$.

Then $X''(t) = s''(t)T(t) + s'(t)T'(t)$ where $T'(t)$ is perpendicular to $T(t)$. If $T'(t)$ is not equal to zero, we define the *Principal Normal* $P(t)$ and the *curvature* $\kappa(t)$ by the condition $T'(t) = s'(t)\kappa(t)P(t)$. If $T'(t) = 0$, we define the curvature $\kappa(t)$ to be 0.

Consider now a curve $X(t)$ for which the curvature $\kappa(t)$ is never zero. In this case, we define the *radius of curvature* to be $r(t) = 1/\kappa(t)$. For each t, we define the *osculating circle* to be the circle in the plane of $T(t)$ and $P(t)$ with radius $r(t)$ centered at $X(t) + r(t)P(t)$. The collection of all of these osculating circles forms the *osculating tube*

$$O(t, v) = X(t) + r(t)P(t) + r(t)(\cos(v)T(t) + \sin(v)P(t)).$$

A straightforward computation shows that the partial derivative vectors $O_t(t, v)$ and $O_v(t, v)$ are linearly dependent when $\sin(v) = -1$ (where $O(t, v) = X(t)$) and also when $r'(t) = 0$ and $P'(t)$ is a multiple of $T(t)$.

Osculating Tubes for Plane Curves

In the case where $X(t)$ is a plane curve, $P'(t)$ will always be a multiple of $T(t)$ so the singularities of the osculating tube occur on the curve itself, where $\sin(v) = -1$ and along the entire osculating circle at a point where $r'(t) = 0$. Points with this last property are called *vertices* of the curve.

Corollary: If $r'(t) \neq 0$ over an interval, then the osculating tube is an immersion of the cylinder into the plane, and the osculating circles are therefore nested. A smaller osculating circle is completely contained in any larger one.

Figure 1 illustrates this phenomenon for an arc of an Archimedean spiral.

For a curve with a finite number of vertices such that the sign of $r'(t)$ changes for each one, the osculating tube can be expressed as the union of embedded cylinders bounded by osculating circles at vertices. Figure 2 shows one of the four pieces that fit together to make the osculating tube of a non-circular ellipse.

A limiting case of this phenomenon is given by the class of *piecewise circular* plane curves, called *PC Curves* in the papers of the author and Peter Giblin [**3**], [**4**]. Such a curve is given by a sequence of circular arcs, with the tangent line of each arc at its endpoint coinciding with the tangent line of the succeeding arc at its beginning point. The osculating circle at an interior point of an arc is the circle containing the arc, while the osculating circles at a *node* where two arcs meet are

FIGURE 1.

the circles tangent to the tangent line at the node lying between the circles of the two arcs at the node. In this case the osculating tube is a union of pieces, one for each node, consisting of the region between the two circles at the node. The author and Peter Giblin will treat PC space curves in a forthcoming article.

Osculating Tubes for Curves in Three-Space

For a smooth curve $X(t)$ in three-space, we may define the *Binormal* $B(t) = T(t) \times P(t)$. Then a standard calculation shows that $P'(t) = -s'(t)\kappa(t)T(t) + s'(t)\tau(t)B(t)$, where $\tau(t)$ is the *torsion* of the curve at $X(t)$.

The previous calculations then show that the osculating tube of a space curve is singular precisely along the curve itself and when simultaneously $r'(t) = 0$ and $\tau(t) = 0$. The latter condition will not occur generically for a single curve, even for a one-parameter family of curves.

Figure 3 below shows a portion of the osculating tube of a circular helix.

Osculating Tubes for Curves on the Two-Sphere

One particular class of interesting curves in three-space is the *spherical curves* $X(t)$ with $|X(t)| = 1$, on the two-sphere of radius 1 about the origin. It is a straightforward calculation to show that the osculating circles of a spherical curve lie on the same sphere. This can be seen as well by projecting the curve stereographically from a point N not on the curve $X(t)$ to a plane curve $\pi_N X(t)$. The osculating circles to this plane curve lie on the plane, and they are defined as the circles that have order of contact at least three with the curve. The images of these circles

FIGURE 2.

under inverse stereographic projection go to circles on the sphere with the same order of contact with the original curve $X(t)$.

The points of the spherical curve $X(t)$ that have torsion equal to zero are precisely the vertices of the plane curve $\pi_N X(t)$, where the osculating circle has contact higher than three. The osculating tube of $X(t)$ will then consist of a union of annular regions on the sphere bounded by osculating circles at torsion zeros of the curve, where each such region is the image under inverse stereographic projection of the analogous regions in the plane.

Note that if an osculating circle of $X(t)$ passes through the point N, then the image of that circle is a straight line with higher order of contact with the curve $\pi_N X(t)$, yielding an inflection point of the projected curve.

Self-Linking Numbers for Curves in Three-Space

For a space curve $X(t)$ with curvature never zero, the *self-linking number* $\mathrm{SL}(X)$ of $X(t)$ is defined to be the linking number of $X(t)$ with the *normal push-off* $X(t)+\epsilon P(t)$, where ϵ is chosen sufficiently small that the linking number is stable, not changing for any smaller positive value. This notion was first developed by Calgareanu [**5**] in 1959, and independently discovered and significantly extended in 1968 by William Pohl [**9**] and his student James White. In 1974, James White and the author in [**1**] established a relationship between the self-linking numbers of the inverted images $I_Q(X(t))$ of $X(t)$, where I_Q represents inversion with respect to a sphere centered at Q. Specifically, if $Q(t)$ is a path that meets the osculating tube of X transversely, then $\mathrm{SL}(I_Q(1)X) - \mathrm{SL}(I_Q(0)X)$ is the algebraic number of times

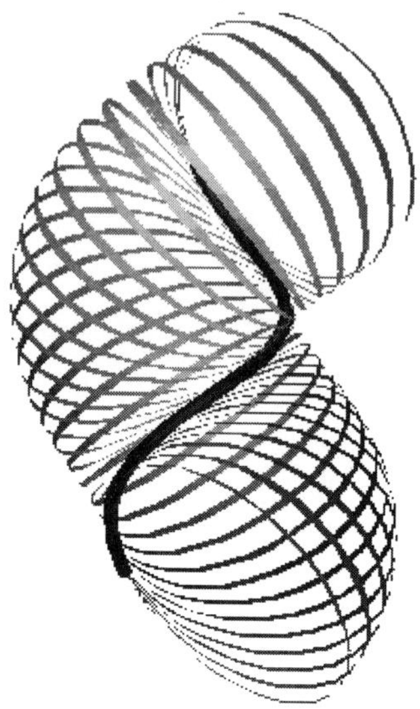

FIGURE 3.

that $Q(t)$ intersects the (oriented) osculating tube of X. In this article, we will use an analogous argument to investigate self-linking numbers of curves projected stereographically into three-space from various points on the three-sphere.

Self-Linking Numbers for Curves on the Three-Sphere

Let $X(t)$ be a curve on the unit three-sphere in four-space, so $|X(t)| = 1$. Then as before, if $T'(t) \neq 0$, we have $T'(t) = \kappa(t)P(t)$. We may resolve this curvature vector as a multiple of the outward unit normal vector $N(t)$ to the three-sphere, namely $X(t)$ itself, plus a vector tangent to the three-sphere. If $P(t)$ is not a multiple of $X(t)$, then we may define the *geodesic curvature* $\kappa_g(t)$, a non-negative quantity, the *normal curvature* $\kappa_N(t)$, and a unit vector $U(t)$ in the tangent hyperplane to the three-sphere by the expression

$$\kappa(t)P(t) = \kappa_g(t)U(t) + \kappa_N(t)N(t).$$

Note that as in the case of a curve on the unit sphere in three-space, the normal curvature of a curve on the four-sphere is -1. This follows since

$$X''(t) = s''(t)T(t) + s'(t)^2\kappa_g(t)U(t) + s'(t)^2\kappa_N(t)N(t)$$

so

$$s'(t)^2\kappa_N = X''(t)\cdot N(t) = X''(t) \cdot X(t).$$

But $X(t) \cdot X'(t) = 0$ yields $X(t) \cdot X''(t) + X'(t) \cdot X'(t) = 0$ so $s'(t)^2\kappa_N = -s'(t)^2$. Therefore

$$\kappa(t)P(t) = \kappa_g(t)U(t) - X(t).$$

The geodesic curvature will be zero when $P(t) = -X(t)$. In this case, the plane through $X(t)$ spanned by $T(t)$ and $P(t)$ will pass through the origin, cutting out a great circle on the three-sphere which is the osculating circle at $X(t)$.

If the geodesic curvature is non-zero for all t, then we have a well-defined normal vector $U(t)$ at each point of $X(t)$ and we can define a push-off curve on the three-sphere by normalizing the push-off curve $X(t) + \epsilon U(t)$.

As in the case of a pair of curves in three-space , two disjoint oriented curves on the three-sphere have a well-defined *linking number* given by the algebraic number of times that one of them intersects any oriented surface bounded by the other. We may compute this linking number by using stereographic projection from a point not on either curve to obtain a pair of curves in three-space with the same linking number as the curves on the sphere.

In analogy with the definition for a curve in three-space, we define the *self-linking number* $\mathrm{SL}(X)$ of a curve $X(t)$ on the three-sphere to be the linking number of $X(t)$ with the normalized push-off of $X(t)+\epsilon U(t)$ where ϵ is chosen small enough that the linking number will not change for smaller values. We can then compute $\mathrm{SL}(X)$ by stereographically projecting both curves into three-space and calculating the linking number there.

Our first observation is that the self-linking number of a curve $X(t)$ on the three-sphere is not the same as the self-linking number of the curve $\pi_Q X(t)$ for an arbitrary Q on the three-sphere not on $X(t)$. The easiest way to see this is to consider a family of examples.

Torus Knots on the Three-Sphere

A (m, n)-*Torus Knot* on the three-sphere is defined by

$$X(t) = (\cos(a)\cos(mt), \cos(a)\sin(mt), \sin(a)\cos(nt), \sin(a)\sin(nt))$$

where $-\pi \leq t \leq \pi$ and a is an angle between 0 and $\pi/4$. Then

$$X'(t) = (-m\cos(a)\sin(mt), m\cos(a)\cos(mt), \\ -n\sin(a)\sin(nt), n\sin(a)\cos(nt))$$

with $s'(t) = \sqrt{m^2\cos^2(a) + n^2\sin^2(a)}$ and

$$X''(t) = (-m^2\cos(a)\cos(mt), -m^2\cos(a)\sin(mt), \\ -n^2\sin(a)\cos(nt), -n^2\sin(a)\sin(nt)).$$

Then $|X''(t)| = \sqrt{m^4\cos^2(a) + n^4\sin^2(a)} = s'(t)^2\kappa(t)$ so the curvature

$$\kappa(t) = \frac{\sqrt{m^4\cos^2(a) + n^4\sin^2(a)}}{m^2\cos^2(a) + n^2\sin^2(a)}$$

is constant.

The *normal curvature* $\kappa_N(t)$ of a curve on a unit sphere is -1, so the only time that all of the curvature is in the normal direction is when $\kappa(t) = 1$, i.e. when $m^4\cos^2(a) + n^4\sin^2(a) = (m^2\cos^2(a) + n^2\sin^2(a))^2$.

Note that $X''(t)$ has no $T(t)$-component since $s''(t) = 0$. The geodesic curvature of $X(t)$ will be zero if $X''(t)$ lies in the same direction as $X(t)$, and this occurs exactly when $m^2 = n^2$. We wish to calculate the self-linking numbers of the curves for which the geodesic curvature is never zero, so from now on we will deal only with the case where $m^2 \neq n^2$.

Self-Linking for Torus Knots on the Three-Sphere

Since $X''(t)$ has constant length for an (m, n)-Torus Knot, we may obtain the push-off $Y(t, s)$ of $X(t)$in the direction of $P(t)$ by normalizing the vector $X(t) + sX''(t)$ for a small value of s. We thus obtain a curve

$$Y(t,s) = \tfrac{1}{c}((1 - sm^2)\cos(a)\cos(mt), (1 - sm^2)\cos(a)\sin(mt),$$
$$(1 - sn^2)\sin(a)\cos(nt), (1 - sn^2)\sin(a)\sin(nt)).$$

where $c = \sqrt{(1 - sm^2)^2\cos^2(a) + (1 - sn^2)\sin^2(a)}$.

Note that

$$Y(t,s) = (\cos(b(s))\cos(mt), \cos(b(s))\sin(mt),$$
$$\sin(b(s))\cos(nt), \sin(b(s))\sin(nt))$$

for some angle $b(s)$ which is small if s is small. The problem of computing the self-linking number of the original torus knot is then seen to be equivalent to finding the linking number of two curves $X(t)$ for two different values of a, namely a and $a + h$ for some small h.

To compute the linking numbers of these two curves, we may stereographically project the curves into three-space from the point $Q = (0, 0, 0, 1)$ to get $\pi_Q X(a, t)$ and $\pi_Q X(a + h, t)$. Figure 4 shows the stereographic projection of a (3,2)-torus knot (as well as the image of its pushoff in the normal direction, defined below). Figure 5 shows the analogous illustration for a (2,3)-torus knot.

Observe first of all that the projection $\pi_z(\pi_Q X(a,t))$ of $\pi_Q X(a,t)$ into the horizontal plane is a locally convex curve, and for such a curve, the self-linking number is the algebraic number of self-crossings, in this case m. However the linking number of $\pi_z(\pi_Q X(a,t))$ and $\pi_z(\pi_Q X(a + h, t))$ is m plus the algebraic number of values of t for which

$$\lim_{h \to 0} \pi_z(\pi_Q X(a + h, t) - \pi_Q X(a, t)) = 0.$$

The formula for stereographic projection from the point $Q = (0, 0, 0, 1)$ is

$$\pi_Q X(a,t) = \tfrac{1}{1 - \sin(a)\sin(nt)}(\cos(a)\cos(mt), \cos(a)\sin(mt), \sin(a)\cos(nt)).$$

The subsequent projection to the x-y-plane yields the curve

$$Y(a,t) = \pi_z \pi_Q X(a,t) = \tfrac{1}{1 - \sin(a)\sin(nt)}(\cos(a)\cos(mt), \cos(a)\sin(mt)),$$

a locally convex plane curve with $n(m - 1)$ crossings. This curve will intersect the curve $Y(a + h, t)$ twice in the neighborhood of each of these crossings, for a total of $2n(m - 1)$ crossings.

There will also be crossings when $Y(a + h, t) = Y(a, t)$, namely when

$$\frac{\cos(a)}{1 - \sin(a)\sin(nt)} = \frac{\cos(a + h)}{1 - \sin(a + h)\sin(nt)}$$

so

$$\cos(a)(1 - \sin(a + h)\sin(nt)) = \cos(a + h)(1 - \sin(a)\sin(nt)).$$

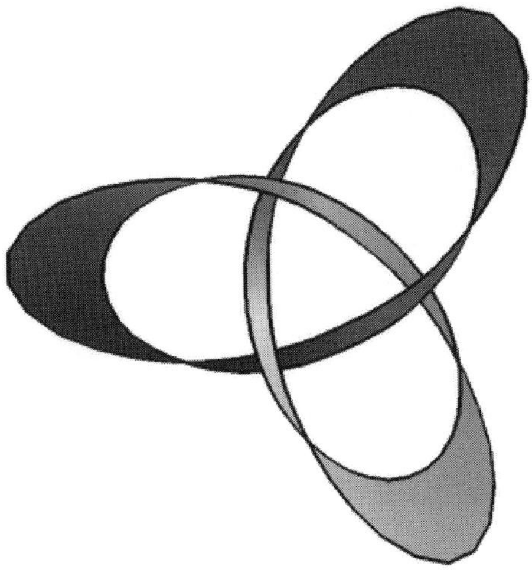

FIGURE 4.

This simplifies to

$$\begin{aligned}
&\cos(a+h) - \cos(a) \\
&\qquad = -\cos(a+h)\sin(a)\sin(nt) + \sin(a+h)\cos(a)\sin(nt) \\
&\qquad = \sin(a)\sin(nt).
\end{aligned}$$

As h goes to 0, the condition for the point of intersection is $0 = \sin(a)\sin(nt)$ and, for non-zero a, this gives $2n$ values for t between 0 and 2π. The total of crossings is then $2n(m-1) + 2n = 2nm$, and at exactly nm of them, the curve $Y(a+h,t)$ crosses the curve $Y(a,t)$, always in a positive direction. The self-linking number of X on the 3-sphere is therefore nm, completing the proof of the Theorem.

It follows that the self-linking number of an (m,n)-torus knot on the three-sphere is nm, as predicted by Kuiper, and this is the major result of this paper.

Self-Linking of a Torus Knot on the Three-Sphere and Self-Linking Numbers of Its Stereographic Projections

We have observed that stereographic projection of an (m,n)-torus knot from the point $(0,0,0,1)$ on the three-sphere, subsequently projected to a plane, gives a locally convex curve with $n(m-1)$ crossings, all of the same sign, leading to a curve in three-space with self-linking number $n(m-1)$. Similarly stereographic projection from $(1,0,0,0)$ on the three-sphere leads to a curve with self-linking number $m(n-1)$. The first observation is that the self-linking number of the

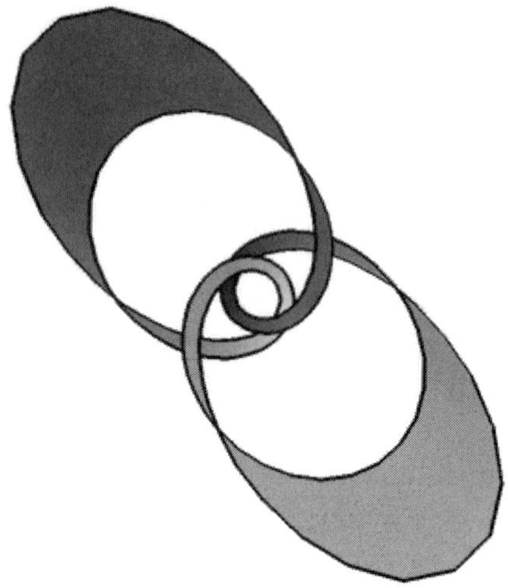

FIGURE 5.

projection depends on the North Pole of the projection, and neither projection has self-linking number equal to the actual self-linking number on the three-sphere, which is mn. The relationship between the point of projection and the self-linking number of the projected curve is related to the geometry of the osculating tube of the torus knot on the three-sphere.

The Osculating Tube for a Curve on the Three-Sphere

Just as the osculating tube of a spherical curve $X(t)$ is given as the image of the osculating tube of $\pi_N X(t)$ under inverse stereographic projection, we may obtain the osculating tubes of a curve $X(t)$ on the three-sphere as the inverse stereographic images of the osculating tube of $\pi_N X(t)$. Note that if N is not on any osculating circle of $X(t)$, then the projected curve $\pi_N X(t)$ will have curvature non-zero for all t.

If we take a path $Q(t)$ on the three-sphere from $Q(0) = (0,0,0,1)$ to $Q(1) = (1,0,0,0)$ then the self-linking number of the projected curve $\pi_Q(t)X(t)$ will change precisely when a point of this curve passes through the osculating tube of $X(t)$ on the three-sphere, and this is the second and final major result of this paper.

References

[1] Thomas F. Banchoff and James White, The Behavior of the Total Twist and Self-Linking Number of a Closed Space Curve under Inversions, *Mathematica Scandinavica* **36** (1975), 254–262.

[2] Thomas F. Banchoff, Self-Linking Numbers of Space Polygons, *Indiana University Mathematics Journal* **25** (1976), 1171–1188.
[3] Thomas F. Banchoff and Peter Giblin, Symmetry Sets of Piecewise Circular Curves, *Proc. Royal Soc. Edinburgh* **123A** (1993), 1–15.
[4] Thomas F. Banchoff and Peter Giblin, On the Geometry of Piecewise Circular Curves, *American Mathematical Monthly* **101** No. 5 (1994), 403–416.
[5] G. Calgareanu, L'Integrale de Gauss et l'analyse des noeuds tridimensionnels, *Revue de Math. Pure et Appl.* **4** (1959), 5–20.
[6] Alfred Gray, Modern Differential Geometry of Curves and Surfaces with Mathematica, *CRC Press*, (1998).
[7] M. Gromov, H. B. Lawson, and W. Thurston, Hyperbolic 4-Manifolds and Conformally Flat 3-Manifolds, *Pub. Math. IHES*, **68** (1988), 27–45.
[8] Nicolaas H. Kuiper, Hyperbolic 4-Manifolds and Tesselations, *Pub. Math. IHES*, **68** (1989), 47–76.
[9] William Pohl, The self-linking of closed space curves, *J. Math. Mech.* **17** (1968), 975–985.

Department of Mathematics, Brown University, Providence, Rhode Island 02912
E-mail address: `Thomas_Banchoff@brown.edu`

Contemporary Mathematics
Volume **288**, 2001

Isoparametric Functions and Harmonic and Minimal Unit Vector Fields

E. Boeckx and L. Vanhecke

Abstract. We prove that any isoparametric function on an Einstein manifold gives rise to a harmonic unit vector field and provide several explicit examples. Also, it is shown that these vector fields are minimal for a number of specific isoparametric functions on real, complex and quaternionic space forms. Furthermore, using the same methods, we construct examples of minimal and harmonic unit vector fields on generalized Heisenberg groups, which are not Einstein spaces.

Introduction

Let (M, g) be a Riemannian manifold and (T_1M, g_S) its unit tangent sphere bundle equipped with the Sasaki metric. A smooth unit vector field ξ on M can be thought of as a mapping between (M, g) and (T_1M, g_S), embedding M into its unit tangent sphere bundle T_1M. When M is compact and orientable, we can define the *energy* of ξ as the energy of the corresponding map, and the *volume* of ξ as the volume of the corresponding embedding. This gives rise to two functionals, energy and volume, on the set of all unit vector fields on M (which we suppose to be non-empty). In analogy with the notions of harmonic maps and minimal immersions, we say that a unit vector field is *harmonic* if it is critical for the energy functional, and *minimal* if it is critical for the volume functional. In the case when the manifold is non-compact or non-orientable, the critical conditions still make sense and the notions of harmonic and minimal unit vector fields can be extended to arbitrary Riemannian manifolds.

The study of the volume of unit vector fields started with work by H. Gluck and W. Ziller ([**20**]) who showed that Hopf vector fields on S^3 have absolute minimal volume. Even though this no longer holds true for higher-dimensional spheres S^{2n+1}, $n \geq 2$, the Hopf vector fields are still critical points for the volume functional ([**24**], [**32**]). In [**19**], it is proved that minimal unit vector fields correspond to minimal submanifolds of the unit tangent sphere bundle (T_1M, g_S).

2000 *Mathematics Subject Classification.* 53B20, 53B25, 53C12, 53C25, 53C40, 58E20.

Key words and phrases. harmonic and minimal unit vector fields, isoparametric functions, isoparametric foliations.

The first author is a Postdoctoral Researcher of the Fund for Scientific Research - Flanders (FWO-Vlaanderen).

The energy of a unit vector field ξ is equal, up to some constants, to the *total bending* of ξ. G. Wiegmink ([**42**]) derived critical point conditions for this functional and showed that harmonic unit vector fields need not correspond to harmonic maps from (M,g) to (T_1M, g_S), i.e., there may exist maps from (M,g) to (T_1M, g_S) which do not come from unit vector fields, but have smaller energy.

In a string of recent papers, large classes of minimal and harmonic unit vector fields have been found ([**8**]–[**10**], [**18**], [**19**], [**21**]–[**23**], [**37**]–[**39**]). In [**9**] and [**10**], we considered radial vector fields on normal neighbourhoods around points and tubular neighbourhoods about submanifolds. The critical conditions for energy and volume take an easy to handle form for such vector fields (see (1)–(3)). In this way, we constructed whole classes of new harmonic and minimal unit vector fields. In particular, we characterized harmonic spaces using harmonic radial vector fields around points (Theorem 1.1). In this paper, we put these results in a larger framework. Starting from an isoparametric function on an Einstein space, we find a very natural harmonic unit vector field (Theorem 2.1). As many examples of isoparametric functions are known (see the references in Section 2), this gives us a wealth of new examples of harmonic unit vector fields. Determining whether these are also minimal or whether they determine harmonic maps is more complicated. We do this for a number of specific isoparametric functions on different types of space forms and on Damek-Ricci spaces. In the final section, using the same formulas, we give examples of harmonic and minimal unit vector fields on generalized Heisenberg groups, which are *not* Einstein spaces. These vector fields occur in the context of transversely oriented Riemannian foliations, which is a further generalization of the framework with isoparametric functions.

1. Harmonic and minimal unit vector fields

In this paper, (M,g) denotes an n-dimensional smooth and connected Riemannian manifold with Levi Civita connection ∇ and Riemann curvature tensor R. We use the sign convention $R(X,Y) = [\nabla_X, \nabla_Y] - \nabla_{[X,Y]}$. Let ξ be a unit vector field on M. We can interpret ξ as the immersion $\xi\colon M \to T_1M\colon x \mapsto \xi_x$ of M into its unit tangent sphere bundle T_1M. The pull-back of the Sasaki metric g_S on T_1M is given by

$$(\xi^* g_S)(X,Y) = g(X,Y) + g(\nabla_X\xi, \nabla_Y\xi).$$

It follows that ξ is an isometric embedding if and only if ξ is parallel.

Now we define operators A_ξ and L_ξ as

$$A_\xi X = -\nabla_X \xi, \qquad L_\xi X = X + {}^tA_\xi(A_\xi X)$$

for $X \in TM$. Then, for a compact orientable manifold M, the energy $E(\xi)$ and the volume $\mathrm{Vol}(\xi)$ of ξ are given by

$$\begin{aligned} E(\xi) &= \frac{1}{2}\int_M \mathrm{tr} L_\xi \, \mu_g = \frac{n}{2}\mathrm{Vol}(M) + \frac{1}{2}\int_M |A_\xi|^2 \, \mu_g, \\ \mathrm{Vol}(\xi) &= \int_M \sqrt{\det L_\xi}\, \mu_g \end{aligned}$$

where μ_g is the volume form on (M,g). Note that the energy is equal, up to constants, to the quantity $\int_M |\nabla\xi|^2 \, \mu_g$, which is the *total bending* of ξ ([**42**]). The critical point conditions for the functionals E and Vol on the set of all unit vector fields have been derived in [**42**] and [**19**], respectively. (See [**17**] for a unified treatment.) As we do not need these conditions in their most general form, we do not write them down here. We do point out though, that these conditions make sense also in the case when the manifold is not compact or not orientable. This allows to define *harmonic* and *minimal* unit vector fields on any Riemannian manifold as those unit vector fields satisfying the critical condition for energy and volume, respectively.

Now we consider a unit vector field on (M, g) satisfying two additional assumptions:

- ξ is geodesic,
- the distribution $\xi^\perp$ is integrable.

In this situation, the operator $A_\xi = -\nabla\xi$ at a point $p \in M$ corresponds to the shape operator for the integral manifold through p of $\xi^\perp$ and vanishes on ξ. In particular, A_ξ is symmetric. As a consequence, the critical conditions for the energy and volume functional take an easier form. This was derived in [**9**]. In order to state them, we denote with λ_i, $i = 1, \dots, n-1$, the eigenvalue functions of A_ξ on $\xi^\perp$ and by $E_1, \dots, E_{n-1}$ a local orthonormal frame of $\xi^\perp$ consisting of corresponding eigenvectors. Then we have:

- ξ is a harmonic unit vector field if and only if

$$X(h) = \rho(X, \xi) \text{ for all } X \perp \xi \tag{1}$$

  where ρ is the Ricci tensor of (M, g) and h is the mean curvature function of the foliation determined by $\xi^\perp$;
- ξ determines a harmonic map if and only if it is a harmonic unit vector field and, additionally,

$$\sum_{i=1}^{n-1} \lambda_i R(\xi, E_i)E_i = 0; \tag{2}$$

- ξ is a minimal unit vector field if and only if for $j = 1, \dots, n-1$ it holds

$$\sum_{i=1}^{n-1} \frac{1}{1+\lambda_i^2}\big(E_j(\lambda_i) - (1 - \lambda_i\lambda_j)g(R(\xi, E_i)E_i, E_j)\big) = 0. \tag{3}$$

In [**9**] and [**10**], we applied these formulas to the case of radial vector fields around points and submanifolds to obtain the following results.

THEOREM 1.1. *A Riemannian manifold is harmonic if and only if all radial vector fields on pointed normal neighbourhoods of arbitrary points are harmonic vector fields. They determine harmonic maps only if the manifold is flat.*

PROPOSITION 1.2. *In a two-point homogeneous space, the radial vector fields on pointed normal neighbourhoods of arbitrary points are minimal vector fields.*

PROPOSITION 1.3. *In a two-point homogeneous space, the radial vector fields on tubular neighbourhoods of invariant totally geodesic submanifolds are both minimal and harmonic. They determine harmonic maps only if the manifold is flat.*

Recall that a submanifold P in a two-point homogeneous space is *invariant* if $J(TP) \subset TP$ where J is the complex structure, or belongs to the quaternionic Kähler or to the Cayley structure, depending on the space we work in. For real space forms, invariance is no additional condition. A final result concerns anti-invariant submanifolds P, i.e., $J(TP) \subset (TP)^\perp$.

PROPOSITION 1.4. *Let (M,g) be a Kähler manifold with constant holomorphic sectional curvature of real dimension $2m$ and P an anti-invariant totally geodesic submanifold of dimension m. Then the radial vector field in a tubular neighbourhood of P is both minimal and harmonic. It determines a harmonic map only if the manifold is flat.*

In the following sections, we aim to extend these results to a larger framework, and to come up with a wealth of new examples of harmonic and minimal unit vector fields.

REMARK 1. In the paper [**39**], K. Tsukada and the second author give examples of radial vector fields on tubular neighbourhoods of special real hypersurfaces of $G_2(\mathbb{C}^{m+2})$ and its dual space which are minimal and harmonic. These spaces are rank two symmetric spaces with a very rich geometrical structure. See [**39**] for further references.

2. Isoparametric functions

A smooth non-constant function $f\colon M \to \mathbb{R}$ defined on a Riemannian manifold (M,g) is called *isoparametric* if

$$|df|^2 = b(f), \qquad \Delta f = a(f) \tag{4}$$

for some smooth function b and some continuous function a. Regular level sets $f^{-1}(c)$ for $c \in \mathbb{R}$ are called *isoparametric hypersurfaces.* In [**41**], it is shown that f has at most two critical values, that the regular level hypersurfaces are parallel and that they are tubes about the singular level sets (i.e., the *focal varieties*). Moreover, the integral curves of $\nabla f/|\nabla f|$ are geodesics. The second condition of (4) means that the associated isoparametric hypersurfaces have constant mean curvature. From (1) we then conclude

THEOREM 2.1. *Let (M,g) be an Einstein space and f an isoparametric function on M. On $U = \{x \in M \mid (df)_x \neq 0\}$, the unit vector field $\xi = \nabla f/|\nabla f|$ is a harmonic vector field.*

As isoparametric functions have attracted a lot of attention, this theorem gives quite some new examples of harmonic vector fields. We refer to [**5**] for an overview and for the relation with homogeneous hypersurfaces and hypersurfaces with constant principal curvatures. (See also further in this section.) Unfortunately, in order to determine whether $\xi = \nabla f/|\nabla f|$ determines a harmonic map or is a minimal unit vector field, we need more information about the principal curvatures of the isoparametric hypersurfaces and about the curvature tensor of (M,g). In the rest of this section, we will look at some of the explicit examples of isoparametric functions in the literature.

2.1. Real space forms. In a real space form, a hypersurface is isoparametric if and only if it has constant principal curvatures. The classification of isoparametric functions and hypersurfaces in $\mathbb{R}^n$ was given by T. Levi Civita ([**28**]) for $n = 3$ and by B. Segre ([**33**]) in the general case.

THEOREM 2.2. *An isoparametric hypersurface in $\mathbb{R}^n$, $n \geq 3$, is either*

- *an affine hyperplane in $\mathbb{R}^n$, or*
- *a tube around a k-dimensional totally geodesic submanifold of $\mathbb{R}^n$ for $0 \leq k \leq n-2$.*

Note that these are precisely the *homogeneous hypersurfaces* of $\mathbb{R}^n$, i.e., the regular orbits of a closed cohomogeneity one subgroup of $I(\mathbb{R}^n)$.

A classification of isoparametric hypersurfaces in the real hyperbolic space $\mathbb{R}H^n$ was realized by E. Cartan ([**11**]). Again, we only have homogeneous ones.

THEOREM 2.3. *An isoparametric hypersurface in $\mathbb{R}H^n$, $n \geq 3$, is either*

- *a totally geodesic $\mathbb{R}H^{n-1}$ or an equidistant hypersurface to it, or*
- *a tube around a k-dimensional totally geodesic submanifold of $\mathbb{R}H^n$ for some $0 \leq k \leq n-2$, or*
- *a horosphere.*

In contrast, no classification for isoparametric hypersurfaces of S^n is known despite the efforts of several decades of mathematicians, starting with E. Cartan ([**11**]–[**13**]) up to H.-F. Münzner ([**30**]) who turned the problem into an algebraic one. Here, an isoparametric hypersurface need no longer be homogeneous, as shown by the examples by H. Ozeki and M. Takeuchi ([**31**]) and by D. Ferus, M. Karcher and H.-F. Münzner ([**16**]).

Still, we have the following general result concerning isoparametric functions in spaces of constant curvature.

THEOREM 2.4. *Let (M, g) be a space of constant curvature and f an isoparametric function on M. Then $\xi = \nabla f/|\nabla f|$ is a minimal unit vector field. It determines a harmonic map only when (M, g) is flat.*

PROOF. The curvature tensor R of a space of constant curvature c is given explicitly by $R(X,Y)Z = c(g(Y,Z)X - g(X,Z)Y)$. In particular, $R(\xi, X)X = c|X|^2\xi$ for X orthogonal to ξ. Moreover, as we mentioned before, isoparametric hypersurfaces of real space forms have constant principal curvatures. So, condition (3) for minimality is easily seen to be satisfied.

The harmonicity condition (2) reads in this case:

$$0 = \sum_{i=1}^{n-1} c\lambda_i \, \xi = ch \, \xi.$$

Hence, if the space is non-flat, *all* the isoparametric hypersurfaces in the isoparametric family must be minimal. It follows easily from the well-known formulas for the mean curvature of parallel hypersurfaces (see, e.g., [**2**]) that this is never the case. □

If we compare the above results with Theorem 1.1, Proposition 1.2 and Proposition 1.3, we see that we have obtained many new examples of minimal and harmonic unit vector fields on (an open part of) S^n, as well as a new minimal and harmonic unit vector field on $\mathbb{R}H^n$, orthogonal to the horosphere foliation.

2.2. Complex space forms. Now we consider complex space forms of non-zero constant holomorphic sectional curvature. A major difference with the case of real space forms is that isoparametric hypersurfaces in the complex setting do not necessarily have constant principal curvatures. Q.-M. Wang ([**40**]) has given examples of certain non-homogeneous isoparametric hypersurfaces in spheres which project to isoparametric hypersurfaces in complex projective spaces with non-constant principal curvatures. (Further examples can be found in [**43**].) Later, M. Kimura ([**26**]) proved that an isoparametric hypersurface in complex projective space has constant principal curvatures if and only if it is homogeneous. Of course, because of Theorem 2.1, also the non-homogeneous examples give rise to harmonic unit vector fields, but checking whether these are also minimal or determine harmonic maps will be more difficult in view of the formulas (2) and (3).

For this reason, we consider only homogeneous hypersurfaces here. We further suppose that they are also *curvature-adapted.* This means that the shape operator and the normal Jacobi operator of these hypersurfaces commute. In a complex space form, this property is equivalent to saying that JN is a principal curvature vector field for the hypersurface, where N is a local unit normal vector field and J is the complex structure. Such hypersurfaces are also known as *Hopf hypersurfaces.* For the complex hyperbolic space, we have the following classification by J. Berndt ([**3**]).

THEOREM 2.5. *A curvature-adapted homogeneous hypersurface in* $\mathbb{C}H^n$, $n \geq 2$, *is either*

- *a tube around a* $(2k)$*-dimensional invariant totally geodesic submanifold of* $\mathbb{C}H^n$ *for some* $0 \leq k \leq n-1$, *or*
- *a tube around an* n*-dimensional anti-invariant totally geodesic submanifold of* $\mathbb{C}H^n$, *or*
- *a horosphere.*

For the complex projective space, R. Takagi ([**34**]) has given the following classification.

THEOREM 2.6. *A homogeneous hypersurface in* $\mathbb{C}P^n$, $n \geq 2$, *is either*

- *a tube around a* $(2k)$*-dimensional invariant totally geodesic submanifold of* $\mathbb{C}P^n$ *for some* $0 \leq k \leq n-1$, *or*
- *a tube around an* n*-dimensional anti-invariant totally geodesic submanifold of* $\mathbb{C}P^n$, *or*
- *a tube around the Segre embedding of* $\mathbb{C}P^1 \times \mathbb{C}P^k$ *into* $\mathbb{C}P^{2k+1}$, *or*
- *a tube around the Plücker embedding of the complex Grassmann manifold* $G_2(\mathbb{C}^2)$ *into* $\mathbb{C}P^9$, *or*
- *a tube around the half spin embedding of the Hermitian symmetric space* $SO(10)/U(5)$ *into* $\mathbb{C}P^{15}$.

All of these are curvature-adapted.

REMARK 2. We see that curvature-adaptedness for hypersurfaces in complex projective space is a consequence of homogeneity. This does not hold in complex hyperbolic space. M. Lohnherr and H. Reckziegel ([**29**]) have recently given an example of a ruled hypersurface in $\mathbb{C}H^n$ which is homogeneous but not curvature-adapted. See also [**5**], [**6**] where similar examples are provided for $\mathbb{H}H^n$.

All of the above homogeneous hypersurfaces are level hypersurfaces for some isoparameteric function defining a Riemannian foliation. We now prove

THEOREM 2.7. *The unit normal vector field* ξ *to any of the foliations provided by the homogeneous hypersurfaces in the two theorems above is a harmonic and minimal vector field, but never determines a harmonic map.*

PROOF. The curvature tensor R of a Kähler space of constant holomorphic sectional curvature is given explicitly by

$$R(X,Y)Z = \frac{c}{4}\big(g(Y,Z)X - g(X,Z)Y - g(JX,Z)JY + g(JY,Z)JX - 2g(JX,Y)JZ\big).$$

In particular, $R(\xi,X)X = \frac{c}{4}|X|^2\xi + \frac{3c}{4}g(JX,\xi)JX$ for X orthogonal to ξ.

The unit normal vector fields are surely harmonic by Theorem 2.1. The additional condition (2) to determine a harmonic map can be rewritten, using the explicit expression for the curvature, as

$$0 = \sum \lambda_i R(\xi, E_i)E_i = \frac{c}{4}\, h\,\xi - \frac{3c}{4} JA_\xi J\xi,$$

or equivalently, $A_\xi J\xi = -(h/3)\,J\xi$. This means that all integral manifolds of $\xi^\perp$ must be Hopf hypersurfaces and that the corresponding principal curvature function is given by $-h/3$. Now we refer to [**2**, p. 67–69], where the principal curvatures and their multiplicities for the hypersurfaces in Theorem 2.5 and Theorem 2.6 are listed. An easy calculation shows that the principal curvature function corresponding to $J\xi$ is never equal to $-h/3$. Hence, ξ never determines a harmonic map.

For minimality, we note first that all principal curvatures are constant along the integral manifolds of $\xi^\perp$, because we restrict to homogeneous isoparametric hypersurfaces. So, we have to check whether

$$\begin{aligned} 0 &= \sum \frac{1-\lambda_i\lambda_j}{1+\lambda_i^2}\, g(R(\xi,E_i)E_i, E_j) \\ &= \frac{3c}{4}\sum \frac{1-\lambda_i\lambda_j}{1+\lambda_i^2}\, g(J\xi,E_i)\, g(JE_j,E_i). \end{aligned}$$

Here the E_i are the principal curvature vectors for the integral manifolds of $\xi^\perp$. As we consider only curvature-adapted hypersurfaces, $g(J\xi, E_i)$ vanishes unless $E_i = J\xi$. The above sum then reduces to the single term

$$\frac{3c}{4}\,\frac{1-\lambda_\xi\lambda_j}{1+\lambda_\xi^2}\, g(JE_j, J\xi)$$

which vanishes as $g(JE_j, J\xi) = g(E_j,\xi) = 0$. Hence, ξ is minimal. □

2.3. Quaternionic space forms. As in the case of complex space forms, we restrict our attention to isoparametric functions for which the associated foliation consists of homogeneous hypersurfaces which are moreover curvature-adapted. This last condition now means that the three-dimensional distribution along the hypersurface spanned by all vectors JN, where N is again a local unit normal vector field and J belongs to the quaternionic structure, is invariant under the shape operator. J. Berndt ([**4**]) and J.E. D'Atri ([**15**]) have given the following classifications for quaternionic hyperbolic space and quaternionic projective space, respectively.

THEOREM 2.8. *A curvature-adapted, homogeneous hypersurface in* $\mathbb{H}H^n$, $n \geq 2$, *is either*

- *a tube around a* $(4k)$*-dimensional invariant totally geodesic submanifold of* $\mathbb{H}H^n$ *for some* $0 \leq k \leq n-1$, *or*
- *a tube around some totally geodesic* $\mathbb{C}H^n$ *in* $\mathbb{H}H^n$, *or*
- *a horosphere.*

THEOREM 2.9. *A homogeneous hypersurface in* $\mathbb{H}P^n$, $n \geq 2$, *is either*

- *a tube around a* $(4k)$*-dimensional invariant totally geodesic submanifold of* $\mathbb{H}P^n$ *for some* $0 \leq k \leq n-1$, *or*
- *a tube around some totally geodesic* $\mathbb{C}P^n$ *in* $\mathbb{H}H^n$.

All of these are curvature-adapted.

We have in this case

THEOREM 2.10. *The unit normal vector field* ξ *to any of the foliations provided by the homogeneous hypersurfaces in the two theorems above is a harmonic and minimal vector field, but never determines a harmonic map.*

PROOF. The proof is completely along the same lines as that of Theorem 2.7. Now we use the explicit expression for the curvature tensor of a quaternionic space form and the formulas for the principal curvatures of the hypersurfaces in Theorem 2.8 and Theorem 2.9 as given in [**4**]. $\square$

REMARK 3. For most isoparametric functions, the question whether the corresponding unit vector field is minimal remains open. In particular, this is the case for those isoparametric functions with non-homogeneous isoparametric hypersurfaces (except in the case of S^n). As we said before, the condition (3) is more complicated in this case. For examples of such hypersurfaces, see the articles [**40**] and [**43**] mentioned earlier and the recent article [**44**] by L. Xiao and the references therein. It would be interesting to find examples where the vector field is not minimal, supposing they exist.

REMARK 4. A particular case of harmonic unit vector fields arising from isoparametric functions is given by the unit normal vector field to the regular orbits of a cohomogeneity one action on an Einstein space. For irreducible symmetric spaces, a lot of work has been done on the classification of such actions. We refer once more to the excellent survey [**6**] and in particular to the work of A. Kollross ([**27**]) who completely describes the situation for irreducible simply connected Riemannian symmetric spaces of compact type. It is worthwhile to note that the construction of cohomogeneity one Einstein metrics has received quite some attention. We refer to [**1**] and [**14**] for some results and for further references on this topic. This provides us with many more new examples of harmonic unit vector fields.

REMARK 5. The hyperbolic complex and quaternionic space forms are special examples of *Damek-Ricci spaces.* We refer to [**7**] for more details about this interesting class of Einstein spaces. The result about the unit vector fields orthogonal to the horosphere foliations in Theorem 2.7 and Theorem 2.10 also holds in an *arbitrary* Damek-Ricci space for the horosphere foliation containing the defining generalized Heisenberg group as a horosphere ([**5**], [**7**]). The harmonicity of the normal unit vector field has already been found in [**23**] where it is also shown that this field does not determine a harmonic map. The minimality has been proved in [**21**]. This horosphere foliation provides an example for the situation described in Remark 4 but, fortunately, the knowledge of the Riemannian curvature tensor of a Damek-Ricci space allows to decide about the conditions (2) and (3) as well.

3. Isoparametric Riemannian foliations

Another situation in which the formula (1) is useful is that of transversely oriented Riemannian foliations of codimension one. First, the transversal orientation makes it possible to choose a global unit normal vector field ξ to the foliation. Second, because the foliation is Riemannian, i.e., the leaves are locally equidistant submanifolds, this vector field ξ is geodesic. So, we are precisely in a situation where the formulas (1)–(3) for the minimality and the harmonicity of ξ hold.

In what follows, we consider an example of an *isoparametric* Riemannian foliation of codimension one, i.e., all leaves have constant mean curvature. Note that in such a case the harmonicity condition (1) simply says that the normal unit vector field must be an eigenvector field of the Ricci tensor. Our example is a left-invariant foliation on a generalized Heisenberg group introduced in [**25**]. We recall some basic facts about this class of spaces and refer to [**7**] for more details.

Let $\mathfrak{v}$ and $\mathfrak{z}$ be real vector spaces of dimensions n and m, respectively. On the direct sum $\mathfrak{n} = \mathfrak{v} \oplus \mathfrak{z}$ we consider an inner product $\langle\, , \rangle$ such that $\mathfrak{v}$ and $\mathfrak{z}$ are orthogonal. Suppose also that a linear mapping $J\colon \mathfrak{z} \to \mathrm{End}(\mathfrak{v})\colon Z \mapsto J_Z$ exists which satisfies

$$\langle J_Z U, V\rangle + \langle U, J_Z V\rangle = 0, \qquad {J_Z}^2 = -\langle Z, Z\rangle\, \mathrm{id}_{\mathfrak{v}}$$

for all $U, V \in \mathfrak{v}$ and $Z \in \mathfrak{z}$. Define a Lie algebra structure on $\mathfrak{n}$ by $[U+X, V+Y] = [U, V]$ where $[U, V] \in \mathfrak{z}$ is completely determined by $\langle [U, V], Z\rangle = \langle J_Z U, V\rangle$. With this Lie bracket, $\mathfrak{n}$ becomes a two-step nilpotent Lie algebra with center $\mathfrak{z}$. $\mathfrak{n}$ is called a *generalized Heisenberg algebra* and the associated connected and simply connected Lie group N equipped with the left-invariant metric g induced by $\langle\, , \rangle$ is a *generalized Heisenberg group.*

In the sequel, U, V, W (resp. X, Y, Z) always denote vectors (or left-invariant vector fields) in $\mathfrak{v}$ (respectively $\mathfrak{z}$). The Levi Civita connection is given by

$$\nabla_{V+Y}(U+X) = -\frac{1}{2}\, J_X V - \frac{1}{2}\, J_Y U - \frac{1}{2}\, [U, V]. \tag{5}$$

Now we fix a unit left-invariant vector field $U \in \mathfrak{v}$. Then, from (5), we see that U is geodesic and also integrable, as $[U^\perp, U^\perp] \subset \mathfrak{z} \subset U^\perp$. So the formulas (1)–(3) apply.

Using U, we can decompose $\mathfrak{v}$ as follows ([**7**])

$$\mathfrak{v} = \ker \mathrm{ad}_{U|\mathfrak{v}} \oplus J_{\mathfrak{z}} U,$$

and we calculate the shape operator A for the leaves of $U^\perp$. From (5), we get

$$\begin{aligned} A_U V &= -\nabla_V U = \frac{1}{2}\, [U, V] = 0, && V \in \ker \mathrm{ad}_U \cap \mathfrak{v}, \\ A_U(J_X U) &= \frac{1}{2}\, [U, J_X U] = \frac{1}{2}\, X, && X \in \mathfrak{z}, \\ A_U X &= \frac{1}{2} J_X U, && X \in \mathfrak{z}. \end{aligned} \tag{6}$$

Hence A_U has constant eigenvalues 0, 1/2 and $-1/2$ with corresponding eigenspaces $\ker \mathrm{ad}_U \cap \mathfrak{v}$, $\{X_+ = X + J_X U \mid X \in \mathfrak{z}\}$ and $\{X_- = X - J_X U \mid X \in \mathfrak{z}\}$. We can now prove easily

PROPOSITION 3.1. *Every unit vector field $U \in \mathfrak{v}$ is harmonic and minimal and determines a harmonic map.*

PROOF. From (6) is follows at once that the mean curvature function h is identically zero. Moreover, the Ricci tensor of a generalized Heisenberg group is given by ([**7**])

$$\rho(V+Y,W+Z) = -\frac{m}{2}\langle V,W\rangle + \frac{n}{4}\langle Y,Z\rangle.$$

So, N is not an Einstein space, but U is an eigenvector of the Ricci operator. Hence, condition (1) is fulfilled and U is a harmonic unit vector field.

Next, we calculate $R(U,E)E$ for eigenvectors E of A_U using the formulas for the curvature tensor of a generalized Heisenberg group ([**7**]) to obtain

$$\begin{aligned} R(U,V)V &= 0, && V\in \ker \mathrm{ad}_U \cap \mathfrak{v}, \\ R(U,X_+)X_+ &= -\frac{1}{2}|X|^2U, && X\in\mathfrak{z}, \\ R(U,X_-)X_- &= -\frac{1}{2}|X|^2U, && X\in\mathfrak{z}. \end{aligned}$$

We see that the two conditions (2) and (3) are satisfied. So, U is also minimal and determines a harmonic map. □

REMARK 6. Examples of minimal and harmonic vector fields on generalized Heisenberg groups have been found before in [**21**], [**23**] and [**37**] using different techniques. Proposition 3.1 agrees with the results obtained there and complements them. We also note that the only left-invariant unit vector fields on generalized Heisenberg groups which are geodesic and have integrable orthogonal distribution are precisely those in $\mathfrak{v}$.

REMARK 7. Even though the example of a Riemannian isoparametric foliation given here is actually a homogeneous one (its leaves are the orbits under the action by H_U, the connected, simply connected subgroup of N having $U^\perp$ as its Lie algebra), it is worthwhile as it uses the formula (1) on a *non-Einstein* space. Additionally, because of the explicit expression for the curvature tensor, we can decide about both minimality and the fact that it determines a harmonic map.

We also note that *harmonic* (i.e., with minimal leaves) and *totally geodesic* Riemannian foliations of codimension one are particular examples of isoparametric Riemannian foliations. We refer to [**35**] and [**36**] and the references therein for more information, existence results and examples.

References

[1] L. Bérard Bergery, *Sur des nouvelles variétés d'Einstein*, Publications de l'Institut Elie Cartan, Nancy, 1982.

[2] J. Berndt, *Über Untermannigfaltigkeiten von komplexen Raumformen,* Thesis, University of Cologne, 1989.

[3] J. Berndt, *Real hypersurfaces with constant principal curvatures in complex hyperbolic space*, J. Reine Angew. Math. **395** (1989), 132–141.

[4] J. Berndt, *Real hypersurfaces in quaternionic space forms*, J. Reine Angew. Math. **419** (1991), 9–26.

[5] J. Berndt, *Homogeneous hypersurfaces in hyperbolic spaces*, Math. Z. **229** (1998), 589–600.

[6] J. Berndt, *On homogeneous hypersurfaces in Riemannian symmetric spaces*, Proc. Second Internat. Workshop Diff. Geom. **2** (1998), 17–34.

[7] J. Berndt, F. Tricerri and L. Vanhecke, *Generalized Heisenberg groups and Damek-Ricci harmonic spaces,* Lecture Notes in Math. 1598, Springer-Verlag, Berlin, Heidelberg, New York, 1995.

[8] E. Boeckx and L. Vanhecke, *Harmonic and minimal vector fields on tangent and unit tangent bundles*, Differential Geom. Appl. **13** (2000), 77–93.

[9] E. Boeckx and L. Vanhecke, *Harmonic and minimal radial vector fields*, Acta Math. Hung. **90** (2001), 317–331.

[10] E. Boeckx and L. Vanhecke, *Radial vector fields on harmonic manifolds*, Bull. Math. Soc. Sc. Math. Roumanie **43** (2000), 181–185.

[11] E. Cartan, *Familles de surfaces isoparamétriques dans les espaces à courbure constante*, Ann. Mat. Pura Appl. **17** (1938), 177-191.

[12] E. Cartan, *Sur des familles remarquables d'hypersurfaces isoparamétriques dans les espaces sphériques*, Math. Z. **45** (1939), 335–367.

[13] E. Cartan, *Sur quelques familles remarquables d'hypersurfaces*, C.R. Congr. Sci. Math. (1939), 30–41.

[14] A. Dancer and M. Wang, *The cohomogeneity one Einstein equations from the Hamiltonian viewpoint*, J. Reine Angew. Math. **524** (2000), 97–128.

[15] J.E. D'Atri, *Certain isoparametric families of hypersurfaces in symmetric spaces*, J. Differential Geom. **14** (1979), 21–40.

[16] D. Ferus, H. Karcher and H.-F. Münzner, *Cliffordalgebren und neue isparametrische Hyperflächen*, Math. Z. **177** (1981), 479–502.

[17] O. Gil-Medrano, *Relationship between volume and energy of unit vector fields*, Differential Geom. Appl., to appear.

[18] O. Gil-Medrano, J.C. González-Dávila and L. Vanhecke, *Harmonic and minimal invariant unit vector fields on homogeneous Riemannian manifolds*, Houston J. Math. **26** (2000), to appear.

[19] O. Gil-Medrano and E. Llinares-Fuster, *Minimal unit vector fields*, Tôhoku Math. J., to appear.

[20] H. Gluck and W. Ziller, *On the volume of a unit vector field on the three-sphere*, Comment. Math. Helv. **61** (1986), 177–192.

[21] J.C. González-Dávila and L. Vanhecke, *Examples of minimal unit vector fields*, Ann. Global Anal. Geom. **18** (2000), 385–404.

[22] J.C. González-Dávila and L. Vanhecke, *Minimal and harmonic characteristic vector fields on three-dimensional contact metric manifolds*, J. Geom., to appear.

[23] J.C. González-Dávila and L. Vanhecke, *Invariant harmonic unit vector fields on Lie groups*, Boll. Un. Mat. Ital., to appear.

[24] D.L. Johnson, *Volume of flows*, Proc. Amer. Math. Soc. **104** (1988), 923–931.

[25] A. Kaplan, *Fundamental solutions for a class of hypoelliptic PDE generated by composition of quadratic forms*, Trans. Amer. Math. Soc. **258** (1980), 147–153.

[26] M. Kimura, *Real hypersurfaces and complex submanifolds in complex projective space*, Trans. Amer. Math. Soc. **296** (1986), 137–149.

[27] A. Kollross, *A classification of hyperpolar and cohomogeneity one actions,* Thesis, University of Augsburg, 1998.

[28] T. Levi Civita, *Famiglie di superficie isoparametriche nell'ordinario spazio euclideo*, Atti Accad. Naz. Lincei Rend. Cl. Sci. Fis. Mat. Natur. **26** (1937), 355–362.

[29] M. Lohnherr and H. Reckziegel, *On ruled real hypersurfaces in complex space forms*, Geom. Dedicata **74** (1999), 267–286.

[30] H.-F. Münzner, *Isoparametrische Hyperflächen in Sphären*, Math. Ann. **251** (1980), 57–71.

[31] H. Ozeki and M. Takeuchi, *On some types of isoparametric hypersurfaces in spheres, I*, Tôhoku Math. J. **27** (1975), 515–559.

[32] S. Pedersen, *Volumes of vector fields on spheres*, Trans. Amer. Math. Soc. **336** (1993), 69–78.

[33] B. Segre, *Famiglie di ipersurficie isoparametriche negli spazi euclidei ad un qualunque numero di dimensioni*, Atti Accad. Naz. Lincei Rend. Cl. Sci. Fis. Mat. Natur. **27** (1938), 203–207.

[34] R. Takagi, *On homogeneous real hypersurfaces in a complex projective space*, Osaka J. Math. **10** (1973), 495–506.

[35] Ph. Tondeur, *Foliations on Riemannian manifolds,* Universitext, Springer-Verlag, Berlin, Heidelberg, New York, 1988.

[36] Ph. Tondeur, *Geometry of foliations,* Monographs in Mathematics, Birkhäuser, Basel, Boston, Berlin, 1997.

[37] K. Tsukada and L. Vanhecke, *Invariant minimal unit vector fields on Lie groups*, Period. Math. Hungar. **40** (2000), 123–133.

[38] K. Tsukada and L. Vanhecke, *Minimality and harmonicity for Hopf vector fields*, Illinois J. Math., to appear.

[39] K. Tsukada and L. Vanhecke, *Minimal and harmonic unit vector fields in* $G_2(\mathbb{C}^{m+2})$ *and its dual space*, Monatsh. Math. **130** (2000), 143–154.

[40] Q.-M. Wang, *Isoparametric hypersurfaces in complex projective spaces*, Differential Geometry and Differential Equations, Proc. 1980 Beijing Sympos. **3** (1982), 1509–1523.
[41] Q.-M. Wang, *Isoparametric functions on Riemannian manifolds, I*, Math. Ann. **277** (1987), 639–646.
[42] G. Wiegmink, *Total bending of vector fields on Riemannian manifolds*, Math. Ann. **303** (1995), 325–344.
[43] L.Xiao, *Isoparametric submanifolds in complex projective spaces*, Acta Math. Sinica **38** (1995), 845–856.
[44] L. Xiao, *Principal curvatures of isoparametric hypersurfaces in $\mathbb{C}P^n$*, Proc. Amer. Math. Soc. **352** (2000), 4487–4499.

Department of Mathematics, Katholieke Universiteit Leuven, Celestijnenlaan 200B, 3001 Leuven, Belgium
E-mail address: eric.boeckx@wis.kuleuven.ac.be

Department of Mathematics, Katholieke Universiteit Leuven, Celestijnenlaan 200B, 3001 Leuven, Belgium
E-mail address: lieven.vanhecke@wis.kuleuven.ac.be

Contemporary Mathematics
Volume **288**, 2001

The Seiberg-Witten Equations and Almost-Hermitian Geometry

S. K. Donaldson

Dedicated to the memory of Alfred Gray

ABSTRACT. This note gives an interpretation of the Seiberg-Witten equations in terms of a second order PDE for an almost-complex structure on a 4-manifold.

Let (M, g_0) be an oriented, 4-dimensional Riemannian spin-manifold. The Seiberg-Witten equations over M are equations for data consisting of a unitary connection A on a complex line bundle L over M and a section Φ of the twisted spin bundle $S^+ \otimes L$. The equations are

$$D_A\Phi = 0 \ , \ F_A^+ = i\tau(\Phi),$$

where $\tau : S^+ \otimes L \to \Lambda^+$ is the standard quadratic map (defined explicitly below) and F_A^+ denotes the self-dual part of the curvature of A. (In fact one does not need to restrict to spin manifolds, since the relevant data is really a $\mathbf{Spin}^c$ structure, but to simplify notation we will ignore this point.) The purpose of this note is to point out that these equations, away from the zero set of Φ, can be interpreted in terms of almost-complex geometry: as equations for a compatible almost-complex structure and a real function related to a conformal factor. There are well-established and profound global links [**3**],[**4**] between on the one hand the Seiberg-Witten equations, and the 4-manifold invariants they define, and on the other hand the Kähler and almost-Kähler structures on the 4-manifold. The observation in this note fits in a similar mould but is by contrast elementary and local, hingeing merely on a re-arrangement of the variables in the equations.

Turning to almost-complex geometry, recall that if g is a Riemannian metric on M the compatible almost-complex structures are parametrised by sections of the "twistor space" Z; the sphere bundle in Λ^+. Thus we can think of such structures as being given by self-dual 2-forms ω on M with $|\omega|^2 = 2$ everywhere (the factor 2 being chosen to fit in with standard conventions). The structure is Kähler if and only if $\nabla\omega = 0$, where ∇ is the Levi-Civita covariant derivative. In the general case we can think of $\nabla\omega$ as a measure of the deviation from Kählerity. This tensor splits

2000 *Mathematics Subject Classification.* 53B35.

naturally into two components. To see this recall that on $\mathbf{C}^2$, with its standard Hermitian metric and associated form ω_0, the self-dual forms decompose as

$$\Lambda^+ = \mathbf{R}\omega_0 \oplus \Lambda^{0,2},$$

where $\Lambda^{0,2}$ is the complex line arising from the usual (p,q) decomposition. (That is, we map a complex form $\theta \in \Lambda^{0,2}$ to the real form $1/2(\theta + \overline{\theta})$.) Thus, given an almost-complex structure ω on M we can split the bundle Λ^+_M into the sum of a trivial bundle, spanned by ω, and its orthogonal complement $\overline{K}$, where K is a complex line bundle—the canonical bundle in the familar complex case. Now, since $|\omega| = 1$, the covariant derivative of ω is a section of $T^*M \otimes_{\mathbf{R}} \overline{K}$. Using the almost complex structure on M we can identify this with

$$T^*M \otimes_{\mathbf{C}} \overline{K} \oplus T^*M \otimes_{\mathbf{C}} K.$$

Now the wedge product gives a complex bilinear map

$$T^*M \times T^*M \to \Lambda^2 T^*M = K,$$

which can be regarded as an isomorphism

$$TM \equiv T^*M \otimes_{\mathbf{C}} \overline{K},$$

(using the metric to identify $\overline{K}$ with K^*). Thus $\nabla\omega$ has two components, one in $T^*M \otimes_{\mathbf{C}} K$ and one in TM. It is well-known that the first of these corresponds to the *Nijenhius tensor* N_ω of the almost-complex stucture and the second to the exterior derivative $d\omega$ (where we use the volume form to identify TM with $\Lambda^3 T^*M$).

One natural condition to consider for such an almost complex structure ω is given by the notion of a "harmonic section" of the twistor fibration: that is we consider the energy functional

$$\int_M |\nabla\omega|^2,$$

and its associated Euler-Lagrange equation [**5**]. Clearly this can be written as

$$(\nabla^*\nabla\omega)^\perp = 0,$$

where $\nabla^*\nabla\omega$ is the usual "covariant Laplacian" and $\perp$ denotes the component orthogonal to ω: i.e. the component in $\overline{K}$. One might also look at the Euler-Lagrange equations arising from the L^2 norms of the components of $\nabla\omega$: for example the functional

$$\int_M |d\omega|^2,$$

yields the Euler-Lagrange equation

$$(\Delta\omega)^\perp = 0,$$

where Δ is the Hodge Laplacian on 2-forms. These equations differ by a term involving the self-dual Weyl tensor W^+. In particular the equations are the same in the case when M is self-dual and the twistor space Z is complex: this is similar to the usual discussion of the harmonic maps energy functional for maps between Kähler manifolds.

The condition we want to consider can be thought of as a deformation of the harmonic section equation above. We consider the obvious $\overline{K}$-valued pairing between TM and $T^*M \otimes \overline{K}$. We apply this to $d\omega$ and N_ω to define a section $\langle d\omega, N_\omega\rangle$

of $\overline{K}$. Thus we can write down a 1-parameter family of elliptic equations for an almost complex structure ω:

$$(\nabla^*\nabla\omega)^\perp + \mu\langle d\omega, N_\omega\rangle = 0, \tag{0.1}$$

with a parameter $\mu \in \mathbf{C}$. This is somewhat reminiscent of the "Wess-Zumino-Witten" deformation of the harmonic mapping equation from a surface into a Lie group.

THEOREM 0.1. *There is a one-to one correspodence between*

- *pairs consisting of a metric g' conformally equivalent to g and a solution of the Seiberg-Witten equation over M with respect to the metric g' such that Φ is nowhere zero;*
- *pairs consisting of a metric g conformally equivalent to g_0 and solutions ω of Equation 0.1 on (M, g), with parameter $\mu = 2$ such that*

$$\frac{3}{2}|N_\omega|^2 + \frac{1}{2}|d\omega|^2 + S(g) < 0$$

everywhere on M, where $S(g)$ is the scalar curvature of g.

The proof of this Theorem involves two parts: first matching up the relevant sets of data and second identifying the equation involved. Recall first that the Seiberg-Witten equations are not conformally invariant: if $g' = e^{2s}g_0$ there is no way of transforming a solution (A, Φ) of the equations with respect to the metric g_0 to one with respect to the metric g'. However the spinors themselves are conformally invariant and if (in an obvious notation) $D_{A,g_0}\Phi = 0$ then $D_{A,g'}(e^{-3s/2}\Phi) = 0$. It follows that there is one-to-one correspondence between solutions of the g'-Seiberg-Witten equations, for metrics g' in the conformal class of g_0, with nowhere-vanishing spinor field Φ, and triples (g, A, Ψ) where g is conformally equivalent to g_0, the spinor field Ψ has $\|\Psi\| = 1$ everywhere (measured in the metric g) and

$$D_{A,g}\Psi = 0\ ,\ F_A^+ = i\lambda\tau(\Psi), \tag{0.2}$$

for some strictly positive function λ on M. To make this correspondence, we just take $g = |\Phi|^{2/3}g'$. Suppose we are given such a triple (g, A, Ψ). Then $\omega = \tau(\Psi)$ has $|\omega| = 2$ everywhere and defines a compatible almost-complex structure on M. Recall now that on an an almost complex manifold there is a canonical isomorphism between the positive spin bundle S^+ and the sum $K^{1/2} \oplus K^{-1/2}$, where $K^{1/2}$ is the square root of K (the ambiguity in forming the square root corresponding to the ambiguity in the choce of spin structure, which we will largely ignore). Hence $S^+ \otimes L$ is a sum $K^{1/2} \otimes L \oplus K^{-1/2} \otimes L$. In this decomposition, Φ lies in the factor $K^{1/2} \otimes L$, so Φ defines a unitary isomorphism between L and $K^{-1/2}$, so our connection A can be thought of as a connection on $K^{-1/2}$, which in turn corresponds to a connection on the line bundle K. Now these line bundles have standard unitary connections, which we denote by A_0, induced from the Levi-Civita connection of (M, g) in the usual way. Thus we can write $A = A_0 + ia$ for a real 1-form a on M. Going backwards, if we start with an almost-complex stucture ω we form the corresponding line bundle $K^{-1/2}$ and there is canonical unit-length section Φ of $S^+ \otimes K^{-1/2}$. We may then consider connections $A = A_0 + ia$ on this line bundle. The following lemma is now the crux of the argument

LEMMA 0.2. *Given a line bundle L over M with a connection A_0 and given a non-vanishing spinor field $\Psi \in \Gamma(S^+ \otimes L)$ there is a unique 1-form a over M such that $D_A\Psi = 0$, where D_A is the Dirc operator of the connection $A = A_0 + ia$.*

To see this we write $D_A\Psi = D_{A_0}\Psi + ia * \Psi$, where $*$ denotes the Clifford multiplication

$$T^*M \times S^+ \otimes L \to S^- \otimes L.$$

We merely need to observe that for any fixed non-zero $\psi \in S^+$ the multiplication

$$a \mapsto a * \psi$$

defines an isomorphism from T^*M to S^-. This can be seen in many ways. From one point of view we may choose identifications of each of the spaces T^*M, S^+, S^- with the quaternions in such a way that the Clifford multiplication is represented by multiplication of quaternions. From an other point of view, the map in question is essentially the usual (complex antilinear) identification between S^- and $T^*M \otimes_{\mathbf{C}} K^{-1/2}$.

Given this Lemma we see that we can use the Dirac equation to eliminate the variable a: it is entirely determined by the canonical section Ψ and the connection A_0 hence by the almost complex structure ω. So we can write $a = a(\omega)$. In fact one can readily show that $a = *d\omega$, although we will not use this explicitly in our calculations below. Then $F_A^+ = F_{A_0}^+ + id^+a$ and the equation 0.2 can be written

$$d^+a(\omega) = -iF_{A_0}^+ + \lambda\omega,$$

for some positive function λ. In other words, if we use the decomposition $\Lambda^+ = \mathbf{R}\omega \oplus \overline{K}$ we get two conditions: a differential equation

$$d_{0,2}^+(a(\omega)) = -iF_{A_0}^{0,2}, \tag{0.3}$$

where, for example, $d_{0,2}^+$ denotes the $\overline{K}$ component of d^+, together with an open condition

$$\langle d^+a(\omega), \omega\rangle > 0. \tag{0.4}$$

We see then without more ado that the Seiberg-Witten equations, with non-vanishing spinor field, in the given conformal class can be expressed in terms of almost-complex structures, and it remains only to identify the terms appearing in equation 0.3 and inequality 0.4 more explicitly. Theorem 1 is a consequence of the the next Lemma.

LEMMA 0.3. *In the situation above, the following formulae hold:*

- $$d_{0,2}^+(a(\omega)) + iF_{A_0}^{0,2} = \frac{1}{2}((\nabla^*\nabla\omega)^\perp + \langle d\omega, N_\omega\rangle);$$
- $$8\langle d^+a(\omega), \omega\rangle = -(S(g) + \frac{3}{2}|N_\omega|^2 + \frac{1}{2}|d\omega|^2).$$

The proof of this Lemma is a differential geometric calculation which can be done in various ways. The approach we will use goes via the Lichnerowicz-Weitzenbock formula for the Dirac equation. Let

$$B : S^+ \times S^+ \to \Lambda^+$$

be the symmetric bilinear map associated to the quadratic map τ and let $\tilde{B}$ be the induced bilinear form on $T^*M\otimes^+$ obtained using the contraction on T^*M furnished by the Riemannian metric. Suppose as before that (A, Ψ) satisfies the equations 0.2 and that $|\Psi| = 1$ everywhere and set $\omega = \tau(\Psi) = B(\Psi, \Psi)$. Then

$$\nabla^*\nabla\omega = \nabla^*\nabla B(\Psi, \Psi) = 2(B(\Psi, \nabla_A^*\nabla_A\Psi, \Psi) - \tilde{B}(\nabla_A\Psi, \nabla_A\Psi).$$

Now the Lichnerowicz-Weitzenbock formula reads

$$D_A^* D_A \Psi = \nabla_A^* \nabla_A \Psi + \rho(F_A^+)(\Psi) + \frac{S}{4}\Psi,$$

where

$$\rho : \Lambda^+ \to \mathbf{su}(S^+)$$

is the standard isomorphism (which again we write down explicitly below). Now since $D_A\Psi = 0$ and $F_A^+ = i\lambda\tau(\Psi)$, we get

$$\nabla_A^* \nabla_A \Psi = (2\lambda + \frac{S}{4})\Psi.$$

Thus

$$\nabla^* \nabla \omega = -(4\lambda + \frac{S}{2})\omega - 2\tilde{B}(\nabla_A \Psi, \nabla_A \Psi).$$

We decompose this equation into the components along ω and perpendicular to ω. For the first of these we use the fact that ω has constant length so

$$0 = \Delta|\omega|^2 = 2(\langle \omega, \nabla^* \nabla \omega\rangle - |\nabla \omega|^2.$$

This gives

$$8\lambda = -S - |\nabla\omega|^2 - 2\langle \tilde{B}(\nabla_A\Psi, \nabla_A\Psi), \omega\rangle.$$

The perpendicular component gives

$$(\nabla^*\nabla\omega)^\perp + 2\tilde{B}(\nabla_A\Psi, \nabla_A\Psi)^\perp = 0,$$

using the obvious notation. It is quite easy to see that these are precisely the formulae required in Lemma 2. Thus to complete the calculation we need to show that

$$(0.5) \quad \tilde{B}(\nabla_A\Psi, \nabla_A\Psi)^\perp = -2\langle N, d\omega\rangle \ , \ \langle \tilde{B}(\nabla_A\Psi, \nabla_A\Psi), \omega\rangle = \frac{1}{4}(|N_\omega|^2 - |d\omega|^2).$$

This is straightforward, since it only involves the first derivativess of the section Ψ. From the point of view of representation theory the issue is to identify the map $\tilde{B}$ on the kernel of the Clifford multiplication map

$$c : S^+ \otimes T^*M \to S^-.$$

To effect the calculation we may reduce to the standard model when M is $\mathbf{R}^4$, and the connection A is trivial. (This is because we may choose a trivialisation in which the connection matrix vanishes at a given point.) Let $(x_0, \dots, x_3)$ be standard co-ordinates on $\mathbf{R}^4$ and s_0, s_1 be a standard basis for the spinors. So our maps τ is given by

$$\tau(fs_0 + gs_1) = \frac{|f|^2 - |g|^2}{2}\omega_0 + Re(f\overline{g})\omega_1 + Im(f\overline{g})\omega_2,$$

where

$$\omega_0 = dx_0dx_1 + dx_2dx_3\omega_1 = dx_0dx_2 + dx_3dx_1\omega_2 = dx_0dx_3 + dx_1dx_2.$$

We compute at the origin and we suppose that $\Psi = fs_0 + gs_1$ with $f = 1, g = 0$ at the origin. This unit spinor determines the standard complex structure on $\mathbf{R}^4$, so we use complex co-ordinates

$$z = x_0 + ix_1 \ , \ w = x_2 + ix_3.$$

The Dirac equation $D\Psi = 0$ becomes

$$f_{\overline{z}} = g_w \ , f_{\overline{w}} = -g_z,$$

where suffixes f_z denote $\frac{\partial f}{\partial z}$ etc. The condition that $|\Psi| = 1$ imples that the partial derivatives of f with respect to the x_λ are pure imaginary at the origin so, at this point,

$$f_w = -\overline{f_{\overline{w}}} \ , \ f_z = -\overline{f_{\overline{z}}}.$$

Now the ω_0 component of $\tilde{B}(\nabla\Psi, \nabla\Psi)$ is

$$\sum_\lambda |\frac{\partial f}{\partial x_\lambda}|^2 - |\frac{\partial g}{\partial x_\lambda}|^2.$$

We can write this as the difference

$$2(|f_z|^2 + |f_{\overline{z}}|^2 + |f_w|^2 + |f_{\overline{w}}|^2) - 2(|g_z|^2 + |g_{\overline{z}}|^2 + |g_w|^2 + |g_{\overline{w}}|^2).$$

Using the relations above we obtain

$$\langle \tilde{B}(\nabla\Psi, \nabla\Psi), \omega_0\rangle = 8(|g_z|^2 + |g_w|^2) - 8(|g_{\overline{z}}|^2 + |g_{\overline{w}}|^2).$$

Similarly, the perpendicular component of $\tilde{B}(\nabla\Psi, \nabla\Psi)$ is represented, using the standard basis element

$$\theta = d\overline{z}d\overline{w} = \omega_2 + i\omega_3,$$

by the complex number

$$\sum_\lambda \frac{\partial f}{\partial x_\lambda}\frac{\partial \overline{g}}{\partial x_\lambda}.$$

In complex co-ordinates this is

$$f_z\overline{g_z} + f_z\overline{g_{\overline{z}}} + f_w\overline{g_w} + f_w\overline{g_{\overline{w}}}.$$

Using the relations above, we see that two terms cancel so we get

$$\tilde{B}(\nabla\Psi, \nabla\Psi)^\perp = (g_z\overline{g_{\overline{w}}} + g_w\overline{g_{\overline{z}}})\ \theta.$$

Now the derivative of the section ω of the twistor space is given at the origin by the derivative of $\overline{f}g$, which is the same as the derivative of g. Thus $d\omega$ corresponds to the anti-holomorphic component $(g_{\overline{z}}, g_{\overline{w}})$ and N_ω to the holomorphic component (g_z, g_w). On the one hand, the natural pairing $\langle N, d\omega\rangle$ is just $g_z\overline{g_{\overline{w}}} + g_w\overline{g_{\overline{z}}}\theta$. On teh other hand, with the standard metrics on the relevant spaces, we have

$$|d\omega|^2 = 16(|g_{\overline{z}}|^2 + |g_{\overline{w}}|^2)\ , |N_\omega|^2 = 16(|g_z|^2 + |g_w|^2).$$

Putting this together we complete the proof of the Lemma.

Remarks

1. The formulation of our main result above does not explicitly give a description of the Seiberg-Witten solutions for a fixed metric, because of the variable conformal factor. If we fix the metric g_0 one can show, using arguments similar to those above, that the Seiberg-Witten equations (assuming that the spinor field does not vanish) can be expressed in terms of the self-dual form $\Omega = \tau(\phi)$. The equation for Ω is

(0.6)

$$\nabla^*\nabla\Omega = -(\frac{S}{2} + |\Omega|^2)\Omega$$
$$- 2\langle d\Omega + *d|\Omega|, N_\Omega\rangle + \frac{1}{2}\left(\frac{|d\Omega|^2}{|\Omega|^2} - |N_\Omega|^2\right)\Omega + (d|\Omega||^2 + 2\langle d|\Omega|, *d\Omega\rangle\frac{\Omega}{|\Omega|^2}.$$

This way of writing the equations is somewhat similar to the usual treatment of the vortex equation in two dimensions [**1**], the dimension reduction of the

Seiberg-Witten equations, which one reduces to a second order equation for a single function.

2. Over a compact 4-manifold M, the existence of any nowhere vanishing section of $S^+ \otimes L$ requires that the second Chern class of this bundle vanishes. This is just the familiar condition, in terms of the Euler chracteristic χ and signature τ of M,
$$c_1(L)^2 = \frac{1}{4}(3\tau + 2\chi),$$
for L^2 to be the canonical bundle of an almost complex sructure on M. In the theory of the Seiberg-Witten invariants one is primarily interested in this case, so it is reasonable to hope that the condition that the spinor fields of the actual solutions have no zeros is satisfied quite often. One thing that our formulation does make transparent is the index formula for the Seiberg-Witten equations in this situation: equation above is precisely the condition under which the expected dimension of the moduli space of solutions is zero. One can see this, assuming there is a zero-free solution, because the symbol of the equivalent, Laplace-type, second order equation is homotopically trivial.

3. Part of the author's motivation for this note comes from unpublished work of V. da Silva [**2**], which took a rather similar point of view. da Silva'a approach was in a sense complementary to that above: starting with any connection A such that $F^+(A)$ is nowhere zero he chooses a spinor field Φ such that $\tau(\Phi) = F^+(A)$, and shows that the Seiberg-Witten equations can be expressed as a second order equation for the connection A. As da Silva points out this raises an interesting question. A generic self-dual form vanishes on a curve in M, whereas a generic spinor field vanishes at isolated points: what happens for a generic solution of the Seiberg-Witten equations?

References

[1] A. Jaffe and C. H. Taubes, *Vortices and Monopoles*, Progress in Physics 2, Birkhauser 1980

[2] V. da Silva, Unpublished Oxford dissertation (1995)

[3] C.H. Taubes, *The Seiberg-Witten and the Gromov invariants*, Math. Res. Letters 2 (1995), 221-238

[4] E. Witten, *Monopoles and four-manifolds*, Math. Res. Letters 1 (1994), 769-96

[5] C. Wood, *Harmonic almost-complex structures*, Compositio Mathematica 99 (1995), 183-212

IMPERIAL COLLEGE, LONDON
E-mail address: s.donaldson@ic.ac.uk

Contemporary Mathematics
Volume **288**, 2001

Sculpture Inspired by Work with Alfred Gray: Kepler Elliptic Curves and Minimal Surface Sculptures of the Planets

by
Helaman Ferguson
Sculpture Studio in Bowie, Maryland, U. S. A.
21 January 2001

Abstract

This paper continues an earlier collaboration of Helaman Ferguson with Alfred Gray on surfaces with negative Gaussian curvature and their sculptural implications. Due to Alfred's unfortunate death by a burst aneurism, Helaman was left to finish the mathematical work as well as carve the stone. The many things Alfred taught Helaman in their brief collaboration were cherished; without them this new work would not have been possible. A photograph of the Alfred Gray Memorial Bronze appears on the title page of this book. This paper includes a photograph of Helaman Ferguson's Carnelian Granite elliptic curve carving corresponding to a seventh root of unity together with nine drawings of Kepler elliptic curve minimal surfaces, one for each of the nine planets of our solar system.

Event

Alfred Gray Memorial Congress on Homogeneous Spaces, Riemannian Geometry, Special Metrics, Symplectic Manifolds and Topology, 18-23 September 2000, Bilbao, Viscaya, Spain. Helaman Ferguson created a special sculpture for this conference, a Costa type surface which was cast in a bronze limited edition of fifty by Alfa Arte Foundry of Eibar, Guipúzcoa, Spain. Elements of this edition were made available to conference participants and other interested persons.

Key Words

negative gaussian curvature, minimal surfaces, elliptic functions, Weierstrass representation, period parallelogram, complex elliptic curves, mathematical sculpture

Mathematics Subject Classification 2000
53A10, 49Q65, 57R42, 58D27, 33E05, 33F05, 14H52, 70F05, 01A10

Figure 0. Photograph of Carnelian Dakota Granite carving: the torus with periods $< 1, e^{2\pi\sqrt{-1}/7} >$ as a negative gaussian curvature surface,

$$Z + Z\,e^{2\pi\sqrt{-1}/7} \hookrightarrow C \Longrightarrow C/(Z + Z\,e^{2\pi\sqrt{-1}/7})$$

This sculpture was commissioned in Fall 2000 by the family of mathematician Kenneth Hoffman for a surprise celebration of his seventh decade.

Introduction

This paper is mostly expository in which the mathematics, which is now classical, has been summarized for my application, viz., sculpture that celebrates Mathematics. The origins of the mathematics itself began with the work of Euler and culminated in fundamental work of Costa for genus one, David Hoffman and Weeks for genus one and higher on embedding. As a sculptor, the starting concept of minimal surface is important to me for structural as well as aesthetic reasons. From my sculptural point of view, an immersion which is not an embedding may have advantages over an embedding – self-intersections may make the structure stronger. But the structural and aesthetic strength I am interested in comes from the negative gaussian curvature, so I emphasize negative gaussian curvature, and let the embedding/immersion issue fall where it may. Most of the (genus one) elliptic curves for my planetary application are immersions. Minimal surfaces are a special case of negative gaussian curvature, additionally zero mean curvature. Again, in my sculpture, specifically carving stone where stone has great compressive strength and not so great tensile strength, I regard the important feature to be negative gaussian curvature.

Elliptic curves over the complex numbers tend to spread themselves out over $C \times C$ or R^4. A central purpose here is to find images that confine themselves to R^3. For each complex elliptic curve I construct a conformal image of a minimal triply punctured tori. Each image is located wholly in R^3, has negative Gaussian curvature of a special kind, principle curvatures equal and opposite in sign, hence mean curvature is zero. The images are parametrized by the usual $SL_2(Z)$ moduli space in the upper half plane. In general the images are immersions, only a line of them (measure zero) are embeddings, this result appears in a 1991 paper of Celso Costa [Co3]. I pay special attention to preparing for the numerical analysis of explicit evaluation of the Weierstrass representation integrals in closed form. I give this attention to the numerical analysis and its reduction to C code is because in my studio I have computer equipment for 'projecting' a virtual mathematical image into a block of stone. This is part of my CRADA (Cooperative Research and Development Agreement) with the Intelligent Systems Group at NIST (National Institute of Standards and Technology), cf., [AF].

My Costa series of sculptures includes, Costa I, $5''$ diameter bronze; Costa II, $16''$ diameter bronze and aluminum (completed); Costa III, $36''$ diameter bronze (wip); Costa IV, $14''$ diameter Carrara marble (completed); Costa V, $30'' \times 24'' \times 16''$, Carrara marble (completed), bronze (edition); Costa VI, D_4 Symmetry, $4'' \times 4'' \times 12''$, nine part bronze (edition); Costa VII, $4'' \times 4'' \times 4''$, serpentine (completed); Costa VIII, $4'' \times 4'' \times 6''$, serpentine (completed); Costa IX, $12' \times 10' \times 10'$, 20 tons, artificial snow, Breckenridge, Colorado; Costa X, $6.5'' \times 5'' \times 4''$, Gray Memorial Conference, bronze (edition of 48); Costa XI, $6' \times 5' \times 4'$, 11 tons, Academy Black California Granite (wip); Costa XII, $7' \times 5' \times 4.5'$, 13 tons, Sunset Red Texas Granite (in progress); Costa XIII, $8' \times 6' \times 5'$, 24 tons, Impala Black South Africa Granite (wip); cf., [F], [FGM], [FFNSW].

Figure 1. Mercury, ☿, $y^2 = x^3 + 15/64$.

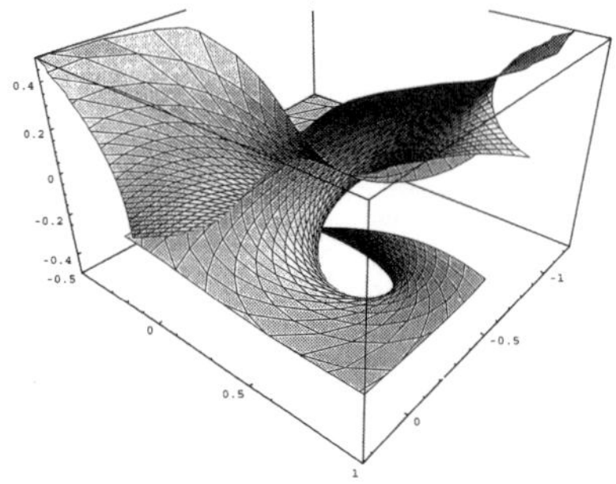

The now classical Costa minimal surface is a triply punctured torus image, the R^3 image of the lemniscatic elliptic curve. This has many elegant sculptural features, primarily non-intersecting negative Gaussian curvature. Although all sculptures in this series share the designation 'Costa' (a conveniently elegant name!) and no self-intersection, my sculptural emphasis has always been negative Gaussian curvature. The Costa surface, qua minimal surface, can be recognized in these sculptures, but none of them are models in any narrow mathematical sense. My creative process is interpretive and sensitive to natural material and other physical considerations. My creative process convolves pure thought and reason with physical materials. Even so, I carved some of these 'Costas' with special computer driven or computer informed devices, partly so that I could learn the mathematical forms, to make them part of my personal repertoire of points, curves, and surfaces. These 'Costas' are special cases of what I develop here to creatively express important elements of the mathematics and physics of each planet in our solar system. As it turns out, some of the planets are pretty close to the Costa surface. This paper represents one possible expression of Kepler's third law and Newton's inverse square law. I use mathematical language elements which have historically been important to our understanding of the physical universe in general and our little solar system in particular. For example, series expansions are such a mathematical language element. One of my aesthetic interests in elliptic curves and cubic curves in general is that Newton's inverse square law and Kepler's third law are polynomial cubics. These laws are usually stated in terms of two bodies and are only valid for two bodies. Of course, there never are just two bodies, there are always perturbations of other bodies (and relativistic effects). These elliptic curves are to be understood to be representative approximations to the facts. I am after all doing sculpture, not rebuilding the solar system!

Newton's inverse square law is the cubic expression,

$$F = G\frac{m_1 m_2}{r^2}, \quad Fr^2 = G\, m_1\, m_2$$

for gravitational constant G, force F, m_1, m_2 the masses and r the distance between them. Kepler's third law

$$P^2 \propto R^3$$

for P the planetary period in earth years and average semi-axis R in astronomical (Sun to Earth) units. Kepler's third law is already a cubic polynomial law, practically an elliptic curve in the form,

$$y^2 = x^3 + ax + b.$$

Jupiter has a period of 11 earth years and orbital radius of 5 earth distances to the sun to a reasonable integer approximation. This (period, distance) pair of whole numbers corresponds to the integral point $(x, y) = (5, 11)$ satisfying,

$$11^2 = 5^3 - 5 + 1, \quad y^2 = x^3 - x + 1,$$

which is an integral point on the elliptic curve $y^2 = x^3 + ax + b$ with $a = -1$ and $b = 1$.

Cubic curves over the complex numbers are generically triply punctured tori. An elegant way to look at the topological form of triply punctured tori is as complete minimal surfaces of genus one. My aesthetic point of view here is that Kepler's third law, stated as, say $P^2 \propto R^3$ or $y^2 = x^3$, a degenerate cubic, is certainly not exact in the real world of multiple bodies and much else. Neither for that matter is Newton's, usually stated for two bodies, similarly for Einstein's general relativity versions. I think it is a good thing that our physical universe seems always to be more interesting than our mathematical formulations of it. I hope that always remains true. But our creative mathematical response is part of nature too.

I have sandwiched each of the planets between elliptic curves. Each planet then has a minimal image as a triply punctured torus wholly in real three space: this negative curvature image is my planetary sculpture image, a kind of image of where the planet is and why.

The Sun actually has its own orbit about the center of gravity of the solar system. The planets do not revolve around the Sun, they revolve around this center of gravity which is located about a Sun radius from the center of the Sun. Kepler's third law is only approximate relative to the Sun. For these elliptic curves, I choose somewhat arbitrary reasonable approximations.

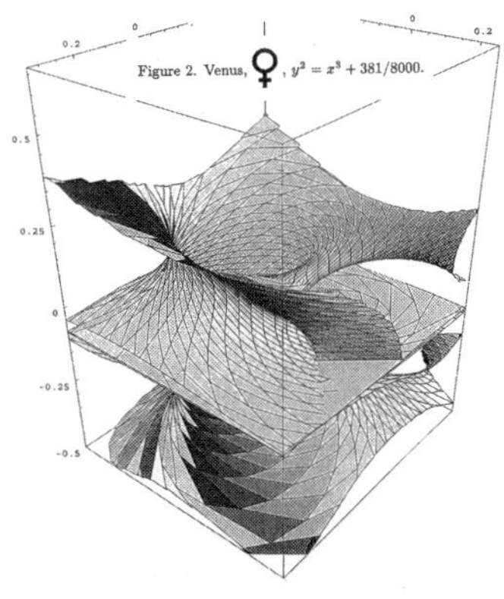

Figure 2. Venus, ♀, $y^2 = x^3 + 381/8000$.

For little and close planets, Mercury, Venus, Earth, and Mars, we will take rational points on elliptic curves for these give more reasonable approximations. For the larger or far out planets we will be able to take elliptic curves with integer points. The idea is to invert these elliptic curves and find the lattices associated with them and hence the lattices associated with the planets. To develop the sandwich, I will designate the rational point pair $(R, P) = (\mathit{distance}, \mathit{period}) = (x, y)$ as (less, more), (less, less), (more, more), or (more, less) depending upon the rational approximations to the known periods and distances. Different scientific sources give different knowns by the way, even best knowns have to be bracketed. Various choices of curves are possible for a given planet, they are more or less equivalent for our purposes, so we get to make an aesthetic choice, which may depend upon the size of the coefficients, the discriminant or the size of the discriminant, and possibly the periods of the elliptic curve. The curve may be chosen so the periods are from an interesting number field. I state the planets in terms of $y^2 = x^3 + ax + b$.

Mercury: $(x, y) = (5/8, 5/8)$ is (less, more) and $(x, y) = (2/5, 6/25)$ is (more, less) yielding

$$y^2 = x^3 + \frac{75}{512} \quad \text{and} \quad y^2 = x^3 - \frac{4}{625},$$

or possibly

$$y^2 = x^3 + \frac{15}{64}x \quad \text{and} \quad y^2 = x^3 - \frac{2}{125}x.$$

Venus: $(x, y) = (7/10, 5/8)$ is (less, more) and $(x, y) = (3/4, 8/13)$ is (more, less),

$$y^2 = x^3 + \frac{381}{8000} \quad \text{and} \quad y^2 = x^3 - \frac{467}{10816},$$

or possibly

$$y^2 = x^3 + \frac{27}{400}\,x + \frac{3}{8000} \quad \text{and} \quad y^2 = x^3 - \frac{10}{169}\,x + \frac{1}{832}.$$

Earth: $(x, y) = (1, 1)$ is (same, same) of course, just as the Sun is $(x, y) = (0, 0)$ and share the same 'degenerate' cubic, $y^2 = x^3$. This Sun-Earth relationship is purely an artifact of our geocentrism, because these Kepler law approximations are in astronomical units of Sun-Earth distances and Earth years. Instead of using this degenerate cubic, I will take the Earth-Moon relationship for the elliptic curve I will associate with the Earth. Again,

Earth/Moon: $(x, y) = (9, 28)$ is (less, less) and $(x, y) = (10, 29)$ is (more, more),

$$y^2 = x^3 + 6\,x + 1 \quad \text{and} \quad y^2 = x^3 + 16\,x + 1.$$

Mars: $(x, y) = (7/5, 5/3)$ is (less, less) and $(x, y) = (2, 3)$ is (more, more),

$$y^2 = x^3 + \frac{x}{45} + \frac{3}{1125} \quad \text{and} \quad y^2 = x^3 + 1.$$

Jupiter: $(x, y) = (5, 11)$ is (less, less) and $(x, y) = (6, 15)$ is (more, more),

$$y^2 = x^3 - x + 1 \quad \text{and} \quad y^2 = x^3 + x + 3.$$

Saturn: $(x, y) = (9, 27)$ is (less, less) and $(x, y) = (10, 32)$ is (more, more),

$$y^2 = x^3 \quad \text{and} \quad y^2 = x^3 + 2\,x + 4.$$

Uranus: $(x, y) = (19, 83)$ is (less, less) and $(x, y) = (20, 89)$ is (more, more),

$$y^2 = x^3 + 2\,x - 8 \quad \text{and} \quad y^2 = x^3 - 4\,x + 1.$$

Neptune: $(x, y) = (30, 164)$ is (less, less) and $(x, y) = (31, 173)$ is (more, more),

$$y^2 = x^3 - 3\,x - 14 \quad \text{and} \quad y^2 = x^3 + 4\,x + 14.$$

Pluto/Charon: $(x, y) = (40, 253)$ is (less, more) and $(x, y) = (39, 244)$ is (more, less),

$$y^2 = x^3 + 9 \quad \text{and} \quad y^2 = x^3 + 6\,x - 17.$$

I perform simple rescalings of the variables to get canonical Weierstrass forms; I do an inversion described below to calculate the periods. I only need one elliptic curve for a sculpture; I get to pick the most expressive.

Complex Lattices

I follow Alfred Gray [G] for notation, for the Costa example, and for his references there, particularly Barbosa-Colares [BC]. Cf., [Ch], [Sh], [Si]. Let $\omega_1, \omega_2 \in C$ complex numbers with $\Im(\omega_2/\omega_1) > 0$, or $\omega_2/\omega_1 \in H^2$ upper half plane, and $0 < \arctan(\omega_2/\omega_1) < \pi/2$. Define the lattice

$$L = Z\omega_1 + Z\omega_2 \in C.$$

Set

$$L^{\times} = L \setminus \{0\} = \{\omega = m\omega_1 + n\omega_2 \in L \quad | \quad \omega \neq 0, \quad m^2 + n^2 > 0\}.$$

Put

$$G_k = \sum_{\omega \in L^{\times}} \frac{1}{\omega^{2k}}.$$

Then

$$g_2 = g_2(\omega_1, \omega_2) = 60G_2 = 60 \sum_{\omega \in L^{\times}} \frac{1}{\omega^4},$$

and

$$g_3 = g_3(\omega_1, \omega_2) = 140G_3 = 140 \sum_{\omega \in L^{\times}} \frac{1}{\omega^6},$$

and $g_k \in Q[g_2, g_3]$.

For example, the first few G_k polynomials in g_2 and g_3 are

$$G_2 = \frac{g_2}{60}, \quad G_3 = \frac{g_3}{140}, \quad G_4 = \frac{{g_2}^2}{8400},$$

$$G_5 = \frac{g_2\, g_3}{18480}, \quad G_6 = \frac{49\, g_2^3 + 750\, g_3^2}{84084000}, \quad G_7 = \frac{{g_2}^2\, g_3}{2402400},$$

$$G_8 = \frac{539\, g_2^4 + 18000\, g_2\, g_3^2}{171531360000}, \quad G_9 = \frac{1421\, g_2^3\, g_3 + 5500\, g_3^3}{506970464000},$$

$$G_{10} = \frac{5929\, g_2^5 + 363750\, g_2^2\, g_3^2}{358500542400000}, \quad G_{11} = \frac{19061\, g_2^4\, g_3 + 184500\, g_2\, g_3^3}{1049428860480000},$$

$$G_{12} = \frac{5519899\, g_2^6 + 529200000\, g_2^3\, g_3^2 + 833085000\, g_3^4}{6302869736042880000}.$$

It is helpful to introduce two auxiliary functions of these two invariants g_2, g_3, viz.,

$$\Delta(\omega_1, \omega_2) = g_2^3 - 27g_3^2, \quad J(\omega_1, \omega_2) = 12^3 \frac{g_2^3}{g_2^3 - 27g_3^2},$$

where the latter denominator, the discriminant of the elliptic curve as discussed below, is always non-zero, $g_2^3 - 27g_3^2 \neq 0$.

The Weierstrass $\wp$ and $\wp'$ functions

The Weierstrass $\wp = \wp_L$ function for the lattice L is defined by the convergent series

$$\wp(z) = \wp(z, \{g_2, g_3\}) = \frac{1}{z^2} + \sum_{\omega \in L^{\times}} \left(\frac{1}{(z - \omega)^2} - \frac{1}{\omega^2}\right)$$

which has the Laurent series at $z = 0$

$$\wp(z) = \frac{1}{z^2} + \sum_{k \geq 1} (2k+1) G_k z^{2k}.$$

Similarly for the derivative of the Weierstrass $\wp$ function,

$$\wp'(z) = \wp'(z, \{g_2, g_3\}) = -\frac{2}{z^3} - 2 \sum_{\omega \in L^{\times}} \frac{1}{(z - \omega)^3}$$

which has the Laurent series at $z = 0$

$$\wp'(z) = -\frac{2}{z^2} + \sum_{k\geq 1}(2k+1)(2k)G_k z^{2k-1}.$$

The $\wp$ is an even function $z \mapsto \wp(-z) = \wp(z)$ which is doubly periodic

$$\wp(z+\omega) = \wp(z + m\omega_1 + n\omega_2) = \wp(z), \quad \omega = m\omega_1 + n\omega_2 \in L.$$

The derivative $\wp'$ is an odd function $z \mapsto \wp(-z) = -\wp(z)$ which is doubly periodic

$$\wp'(z+\omega) = \wp'(z + m\omega_1 + n\omega_2) = \wp'(z), \quad \omega = m\omega_1 + n\omega_2 \in L.$$

Figure 3. Earth, ⊕, $y^2 = x^3 + 6x + 1$.

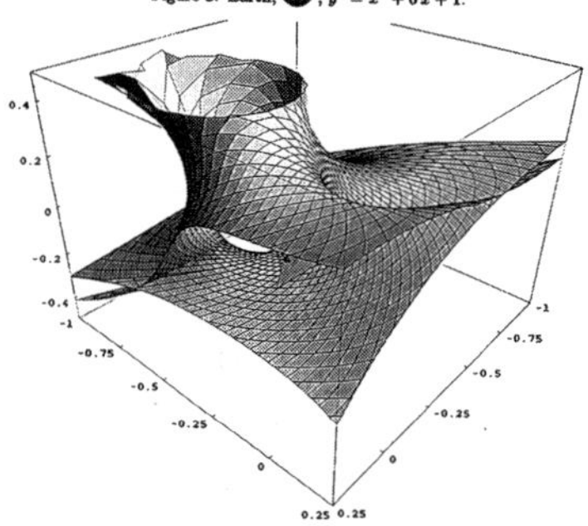

Define the fundamental period parallelogram, assuming the identification of opposite sides, a torus,

$$F = F(\omega_1, \omega_2) = I\omega_1 + I\omega_2$$

for the real interval $I = [0, 1)$. Then

$$L \hookrightarrow C \Rightarrow F \approx C/L$$

$$Z^2 \hookrightarrow C \Rightarrow F \approx R^2/Z^2$$

are exact in the sense that $C = F + L$. $\wp$ and $\wp'$ are functions on the torus F. The relationship between $\wp$ and $\wp'$ and the torus can be made more precise. One can verify directly that the pair satisfies the differential equation

$$(\wp')^2 = 4\wp^3 - g_2\wp - g_3.$$

The map $z \mapsto (\wp(z), \wp'(z))$ gives an isomorphism of C/L onto the elliptic curve

$$E = \{(x, y) \in C \times C \quad | \quad y^2 = 4x^3 - g_2 x - g_3\}.$$

In real variables this complex curve can be thought of as a surface in $R^2 \times R^2$ or in R^4. Now R^4 per se is not a very convenient space in which to do sculpture.

Returning to the elliptic curve formulation, we identify exactly the zeros of the cubic on the right hand side as

$$(\wp')^2 = 4\wp^3 - g_2\wp - g_3 = 4(\wp - e_1)(\wp - e_2)(\wp - e_3).$$

It suffices to show from the fact that $\wp'$ is odd and doubly periodic that it is zero at the three half periods

$$\frac{\omega_1}{2}, \quad \frac{\omega_2}{2}, \quad \frac{\omega_3}{2} \quad \text{where} \quad \omega_3 = \omega_1 + \omega_2.$$

Consequently,

$$e_1 = \wp(\frac{\omega_1}{2}), \quad e_2 = \wp(\frac{\omega_2}{2}), \quad e_3 = \wp(\frac{\omega_3}{2}),$$

are exactly the zeros. These three evaluations of the $\wp$ function are distinct.

The coefficients of the cubic on the right hand side will come up again in the forms

$$0 = e_1 + e_2 + e_3, \quad -\frac{g_2}{4} = e_1 e_2 + e_2 e_3 + e_1 e_3, \quad \frac{g_3}{4} = e_1 e_2 e_3.$$

Because these three roots are always distinct, the discriminant of this cubic

$$g_2^3 - 27 g_3^2 = 4^2 (e_1 - e_2)^2 (e_1 - e_3)^2 (e_2 - e_3)^2 .$$

is always non-zero.

$\wp'$ is doubly periodic so it has the same number of poles as zeros on the (closure of) the fundamental parallelogram. From the partial fraction decomposition for $\wp'$ the non-zero poles are at the three periods, $\omega_1, \omega_2, \omega_3 = \omega_1 + \omega_2$. On the other hand, where are the zeros and poles of $\wp$? Again, from the partial fraction decomposition of $\wp$, the non-zero poles are again at the three periods, $\omega_1, \omega_2, \omega_3 = \omega_1 + \omega_2$. Where the zeros of $\wp$ are will depend upon the lattice L, its generators $\langle \omega_1, \omega_2 \rangle$, and the specific choice of $\Im(\omega_2/\omega_1) > 0$ and finally the g_2 and g_3. If $g_3 = 0$, then one of the roots e_1, e_2, e_3 has to be zero, say $e_1 = 0$ so that $\omega_1/2$ is a root of $\wp$. As long as $g_2 \neq 0$ the other two roots can be neither ω_2 nor ω_3.

The addition formula for the Weierstrass $\wp$ function is

$$\wp(z) + \wp(z+w) + \wp(w) = \frac{1}{4} \left(\frac{\wp'(z) - \wp'(w)}{\wp(z) - \wp(w)} \right)^2 .$$

This addition formula will be crucial in integrating in closed form the Weierstrass minimal surface representation. The integrals become essentially phase shifts of the arguments of Weierstrass ζ functions. The consequence of the addition formula which we will use in the integration will be the following formula which relates a half period phase shift of the argument of $\wp$ to a reciprocal of a shift of the value of $\wp$.

Figure 4. Mars, ♂, $y^2 = x^3 + 1$.

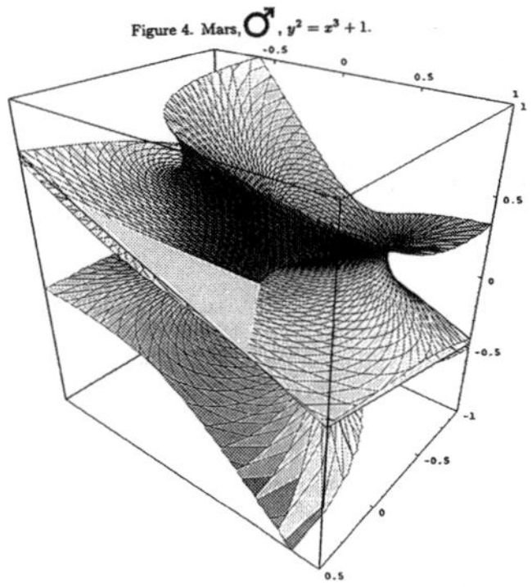

Let the following two unordered sets be the same, $\{i, j, k\} = \{1, 2, 3\}$. Then set

$$E_j = 2\, e_j^2 + e_i\, e_k,$$

so that

$$\wp(z - \frac{\omega_j}{2}) = e_j + \frac{E_j}{\wp(z) - e_j},$$

and

$$\frac{1}{\wp(z) - e_j} = \frac{1}{E_j} \cdot \left(\wp(z - \frac{\omega_j}{2}) - e_j \right)$$

It is helpful to sketch the proof of this. From the above definitions of the e_j and the half periods, we have for $j = 1, 2, 3$,

$$\wp'(\frac{\omega_j}{2}) = 0, \quad \wp(\frac{\omega_j}{2}) = e_j.$$

Making the appropriate substitution in the addition formula gives

$$\wp(z - \frac{\omega_j}{2}) = \frac{1}{4} \frac{(\wp'(z))^2}{(\wp(z) - e_j)^2} - \wp(z) - e_j$$

But $(\wp')^2 = 4(\wp - e_1)(\wp - e_2)(\wp - e_3)$ is a cubic, so

$$\wp(z - \frac{\omega_j}{2}) = \frac{(\wp(z) - e_i)(\wp(z) - e_k)}{\wp(z) - e_j} - (\wp(z) + e_j).$$

Noting that $e_i + e_k = -e_j$ in the simplification, we get the desired phase shift and reciprocal relations.

We use the following fact frequently for the integration of the Weierstrass representation in closed form,

$$\frac{p}{(p-a)(p-b)(p-c)} =$$

$$\frac{a}{(p-a)(a-b)(a-c)} + \frac{b}{(b-a)(p-b)(b-c)} + \frac{c}{(c-a)(c-b)(p-c)}.$$

The Weierstrass ζ and σ functions:

We define the Weierstrass ζ function by the lattice sum over reciprocals adjusted for convergence by

$$\zeta(z) = \zeta(z, \{g_2, g_3\}) = \frac{1}{z} + \sum_{\omega \in L^\times} (\frac{1}{z-\omega} + \frac{1}{\omega} + \frac{z}{\omega^2}).$$

The essential relation between ζ and $\wp$ is that the derivative of ζ is the negative of $\wp$,

$$\zeta'(z) = -\wp(z).$$

Therefore ζ has the Laurent series at $z = 0$

$$\zeta(z) = \zeta(z, \{g_2, g_3\}) = \frac{1}{z} - \sum_{k \geq 2} G_k z^{2k+1}.$$

This odd function $z \mapsto \zeta(-z) = -\zeta(z)$ is not doubly periodic, but nearly doubly periodic up to the addition of even multiples of values at half periods,

$$\zeta(z + m\omega_1 + n\omega_2) = \zeta(z) + 2m\zeta(\frac{\omega_1}{2}) + 2n\zeta(\frac{\omega_2}{2}), \quad m\omega_1 + n\omega_2 \in L.$$

An immediate consequence of this is that

$$\zeta(\frac{\omega_3}{2}) = \zeta(\frac{\omega_1}{2}) + \zeta(\frac{\omega_2}{2}), \quad \omega_3 = \omega_1 + \omega_2.$$

A special case of this nearly periodic ζ is the first place the half-periods appear in a form helpful to us later when we need them to prove the path independence, viz., for $j = 1, 2, 3$, note $j = 3$,

$$\frac{1}{\omega_j}(\zeta(z + \omega_j) - \zeta(z)) = \frac{2}{\omega_j}\zeta(\frac{\omega_j}{2}).$$

Figure 5. Jupiter, ♃, $y^2 = x^3 + x + 3$.

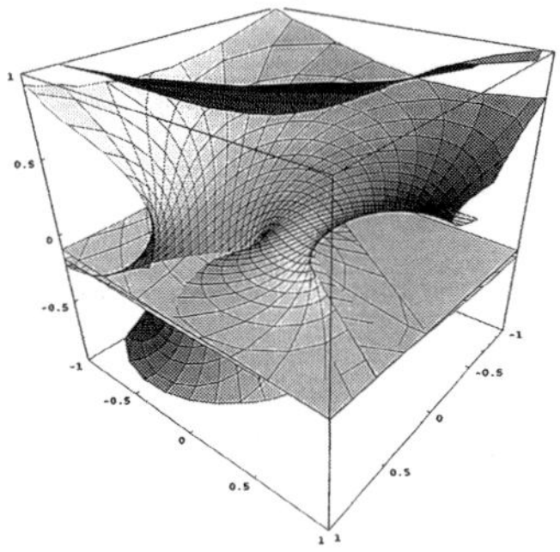

We record the remarkable Legendre relation

$$\frac{\omega_2}{2}\zeta(\frac{\omega_1}{2}) - \frac{\omega_1}{2}\zeta(\frac{\omega_2}{2}) = \frac{\pi}{2},$$

or alternatively,

$$\frac{2}{\omega_1}\zeta(\frac{\omega_1}{2}) - \frac{2}{\omega_2}\zeta(\frac{\omega_2}{2}) = \frac{2\,\pi}{\omega_1\,\omega_2},$$

where recall that we have already chosen $\Im(\omega_2/\omega_1) > 0$, which accounts for the lack of symmetry between ω_1 and ω_2. We wish to develop numerically fast algorithms for computing elliptic function expressions in $\wp$, $\wp'$, and ζ which give minimal surface images in R^3. It is helpful to have on hand four more functions, the Weierstrass σ function and the three theta functions. The logarithmic derivative of the σ function is the ζ function. The Weierstrass $\sigma(z)$ function is an odd entire function of z with its set of zeros (all simple) exactly the discrete lattice L, exactly the periods of $\wp$. Define the σ function by

$$\sigma(z) = z \prod_{\omega \in L^\times} (1 - \frac{z}{\omega}) e^{\frac{z}{\omega} + \frac{z^2}{2\omega^2}}.$$

Since $\sigma(z)$ is an entire function of z, we include a number of terms of its Taylor series, convergent for all $z \in C$. We list enough terms to do computation accurate enough to do specific sculptures. Of course, the number of terms required in general will depend upon the lattice L and its g_2 and g_3 and the polynomials in them.

$$\frac{\sigma(z)}{z} = 1 - \frac{g_2\, z^4}{2^4 \cdot 3 \cdot 5} - \frac{g_3\, z^6}{2^3 \cdot 3 \cdot 5 \cdot 7} - \frac{g_2^2\, z^8}{2^9 \cdot 3^2 \cdot 5 \cdot 7}$$
$$- \frac{g_2\, g_3\, z^{10}}{2^7 \cdot 3^2 \cdot 5^2 \cdot 7 \cdot 11} + \frac{(23\, g_2^3 - 2^6 \cdot 3^2\, g_3^2)\, z^{12}}{2^{13} \cdot 3^4 \cdot 5^2 \cdot 7 \cdot 11 \cdot 13} + \frac{19\, g_2^2\, g_3\, z^{14}}{2^{12} \cdot 3^3 \cdot 5^3 \cdot 7^2 \cdot 11 \cdot 13}$$
$$+ \frac{(107\, g_2^4 + 2^8 \cdot 3^2 \cdot 23\, g_2\, g_3^2)\, z^{16}}{2^{19} \cdot 3^5 \cdot 5^3 \cdot 7^2 \cdot 11 \cdot 13 \cdot 17} + \frac{(311\, g_2^3\, g_3 + 2^6 \cdot 3 \cdot 23\, g_3^3)\, z^{18}}{2^{16} \cdot 3^5 \cdot 5^3 \cdot 7^2 \cdot 11 \cdot 13 \cdot 17 \cdot 19}$$
$$+ \frac{(7 \cdot 23 \cdot 37\, g_2^5 + 2^7 \cdot 3^2 \cdot 5 \cdot 53\, g_2^2\, g_3^2)\, z^{20}}{2^{23} \cdot 3^6 \cdot 5^4 \cdot 7^3 \cdot 11 \cdot 13 \cdot 17 \cdot 19} \cdots .$$

Note that the polynomials in the g_2, g_3 appearing in the $2n^{th}$ numerator have virtual degree $n = i\nu_i + \nu_j$ for each monomial $g_i^{\nu_i} \cdot g_j^{\nu_j}$.

It follows directly from the product expansion for σ that

$$\zeta(z) = \frac{d}{dz} \log \sigma(z) = \frac{\sigma'(z)}{\sigma(z)}.$$

One could reverse the order in which we have presented these functions, beginning with σ, and differentiating variously to get ζ, $\wp$, $\wp'$, even realizing

$$\wp'(z) = -\frac{\sigma(2z)}{\sigma(z)^4}, \quad \wp(z) - \wp(w) = -\frac{\sigma(z+w)\sigma(z-w)}{\sigma(z)^2\sigma(z)^2},$$

and

$$\zeta(z+w) = \zeta(z) + \zeta(w) + \frac{1}{2}\frac{\wp'(z) - \wp'(w)}{\wp(z) - \wp(w)}.$$

σ is called the cardinal interpolating function for elliptic functions because every elliptic function has the same number of poles as zeros and can be written as a zero-pole quotient of products of phase shifts of σ functions.

Inverting the Invariants to get the Periods

Figure 6. Saturn, ♄, $y^2 = x^3 + 2x + 4$.

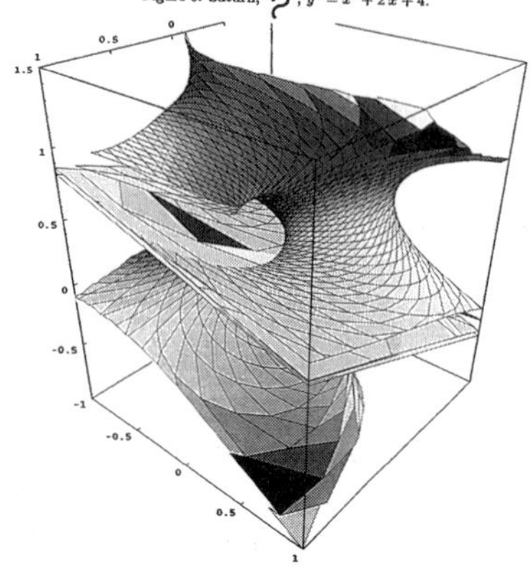

We want to go from the identifiable invariants g_2 and g_3 to the periods ω_1 and ω_2 of the elliptic curve. It is straightforward to go from the periods ω_1 and ω_2 to the invariants g_2, g_3. Set $e_j = \wp(w_j/2)$ for $j = 1, 2, 3$ to get $g_2 = -4(e_1e_2 + e_1e_3 + e_2e_3)$ and $g_3 = 4e_1e_2e_3$. Again, the elliptic curve in Weierstrass form is $\eta^2 = 4\xi^3 - g_2\xi - g_3$, and the Weierstrass function $\wp(z)$ is the inverse of an elliptic integral of the first kind,

$$z = \int_{\wp(z)}^{\infty} \frac{d\xi}{\sqrt{4\xi^3 - g_2\xi - g_3}},$$

where the integral can be taken over a suitable curve from $\wp(z)$ to ∞ which does not pass through a zero of $4\xi^3 - g_2\xi - g_3$. It is always the case for all our lattices L that the zeros are distinct since the cubic in the denominator has discriminant non-zero, $g_2^3 - 27g_3^2 \neq 0$. From this inverse integral formula for $\wp$ we have the half periods as integrals

$$\frac{\omega_j}{2} = \int_{e_j}^{\infty} \frac{d\xi}{\sqrt{4\xi^3 - g_2\xi - g_3}},$$

where recall $e_j = \wp(w_j/2)$ and the path of integration is chosen appropriately. The first few terms of the two variable Taylor series in g_2, g_3 for these integrals is given by

$$\frac{\omega_j}{2} = \frac{1}{e_j{}^{1/2}} + \frac{g_2}{40e_j^{5/2}} + \frac{g_3}{56e_j^{7/2}} +$$

$$\frac{g_2^2}{384e_j^{9/2}} + \frac{3g_2g_3}{704e_j^{11/2}} + \frac{5g_2^3 + 24g_3^2}{13312e_j^{13/2}} + + \cdots.$$

Alternatively, we can observe that the integral for the periods

$$\int_{e_j}^{\infty} \frac{d\xi}{\sqrt{4\xi^3 - g_2\xi - g_3}} = \int_{-a}^{a} \frac{dt}{\sqrt{(a^2 - t^2)(b^2 + t^2)}},$$

where the latter integral can be computed by arithmetic-geometric means. For the identity above cf., [WW, page 516] and [BB, pages 27-28] for the identity again and the arithmetic-geometric mean context for a, $b > 0$. The first problem to solve is the relationship between the a, b and the g_2, g_3. In one direction it is given by the set of two equations

$$12g_2 = (a^2 - b^2)^2 - 12a^2b^2,$$

$$-216g_3 = (a^2 - b^2)[(a^2 - b^2)^2 + 36a^2b^2].$$

Note that the discriminant for the elliptic curve in terms of a and b is given by

$$g_2^3 - 27g_3^2 = -\frac{a^2b^2(a^2 + b^2)^4}{16}.$$

We solve for a^2 and b^2 by introducing polynomials $p(t)$, $q(t)$ in the rational field $Q(t, g_2, g_3)$. We define p by setting $3 \cdot (17^2 g_2^3 + 13^3 g_3^2) \cdot p(t) =$

$$-2^5 5^4 g_2^2 g_3 - 29257 g_2^3 t + 91767 g_3^2 t - 228514 g_2 g_3 t^2 - 26588 g_2^2 t^3 - 2^3 13^2 g_3 t^4 + 2^4 221 g_2 t^5,$$

$$q(t) = g_2^3 - 3^3 g_3^2 - 2 \cdot 3^4 g_2 g_3 t - 3^3 5 g_2^2 t^2 - 2^3 3^3 5 g_3 t^3 - 2^3 \cdot 3 \cdot 5 g_2 t^4 + 2^4 t^6.$$

Then we have explicit solutions for a^2 and b^2,

$$(a^2, b^2) = (p(\tau), \tau),$$

where τ is a root of q, i.e., $q(\tau) = 0$.

The period integral as an elliptic integral of the first kind as given above can now be evaluated by the arithmetic-geometric means,

$$\int_{-a}^{a} \frac{dt}{\sqrt{(a^2 - t^2)(b^2 + t^2)}} = \frac{\pi}{M(b, \sqrt{a^2 + b^2})},$$

where M is the arithmetic-geometric mean limit of the transformation

$$(u, v) \mapsto (\frac{u + v}{2}, (u v)^{\frac{1}{2}}),$$

beginning with $(u, v) = (b, (a^2 + b^2)^{\frac{1}{2}})$.

This arithmetic-geometric mean limit converges quadratically fast for positive real numbers. For complex numbers u and v things are more interesting including quadratically fast convergence if the initial values are not π radians apart. Note that for complex numbers the geometric mean translates into the arithmetic mean on the exponents. For the arguments of u and v sufficiently close (usually eventually satisfied) then both the arithmetic and geometric mean lie near the intersection of an annulus defined by the radii $|u|$, $|v|$ and the addition parallelogram with vertices 0, u, v, $u+v$. The arithmetic-geometric mean for complex numbers clearly has value for the present context of computing periods of elliptic curves. A quite complete discussion of the positive real case of the arithmetic-geometric mean can be found in [BB].

Alternatively, we can go another way from the identifiable invariants g_2 and g_3 to the periods ω_1 and ω_2 of the elliptic curve, cf., [Ch, VI, §, pages 90-92]. Define the variable

$$t = \frac{\omega_2}{\omega_1},$$

which has positive imaginary part and is in the upper half plane. At this point for simplicity of exposition we will abuse the notation slightly and regard g_2 and g_3 as fixed given constants and $g_2(t)$ and $g_3(t)$ as functions of the period variable t in the upper half plane. The constant invariants $g_2 = 60G_2$ and $g_3 = 140G_3$ are sums over the lattice given by the periods ω_1 and ω_2. On the other hand, the functions $g_2(t)$ and $g_3(t)$ have power series in the variable $q = e^{2\pi t\sqrt{-1}}$,

$$g_2(t) = (2\pi)^4 \left(\frac{1}{2^2 \cdot 3} + 2^2 \cdot 5 \cdot \sum_{n \geq 1} \sigma_3(n) q^n \right),$$

$$g_3(t) = (2\pi)^6 \left(\frac{1}{2^3 \cdot 3^3} - \frac{7}{3} \cdot \sum_{n \geq 1} \sigma_5(n) q^n \right),$$

where $\sigma_k(n) = \sum_{d|n} d^k$ is over divisors of n.

The Klein invariant J of the elliptic curve is a function of the ratio t, $q = e^{2\pi t\sqrt{-1}}$,

$$J(t) = \frac{g_2(t)^3}{g_2(t)^3 - 27g_3(t)^2} =$$

$$= \frac{1}{q} + 744 + 196884\,q + 21493760\,q^2 + 864299970\,q^3 +$$

$$20245856256\,q^4 + 333202640600\,q^5 + 4252023300096\,q^6 +$$

$$44656994071935\,q^7 + 401490886656000\,q^8 + 3176440229784420\,q^{18} +$$

$$22567393309593600\,q^{20} + \ldots.$$

For any $z \in C$, the equation $J(t) = z$ has a unique solution t in the fundamental domain for the group $SL(2, Z)$ acting on the upper half plane in the usual way by fractional linear transformations.

If both $g_2 \neq 0$ and $g_3 \neq 0$ then $J(t) = g_2^3/(g_2^3 - 27g_3^2)$ has a unique solution $t = t_\omega = \omega_2/\omega_1$. But then from the lattice sum, $\omega_1^2 g_2(t_\omega)/g_3(t_\omega) = g_2/g_3$ determines ω_1 and hence $\omega_2 = \omega_1 \cdot t_\omega$.

If $g_2 = 0$ then $g_3 \neq 0$. Then $t_\omega = e^{2\pi\sqrt{-1}/3} = \rho$ and $\omega_1^{-6} \cdot g_3(\rho) = g_3$ determines ω_1 and again $\omega_2 = \omega_1 \cdot \rho$.

If $g_3 = 0$ then $g_2 \neq 0$. Then $t_\omega = e^{2\pi\sqrt{-1}/4} = i$ and $\omega_1^{-4} \cdot g_2(i) = g_2$ determines ω_1 and again $\omega_2 = \omega_1 \cdot i$.

Weierstrass Integral Representation for Minimal Surfaces

The Weierstrass integral representation formula for minimal surfaces gives essentially all minimal surface immersions in R^3. The formulae are parametrized by two meromorphic functions f and g and a base point z_0. The immersion is given by

$$z \mapsto (\Re\frac{1}{2}\int_{z_0}^{z} f(1+g^2)dw, \Re\frac{i}{2}\int_{z_0}^{z} f(1-g^2)dw, \Re\int_{z_0}^{z} fgdw),$$

where f, g are functions of the complex variable w. $\Re(z) = x$ denotes the real part of $z = x + iy$ and $\Im(z) = y$ denotes the imaginary part of $z = x + iy$. There are continuously many parameters to work with. Conditions will be given rendering the base point irrelevant in certain cases. Even with the base point the result is a surface immersion.

Figure 7. Uranus, ⛢, $y^2 = x^3 - 4x + 1$.

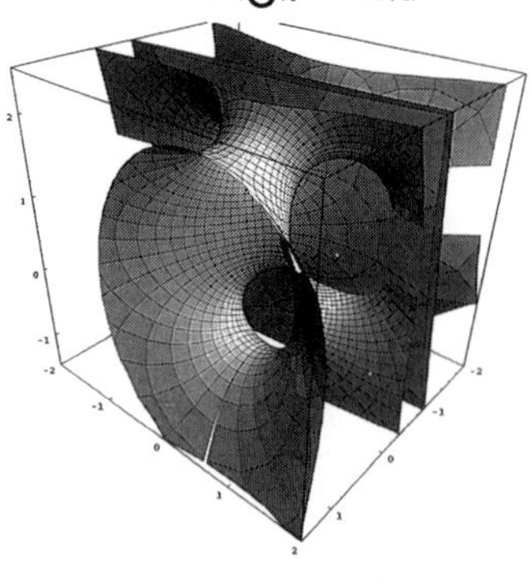

The classical minimal surfaces appear by making specific choices of f and g and a domain $D \subseteq C$ for them. For examples, $f(z) = 1$ and $g(z) = z$ with $D = C$ gives Enneper's surface; $f(z) = 1/z^2$ and $g(z) = z$ with $D = C^\times$ or $f(z) = -e^{-z}$ and $g(z) = -e^z$ with $D = C$ gives the catenoid; $f(z) = e^{-z}$ and $g(z) = -ie^z$ with $D = C$ gives the helicoid; $f(z) = 4/(1-z^4)$ and $g(z) = z$ with $D = \{z \,|\, |z| < 1\}$ gives Sherk's surface; $f(z) = 2/(1-z^4)$ and $g(z) = z$ gives Henneberg's incomplete minimal Möbius strip, etc.

These surfaces are all minimal immersions, but only the catenoid and helicoid are embeddings. Up to path independence, any f and g gives an immersion. The genus one minimal surface discovered by Celso Costa in 1983 [C1, C2, C3], [DH], [BC], [G], [FGM], appears in this context by choosing f and g to be certain lemniscatic elliptic functions for the square period parallelogram given by the choice of Gaussian integer lattice $L = \langle \omega_1, \omega_2 \rangle = \langle 1, \sqrt{-1} \rangle$. For $f = \wp$ and $g = c/\wp'$ and the square domain $D = [0, 1) + [0, 1)\sqrt{-1}$ and a special choice of constant c, then the resulting surface is not only an immersion but also an embedding. One problem we are solving here is to carry the Costa program through for arbitrary lattices, i.e., any period parallelogram. The first problem to solve is what should the corresponding constant c be? It turns out that the case of a general lattice is interesting because there are several possibilities.

A difficult global problem, it would appear, is to choose f and g to get an embedding. Embedding means that the image surface has no self-intersections in R^3. Celso Costa [C3] has shown in 1991 that for the general lattice, those and only those with rectangular period parallelograms are embedded. This is a set of measure zero among all lattices.

The metrics, normal vectors, and curvatures of these surfaces can all be given in terms of the functions f and g. We state them at the outset.

The metric ds on the surface is

$$ds^2 = \frac{1}{2}|f|^2(1 + |g|^2)^2\,|dz|^2.$$

This metric is complete if every divergent path on the surface has infinite length.

The normal vector N on the surface is

$$N = \frac{(2\Re(g), 2\Im(g), |g|^2 - 1)}{|g|^2 + 1}.$$

The Gaussian curvature κ of the surface is

$$\kappa = -\Big(\frac{4\,|g'|}{|f|\,(|g|^2 + 1)^2}\Big)^2.$$

Denote the principle curvatures by ν_+ and ν_-. The sum of the principal curvatures $\nu_+ + \nu_-$ is zero and their product $\nu_+ \cdot \nu_-$ is κ. The Gauss map is given by g.

The Weierstrass integral representation formula calls for integrating three differentials defined in terms of the meromorphic functions f, g and the base point z_0. An auxiliary condition, path independence of the integrals, eliminates any specific base point and suffices to give a complete minimal surface immersion in R^3. The resulting coordinate functions are all harmonic and are hence solutions of Laplace's equation.

Define the vector valued meromorphic differential α by

$$\alpha = (\alpha_1, \alpha_2, \alpha_3) = \Big(\frac{f}{2}\,(1 - g^2)\,dw,\ \frac{i\,f}{2}\,(1 + g^2)\,dw,\ f\,g\,dw\Big).$$

Then

$$\sum \alpha_j^2 = 0, \quad \sum |\alpha_j|^2 > \frac{1}{2} f^2\,(|g|^2 + 1)^2 > 0,$$

and

$$f(w)\,dw = \alpha_1 - i\,\alpha_2, \quad g(w) = \frac{\alpha_3}{\alpha_1 - i\,\alpha_2},$$

so that the functions f, g are equivalent to the α_j.

The Weierstrass representation formula gives the harmonic minimal immersion and the conjugate minimal immersion

$$x_j = \Re \int^z \alpha_j, \quad x_j = \Im \int^z \alpha_j,$$

respectively, assuming the integrals are path independent.

Lattice Minimal Surfaces

We do the three integrations implicit in the Weierstrass representation. Closed form integrations seem to be a small miracle in itself. We take the three differentials α_j in turn. Take

Figure 8. Neptune, Ψ, $y^2 = x^3 + 4x + 14$.

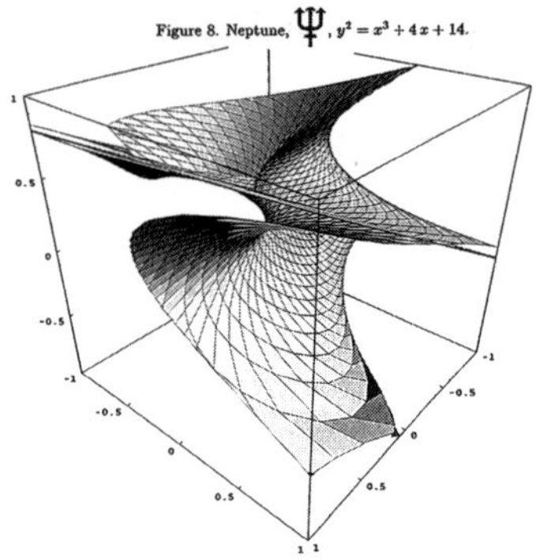

$$f(z) = \wp(z),$$

and

$$g(z) = \frac{c}{\wp'(z)},$$

where c is a constant to be chosen later. It is technically convenient to do $j = 1, 2$ simultaneously, writing the f, g form as

$$\alpha_j = (-1)^{\frac{j-1}{2}} \cdot \frac{f}{2} (1 + (-1)^j g^2).$$

Then

$$\alpha_j = (-1)^{\frac{j-1}{2}} \cdot \frac{\wp}{2} (1 + (-1)^j \frac{c^2}{\wp'})$$

$$\alpha_j = \frac{(-1)^{\frac{j-1}{2}}}{2} \cdot (\wp + (-1)^j \frac{c^2}{4} \cdot \{\frac{\wp}{(\wp - e_1)(\wp - e_2)(\wp - e_3)}\}).$$

Now use the partial fraction decomposition of cubic in the denominator to a sum of reciprocal linears. Set

$$E_l = 2\, e_l^2 + e_m\, e_n, \quad f_l = \frac{e_l}{(e_l - e_m)\,(e_l - e_n)},$$

where again $\{l, m, n\} = \{1, 2, 3\}$, and get

$$\alpha_j(w) = \frac{(-1)^{\frac{j-1}{2}}}{2} \cdot \left(\wp(w) + (-1)^j \frac{c^2}{4} \cdot \{\sum_l \frac{f_l}{E_l} \cdot (\wp(w - \frac{\omega_l}{2}) - e_l)\}\right).$$

Recall that $\zeta' = -\wp$ and integrate on a path between some base point z_0 and z to get $(j = 1, 2)$

$$\beta_j = \beta_j(z, z_0) = \int_{z_0}^{z} \alpha_j = \frac{(-1)^{\frac{j-1}{2}}}{2} \cdot \Big(\zeta(z_0) - \zeta(z) +$$

$$+ (-1)^j \frac{c^2}{4} \cdot \{\sum_l \frac{f_l}{E_l} \cdot (\zeta(z_0 - \frac{\omega_l}{2}) + e_l\, z_0 - \zeta(z - \frac{\omega_l}{2}) - e_l\, z)\}\Big).$$

We now do the $j = 3$ case, where the third coordinate differential is

$$\alpha_3 = f\,g\,dz = \frac{c\wp}{\wp'}dz = \frac{c\wp\wp'}{(\wp')^2}dz = \frac{c\,\wp'}{4} \cdot \frac{\wp}{(\wp - e_1)(\wp - e_2)(\wp - e_3)}.$$

Using the partial fraction for the reciprocal cubic again and the definition of f_l given above

$$\alpha_3 = \frac{c}{4} \cdot \{\sum_l f_l \frac{\wp'}{\wp - e_l}\}.$$

We do not need the half period phase shift this time since we can integrate directly with logarithms,

$$\beta_3 = \beta_3(z, z_0) = \int_{z_0}^{z} \alpha_3 = \frac{c}{4} \cdot \{\sum_l f_l \log\left(\frac{\wp(z) - e_l}{\wp(z_0) - e_l}\right)\}.$$

This completes the integrations in closed form.

Residue Computation and the Costa Constant c

We now wish to compute the residues for the three coordinate functions α_j for $j = 1, 2, 3$ at the origin 0 and the three poles

$$\frac{\omega_1}{2}, \quad \frac{\omega_2}{2}, \quad \frac{\omega_3}{2}.$$

Thus we have nine separate residue computations to look at. The three poles are the punctures of the triply punctured torus which is the immersed image of the elliptic curve in R^3. First we compute the residues for α_3. We have written this differential as a sum of terms,

$$\frac{\wp'(z)}{\wp(z) - e_j} = -2\,c\,z + O(z^2).$$

From this Laurent series we have that

$$Res(\alpha_3)_{z=0} = -\frac{1}{2}c \sum_l f_l = -\frac{1}{2}c \cdot 0 = 0.$$

Alternatively, directly from the Laurent series,

$$\frac{c\wp(z)}{\wp'(z)} = -2\,c\,z + O(z^2).$$

By the residue theorem, taking a small circle around $\frac{\omega_j}{2}$,

$$Res(\alpha_3)_{\{z=\omega_j/2\}} = \frac{c}{2} \cdot f_j$$

for $j = 1, 2, 3$.

We have already written expressions for the α_j for $j = 1, 2$ as linear combinations of the $\wp$ functions at various arguments which yield $Res(\alpha_j)_{\{z=\omega_k/2\}}$.

Closed Paths

Figure 9. Pluto/Charon, $\wp$, $y^2 = x^3 + 6x - 17$.

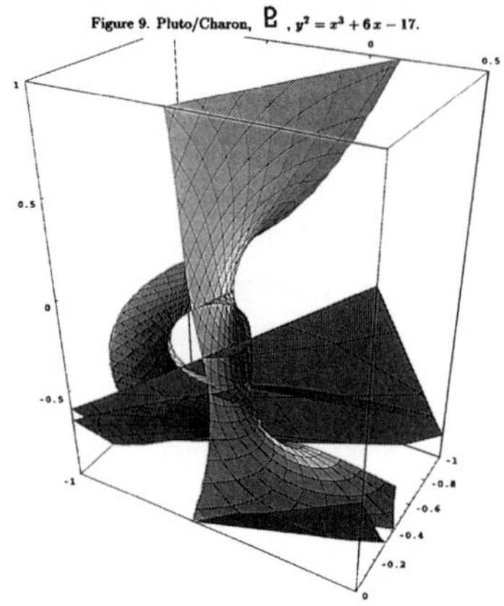

We study closed paths that give the generators of the fundamental group for the torus, i.e., the two cycles for $\{j, k\} = \{1, 2\}$ given by $w = t_1\omega_1 + t_2\omega_2$

$$\Gamma_j = \{t_1\omega_1 + t_2\omega_2 \,|\, t_j \neq \frac{1}{2}, \, 0 \leq t_k \leq 1\}.$$

Using the fact that ζ is the anti-derivative of $-\wp$ and the half period identities from the nearly periodic ζ, we can evaluate the two integrals, $\{j, k\} = \{1, 2\}$,

$$\int_{\Gamma_j} \wp(w - \frac{\omega_k}{2}) dt_j = -\frac{2}{\omega_k}\zeta(\frac{\omega_k}{2}).$$

These are used to evaluate the residues of α_j for $j = 1, 2$. The fact that α_3 is path independent follows from the usual properties of the logarithm.

For $j = 1, 2$ and $k = 1, 2$ we will have

$$c_{j,k} = \int_{\Gamma_j} \alpha_k =$$

$$\frac{(-1)^{\frac{j-1}{2}}}{2} \cdot \Big(-\frac{2}{\omega_k} \cdot \zeta(\frac{\omega_k}{2}) + (-1)^j \frac{c^2}{4} \cdot \{\sum_l \frac{f_l}{E_l} \cdot (\frac{-2}{\omega_k} \zeta(\frac{\omega_k}{2}) - e_l)\} \Big).$$

Acknowledgement

I would like to thank algebraic geometer James Carlson for asking me the question, *since general cubic curves are triply punctured tori does the Costa surface have a moduli space?* His question stimulated my research for this paper and this new series of negative gaussian curvature sculptures.

References

[AF] James Albus, Helaman Ferguson, et al, String-Potentiometer Measurement System SP-2, hardware/software, Draft document, NIST, 3 March 1993; see also, Helaman Ferguson, Stone Sculpture from Equations, in J. Albus (ed.), Intelligent Systems and Semiotics, National Institute of Standards and Technology, Gaithersburg, Maryland, 14-17 September 1998, Proceedings 1998 IEEE ISIC CIRA ISAS Joint Conference, Section VI: Autonomy, pages 765-770.

[BC] J. Lucas M. Barbosa and A. Gervasio Colares, Minimal Surfaces in R^3, Lecture Notes in Mathematics, Volume 1195, Chapter 7, pages 84-90, Springer-Verlag, Berlin-New York, 1986, x + 124 pages.

[BB] Jonathan M. Borwein and Peter B. Borwein, Pi and the AGM: A Study in Analytic Number Theory and Computational Complexity, Canadian Mathematical Society, John Wiley & Sons, New York, 1987, xv + 414 pages.

[Ch] K. Chandrasekharan, Elliptic Functions, Grundlehren der Mathematischen Wissenshaften 281, Springer-Verlag, Berlin-New York, 1985, xii + 189 pages.

[Co1] Celso J. Costa, Example of a complete minimal immersion of genus one and three embedded ends. Bol. Soc. Bras. Mat. volume 15, 1984, pages 47-54.

[Co2] Celso J. Costa, Uniqueness of complete minimal surfaces in R^3 of genus one and three ends, Journal of Diff. Geometry, vol. 30, 1989, pages 597-618.

[Co3] Celso J. Costa, Classification of complete minimal surfaces in R^3 with total curvature 12π, Inventiones Mathematicae, volume 105, 1991, pages 273-303.

[DH] David Hoffman, The computer-aided discovery of new embedded minimal surfaces, The Mathematical Intelligencer, volume 9, number 3, 1987, pages 8-21.

[FFNSW] Helaman Ferguson, Claire Ferguson, Tomas Nemeth, Dan Schwalbe, Stan Wagon, "Invisible Handshake", cover article, The Mathematical Intelligencer, Volume 21, Number 4, Fall 1999, pages 30-35.

[F] H. Ferguson, Sculpture Inspired by Alfred Gray, Annals of Global Analysis and Geometry, volume 18, numbers 3-4, August 2000, pages 341-345.

[FGM] H. Ferguson, A. Gray, S. Markvorsen, Costa's Minimal Surface via Mathematica, Mathematica in Research and Education, vol. 5, 1996, pages 5-10.

[G] Alfred Gray, Modern Differential Geometry of Curves and Surfaces with Mathematica, CRC Press, Boca Raton, 1998, xxiv + 1053 pages. Chapter 32, Minimal Surfaces via Weierstrass, 32.5, Costa's Minimal Surface, pages 747-759.

[Sh] Goro Shimura, Introduction to the Arithmetic Theory of Automorphic Functions, Chapter 4.2 Elliptic Curves over C, Kano Memorial Lectures I, Publications of the Mathematical Society of Japan, Volume 11, Iwanami Shoten and Princeton University Press, 1971, xiv + 267 pages.

[Si] Carl L. Siegel, Topics in Complex Function Theory, Volume 1: Elliptic Functions and Uniformization Theory, Chapter 1, paragraphs 11-15 on the $\wp$ function, pages 56-89, translated by A. Shenitzer and D. Solitar, John Wiley & Sons, 1988, xii +186.

[WW] E. T. Whittaker and G. N. Watson, A Course of Modern Analysis, Chapter XX: Elliptic Functions. General theorems and the Weierstrassian Functions; Chapter XXI: The Theta Functions, First Edition, 1902, Fourth Edition, Reprinted, Cambridge University Press, 1965, viii + 608

Contemporary Mathematics
Volume **288**, 2001

Mesoscopic curvature and hyperbolicity

Misha Gromov

Dedicated to the memory of Alfred Gray

We introduce in this paper a notion of $\mathrm{CAT}_\delta(\kappa)$-*spaces* mediating between $\mathrm{CAT}(\kappa)=\mathrm{CAT}_0(\kappa)$ and δ-hyperbolicity (essentially) corresponding to $\mathrm{CAT}_\delta(-\infty)$. This allows the local-to-global passage (in the spirit of Cartan-Hadamard) most convenient for applications to combinatorial group theory. There is little new about the proofs : everything follows by "δ-perturbing" similar arguments in the $\mathrm{CAT}(\kappa)$-case. This has been done in $[\mathrm{Gro}]_{\mathrm{HG}}$ for $\kappa = -\infty$ and we do more or less the same for finite κ with minor technical adjustments. We shall see, in particular, that $\mathrm{CAT}_\delta(\kappa)$-property of the σ-balls for a given σ in a *simply connected* space X implies δ_1-hyperbolicity of X with $\delta_1 < 2/\sqrt{-\kappa}$ for all $\kappa < 0$, provided $\kappa > 0$, $\sigma > 0$ and $\delta \leq \delta_0(-\kappa, \sigma) > 0$ for some universal *strictly* positive function δ_0 in two strictly positive variables. (This is standard for $\delta = 0$).

0. Preliminary notations and definitions.

Given a metric space X we write

$$|x - y| = |x - y|_X = \mathrm{dist}_X(x, y)$$

(compare $[\mathrm{Gro}]_{\mathrm{HG}}$ and $[\mathrm{Gro}]_{\mathrm{CAT}}$). X is called a *path metric* space (see [G-L-P]) if $|x - y|$ equals the infimum of length of curves between x and y for all $x, y \in X$. We say that X is *geodesic* if this infimum is assumed where the minimal curve is called a *geodesic segment* between x and y, denoted $[x, y] \subset X$.

Recall (see [Bri-Hae] and references therein) that $\mathrm{CAT}(\kappa)$-spaces are introduced by improving certain inequalities on the six distances between four points $x_i \in X$, $i = 1, \ldots, 4$, similar to those satisfied by (comparison) quadruples $\{x'_i\}$ in the *model* space X' that is the simply connected space X' with $\dim X' = 2$ of constant curvature κ. We are mostly concerned here with negative curvature $\kappa < 0$, where X' equals the hyperbolic plane H^2_κ. If $\kappa \to -\infty$, then H^2_κ (Hausdorff) converges to an infinitely branched tree (see [G-L-P], $[\mathrm{Gro}]_{\mathrm{AI}}$ serving as the model space for $\kappa = -\infty$.

2000 *Mathematics Subject Classification.* 53C23, 53C24.
Key words and phrases. Negative curvature, hyperbolic groups.

The comparison inequality can be expressed by saying that for each quadruple $\{x_i\}$ in X there is a quadruple $\{x_i'\} \subset X'$ making a convex 4-gone for the cyclic order on x_i, such that the four edges of this 4-gone equal the corresponding 4 distances between x_i,

$$|x_i' - x_{i+1}'|_{X'} = |x_i - x_{i+1}|_X\,, \; i = 1, \ldots, 4,$$

(with the convention $4+1=1$), while the remaining two distances satisfy

$$|x_1' - x_3'|_{X'} \geq |x_1 - x_3|_X$$

and

$$|x_2' - x_4'|_{X'} \geq |x_2 - x_4|_X\,.$$

This property for a quadruple $\{x_i\}$ is called $\mathrm{Cycl}_4^{(\kappa)}$-*inequality*. Similarly, given a cyclicly ordered k-tuple of points $\{x_i\} \subset X$, $i = 1, \ldots k$, $k+1 = 1$, one says that $\{x_i\}$ satisfies $\mathrm{Cycl}_k(\kappa)$ if there exists $\{x_i'\} \subset X'(= H_\kappa^2)$, such that

$$|x_i' - x_j'|_{X'} \geq |x_i - x_j|_X$$

for all $i, j = 1, \ldots, k$ and

$$|x_i' - x_{i+1}'|_{X'} \leq |x_i - x_{i+1}|_X$$

for $i = 1, \ldots, k$, and where one additionally requires that $\{x_i'\}$ makes a *convex* k-gone in X'. One knows (Resketnyak's theorem) that

$$\mathrm{Cycl}_4(\kappa) \Rightarrow \mathrm{Cycl}_k(\kappa), \; k = 5, 6, \ldots \text{ for all k},$$

for the *geodesic* spaces X (compare [Gro]$_{\mathrm{CAT}}$).

One can express the inequality $\mathrm{Cycl}_4(\kappa)$ as

$$|x_1 - x_4| \leq F_\kappa(|x_i - x_j|), \tag{$(\kappa)_4$}$$

where F_κ is a certain function in 5 variables $|x_i - x_j|$, $1 \leq i < j \leq 4$, except $i = 1$, $j = 4$, that is monotone increasing in $|x_i - x_{i+1}|$ and decreasing in $|x_2 - x_4|$. Namely, one takes F_κ, such that $(\kappa)_4$ become *equality* for the model space X_κ'. This F_κ is defined for all *non-negative* values of the five arguments satisfying two *triangle inequalities* (for $\Delta(x_1, x_2, x)$ and $\Delta(x_1, x_3, x_4)$. Thus the domain of definition of F_κ is a convex subset, denoted $\Delta_5 \subset \mathbb{R}_+^5$. The function $F_\kappa : \Delta_5 \to \mathbb{R}_5$ is smooth in the interior of Δ_5 and 1/2-*Hölder* on all of Δ_5; but it is *not Lipschitz*. This causes some technical problems when we perturb this inequality by a small $\delta > 0$.

If $\kappa = -\infty$, then $(\kappa)_4$ becomes an *equality* and its δ-perturbation leads to the *δ-hyperbolicity* of X expressed by the following $\mathrm{CAT}_\delta(-\infty)$-*relation* between the three numbers

$$m_i = |x_1 - x_i| + |x_j - x_k|,$$

for $i \neq j \neq k \neq i = 2, 3, 4$, as follows

$$m_2 \leq \max(m_3, m_4) + 2\delta \tag{$\boxtimes_\delta$}$$

for all quadruples $\{x_i\} \subset X$. (See [Bri-Hae] and references therein). Similarly, one can introduce $\mathrm{CAT}_\delta(\kappa)$-inequality to be

$$|x_1 - x_4| \le F_x(|x_i - x_j|) + \delta$$

expressing the idea that the quadruple $\{x_i\}$ is δ-close (in the Hausdorff sense) to $\{x'_i\}$ satisfying $(\kappa)_4$. Unfortunately this does not work as well for $\kappa > -\infty$ as for $\kappa = -\infty$ due to non-Lipschitz behaviour of F_κ on the boundary of Δ_5 and leads to somewhat heavier notations and inequalities than those for $\kappa = -\infty$.

COMPARISON MAPS IN GEODESIC SPACES. If X is a geodesic space, then one can express the CAT(κ)-property in terms of the *comparison maps*. Such a map c is defined with a triple $\{x_i\} \subset X$ and sends the *comparison triangle* $\Delta' \subset X' = H^2_\kappa$, spanned by $\{x'_i\} \subset H^2_\kappa$ with $|x'_i - x'_j|_{X'} = |x_i - x_j|_X$, to X, such that $x'_i \mapsto x_i$, $i = 1, 2, 3$ and the geodesic segments $[x'_i, x'_j]$ *isometricly* go to (possibly non-unique) segments $[x_i, x_j] \subset X$. The Alexandrov comparison inequality reads

$$|c(x') - c(y')|_X \le |x' - y'|_{X'}$$

for all x', y' on the perimeter of Δ' (and it is not hard to extend this c from the boundary of Δ to a 1-*Lipschitz* map on all of Δ, see [La-Sch]).

1. Broken Triangles and CAT$_\delta(\kappa)$.

It is slightly more convenient to depart from Cycl$_5$ rather than from Cycl$_4$ and work with 5-tuples of points in X, denoted by $\{x_1, y_2, x_2, x_3, y_3\}$, thought of as a broken (geodesic) triangle $\Delta(x_1, x_2, x_3)$ with the "breaks" y_2 and y_3, see Fig. 1.

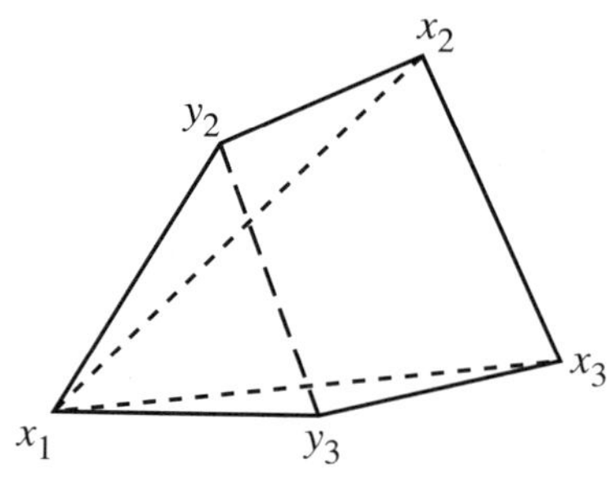

FIGURE 1

where we are concerned with an upper bound on $|y_2 - y_3|_X$ in terms of the five boundary edges and $|x_1 - x_i|$, $i = 2, 3$. We consider the comparison triangle (x'_1, x'_2, x'_3) in the model space $X' = H^2_\kappa$, where

$$|x'_1 - x'_i|_{X'} = |x_1 - y_i|_X + |y_i - x_i|_X \text{ for } i = 2, 3. \qquad (*)$$

and

$$|x'_2 - x'_3|_{X'} = \max(|x_2 - x_3|_X\,,\ |x'_1 - x'_2| - |x'_1 - x'_3|) \qquad \binom{*}{*}$$

and where $|x'_1 - x'_i|$ are computed according to $(*)$. (In other words, if the numbers $|x'_i - x'_j|$ violate the triangle inequality, we use the degenerate triangle in X' with $|x'_2 - x'_3| = ||x'_1 - x'_2|_{X'} - |x'_1 - x'_3|_{X'}|$). We measure non-straightness of the "broken edges" $[x_1, x_i]$, $i = 2, 3$ by the errors $\beta_i = |x_1 - y_i| + |y_i - x_i| - |x_1 - x_i|$ for $i = 2, 3$, where the equalities $\beta_i = 0$ signify that there is no breaks in these "edges".

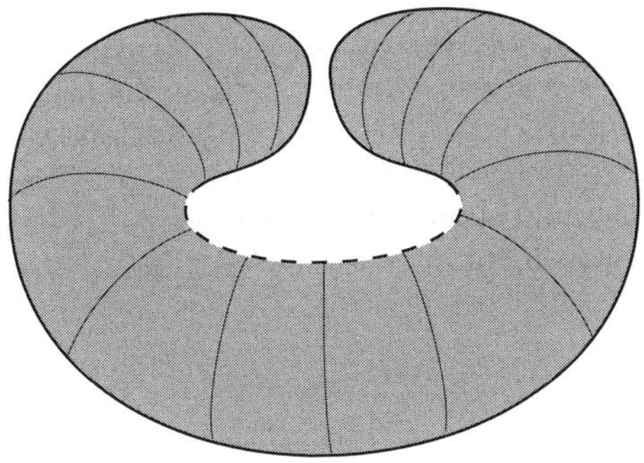

FIGURE 2

4. $\text{CAT}_\delta(\kappa|\sigma)$-spaces.

The above argument needed the $\text{CAT}_\delta(\kappa)$-properly only for subsets of the size $\leq \sigma$. With this in mind, we say that Y *is* $\text{CAT}_\delta(\kappa)$ *on the scale* $\leq \sigma$ and denote the corresponding class of spaces by $\text{CAT}_\delta(\kappa|\sigma)$ if the comparison inequality $(*_\delta)$ is satisfied by the triangles of diameter $\leq 10\sigma$. Then we have the following

4.A. CAT_δ HADAMARD-CARTAN THEOREM. *Let X be $\text{CAT}_\delta(\kappa|\sigma)$ with $\sigma \geq 10^5(-\kappa)^{\frac{1}{2}} \max(\delta\sqrt{-\kappa}, (\delta\sqrt{-\kappa})^{\frac{1}{4}})$. If X is locally and globally simply connected, then the metric balls in X are geodesicly ϱ-convex for some $\varrho \leq 1000(-\kappa)^{\frac{1}{2}} \max \delta\sqrt{-\kappa}, (\delta\sqrt{-\kappa})^4)$. Furthermore, X is δ'-hyperbolic for δ' depending only on $\kappa < 0$ and δ.*

PROOF. The only point requiring an explanation is the δ'-hyperbolicity. But this follows from the contraction property of the normal "projections" of X to balls (compare [Gro]$_{\text{HG}}$).

4.B. REMARK. It is not hard to show that $\delta' \leq 10^5((-\kappa)^{\frac{1}{2}} + \delta)$ and that X is $\text{CAT}_{\delta'}(\kappa)$ with the same bound on δ'.

5. Definition of $K_\delta(X|\sigma)$.

One could define $K_\delta(X|\sigma| \leq \kappa)$ as the $\text{CAT}_\delta(\kappa|\sigma)$-property but then one would loose, for example spaces of negative curvature with cusps. To take care of this, we have to require $\text{CAT}_\delta(\kappa|\sigma)$ not for balls but for suitable *multi-balls* $\widetilde{B}(R) \to X$. So we say that X has *curvature $\leq \kappa$ with error $\leq \delta$ in the σ-multivicinity of $x_0 \in X$*, and write $K_\delta(X, x_0|\sigma) \leq -\kappa$ if there exists a multidomain $(\widetilde{U}, \widetilde{x}_0) \to X$ with $\widetilde{x}_0 \mapsto x_0$ (compare 3.A.) such that

(i) $\text{dist}(\widetilde{x}_0, \partial\widetilde{U}) \geq 30\sigma$, where the distance in $\widetilde{U}$ is defined via the *path* metric induced from X and where $\partial\widetilde{U} \subset \widetilde{U}$ is defined as the subset where the (locally isometric!) map $\widetilde{U} \to X$ fails to be locally homeomorphic.

(ii) Every geodesic triangle in $\widetilde{U}$ of diameter $\leq 10\sigma$ and lying 10σ far from the boundary $\partial\widetilde{U} \subset \widetilde{U}$ satisfies the comparison inequality $(*_\delta)$.

(iii) $\widetilde{U}$ is (δ, σ) *regular* in the sense that every 100δ-small in diameter closed curve in U staying 10σ-from ∂U bounds a disk of diameter $\leq \sigma/100$ in $\widetilde{U}$ (compare [Gro]$_{\text{HG}}$).

5.A. K_δ-CARTAN-HADAMARD THEOREM. *If $K_\delta(X,x|\sigma) \leq -\kappa$ at all point $x \in X$, the space X is simply connected and $\sigma \geq 10^5(-\kappa)^{\frac{1}{2}} \max(\delta\sqrt{-\kappa}, (\delta\sqrt{-\kappa^{\frac{1}{4}}})$, then X is δ'-hyperbolic for some $\delta' = \delta'(\kappa,\delta) < \infty$. Moreover, $\delta' \leq 10^5((-\kappa)^{\frac{1}{2}} + \delta)$ and X is $\mathrm{CAT}_{\delta_\bullet}(\kappa_\bullet)$ for $\delta_\bullet \leq 10^5\delta$.*

The proof is quite similar to the case of $K(X) \leq \kappa$ as well as of the the δ-hyperbolic case (see $[\mathrm{Gro}]_{\mathrm{HG}}$, $[\mathrm{Gro}]_{\mathrm{CAT}}$ and [Bri-Hae].

5.B. REMARK. One can define K_δ with suitable fillings of closed contractible curves $S = S_\ell$ in X by κ-disks (D_μ) with length $(\partial D) = \ell$ (compare $[\mathrm{Gro}]_{\mathrm{CAT}}$. Here the relevant maps : $(D,\mu) \to X$ are required to be distance decreasing up-to an additive error δ, i.e.

$$|\varphi(x) - \varphi(y)|_X \leq |x-y|_\mu + \delta,$$

where we additionally insist on the continuity of φ (which is unnecessary under the above regularity assumption). In fact, one can localize this definition by applying it only to short curves S, i.e. of length $\ell \leq \sigma$ and show that this suffices for the conclusion of the K_δ-Cartan-Hadamard theorem to hold true.

6. Evaluation of K_δ for coned spaces.

The *hyperbolic cone* $C_{\kappa,r}(U)$ over a path metric space U is defined as the topological cone $U \times [0,r]/U \times 0$ with the path metric, such that for each (short) curve $S \subset U$ the metric on $S \times [0,r]/S \times 0 \subset \mathrm{Cone}_{\kappa,r}(U)$ equals that of the segment in C' the r-disk $D(r)$ in H^2_κ over an ark $S' \subset \partial D(r)$ with $\mathrm{length}(S') = \mathrm{length}(S)$, where S' is (obviously) identified with S and C' with $S \times [0,r]/S \times 0$.

6.A. LEMMA. *If U is $\mathrm{CAT}(\kappa_0 \leq 0)$ then $C_{\kappa,r}(U)$ is $\mathrm{CAT}(\kappa)$.*

This is obvious (compare [Bri-Hae]). In fact, it suffices to assume that U is $\mathrm{CAT}(\kappa_+)$ where κ_+ is the intrinsic curvature of the r-sphere in H^3_κ.

6.A′. COROLLARY. *If U is $\mathrm{CAT}_\delta(\kappa_0 \leq 0)$ then $C_{\kappa,r}(U)$ is $\mathrm{CAT}_\delta(\kappa)$.*

REMARK. All of the above generalizes to *non*-path metric spaces.

6.B. Given subsets U_i in a geodesic space X we consider the cones $C_{\kappa,r}(U_i)$, attach each $C_{\kappa,r}(U_i)$ to X accross $U_i = U_i \times r \subset C_{\kappa,r}$, and let $X^\bullet \supset X$ denote so multi-coned spaced X with the natural path metric (compare $[\mathrm{Gro}]_{\mathrm{CAT}}$).

We want to evaluate $K_\delta(X^\bullet)$ for the resulting space $X^\bullet$, where we are concerned with the points away from the apices of the cones as the latter has been taken care of in the above lemmas. Start with the case where X is a tree, i.e. a $\mathrm{CAT}(-\infty)$-space and U_i, $i \in I$ are subtrees. Here an important characteristics of $\{U_i\}$ is the *(maximal) overlap* between U_i, i.e. $\ell = \ell\{U_i\} = \mathrm{supdiam}(U_i \cap U_i)$.

6.C. CONING LEMMA. *Let $\sigma \leq r/2$ and $\ell \leq 0.01\min(1/\sqrt{-\kappa}, \sigma)$. Set*

$$\delta = \delta(\beta) = \delta_\kappa(\beta + \epsilon) + \epsilon$$

for δ_κ defined in §1 and

$$\epsilon = 100(\ell\sqrt{-\kappa})^{\frac{2}{3}}\sigma\sqrt{-\kappa}.$$

Then $X^\bullet$ is $\mathrm{CAT}_\delta(\kappa)$ within the σ ball around each point $x \in X \subset X^\bullet$. Notice that we returned to *functions* $\delta(\beta)$ and $\delta_\kappa(\beta)$ from §1).

PROOF. To grasp the idea, let first $\ell = 0$. In this case $X^\bullet$ is obviously (compare [Gro]$_{\mathrm{HG}}$, [Gro]$_{\mathrm{CAT}}$ and [Bri-Hae]) CAT(κ). Next we observe that CAT$_\delta(\kappa)$ (including CAT$_\delta(-\infty)$ corresponding to hyperbolicity) is stable under perturbations of spaces in the Hausdorff metric (with $\frac{1}{2}$-Hölder dependence of δ on the Hausdorff distance). This, and the essential locality of CAT$_\delta(\kappa)$ (by Hadamard-Cartan) imply that $X^\bullet$ is CAT$_\delta(\kappa)$, where $\delta = \delta(\ell) \to 0$ for $\ell \to 0$.

To get a feeling for an actial bound on δ, one may look at the simplest tree, namely $X = \mathbb{R}$ where U_i are just subsegments $\subset \mathbb{R}$. The major problems in understanding the geometry of the corresponding $X^\bullet$ consists in evaluating by how much the distance $|x-y|_{X^\bullet}$ is smaller than $|x-y|_{\mathbb{R}}$ for $x, y \in \mathbb{R}$. The small (i.e. $\ll \ell$) segments $U_i \subset \mathbb{R}$ do not have a significant effect on the shortening $|x-y|_{\mathbb{R}} \rightsquigarrow |x-y|_{X^\bullet}$ and one may concentrate on the large ($\gg \ell$) U_i's. These can be assumed to have only double intersections, where the combinatorics is transparent and $|x-y|_{X^\bullet}$ is easily computable.

For general trees X, the problem is complicated by the presence of brance points, but the number of these can be, a priori, bounded. Thus one essentially reduces the general case to that of the Y-shaped tree, where the branched point is treated as if it is covered by a large U_i.

Keeping all this in mind we start by observing, now for a general tree X, that the radical projections from the apices of the cones send every geodesic segment $[x_1^\bullet, x_2^\bullet]$ in $X^\bullet$ missing the apices to *geodesic* segments in X, i.e. these projection *biject* $[x_1^\bullet, x_2^\bullet]$ to a segment $[x_1, x_2]$ in the tree X, as an obvious argument shows. Since the CAT$_\delta$-property concerns 5-tuples of points $\{x_i^\bullet\} \subset X^\bullet$, $i = 1, \dots, 5$, we may assume that our X is spanned by five extreme points (leaves) $x_i \in X$. Now X contains at most three interior branch points of valency (degree) at most five and to simplify the computation, we also assume $\kappa = -1$. (This can always be achieved by scaling $X^\bullet$ to $\sqrt{-\kappa X^\bullet}$). We divide U_i into two groups : those with $\operatorname{diam} U_i \geq \ell^{\frac{1}{3}}$ and the remaining short ones with $\operatorname{diam} U_i < \ell^{\frac{1}{2}}$. The intersection between two large U_i, say $U_i \cap U_j$, consists of at most $14 = 7 + 7$ disjoint subtrees of diameters $\geq \ell$, where each of these meets a vertex (i.e. an extreme or a branch point) either in U_i or in U_j and no more than five large U_i ever meet at a single point. In fact, the union of all these double intersections may consist of at most three subtrees meeting the branch points of X and having diameters $\leq 3\ell$ and several disjoint segments of diameters $\leq \ell$, where each of the segments meets exteme points of exactly two U_i.

Now, we perform two operations over $X^\bullet$: first we remove the cones over all short U_i with the exception of those containing our points $x_i^\bullet$, $i = 1, \dots, 5$. Clearly the distances between x_i' in the resulting space may be greater than they were in $X^\bullet$ but not by face. In fact, as a geodesic in $X^\bullet$ passes through a cone C_j with short base U_j, it remains within distance $\leq 5(\ell^{\frac{1}{3}})^2$ from U_j (which was *not* at all removed along with C_j by our construction) and its length is only slightly shorter at most by the factor $1 - 5\ell^{\frac{2}{3}}$. Thus the total enlargement in the distances does not exceed $\sigma(20\ell^{\frac{2}{3}} + 30(\ell^{\frac{1}{3}})^2) \leq \frac{\epsilon}{2}$. Next, after removing the above C_j we shrink to points the connected components of the pairwise intersections of the remaining U_i in X (i.e. the great ones and those containing the ends) and conicly extend this shrinking to the cones C_i. Dispite the fact that the total length of U_i can be quite large compared to σ (since the apex projections shorten *long* geodesic segments in C_i near their ends where they meet U_i, while the total length of these segments strongly increases under these projections), the shortening effect to each C_i is bounded by 2ℓ since our shrinking is conical. On the other hand, the total length of these segments does not exceed 10σ and so the enlargement is bounded by $\epsilon/2 = 50\sigma\ell/\ell^{\frac{1}{3}}$. Now the lemma follows easily with the local-to-global transition provided by the Hadamard-Cartan theorem.

COROLLARY. *If $\ell \geq 10^{-11}/\sqrt{-\kappa}$ and $r \geq 0.2/\sqrt{-\kappa}$, then $X^\bullet$ is $(10^3/\sqrt{-\kappa})$-hyperbolic.*

6.D. ON THE ROLE OF CAT_δ AND K_δ. One could prove the Corollary within a purely δ-hyperbolic framework and essentially by the same argument without bothering with the notions of CAT_δ and K_δ. In fact, as far as applications to random groups are concerned, one could go without these spaces. However, the lemma carries extra geometric information, such a (better than just δ-hyperbolic) bound on the distance between "parallel" geodesics that is needed in the combinatorial group theory. Actually, one needs this mesoscopic curvature not so much for $X^\bullet$ itself, but rather for quotient spaces $X^\bullet/\Gamma$ for (rotation) isometry groups Γ acting on $X^\bullet$ where the action is free apart from the apices of the cones. If the curvature remains $\leq \kappa$ at the appices, then one can prove (compare $[\mathrm{Gro}]_{\mathrm{CAT}}$) that $X^\bullet/\Gamma$ is $\mathrm{CAT}_\delta(\kappa)$ which leads to the $\mathrm{CAT}_\delta(\kappa)$-refinement of the hyperbolicity (i.e. $\mathrm{CAT}_{\delta'}(-\infty)$ with $\delta' \gg \delta$) of small cancellation groups. Furthermore, this allows a similar sharpening of the *hyperbolic small cancellation* theory as we shall see below.

REMARK. If the subtrees U_i in the tree X are homeomorphic to $\mathbb{R}$, then $X^\bullet$ can be (canonicly) Hausdorff approximated by a $\mathrm{CAT}(\kappa)$ space (see $[\mathrm{GRO}]_{\mathrm{CAT}}$).

7. Cones over hyperbolic spaces.

Let X be a δ_0-hyperbolic space and let us look first at an individual cone $C = C_{\kappa,r}(X)$, where the important case is where $|\kappa|^{\frac{1}{2}}$ is small compared to $1/\delta_0$. Since X behaves like a tree on the scale $\gg \delta_0$, we see that C has curvature $\leq \kappa$ up to a δ_0-error; more precisely we have the following

7.A. LEMMA. *The cone C is $\mathrm{CAT}_\delta(\kappa)$ for*

$$\delta = \delta(\beta) = \delta_\kappa(\beta) + 10\delta_0.$$

In particular, if $100\sqrt{-\kappa} \leq \delta_0$, then C is δ_1-hyperbolic for $\delta_1 = 10/\sqrt{-\kappa}$.

Notice that this hyperbolicity less informative than the $\mathrm{CAT}_\delta(\kappa)$-property since the former (non-trivially) controls the geometry of the balls of radii $\gamma\sigma$ for σ starting from something like $100\sqrt{-\delta_0/\kappa}$ while the hypbolicity of C emerges only on the scale of $\sigma \approx 10/\sqrt{-\kappa}$.

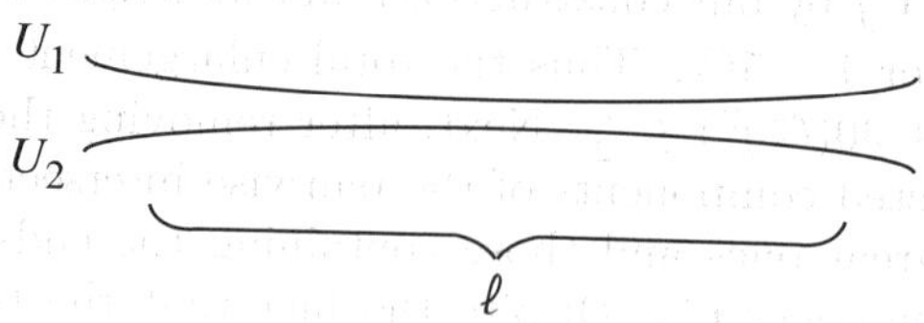

FIGURE 3

Take ϵ-convex (see [Gro]$_{\text{HG}}$) subsets $U_i \subset X$ and obtain $X^\bullet$ by attracting the cones $C_{\kappa,r}(U_i)$ to X as earlier. The definition of the overlap ℓ defined via the intersections $U_i \cap U_j$ is no good anymore. For example, if we have two distinct geodesics U_1 and U_2 in H^2_κ going close to each other along segments of length $\approx \ell$ and then going far apart, we should assign the overlap ℓ to them rather than zero, see Fig. 3.

This can be taken care of by either considering $(2\epsilon + 10\delta)$-neighbourhoods of U_i and taking

$$\ell_1 = \sup_{i \neq j} \operatorname{Diam}((U_i + 2\epsilon + 10\delta) \cap (U_j + 2\epsilon + 10\delta))$$

or by using the supremum ℓ_2 of the diameters of the normal (i.e. nearest point(s)) projections $U_i \to U_j$ for all i and $j \neq i$ with either of these ℓ's one has a generalization of the tree coning lemma, where we stick below to $\ell = \ell_1$.

7.B. Hyperbolic Coning Lemma. *If* $1/\sqrt{-\kappa} \geq 1000\delta_0$ *and* $\ell \leq 10^{-3}/\sqrt{-\kappa}$ *then* $X^\bullet$ *is* $\mathrm{CAT}_\delta(\kappa)$ *on the* σ*-balls in* $X^\bullet$ *for*

$$\delta = \delta(\beta) = \delta_\kappa(\beta - \epsilon + 10\delta_0) + \epsilon_0 + 10^4\delta_0$$

with

$$\epsilon_\bullet = 10^{10}(\ell\sqrt{-\kappa})^{\frac{2}{3}} \min(1, \sigma\sqrt{-\kappa}),$$

where δ_κ *is from §1. In particular, if* $r \geq 1/\sqrt{-\kappa} \geq 10^5\delta_0$ *and* $\ell \leq 10^{-50}/\sqrt{-\kappa}$, *then* $X^\bullet$ *is* δ_1*-hyperbolic for* $\delta_1 = 10^{20}/\sqrt{-\kappa}$.

Proof. The idea is that the space X is tree-like on δ_0-scale and so 7.A applies with a δ_0-error. Actually, one could go to the limit for $\delta_0 \to 0$, where X converges to a tree, and thus derive the qualitative form of the present lemma from 7.A. Since it is convenient (at least notationally) to keep track of the constants, we do it differently. Take five points x_i in $X^\bullet$, join them by ten segments in $X^\bullet$ and project these to X from the apices of the cones in $X^\bullet$. (This applies only to points within σ-balls around points $x \in X \subset X^\bullet$ with $\sigma \leq \min(r/4,\ 1/\sqrt{-\kappa})$, while the balls of radii $\leq r/2$ away from $X \subset X^\bullet$ are already taken care of. Then the properties of larger balls follow with our generous constants). Segments project to segments and the union of the ten projected segments can be δ_0'-approximated by a tree Y in X built of seven segments with $\delta_0' \leq 1000\delta$ (see [Gro]$_{\text{HG}}$). Now we apply the previous proof to this subtree with minor adjustments related to the position of Y relative to U_i. A little difficulty comes from the fact, that U_i are *not* geodesicly convex in the strict sense and so some edges of Y with vertices in some U_i may go outside U_i. The easiest way to help it is to stabilize U_i by taking their $4\delta_0$-neighbourhoods. This only slightly changes the geometry of $X^\bullet$ and brings a minor (by far less than we presented) modification of the constants. (Recall that ℓ refers to the geometric overlaps of neighbourhoods of U_i anyway and with this convention the constants hardly change at all).

7.C. Applications. The main case of interest is where X is acted upon an isometry group Γ generated by Γ_i where each Γ_i keeps U_i invariant and has "large" displacements outside U_i. Then Γ naturally acts on $X^\bullet$ with the apices of the cones kept fixed by Γ_i and, under suitable (and easily verifiable) conditions each quotient space $X^\bullet/\Gamma_i$ is $\mathrm{CAT}_\delta(\kappa)$. Then by a straightforward generalization of the argument in [Gro]$_{\mathrm{CAT}}$ one obtains the $\mathrm{CAT}_{\delta'}(\kappa)$ property for $X^\bullet/\Gamma$ with something like $\delta' = 4\delta$ (or slightly worse but yet controlled on large scale σ). This shows, that the hyperbolic small cancellation groups are not only hyperbolic, but behave as if they were acting on spaces of negative curvature $\kappa < 0$ where "as if" corresponds to δ in $\mathrm{CAT}_\delta(\kappa)$ that can be made arbitrarily small (for a fixed $\kappa < 0$) in many cases of interest.

8. Coning $\mathrm{CAT}_\delta^+(\kappa, r)$-spaces.

When we compare a space X_+ with the sphere $X'_+ = S^2_{\kappa_+}$ of radius $1/\sqrt{x_+}$ we must be aware of the fact that some configurations of points in X_+ have no counterparts in X'_+ as the distances between these points may exceed $\pi/\sqrt{\kappa_+} = \operatorname{diam} X'_+$; Then, defining $\mathrm{CAT}_\delta(\kappa_+)$-property, we require the relevant inequality between distances only from those 5-tuples $\{x_1, \dots, y_3\} \subset X_+$ which are suitably realizable in X'. As we are eventually concerned with $C_{r,\kappa}$-cones over these spaces we just *define* the relevant property via these cones as follows.

A space X_+ is *called* $\mathrm{CAT}_\delta^+(\kappa, r)$ for some $\kappa < 0$ and $\delta \geq 0$ (which may be understood either as a function $\delta(\beta)$ or just a single number $\delta = \delta(0)$) if the cone $C_{\kappa,r}(X_+)$ is $\mathrm{CAT}_\delta(\kappa)$.

For example, $\mathrm{CAT}_{\delta=0}^+(0,1)$ is obviously equivalent to $\mathrm{CAT}(1)$. It is easy to show that

$$\mathrm{CAT}_\delta^+(\kappa, r) \Rightarrow \mathrm{CAT}_\delta^+(\kappa', r')$$

for all $\kappa' \leq \kappa$ and $r' \leq r$.

What we shall actually need is the following obvious

8.A. LEMMA. *Let X_+ be a geodesic metric space that is δ_0-hyperbolic in every ball of radius R in X_+. Then X_+ is $\mathrm{CAT}_{\delta_1}^+(\kappa, r)$, for $\delta_1 = \delta_1(\beta) = \delta_\kappa(\beta) + 10\delta_0$,*

$$r \leq 0.1(\log\sqrt{-\kappa R}/\sqrt{-\kappa}.$$

Notice, that this inequality needs to be confronted with the lower bound $r \geq 1/\sqrt{-\kappa}$. Also observe, that if $X_+ = \widetilde{X}_+/\Gamma_+$ for a (now globally) δ_0-hyperbolic space $\widetilde{X}_+$ and a free isometry group Γ acting on $\widetilde{X}_+$ with $|x - \gamma(x)|_{\widetilde{X}_+} \geq 3R \geq 100\delta_0$ for all $x \in \widetilde{X}_+$ and all non-identity elements $\gamma \in \Gamma_+$, then X_+ is δ_0-hyperbolic on the R-balls in X_+. This is exactly where the above Lemma becomes useful.

9. Final Remarks and open questions.

(a) It is unclear if the $\mathrm{CAT}_\delta(\kappa)$-spaces can be Hausdorff δ'-approximated by $\mathrm{CAT}(\kappa)$-spaces with $\delta' = \delta'(\delta) \to 0$ for $\delta \to 0$. Probably, to have such approximation, one should strengthen the definition of $\mathrm{CAT}_\delta(\kappa)$ by incorporating more (δ-perturbed) inequalities, such as all $\mathrm{Cycl}_k(\kappa)$, for example.

(b) There is an obvious δ-perturbed version of the inequality defining Alexandrov's spaces with $K > 0$, and it may be interesting to look at the corresponding spaces with $K > \kappa$ up-to an error δ.

(c) The above mesoscopic extension (δ-perturbation) of locally defined classes of spaces goes along with the Hausdorff distance between metric spaces. If we look from the angle of *metric measure* spaces (see [G-L-P]) we come up with other classes where the relevant (e.g. 4-point) inequalities are satisfied up-to δ-error understood in some *measure theoretic* (integral) sense. What happens, for example, to Ricci > 0 in this setting ?

References

[Bri-Hae] Bridson, Martin R., Haefliger, A., *Metric Spaces of Non-Positive Curvature*, Grundlehren Der Mathematischen Wissenschaften, Vol **319**.

[Gro_{AI}] Gromov, M., *Asymptotic invariants of infinite groups. Geometric group theory*, Volume 2, Proc. Symp. Sussex Univ., Brighton, July 14–19, 1991; Lond. Math. Soc. Lecture Notes **182**, Niblo and Roller ed., Cambridge Univ. Press, Cambridge (1993), 1–295.

[Gro_{CAT}] Gromov, M., CAT(κ)-*spaces: construction and concentration* (to Victor Abramovich Zalgaller), January 2001.

[Gro_{HG}] Gromov, M., *Hyperbolic groups*, in *Essays in Group Theory*, Mathematical Sciences Research Institute Publications **8** (1978), 75–263, Springer-Verlag.

[G-L-P] Gromov M., *Metric Structures for Riemannian and Non-Riemannian Spaces*, based on: *Structures Métriques des Variétés Riemanniennes* (with Appendices by M. Katz, P. Pansu, and S. Semmes), edited by J. LaFontaine and P. Pansu, English Translation by Sean M. Bates, Birkhäuser, Boston - Basel - Berlin (1999).

[La-Sch] Lang, U., Schroeder, V., *Kirszbraun's theorem and metric spaces of bounded curvature*, Geom. Funct. Anal. **7** (1997), no. 3, 535–560.

Current address: Institut des Hautes Etudes Scientifiques, Bois-Marie, 35 route de Chartres, F-91440 Bures-sur-Yvette, and New York University, Courant Institute of Mathematical Sciences, 251 Mercer St, New York, NY 10012-1110, USA.

E-mail address: gromov@ihes.fr

Contemporary Mathematics
Volume **288**, 2001

Stable Forms and Special Metrics

Nigel Hitchin

Dedicated to the memory of Alfred Gray.

ABSTRACT. We show how certain diffeomorphism-invariant functionals in dimensions $6, 7$ and 8 generate in a natural way special geometrical structures in these dimensions: metrics of holonomy G_2 and $Spin(7)$, metrics with weak holonomy $SU(3)$ and G_2, and a new and unexplored example in dimension 8. The general formalism becomes a practical tool for calculating homogeneous or cohomogeneity one examples, and we illustrate this with some newly discovered examples of $Spin(7)$ and G_2 metrics.

1. Introduction

One of Alfred Gray's most original concepts was that of weak holonomy [**G**]. This was an idea clearly ahead of its time, as became evident in the later work on Killing spinors of Baum et al. [**B**], setting it in a natural context. We shall give here another natural approach to both weak holonomy and special holonomy in low dimensions through the use of certain invariant functionals of differential forms. This provides both a canonical setting for these structures and sometimes an effective means of finding them.

Our starting point is the question: "What is a non-degenerate form?". Symplectic geometry is the geometry determined by a closed non-degenerate 2-form where non-degeneracy means $\omega^m \neq 0$. Another way to describe such genericity at each point is to note that the orbit of a nondegenerate $\omega \in \Lambda^2 V^*$ under the natural action of $GL(V)$ on $\Lambda^2 V^*$ is open. We shall say in general that $\rho \in \Lambda^p V^*$ is *stable* if it lies in an open orbit. (The use of this word rather than "non-degenerate" avoids possible confusion: a 2-form in odd dimensions may be stable but is always degenerate as a bilinear form. Also, ρ is stable in the sense of deformation invariance: all forms in a neighbourhood of ρ are $GL(V)$-equivalent to ρ).

When does stability occur? Clearly not very often since $\dim GL(V) = n^2$ is usually much smaller than $\dim \Lambda^p V^* = n!/p!(n-p)!$. The result (over the complex numbers) is classical [**R**], [**S**], [**Gu1**], [**Gu2**] but Robert Bryant explained to me all

1991 *Mathematics Subject Classification.* Primary 53C25, 53C10; Secondary 58E11, 58A10.
Key words and phrases. Special holonomy, nearly Kähler, weak holonomy.
The author wishes to thank the Universidad Autónoma, Madrid and the Programa Catèdra Fundación Banco de Bilbao y Vizcaya for support during the preparation of this paper.

the real cases. Apart from the obvious case of $p = 1, 2$ there are essentially only three more, where $p = 3$ and $n = 6, 7, 8$. In these three cases the stabilizer subgroup of $\rho \in \Lambda^p V^*$ in $GL(V)$ is a real form of one of the complex groups

$$SL(3) \times SL(3), \quad G_2, \quad PSL(3)$$

respectively. We shall be concerned here only with the real forms $SL(3, \mathbf{C})$ and the compact groups G_2 and $PSU(3)$. Thus if a manifold M^n admits a global stable p-form, it has a G-structure where G is one of these groups.

Note that if $GL(V)$ has an open orbit in $\Lambda^p V^*$ then it also does on the dual space $\Lambda^p V \cong \Lambda^{n-p} V^* \otimes \Lambda^n V$. Since for $p \neq 0$ the scalars act non-trivially, there is then an open orbit on $\Lambda^{n-p} V^*$. Thus if we are in a dimension n where stable p-forms exist, we can also consider stable $(n - p)$-forms.

The three stabilizers of ρ above, as well as the symplectic group, each preserve a volume element $\phi(\rho) \in \Lambda^n V^*$. Thus if we have a compact oriented manifold M^n and a p-form ρ which is everywhere stable, we can integrate $\phi(\rho)$ to obtain a volume $V(\rho)$. Openness of the orbit implies that nearby forms are also stable, so that the volume functional is defined and smooth on an open set of forms.

We now set up a number of variational problems involving this functional. First we consider a critical point of $V(\rho)$ restricted to a fixed cohomology class of closed forms in $H^p(M, \mathbf{R})$, performing a non-linear version of Hodge theory. If ρ is everywhere stable we find the following structures:

- for $n = 2m$ and $p = 2$ or $2m - 2$, a symplectic manifold,
- for $n = 6$ the structure of a complex 3-manifold with trivial canonical bundle (this is described in some detail in [**H**]),
- for $n = 7$ and $p = 3$ or 4 a Riemannian manifold with holonomy G_2.

The critical points for $n = 8$ and $p = 3$ or 5 form a class of geometric structures which is largely unexplored (though I have benefited from Robert Bryant's thoughts on these). We study them briefly in Section 4, showing that they admit a solution to the Rarita-Schwinger equations, which imposes constraints on the Ricci tensor. The lack of concrete compact examples beyond the 8-manifold $SU(3)$ itself is currently a stumbling block in taking the analysis of these further, but the fact that they arise from the same variational origins as G_2-manifolds suggests that they ought to exist in abundance.

To continue with other variational characterizations, we note that there is a canonically defined indefinite quadratic form

$$Q(d\alpha) = \int_M \alpha \wedge d\alpha$$

on the space of exact 4-forms $d\alpha$ on a 7-manifold M. Thus Q defines an indefinite metric on the closed 4-forms in a fixed cohomology class. We then find:

- the stable critical points of $V(\rho)$ on the trivial cohomology class, subject to the constraint $Q(d\alpha) = const$ define a 7-manifold with a weak holonomy G_2 structure,
- the gradient flow of $V(\rho)$ on a fixed degree 4 cohomology class yields a Riemannian metric of holonomy $Spin(7)$ on $M \times \mathbf{R}$.

Finally, consider a 6-manifold M and a stable closed 3-form ρ together with a stable closed 4-form σ. We note that the spaces of exact 3-forms $d\alpha$ and exact 4-forms $d\beta$ on a 6-manifold are formally dual to each other via the pairing

$$\langle d\alpha, d\beta\rangle = \int_M \alpha \wedge d\beta.$$

We find

- for ρ and σ exact, the stable critical points of $3V(\rho)+8V(\sigma)$ subject to the constraint $\langle \sigma, \rho\rangle = const$ define a manifold with weak holonomy $SU(3)$,
- the pairing defines a formal symplectic structure on the product of a degree 3 and degree 4 cohomology class and the Hamiltonian flow of the functional $V(\rho)-2V(\sigma)$, with an appropriate initial condition, generates a Riemannian metric with holonomy G_2 on $M \times \mathbf{R}$.

(The particular coefficients here are a computational convenience and have no geometrical meaning.)

The formal variational setting of all these special geometries is perfectly general – any such structure appears in this way – but it also has a practical value. In the homogeneous or cohomogeneity one situation we merely set up the same variational equations using *invariant* forms and easily derive the corresponding equations. We demonstrate this in the case of S^7 by deriving the equations recently used by Čvetic et al. [**C**] to find a new $Spin(7)$ metric on $\mathbf{R}^8$, and in the work of Brandhuber et al. [**Br**] for a new example of a metric of holonomy G_2.

2. The linear algebra of stable forms

Let $\rho \in \Lambda^p V^*$ be stable, in the sense described above, i.e. it lies in an open orbit U of $GL(V)$. We consider the cases:

1. $\dim V = 2m$, $p = 2$ or $2m-2$: stabilizer $Sp(2m, \mathbf{R})$
2. $\dim V = 6$, $p = 3$: stabilizer $SL(3, \mathbf{C})$
3. $\dim V = 7$, $p = 3$ or 4: stabilizer G_2
4. $\dim V = 8$, $p = 3$ or 5: stabilizer $PSU(3)$

We see that each stabilizer preserves a volume form: G_2 and $PSU(3)$ are compact so this is the volume form of an invariant positive definite inner product on V. The symplectic group fixes the Liouville volume and $SL(3, \mathbf{C})$ preserves a complex 3-form Ω and hence the real 6-form $i\Omega \wedge \bar{\Omega}$. In the appendix we give the concrete expression of this form in each case. It is algebraically determined by ρ and smooth on U.

The volume form $\phi(\rho)$ associated to ρ defines a $GL(V)$-invariant map

$$\phi : U \to \Lambda^n V^*.$$

Applying invariance to the action of the scalar matrices, we see that

$$\phi(\lambda^p \rho) = \lambda^n \phi(\rho)$$

so that ϕ is homogeneous of degree n/p.

The derivative of ϕ at ρ is an invariantly defined element of $(\Lambda^p V^*)^* \otimes \Lambda^n V^*$. Since $(\Lambda^p V^*)^* \cong \Lambda^{n-p} V^* \otimes \Lambda^n V$, the derivative lies in $\Lambda^{n-p} V^*$ so there is a unique element $\hat{\rho} \in \Lambda^{n-p}V^*$ for which

$$D\phi(\dot{\rho}) = \hat{\rho} \wedge \dot{\rho} \tag{2.1}$$

Taking $\dot{\rho} = \rho$, Euler's formula for a homogeneous function gives

$$\hat{\rho} \wedge \rho = \frac{n}{p}\phi(\rho) \tag{2.2}$$

EXAMPLE 2.1. If $\dim V = 2m$ and $\omega \in \Lambda^2 V^*$ is stable (meaning non-degenerate here), then we take the Liouville volume form

$$\phi(\omega) = \frac{1}{m!}\omega^m \in \Lambda^{2m}V^*$$

This is clearly homogeneous in ω of degree $m = 2m/2 = n/p$. Differentiating, we see that

$$D\phi(\dot{\omega}) = \frac{1}{(m-1)!}\dot{\omega} \wedge \omega^{m-1}$$

so that

$$\hat{\omega} = \frac{1}{(m-1)!}\omega^{m-1}.$$

We may also consider a stable $\rho \in \Lambda^{2m-2}V^*$, which is in the open orbit U consisting of forms $\rho = \omega^m/(m-1)!$ for a non-degenerate ω. In this case we have $\hat{\rho} = \omega/(m-1) \in \Lambda^2 V^*$.

The precise form of $\hat{\rho}$ is determined by seeing which elements of $\Lambda^{n-p}V^*$ are fixed by the stabilizer. The symplectic case is done in the example above. For the others we see easily that:

- for $n = 6$, $p = 3$, $\hat{\rho}$ is determined by the property that $\Omega = \rho + i\hat{\rho}$ is a complex $(3,0)$-form preserved by $SL(3, \mathbf{C})$,
- for $n = 7$, $p = 3$ or 4, $\hat{\rho} = *\rho$, where $*$ is the Hodge star operator for the inner product on V
- for $n = 8$, $p = 3$ or $p = 5$, $\hat{\rho} = - * \rho$.

REMARK 2.2. There is clearly a choice in what we call the volume in each case. There are conventions – the Liouville volume $\omega^m/m!$ in the symplectic case for example. For most purposes it makes no difference to the results that follow, but when we need to find a metric, as in Sections 5 and 6, we shall make a more explicit choice to aid the calculations.

3. Critical points

Suppose now that M is a closed, oriented n-manifold. If $\rho \in \Omega^p(M)$ is a global p-form, then it is a section of $\Lambda^p T^*$. Suppose its value is stable at each point (such an assumption requires of course the reduction of the structure group of the tangent bundle to one of the stabilizers above). Then we can define a functional by taking the total volume:

$$V(\rho) = \int_M \phi(\rho).$$

By definition of stability, nearby forms will be stable and so we can differentiate the functional. We shall set up a variational problem by considering the volume restricted to *closed* stable p-forms in a given cohomology class.

THEOREM 3.1. *A closed stable form $\rho \in \Omega^p(M)$ is a critical point of $V(\rho)$ in its cohomology class if and only if $d\hat{\rho} = 0$.*

PROOF. Take the first variation of $V(\rho)$:

$$\delta V(\dot{\rho}) = \int_M D\phi(\dot{\rho}) = \int_M \hat{\rho} \wedge \dot{\rho}$$

from (2.1). But the variation is within a fixed cohomology class so $\dot{\rho} = d\alpha$. Thus

$$\delta V(\dot{\rho}) = \int_M \hat{\rho} \wedge d\alpha = \pm \int_M d\hat{\rho} \wedge \alpha$$

and the variation vanishes for all $d\alpha$ if and only if

$$d\hat{\rho} = 0.$$

□

EXAMPLE 3.2. A closed stable 2-form ω on an even-dimensional manifold is a symplectic form. The volume $V(\omega)$ is then *constant* on a fixed cohomology class $[\omega]$ since it is just the evaluation of the cup product:

$$V(\omega) = \frac{1}{m!}[\omega]^m[M].$$

Rather trivially a symplectic manifold appears as a critical point here.

On the other hand if we take $\rho = \omega^{m-1}/(m-1)!$ to be the closed form then the condition $d\rho = 0$ is (for $m > 2$) weaker than $d\omega = 0$. The functional genuinely varies and the critical points are where

$$(m-1)d\hat{\rho} = d\omega = 0.$$

This is an alternative way of obtaining a symplectic manifold as a critical point.

EXAMPLE 3.3. A 7-manifold M which has either a closed stable 3-form or a closed stable 4-form which is a critical point for V in its cohomology class has the structure of a Riemannian manifold with holonomy G_2. This follows from the theorem of M. Fernández and A. Gray [**F**] that the holonomy reduces to G_2 if and only if

$$d\rho = d * \rho = 0.$$

EXAMPLE 3.4. The case of $n = 6$ means that we have a complex closed locally decomposable 3-form $\rho + i\hat{\rho}$. This yields the structure of a complex 3-manifold with trivial canonical bundle as shown in [**H**].

The 8-dimensional case will occupy us next.

4. Eight-manifolds with $PSU(3)$ structure

Suppose that M is a compact 8-manifold with a stable 3-form ρ such that, with respect to the metric determined by ρ,

$$d\rho = d * \rho = 0.$$

An example is $SU(3)$ itself, where ρ is covariant constant and is a multiple of the standard bi-invariant form

$$\rho = \mathrm{tr}(g^{-1}dg)^3.$$

We shall never find a metric on a compact simply-connected M^8 whose holonomy is $PSU(3)$ other than this example, from Berger's classification of Riemannian holonomy groups. This is not the case of weak holonomy either – there are no Killing spinors. We do however have one interesting object: a Rarita-Schwinger field. In physics terminology this is a spin 3/2 field. We take one of the two spinor bundles (say S^+) and consider a spinor-valued 1-form – a section γ of $S^+ \otimes \Lambda^1$. This satisfies the *Rarita-Schwinger equation* if

$$D\gamma = 0$$

and

$$d^*\gamma = 0.$$

Here

$$D : C^\infty(S^+ \otimes \Lambda^1) \to C^\infty(S^- \otimes \Lambda^1)$$

is the Dirac operator with coefficients in the bundle of 1-forms Λ^1 and

$$d^* : C^\infty(S^+ \otimes \Lambda^1) \to C^\infty(S^+)$$

is the covariant d^* operator on 1-forms with coefficients in the spinor bundle S^+.

First we describe γ as a $PSU(3)$-invariant object.

LEMMA 4.1. *Let S^+, S^- be the two spin representations and Λ^1 the standard vector representation of $Spin(8)$. Then restricted to the lift $PSU(3) \subset Spin(8)$ of the adjoint representation, these three representations are equivalent.*

PROOF. If $\pm x_1, \dots, \pm x_4$ are the weights of the 8-dimensional vector representation of $Spin(8)$, the weights of the spin representations $S^\pm$ are

$$\frac{1}{2}(\pm x_1 \pm x_2 \cdots \pm x_4) \tag{4.1}$$

where there is an even number of minus signs for S^+ and an odd number for S^-.

If $\alpha, \beta, \alpha + \beta$ are the positive roots of $SU(3)$ then substituting

$$x_1 = 0, \quad x_2 = \alpha, \quad x_3 = \beta, \quad x_4 = \alpha + \beta$$

we have from (4.1) for S^+ and S^- the same weights $0, \pm\alpha, \pm\beta, \pm\alpha+\beta$ as the adjoint representation. □

REMARK 4.2. The lemma implies a rather interesting property of the Lie algebra of $SU(3)$. Recall that Clifford multiplication of vectors on spinors is skew adjoint and satisfies $x^2 = -(x,x)1$. In eight dimensions the three representations S^+, S^- and Λ^1 are all real and 8-dimensional (this is triality). If $\varphi \in S^\pm$ and $x \in \Lambda^1$ then

$$(x\varphi, x\varphi) = (-x^2\varphi, \varphi) = (x,x)(\varphi,\varphi)$$

so that Clifford multiplication

$$\Lambda^1 \otimes S^+ \to S^-$$

is an orthogonal product. From the lemma all three representations are equivalent under $SU(3)$ so we must have an $SU(3)$-invariant orthogonal multiplication on the Lie algebra of 3×3 skew-hermitian matrices with trace zero. Here it is:

$$A \times B = \omega AB - \bar{\omega} BA - \frac{i}{\sqrt{3}} \operatorname{tr}(AB) I$$

where $\omega = (1 + i\sqrt{3})/2$.

The $SU(3)$-invariant isomorphism $S^+ \cong \Lambda^1$ defines a section of $(S^+)^* \otimes \Lambda^1 \cong S^+ \otimes \Lambda^1$ and this is what we take to be γ. There is of course an equivalent section of $S^- \otimes \Lambda^1$. We now prove:

THEOREM 4.3. *If M is an 8-manifold with a $PSU(3)$ structure defined by a 3-form ρ with $d\rho = d * \rho = 0$, then the section $\gamma \in C^\infty(S^+ \otimes \Lambda^1)$ satisfies the Rarita-Schwinger equations.*

PROOF. The covariant derivative of the 3-form ρ at any point can be written

$$\nabla \rho = A(\rho)$$

where $A \in \Lambda^2 \otimes \Lambda^1$ and acts on ρ by identifying Λ^2 with the Lie algebra of $SO(8)$. Since ρ is fixed by $SU(3)$ we may as well assume that $A \in \Lambda^2_0 \otimes \Lambda^1$ where Λ^2_0 is the orthogonal complement in Λ^2 to the Lie algebra of $SU(3) \cong \Lambda^1$. Since ρ is not covariant constant, A will not vanish. However, the harmonicity condition on ρ will force many of its components (as representations of $SU(3)$) to vanish.

We shall index representations by their highest weight – the adjoint representation Λ^1 of $SU(3)$ has highest weight $\alpha + \beta$. If we decompose the tensor product $\Lambda^1 \otimes \Lambda^1$ into irreducible representations as in Chapter 6 of Salamon's book [**Sal**], we find that Λ^2_0 is the direct sum of two 10-dimensional irreducibles Λ^2_+, Λ^2_- with highest weight $2\alpha + \beta$ and $\alpha + 2\beta$ respectively. These are interchanged under a change of orientation, just like self-dual and anti-self-dual forms in four dimensions. Similarly $V(2\alpha + \beta) \otimes V(\alpha + \beta)$ breaks up into irreducibles with highest weights

$$3\alpha + 2\beta, \quad 2\alpha + 2\beta, \quad 2\alpha + \beta, \quad \alpha + \beta$$

and each with multiplicity one. Robert Bryant informed me of the following result:

LEMMA 4.4. *If $d\rho = d * \rho = 0$, the components of A with highest weight $\alpha + \beta$, $2\alpha + \beta$, $\alpha + 2\beta$ and $2\alpha + 2\beta$ all vanish.*

PROOF. Since we are considering exterior powers of the Lie algebra it is convenient to think of these as spaces of left-invariant forms on the group. Then, for example, Λ^2_0 is the space of coclosed invariant 2-forms. The 3-form ρ is built out of the structure constants of the Lie algebra $\mathfrak{su}(3)$ and as a consequence of this, the action of $a \in \Lambda^2$ on ρ can be rewritten as da if we consider a as a 2-form on the group. Thus, if

$$A = \sum_i a_i \otimes e_i \in \Lambda^2_0 \otimes \Lambda^1$$

then

$$d\rho = \sum_i e_i \wedge da_i$$

and

$$d^* \rho = \sum_i \iota(e_i) da_i.$$

Now if $x_\gamma \in \mathfrak{su}(3) \otimes \mathbf{C}$ denotes a root vector for the root γ, the vector

$$A = x_\alpha \wedge x_{\alpha+\beta} \otimes x_\beta$$

lies in $V(2\alpha + 2\beta) \subset V(2\alpha + \beta) \otimes V(\alpha + \beta)$. Let h be an element of the Cartan subalgebra, then because $\iota(h) dx_\alpha = \mathcal{L}_h x_\alpha = \alpha(h) x_\alpha$ etc.

$$\iota(h) d\rho = \iota(h)(x_\beta \wedge d(x_\alpha \wedge x_{\alpha+\beta})) = \beta(h) x_\beta \wedge x_{\alpha+\beta} \wedge x_\alpha.$$

Since we can find h for which $\beta(h) \neq 0$ this shows that the irreducible representation $V(2\alpha+2\beta) \subset \Lambda^2_+ \otimes \Lambda^1$ maps non-trivially into Λ^3 under $A(\rho)$. That representation is also contained in $\Lambda^2_- \otimes \Lambda^1$, interchanging the roles of α and β. However, since we can choose h such that $\alpha(h) = 0$ and $\beta(h) \neq 0$ and vice-versa, $V(2\alpha+2\beta)$ appears twice in Λ^3, and so if $d\rho = 0$, then both of these components in A vanish.

We now work similarly with $V(\alpha+\beta) \subset V(2\alpha+\beta) \otimes V(\alpha+\beta)$ using the vector

$$A = x_\alpha \wedge x_{\alpha+\beta} \otimes x_{-\alpha}.$$

We deduce that if $d\rho = 0$, then both of the $V(\alpha+\beta)$ components in A vanish.

Now consider $V(2\alpha+\beta) = \Lambda^2_+ \subset \Lambda^2_+ \otimes \Lambda^1$. The inclusion is unique and is given by

$$a \mapsto \sum_i \mathcal{L}_{e_i} a \otimes e_i$$

for an orthonormal basis $\{e_i\}$ of the Lie algebra $\mathfrak{su}(3) \cong \Lambda^1$. In this case

$$d^*\rho = \sum_i \iota(e_i) d(\mathcal{L}_{e_i} a) = \sum_i \mathcal{L}^2_{e_i} a - d(\iota(e_i)\mathcal{L}_{e_i} a)$$

and $a \mapsto \iota(e_i)\mathcal{L}_{e_i} a$ is an invariant map from the irreducible Λ^2_+ to the irreducible Λ^1 and thus must vanish. Hence $d^*\rho$ is the Casimir $\sum_i \mathcal{L}^2_{e_i} a$ which is a non-zero scalar multiple of a. It follows that if $d^*\rho = 0$, the component $V(2\alpha+\beta)$ (and similarly $V(\alpha+2\beta)$) in A vanishes. This completes the proof of the lemma. □

To return to the theorem, consider $D\gamma \in C^\infty(S^- \otimes \Lambda^1)$. The right hand side, since $S^- \cong \Lambda^1$, is a vector bundle associated to the representation $V(\alpha+\beta) \otimes V(\alpha+\beta)$. The skew part we calculated to have highest weights $2\alpha+\beta, \alpha+2\beta$ and $\alpha+\beta$. The calculation gives at the same time the symmetric part to have weights 0, $\alpha+\beta$ and $2\alpha+2\beta$. But $D\gamma$ is the image under an invariant map of A which from the lemma only has components with highest weights $3\alpha+2\beta$ and $2\alpha+3\beta$. Since these do not occur in $S^- \otimes \Lambda^1$ we deduce that $D\gamma = 0$. Similarly, since $S^+ = V(\alpha+\beta)$, we see that $d^*\gamma = 0$. □

The existence of a Rarita-Schwinger field implies the vanishing of certain components of the Ricci tensor. In a sense the Einstein equations are the integrability condition for a linear system given by the Rarita-Schwinger operator [**J**]. More concretely, given $\gamma \in C^\infty(S^+ \otimes \Lambda^1)$ we form the second covariant derivative

$$\nabla^2\gamma \in C^\infty(S^+ \otimes \Lambda^1 \otimes \Lambda^1 \otimes \Lambda^1)$$

with components $\gamma_{i;jk}$ in an orthonormal basis. Covariantly differentiating the equation $D\gamma = 0$ gives

$$\sum_j e_j \gamma_{i;jk} = 0$$

and by contraction

$$\sum_{i,j} e_j \gamma_{i;ji} = 0. \tag{4.2}$$

But differentiating the equation $d^*\gamma = 0$ gives

$$\sum_{i,j} \gamma_{i;ij} = 0$$

and hence also

$$\sum_{i,j} e_j \gamma_{i;ij} = 0. \tag{4.3}$$

The difference of (4.2) and (4.3) is skew-symmetric in the last two indices and can therefore be rewritten in terms of the action of the curvature tensor on $S^+ \otimes \Lambda^1$. This only involves the Ricci tensor:

$$\sum_{i,j} R_{ij} e_i \gamma_j = 0 \tag{4.4}$$

The expression $\sum_{i,j} R_{ij} e_i \gamma_j$ in (4.4) defines an $SU(3)$-invariant map from the space of symmetric tensors R_{ij} to S^-. We have seen that $S^- \cong V(\alpha + \beta)$ and this representation occurs with multiplicity one in $Sym^2\Lambda^1$. Thus equation (4.4) implies the vanishing of 8 of the 36 components of the Ricci tensor.

REMARK 4.5. One consequence of the theorem is that the operator

$$\mathcal{D} : \Omega^1(M) \to \Omega^1(M)$$

defined by the orthogonal multiplication

$$\mathcal{D}\alpha = \sum_i e_i \times \nabla_i \alpha$$

is equivalent (under the isomorphisms γ for S^+ and S^-) to the Dirac operator. The multiplication $\alpha \times \beta$ is of course Clifford multiplication under these isomorphisms, but the Levi-Civita connections on Λ^1 ond on $S^\pm$ are different. Nevertheless, $D\gamma = 0$ implies that the Dirac operators correspond.

5. Constrained critical points

On a compact oriented manifold M^n there is a non-degenerate pairing between the spaces of forms $\Omega^p(M)$ and $\Omega^{n-p}(M)$ defined by

$$\int_M \alpha \wedge \beta.$$

If $\alpha = d\gamma \in \Omega^p(M)$ is exact, then by Stokes'theorem

$$\int_M d\gamma \wedge \beta = (-1)^p \int_M \gamma \wedge d\beta$$

which vanishes for all γ if and only if β is closed. We thus have a non-degenerate pairing between $\Omega^p_{exact}(M)$ and

$$\Omega^{n-p}(M)/\Omega^{n-p}_{closed}(M)$$

Since the exterior derivative d maps this latter space isomorphically onto $\Omega^{n-p+1}_{exact}(M)$, then formally (in the sense of non-degenerate pairings) we can say that

$$\Omega^p_{exact}(M)^* \cong \Omega^{n-p+1}_{exact}(M) \tag{5.1}$$

Now consider $n = 7$ and $p = 4$. We have seen that a metric of holonomy G_2 arises from a stable critical point for the functional $V(\rho)$ on a fixed cohomology class $[\rho]$ of closed 4-forms. On the other hand since $*\rho \wedge \rho = 7\phi(\rho)/4$ and ρ is both closed and coclosed, we must have $[*\rho] \cup [\rho][M] = V(\rho) \neq 0$ and in particular

$[\rho] \neq 0$. For the trivial cohomology class it follows that there are no stable critical points for this functional. However, we have from (5.1)

$$\Omega^4_{exact}(M)^* \cong \Omega^4_{exact}(M)$$

and hence a non-degenerate quadratic form on $\Omega^4_{exact}(M)$ given by

$$Q(d\gamma) = \int_M \gamma \wedge d\gamma.$$

We now have two natural functionals on the space of exact 4-forms: V and Q.

THEOREM 5.1. *An exact stable 4-form ρ on a compact 7-manifold is a critical point of $V(\rho)$ subject to the constraint $Q(\rho) = const.$ if and only if ρ defines a metric with weak holonomy G_2.*

PROOF. From Theorem (3.1), the first variation of V at $\rho = d\gamma$ is

$$\delta V(d\dot\gamma) = \int_M *\rho \wedge d\dot\gamma$$

and the first variation of the quadratic form Q is

$$\delta Q(d\dot\gamma) = 2\int_M \dot\gamma \wedge \rho.$$

Thus, introducing a Lagrangian multiplier, the constrained critical point is given by

$$d(*\rho) = \lambda\rho$$

and from [**Fr**], this is equivalent to the structure of a manifold with weak holonomy G_2 (sometimes called a *nearly parallel* G_2 structure). □

REMARK 5.2. A 7-manifold with weak holonomy G_2 is an Einstein manifold with positive scalar curvature. There are many examples, such as 3-Sasakian manifolds (see [**Bo**]) and their squashed versions (see [**Fr**]).

Next consider the case of $n = 6$ and $p = 3$. Then (5.1) tells us

$$\Omega^3_{exact}(M)^* \cong \Omega^4_{exact}(M)$$

so that the spaces of exact 3-forms and exact 4-forms are formally dual to each other. The pairing for an exact 3-form $\rho = d\alpha$ and an exact 4-form $\sigma = d\beta$ is

$$\langle \rho, \sigma \rangle = \int_M \alpha \wedge \sigma = -\int_M \rho \wedge \beta \tag{5.2}$$

We shall consider in 6 dimensions a variational problem involving the three functionals $V(\rho), V(\sigma)$ and $\langle \rho, \sigma \rangle$, but before proving a theorem, let us see how ρ and σ define a reduction of the structure group of the tangent bundle of M^6 to $SU(3)$. The 3-form ρ, being stable, provides a reduction to $SL(3, \mathbf{C})$ and $\sigma = \omega^2/2$ to $Sp(6, \mathbf{R})$. The group $SU(3)$ is an intersection of these two but there are two compatibility conditions for ρ and σ to achieve this. The first is

$$\omega \wedge \rho = 0 \tag{5.3}$$

From the point of view of ρ and the complex structure it describes, this says that ω is of type $(1,1)$. Since $\rho + i\hat\rho$ is a $(3,0)$ form the property $\omega \wedge \hat\rho = 0$ follows from (5.3). From the point of view of the symplectic form ω, the equation says that ρ is primitive. For a stable 3-form ρ and a stable 4-form $\sigma = \omega^2/2$, we shall say that the pair (ρ, σ) is of positive type if the almost complex structure I determined by ρ (see

Appendix (8.2)) makes $\omega(X, IX)$ a positive definite form. This is clearly an open condition. If $\omega \wedge \rho = 0$, the condition of positive type means that the Hermitian form defined by ω is positive definite.

The second condition is

$$\phi(\rho) = c\phi(\sigma) \tag{5.4}$$

For a constant c. This says that the complex 3-form $\rho + i\hat{\rho}$ has constant length relative to the Hermitian metric.

Since ρ and σ satisfying these conditions define a metric, it is natural to normalize the constant c above so that the volume forms $V(\rho)$ and $V(\sigma)$ are fixed multiples of the metric volume form. The most convenient way to do this is to see the normal form of the G_2 3-form φ and its dual $*\varphi$ in terms of an orthonormal basis as in [**Sal**]:

$$\varphi = e_7(e_5e_6 + e_1e_4 + e_3e_2) + (e_1e_2e_3 - e_3e_4e_5 + e_1e_3e_6 - e_4e_2e_6)$$

$$*\varphi = e_7(e_3e_4e_6 - e_1e_2e_6 + e_1e_3e_5 - e_4e_2e_5) + (e_1e_3e_4e_2 + e_5e_6e_2e_3 + e_5e_6e_4e_1).$$

If $\omega = e_5e_6 + e_1e_4 + e_3e_2$ and $\rho = e_1e_2e_3 - e_3e_4e_5 + e_1e_3e_6 - e_4e_2e_6$ then $\hat{\rho} = e_3e_4e_6 - e_1e_2e_6 + e_1e_3e_5 - e_4e_2e_5$. We then have from the above expressions

$$\phi(\sigma) = \frac{1}{6}\omega^3 = \frac{1}{4}\hat{\rho} \wedge \rho = \frac{1}{2}\phi(\rho) \tag{5.5}$$

and

$$\varphi = e_7 \wedge \omega + \rho, \qquad *\varphi = e_7 \wedge \hat{\rho} - \sigma \tag{5.6}$$

We now prove the theorem.

THEOREM 5.3. *A pair (ρ, σ) of exact, stable forms of positive type on a compact 6-manifold forms a critical point of $3V(\rho)+8V(\sigma)$ subject to the constraint $\langle \rho, \sigma \rangle =$ const. if and only if ρ and σ define a metric with weak holonomy $SU(3)$.*

PROOF. From Theorem (3.1), the first variation of $3V(\rho) + 8V(\sigma)$ is

$$3\int_M \hat{\rho} \wedge \dot{\rho} + 4\int_M \omega \wedge \dot{\sigma}.$$

The first variation of $P = \langle \rho, \sigma \rangle$ is

$$\delta P(\dot{\rho}, \dot{\sigma}) = \int_M \dot{\rho} \wedge \beta + \int_M \dot{\sigma} \wedge \alpha$$

where $\rho = d\alpha$ and $\sigma = d\beta$. Using Stokes' theorem and a Lagrange multiplier 12λ we find the constrained critical point to be given by the equations $3d\hat{\rho} = -12\lambda\sigma = -6\lambda\omega^2$ and $4d\omega = 12\lambda\rho$, i.e.

$$d\hat{\rho} = -2\lambda\omega^2 \tag{5.7}$$

$$d\omega = 3\lambda\rho \tag{5.8}$$

The compatibility conditions (5.3), (5.4) actually follow from these equations. From (5.8) we have

$$\lambda\omega \wedge \rho = \frac{1}{3}\omega \wedge d\omega = \frac{1}{6}d(\omega^2) = 0$$

since $\sigma = \omega^2/2$ is closed. Moreover, from (5.7)

$$2\lambda\omega^3 = -\omega \wedge d\hat{\rho} = -d(\omega \wedge \hat{\rho}) + d\omega \wedge \hat{\rho} = 3\lambda\rho \wedge \hat{\rho}$$

using (5.8) and the fact that $\omega \wedge \hat{\rho} = 0$, which, as we have seen, follows from $\omega \wedge \rho = 0$. Thus $\omega^3 = 3\hat{\rho} \wedge \rho/2$ as in (5.5).

The equations (5.7),(5.8) give a metric of weak holonomy $SU(3)$, sometimes called a *nearly Kähler* metric. To see this, consider the 3-form on $M \times \mathbf{R}$

$$\varphi = \frac{r^2}{\lambda} dr \wedge \omega + r^3 \rho.$$

This is stable, and from (5.8) closed. Moreover, by comparing with the normal forms above, we see that the G_2 metric it defines is

$$(dr/\lambda)^2 + r^2 g$$

where g is the $SU(3)$ metric. From (5.6) we have

$$*\varphi = \frac{r^3}{\lambda} dr \wedge \hat{\rho} - \frac{r^4}{2}\omega^2$$

and then from (5.7), $d{*\varphi} = 0$. It follows that the cone metric above is a Riemannian metric of holonomy G_2. However, from [**Ba**], this implies that M^6 is nearly Kähler. □

REMARK 5.4. The particular coefficients of $V(\rho)$ and $V(\sigma)$ in the theorem are not crucial. Any two positive numbers will, after a rescaling of metrics, give the same result.

REMARK 5.5. Manifolds with weak holonomy $SU(3)$ structure are (currently) far less plentiful than their 7-dimensional counterparts with weak holonomy G_2. They are again Einstein manifolds with positive scalar curvature, but the only known examples are S^6, the twistor spaces of S^4 and $\mathbf{C}P^2$ and $S^3 \times S^3$.

6. Evolution equations

We return to a 7-manifold and consider a fixed cohomology class $\mathcal{A}$ of closed 4-forms. This is an infinite dimensional affine space, whose tangent space at each point is naturally isomorphic to $\Omega^4_{exact}(M)$. The quadratic form Q defined above provides an indefinite metric on $\mathcal{A}$:

$$(\dot{\rho}, \dot{\rho}) = \int_M \gamma \wedge d\gamma$$

where $\dot{\rho} = d\gamma$.

In this setting the functional $V(\rho)$ defines a gradient vector field X on $\mathcal{A}$, and the critical points of $V(\rho)$ are the zeros of X. More generally, we can consider the gradient flow, and we find:

THEOREM 6.1. *Let M be a closed 7-manifold and suppose $\rho(t)$ is a closed stable 4-form which evolves via the gradient flow of the functional $V(\rho)$ restricted to a cohomology class in $H^4(M, \mathbf{R})$. Then the 4-form*

$$\varphi = dt \wedge *\rho + \rho$$

defines a metric with holonomy $Spin(7)$ on the 8-manifold $N = M \times (a, b)$ for an interval (a, b). Conversely, if N is an 8-manifold with holonomy $Spin(7)$, foliated by equidistant compact hypersurfaces diffeomorphic to M, the restriction of the defining 4-form to each hypersurface evolves as the gradient flow of $V(\rho)$.

PROOF. The gradient flow is the solution of the equation

$$\frac{dx}{dt} = X$$

where $(X, Y) = D\phi(Y)$ for any vector field Y. In our case we have x described by a 4-form $\rho(t)$ so using the inner product above with $Y = d\gamma$,

$$\left(\frac{dx}{dt}, Y\right) = \int_M \frac{\partial \rho}{\partial t} \wedge \gamma = D\phi(Y) = \int_M *\rho \wedge d\gamma$$

since $D\phi = *\rho$. But this equation holds for all 3-forms γ and thus yields the gradient flow equation

$$\frac{\partial \rho}{\partial t} = d(*\rho) \tag{6.1}$$

But $d\rho = 0$, so we obtain

$$d(dt \wedge *\rho + \rho) = 0$$

and from **[Sal]** we see that the 4-form $dt \wedge *\rho + \rho$ defines a metric with holonomy $Spin(7)$. In this metric $\|dt\|^2 = 1$.

Conversely, if a $Spin(7)$ manifold is foliated by equidistant hypersurfaces, defining the function t to be the distance to a fixed hypersurface M, we can write the defining 4-form in the form $dt \wedge *\rho + \rho$ and the statement that this is closed is equivalent to the gradient flow equation (6.1). □

We can do something similar with the 6-dimensional case. Here we take $\mathcal{A}$ to be a cohomology class in $H^3(M, \mathbf{R})$ and $\mathcal{B}$ to be a class in $H^4(M, \mathbf{R})$. Then the tangent space at each point of the product of affine spaces $\mathcal{A} \times \mathcal{B}$ is naturally isomorphic to

$$\Omega^3_{exact}(M) \times \Omega^4_{exact}(M)$$

and (5.1) shows that the pairing $\langle \rho, \sigma \rangle$ defines formally a symplectic structure on $\mathcal{A} \times \mathcal{B}$:

$$\omega((\rho_1, \sigma_1), (\rho_2, \sigma_2)) = \langle \rho_1, \sigma_2 \rangle - \langle \rho_2, \sigma_1 \rangle.$$

REMARK 6.2. The pairing between V and V^* to define a symplectic structure on $V \times V^*$ can also be used to define an indefinite metric:

$$((\rho_1, \sigma_1), (\rho_2, \sigma_2)) = \langle \rho_1, \sigma_2 \rangle + \langle \rho_2, \sigma_1 \rangle.$$

The gradient flow of a function of the form $f(\rho) + g(\sigma)$ is then equivalent to the Hamiltonian flow of $f(\rho) - g(\sigma)$. With $V = \Omega^3_{exact}(M)$, the gradient flow shows, as a particular consequence, the relationship between weak holonomy $SU(3)$ and conical G_2 metrics as in **[Ba]**. Nevertheless, the Hamiltonian interpretation of the equations has certain advantages over the gradient viewpoint, which is why we adopt it in the next theorem.

THEOREM 6.3. *Let $\mathcal{A} \in H^3(M, \mathbf{R})$ and $\mathcal{B} \in H^4(M, \mathbf{R})$ be cohomology classes and $(\rho, \sigma) \in \mathcal{A} \times \mathcal{B}$ be stable forms of positive type which evolve via the Hamiltonian flow of the functional $H = V(\rho) - 2V(\sigma)$. If for time $t = t_0$, ρ and σ satisfy the compatibility conditions $\omega \wedge \rho = 0$ and $\phi(\rho) = 2\phi(\sigma)$ then the 3-form*

$$\varphi = dt \wedge \omega + \rho$$

(where $\sigma = \omega^2/2$) defines a metric with holonomy G_2 on the 7-manifold $M \times (a, b)$. Conversely, if N is a 7-manifold with holonomy G_2, foliated by equidistant compact

hypersurfaces diffeomorphic to M, the restriction of the defining closed forms ρ, σ to each hypersurface evolves as the Hamiltonian flow of H.

PROOF. As before, we have the derivative of the Hamiltonian $H = V(\rho) - 2V(\sigma)$ given by

$$DH(d\alpha, d\beta) = \int_M \hat{\rho} \wedge d\alpha - \int_M \omega \wedge d\beta$$

and the Hamiltonian vector field X is defined by $\iota(X)\omega = dH$ so in our case we have

$$\int_M \dot{\rho} \wedge \beta - \int_M \dot{\sigma} \wedge \alpha = \int_M \hat{\rho} \wedge d\alpha - \int_M \omega \wedge d\beta$$

and this gives the equations:

$$\frac{\partial \rho}{\partial t} = d\omega \tag{6.2}$$

$$\frac{\partial \sigma}{\partial t} = \omega \wedge \frac{\partial \omega}{\partial t} = -d\hat{\rho} \tag{6.3}$$

We shall see that if the compatibility conditions (5.3), (5.4) between ω and ρ hold for $t = t_0$, then they hold for all subsequent time.

First consider the condition $\omega \wedge \rho = 0$. This can be viewed as the vanishing of the moment map for the natural action of $\mathrm{Diff}(M)$ on the symplectic manifold $\mathcal{A} \times \mathcal{B}$. To see this, for a vector field X on M, consider the function

$$\mu_X(\rho, \sigma) = \int_M \iota(X)\sigma \wedge \rho = \int_M \iota(X)\omega \wedge \omega \wedge \rho \tag{6.4}$$

We have

$$d\mu_X(\dot{\rho}, \dot{\sigma}) = \int_M \iota(X)\dot{\sigma} \wedge \rho + \int_M \iota(X)\sigma \wedge \dot{\rho}$$

But the first term is

$$-\int_M \dot{\sigma} \wedge \iota(X)\rho$$

since $\dot{\sigma} \wedge \rho = 0$. Thus

$$d\mu_X(\dot{\rho}, \dot{\sigma}) = \int_M \iota(X)\rho \wedge \dot{\sigma} + \int_M \iota(X)\sigma \wedge \dot{\rho} \tag{6.5}$$

Now since ρ and σ are closed,

$$\mathcal{L}_X(\rho, \sigma) = (d\iota(X)\rho, d\iota(X)\sigma)$$

so (6.5) can be written

$$d\mu_X(\dot{\rho}, \dot{\sigma}) = \langle \mathcal{L}_X(\rho, \sigma), (\dot{\rho}, \dot{\sigma}) \rangle$$

using the definition of the symplectic form. We deduce that μ_X is the moment map for $\mathrm{Diff}(M)$, evaluated on X. Since ω is non-degenerate, from (6.4) μ_X vanishes for all X if and only if $\omega \wedge \rho = 0$.

Now since the functional H is diffeomorphism invariant, H Poisson commutes with all the functions μ_X. Hence the Hamiltonian flow of H is tangential to the zero set of all the functions μ_X, i.e. the space of pairs $(\rho, \sigma) \in \mathcal{A} \times \mathcal{B}$ such that $\omega \wedge \rho = 0$. Thus if $\omega \wedge \rho = 0$ holds for $t = t_0$, it holds for all time. Note that it follows then that $\omega \wedge \hat{\rho} = 0$ for all time too.

Next consider the second compatibility condition (5.4). The form $\hat{\rho}$ is defined by the derivative of ϕ, $D\phi(\dot{\rho}) = \hat{\rho} \wedge \dot{\rho}$, so the derivative of $\hat{\rho}$ can be expressed via

$D^2\phi$. The volume $\phi(\rho)$ is homogeneous of degree 2, so its derivative is homogeneous of degree one and hence

$$D^2\phi(\rho, \dot{\rho}) = D\phi(\dot{\rho}) = \hat{\rho} \wedge \dot{\rho}$$

We have

$$\frac{\partial \hat{\rho}}{\partial t} \wedge \rho = D^2\phi(\rho, \frac{\partial \rho}{\partial t}) = D\phi(\frac{\partial \rho}{\partial t}) = \hat{\rho} \wedge \frac{\partial \rho}{\partial t} \tag{6.6}$$

So from (6.6)

$$\frac{\partial}{\partial t}(\hat{\rho} \wedge \rho) = 2\hat{\rho} \wedge \frac{\partial \rho}{\partial t} = 2\hat{\rho} \wedge d\omega$$

from (6.2). But $\omega \wedge \hat{\rho} = 0$, and hence $\hat{\rho} \wedge d\omega = -d\hat{\rho} \wedge \omega$ Thus

$$\frac{\partial}{\partial t}(\hat{\rho} \wedge \rho) = -2d\hat{\rho} \wedge \omega = 2\omega^2 \frac{\partial \omega}{\partial t} = \frac{2}{3}\frac{\partial}{\partial t}\omega^3$$

and so if $\phi(\rho) = 2\phi(\sigma)$ at $t = t_0$ then it holds for all t.

The evolution equation thus preserves the $SU(3)$ geometry on M, and we define the 3-form

$$\varphi = dt \wedge \omega + \rho$$

From (6.2), $d\varphi = 0$. From (5.6), using the G_2 metric defined by φ, we have

$$*\varphi = dt \wedge \hat{\rho} - \sigma$$

and from (6.3), $d * \varphi = 0$. Thus φ defines a metric of holonomy G_2 on $M \times (a, b)$ for some interval. □

7. Examples

It is a well-known principle that if we are looking for a G-invariant critical point of an invariant function, we need only consider critical points of the same function restricted to the fixed point set of G. Thus if G acts on M, a G-invariant special metric of any of the types we are considering can be found by restricting the volume functionals to G-invariant forms. This applies in particular when M is homogeneous under G, in which case we are reduced to a finite-dimensional variational problem.

Since metrics of holonomy G_2 and $Spin(7)$ have zero Ricci tensor, we have no interesting compact homogeneous examples. We can nevertheless apply the principle to compact spaces for weak holonomy G_2 and $SU(3)$, and also to the evolution equation to find non-compact examples with holonomy $Spin(7)$ and G_2.

7.1. A 7-dimensional example. We consider $M^7 = S^7$ as a principal $SU(2)$ bundle over S^4, and $G = SO(5)$ acting transitively on it with stabilizer $SU(2)$. All the relevant cohomology classes are trivial here, so we are dealing with exact forms, which must be built out of the three invariant 1-forms $\alpha_1, \alpha_2, \alpha_3$ which are the components of the $SU(2)$-connection form, and the components $\omega_1, \omega_2, \omega_3$ of the curvature. These are defined by:

$$d\alpha_1 + 2\alpha_2\alpha_3 = \omega_1 \tag{7.1}$$

and similar expressions for α_2, α_3. The curvature forms themselves satisfy the Bianchi identity

$$d\omega_1 = 2(\omega_2\alpha_3 - \omega_3\alpha_2) \ etc. \tag{7.2}$$

A basis for the exact invariant 4-forms is provided by

$$u_1 = d(\alpha_1\omega_1), u_2 = d(\alpha_2\omega_2), u_3 = d(\alpha_3\omega_3), u_4 = d(\alpha_1\alpha_2\alpha_3)$$

and the quadratic form

$$Q(d\gamma) = (d\gamma, d\gamma) = \int_M \gamma d\gamma$$

defines a metric on this space. We have, for example,

$$(u_1, u_1) = \int_M \alpha_1\omega_1 d(\alpha_1\omega_1) = \int_M \alpha_1\omega_1(-2\alpha_2\alpha_3)\omega_1 = -2\int_M \alpha_1\alpha_2\alpha_3\omega_1^2$$

using (7.1) and (7.2). Here, because the connection is anti-self-dual, we have

$$\omega_1^2 = \omega_2^2 = \omega_3^2, \quad \omega_1\omega_2 = \omega_2\omega_3 = \omega_3\omega_1 = 0.$$

Continuing this way, and normalizing the volume of $\alpha_1\alpha_2\alpha_3\omega_1^2$, we find the metric with respect to this basis to be

$$(u_1, u_1) = (u_2, u_2) = (u_3, u_3) = 2, \quad (u_1, u_2) = (u_2, u_3) = (u_3, u_1) = -2$$

$$(u_1, u_4) = (u_2, u_4) = (u_3, u_4) = 1, \quad (u_4, u_4) = 0.$$

We now need the volume $V(\rho)$ for a general exact invariant 4-form

$$\rho = \sum_1^3 x_i d(\alpha_i\omega_i) + 2x_4 d(\alpha_1\alpha_2\alpha_3)$$

which from (7.1) and (7.2) can be written

$$\rho = (x_1 + x_2 + x_3)\nu + k_1\alpha_2\alpha_3\omega_1 + k_2\alpha_3\alpha_1\omega_2 + k_3\alpha_1\alpha_2\omega_3$$

where $k_1 = 2(x_4 - x_1 + x_2 + x_3)$ etc. Instead of working out the volume as in (8.4), it is easier to transform ρ to the standard form (5.6). In fact we can rewrite that formula as

$$*\varphi = e_1e_2e_3e_4 + e_5e_6(e_2e_3 + e_4e_1) + e_6e_7(e_4e_3 + e_1e_2) + e_7e_5(e_1e_3 + e_2e_4)$$

Then putting $e_5 = y_1\alpha_1, e_6 = y_2\alpha_2, e_7 = y_3\alpha_3$ and rescaling $e_i = y_4v_i$ for $1 \le i \le 4$, where v_i is a local orthonormal basis such that $\omega_1 = (v_4v_3 + v_1v_2)$ etc. we can transform ρ into $*\varphi$ if

$$k_1 = y_2y_3y_4^2, \quad k_2 = y_3y_1y_4^2, \quad k_3 = y_1y_2y_4^2, \quad y_4^4 = -2(x_1 + x_2 + x_3)$$

The Riemannian volume is then

$$V(\rho) = y_1y_2y_3y_4^4 \tag{7.3}$$

Using the expression for the inner products above, we find the gradient flow equation

$$2(-\dot{x}_1 + \dot{x}_2 + \dot{x}_3 + \dot{x}_4) = \frac{\partial V}{\partial x_1}, \; etc.$$

$$2(\dot{x}_1 + \dot{x}_2 + \dot{x}_3) = \frac{\partial V}{\partial x_4}$$

Converting to coordinates y_i defined by

$$y_1y_2y_4^2 = 2(x_1 + x_2 - x_3 + x_4) \; etc., \quad y_4^4 = -2(x_1 + x_2 + x_3),$$

we find the equations

$$\dot{y}_1 = -1 + \frac{1}{2y_2y_3}(y_2^2 + y_3^2 - y_1^2) + \frac{y_1^2}{2y_4^2}$$

$$\dot{y}_2 = -1 + \frac{1}{2y_3y_1}(y_3^2 + y_1^2 - y_2^2) + \frac{y_2^2}{2y_4^2}$$

$$\dot{y}_3 = -1 + \frac{1}{2y_1y_2}(y_1^2 + y_2^2 - y_3^2) + \frac{y_3^2}{2y_4^2}$$

$$4y_4\dot{y}_4 = -(y_1 + y_2 + y_3)$$

Putting $-\alpha = y_2 = y_3, \beta = -y_1, \gamma = y_4$ gives the equations solved by Čvetic et al in [**C**]. Their solution represents a $Spin(7)$ manifold with an extra S^1 symmetry.

The constrained variational problem to get a weak holonomy G_2 structure consists of finding critical points of V subject to the condition that the quadratic form is constant. This is precisely when the gradient field is parallel to the position vector. As an example, if $y_1 = y_2 = y_3 = y$ in the equations above, then we obtain

$$\lambda y = -\frac{1}{2} + \frac{y^2}{2y_4^2}, \quad 4\lambda y_4^2 = -3y$$

which yields $y = -3/10\lambda$, and this is the squashed 7-sphere.

In this same symmetric situation the gradient flow equations become

$$\dot{y} = -\frac{1}{2} + \frac{y^2}{y_4^2}, \quad 2y_4\dot{y}_4 = -\frac{3}{2}y$$

Putting $s = y_4^2$ the equation becomes

$$y\frac{dy}{ds} + \frac{y^2}{3s} = \frac{1}{3}$$

which is solved easily:

$$y^2 = \frac{2}{5}s + cs^{-2/3}.$$

When $c = 0$ we have the cone on the squashed S^7 and for $c \neq 0$, the complete Bryant-Salamon metric on the spin bundle over S^4 [**Sal**].

7.2. A 6-dimensional example. We take now $M^6 = S^3 \times S^3$ and the forms to be invariant under G given by the left action of $S^3 \times S^3$ and *anti-invariant* under the $\mathbf{Z}/2$ action which interchanges the two factors. We have left-invariant 1-forms

$$\sigma_1, \sigma_2, \sigma_3, \quad \Sigma_1, \Sigma_2, \Sigma_3$$

on the two factors, satisfying

$$d\sigma_1 = -\sigma_2\sigma_3, \ etc.$$

We shall take $\mathcal{A} \in H^3(S^3 \times S^3)$ to be $\nu_1 - \nu_2$ where ν_1 is the pull-back of a generator of $H^3(S^3)$ on the first factor and ν_2 the pull-back from the second. This is clearly anti-invariant under the $\mathbf{Z}/2$ action. Since $H^4(S^3 \times S^3) = 0$, $\mathcal{B}$ must be the trivial class.

We find the following two three-dimensional spaces of closed invariant 3-forms ρ and 4-forms σ:

$$\rho = \sigma_1\sigma_2\sigma_3 - \Sigma_1\Sigma_2\Sigma_3 + x_1(\sigma_1\Sigma_2\Sigma_3 - \sigma_2\sigma_3\Sigma_1) + \cdots \tag{7.4}$$

$$\sigma = y_1\sigma_2\Sigma_2\sigma_3\Sigma_3 + y_2\sigma_3\Sigma_3\sigma_1\Sigma_1 + y_3\sigma_1\Sigma_1\sigma_2\Sigma_2 \tag{7.5}$$

Since $d(\sigma_1\sigma_1) = \sigma_1\Sigma_2\Sigma_3 - \sigma_2\sigma_3\Sigma_1$ we see that ρ lies in the fixed cohomology class $\mathcal{A}$. The 4-form σ is clearly closed. We can work out the pairing between invariant exact 3-forms and invariant exact 4-forms in a straightforward manner, e.g.

$$\langle \sigma_2\Sigma_2\sigma_3\Sigma_3, d(\sigma_1\Sigma_1)\rangle = \int_{S^3\times S^3} \sigma_2\Sigma_2\sigma_3\Sigma_3\sigma_1\Sigma_1 = \nu_1[S^3]\nu_2[S^3] \neq 0.$$

It turns out that the symplectic form is a multiple of

$$dx_1 \wedge dy_1 + dx_2 \wedge dy_2 + dx_3 \wedge dy_3$$

We can use (8.2) to find $V(\rho)$ and then we find

$$V(\rho)^2 = (1 + x_1 + x_2 + x_3)(x_2 + x_3 - x_1 - 1)(x_3 + x_1 - x_2 - 1)(x_1 + x_2 - x_3 - 1)$$

$$V(\sigma)^2 = y_1y_2y_3$$

Since we are only interested in solutions where $V(\rho) = 2V(\sigma)$, after a change of parameter t, it is equivalent to consider the Hamiltonian flow of $4V^2(\sigma) - V^2(\rho)$ i.e. for the Hamiltonian

$$H = 4y_1y_2y_3 - (1+x_1+x_2+x_3)(x_2+x_3-x_1-1)(x_3+x_1-x_2-1)(x_1+x_2-x_3-1)$$

The six first order equations arising from this are equivalent to those in [**Br**], where the authors produce an explicit complete solution with $x_2 = x_3$ and $y_2 = y_3$. They express their solution using an orthonormal basis $A_j(\sigma_j - \Sigma_j), B_j(\sigma_j + \Sigma_j)$ for the metric and our coordinates relate to theirs by $x_1 = A_1A_2A_3 + A_1A_2B_3 + A_3B_1B_2 - A_1B_2B_3$ etc. and $y_1 = 4A_2B_2A_3B_3$ etc.

When $x_1 = x_2 = x_3 = x$ and $y_1 = y_2 = y_3 = y$, we don't need to solve the equations since the solutions are given by the vanishing of the Hamiltonian:

$$4y^3 - (1 + 3x)(x - 1)^3 = 0$$

This is the metric originally found by Bryant and Salamon [**Sal**] on the spin bundle over S^3.

For a weak holonomy $SU(3)$ metric we need the cohomology classes $\mathcal{A}, \mathcal{B}$ to be trivial, and then the "1" terms disappear in the functional. In the fully symmetric situation, we need then to find critical points $8y^{3/2}+3\sqrt{3}x^2$ subject to the condition $xy = c$. The solution $y^{7/2} = \sqrt{3}c^2/2$ gives the weak holonomy $SU(3)$ metric on $S^3 \times S^3$.

8. Appendix: Definition of the volumes

We assume V to be oriented and then make all volume forms positive.

8.1. $\mathbf{p = 2m - 2, n = 2m}$. The vector space V is $2m$-dimensional and $\rho \in \Lambda^{2m-2}V^*$. Use the isomorphism

$$\Lambda^{2m-2}V^* \cong \Lambda^2 V \otimes \Lambda^{2m}V^*$$

to write ρ as $\sigma \in \Lambda^2 V \otimes \Lambda^{2m}V^*$. Then we have using the exterior product

$$\sigma^{2m} \in \Lambda^{2m}V \otimes (\Lambda^{2m}V^*)^m \cong (\Lambda^{2m}V^*)^{m-1}$$

We define

$$\phi(\rho) = |\sigma^{2m}|^{1/(m-1)}.$$

8.2. $\mathbf{n = 6, p = 3}$. We have a 6-dimensional space V and $\rho \in \Lambda^3 V^*$. For $v \in V$ define

$$K(v) = \iota(v)\rho \wedge \rho \in \Lambda^5 V^* \cong V \otimes \Lambda^6 V^*.$$

Then

$$\operatorname{tr}(K^2) \in (\Lambda^6 V^*)^2.$$

The stable forms with stabilizer $SL(3, \mathbf{C})$ are characterized by $\operatorname{tr}(K)^2 < 0$ and there

$$\phi(\rho) = |\sqrt{-\operatorname{tr} K^2}| \in \Lambda^6 V^*$$

8.3. $\mathbf{n = 7, p = 3}$. The space V is 7-dimensional and $\rho \in \Lambda^3 V^*$. Given $v, w \in V$ form

$$\iota(v)\rho \wedge \iota(w)\rho \wedge \rho \in \Lambda^7 V^*.$$

This is a symmetric bilinear form on V with values in $\Lambda^7 V^*$, and so a linear map

$$G : V \to V^* \otimes \Lambda^7 V^*$$

We have

$$\det G \in (\Lambda^7 V^*)^9$$

and we define

$$\phi(\rho) = |\det G|^{1/9}.$$

8.4. $\mathbf{n = 7, p = 4}$. Take $\rho \in \Lambda^4 V^*$ but identify with an element $\sigma \in \Lambda^3 V \otimes \Lambda^7 V^*$. Now apply the construction of the previous case and we find a linear map

$$H : V^* \to V \otimes (\Lambda^7 V^*)^2$$

This gives

$$\det H \in (\Lambda^7 V^*)^{12}$$

and we define

$$\phi(\rho) = |\det H|^{1/12}$$

8.5. $\mathbf{n = 8, p = 3}$ and $\mathbf{p = 5}$. The space V is 8-dimensional and $\rho \in \Lambda^3 V^*$. For $v, w \in V$ form

$$\iota(v)\rho \wedge \iota(w)\rho \wedge \rho \in \Lambda^7 V^* \cong V \otimes \Lambda^8 V^*$$

This gives an element

$$d \in V^* \otimes V^* \otimes V \otimes \Lambda^8 V^*$$

Take $d \otimes d$ and contract to get

$$G \in V^* \otimes V^* \otimes (\Lambda^8 V^*)^2$$

or equivalently

$$G : V \to V^* \otimes (\Lambda^8 V^*)^2.$$

Then we have

$$\det G \in (\Lambda^8 V^*)^{18}$$

and define

$$\phi(\rho) = |\det G|^{1/18}.$$

For $p = 5$ we use $\Lambda^5 V^* \cong \Lambda^3 V \otimes \Lambda^8 V^*$ and proceed similarly.

References

[Ba] C. Bär, *Real Killing spinors and holonomy*, Comm. Math. Phys. **154** (1993), 509–521.

[B] H. Baum, Th. Friedrich, R. Grunewald, I. Kath, *Twistor and Killing spinors on Riemannian manifolds*, Teubner-Texte zur Mathematik, **124**, B. G. Teubner, Stuttgart, 1991.

[Bo] C. P. Boyer, K. Galicki, B. M. Mann and E. G. Rees, *Compact 3-Sasakian 7-manifolds with arbitrary second Betti number*, Inventiones Mathematicae, **131**, (1998), 321-344.

[Br] A. Brandhuber, J. Gomis, S. Gubser and S. Gukov, *Gauge theory at large N and new G_2 holonomy metrics*, **hep-th/0106034**

[C] M. Čvetic, G. W. Gibbons, H. Lü, C. N. Pope, *New cohomogeneity one metrics with $Spin(7)$ holonomy*, **math.DG/0105119**

[F] M. Fernández and A. Gray, *Riemannian manifolds with structure group G_2*, Ann. Mat. Pura Appl. **132** (1982), 19–45 (1983).

[Fr] Th. Friedrich, I. Kath, A. Moroianu and U. Semmelmann, *On nearly parallel G_2-structures*, J. Geom. Phys. **23** (1997), 259–286.

[G] A. Gray, *Weak holonomy groups*, Math. Z. **123** (1971), 290–300.

[Gu1] G. B. Gurevich, *Classification of trivectors of rank eight* (in Russian), Dokl. Akad. Nauk SSSR **2** (1935), 353-355.

[Gu2] G. B. Gurevich, *Algebra of trivectors II.* (in Russian), Trudy Sem. Vektor. Tenzor. Anal. **6** (1948), 28-124.

[H] N. J. Hitchin, *The geometry of three-forms in six dimensions*, J.Differential Geometry **56** (2001), see also **math.DG/0010054**

[J] B. Julia, *Système linéaire associé aux équations d'Einstein*, C. R. Acad. Sci. Paris Sér. II **295** (1982), 113–116.

[R] W. Reichel, *Über die Trilinearen Alternierenden Formen in 6 und 7 Veränderlichen*, Dissertation, Greifswald, (1907).

[Sal] S. Salamon, *Riemannian geometry and holonomy groups*, Pitman Research Notes in Mathematics **201**, Longman, Harlow 1989.

[S] J.A.Schouten, *Klassifizierung der alternierenden Grössen dritten Grades in 7 Dimensionen*, Rend. Circ. Mat. Palermo, 55 (1931), 137–156.

Mathematical Institute, 24-29 St Giles, Oxford OX1 3LB, UK

E-mail address: `hitchin@maths.ox.ac.uk`

Contemporary Mathematics
Volume **288**, 2001

A CR–Momentum Ansatz for Nilpotent Groups

A. Huckleberry and M. Völler

Dedicated to the memory of Alfred Gray

Abstract. A program is described for connecting the well-established representation theory for nilpotent Lie groups G and with the complex geometry of CR-manifolds which arise in a certain way as G-orbits in $G^{\mathbb{C}}$-homogeneous spaces. On one hand the Ansatz gives a complex analytic view of representations associated to coadjoint strata and on the other provides a natural method for constructing CR-manifolds with interesting extension properties. This is carried out in various concrete examples.

1. Introduction

The title is an indication of our approach to associating a Cauchy-Riemann manifold M and its envelope of holomorphy $\widehat{M}$ to an isotropy type stratum $\mathcal{S}$ of the coadjoint representation of a nilpotent group G. Although this has only been carried out in a number of examples, some of which will be explained in the sequel, it is hoped that it will prove to be a general approach which will link the direct integral representation defined by $\mathcal{S}$ to representations of G on spaces $\mathcal{H} \subset \mathcal{O}(M) = \mathcal{O}(\widehat{M})$. In special cases involving conal Heisenberg algebras this latter point has been confirmed in ([**HNO**])

The details involve Levi-extension phenomena, explicit description of Cauchy-Riemann envelopes of holomorphy, computation of generators of certain algebras of invariant smooth functions, associated Kähler or pseudo-Kähler forms and moment maps. This method generates in a systematic way interesting high codimensional CR-manifolds M with computable hulls $\widehat{M}$ which are not necessarily complex manifolds.

This paper is organized as follows. After outlining Kirillov's method in 2, in 3 we explain the idea of our approach in the context of the simplest case of the

1991 *Mathematics Subject Classification.* Primary: 43A65,32V40; Secondary: 22E25, 32V05, 32V10, 32V25.

Key words and phrases. Nilpotent groups, CR-manifolds, representations.

3-dimensional Heisenberg group. In 5 and 6 we explain some details for the infinite series K_n and the group N_4 of 4×4-unipotent upper-triangular matrices. Computations are carried out in complete detail in ([**V**])

2. Kirillov's Method

Let G be a connected Lie goup and recall that a symplectic manifold $N = G/H$ which is homogeneous under the action of G by symplectic diffeomorphisms can for all practical purposes be regarded as an orbit $O = G.\ell$ of a function $\ell \in \mathfrak{g}^*$ via the coadjoint representation $Ad^* : G \to Gl(\mathfrak{g}^*)$. The symplectic form on N is then given by the canonical orbit form ω_o which is in turn defined by the bilinear form

$$B_\ell : \mathfrak{g} \times \mathfrak{g} \to \mathbb{R}, (\xi, \eta) \mapsto \ell([\xi, \eta]).$$

The degeneracy $\mathfrak{g}^\perp$ of B_ℓ is exactly the isotropy algebra $\mathfrak{g}_\ell$ and it defines ω_o on $O = G.\ell$.

From the point of view of this paper it is important to note that ω_o arises as the restriction of the standard Poisson structure

$$\{\ ,\ \} : \mathcal{E}(\mathfrak{g}^*) \times \mathcal{E}(\mathfrak{g}^*) \to \mathcal{E}(\mathfrak{g}^*),\ \{f, g\}(\ell) := \ell([df(\ell), dg(\ell)]).$$

The restriction $\{\ \}_o$ is well-defined in the obvious way and it defines ω_o by $\{f, g\}_o = \omega_o(V_f, V_g)$, where V_f and V_g are the Hamiltonian fields associated to f and g by $df = i_{V_f}\omega_o$.

Kirillov's method follows the line of *geometric quantization* which requires a polarization on the underlying symplectic manifold, i.e. a Lagrangian foliation. In the case where G is nilpotent he shows that such can always be defined by a homogeneous fibration

$$O = G.\ell = G/G_\ell \to G/\mathcal{P},$$

where the foliation is given by the fibers of the bundle.

There are of course numerous settings where geometric quantization has been successful. However, here we now restrict to G being a simply-connected nilpotent Lie group. Via its polynomial exponential map, such is an $\mathbb{R}$-algebraic group acting algebraically on $\mathfrak{g}^*$. Its orbits are quite simple: $O = G.\ell = \mathbb{R}^{n(\ell)}$.

The polarizing subgroup is also connected, $\mathcal{P} = exp(\mathfrak{p})$, and the *polarization* $\mathfrak{p}$ is simply a subalgebra of $\mathfrak{g}$ which is B_ℓ-isotropic and maximal with respect to this property.

The fact that $\mathfrak{p}$ is isotropic, i.e. $\ell([\mathfrak{p}, \mathfrak{p}]) = 0$, implies that

$$\chi_\ell := e^{2\pi i \ell} : \mathcal{P} \to S^1$$

is a character on $\mathcal{P}$ which defines the G-linearized line bundle $L_\ell := G \times_{\chi_\ell} \mathbb{C} = G \times \mathbb{C}/\mathcal{P}$, where the $\mathcal{P}$-action on $G \times \mathbb{C}$ is defined by $p(g, v) = (gp^{-1}, \chi_\ell(p).v)$.

The space of sections $\Gamma(G/\mathcal{P}, L_\ell)$ can be identified as usual with the $\mathcal{P}$-equivariant maps $f : G \to \mathbb{C}$, i.e. $f(gp^{-1}) = \chi_\ell(p)f(g)$. Since G is unimodular, $G/\mathcal{P}$ possesses a natural invariant measure which yields a unitary representation $\pi_{\ell,\mathfrak{p}}$ of G on $\Gamma_{L^2}(G/\mathcal{P}, L_\ell)$.

Kirillov's results in this context can be formulated as follows ([**K**], see e.g. also [**CG**]). The representation $\pi_{\ell,\mathfrak{p}}$ is irreducible and its equivalent class in the space

$\widehat{G}$ of irreducible representations only depends on the orbit $O = G.\ell$. Thus there exists a map

$$Rep : \mathfrak{g}^*/G \to \widehat{G}, O \mapsto \pi_o,$$

which Kirillov shows to be bijective

Before moving on to an explanation of the CR-Momentum Ansatz, we would like to make a few remarks which indicate the origin of this project.

REMARKS. 1. The Poisson structure $\{\ ,\ \}$ on a symplectic manifold has only the locally constant functions as universal constants of motion. It is fair to say that many physical systems have non-trivial conservation laws, i.e., in the Poisson setting, non-constant functions $F \in \mathcal{E}(M)$ such that $\{H, F\} = 0$ for all Hamiltonians H.

2. If $\mathcal{S}$ is a G-invariant submanifold of $\mathfrak{g}^*$, then $(\mathcal{S}, \{\ ,\ \})$ is Poisson.

3. It is a reasonable goal to produce representations on algebras of global functions arising in e.g. complex geometry such as $L^2_{\mathcal{O}}(M)$, where M has some sort of G-invariant complex structure. The multiplicative structure of such algebras often destroys the possibility of an irreducible representation where automorphic behavior, e.g. the equivariance resulting from induction, is required.

4. It is of interest to consider representations where the Hilbert space at hand is invariant under some (perhaps restricted) form of tensor product construction. Geometrically speaking, it is on the one hand desireable that the *semi-classic* is visible, i.e. going to infinity in some way in $\mathcal{S}$, as well as *localizing* at an orbit.

3. The Ansatz in the Heisenberg context

Here we sketch the CR-Momentum Ansatz with reference to the 3-dimensional Heisenberg group. For this let $\mathfrak{g} = ((X, Y, Z))$ be the Lie algebra with center $\mathfrak{z} = \mathfrak{g}' = ((Z))$ and structure given by $[X, Y] = Z$. As is the case with every simply-connected nilpotent Lie group, the exponential map $exp : \mathfrak{g} \to G$ is a polynomial isomorphism. Letting $g = exp(xX + yY + zZ)$ have the coordinates (x, y, z) and $v = (x, y)$, the group multiplication law is given by $(v, z) \cdot (v', z') = (v + v', z + z' + \frac{1}{2}\omega_{std}(v, v'))$, where ω_{std} is the standard sympletic form on $\mathbb{R}^2$ defined by the determinant.

Let $< X^*, Y^*, Z^* >$ be the dual basis and represent $\ell \in \mathfrak{g}^*$ by $\ell = \ell_1 X^* + \ell_2 Y^* + \ell_3 Z^* = (\ell_1, \ell_2, \ell_3)$. Then

$$Ad^*(g)(\ell) = (x, y, z) \cdot (\ell_1, \ell_2, \ell_3) = (\ell_1 + \ell_3 y, \ell_2 - \ell_3 x, \ell_3).$$

Therefore, $Fix(G) = ((X^*, Y^*))$ is the (ℓ_1, ℓ_2)-plane and otherwise $G.\ell$ is the parallel plane determined by the coordinate ℓ_3.

The fixed points of G correspond to its irreducible representations as an Abelian group. Here this might be interpreted as the points where $\hbar = 0$.

From now on let $\ell = (0, 0, \ell_3) = \ell_3 Z^*$ with $\ell_3 \neq 0$. If $O = G.\ell$, then $\omega_o = \ell_3^{-1} d\ell_1 \wedge d\ell_2$.

The coadjoint representation of a Heisenberg group is very special in a number of ways. In particular, an isotropy group G_ℓ is either the entire group G or the center, i.e. $G_\ell = exp((Z))$. In general the decomposition of $\mathfrak{g}^*$ by isotropy-type, the partition defined by the equivalence relation $\ell \sim \ell'$ iff G_ℓ and G'_ℓ are conjugate, is much more complicated.

If $\{\mathcal{S}_\alpha\}$ is this partition, then the $\mathcal{S}_\alpha$ are locally closed submanifolds and it is indeed a stratification, but it is in general not locally finite (see 6 for examples). In the example at hand there are only two strata, $\mathcal{S}_1 = \{\ell_3 \neq 0\}$ and $\mathcal{S}_2 = \{\ell_3 = 0\}$. Actually we require that strata are connected and hence there are actually three in this case.

Consider $\mathcal{S} = \{\ell_3 > 0\}$ and let $\Lambda = \{(0,0,\ell_3) : \ell_3 > 0\}$. Then Λ is a slice for the G-action of $\mathcal{S}$ in the sense that the natural map $G \times \Lambda \to \mathcal{S}$ induces an equivariant isomorphism $G/G_\Lambda \times \Lambda \cong \mathcal{S}$, where G_Λ is the fixed isotropy group at every point of Λ.

Of course in this example the stratification is very simple. But in general this is not the case and therefore we underline this difficulty by regarding it as the first general step, GS_1, in the Ansatz.

GS_1 Determine the orbit type stratification with slices $\{(\mathcal{S}_\alpha, \Lambda_\alpha)\}$ of the coadjoint representation.

If G_α is the isotropy subgroup at points of Λ_α, then Kirillov's theory yields a polarizing subgroup $\mathcal{P}_\alpha$ (resp. polarization $\mathfrak{p}_\alpha$) containing G_α. This subgroup (resp. algebra) is in general not unique. For example, for the stratum-slice $(\mathcal{S}, \Lambda)$ under discussion, both both $\mathfrak{p}_1 = ((Y, Z))$ and $\mathfrak{p}_2 = ((X, Z))$ do the job. We choose $\mathfrak{p} = \mathfrak{p}_1$ and have a triple $(\mathcal{S}, \Lambda, \mathcal{P})$, where $\mathcal{P} = exp(\mathfrak{p})$.

Let us concretely compute the representations π_ℓ for $\ell \in \Lambda$ via the polarization $\mathcal{P}$. In this case $\mathcal{P} = \{x = 0\}$ and for $\ell = \ell_3 Z^* \in \Lambda$ it follows tht $\chi_\ell(0, y, z) = e^{2\pi i \ell_3 z}$.

Here it is a simple matter to compute the transformation rule for the functions $f : G \to \mathbb{C}$ which arise as sections $\sigma \in \Gamma(G/\mathcal{P}, L_\ell)$:

$$f(a, b - y, c - z - \frac{ay}{2}) = e^{2\pi i \ell_3 z} f(a, b, c)$$

Now the bundle $G \to G/\mathcal{P}$ is trivialized by the section $\Sigma = exp((X))$ and furthermore the standard uniform density dx on Σ is the measure induced on $G/\mathcal{P}$ from G. Given $f = f(t)$ in $L^2(\Sigma, dx)$ it extends canonically by the transformation rule to G:

$$f(x, y, z + \frac{1}{2}xy) = e^{-2\pi i \ell_3 z} f(x, 0, 0)$$

In fact the unit function 1_Σ extends to a section which trivializes L_ℓ so that

$$\Gamma_{L^2}(G/\mathcal{P}, L_\ell) = L^2(\Sigma, dx) \cong L^2(\mathbb{R}).$$

In order to compute the G-representation on this space it is only necessary to do so for the obvious generating 1-parameter groups:

$$g = (x, 0, 0) \Rightarrow g(f)(t) = f(t - x)$$

$$g = (0, y, 0) \Rightarrow g(f)(t) = e^{2\pi i \ell_3 t y} f(t)$$
$$g = (0, 0, z) \Rightarrow g(f)(t) = e^{2\pi i \ell_3 z} f(t)$$

This is done as follows. For $t = (t, 0, 0)$ the point $g^{-1}t$ is in general not in Σ and must be transformed back to Σ by right-multiplication with a unique element $p \in \mathcal{P}$. In this way $g(f)(t) = f(g^{-1}t) = \chi_\ell^{-1}(p) f(g^{-1}tp^{-1})$.

REMARKS. 1. Dependence on $\ell \in \Lambda$ is very simple; in particular $\{\pi_o : O \subset \mathcal{S}\}$ is stable under multiplication.
2. At the operator level $(x,0,0)$ corresponds to $-\frac{d}{dt}$ and $(0,y,0)$ to the multiplication operator $2\pi i\ell_3 t$. Introducing the appropriate factor to achieve self-adjointness, $p \mapsto \Phi(p) = i\frac{d}{dt}$ and $q \mapsto \Phi(q) = 2\pi\ell_3 t$. Hence $[p,q] = 2\pi i\ell_3 = i\hbar$ and therefore $\ell_3 = \frac{\hbar}{2\pi}$. Hence $\mathcal{S}$ is suitable for discussing semi-classical limits. In this case $\hbar \to 0$ corresponds to going toward $Fix(G)$.

In the Heisenberg example above it is obvious that the representational ineffectivity of $\mathcal{S}$, i.e.

$$I = \{g : \pi_o(g) = Id \;\forall\; O \subset \mathcal{S}\},$$

is trivial. In general this is not the case, but it is a simple matter to compute it. So we reformulate the first general step as follows.

GS_1 Determine the data $\{(\mathcal{S}_\alpha, \Lambda_\alpha, \mathcal{P}_\alpha, I_\alpha)\}$, where $\{(\mathcal{S}_\alpha, \Lambda_\alpha)\}$ is the orbit type stratification with each stratum accompanied by a slice, $\mathcal{P}_\alpha$ is a polarization containing the fixed isotropy group G_α and I_α is the representational ineffectivity of $\mathcal{S}_\alpha$.

In fact it is difficult to separaate this from the steps on the complex analytic side, because the determination of the CR-manifold, its envelope of holomrophy and an appropriate moment map actually determines data of the type (S, Λ, P, I).

Now we come to the connection to complex analysis. As above we describe this here with reference to the 3-dimensional Heisenberg example. Let us begin with some ad hoc remarks.

Recall that, since G is the N of the standard KAN-decomposition of $SU(1,2)$, it acts on the non-compact model of the 2-dimensional Euclidean ball $\mathbb{B}_2$ in $\mathbb{C}^2$. This is in fact the restriction of an action on $X := \mathbb{C}^2$. For $(x,y,z) = g \in G$ and $(z_0, w) \in X$ we have

$$(x,y,z)\cdot(z_0,w) = (z_0 + z + xw + \frac{1}{2}ix^2, w + x + iy)$$

Note that this action is free and that $\Sigma = \{(ir,0) : r \in \mathbb{R}\}$ is a slice.

Let $x_0 = (i,0)$ be the neutral point and $M = G.x_0$. Then M, which is identified with G by the action, is the real hypersurface

$$M = \{(z_0,w) \in X : \frac{1}{2}|w|^2 - Im(z_0) + 1 = 0\}.$$

The complex manifold X should be regarded as an orbit of the complexification, $X = G^{\mathbb{C}}.x_0$.

Being a real hypersurface in X, the manifold M has natural Cauchy-Riemann structure. In this case this means that $T_pM \cap J_pT_pM$ is a 1-dimensional complex vector space with respect to the structure J_p. It is more convenient to consider this in the complexification:

$$(T_pM \cap J_pT_pM)^{\mathbb{C}} = T_p^{1,0}M \oplus T_p^{0,1}M \subset T_pM^{\mathbb{C}} = \mathfrak{g}^{\mathbb{C}},$$

where the last equality comes from the identification by the action.

The fact that the CR-structure is an invariant structure on G translates to $T_p^{1,0}M = \mathfrak{a}$ being a complex subalgebra of $\mathfrak{g}^{\mathbb{C}}$. In the case of the Heisenberg group, $\mathfrak{a} = ((X - iY))_{\mathbb{C}}$ and the homogeneous space X is simply $G^{\mathbb{C}}/\bar{A}$, where $\bar{A} = exp(\bar{\mathfrak{a}})$.

It is a general principle that a subalgebra $\mathfrak{a} < \mathfrak{g}^{\mathbb{C}}$ determines an invariant CR-structure on $M = G$ and that this is induced from the natural embedding of M in $X = G^{\mathbb{C}}/\bar{A}$ as the G-orbit of the neutral point.

GS_2 *Determine a canonical CR-structure $\mathfrak{a}_\alpha$ associated to given data $(\mathcal{S}_\alpha, \Lambda_\alpha, \mathcal{P}_\alpha, I_\alpha)$ from GS_1.*

This is of course an essential step in our Ansatz. In 4 we give details on the method of carrying this out in the general case.

In the example at hand I_α is trivial, but in general one must take it into consideration. Thus the appropriate complex analytic setting is

$$M_\alpha =: G/I_\alpha \hookrightarrow G^{\mathbb{C}}/\bar{A}I_\alpha^{\mathbb{C}} =: X_\alpha.$$

Let $\mathcal{O}(M_\alpha)$ denote the holomorphic functions on M_α, i.e. those defined by the $\bar{\partial}$-equation given by the CR-structure (see e.g.[**G**],[**B**]).

One of the essential points is that functions in $\mathcal{O}(M_\alpha)$ extend to holomorphic functions on a much larger manifold $\widehat{M}_\alpha$ which contains a continuum of G-orbits corresponding to the G-orbits in the given stratum $\mathcal{S}_\alpha$ (see e.g. [**BP**],[**BR**], [**G**] and [**T**] for relevant general results). In the present example $\widehat{M} = \{\frac{1}{2}|w|^2 - Im(z_0) + 1 \leq 0\}$ with boundary M. This is of course a well-known situation, but in general $\widehat{M}$ might very well be a high-codimensional CR-manifold with M being a small piece of its boundary (see 5 and 6 for examples).

It should be emphasized that the conditions imposed on $\mathfrak{a}_\alpha$ are in particular such that $\widehat{M}_\alpha$ is maximal dimensional and non-trivial in the sense that it lies on "one side" of M. For example, it would not be of interest if $\widehat{M} = X$ where $G^{\mathbb{C}}$ acts!

GS_3 *Compute the envelope of holomorphy $\widehat{M}$.*

The above general steps are carried out having a goal of a canonical (multiplicity-free) geometric relation between the manifolds $\widehat{M}_\alpha$ and open sets in the strata $\mathcal{S}_\alpha$. This is realized by a moment map associated to a well-chosen invariant symplectic structure.

GS_4 *Determine a G-invariant Kähler or pseudo-Kähler structure ω_α on a neighborhood X_α^+ of $\widehat{M}_\alpha$ and an associated moment map $\mu_\alpha : \widehat{M}_\alpha \to \mathcal{S}_\alpha$ which is an open immersion separating G-orbits in such a way that $\mu_\alpha|\Sigma_\alpha : \Sigma_\alpha \to \Lambda_\alpha$ is a diffeomorphism onto its open image.*

For simplicity we have intensionally supressed the discussion of the slice Σ_α for the G-action on an appropriate open set containing $\widehat{M}_\alpha$. It suffices here to say that its construction is an aspect of the definition of $\mathfrak{a}_\alpha$.

A G-action by symplectic diffeomorphisms is geometrically multiplicity-free if its generic orbits are coisotropic. In our case, where the moment map is in fact a bundle with isotropic fibers all of which are contained in orbits, the action is multiplicity-free in the strongest possible sense. This has been set up in this way so that the representation theory at the level of M (resp. $\widehat{M}$) can be related to the direct integral representation on $\mathcal{S}$ which is multiplicity-free by definition.

Let us return to the example. Here $\omega = dd^c\varphi$, where $\varphi = e^{c\rho}$ and $\rho = \frac{1}{2}|w|^2 - Im(z_0)$ is a canonical generator of $\mathcal{E}(X)^G$ determined by the slice Σ. Here $c \in \mathbb{R}^*$ is some constant.

REMARK. In this case Σ is a global slice for the G-action on X, but in general no such will exist and consequently one discusses the invariants $\mathcal{E}(X^+_\alpha)^G$, where X^+ is some possibly smaller open set where the action can be better controlled.

There will always be some freedom of signs, but these can be normalized by requiring that the image $\mu(\widehat{M})$ be contained in $\mathcal{S}$ and the restriction of ω to the complex tangent bundle of M to be positive definite.

Here this amounts to $\omega = dd^c\varphi$, $c < 0$ and $\mu = (0, 0, -\frac{1}{4}ce^{-cIm(z_0)})$ on Σ. Therefore $\mu(\widehat{M})$ is a half-space $\{\ell_3 \geq const.\}$ in $\mathcal{S}$.

In the cases which we have considered, i.e. those in 5 and 6 and several other special cases, the invariant pseudo-Kähler forms ω_α are constructed in a rather obvious way from canonical bases of $\mathcal{E}(X^+_\alpha)^G$ defined by Σ_α. In most cases the moment map is directly related to a potential φ_α by

$$\xi \circ \mu_\alpha = J\xi_X(\varphi_\alpha).$$

Even in the simple cases treated in 5 and 6, this posed interesting challenges and therefore we would like to mention it separately.

GS_4' Compute the natural generators of $\mathcal{E}(X^+_\alpha)^G$ and determine the associated invariant psuedo-Kähler structure.

The final step is to prove that there is the desired connection to the representation theory. In the case of conal Heisenberg algebras, where, as in the present example, there is a natural relationship to Siegel domains, this has been carried out in ([**HNO**]).

GS_5 Determine the relationship between the representation theory in subspaces of $\mathcal{O}(\widehat{M_\alpha})$ and the direct integral $\int_{\ell\in\Lambda_\alpha}\pi_\ell d\sigma_\alpha$ for the appropriate measure $d\sigma_\alpha$ on the associated piece of $\widehat{G}$ interpreted as the stratum $\mathcal{S}_\alpha$.

This program can be carried out in the Heisenberg case in an obvious way (see [**V**]). It would be very interesting to test it in some general setting, e.g. that of flat coadjoint orbits (see [**CG**] [**MW**], [**RW**], [**W**]). Sections 5 and 6 of the present work are devoted to outlining the geometry, i.e. $GS_1 - GS_4$, in the cases K_n and $N4$ where the orbits are not all flat.

4. Generalities on associated CR-Structures

Given a coadjoint orbit $O = G.\ell$ we look for a CR-manifold M which is homogeneous under a G-action of holomorphic transformations and which is closely related to O and the representation theory associated to the relevant stratum in which it is contained. The theory is in a state of flux and we are still not completely certain that we have determined all the conditions which should be imposed on the CR-structure. Here we present several requirements whcih have led to the determination of what would seem to be the appropriate CR-structures in all examples computed thus far.

4.1. Invariant CR-structures on Lie groups. Let G be a connected real Lie group which is embedded in its complexification $G^{\mathbb{C}}$ and $X = G^{\mathbb{C}}/H$ be a $G^{\mathbb{C}}$-homogeneous complex manifold, where H denotes the $G^{\mathbb{C}}$-isotropy group at a base point $x_0 \in X$.

The manifold $M = G.x_0$ is equipped with a canonical G-invariant CR-structure which comes from the its embedding in X. For our purposes it is essentially enough to consider the case where $M = G.x_0$ can be identified with the group G itself, i.e., the isotropy group $G_{x_0} = \{e\}$ is trivial.

Under this or the slightly weaker assumption that G_{x_0} is discrete, let $\pi : \mathfrak{g}^{\mathbb{C}} \to \mathfrak{g}^{\mathbb{C}}/\mathfrak{h}$ be the natural projection and regard the CR-tangent space $T^{CR}_{x_0}M = T_{x_0} \cap iT_{x_0}M =: \mathfrak{m}$ as a subspace of $\mathfrak{g}^{\mathbb{C}}/\mathfrak{h}$. Since $\pi^{-1}(\mathfrak{m})$ is complex and defined over $\mathbb{R}$, i.e. invariant under complex conjugation, it follows that $\pi^{-1}(\mathfrak{m}) \supset \mathfrak{h} \oplus \bar{\mathfrak{h}}$. Note that this sum is direct, because G_{x_0} is assumed to be trivial.

Conversely, if $v \in \pi^{-1}(\mathfrak{m})$, then $\pi(v)$, $\pi(iv) \in \pi(\mathfrak{g})$. Thus there exit g_1,$g_2 \in \mathfrak{g}$ and h_1,$h_2 \in \mathfrak{h}$ so that $v = g_1 + h_1$ and $iv = g_2 + h_2$. Consequently $g_1 + ig_2 \in \mathfrak{h}$ and this in turn implies that g_1,$g_2 \in \mathfrak{h} \oplus \bar{\mathfrak{h}}$. Hence $v \in \mathfrak{h} \oplus \bar{\mathfrak{h}}$.

It is convenient for us to identify $\mathfrak{m}^{1,0}$ with $\bar{\mathfrak{h}}$. So, for notational clarity we let $X = G^{\mathbb{C}}/\bar{A}$, where $\bar{A} = exp(\bar{\mathfrak{a}})$ and $\mathfrak{a}$ is a complex Lie subalgebra of $\mathfrak{g}^{\mathbb{C}}$ with $\mathfrak{a} \cap \bar{\mathfrak{a}} = \{0\}$. We then have $\mathfrak{m}^{1,0} = \mathfrak{a}$ and $\mathfrak{m}^{0,1} = \bar{\mathfrak{a}}$ in the complexified tangent space $(T_{x_0}M)^{\mathbb{C}} = T_{x_0}M \otimes_{\mathbb{R}} \mathbb{C} \cong \mathfrak{g}^{\mathbb{C}}$. In the embedding $M \hookrightarrow X$ we make the identification $T^{CR}_{x_0}M \cong \mathfrak{a}$.

4.2. The basic decomposition of $T_{x_0}M$. We wish to relate M to a given coadjoint orbit $O = G.\ell = G/G_\ell$ by a moment map μ in a geometrically multiplicity-free setting. Here this means that, up to connected components, the μ-fiber at x_0 is exactly the orbit of the isotropy group $G_\ell.x_0$.

For this the required decomposition of $T_{x_0}M$ is most conveniently described at the level of the complexification:

$$(T_{x_0}M)^{\mathbb{C}} = \mathfrak{g}_\ell^{\mathbb{C}} \oplus \mathfrak{a} \oplus \bar{\mathfrak{a}}$$

This implies that $T_{x_0}M = \mathfrak{g}_\ell \oplus \mathfrak{a}$ and $T_{x_0}X = \mathfrak{g} \oplus \mathfrak{a} \oplus i\mathfrak{g}_\ell$, where the space $i\mathfrak{g}_\ell$ can be regarded as defining the normal bundle of M in X.

4.3. Levi-extension phenomena. Let $\mathcal{O}(M)$ denote the holomorphic functions (sometimes called CR-functions) on the CR-manifold M. Most simply stated, these are the maps $f : M \to \mathbb{C}$ which respect the CR-structure: $df(iv) = idf(v)$ for all $v \in T^{CR}M$.

Just as in the case of holomorphic functions on complex manifolds, it is quite possible that the functions $f \in \mathcal{O}(M)$ extend as CR-functions to some larger CR-manifold $\widehat{M}$. An early theory for this type of extension phenomenon was developed by S. Greenfield ([**G**]).

For this it is essential to understand the lack of integrability of the bundle $T^{1,0}M \oplus T^{0,1}M$.–the individual summands are of course integrable. For $p \in M$ consider the subspace of $(T_pM)^{\mathbb{C}}$ generated by Lie brackets of all orders of fields from $T^{1,0}M$ and $T^{0,1}M$. In general this varies from point to point, but in the case at hand the homogeneity allows us to restrict our attention to the neutral point x_0 and compute this in terms of the algebra $\mathfrak{a}$.

Let $\mathcal{L}(\mathfrak{a})$ be the Lie subalgebra of $\mathfrak{g}^{\mathbb{C}}$ which is generated by $\mathfrak{a}$ and $\bar{\mathfrak{a}}$. We refer to it as the Levi-algebra of the CR-homogeneous space M. The Levi-extension dimension is then defined as

$$e := exdim_{\mathbb{C}}M = dim_{\mathbb{C}}\mathcal{L}(\mathfrak{a}/(\mathfrak{a} \oplus \bar{\mathfrak{a}}).$$

Using the basic theorem of Greenfield along with the homogeneity, it is possible to prove an extension theorem of a somewhat global nature (see [**V**] p.12-15)

PROPOSITION 4.3.1. *There exists a generic locally closed G-invariant CR-submanifold $\widehat{M}$ in X with $dim_{\mathbb{R}}\widehat{M} = dim_{\mathbb{R}}M + e$ and with M in its boundary so that every CR-function $f \in \mathcal{O}(M)$ extends to a function $f \in \mathcal{O}(\widehat{M})$ which is continuous on the closure of $\widehat{M}$.*

In all examples of the CR-Momentum Ansatz which are know to us, the manifolds $\widehat{M}$ can be chosen to be maximal in the sense that if $\{p_n\} \subset \widehat{M}$ is a divergent sequence, the there exists $f \in \mathcal{O}(\widehat{M})$ such that $\overline{lim}|f(p_n)| = \infty$. Since in these cases $\widehat{M}$ is contained in X, i.e. there is no problem with "Schlichtheit", and the latter is Stein, it is quite reasonable to regard such CR-manifolds as the "envelopes of holomorphy" of the manifolds M which arise in our procedure.

We say that M has Levi-maximal CR-structure if $\mathcal{L} = \mathfrak{g}^{\mathbb{C}}$. In this case $\widehat{M}$ is an G-invariant domain in X.

An extreme case of analytic continuation would be that where the functions in $\mathcal{O}(M)$ extend to functions which are holomorphic on all of X. Although there could be interesting phenomena related to the topologies of $\mathcal{O}(M)$ and $\mathcal{O}(X)$, at least formally this would mean that we are discussing a $G^{\mathbb{C}}$-representation. This is not desireable and therefore we impose a natural first order Levi-extension condition.

The "first Levi-form" is the Hermitian vector valued map

$$L_1 : \mathfrak{a} \times \mathfrak{a} \to (T_{x_0})^{\mathbb{C}}/(\mathfrak{a} \oplus \bar{\mathfrak{a}}) \cong \mathfrak{g}_{\ell}^{\mathbb{C}}$$

which is defined by $(v, w) \mapsto \frac{i}{2}\pi([v, \bar{w}])$, where $\pi : \mathfrak{g}^{\mathbb{C}} \to \mathfrak{g}^{\mathbb{C}}/(\mathfrak{a} \oplus \bar{\mathfrak{a}})$ is the natural projection.

For $M = \{\rho = 0\}$ a real hypersurface this is defined by the restriction of the complex Hessian $\frac{\partial^2 \rho}{\partial z_i \partial \bar{z}_j}$ to $\mathfrak{a}$. In this case it is classical that the positive-definiteness of this form implies holomorphic extension to the side of M defined by $\{\rho < 0\}$. In our possibly higher codimensional situation we require a similar condition.

The first Levi-form L_1 is said to be semi-definite and non-trivial if there exists $\lambda \in (\mathfrak{a} \oplus \bar{\mathfrak{a}})^0 \subset \mathfrak{g}^*$ so that $\lambda \circ L_1(v, v) \geq 0$ for all $v \in \mathfrak{a}$ and if for some such v there is strict positivity. The envelope of holomorphy $\widehat{M}$ will then lie on a "side" of $\widehat{M}$ (see e.g. [**BR**], [**BP**], [**T**]) and in particular $G^{\mathbb{C}}$ will not act on $\mathcal{O}(\widehat{M})$.

4.4. The scalar Levi-Kirillov-form. For our Ansatz it is appropriate to line up the Levi-geometry of M with the symplectic geometry of the initial coadjoint orbit $O = G.\ell$

Recall that the Kirillov-form B_ℓ is defined by $B_\ell(v, w) = \ell([v, w])$. By complexifying ℓ we have the Hermitian form $H_\ell : \mathfrak{a} \times \mathfrak{a} \to \mathbb{C}$ defined by $(v, w) \mapsto \ell^{\mathbb{C}}([v, \bar{w}])$. If

$$(\mathfrak{a} \oplus \bar{\mathfrak{a}}) \cap [\mathfrak{g}^{\mathbb{C}}, \mathfrak{g}^{\mathbb{C}}] \subset ker(\ell^{\mathbb{C}}),$$

then this factors through the first Levi-form, i.e. the associated "scalar Levi-form" $S_\ell := \ell^{\mathbb{C}} \circ L_1$ is well-defined. We also require this of the CR-structure $\mathfrak{a}$.

In summary, we refer to a CR-structure $\mathfrak{a}$ as being "associated to $\ell \in \mathfrak{g}^*$" if $\mathfrak{g}^{\mathbb{C}} = \mathfrak{g}_{\ell}^{\mathbb{C}} \oplus \mathfrak{a} \oplus \bar{\mathfrak{a}}$, the first Levi-form is non-trivial and semi-definite and the scalar Levi-Kirillov-form is well-defined.

It would seem that maximal extension dimension would also be desireable and we demand this whenever possible. However, in general it seems that there will be substrata where this can not be attained.

4.5. Ineffectivity. Observe that if I is a closed normal Lie subgroup of G, then the representation theory of G/I corresponds to a certain set of coadjoint orbits of G, i.e., those where I acts trivially on the associated Hilbert space.

Given an orbit $O = G.\ell$, the subgroup $I_O := ker(\pi_O)$ will be called the (representation theoretical) ineffectivity associated to O. Although it is defined abstractly, it may be most conveniently computed via a polarization.

If P is a polarizing subgroup of ℓ, i.e. $G/G_\ell \to G/P$ is a Lagrangian fibration, then clearly I_O is contained in the ineffectivity I_P of the G-action on the base manifold G/P. Recall how I_P acts on the functions $f \in \mathcal{E}(\Sigma)$, where Σ is a trivializing section for the fibration $G \to G/P$:

$$f(p^{-1}g) = f(g(g^{-1}pg)^{-1}) = f(g\tilde{p}^{-1}) = \chi_\ell(\tilde{p})f(g).$$

So $I_O = \{p \in I_O : gpg^{-1} \in ker(\chi_\ell) \ \ \forall g \in G\}$.

The roughest stratification of $\mathfrak{g}^*$ is defined by this ineffectivity. Examples show that it is not necessarily locally finite (see 6). However, it seems to be rather easy to compute and could be related in an interesting way to $\mathbb{R}$-algebraic geometric invariants of the G-action on $\mathfrak{g}^*$.

If $I_O = (e)$, then the appropriate CR-manifold is $M = G.x_0 \hookrightarrow X$ as above with $G_{x_0} = (e)$. The structure is that of a left-invariant CR-structure on G. However, in general the appropriate complex manifold is $X_{red} = X/I_O^{\mathbb{C}} = G^{\mathbb{C}}/\bar{A}I_O^{\mathbb{C}}$. The CR-manifold is again the orbit of the neutral point. However, in this case $M = G/I_O$.

Since I_O is a normal subgroup of G and acts trivially on M, the above procedure essentially leads to nothing new. It is just a way to systematically handle all G-orbits in $\mathfrak{g}^*$. There might be a slight advantage in that before reduction the structure $\mathfrak{a}$ might have a particularly simple form and might be more visiable than that for the corresponding orbit/stratum in $(\mathfrak{g}/\mathfrak{i}_O)^*$.

4.6. Appropriate domains in X. Let us recall that we wish to associate the envelope of holomorphy $\widehat{M}$ with a stratum, or at least an essential parat of a stratum, which contains the intial point $\ell \in \mathfrak{g}^*$. This should be realized by a moment map $\mu : X^+ \to \mathfrak{g}^*$, where X^+ is an appropriate G-invariant domain in X which contains $\widehat{M}$.

For us it is of essential importance that there exists a natural G-invariant Kähler or pseudo-Kähler structure on X^+. We even wish to normalized this to be positive-definite on $T^{1,0}M$ and of course it should be closely related to the scalar valued Levi-Kirillov-form.

Now the embedding $M \hookrightarrow X$ has been set up so that $\Sigma := exp(i\mathfrak{g}_\ell)$ is transversal to $M = G.x_0$ and is a slice for the G-action in some neighborhood of M. It may be necessary to shrink Σ to an open subset Σ^+, but after doing so $X^+ := G.\Sigma^+ \cong G \times \Sigma^+$ is an open neighborhood of M where the G-action is proper.

In such a situation it is a relatively simple matter to compute natural algebraic generators for the G-invariant functions. They are rational, possibly having poles in the complement $X \setminus X^+$.

In all examples considered thus far it has been possible to combine these generators in an obvious manner to obtain G-invariant functions φ with $\omega = dd^c\varphi$ non-degenerate, i.e., to obtain an invariant Kähler or pseudo-Kähler structure on X^+. Furthermore, X^+ always contains $\widehat{M}$ and, even if $\widehat{M}$ is a high-codimensional submanifold, it is possible to discuss its momentum geometry in terms of its embedding in (X^+, ω). In fact, due to the multiplicity-free nature of the situation and the fact that the restriction of ω to $T^{1,0}M$ is symplectic, the restricted moment map $\mu|M : M \to O = G.\ell$ is just a form of symplectic reduction.

5. The series K_n

In ([**V**]) the second author gives a detailed account of the CR-Momentum Ansatz for the infinite series H_n and K_n and the group N_4 of (4×4)-uppertriangular matrices.

The results for the Heisenberg series H_n are just the obvious generalizations of the remarks in 3 and will therefore not be discussed in the sequel. In the final two sections, avoiding as much of the technical work as is possible, we indicate the results for K_n and for N_4. Of course we refer to ([**V**]) for details.

5.1. The coadjoint representation of K_n. The $(n+1)$-dimensional Lie algebra $\mathfrak{g} = \mathfrak{k}_n = ((e_1, \dots e_n))_{\mathbb{R}}$ is defined by $[e_{n+1}, e_{j+1}] = e_j$, $j = 1, \dots, n$, with all other brackets being zero. Identifying the group $G = K_n$ with $\mathfrak{g} = \mathfrak{k}_n$ by the polynomial diffeomorphism $exp : \mathfrak{g} \to G$, one computes the coadjoint representation as an $\mathbb{R}$-algebraic G-action on $\mathfrak{g}^*$ (see [**V**], p.38). Except for the fixed points, which we will neglect throughout the discussion, all orbits are 2-dimensional.

The roughest decomposition of this action is by representation theoretical ineffectivity. For this let $(\alpha_1, \dots, \alpha_{n+1})$ be the coordinates of $\ell = \sum \alpha_j e_j^* \in \mathfrak{g}^*$ and $\mathcal{S}_m = \{\alpha : \alpha_1 = \dots \alpha_{m-1} = 0, \alpha_m \neq 0\}$.
For $\ell \in \mathcal{S}_m$ the orbit $G.\ell \cong \mathbb{R}^2$ can be parameterized by $(s,t) \mapsto (0, \dots, 0, \alpha_m, t, P_2(t), \dots, P_{n-m}(t), s)$, where

$$P_j(t) = \sum_{k=m}^{m+j} \frac{1}{(m+j-k)!} \frac{(t-\alpha_{m+1})^{m+j-k}}{{\alpha_m}^{m+j-k}} \alpha_k$$

The linear set $\Lambda_m := \{\alpha \in \mathcal{S}_m : \alpha_{m+1} = \alpha_{n+1} = 0\}$ is an appropriate slice for the G-action on $\mathcal{S}_m$ and for $\ell \in \mathcal{S}_m$ one computes that

$$\mathfrak{g}_\ell = ((e_1, \dots, e_m, e_{m+2}, e_{m+3} - \frac{\alpha_{m+2}}{\alpha_m} e_{m+1}, \dots, e_n - \frac{\alpha_{n-1}}{\alpha_m} e_{m+1}))_{\mathbb{R}} .$$

The stratification $\{\mathcal{S}_m\}$ is indeed defined by ineffectivity: $\mathfrak{i}_O = ((e_1, \dots, e_{m-1}))$ for $O \subset \mathcal{S}_m$. The orbit type stratification is even finer (see [**V**], p.39), but this turns out to play only a minor role in the sequel.

5.2. Associated CR-structures in the K_n-case. The appropriate CR-structures in this case are 1-dimensional.

PROPOSITION 5.2.1. *Up to isomorphism, the CR-structures associated to coadjoint orbits $G = K_n$ are given by $\mathfrak{a}_j = ((e_{n+1} + ie_j))_{\mathbb{C}}$ with Levi-algebra $\mathcal{L}(\mathfrak{a}_j) = ((e_1, \dots, e_j, e_{n+1}))_{\mathbb{C}}$, $j = 1, \dots, n$.*

For a proof see ([**V**], p.41-42)

In the case of $G = K_n$ there are essentially no interesting new quotients G/I_O and consequently the higher codimensional strata determined by non-trivial ineffectivity are not of particular interest. Thus we restrict our attention to the open stratum $\mathcal{S} = \mathcal{S}_1$. Let us now check which $\mathfrak{a}_j$ are associated to which points $\ell \in \mathcal{S}$.

First, consider $\mathfrak{a} = \mathfrak{a}_n = ((e_{n+1} + ie_n))$ which is Levi-maximal, i.e., $\mathcal{L}(\mathfrak{a}) = \mathfrak{g}^{\mathbb{C}}$ and $\widehat{M}$ is open in X. In this case $\mathfrak{a} \oplus \bar{\mathfrak{a}} = ((e_n, e_{n+1}))$ has trivial intersection with $[\mathfrak{g}^{\mathbb{C}}, \mathfrak{g}^{\mathbb{C}}]$ and consequently the scalar Levi-Kirillov-form is well-defined for all $\ell \in \mathfrak{g}^*$.

For $v = z(e_{n+1} + ie_n)$ one computes that $[v, \bar{v}] = (-2ie_{n-1})|z|^2$. Applying $\lambda = e^*_{n-1}$ we have $\lambda \circ L_1(v, v) = |z|^2$. Consequently the first Levi-form is semi-definite and non-trivial.

The condition $\mathfrak{g}^{\mathbb{C}} = \mathfrak{g}^{\mathbb{C}}_{\ell} \oplus \mathfrak{a} \oplus \bar{\mathfrak{a}}$ is fulfilled for a generic point $\ell \in \mathcal{S}$, but not for all points. As a result we are forced to use all of the $\mathfrak{a}_j$ to obtain CR-structures associated to every such ℓ. Of course, given a choice, we always choose the $\mathfrak{a}_j$ with maximal extension dimension.

Beginning at the generic level, the structures associated to points in $\mathcal{S}$ are determined as follows. If $\ell \in \Lambda$ has coordinates $(\alpha_1, 0, \alpha_2, \ldots, \alpha_n, 0)$ and $\alpha_{n-1} \neq 0$, then the condition $\mathfrak{g}^{\mathbb{C}} = \mathfrak{g}^{\mathbb{C}}_{\ell} \oplus \mathfrak{a} \oplus \bar{\mathfrak{a}}$ is satisfied.

Otherwise, we go to the next level. In this way $\mathfrak{a}_{n-j}$ is naturally associated to the substratum determined by the slice

$$\Lambda^{n-j} := \{\alpha_{n-1} = \ldots = \alpha_{n-j} = 0, \alpha_{n-j-1} \neq 0\}.$$

The smallest substratum is given by the condition $\alpha_3 = \ldots \alpha_{n-1} = 0$. In this case $\mathfrak{g}_\ell = ((e_1, e_3, e_4, \ldots, e_n))$ and only $\mathfrak{a}_2 = ((e_{n+1} - ie_2))$ does the job.

At this stage it is prudent to compute dimensions in order to make sure that the moment maps which are to be constructed have at least a chance of mapping the associated $\widehat{M}$'s to the relevant strata.

Now $dim_{\mathbb{C}} X = n$ and $\mathcal{L}(\mathfrak{a}_{n-j}) = ((e_1, \ldots, e_{n-j}, e_{n+1}))$. Thus the extension dimension is $e = n - j - 1$ and consequently the expected dimension is $dim_{\mathbb{R}} \widehat{M} = 2n - j$. Since the Kählerian structure should be multiplicity-free, the dimension of the fiber of the moment map μ should be $(n + 1 - 2)$.

Consequently $dim_{\mathbb{R}} \mu(\widehat{M}) = (2n - j) + (n - 1) = n - j + 1$. This is indeed the same as $dim_{\mathbb{R}} G.\Lambda^{n-j}$, because Λ^{n-j} is defined by the equations $\alpha_{n-1} = \ldots = \alpha_{n-j} = 0$ in Λ.

5.3. Invariant functions and Kählerian structures. Here we devote most of our attention the the set of generic points which is defined by $\Lambda^0 = \{\alpha \in \Lambda : \alpha_{n-1} \neq 0\}$. For this let $X = G^{\mathbb{C}}/\bar{A}$, where $\mathfrak{a} = \mathfrak{a}_n$ and $\bar{A} := exp(\bar{\mathfrak{a}})$. In order to determine natural generators of the space $\mathcal{E}(X^+)^G$ of invariants it is convenient to introduce coordinates on X and explicitly compute the G-action (see [**V**], p.43).

In the coordinates $(z_1, \ldots, z_{n-1}, w_1)$, the base point x_0 corresponds to $0 \in \mathbb{C}^n \cong X$ and the complex tangent line to $M = G.x_0$ at x_0 corresponding to $\mathfrak{a} = T^{CR}_{x_0} M \cong T^{1,0}_{x_0} M$ is the last coordinate axis $\{(0, \ldots, o, w_n)\}$.

The most convenient slice for the G-action in a neighborhood of M is given by $\Sigma := \{(iIm\, z_1, \ldots, iIm\, z_{n-1}, 0) \in X\}$ (see [**V**], p.43). In fact in this case $X =$

$G.\Sigma \cong G \times X$. Extending linear coordinate functions on Σ to G-invariant functions on $X = X^+$, one obtains a natural generating system for $\mathcal{E}(X)^G$.

PROPOSITION 5.3.1. *The G-invariant functions on X which are defined by*

$$\rho_l = (-1)^{l+1} \frac{l}{(l+1)!} (\operatorname{Im} w_n)^{l+1} - \sum_{j=0}^{l-1} (-1)^j \frac{1}{j!} (\operatorname{Im} w_n)^j \operatorname{Im} z_{n-(l-j)}$$

for $l = 1, \dots, n-1$ satisfy the following conditions:

(i) $\{z \in X : (d\rho_1 \wedge \dots \wedge d\rho_{n-1})(z) = 0\} = \emptyset$

(ii) *For $F : X \to \mathbb{R}^{n-1}$ with $F := (\rho_1, \dots, \rho_{n-1})$ it follows that for all $y \in \mathbb{R}^{n-1}$ the fiber $F^{-1}(\{y\})$ is a G-orbit.*

Due to the linear nature of the ρ_j it is natural to attempt to find constants c_j so that $\varphi = \sum e^{c_j \rho_j}$ is a Kähler or pseudo-Kähler potential. This turns out to be possible, although the verification of non-degeneracy is somewhat tedious (see **[V]** p.51-55).

THEOREM 5.3.2. *For $c_j = -1$, $j = 1, \dots, n-1$, the G-invariant $(1,1)$-form $\omega = dd^c\varphi$ is a pseudo-Kähler form on $X^+ = G.\Sigma^+$, where*

$$\Sigma^+ := \{(i\operatorname{Im} z_1, \dots, i\operatorname{Im} z_{n-1}, 0) : \operatorname{Im} z_{n-1-2k} \geq 0 \ \ \forall k = 0, \dots, [\frac{n}{2}] - 1\}.$$

It should be underlined that for functions of the type φ it is necessary to restrict to the smaller open set X^+ to obtain the pseudo-Kähler property. It would be interesting to know if this is only a technical matter or if there exists a complex geometric explanation.

While the proof of the above theorem requires some technical work, the following goes by straightforword verification (see **[V]** p.55-56).

PROPOSITION 5.3.3. *The restriction of $\omega = dd^c\varphi$ to $T^{1,0}M$ is Kählerian.*

It is perhaps worth noting that there are points $\zeta_0 \in \Sigma^+$ such that the restriction of ω to $T^{1,0}G.\zeta_0$ is degnerate. This is reflects the typical problem that arises when working with non-positive-definite structures on the ambient space X.

The technical discussion in the cases of the $\mathfrak{a}_j$, $j < n$, i.e., in the situations where the $\widehat{M}$ are not open, is in fact a bit more complicated (see **[V]** p.62-69). However, conceptually speaking it essentially goes along the same lines as those above. We go into some detail on this in 5.5.

5.4. The envelope of holomorphy. Here we also devote most of our attention to the case $\mathfrak{a} = \mathfrak{a}_n$.

Note that the Abelian subgroup $(\mathbb{R}^n, +)$ is naturally contained in G and acts via translation: $(x_1, \dots, x_n, 0)(z_1, \dots, z_{n-1}, w_n) = (z_1 + x_1, \dots, z_{n-1} + x_{n-1}, w_n + x_n)$. Thus the envelope of holomorphy $\widehat{M}$ is a tube domain in $X \cong \mathbb{C}^n$ (see **[V]** p.15). One would think that it would therefore be a simple matter to describe $\widehat{M}$, but this seems nevertheless not to be the case. As a first remark we underline the following rough statement which is of basic importance for our Ansatz.

PROPOSITION 5.4.1. *The envelope of holomorphy $\widehat{M}$ is contained in the domain of definition X^+ of the invariant pseudo-Kähler structure $\omega = dd^c\varphi$.*

Sketch of Proof. The fact that $(\mathbb{R}^n, +)$ acts by real translations on M implies that the envelope of holomorphy $\widehat{M}$ is defined as the tube domain over the geometric convex hull of the "imaginary part" $Im(M) \cong M/\mathbb{R}^n \subset i\mathbb{R}^n$. In the case at hand $Im(M)$ is simply a curve which is parameterized by $\gamma : \mathbb{R} \to i\mathbb{R}^n$, $t \mapsto i(\frac{t^n}{n!}, \ldots, \frac{t^2}{2!}, t)$.
The image $Im(\gamma)$ is contained in the region where $Im(z_{n-1-2k}) =: y_{n-1-2k} > 0$ for $k = 0, \ldots, [\frac{n}{2}] - 1$. Thus the geometric convex hull of $Im(M)$ is contained in the intersection of the half-spaces $\{y_{n-1-k} > 0\}$. These are exactly the inequalities which define Σ^+! $\square$

We know of no simple description of the geometric convex hull of the above curve. It is however possible to develop some feeling for $\widehat{M}$ via an inductive process.
For this let $Z_n := exp((e_1))_{\mathbb{C}}$ be the center of $G^{\mathbb{C}} = K_n^{\mathbb{C}}$ and consider the fibration

$$K_n^{\mathbb{C}}/\bar{A}_n = X_n \to K_n^{\mathbb{C}}/\bar{A}_n Z_n = K_{n-1}^{\mathbb{C}}/\bar{A}_{n-1} = X_{n-1} = \mathbb{C}^{n-1}.$$

This is in fact the coordinate projection $(z_1, \ldots, z_{n-1}, w_n) \to (z_2, \ldots, z_{n-1}, w_n)$ which can be repeated until reaching X_2. Recall that K_2 is the 3-dimensional Heisenberg group and thus the situation $M_3 \subset X_2$ is well-understood.

On shows that the above tower of central extensions is compatible with the envelope of holomorphy construction (see **[V]** p.49).

PROPOSITION 5.4.2. *The induced map* $\widehat{M_n} \to \widehat{M_{n-1}}$ *is surjective.*

The fibers of the map $\widehat{M_n} \to \widehat{M_{n-1}}$ can be regarded as strips in the complex plane $\{(z_1, 0, \ldots, 0)\}$. The reason that it is difficult to get an explicit hold on $\widehat{M}$ is that the width of these strips varies as a function on the base.

For $3 \le j \le n$ one can obtain a bit better estimate on $\widehat{M_{j+1}}$ than the trivial linear one mentioned above (see **[V]** p.49):

$$\widehat{M_{j+1}} \subset \cap_{k=0}^{[\frac{j}{2}]-1} \left\{ \operatorname{Im} z_{n-1-2k} \ge \frac{1}{(2k+2)!} (\operatorname{Im} w_n)^{2k+2} \right\}.$$

The computation of the envelope of holomorphy in the higher codimensional case goes in exactly the same way, because $\widehat{M}$ is contained in an affine subspace where the above discussion applies (see **[V]** p.64).

5.5. Momentum reduction. Since the symplectic structure $\omega = dd^c\varphi$ is defined by an invariant potential φ, an equivariant moment map $\mu : X^+ \to \mathfrak{g}^*$ can be defined by $\xi \circ \mu = \mu_\xi := J\xi_X(\varphi)$. As ususal, for $\xi \in \mathfrak{g}$ this is the ξ-th coordinate of μ. To emphasize the role of the potential we sometimes write $\mu = \mu^\varphi$. In the case where $\widehat{M}$ is not open in X^+ we define $\mu := \widehat{M} \to \mathfrak{g}^*$ by restriction.
In 5.2 it was shown that if all substrata of the open stratum $\mathcal{S}$ are to be taken into consideration, then we must make use of all associated CR-structures $\mathfrak{a}_j$, $j = 2, \ldots, n$.
Let us begin with the set of generic points defined by the condition $\{\alpha_{n-1} \neq 0\}$ in Λ. For this we use $\mathfrak{a} = \mathfrak{a}_n = ((e_{n+1} + ie_n))_{\mathbb{C}}$ which defines a CR-manifold M in $X = G^{\mathbb{C}}/\bar{A} \cong \mathbb{C}^n$. This has an open envelope of holomorphy which was discussed in some detail in 5.4 above. For our purposes here it is only important that $\widehat{M}$ is contained in X^+ which possesses the pseudo-Kähler structure $\omega = dd^c\varphi$ discussed in 5.3.

By a direct calculation (see [**V**] p.57-59) one proves that μ^φ connects $\widehat{M}$ with the desired stratum.

THEOREM 5.5.1. *In the case of the open domain $\widehat{M}$ the moment map μ^φ is given on the slice Σ^+ by*

$$\mu(i\mathrm{Im}\, z_1, \ldots, i\mathrm{Im}\, z_{n-1}, 0) = \sum_{j=1}^{n-1} e^{\mathrm{Im}\, z_j} e_j^* - \sum_{j=2}^{n-1} e^{\mathrm{Im}\, z_{j-1}} \mathrm{Im}\, z_j e_n^* \,.$$

In particular, $\mu^\varphi|\Sigma^+ : \Sigma^+ \to \Lambda$ is a diffeomorphism onto its open image.

Since the G-action on X^+ is free and the isotropy along the image slice is exactly complementary to the CR-structure, μ^φ is geometrically multiplicity free as required by the Ansatz. Note that the substrata $\Lambda^j = \{\alpha_{n-1} = \ldots = \alpha_{n-j} = 0\}$, $j = 1, \ldots, n-1$, are not in the image of μ.

These substrata are filled out by the images of the moment maps of the $\widehat{M}_j$ which are the the envelopes of holomorphy of the CR-manifolds M_j defined by $\mathfrak{a}_j = ((e_{n+1} + ie_j))$, $j = 1, \ldots, n-1$.

For these $\mathfrak{a}_j$ the theory is very much the same as that for $\mathfrak{a}_n$. This is reflected by the natural choice of generators for $\mathcal{E}(X^+)^G$ (see [**V**] p.64):

$$\rho_j = (-1)^{j+1} \frac{j}{(j+1)!} \frac{(\mathrm{Im}\, w_l)^{j+1}}{(1 + \mathrm{Im}\, z_{l+1})^j} - \sum_{k=0}^{j-1} (-1)^k \frac{1}{k!} \frac{(\mathrm{Im}\, w_l)^k}{(1 + \mathrm{Im}\, z_{l+1})^k} \mathrm{Im}\, z_{l-(j-k)}$$

$$\rho_j = \mathrm{Im}\, z_{j+1} \quad \text{for } j = l, \ldots, n-1.$$

Now, in order to obtain the appropriate moment map, we must define $\varphi = \sum_{i=1}^{j-1} e^{c_i \rho_i} + \sum_{i=j}^{n-1} (\rho_i)^2$. In [**V**] another choice was made, but the calculations of $dd^c\varphi = \omega$, which are carried out to show that ω is a pseudo-Kähler form on X^+ and that its restriction to $T^{1,0}M$ is Kählerian, remains the same (see [**V**], p.65-67). The moment map is computed directly in slice coordinates $(iIm\, z_1, \ldots, iIm\, z_{j-1}, 0, iIm\,_{j+1}, 0, \ldots, 0)$:

$$\mu^\varphi = \sum_{i=1}^{j-3} e^{\mathrm{Im}\, z_i} e_i^* - \sum_{i=2}^{j-3} e^{\mathrm{Im}\, z_{j-i-2}} \frac{\mathrm{Im}\, z_{(j-i-1)}}{1 + \mathrm{Im}\, z_{j-1}} e_{j-2}^* + e^{\mathrm{Im}\, z_{j-1}} e_{j-1}^*$$

for $j = 2, \ldots, n$.
As above $\mu^\varphi|\Sigma_j$ maps Σ_j diffeomorphically onto its open image in Λ_j and the condition of multiplicity-free is satisfied.

In summary, for the example of $G = K_n$ the CR-Momentum Ansatz produces CR-manifolds M_j with envelopes of holomorphy $\widehat{M_j}$ with naturally associated multiplicity-free moment maps $\mu_j : \widehat{M_j} \to \mathfrak{g}^*$ which perfectly aligne slices for the various strata of representation theoretically effective orbits.

Although nothing here is canonical, at every step of the procedure we had virtually no choices. The end results fit so tightly that we strongly believe that the representation theory associated to the G-actions on the $\mathcal{O}(M_j)$ must be a strong background force.

6. The example N_4

In this section $G = N_4$ denotes the group of (4×4)-uppertriangular matrices. A typical element of its Lie algebra is

$$\begin{pmatrix} 0 & x_4 & x_2 & x_1 \\ 0 & 0 & x_5 & x_3 \\ 0 & 0 & 0 & x_6 \\ 0 & 0 & 0 & 0 \end{pmatrix}$$

and the structure constants in the corresponding basis $\mathfrak{g} = ((e_1, \ldots, e_6))$ are

$$\begin{aligned} e_1 &= [e_2, e_6] = [e_4, e_3] \\ e_2 &= [e_4, e_5] \\ e_3 &= [e_5, e_6]. \end{aligned} \tag{1}$$

6.1. The stratification of the coadjoint representation. The open stratum $\mathcal{S} \subset \mathfrak{g}^*$ consists of those $\ell = \sum \alpha_i e_i^*$ such that $\alpha_1 \neq 0$. This is in fact characterized by maximal orbit-dimension: $\ell \in \mathcal{S} \iff \mathit{dim}\, G.\ell = 4$. For $\ell \in \mathcal{S}$ the orbit $O = G.\ell$ can be described by the parameterization

$$(s, t, u, v) \mapsto (\alpha_1, s, t, u, (\alpha_5 - \frac{\alpha_2 \alpha_3}{\alpha_1}) + \frac{1}{\alpha_1} st, v).$$

The appropriate slice is

$$\Lambda := \{(\alpha_1, 0, 0, 0, \alpha_5, 0) : \alpha_1 \neq 0\}.$$

At $\ell \in \Lambda$ the isotropy algebra has the particularly simple form $\mathfrak{g}_\ell = ((e_1, e_5))_{\mathbb{R}}$ and at such an orbit $O = G.\ell$ the ineffectivity I_O is trivial.

The second type of coadjoint orbit defines a stratum $\mathcal{S}_2 = \{\alpha_1 = 0, \alpha_2 \alpha_3 \neq 0\}$ with a slice being given by $\Sigma_2 := \{(0, \alpha_2, \alpha_3, \alpha_4, 0, \alpha_6 : \alpha_2 \alpha_3 \neq 0\}$.

In this case the orbit-type stratification requires a further decomposition. It is most easily defined at the slice level, where for $\Sigma_{2,\lambda} := \{(0, \alpha_2, \alpha_2 \cdot \lambda, \alpha_4, 0, \alpha_6) : \alpha_2 \neq 0\}$, $\lambda \in \mathbb{R}^*$, the isotropy algebras is $\mathfrak{g}_\lambda = ((e_1, e_2, e_3, e_6 + \lambda e_4))$. This decomposition indeed defines the orbit-type stratification in $\mathcal{S}_2$.

The ineffectivity associated to an orbit $O = G.\ell$, $\ell \in \Lambda_{2,\lambda}$, is $\mathfrak{i}_O = ((e_1, \lambda e_2 - e_3))$. Unlike the case of $G = K_n$, where the ineffectivity leads to nothing new, the group G/I_O is not of the same form as G. Thus it is of interest to carry out the Ansatz for these orbits as well. Although a number of new features arise, we do not do this here and instead only refer to ([**V**]) for details. For the sake of completeness of this section let us however describe the remaining coadjoint orbits.

The third type of orbit is defined by the slice $\{\alpha_1 = \alpha_2 \alpha_3 = 0, \alpha_2^2 + \alpha_3^2 \neq 0\}$. These correspond to the values $\lambda = 0, \infty$ on the boundaries of the slices $\Sigma_{2,\lambda}$. As in the case of orbits of the second type, these orbits are (flat) affine planes in $\mathfrak{g}^*$

For $\lambda = 0$ it follows that $\mathfrak{g}_\ell = ((e_1, e_2, e_3, e_6))$ and for $\lambda = \infty$ we have $\mathfrak{g}_\ell = ((e_1, e_2, e_3, e_4))$. These algebras are not conjugate by inner automorphisms.

The ineffectivity also varies depending on λ. For $\lambda = 0$ it is $\mathfrak{i}_O = ((e_1, e_3))$ and for $\lambda = \infty$ we have $\mathfrak{i}_O = ((e_1, e_2))$.

6.2. Associated CR-structures. In the case of $G = N_4$ there are no associated CR-structures with $\mathcal{L}(\mathfrak{a}) = \mathfrak{g}^{\mathbb{C}}$. As we will later explain, this can be understood as a consequence of the nature of the representation theory associated to the generic coadjoint orbits.

It can be shown that up to isomorphism ther is a unique structure $\mathfrak{a} := ((e_2 + ie_6, e_3 - ie_4))$ which satisfies our usual requirements and is associated to orbits $O = G.\ell$, $\ell \in \Lambda$ (see [**V**] p.82). One easily computes its 1-codimensional Levi-algebra: $\mathcal{L}(\mathfrak{a}) = ((e_1, e_2, e_3, e_4, e_6))_{\mathbb{C}}$.

Elementary Lie algebra considerations show that the 1-dimensional structures $\mathfrak{a}$ all satisfy $dim_{\mathbb{C}}\mathcal{L}(\mathfrak{a}) \leq 4$ and for $dim_{\mathbb{C}}(\mathcal{L}(\mathfrak{a}) = 4$ there is effectively only one of interest. Written as being dependent on a parameter in order to emphasize its relation with $\ell \in \Sigma_{2,\lambda}$, we have $\mathfrak{a}_\lambda = ((\lambda_4 + e_6, ie_5))$, $\lambda \in \mathbb{R}^*$.

Finally, for $dim_{\mathcal{L}}(\mathfrak{a}) = 3$, corresponding to the orbits in the smallest strata, we have $\mathfrak{a}_0 = ((e_4 + ie_5))$ for $\lambda = 0$ and $\mathfrak{a}_\infty = ((e_5 + ie_6))$ for $\lambda = \infty$.

Let us now note some details for the structure $\mathfrak{a} = ((e_2 + ie_6, e_3 - ie_4))$ associated to $\ell \in \Lambda = \{(\alpha_1, 0, 0, 0, \alpha_5, 0) : \alpha_1 \neq 0\}$. As we have already mentioned, $\mathcal{L}(\mathfrak{a}) = ((e_1, e_2, e_3, e_4, e_6))$ is 1-codimensional and consequently $\widehat{M}$ can be expected to be a real hypersurface in $X \cong \mathbb{C}^4$.

Furthermore $\mathfrak{a} \oplus \bar{\mathfrak{a}} = ((e_2, e_3, e_4, e_6))$ and $\mathfrak{g}_\ell^{\mathbb{C}} = ((e_1, e_5))$. So the condition $\mathfrak{g}^{\mathbb{C}} = \mathfrak{g}_\ell^{\mathbb{C}} \oplus \mathfrak{a} \oplus \bar{\mathfrak{a}}$ is clearly satisfied.

Finally, we come to the computation of the first Levi-form L_1. For $Z = z_1(e_2 + ie_6) + z_2(e_3 - ie_4)$ and W correspondingly defined, one computes directly that $L_1(Z, W) = \frac{i}{2}[Z, \bar{W}] = (Z\bar{W}^T)e_1$, where $Z(z_1, z_2)$ and $W = (w_1, w_2)$ are now regarded in coordinates. Thus, for $\lambda = e_1^*$, the scalar valued form $\lambda \circ L_1$ is the standard positive-definite Hermitian form on $\mathbb{C}^2$.

Hence we explicitly see that $\mathfrak{a}$ satisfies all of our requirements for a CR-structure associated to $\ell \in \Lambda$.

6.3. Invariant structures. Here we briefly discuss generators of the invariant functions and associated pseudo-Kähler structures in the case of $\mathfrak{a} = ((e_2 + ie_6, e_3 - ie_4))_{\mathbb{C}}$. For this we coordinatize $X = G^{\mathbb{C}}/\bar{A} \cong \mathbb{C}^4$ in the usual way with the w-variables in (z_1, w_2, w_3, z_4) parameterizing the tangent space $T^{1,0}_{x_0}M$ at the neutral point $x_0 = 0 \in M$.

For $(x_1, \ldots, x_6)$ in Lie algebra coordinates the action is given by

$$\begin{aligned}(x_1, \ldots, x_6) \cdot (z_1, w_2, w_3, z_4) &= (z_1 + x_4 w_3 - x_6 w_2 - x_4 x_6 z_4 + \frac{1}{2} i(x_4^2 + x_6^2) \\ &+ x_1 + \frac{1}{2}(x_3 x_4 - x_2 x_6) - \frac{1}{3} x_4 x_5 x_6, w_2 + x_4 z_4 + x_2 - ix_6 + \frac{1}{2} x_4 x_5, \\ &w_3 - x_6 z_4 + x_3 + ix_4 - \frac{1}{2} x_5 x_6, z_4 + x_5)\end{aligned}$$

(2)

Note that the Abelian group $exp((e_1, e_2, e_3, e_5))$ acts as the group $(\mathbb{R}^4, +)$ of real translations. Consequently, just as in the case of 5 much of the underlying complex geometry is real in nature.

The appropriate slice for the G-action in a neighborhood of $G_{x_0} = M$ is

$$\Sigma := \{(iIm\, z_1, 0, 0, iIm\, z_4) : (Im\, z_4)^2 \neq 1\}$$

which is open in $exp(i\mathfrak{g}_\ell).x_0$. The G-action on $X^+ = G.\Sigma = \{z_1, w_2, w_3, z_4) : (Im\, z_4)^2 \neq 1\}$ is free and proper.

Of course X^+ is not connected. However, for all practical purposes one can restrict to the Stein domain defined by $-1 < Im\, z_4 < 1$.

The usual method of invariantly extending coordinate functions from Σ produces the basis

$$\begin{aligned} \rho_1 &= \frac{(\mathrm{Im}\, w_2)^2 + (\mathrm{Im}\, w_3)^2}{2[1 - (\mathrm{Im}\, z_4)^2]} - \frac{\mathrm{Im}\, w_2 \mathrm{Im}\, w_3 \mathrm{Im}\, z_4}{1 - (\mathrm{Im}\, z_4)^2} - \mathrm{Im}\, z_1 \\ \rho_2 &= \mathrm{Im}\, z_4 \end{aligned}$$

of generators of $\mathcal{E}(X^+)^G$. The importance of the hypersurface $(Im\, z_4)^2 = 1$ is transparent!

In this case it is relatively simple to produce invariant symplectic structures (see **[V]** p.87-89). As usual we use $\varphi = e^{c_1 \rho_1} + e^{c_2 \rho_2}$ with $c_1, c_2 \in \mathbb{R}^*$.

PROPOSITION 6.3.1. *For any choice of the constants c_1 and c_2 the form $\omega = dd^c\varphi$ is non-degenerate on X^+. If $c_1 < 0$, then its restriction to $T^{CR}M$ is positive-definite.*

6.4. The envelope of holomorphy and momentum reduction. Since the linear function $\rho_2 = Im\, z_4$ is G-invariant and pluri-harmonic, it is clear that $\widehat{M}$ is contained in the real hyperplane $\{Im\, z_4 = 0\}$.

More generally, we introduce the Levi-fibration $X = G^{\mathbb{C}}/\bar{A} \to G^{\mathbb{C}}/\widehat{J}$, where $\widehat{J} = exp(\mathcal{L}(\mathfrak{a}))$ is defined by the Levi-algebra. In the case at hand the base $G^{\mathbb{C}}/\widehat{J}$ is 1-dimensional and this fibration just amounts to the projection $(z_1, w_1, w_2, z_4) \mapsto z_4$.

It is useful to look more closely at the Lie algebra structures. Recall that $\mathfrak{a} = ((e_2 + ie_6, e_3 - ie_4))_{\mathbb{C}}$ and $\mathcal{L}(\mathfrak{a}) = ((e_1, e_2, e_3, e_4, e_6)) = \widehat{\mathfrak{j}}$.

Now $\mathfrak{j}$ is an ideal in $\mathfrak{g}^{\mathbb{C}}$ and therefore the $G^{\mathbb{C}}$-action on the base is simply by translations. The non-zero structure constants for $\widehat{\mathfrak{j}}$ are given by $[e_2, e_6] = [e_4, e_3] = e_1$. Thus $\widehat{\mathfrak{j}}$ is just the 5-dimensional Heisenberg algebra.

The fiber $F = \widehat{J}.x_0$ over the base point $z_4 = 0$ is naturally isomorphic to $\mathbb{C}^3$ with coordinates (z_1, w_2, w_3) and the real form $\widehat{J} \cap G =: J$ leaves the function

$$\rho_1|F =: \rho^F = \frac{1}{2}[(Im\, w_2)^2 + (Im\, w_3)^2] - Im\, z_1$$

invariant. So clearly $\widehat{M} \cap F$ contains the non-compact model of $\mathbb{B}_3$ defined by $\rho^F < 0$.

As the base point varies in $G/J = \{Im\, z_4 = 0\}$, the fiber defining equation doesn't change. This is something that is not a general phenomenon. Here it is related to the translation invariance. As a consequence $\widehat{M}$ is just the CR-product $\mathbb{B}_3 \times \mathbb{R}$.

In the usual way it is possible to compute the moment map $\mu^\varphi : X^+ \to \mathfrak{g}^*$ (see **[V]** p.89-90). Here we of course are interested in its restriction to $\widehat{M}$.

PROPOSITION 6.4.1. *The restriction of* $\mu^\varphi : X^+ \to \mathfrak{g}^*$ *to the slice* Σ^+ *is given by*

$$\mu^\varphi(iIm\, z_1, 0, 0, iIm\, z_4) = -\frac{1}{4}(c_1 e^{-c_1 Im\, z_1} e_1^* + c_2 e^{-c_2 Im\, z_4} e_5^*).$$

Furthermore, $\mu^\varphi|\Sigma^+ : \Sigma^+ \to \Lambda$ *is a diffeomorphism onto its open image.*

Note that e.g. for $c_1 = c_2 = -1$ it follows that $\mu^\varphi|(\Sigma^+ \cap \widehat{M})$ is given by $\frac{1}{4}(e^{Im\, z_1} e_1^* + e_5^*)$. We see that $\mu^\varphi(\widehat{M})$ is not an open domain in $\mathcal{S}$, but rather a real hyypersurface and that the "missing direction" $((e_5))$ is distinguished in $\mathcal{L}(\mathfrak{a})$.

There are (happily) representation theoretical reasons for this: For $\ell \in \mathcal{S}$ the representation π_O associated to the orbit $O = G.\ell$ is induced from the fibration $M \to M/J$ which is realized above in the holomorphic Levi-fibration $G^{\mathbb{C}}/\bar{A} \to G^{\mathbb{C}}/\widehat{J}$.

Acknowledgements The intial stages of the this project were supported by the Graduiertenkolleg "Geometrie und Mathematische Physik" of the Ruhr-Universität Bochum. The second author's dissertation appears as [**V**] in the Schriftenreihe of the Graduiertenkolleg. The first author was supported for the entire project by the SFB-237 of the Deutsche Forsuchungsgemeinschaft and during its last phase by the Japanese Society for Promotion of Science during his stay at Nagoya University. The authors are thankful for this kind support.

References

[BR] M. S. Baouendi and L. P. Rothschild, Cauchy-Riemann functions on manifolds of higher codimension in complex space. Invent. Math **101**(1990)no.1 42-67

[B] A. Boggess, *CR manifolds and the tangential Cauchy-Riemann complex.* CRC Press, Boca Raton, FL, 1991.

[BP] A. Boggess and J. C. Polking, Holomorphic extension of CR functions. *Duke Math. J.* **49**, 757-784 (1982).

[CG] L. Corwin und F. P. Greenleaf, *Representations of nilpotent Lie groups and their applications, Part 1: Basic theory and examples.* Cambridge University Press, Cambridge, 1990.

[G] S. J. Greenfield, Cauchy-Riemann equations in several variables. *Ann. Scuola Norm. Sup. Pisa* **22**, 275-314 (1968).

[HNO] J. Hilgert, K.H. Neeb and B. Oersted, Conal Heisenberg algebras and associated Hilbert spaces. *J. reine angew. Math.* **474**(1996)67-112

[I] A. Iannuzzi, Characterizations of G-tube domains. *Manuscr. Math.* **98**, 425-445 (1999).

[K] A. A. Kirillov, Unitary representations of nilpotent Lie groups. *Russ. Math. Surveys* **17**, 53-104 (1962), *Uspekhi Mat. Nauk* **17**, 57-110 (1962).

[L] H. Lewy, On the local character of the solutions of an atypical linear differential equation in three variables and a related theorem for regular functions of two complex variables. *Ann Math.* **64**, 514-522 (1956).

[MW] C. C. Moore and J. A. Wolf, Square integrable representations of nilpotent groups. Trans. AMS **185**(1973)445-462

[RW] L. P. Rothschild and J. A. Wolf, Eigendistribution expansions on Heisenberg groups, Indiana U. Math. J., **25** (1976) 753-762

[T] A. E. Tumanov, Extending CR functions on manifolds of finite type to a wedge. *Mat. Sbornik* **136**, 128-139 (1988).

[Va] V. S. Varadarajan, *Lie groups, Lie algebras, and their representations.* Prentice-Hall, Englewood Cliffs, NJ, 1974.

[V] M. Völler, Invariante CR-Strukturen und koadjungierte Geometrie nilpotenter Liegruppen, *Schriftenreihe des Graduiertenkollegs Geometrie und Mathematische Physik der Ruhr-Universität Bochum* **35**(1999)1-126 (Contact the first author for a hard copy or a PS-file)

[W] Joseph A. Wolf, *Classification and Fourier inversion for parabolic subgroups with square integrable nilradical.* Mem. AMS **22** (1979)

Alan Huckleberry, Ruhr-Universität Bochum, D-44780 Bochum, Germany
E-mail address: ahuck@cplx.ruhr-uni-bochum.de

Michaele Völler, Boston Consulting Group, Düsseldorf, Germany
E-mail address: voeller.michaele@bcg.com

Contemporary Mathematics
Volume **288**, 2001

Can Tangent Sphere Bundles over Riemannian Manifolds Have Strictly Positive Sectional Curvature?

O. Kowalski, M. Sekizawa, and Z. Vlášek

This paper is dedicated to the memory of Alfred Gray.

Abstract. We prove, except some particular cases, that for every point x of a Riemannian manifold $(M, g), \dim M \geq 3$, there is a curvature operator $R_x(X, Y)$, $X \wedge Y \neq 0$, with nontrivial kernel. Then we apply our results to the geometry of tangent sphere bundles over Riemannian manifolds equipped with the induced Sasaki metric.

1. Introduction

Let (M, g) be a Riemannian manifold and let $(T_rM, \tilde{g})$ denote the tangent sphere bundle of radius $r > 0$ equipped with the induced Sasaki metric. The first two authors have proved in [**18**, Theorem 4], that, if (M, g) is a locally symmetric space with positive sectional curvature, then $(T_rM, \tilde{g})$ is a space of non-negative sectional curvature for all sufficiently small $r > 0$. (See also [**7**].) We shall prove in Theorem 3.6, Section 3, that the conclusion of Theorem 4 in [**18**] is the best possible for $\dim M \geq 3$, *i.e.*, the strictly positive sectional curvature of $(T_rM, \tilde{g})$ is never attained over a locally symmetric space (M, g). Moreover, we shall prove the following more general theorem:

Theorem 1.1. *Let $(T_rM, \tilde{g})$ be a tangent sphere bundle over a Riemannian manifold (M, g) of dimension $n \geq 3$, $n \neq 8$. Then $(T_rM, \tilde{g})$ is never a space of positive sectional curvature.*

The references in this paper will be limited to a necessary minimum. For more references concerning related topics (geometry of tangent bundles with the Sasaki metric or Cheeger-Gromoll metric) see [**18**] and [**3**]–[**6**].

1991 *Mathematics Subject Classification.* Primary 53C07; Secondary 53C20, 53C25, 53C35.

Key words and phrases. Riemannian manifold, sectional curvature, tangent sphere bundle, Sasaki metric, locally symmetric space.

The first and third authors were supported by the grant GA ČR 201/99/0265 and by the project MSM 113200007.

The second author was supported by the Grant-in-Aid for Scientific Research (C) 11640014.

2. Tangent sphere bundles—a short review

Let M be a smooth and connected n-manifold, $n \geq 2$. Then the tangent bundle TM over M consists of all pairs (x, u), where x is a point of M and u is a vector from the tangent space M_x of M at x. We denote by p the natural projection of TM to M defined by $p(x, u) = x$.

Let g be a Riemannian metric on the manifold M and ∇ its Levi-Civita connection. Then the tangent space $(TM)_{(x,u)}$ of TM at (x, u) splits into the horizontal and vertical subspaces $H_{(x,u)}$ and $V_{(x,u)}$ with respect to ∇:

$$(TM)_{(x,u)} = H_{(x,u)} \oplus V_{(x,u)}.$$

For a vector $X \in M_x$, the *horizontal lift* of X to a point $(x, u) \in TM$ is a unique vector $X^h \in H_{(x,u)}$ such that $p_* X^h = X$. The *vertical lift* of X to (x, u) is a unique vector $X^v \in V_{(x,u)}$ such that $X^v(\mathrm{d}f) = Xf$ for all smooth functions f on M. Here we consider a 1-form $\mathrm{d}f$ on M as a function on TM. The map $X \longmapsto X^h$ is an isomorphism between M_x and $H_{(x,u)}$; and the map $X \longmapsto X^v$ is an isomorphism between M_x and $V_{(x,u)}$. In an obvious way we can define horizontal and vertical lifts of vector fields on M. These are uniquely defined vector fields on TM.

To each system of local coordinates $(x^1, x^2, \ldots, x^n)$ in M, one defines, in the standard way, the system of local coordinates $(x^1, x^2, \ldots, x^n; u^1, u^2, \ldots, u^n)$ in TM. The *canonical vertical vector field* on TM is a vector field $\mathfrak{U}$ defined, in terms of local coordinates, by $\mathfrak{U} = \sum_i u^i \partial/\partial u^i$. Here $\mathfrak{U}$ does not depend on the choice of local coordinates and it is defined globally on TM. For a vector $u = \sum_i u^i (\partial/\partial x^i)_x \in M_x$, we see that $u^h_{(x,u)} = \sum_i u^i (\partial/\partial x^i)^h_{(x,u)}$ and $u^v_{(x,u)} = \sum_i u^i (\partial/\partial x^i)^v_{(x,u)} = \mathfrak{U}_{(x,u)}$.

The *Sasaki metric* on the tangent bundle TM of a Riemannian manifold (M, g) is determined, at each point $(x, u) \in TM$, by the formulas

$$\text{(2.1)} \qquad \begin{cases} \bar{g}_{(x,u)}(X^h, Y^h) = g_x(X, Y), \\ \bar{g}_{(x,u)}(X^h, Y^v) = 0, \\ \bar{g}_{(x,u)}(X^v, Y^v) = g_x(X, Y), \end{cases}$$

where X and Y are arbitrary vectors from M_x. Evidently, we have $\bar{g}_{(x,u)}(X^h, \mathfrak{U}) = 0$ and $\bar{g}_{(x,u)}(X^v, \mathfrak{U}) = g_x(X, u)$. Let $\bar{\nabla}$ be the Levi-Civita connection of $(TM, \bar{g})$, and let X and Y be vector fields on M, then we have at each *fixed* point $(x, u) \in TM$,

$$\text{(2.2)} \qquad \begin{cases} (\bar{\nabla}_{X^h} Y^h)_{(x,u)} = (\nabla_X Y)^h_{(x,u)} - \dfrac{1}{2} (R_x(X, Y)u)^v, \\ (\bar{\nabla}_{X^h} Y^v)_{(x,u)} = \dfrac{1}{2} (R_x(u, Y)X)^h + (\nabla_X Y)^v_{(x,u)}, \\ (\bar{\nabla}_{X^v} Y^h)_{(x,u)} = \dfrac{1}{2} (R_x(u, X)Y)^h, \\ (\bar{\nabla}_{X^v} Y^v)_{(x,u)} = 0, \end{cases}$$

where R is the Riemannian curvature tensor of (M, g) defined by $R(X, Y) = [\nabla_X, \nabla_Y] - \nabla_{[X,Y]}$. As concerns the canonical vertical vector field $\mathfrak{U}$, we have

$$\text{(2.3)} \qquad \begin{cases} \bar{\nabla}_{X^h} \mathfrak{U} = 0, \quad \bar{\nabla}_{X^v} \mathfrak{U} = X^v, \\ \bar{\nabla}_{\mathfrak{U}} X^h = 0, \quad \bar{\nabla}_{\mathfrak{U}} X^v = 0, \\ \bar{\nabla}_{\mathfrak{U}} \mathfrak{U} = \mathfrak{U} \end{cases}$$

for each vector field X on M.

Let r be a positive number. Then the *tangent sphere bundle of radius* r over a Riemannian manifold (M,g) is the hypersurface $T_rM = \{(x,u) \in TM \mid g_x(u,u) = r^2\}$. The canonical vertical vector field $\mathcal{U}$ is normal to T_rM in $(TM,\bar{g})$ at each point $(x,u) \in T_rM$. Also, $\bar{g}(\mathcal{U},\mathcal{U}) = r^2$ along T_rM. For any vector field X tangent to M, the horizontal lift X^h is always tangent to T_rM at each point $(x,u) \in T_rM$. Yet, in general, the vertical lift X^v is not tangent to T_rM at (x,u). The *tangential lift* of X (see [**6**]) is a vector field X^t tangent to T_rM and defined by

$$X^t = X^v - \frac{1}{r^2}\bar{g}(X^v,\mathcal{U})\mathcal{U}.$$

Thus, at each point $(x,u) \in T_rM$, we have

$$X^t_{(x,u)} = X^v_{(x,u)} - \frac{1}{r^2}g_x(X,u)\mathcal{U}_{(x,u)}.$$

Now we endow the hypersurface $T_rM \subset (TM,\bar{g})$ with the induced Riemannian metric $\tilde{g}$, which is uniquely determined by the formulas

$$\begin{cases} \tilde{g}(X^h,Y^h) = \bar{g}(X^h,Y^h), \\ \tilde{g}(X^h,Y^t) = 0, \\ \tilde{g}(X^t,Y^t) = \bar{g}(X^v,Y^v) - \dfrac{1}{r^2}\bar{g}(X^v,\mathcal{U})\bar{g}(Y^v,\mathcal{U}), \end{cases} \tag{2.4}$$

where X and Y are arbitrary vector fields on M. In the following we shall use the symbol $\langle\cdot,\cdot\rangle$ for the scalar product g_x on M_x. Then (2.4) can be rewritten, at each fixed point $(x,u) \in T_rM$, in the form

$$\begin{cases} \tilde{g}_{(x,u)}(X^h,Y^h) = \langle X,Y\rangle, \\ \tilde{g}_{(x,u)}(X^h,Y^t) = 0, \\ \tilde{g}_{(x,u)}(X^t,Y^t) = \langle X,Y\rangle - \dfrac{1}{r^2}\langle X,u\rangle\langle Y,u\rangle, \end{cases} \tag{2.5}$$

where X and Y are arbitrary vectors from M_x.

We notice that $u^t_{(x,u)} = 0$ for $(x,u) \in T_rM$ and hence the tangent space $(T_rM)_{(x,u)}$ coincides with the set $\{X^h + Y^t | X \in M_x, Y \in \{u\}^\perp \subset M_x\}$.

In [**18**] all basic formulas for the *curvature operators* on the tangent sphere bundle T_rM have been derived by calculating first the shape operator and then using the Gauss equation. We shall not reproduce them here. Let us only remark that these formulas are not essentially new. They can be derived from the more general formulas for the Riemannian submersions (see Besse [**2**, Chapter 9], O'Neill [**20**] and Gray [**12**]). They have been also obtained independently by Borisenko and Yampolsky [**7**] in a different form. In [**18**] the authors used the particular form of these formulas as presented by Boeckx and Vanhecke in [**3**]–[**6**] for the special case $r = 1$.

It is obvious that each tangent 2-plane $\widetilde{P} \subset (T_rM)_{(x,u)}$ is spanned by an orthonormal basis of the form $\{{X_1}^h + {Y_1}^t, {X_2}^h + {Y_2}^t\}$. For such a basis we have $\|X_i\|^2 + \|Y_i\|^2 = 1$, $i = 1,2$, and $\langle X_1,X_2\rangle + \langle Y_1,Y_2\rangle = 0$. Moreover, we can assume $\langle X_1,X_2\rangle = \langle Y_1,Y_2\rangle = 0$. This can be reached easily by a convenient rotation of the given basis. As usual, Y_1 and Y_2 are supposed to be orthogonal to u. From the formulas for the curvature operators one obtains as in [**18**] the following formula for the sectional curvature of the 2-plane $\widetilde{P}$:

$$
\begin{aligned}
\widetilde{K}(\widetilde{P}) = {} & \langle R_x(X_1,X_2)X_2,X_1\rangle + 3\langle R_x(X_1,X_2)Y_2,Y_1\rangle + \frac{1}{r^2}\|Y_1\|^2\|Y_2\|^2 \\
& - \frac{3}{4}\|R_x(X_1,X_2)u\|^2 + \frac{1}{4}\|R_x(u,Y_2)X_1\|^2 + \frac{1}{4}\|R_x(u,Y_1)X_2\|^2 \\
& + \frac{1}{2}\langle R_x(u,Y_1)X_2, R_x(u,Y_2)X_1\rangle - \langle R_x(u,Y_1)X_1, R_x(u,Y_2)X_2\rangle \\
& + \langle(\nabla_{X_1}R)_x(u,Y_2)X_2,X_1\rangle + \langle(\nabla_{X_2}R)_x(u,Y_1)X_1,X_2\rangle.
\end{aligned} \tag{2.6}
$$

3. The proof of the main theorem

The proof of the main theorem is based on the following two propositions.

PROPOSITION 3.1. *Let (M,g), $n = \dim M$, be a Riemannian manifold such that $n \geq 3$, $n \neq 4$, 8. Then, at every point $x \in M$, there are unit vectors X, Y and Z in the tangent space M_x such that $\langle X,Y\rangle = 0$ and $R_x(X,Y)Z = 0$.*

FIRST PROOF. Let $x \in M$ be a fixed but arbitrary point. The conclusion of the Proposition is equivalent to the statement that the rank of *some* curvature operator $R_x(X,Y)$, $X \wedge Y \neq 0$, is less than the dimension n. This is obvious for odd n because the rank of $R_x(X,Y)$ is always an even number. Let now n be even and assume the opposite, *i.e.*, that the rank of $R_x(X,Y)$ is always equal to n. This means that the rank of $R_x(X,Y)$ is constant. If now $n \neq 4$ and $n \neq 8$, then a highly nontrivial result by Gilkey, Leahy and Sadofski (see [**10**], Theorem B) asserts that $\operatorname{rank}(R_x(X,Y)) \leq 2$, which gives a contradiction. (The authors are obliged to P.B.Gilkey for the valuable hints and references.) □

SECOND PROOF. Very recently, A.L.Yampolsky [**24**] supplied us a very elegant and short proof of Proposition 3.1. Here is a more rigorous version of his private communication:

Suppose that there is a point $x \in M$ such that $R_x(X,Y)Z \neq 0$ holds for every triplet $\{X,Y,Z\}$ of unit vectors satisfying $\langle X,Y\rangle = 0$. Let $\{E_1,E_2,\dots,E_n\}$ be an orthonormal basis of $(M_x,\langle\cdot,\cdot\rangle)$. It is well-known that, for every vector $Z \in M_x$, there is a canonical isomorphism $T_Z : M_x \longrightarrow (TM)_Z$ defined as follows: for each $W \in M_x$ we put $T_Z(W) = \left.\frac{\mathrm{d}}{\mathrm{d}t}\right|_0 (Z+tW)$. Now, let $S_x \subset M_x$ denote the unit tangent sphere at x. For $i = 1,2,\dots,n-1$ and each $Z \in S_x$, the vector $R_x(E_i,E_n)Z$ is nonzero and orthogonal to Z. Hence the vector $(\mathcal{V}_i)_Z = T_Z(R_x(E_i,E_n)Z) \neq 0$ is always tangent to S_x at the end-point of Z. We obtain $n-1$ nondegenerate vector fields $\mathcal{V}_1,\mathcal{V}_2,\dots,\mathcal{V}_{n-1}$ on the sphere S_x.

Now the vector fields $\mathcal{V}_1,\mathcal{V}_2,\dots,\mathcal{V}_{n-1}$ are linearly independent at each point. Indeed, suppose that $\sum_{i=1}^{n-1}\lambda_i(\mathcal{V}_i)_Z = 0$ for a nontrivial set $\{\lambda_1,\lambda_2,\dots,\lambda_{n-1}\}$ of real numbers and some $Z \in S_x$. This means that $R_x(\sum_{i=1}^{n-1}\lambda_iE_i,E_n)Z = 0$, which is a contradiction. Hence the sphere S_x is parallelizable. But, according to the well-known theorem by Adams [**1**], the only nontrivial parallelizable spheres are those of dimensions 1, 3 and 7. □

This completes the proof.

PROPOSITION 3.2. *Let $\dim M = 4$ and suppose, in addition, that (M,g) is a space of positive sectional curvature. Then the conclusion of Proposition 3.1 still holds.*

PROOF. Let $x \in M$ and choose a so-called *Chern basis* $\{E_1, E_2, E_3, E_4\}$ in M_x (see [**8**], [**16**], [**17**]). We consider the vector equation

$$R_x(E_1, \sum_{i=2}^{4} y^i E_i)(\sum_{j=1}^{4} z^j E_j) = 0, \tag{3.1}$$

where y^2, y^3, y^4 and z^1, z^2, z^3, z^4 are parameters satisfying the conditions $\sum_{i=2}^{4}(y^i)^2 = 1$ and $\sum_{j=1}^{4}(z^j)^2 = 1$. Suppose first that the parameters y^2, y^3, y^4 are fixed. Then we obtain a system of n linear homogeneous equations for the unknowns z^1, z^2, z^3, z^4 which can be written in the form

$$\sum_{j=1}^{4} a_{jk} z^j = 0, \quad k = 1, 2, 3, 4, \tag{3.2}$$

where

$$a_{jk} = -\sum_{i=2}^{4} y^i R_{1ijk}, \quad j, k = 1, 2, 3, 4. \tag{3.3}$$

Obviously $a_{kj} = -a_{jk}$, $j, k = 1, 2, 3, 4$, and the matrix $\boldsymbol{A} = (a_{jk})$ of the system (3.2) is skew-symmetric.

With respect to a Chern basis the curvature components satisfy

$$R_{1213} = R_{1214} = R_{1223} = R_{1224} = 0, \quad R_{1314} = R_{1323} = 0. \tag{3.4}$$

(In the generic case, the Chern basis is uniquely determined by the condition (3.4) up to a finite group of reflections.) There are 14 remaining independent curvature components. We shall need here only eight of them and they will be denoted by short symbols. Namely, we put

$$\begin{aligned} &A_1 = R_{1212}, \quad A_2 = R_{1234}, \\ &B_1 = R_{1313}, \quad B_2 = R_{1324}, \quad B_3 = R_{1334}, \\ &C_1 = R_{1414}, \quad C_2 = R_{1424}, \quad C_3 = R_{1434}. \end{aligned} \tag{3.5}$$

Now, we write (3.1) in the form

$$R_x(E_1, \lambda E_2 + \mu E_3 + \nu E_4)(\sum_{j=1}^{4} z^j E_j) = 0, \tag{3.6}$$

where $\lambda, \mu, \nu \in \mathbb{R}$ stand for y^2, y^3, y^4. We use the same notation for (3.3). Using (3.4) and (3.5), we can calculate the elements a_{ij} of the corresponding matrix $\boldsymbol{A}$:

$$\begin{aligned} &a_{12} = -A_1\lambda, \quad a_{13} = -B_1\mu, \quad a_{14} = -C_1\nu, \\ &a_{23} = (A_2 - B_2)\nu, \quad a_{24} = -B_2\mu - C_2\nu, \\ &a_{34} = -A_2\lambda - B_3\mu - C_3\nu. \end{aligned} \tag{3.7}$$

Now, we check easily that $\det \boldsymbol{A} = Q^2$, where Q is so-called *Pfaffian* of $\boldsymbol{A}$,

$$Q = a_{12}a_{34} - a_{13}a_{24} + a_{14}a_{23}. \tag{3.8}$$

Hence we get

$$\begin{aligned} Q = {} & A_1A_2\lambda^2 - B_1B_2\mu^2 + C_1(B_2 - A_2)\nu^2 \\ & + A_1B_3\lambda\mu + A_1C_3\lambda\nu - B_1C_2\mu\nu. \end{aligned} \tag{3.9}$$

The equation (3.6) has a nontrivial solution for fixed λ, μ, ν, $\lambda^2 + \mu^2 + \nu^2 = 1$, if and only if $Q = Q(\lambda, \mu, \nu) = 0$. We shall show that such triplet (λ, μ, ν) always exists. Because (M, g) is a space of positive sectional curvature, we have

$$A_1 > 0, \quad B_1 > 0, \quad C_1 > 0. \tag{3.10}$$

Now, we shall distinguish four cases. In the first two cases we put $\epsilon = \pm 1$.

1. $\epsilon A_2 > 0$, $\epsilon B_2 < 0$. Then $\epsilon(B_2 - A_2) < 0$. Put $(\lambda, \mu, \nu) = (\cos t, 0, \sin t)$, where $t \in [0, \pi/2]$. We get $\epsilon Q = \epsilon A_1 A_2 > 0$ for $t = 0$ and $\epsilon Q = \epsilon C_1(B_2 - A_2) < 0$ for $t = \pi/2$. Hence $Q = 0$ for some $t \in (0, \pi/2)$.
2. $\epsilon A_2 > 0$, $\epsilon B_2 > 0$. Here we consider a triplet $(\lambda, \mu, \nu) = (\cos t, \sin t, 0)$, $t \in [0, \pi/2]$, and we make the same conclusion.
3. $A_2 = 0$, $B_2 \neq 0$. We put $(\lambda, \mu, \nu) = (0, \cos t, \sin t)$ with the same conclusion.
4. $A_2 = B_2 = 0$. Then we have $Q = 0$ for $(\lambda, \mu, \nu) = (1, 0, 0)$.

This concludes the proof of Proposition 3.2. □

Now, we are going to prove Theorem 1.1. Suppose that the tangent sphere bundle $(T_rM, \tilde{g})$ over a Riemannian manifold (M, g) (with arbitrary fixed $r > 0$) has positive sectional curvature $\widetilde{K}(\widetilde{P})$. Choose $x \in M$ and an orthonormal pair $\{X_1, X_2\}$ in M_x. For any $u \in M_x$ choose a tangent 2-plane $\widetilde{P} \subset (T_rM)_{(x,u)}$ as $\mathrm{span}\{X_1{}^h, X_2{}^h\}_{(x,u)}$. From the formula (2.6) we now see that

$$\widetilde{K}(\widetilde{P}) = \langle R_x(X_1, X_2)X_2, X_1 \rangle - \frac{3}{4}\|R_x(X_1, X_2)u\|^2. \tag{3.11}$$

Hence for $P = \mathrm{span}\{X_1, X_2\}$ we get

$$K(P) = \widetilde{K}(\widetilde{P}) + \frac{3}{4}\|R_x(X_1, X_2)u\|^2 > 0. \tag{3.12}$$

We conclude that (M, g) is a space of positive sectional curvature, as well, and we can apply Proposition 3.1 and also Proposition 3.2. So, for all $n \neq 8$, there are unit vectors $X, Y, Z \in M_x$ such that $\langle X, Y \rangle = 0$ and $R_x(X, Y)Z = 0$. (Recall that, in the case $\dim M = 4$, the vector X is *generically* uniquely determined up to a sign—as the first member E_1 of the Chern basis). Choose $u \in M_x$ as rX, and let Y_2 and X_1 be unit vectors such that $\langle Y_2, u \rangle = 0$ and $R(u, Y_2)X_1 = 0$. Now, choose $\widetilde{P} \subset (T_rM)_{(x,u)}$ such that $\widetilde{P} = \mathrm{span}\{X_1{}^h, Y_2{}^t\}_{(x,u)}$. According to the formula (2.6) in which $X_2 = Y_1 = 0$, the only nontrivial term is $(1/4)\|R_x(u, Y_2)X_1\|^2$, and it is now also equal to zero. Hence $\widetilde{K}(\widetilde{P}) = 0$, a contradiction. □ □

REMARK 3.3. It remains an open problem if Theorem 1.1 and Proposition 3.1 still hold in dimension $n = 8$.

REMARK 3.4. In Proposition 3.2, we cannot remove the assumption about the positive sectional curvature. In fact, Ivanov and Petrova [**14**] found an example of a 4-dimensional algebraic curvature tensor of rank 4. An explicit expression for the components of this algebraic tensor $\hat{R}$ with respect to a Chern basis is given by the following formulas:

$$\begin{aligned} &\hat{R}_{1212} = -2, \quad \hat{R}_{1313} = -2, \quad \hat{R}_{1414} = 1, \\ &\hat{R}_{1234} = 1, \quad \hat{R}_{1324} = -1, \\ &\hat{R}_{2323} = 1, \quad \hat{R}_{2424} = -2, \quad \hat{R}_{3434} = -2, \end{aligned} \tag{3.13}$$

and other *independent* components $\hat{R}_{ijkl}$ are equal to zero. See [**9**] for the reference.

Hence a 4-dimensional Riemannian manifold (M, g) can be constructed for which the conclusion of Proposition 3.1 is false. Indeed, for an arbitrary real analytic Riemannian manifold (M, g), there is a well-known formula expressing the metric components g_{ij} in a normal coordinate system $(x^1, x^2, \dots, x^n)$ centered at a point $o \in M$, namely

$$g_{ij}(x) = \delta_{ij} - \frac{1}{3}\sum_{k,l}(R_{ikjl})_o x^k x^l + \text{higher order terms.} \tag{3.14}$$

(See [**13**] as a reference.) Let now $\hat{R}$ be an *algebraic* curvature tensor defined on a vector space V_n with a scalar product $\langle \cdot, \cdot \rangle$. Let $\hat{R}_{ijkl}$ be the components of $\hat{R}$ with respect to any orthonormal basis. Then let us define on $\mathbb{R}^n$ a set of functions $\{g_{ij} \mid i, j = 1, 2, \dots, n\}$ by the formula

$$g_{ij}(u^1, u^2, \dots, u^n) = \delta_{ij} - \frac{1}{3}\sum_{k,l}\hat{R}_{ikjl}u^k u^l, \tag{3.15}$$

where $(u^1, u^2, \dots, u^n)$ are the Cartesian coordinates. As easy calculation shows that

(a) g_{ij} are components of a Riemannian metric g in a sufficiently small neighborhood of the origin,
(b) the coordinates $(u^1, u^2, \dots, u^n)$ are normal coordinates for the metric g centered at the origin,
(c) $\hat{R}_{ijkl}$ coincide with the corresponding curvature components of g at the origin.

The only nontrivial assertion (b) follows immediately from the standard "criterion of normality" (see [**21**]):

$$\sum_i g_{ik}(u^1, u^2, \dots, u^n)u^i = u^k, \quad k = 1, 2, \dots, n. \tag{3.16}$$

In particular, a metric corresponding to the example by Ivanov and Petrova can be calculated in a routine way, with the metric components being quadratic polynomials. The easy computation is left to the reader.

REMARK 3.5. For $\dim M = 2$, Theorem 1.1 is not more valid. Indeed, according to a result by Yampolsky [**23**], the sectional curvature of a tangent sphere bundle $(T_rM, \tilde{g})$ is positive if and only if the Gaussian curvature K of (M, g) satisfies the inequality

$$\| \operatorname{grad} K \|^2 < K^3(1 - (3/4)r^2 K)$$

at each point of M. This inequality is satisfied, for example, by any constant $K > 0$ and any sufficiently small radius $r > 0$.

Finally, the following result shows that the conclusion of Theorem 4 in [**18**] is the best possible for $n \geq 3$.

THEOREM 3.6. *Let $(T_rM, \tilde{g})$ be a tangent sphere bundle over a locally symmetric Riemannian manifold (M, g) of dimension $n \geq 3$, and r be an arbitrary positive number. Then $(T_rM, \tilde{g})$ is never a space of positive sectional curvature.*

PROOF. Suppose that (M, g) is a locally symmetric space of dimension $n \geq 3$ with positive sectional curvature. As is well known, (M, g) is locally isometric to a rank one symmetric space of compact type. We recall the following theorem, which belongs to Wolf [**26**, Theorem 1]. (The authors are obliged to J.Berndt for the reference.) In the following we denote by $\boldsymbol{P}^k(\mathbb{F})$ a real, complex, quaternionic or Cayley projective space of dimension k according to $\mathbb{F} = \mathbb{R}, \mathbb{C}, \mathbb{H}$ or **Cay**.

THEOREM 3.7. *Let M be a connected compact Riemannian symmetric space of rank one, and let N be a connected totally geodesic submanifold of M. Then*

1. $M = \boldsymbol{S}^k$ *and* $N = \boldsymbol{S}^l$, $l = 1, 2, \dots, k$; *or*
2. $M = \boldsymbol{P}^k(\mathbb{R})$, *and either* $N = \boldsymbol{S}^1 = \boldsymbol{P}^1(\mathbb{R})$, *or* $N = \boldsymbol{P}^l(\mathbb{R})$, $l = 2, 3, \dots, k$; *or*
3. $M = \boldsymbol{P}^k(\mathbb{C})$, *and either* $N = \boldsymbol{S}^l$, $l = 1, 2$, *or* $N = \boldsymbol{P}^l(\mathbb{R}$ *or* $\mathbb{C})$, $l = 2, 3, \dots, k$; *or*
4. $M = \boldsymbol{P}^k(\mathbb{H})$, *and either* $N = \boldsymbol{S}^l$, $l = 1, 2, 3, 4$, *or* $N = \boldsymbol{P}^l(\mathbb{R}, \mathbb{C}$ *or* $\mathbb{H})$, $l = 2, 3, \dots, t$; *or*
5. $M = \boldsymbol{P}^2(\mathbf{Cay})$, *and either* $N = \boldsymbol{S}^l$, $l = 1, 2, \dots, 8$, *or* $N = \boldsymbol{P}^2(\mathbb{R}, \mathbb{C}, \mathbb{H}$ *or* $\mathbf{Cay})$,

These inclusions all exist; they are unique in the sense that, if two connected totally geodesic submanifolds of M are homeomorphic, then they are equivalent under an element of the identity component $\mathbf{I}_0(M)$ of the Lie group $\mathbf{I}(M)$ of all isometries of M.

We see at once that every space M from this list ($\dim M \geq 3$) is either of odd dimension, or it contains one of the following *totally geodesic* submanifolds: $\boldsymbol{S}^4, \boldsymbol{P}^4(\mathbb{R}), \boldsymbol{P}^2(\mathbb{C})$. These 4-dimensional spaces have positive sectional curvature and hence, due to Proposition 3.2, they satisfy the conclusion of Proposition 3.1. The last property must be *inherited* by the space M itself.

The rest of the proof is the same as that of Theorem 1.1. □

We conclude with the following observation: if Theorem 1.1 does not hold for $n = 8$, then the corresponding counter-example cannot be a locally symmetric space.

References

[1] J.F.Adams, On the non-existence of elements of Hopf invariant one, *Ann. Math.* **72**(1960), 20–104.

[2] A.L.Besse, *Einstein manifolds,* Springer-Verlag, Berlin-Heidelberg-New York, 1987.

[3] E.Boeckx and L.Vanhecke, Characteristic reflections on unit tangent sphere bundles, *Houston J. Math.* **23**(1997), 427–448.

[4] E.Boeckx and L.Vanhecke, Geometry of the tangent sphere bundle, *Proceedings of the Workshop on Recent Topics in Differential Geometry,* (Edited by L.A.Cordero and E.García-Río) Santiago de Compostela (Spain) 1997, pp. 5–17.

[5] E.Boeckx and L.Vanhecke, Curvature homogeneous unit tangent sphere bundles, *Publ. Math. Debrecen* **35**(1998), 389–413.

[6] E.Boeckx and L.Vanhecke, Unit tangent sphere bundles with constant scalar curvature, to appear in *Czechoslovak Math. J.*

[7] A.A.Borisenko and A.L.Yampolsky, The sectional curvature of the Sasaki metric of T_rM^n, Ukrainskii Geometricheskií Sbornik, No.30, 1987, pp. 10–17. English translation: Plenum Publishing Corporation, 1990.

[8] S.S.Chern, On the curvature and characteristic classes of a Riemannian manifold, *Abh. Math. Sem. Univ. Hamburg* **20**(1955), 117–126.

[9] P.B.Gilkey,and R.Ivanova, Geometric consequences of some algebraic properties of the curvature tensor, preprint.

[10] P.B.Gilkey, J.V.Leahy and H.Sadofsky, Riemannian manifolds whose skew-symmetrid curvature operator has constant eigenvalues, *Indiana Univ. Math. J.* **48**(1999), 615–634.

[11] P.B.Gilkey and T.Zhang, Algebraic curvature tensors whose skew-symmetric curvature operator has constant rank, preprint.

[12] A.Gray, Pseudo-Riemannian almost product manifolds and submersions, *J. Math. Mech.* **16**(1967), 715–737.

[13] A.Gray, The volume of a small geodesic ball of a Riemannian manifold, *Michigan Math. J.* **20**(1973), 329–344.

[14] S.Ivanov and I.Petrova, Riemannian manifold in which the skew-symmetric curvature operator has pointwise constant eigenvalues, *Geometriae Dedicata* **70**(1998), 269–282.

[15] S.Kobayashi and K.Nomizu, *Foundations of Differential Geometry* II, Interscience Publishers, New York-London-Sydney, 1969.

[16] R.Klinger, A basis that reduces to zero as many curvature components as possible, *Abh. Math. Sem. Univ. Hamburg* **61**(1991), 243–248.

[17] O.Kowalski and F.Prüfer, Curvature tensors in dimension four which do not belong to any curvature homogeneous space, *Archivum math.* **30**(1994), 45–57.

[18] O.Kowalski and M.Sekizawa, On tangent sphere bundles with small or large constant radius, *Ann. Global Anal. Geom.* **18**(2000), 207–219.

[19] P.T.Nagy, On the tangent sphere bundle of a Riemannian 2-manifold, *Tôhoku Math. J.* **29**(1977), 203–208.

[20] B.O'Neill, The fundamental equations of a submersion, *Michigan Math. J.* **13**(1966), 459–469.

[21] H.S.Ruse, A.G.Walker and T.J.Willmore, *Harmonic spaces*, Roma, Cremonese, 1961.

[22] S.Sasaki, On the differential geometry of tangent bundles, *Tôhoku Math. J.* **10**(1958), 338–354.

[23] A.L.Yampolsky, On the geometry of tangent sphere bundles of Riemannian manifolds (Russian), *Ukrain. Geom. Sb.* **24**(1981), 129–132.

[24] A.L.Yampolsky, On Sasaki metric of tangent and normal bundle (Russian), *PhD Thesis (Thesis for the degree "Candidate of Sciences")*, Odessa, 1986.

[25] A.L.Yampolsky, On the strong sphericity of the Sasaki metric of a spherical tangent bundle, *Ukrain. Geom. Sb.* **35**(1992), 150–159.

[26] J.A.Wolf, Elliptic spaces in Grassmann manifolds, *Illinois J. Math.* **7**(1963), 447–462.

Faculty of Mathematics and Physics, Charles University Prague, Sokolovská 83, 186 75 Praha 8, Czech Republic
E-mail address: `kowalski@karlin.mff.cuni.cz`

Tokyo Gakugei University, Koganei-shi Nukuikita-machi 4-1-1, Tokyo 184-8501, Japan
E-mail address: `sekizawa@u-gakugei.ac.jp`

Faculty of Mathematics and Physics, Charles University Prague, Sokolovská 83, 186 75 Praha 8, Czech Republic
E-mail address: `vlasek@karlin.mff.cuni.cz`

Contemporary Mathematics
Volume **288**, 2001

Volumes of Geodesic Balls and Spheres Associated to a Metric Connection with Torsion

Dedicated to the memory of Alfred Gray

Vicente Miquel

ABSTRACT. We give bounds for the volume of a geodesic disk associated to a metric connection with torsion in a surface, and the first terms of the asymptotic expansion of the volume of a geodesic sphere for a riemannian manifold of any dimension.

§1 Introduction

After the writing of [**Mi**], A. Gray posed me the problem of getting bounds for the volume $V_m^D(r)$ of a geodesic ball $B_m^D(r)$ of center m and radius r, associated to a metric connection D on a Riemannian manifold M, with bounded curvature R^D of D. On the other hand, in [**MN**] and [**Mi**], the first terms of the Taylor expansion of $V_m^D(r)$ were computed, but we were unable to compute the corresponding terms of the Taylor expansion of $A_m^D(r)$, the area of the D-geodesic sphere $\partial B_r^D(m)$ of center m and radius r in M.

My recent interest on comparison theorems for riemannian invariants having an homogeneous or even symmetric space as a model, renewed my attention to metric connections with torsion. In fact, in order to handle this problem, it is natural to use Cartan connections (see [**MO**] and [**Ru**]), the curvature of such a connection is the difference between a "curvature of the manifold" and the curvature of the model symmetric space, but this "curvature of the manifold" is usually the curvature of a metric connection with torsion. Then, the study of $V_m^D(r)$ and $A_m^D(r)$ is a natural first step towards a full understanding of comparison theorems with symmetric spaces using Cartan connections.

In this paper I complete the work in [**Mi**] by getting the first terms of the Taylor expansion of $A_m^D(r)$, overcoming the difficulty introduced by the failing of Gauss Lemma for D-geodesic spheres. Moreover, I give a bound for $V_m^D(r)$ for surfaces when some riemannian invariant depending on the curvature and the torsion of D

Key words and phrases. geodesic sphere, geodesic ball, volume, metric connection with torsion, comparison theorem, Cartan connection..

Work partially supported by a DGICYT Grant No. PB97-1425.

2000 Mathematics Subject Classification 53C20, 53C21.

is bounded. Before giving the precise statements of the results, we introduce the notation.

M will be a Riemannian manifold of dimension n, with metric $\langle , \rangle$ and metric connection D with torsion tensor T, $m \in M$, and

$\exp_m$ will denote the exponential map associated to D,
$r_D :=$ D-injectivity radius at m, $r \leq r_D$.
$B_r(0) := \{v \in T_mM, |v| \leq r\}$, $\partial B_r(0) := \{v \in T_mM, |v| = r\}$,
$B_r^D(m) := \exp_m(B_r(0))$, $\partial B_r^D(m) := \exp_m(\partial B_r(0))$,
$V_m^D(r) := n -$ volume$(B_m^D(r))$, $A_m^D(r) := (n-1) -$ volume$(\partial B_m^D(r))$,
ω_e volume element of T_mM, ω volume element of M,
η_e volume element of $\partial B_r(0)$, η_r volume element of $\partial B_r^D(m)$,
du volume element of $S^{n-1} := \partial B_1(0)$,
ξ_r unit vector field normal to $\partial B_r^D(m)$ such that $\eta_r = \iota_{\xi_r}\omega$.

R^D will denote the curvature tensor of D (with the convention sign of [**Gr2**]), and K^D the corresponding sectional curvature.

Given a surface M, if $\mathcal{U}M$ denotes its unit tangent bundle, for every $u \in \mathcal{U}M$, the vector $Ju \in \mathcal{U}M$ orthogonal to u is well defined up to the sign.

Here I will prove:

THEOREM 1. *Let M be a surface, $\lambda \in \mathbb{R}$, and $c = \{\sup |\langle T(u, Ju), Ju\rangle|;$ $u \in \mathcal{U}M\}$.*

(a) If $K^D - \langle D_u(T)(u, Ju), Ju\rangle \geq \lambda$ for every $u \in \mathcal{U}M$, then

$$V_m^D(r) \leq 2\pi \int_0^r e^{\frac{c}{2}t} s_{\lambda - \frac{c^2}{4}}(t)dt.$$

(b) If $K^D - \langle D_u(T)(u, Ju), Ju\rangle \leq \lambda$ for every $u \in \mathcal{U}M$, then, for $t \in$ $]0, \min\{z(co_{\lambda-\frac{c^2}{4}}), r_D\}[$, we have

$$V_m^D(r) \geq 2\pi \int_0^r e^{-\frac{c}{2}t} s_{\lambda - \frac{c^2}{4}}(t)dt.$$

As it will be clearer after reading the proof, the bound cannot be attained for any metric connection on any surface, because the equality will imply $\langle T(u, Ju), Ju\rangle$ $= |\langle T(u, Ju), Ju\rangle| = c$ for every u tangent to a D-geodesic starting from m, which is impossible for points near m, since, at m, $\langle T(u, Ju), Ju\rangle = -\langle T(-u, J(-u)), J(-u)\rangle$. Then, it is an open problem if the bound is the best or not. If so, it will show what different is this situation with the standard Riemannian case (see [**BC, page 256**]). As far as I can understand now, the main difference is introduced by the fact that (if $c \neq 0$) we have no model space with $K^D - DT$ and T constants.

To state the next theorem, we recall that torsion tensor T can be considered as a 3 covariant tensor by $T_{XYZ} = \langle T(X, Y), Z\rangle$. If $V = T_mM$, $T_m \in \wedge^2 V^* \otimes V^*$, and we can decompose this tensor sapce into 3 $O(n)$ irreducible components $V_1 := \wedge^3 V^*$, V_2 and V_3 (see [**Mi**]), and T^i is defined as the V_i-component of T.

THEOREM 2. *For any n-dimensional Riemannian manifold M,*

$$A_m^D(r) = \frac{2\pi^{\frac{n}{2}}}{(\frac{n}{2}-1)!} r^{n-1}$$
$$\times \left\{1 - \frac{1}{6n}\left(\tau^D + 2\partial\overline{T} + \frac{1}{6}||T^1||^2 - \frac{9+n+2}{12(n+2)}||T^2||^2 - \frac{3n+1}{12}||T^3||^2\right) r^2 + O(r^3)\right\}$$

and

$$A_m^D(r) = \frac{2\pi^{\frac{n}{2}}}{(\frac{n}{2}-1)!}\, r^{n-1} \times \left\{1 - \frac{1}{6n}\left(\tau^\nabla + \frac{3(2n+1)}{8(n+2)}||T^2||^2 + \frac{n+2}{8}||T^3||^2\right) r^2 + O(r^3)\right\}.$$

If ∇ is the Levi-Civita connection of M, $B_r^D(m) \subset B_r^\nabla(m)$, however, we have no a priori relation between $A_r^D(m)$ and $A_r^\nabla(m)$. A consequence of Theorem 2 is that, for r small enough, $A_r^D(m) \leq A_r^\nabla(m)$. It would be interesting to find the geometric reasons of this fact.

Moreover, we have

COROLLARY.

$$\frac{d^2 A_m^D}{dr^2}(0) = \frac{d^2 A_m^\nabla}{dr^2}(0) \text{ for every } m \in M$$

if and only if D and ∇ have the same geodesics.

The plan of the paper is the following: In section 2, I collect some facts about D-Jacobi fields that we shall use later. In section 3, I'll stablish the integral formulae for the computation of $V_m^D(r)$ and $A_m^D(r)$ and define the operators which traces give information about the derivative of the corresponding integrands. We shall stablish the differential equations satisfied by these traces for possible future use, although we shall only use (in section 4) that giving information on $V_m^D(r)$. Finally, in sections 4 and 5 I'll prove the theorems.

Two final remarks about the notation: the symbol ($'$) will denote the usual derivative for a function from $\mathbb{R}$ to $\mathbb{R}$, the tangent vector for a curve in M, and the D-covariant derivative for a vector field along a curve;

$$co_\mu(t) = \begin{cases} \sqrt{\mu}\cot(\sqrt{\mu}t) & if \quad \mu > 0 \\ \sqrt{|\mu|}\coth(\sqrt{|\mu|}t) & if \quad \mu < 0 \end{cases}.$$

§2. *D*-Jacobi fields

Although Jacobi fields for connections with torsion are known for a longtime (cfr. [**KN**]), it seems that they are not part of the common background of geometers. Then, for the convenience of the reader, I'll collect here the basic facts that I'll use in this paper.

Given $m \in M$, we shall consider D-Jacobi fields $Y(t)$ along D-geodesics of the form $\exp_m t\, u$, with $|u| = 1$, satisfying the initial conditions $Y(0) = 0$ and $Y'(0)$ orthogonal to u and of norm 1. For such a D-Jacobi fields $Y(t)$, there is a curve $u(s)$ in the sphere of radius 1 in T_mM such that $u(0) = u$,

$$Y(t) = \frac{\partial}{\partial s}|_{s=0} \exp_m t\, u(s), \text{ and } Y'(0) = u'(0), \tag{2.1}$$

or, equivalently, there is a unit vector $e \in T_mM$ orthogonal to u satisfying

$$Y(t) = \exp_{m*tu} t\, e, \text{ and } Y'(0) = e. \tag{2.2}$$

Arguing similarly to the case when D is the Levi-Civita connection, one gets that (2.1) or (2.2) are equivalent to

(2.3) $$Y''(t) - D_N(T(N,Y)) + R^D(N,Y)N = 0, \quad Y(0) = 0, \quad Y'(0) = e,$$

where $N(t) := \gamma_u'(t)$, $\gamma_u(t) := \exp_m t\, u$.

A remark that makes easier the computation of the asymptotic expansion is the following. Given the D-geodesic $\gamma_u(t)$ and the D-Jacobi Field $Y(t)$ along $\gamma_u(t)$, let (U, x) be a normal coordinate system with respect to D (that is, U is the image of an open neighbourhood of 0 by $\exp_m$, and this map is a diffeomorphism on this neighbourhood), and $x(\exp_m u^i e_i) = (u^1, ..., u^n)$, for an orthonormal basis $e_1 := u, e_2, ..., e_i := Y'(0), ..., e_n$. Then

(2.4) $$Y(t) = t\, \frac{\partial}{\partial x^i}|_{\gamma_u(t)},$$

as follows from (2.2) and the fact that $\frac{\partial}{\partial x^i} = \exp_{m*tu}\, e_i$.

§3. The expression for the volumes of D-geodesic balls and spheres

For a D-geodesic ball one has

$$V_m^D(r) = \int_{B_r^D(m)} \omega = \int_{B_r(0)} \exp_m^* \omega.$$

Given $v \in B_r(0) \subset T_m M$, let u be the unit vector in the direccion of v, that is, the vectorsatisfyng $v = tu$, $t = |v| \leq r$. For every orthonormal basis of $T_v T_m M \equiv T_m M$ of the form $\{u, e_2, ..., e_n\}$, we have, by (2.2),

$$\begin{aligned}\exp_m^* \omega_{\exp_m v} &= \exp_m^* \omega(u, e_2, ..., e_n)\, \omega_{ev} \\ &= \omega(\exp_{m*v} u, \exp_{m*v} e_2, ..., \exp_{m*v} e_n)\, \omega_{ev} \\ &= \omega(N, \frac{Y_2}{t}, ..., \frac{Y_n}{t})\, \omega_{ev},\end{aligned}$$

where Y_i are the D-Jacobi fields along $\gamma_u(t)$ satisfying $Y_i(0) = 0$ and $Y_i'(0) = e_i$, $2 \leq i \leq n$.

Since we are interested on computing the derivatives of $V_m^D(r)$ respect to r, it is better to write the expression for the volume using spherical coordinates (t, u) in $T_m M$. In these coordinates, $\omega_e = t^{n-1} dt \wedge du$, and

(3.1) $$\begin{aligned}V_m^D(r) &= \int_0^r \int_{S^{n-1}} t^{n-1} \omega(N, \frac{Y_2}{t}, ..., \frac{Y_n}{t})\, dt \wedge du \\ &= \int_0^r \int_{S^{n-1}} \omega(N, Y_2, ..., Y_n)\, dt \wedge du.\end{aligned}$$

For a D-geodesic sphere,

$$A_m^D(r) = \int_{\partial B_r^D(m)} \eta_r = \int_{\partial B_r(0)} \exp_m^* \eta_r.$$

For every orthonormal basis $\{e_2, ..., e_n\}$ of $T_v \partial B_r(0)$ identified with the vector hyperplane of $T_m M$ orthogonal to $v = ru$, we have, by (2.2),

$$\begin{aligned} \exp_m^* \eta_{rv} &= \exp_m^* \eta_r(e_2, ..., e_n)\ \eta_{ev} \\ &= \eta_r(\exp_{m*v} e_2, ..., \exp_{m*v} e_n)\ \eta_{ev} \\ &= \eta_r(\frac{Y_2}{r}, ..., \frac{Y_n}{r})\ \eta_{ev}, \end{aligned}$$

where Y_i are the D-Jacobi fields along $\gamma_u(t)$ satisfying $Y_i(0) = 0$ and $Y_i'(0) = e_i$, $2 \leq i \leq n$.

Using spherical coordinates (t, u) in $T_m M$, $\eta_e = r^{n-1} du$, and

$$A_m^D(r) = \int_{S^{n-1}} r^{n-1} \eta_r(\frac{Y_2}{r}, ..., \frac{Y_n}{r})\ du = \int_{S^{n-1}} \eta_r(Y_2, ..., Y_n)\ du. \tag{3.2}$$

To get information about the derivatives of V_m and A_m, it will be enough to know the derivatives of the integrands of (3.1) and (3.2). The main point of this section will be to get potentially useful differential equations for these derivatives.

Let us begin with the integrand of (3.1). Since D is metric, we have $D\omega = 0$, then

$$\frac{\partial \omega(N, Y_2, ..., Y_n)}{\partial t} = \sum_{i=2}^{n} \omega(N, Y_2, ..., Y_i' ..., Y_n) = \sum_{i=2}^{n} \theta^i(Y_i') \omega(N, Y_2, ..., Y_n), \tag{3.3}$$

where $\{\theta^1, ..., \theta^n\}$ is the dual reference of $\{N, Y_2, ..., Y_n\}$.

If we consider the map

$$S_N(t) : T_{\gamma_u(t)} \partial B_t(m) \longrightarrow T_{\gamma_u(t)} \partial B_t(m),$$

defined by its action on the basis $\{Y_2, ..., Y_n\}$

$$S_N(t) Y_i(t) = -\pi_N(Y_i'(t)), \tag{3.4}$$

where π_N is the projection on $T_{\gamma_u(t)} \partial B_t(m)$ parallel to N (usually, π_N will be different from the orthogonal projection), then

$$\frac{\partial \omega(N, Y_2, ..., Y_n)}{\partial t} = -\operatorname{tr} S_N(t)\ \omega(N, Y_2, ..., Y_n). \tag{3.5}$$

Then, we may know $\omega(N, Y_2, ..., Y_n)$ if we get $\operatorname{tr} S_N(t)$.

Instead of $\operatorname{tr} S_N(t)$, we shall obtain a differential equation that may be potentially useful to get comparison theorems. First we observe that

$$\theta^{i\prime}(N) = 0, \quad \theta^{i\prime}(Y_j) = -\theta^i(Y_j'),$$

then, from (3.3) and (2.3), we have,

$$\begin{aligned} (\operatorname{tr} S_N)'(t) &= -\frac{\partial}{\partial t}(\sum_{i=2}^{n} \theta^i(Y_i')) = -\sum_{i=2}^{n} \left(-\theta^i(\sum_{j=2}^{n} \theta^j(Y_i') Y_j') + \theta^i(Y_i'') \right) \\ &= \sum_{i=2}^{n} \left(\theta^i(-D_N(T(N, Y_i)) + R^D(N, Y_i)N) + \sum_{j=2}^{n} \theta^i(Y_j')\theta^j(Y_i') \right). \end{aligned}$$

Let us define T_N and R_N^D by

$$T_N(X) = T(N, X), \quad R_N^D(X) = R^D(N, X)N,$$

then we have

$$D_N(T(N, Y_i)) = D_N(T)(N, Y_i) + \sum_{j=2}^{n} T(N, Y_j)\theta^j(Y_i') = D_N(T)(N, Y_i) - T_N \circ S_N(Y_i),$$

and we obtain

$$(\operatorname{tr} S_N)'(t) = \operatorname{tr} R_N^D - \operatorname{tr} D_N(T_N) + \operatorname{tr}(\pi_N \circ T_N \circ S_N) + \operatorname{tr}(S_N^2), \tag{3.6}$$

since $\operatorname{tr} R_N^D = \operatorname{tr}(\pi_N \circ R_N^D \circ i)$ and $\operatorname{tr} D_N(T_N) = \operatorname{tr}(\pi_N \circ D_N(T_N) \circ i)$.

In order to get the derivative of the integrand of (3.2), we have to extend it on an open set of M. First we consider the vector field ξ defined at each point $\exp_m tu$ in the open set $B_{r_D}^D(m) - \{m\}$ as $\xi(\exp_m tu) = \xi_t(\exp_m tu)$. This allows to extend the η_r to a unique $(n-1)$-form η defined on the same open set of M by $\eta = \iota_\xi \omega$.

Now, we compute

$$\begin{aligned}\frac{\partial \eta_t(Y_2, ..., Y_n)}{\partial t} &= \frac{\partial \omega(\xi, Y_2, ..., Y_n)}{\partial t} = \sum_{i=2}^{n} \omega(\xi, Y_2, ..., Y_i' ..., Y_n) \\ &= \sum_{i=2}^{n} \varphi^i(Y_i')\eta_t(Y_2, ..., Y_n),\end{aligned} \tag{3.7}$$

where $\{\varphi^1, ..., \varphi^n\}$ is the dual reference of $\{\xi, Y_2, ..., Y_n\}$, and we have used that $D_N\xi$ is orthogonal to ξ, and, then, a linear combination (for each t) of $Y_2(t), ..., Y_n(t)$.

Now, we consider the map

$$S_\xi(t) : T_{\gamma_u(t)}\partial B_t(m) \longrightarrow T_{\gamma_u(t)}\partial B_t(m),$$

defined by its action on the basis $\{Y_2, ..., Y_n\}$

$$S_\xi(t)Y_i(t) = -\pi_\xi(Y_i'(t)), \tag{3.8}$$

where π_ξ is the orthogonal projection on $T_{\gamma_u(t)}\partial B_t(m)$. Then

$$\frac{\partial \eta_t(Y_2, ..., Y_n)}{\partial t} = -\operatorname{tr} S_\xi(t)\eta_t(Y_2, ..., Y_n). \tag{3.9}$$

As above, we get now a differential equation for $\operatorname{tr} S_\xi(t)$. In this situation we have:

$$\varphi^1(X) = \langle \xi, X\rangle, \quad \varphi^{i\prime}(\xi) = -\varphi^i(\xi'), \varphi^{i\prime}(Y_j) = -\varphi^i(Y_j').$$

From (3.7) and (2.3), we have,

$$\begin{aligned}(\operatorname{tr} S_\xi)'(t) &= -\frac{\partial}{\partial t}(\sum_{i=2}^{n} \varphi^i(Y_i')) = -\sum_{i=2}^{n} \left(-\varphi^i(\sum_{j=2}^{n} \varphi^j(Y_i')Y_j' + \langle \xi, Y_i'\rangle\xi') + \varphi^i(Y_i'') \right) \\ &= \sum_{i=2}^{n} \left(\varphi^i(-D_N(T(N, Y_i)) + R^D(N, Y_i)N) + \sum_{j=2}^{n} \varphi^i(Y_j')\varphi^j(Y_i') + \langle \xi, Y_i'\rangle\varphi^i(\xi') \right),\end{aligned}$$

and now

$$\sum_{i=2}^{n}\varphi^i(-D_N(T(N,Y_i)) = \sum_{i=2}^{n}(-\varphi^i(D_N(T)(N,Y_i) - \langle Y_i',\xi\rangle T(N,\xi)) + \varphi^i(T(N,S_\xi Y_i))$$
$$= -\operatorname{tr}(\pi_\xi \circ D_N T_N \circ i) + \operatorname{tr}(\pi_\xi \circ T_N \circ S_\xi) + \sum_{i=2}^{n}\langle \xi', Y_i\rangle \varphi^i(T(N,\xi)),$$

then we obtain

$$(\operatorname{tr} S_\xi)'(t) = \operatorname{tr}(S_\xi^2) + \operatorname{tr}(\pi_\xi \circ R_N^D \circ i) - \operatorname{tr}(\pi_\xi \circ D_N(T_N) \circ i) + \operatorname{tr}(\pi_\xi \circ T_N \circ S_\xi) - |\xi'|^2 + \langle \xi', T(N,\xi)\rangle. \tag{3.10}$$

§4. Proof of Theorem 1

In dimension 2, $S_N(t)$ and $\pi_N \circ T_N(t)$ are linear operators acting on a 1-dimensional vector space, then

$$S_N(t)Y(t) = k(t)Y(t), \quad \text{and} \quad \pi_N \circ T_N(t)Y(t) = \mathcal{T}(t)Y(t),$$

and the equation (3.6) takes the form

$$k'(t) = k^2(t) + \mathcal{T}(t)k(t) + K^D - \langle D_N(T)(N,E), E\rangle, \tag{4.1}$$

where $E(t)$ is a unit vector normal to N.

LEMMA.
$$\lim_{t\to 0}\frac{k(t)}{1/t} = -1 \quad \textit{and, then} \quad \lim_{t\to 0} k(t) = -\infty.$$

PROOF. The lemma follows from the following remarks:

$$k(t)\frac{Y(t)}{|Y(t)|} = S_N(t)\frac{Y(t)}{|Y(t)|} = -\pi_N\frac{Y'(t)}{|Y(t)|},$$

$$-\lim_{t\to 0}\pi_N\frac{Y'(t)}{|Y(t)|} = -\lim_{t\to 0}\pi_N\frac{1}{t}\frac{Y'(t)}{|\exp_{m*tu} Y'(0)|},$$

$$\lim_{t\to 0}\frac{Y'(t)}{|\exp_{m*tu} Y'(0)|} = \frac{Y'(0)}{|Y'(0)|},$$

and

$$\lim_{t\to 0}\frac{Y(t)}{|Y(t)|} = \lim_{t\to 0}\frac{t\ \exp_{m*tu} Y'(0)}{|t\ \exp_{m*tu} Y'(0)|} = \frac{Y'(0)}{|Y'(0)|}.$$

Now, we proceed to the proof of Theorem 1, (a). From (4.1) and the hypothesis on the bounds, we have

$$k' \geq k^2 + \mathcal{T}k + \lambda.$$

From the Lemma, there is a nonempty maximal interval $]0, t_0[$ where $k(t) < 0$. For this interval we have

$$k' \geq k^2 + ck + \lambda,$$

then

$$(k + \frac{c}{2})' \geq (k + \frac{c}{2})^2 + \lambda - \frac{c^2}{4},$$

and, from Lemma 8.19 [**Gr2, page 175**] (see also [**Gr1**]), we have, for every $t < t_0$,

$$k(t) + \frac{c}{2} \geq -co_{\lambda - \frac{c^2}{4}},$$

and the inequality is obvious for $t < t_0$. Now, part (a) of Theorem 1 follows from this, the equation (3.5) and (3.1).

To prove part (b), we use again the Lemma to obtain, from (4.1) and the hypotheses, the inequality

$$k' \leq k^2 - ck + \lambda,$$

then

$$(k - \frac{c}{2})' \geq (k - \frac{c}{2})^2 + \lambda - \frac{c^2}{4},$$

and, now, for every $t < t_0$,

$$k(t) + \frac{c}{2} \geq -co_{\lambda - \frac{c^2}{4}},$$

but, if this inequality is true for $t \in]0, t_0[$, t_0 being the first zero of $k(t)$ implies that $t_0 \leq z(co_{\lambda - \frac{c^2}{4}})$, and the theorem is proved.

§5. Proof of Theorem 2

In order to obtain the asymptotic expansion of $A_m^D(r)$, we shall also use the following formulae obtained in [**Mi**]:

$$(D_{\partial_1} \partial_i)_m = \frac{1}{2} T(u, e_i), \tag{5.1}$$

$$(D^2_{\partial_1 \partial_1} \partial_i)_m = -\frac{1}{3} R^D(u, e_i)u + \frac{2}{3} D_u(T)(u, e_i) + \frac{1}{3} T(u, T(u, e_i)). \tag{5.2}$$

Moreover, it is better to consider the integrand of the first integral in (3.2). From the remark (2.4), we may write

$$\eta_r(\frac{Y_2}{r}, ..., \frac{Y_n}{r}) = \omega(\xi, \partial_2, ..., \partial_n).$$

Using this expression, the above Lemma and $\omega(\xi, \partial_2, ..., \partial_n)(m) = \omega(\xi, e_2, ..., e_n) = 1$, we obtain

$$\frac{\partial}{\partial t} \eta_t(\frac{Y_2}{t}, ..., \frac{Y_n}{t})(0) = \sum_{i=2}^n \omega(\xi, \partial_2, ..., D_N \partial_i, ..., \partial_n)(m) = \frac{1}{2} \sum_{i=2}^n \langle T(u, e_i), e_i \rangle = \frac{1}{2} \overline{T}(u),$$

with the obvious definition of $\overline{T}$. Moreover,

$$\begin{aligned} \frac{\partial^2}{\partial t^2} \eta_t(\frac{Y_2}{t}, ..., \frac{Y_n}{t})(0) = & \sum_{i=2}^n \omega(D_N \xi, \partial_2, ..., D_N \partial_i, ..., \partial_n)(m) \\ & + \sum_{i,j=2}^n \omega(\xi, \partial_2, ..., D_N \partial_i, ..., D_N \partial_j, ..., \partial_n)(m) \\ & + \sum_{i2}^n \omega(\xi, \partial_2, ..., D^2_N \partial_i, ..., \partial_n)(m). \end{aligned}$$

Now, we compute each summand

$$\omega(D_N\xi,\partial_2,...,D_N\partial_i,...,\partial_n)(m)$$
$$= \omega(\psi^1(D_N\xi)\xi+\psi^i(D_N\xi)\partial_i,\partial_2,...,\psi^1(D_N\partial_i)\xi+\psi^i(D_N\partial_i)\partial_i,...,\partial_n)(m),$$

where $\psi^1,...,\psi^n$ is the dual basis of $\xi,\partial_2,...,\partial_n$. But $\psi^1(D_N\xi)=\langle\xi,D_N\xi\rangle=0$, and, since $g^{1j}=\delta^{1j}$, we have, for $i\neq 1$,

$$\lim_{t\to 0}\psi^i(D_N\xi)=\lim_{t\to 0}\sum_{j=2}^n g^{ij}\langle\partial_j,D_N\xi\rangle=-\sum_{j=2}^n g^{ij}\langle D_N\partial_j,\xi\rangle(0)=-\frac{1}{2}\langle T(u,e_i),u\rangle,$$

then

$$\omega(D_N\xi,\partial_2,...,D_N\partial_i,...,\partial_n)(m)=\frac{1}{4}\langle T(u,e_i),u\rangle^2.$$

For the second summand we have

$$\omega(\xi,\partial_2,...,D_N\partial_i,...,D_N\partial_j,...,\partial_n)(m)$$
$$=\omega(\xi,\partial_2,...,\psi^i(D_N\partial_i)\partial_i+\psi^j(D_N\partial_i)\partial_i,...,\psi^i(D_N\partial_j)\partial_i+\psi^j(D_N\partial_j)\partial_j,...,\partial_n)(m)$$
$$=\left(\psi^i(D_N\partial_i)(m)\psi^j(D_N\partial_j)(m)-\psi^j(D_N\partial_i)(m)\psi^i(D_N\partial_j)(m)\right)\omega(\xi,\partial_2,...,\partial_n)(m)$$
$$=\frac{1}{4}\left\{\langle T(u,e_i),e_i\rangle\langle T(u,ej),e_j\rangle-\langle T(u,e_i),e_j\rangle\langle T(u,e_j),e_i\rangle\right\}.$$

And the third

$$\omega(\xi,\partial_2,...,D_N^2\partial_i,...,\partial_n)(m)=\psi^i(D_{NN}^2\partial_i)\omega(\xi,\partial_2,...,\partial_n)(m)$$
$$=\left\langle-\frac{1}{3}R^D(u,e_i)u+\frac{2}{3}D_u(T)(u,e_i)+\frac{1}{3}T(u,T(u,e_i)),\ e_i\right\rangle.$$

By addition of these summand we obtain

$$\frac{\partial^2}{\partial t^2}\eta_t(\frac{Y_2}{t},...,\frac{Y_n}{t})(0)=\frac{1}{4}\left(\sum_{i=2}^n\langle T(u,e_i),u\rangle^2+\overline{T}(u)^2-\widetilde{T}(u)\right)$$
$$+\frac{1}{3}\left\{-Ric^D(u,u)+2D_u(\overline{T})(u)+\widetilde{T}(u)\right\},$$

where $\overline{T}(u)=\sum_i T_{ue_ie_i}$, and $\widetilde{T}(u)=\sum_{i,j}T_{ue_je_i}T_{ue_ie_j}$.

From here, using the standard methods of integration along S^{n-1} (see [**Gr2, page 213, ss.**] or [**GV**]) we get

$$\int_{S^{n-1}}\sum_{i=2}^n\langle T(u,e_i),u\rangle^2du=\frac{\pi^{\frac{n}{2}}}{(\frac{n}{2})!(n+2)}\left(||\overline{T}||^2+\breve{T}+||T||^2\right),$$

where $\breve{T}:=\sum_{i=1}^n\widetilde{T}(e_i)$,

$$\int_{S^{n-1}}\left\{\frac{1}{4}\left(\overline{T}(u)^2-\widetilde{T}(u)\right)+\frac{1}{3}\left\{-Ric^D(u,u)+2D_u(\overline{T})(u)+\widetilde{T}(u)\right\}\right\}du$$
$$=\frac{\pi^{\frac{n}{2}}}{(\frac{n}{2})!}\left(\frac{1}{4}||\overline{T}||^2+\frac{1}{12}\breve{T}-\frac{1}{3}(\tau^D+2\partial\overline{T})\right),$$

where $\partial\overline{T}=-\sum_{i=1}^n D_{e_i}\overline{T}(e_i)$. On the other hand, the integral of $\overline{T}(u)$ along S^{n-1} is zero because $\overline{T}(u)=-\overline{T}(-u)$.

Now, Theorem 1 follows from these formulae, (3.2) and the following formulae in [**Mi**]:

$$\breve{T} = -||T^1||^2 + \frac{1}{2}||T^2||^2 + \frac{1}{2}||T^3||^2; \quad ||\overline{T}||^2 = \frac{n-1}{2}||T^3||^2;$$
$$\tau^D + 2\partial\overline{T} = \tau^\nabla + \frac{1}{4}||T||^2 + \frac{1}{2}\breve{T} + ||\overline{T}||^2.$$

References

[BC] R.L. Bishop and R.H. Crittenden, *Geometry of manifolds*, Academic Press, New York, 1964.

[Gr1] A. Gray, *Comparison theorems for the volumes of tubes as generalizations of the Weyl tube formula*, Topology **21** (1982), 201–228.

[Gr2] A. Gray, *Tubes*, Addison-Wesley, New York, Reading, 1990.

[GV] A. Gray and L. Vanhecke, *Riemannian geometry as determined by the volume of small geodesic balls*, Acta Math. **142** (1979), 157–198.

[KN] S. Kobayashi and K. Nomizu, *Foundations of Differential Geometry*, vol. II, Interscience, New York, 1969.

[Mi] V. Miquel, *The volumes of small geodesic balls for a metric connection*, Compositio Math. **46** (1982), 121–132.

[MN] V. Miquel and A.M. Naveira, *Sur la relation entre la fonction volume de certaines boules géodésiques et la géométrie d'une variété riemannienne*, C. R. Acad. Sci. Paris **290** (1980), 379–381.

[MO] M. Min-Oo, *Almost Symmetric spaces*, Astérisque **163-164** (1988), 221–246.

[Ru] E. A. Ruh, *Cartan Connections*, Proc. Of Symposia in Pure Mathematics **54, Part 3** (1993), 505–519.

Departamento de Geometría y Topología. Universidad de Valencia. Burjasot (Valencia) Spain

E-mail address: miquel@uv.es

Contemporary Mathematics
Volume **288**, 2001

Special Isothermic Surfaces and Solitons

Emilio Musso and Lorenzo Nicolodi

Abstract. We establish a correspondence between Darboux's special isothermic surfaces of type $(A, 0, C, D)$ and the solutions of the second order p.d.e. $\Phi\Delta\Phi - |\nabla\Phi|^2 + \Phi^4 = s, s \in \mathbb{R}$. We then use the classical Darboux transformation for isothermic surfaces to construct a Bäcklund transformation for this equation and prove a superposition formula for its solutions. As an application we discuss 1 and 2-soliton solutions and the corresponding surfaces.

1. Introduction

The theory of isothermic surfaces in conformal geometry has been the focus for intense research over the past years due to its relation with the theory of integrable systems; see for instance [**6, 9, 10, 14, 15, 19, 20, 21**] and the literature therein.

Among isothermic surfaces Darboux [**17**] distinguished the class of *special isothermic surfaces*. These were introduced in connection with the problem of isometric deformation of quadrics and have been investigated by Bianchi [**3, 4**] and Calapso [**12**]. An isothermic immersion $f : U \subset \mathbb{R}^2 \to \mathbb{R}^3$ admits conformal curvature line coordinates x, y for which its first and second fundamental forms read

$$I = \Theta^2(dx^2 + dy^2), \quad II = \Theta^2(h_1 dx^2 + h_2 dy^2),$$

where Θ is a nowhere vanishing smooth function and h_1, h_2 denote the principal curvatures. The immersion f is called special of type (A, B, C, D) if its mean curvature H satisfies the equation

$$4\Theta^2|\nabla H|^2 + M^2 + 2AM + 2BH + 2CL + D = 0, \tag{1.1}$$

where A, B, C, D are real constants and $M = -HL$ being $L = \Theta^2(h_1 - h_2)$. Examples include surfaces of constant mean curvature (cmc).

2000 *Mathematics Subject Classification.* Primary 53A30, 53A05; Secondary 35Q51, 37K35.

Key words and phrases. Isothermic surfaces, Special isothermic surfaces, Darboux transformation, Bäcklund transformation, Solitons.

This research was partially supported by the MURST project *Proprietà Geometriche delle Varietà Reali e Complesse*, by the GNSAGA of the CNR, and by the European Contract Human Potential Programme, Research Training Network HPRN-CT-2000-00101 (EDGE). Part of this work was done while the first author was visiting the Laboratoire de Mathématiques J.-A. Dieudonné, Université de Nice Sophia-Antipolis. We thank these institutions for their support and/or hospitality.

This article has its origins in the two seminal papers of Bianchi [**3, 4**] on isothermic surfaces and deals with special isothermic surfaces of type $(A, 0, C, D)$, henceforth simply referred to as special isothermic. These surfaces are invariant under the group of conformal transformations. Since the work of Bianchi, it has been known that umbilic free special isothermic surfaces are conformally equivalent to cmc surfaces in 3-dimensional space forms[1]. Examples of special isothermic surfaces with umbilic lines can be constructed by revolving elastic curves in the hyperbolic half-plane about the boundary at infinity (cf. Section 4).

We prove that special isothermic immersions are in correspondence with the solutions $\Phi(x, y)$ of the partial differential equation [2]

$$\Phi\Delta\Phi - |\nabla\Phi|^2 + \Phi^4 = s, \quad s \in \mathbb{R}. \tag{1.2}$$

Further, we show how the geometric properties of the Darboux transformation of special isothermic surfaces apply to obtain analytic results for this equation. As for the study of Darboux transforms of special isothermic surfaces, our work is related to that of Hertrich-Jeromin–Pedit [**20**] which discusses the case of cmc immersions in Euclidean space and that of Hertrich-Jeromin–Musso–Nicolodi [**21**] concerning immersions with cmc H, $|H| = 1$, in hyperbolic 3-space.

The special isothermic surface equation (1.2) can be reformulated as a zero-curvature equation

$$P(m)_y - Q(m)_x - [P(m), Q(m)] = 0 \tag{1.3}$$

involving an auxiliary parameter m. This equation expresses the compatibility condition for a linear differential system

$$V_x = -P(m)V, \quad V_y = -Q(m)V \tag{1.4}$$

in Minkowski 5-space, where the Lax pair matrices $P(m)$, $Q(m)$ take values in $\mathbf{so}(4, 1)$ (cf. Section 3). In this representation, for every solution Φ of (1.2), the spectral parameter m describes a family of special isothermic surfaces which are second order conformal deformations of each other[3] and are obtained by integrating the loop of $\mathbf{so}(4, 1)$-valued 1-forms $P(m)dx + Q(m)dy$.

The linear system (1.4) have the conserved quantities

$$C_1(V) = -2v^0v^4 + \sum_{a=1}^{3}(v^a)^2, \quad C_2(V) = \Phi^2 v^0 - mv^3 + \frac{s}{2}\Phi^{-2}v^4. \tag{1.5}$$

The classical Darboux transforms of a special isothermic immersion with potential Φ are constructed from the solutions V of the linear system (1.4) satisfying $C_1(V) = 0$ (cf. [**3, 4**] and the recent papers [**9, 15, 10, 21, 20**]). Moreover, if V satisfies the additional constraint $C_2(V) = 0$, the Darboux transforms of special isothermic surfaces are still special isothermic. Analytically, this is equivalent to the statement that if Φ is a solution of (1.2) and if V is a solution of the linear system satisfying

[1]Observe that cmc immersions in 3-dimensional space forms are obtained as T-transforms (spectral deformations) of minimal immersions in space forms (Willmore isothermic surfaces). For a recent account of these facts see [**21**], [**13**], and the monograph of Burstall [**9**].

[2]For a special isothermic immersion, the corresponding solution is the Calapso potential Φ defined by $\Phi^2(dx^2 + dy^2) = (1/4)(h_1 - h_2)^2 I$ which only depends on the conformal class of the induced metric.

[3]They amount to a special isothermic immersion together with its T-transforms [**23**] (see also Theorem 4.6 below).

$C_1(V) = C_2(V) = 0$, then $(v^3/v^4)\Phi$ is again a solution which can be regarded as a Bäcklund transformation of (1.2). This description furnishes a procedure for generating new solutions of (1.2) by solving (1.4).

In Section 7 we prove a superposition formula for the solutions of (1.2). Namely, if Φ is a known solution and Φ_1, Φ_2 are solutions respectively generated from Φ by solutions V and W of (1.4) corresponding to different values of the spectral parameter, then

$$\Phi_3 = \Phi - \frac{(h-k)v^3w^3}{v^1w^1 + v^2w^2 + v^3w^3 - v^0w^4 - v^4w^0}\left[\frac{\Phi_2 - \Phi_1}{\Phi_1\Phi_2}\right]$$

represents a new solution of (1.2). This formula, which geometrically amounts to the Bianchi permutability theorem for special isothermic immersions, shows that, after the first step, the procedure for obtaining new solutions can be carried out without the quadratures associated with the Bäcklund transformation; it also provides a method for obtaining the multisoliton solutions of equation (1.2) by algebraic means only. As an application, we compute the 1 and 2-soliton solutions arising from the trivial solution $\Phi = 1$ of (1.2) with $s = 1$.

The Bäcklund transformation for the fourth-order equation defining isothermic immersions — the so-called Calapso equation [4] — as well as its 1-solitons have been previously considered in [**27**] and have been used by Bernstein [**2**] to obtain explicit examples of non-special isothermic tori with spherical curvature lines (cf. examples in Section 6)

2. The conformal compactification of Euclidean space

Let us begin by recalling some basic facts. The one-point conformal compactification $\mathcal{M} = \mathbb{R}^3 \cup \{\infty\}$ of Euclidean space is classically realized as the projectivization of the light cone $\mathcal{L}$ of Minkowski 5-space $\mathbb{R}^5_1$ with Lorentz scalar product $\langle\ ,\ \rangle$:

$$\mathcal{M} \cong \mathbb{P}(\mathcal{L}) = \left\{[X] \in \mathbb{P}^4 : \langle X, X\rangle = 0\right\}.$$

We consider linear coordinates $x^0, \dots, x^4$ such that

$$\langle X, Y\rangle = -\left(x^0y^4 + x^4y^0\right) + x^1y^1 + x^2y^2 + x^3y^3 = g_{ij}x^iy^j, \tag{2.1}$$

and identify $\mathcal{M}$ with $\mathbb{P}[\mathcal{L}]$ by means of the conformal map

$$j : (p^1, p^2, p^3) \in \mathbb{R}^3 \mapsto \left[\left(1, p^1, p^2, p^3, \frac{1}{2}|p|^2\right)\right] \in \mathbb{P}[\mathcal{L}], \quad j(\infty) = [(0,0,0,0,1)].$$

The linear action of the pseudo-orthogonal group $G \cong SO(4,1)$ of (2.1) descends to a transitive action on $\mathcal{M}$ by conformal (Möbius) diffeomorphisms. In this model for $\mathcal{M}$, the de Sitter space $S^4_1 = \left\{Y \in \mathbb{R}^5_1 : \langle Y, Y\rangle = 1\right\}$ parametrizes the 2-spheres in $\mathcal{M}$ by

$$Y \mapsto \mathbb{P}((Y)^\perp \cap \mathcal{L})$$

and G acts transitively on the set of 2-spheres.

A *Möbius frame* is a basis $B = (B_0, \dots, B_4)$ of $\mathbb{R}^5_1$ such that the vectors form the columns of a matrix of G. Geometrically, the unit space-like vectors B_1, B_2, B_3

[4]For more information on the Calapso equation we refer the reader to [**11, 10, 14, 9**]

represent 2-spheres which intersect orthogonally, and $[B_0], [B_4] \in \mathcal{M}$ their intersection points. Regarding $B_0, \dots, B_4$ as $\mathbb{R}^5$-valued functions defined on G, there are unique 1-forms $\{\omega^I_J\}_{0\le I,J\le 4}$ such that

$$dB_I = \omega^J_I B_J, \quad \omega^K_I g_{IJ} + \omega^K_J g_{KI} = 0$$

and satisfying the structure equations

$$d\omega^I_J = -\omega^I_K \wedge \omega^K_J .$$

$(\omega^I_J) = B^{-1}dB$ is the Maurer-Cartan form of G with values in the Lie algebra $\mathcal{G}$ of G.

REMARK 2.1. The set $\dot{\mathcal{L}}$ of all $X \in \mathcal{L}$ such that $x^4 \neq 0$ can be given a Lie group structure. For, let $X \in \dot{\mathcal{L}}$ and define

$$g^+(X) = \begin{pmatrix} 1/x^4 & x^1/x^4 & x^2/x^4 & x^3/x^4 & x^0 \\ 0 & 1 & 0 & 0 & x^1 \\ 0 & 0 & 1 & 0 & x^2 \\ 0 & 0 & 0 & 1 & x^3 \\ 0 & 0 & 0 & 0 & x^4 \end{pmatrix}.$$

The mapping

$$g^+ : X \in \dot{\mathcal{L}} \mapsto g^+(X) \in G$$

is a smooth embedding which induces a Lie group structure on $\dot{\mathcal{L}}$. The group operation is given by

$$X \star Y = \left(x^0y^4 + \frac{1}{x^4}(y^0 + \sum_{j=1}^{3} x^j y^j), x^1 + x^4y^1, x^2 + x^4y^2, x^3 + x^4y^3, x^4y^4 \right)^t .$$

In particular, $X^{-1} = (x^0, \frac{-x^1}{x^4}, \frac{-x^2}{x^4}, \frac{-x^3}{x^4}, \frac{1}{x^4})^t$ and $1 = (0,0,0,0,1)^t$.

3. The Lax pair

DEFINITION 3.1. Throughout the paper a solution Φ of equation (1.2) will be referred to as a *wave potential* with *character* s.

Let $U \subset \mathbb{R}^2$ be a simply connected domain with coordinates (x, y), and let $\Phi : U \to \mathbb{R}$ be a nowhere vanishing smooth function. For $m \in \mathbb{R}$, we define $P(m), Q(m) : U \to \mathcal{G}$ by

$$P(m) = \begin{pmatrix} 2\Phi^{-1}\Phi_x & m\Phi^{-1} - \frac{s}{2}\Phi^{-3} & 0 & 0 & 0 \\ \Phi & 0 & -\Phi^{-1}\Phi_y & -\Phi & m\Phi^{-1} - \frac{s}{2}\Phi^{-3} \\ 0 & \Phi^{-1}\Phi_y & 0 & 0 & 0 \\ 0 & \Phi & 0 & 0 & 0 \\ 0 & \Phi & 0 & 0 & -2\Phi^{-1}\Phi_x \end{pmatrix},$$

$$Q(m) = \begin{pmatrix} 2\Phi^{-1}\Phi_y & 0 & -m\Phi^{-1} - \frac{s}{2}\Phi^{-3} & 0 & 0 \\ 0 & 0 & \Phi^{-1}\Phi_x & 0 & 0 \\ \Phi & -\Phi^{-1}\Phi_x & 0 & \Phi & -m\Phi^{-1} - \frac{s}{2}\Phi^{-3} \\ 0 & 0 & -\Phi & 0 & 0 \\ 0 & 0 & \Phi & 0 & -2\Phi^{-1}\Phi_y \end{pmatrix}.$$

A straightforward calculation gives:

LEMMA 3.2. *Equation (1.2) is equivalent to the matrix Lax equation*

$$P(m)_y - Q(m)_x - [P(m), Q(m)] = 0, \tag{3.1}$$

where $P(m), Q(m)$ are as defined above.

This is the compatibility condition for the linear system

$$V_x = -P(m)V, \quad V_y = -Q(m)V \tag{3.2}$$

in Minkowski 5-space. The linear system (3.2) is referred to as the *D_m-system*[5] associated to Φ. The $\mathcal{G}$-valued one-form $\beta(m) := P(m)dx + Q(m)dy$ satisfies the Maurer-Cartan equation $d\beta(m) + \beta(m) \wedge \beta(m) = 0$ and can consequently be integrated to a frame $B(m) : U \to G$, uniquely determined up to left multiplication by a constant element of G, such that

$$dB(m) = \beta(m)B. \tag{3.3}$$

We call $B(m)$ a *normal frame field* for the wave potential Φ with *spectral parameter* m. Because of (3.3), the components $v^0, \dots, v^4$ of any constant vector $X \in \mathbb{R}^5_1$ with respect to a normal frame $B(m)$ provide a solution V of the D_m-system. From (3.3) it also follows that

$$b(m) = \frac{s}{2}\Phi^{-2}B(m)_0 + mB(m)_3 + \Phi^2 B(m)_4 \tag{3.4}$$

is a constant vector; we call $b(m)$ the *pointing vector* of the normal frame $B(m)$. The *kinetic energy* $C_1(V)$ and the *linear momentum* $C_2(V)$ of a solution V of the linear system (3.2) are defined by

$$C_1(V) = -2v^0v^4 + (v^1)^2 + (v^2)^2 + (v^3)^2, \tag{3.5}$$

$$C_2(V) = \Phi^2 v^0 - mv^3 + \frac{s}{2}\Phi^{-2}v^4. \tag{3.6}$$

If $X \in \mathbb{R}^5_1$ is the initial condition of V with respect to the normal frame $B(m)$, then $C_1(V) = \langle X, X\rangle$ and $C_2(V) = -\langle b(m), X\rangle$. Thus, C_1 and C_2 are two first integrals of the D_m-system.

4. Potentials and special isothermic surfaces

Let $f : U \to \mathbb{R}^3$ be an isothermic immersion with conformal principal coordinates x, y and let

$$I = \Theta^2(dx^2 + dy^2), \quad II = \Theta^2(h_1 dx^2 + h_2 dy^2), \tag{4.1}$$

be its fundamental forms, where Θ is a nowhere vanishing smooth function and h_1 and h_2 are the principal curvatures.

REMARK 4.1. Recall that the notion of an isothermic immersion is conformally invariant, that is, if f is an isothermic immersion and $A \in G$ is a conformal diffeomorphism, then $A \circ f$ is also isothermic [**9, 24**]. In the following we will not make any distinction between isothermic immersions in $\mathbb{R}^3$ or in $\mathcal{M}$.

Following Bianchi [**3**] we set

$$2H = h_1 + h_2, \quad L = \Theta^2(h_1 - h_2), \quad M = -HL.$$

and give the following

---

[5]The linear system (3.2) is gauge equivalent to a special case of the system considered by Darboux and Bianchi for the construction the Darboux transformation [**3, 10**] (see also Section 6)

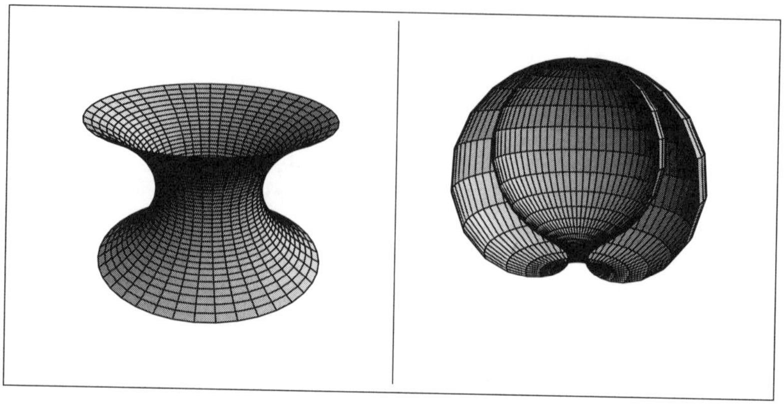

FIGURE 1. The catenoid and a Möbius transform of the catenoid; special isothermic surfaces with Calapso potential $\Psi(x,y) = \sqrt{2}\,\mathrm{Sech}(\sqrt{2}y)$ $(m = s = 0)$.

DEFINITION 4.2. f is called *special isothermic* of type (A, B, C, D) if there exist real constants A, B, C, D such that

$$4\Theta^2|\nabla H|^2 + M^2 + 2AM + 2BH + 2CL + D = 0. \tag{4.2}$$

We shall assume that $f : U \to \mathbb{R}^3$ is *generic*, i.e., either one of the following conditions hold: $L_x L_y \neq 0$, or $L_x \neq 0$ and $L_y = 0$, or $L_y \neq 0$ and $L_x = 0$, or else $L_x = L_y = 0$.

Next, let $\Phi = \frac{1}{2}(h_1 - h_2)\,\Theta$ be the *Calapso potential* [6] of the isothermic immersion f.

We are now in a position to state:

PROPOSITION 4.3. *Let $f : U \to \mathbb{R}^3$ be a special isothermic immersion of type $(A, 0, C, D)$. Then the Calapso potential Φ is a wave potential with character $D/4$.*

PROOF. The Gauss and Codazzi equations

$$\Theta\Delta\Theta - |\nabla\Theta|^2 + h_1 h_2 \Theta^4 = 0, \quad \Theta \neq 0, \tag{4.3}$$

$$(h_1)_y = -\frac{\Theta_y}{\Theta}(h_1 - h_2), \quad (h_2)_x = \frac{\Theta_x}{\Theta}(h_1 - h_2), \tag{4.4}$$

imply

$$L_x = 2\Theta^2 H_x, \quad L_y = -2\Theta^2 H_y, \quad L_{xy} = \Theta^{-1}\left(\Theta_y L_x + \Theta_x L_y\right),$$

and

$$M_x = -h_1 L_x, \quad M_y = -h_2 L_y.$$

Then, (4.2) becomes [7]

$$\Theta^{-2}|\nabla L|^2 + M^2 + 2AM + 2CL + D = 0. \tag{4.5}$$

[6] For more information on the Calapso potential see [**11, 2, 9, 10, 24**]

[7] Note that the functions $\Theta^{-1}, -h_1\Theta^2, h_2\Theta^2$ are related to the Christoffel transformation of f. Thus, from (4.5) it follows that the Christoffel transform of an isothermic immersion of type $(A, 0, C, D)$ is special of type $(A, C, 0, D)$, see also [**3**].

Since

$$2\Phi = (h_1 - h_2)\Theta, \quad L = 2\Phi\Theta, \quad M = -2\Phi H\Theta, \tag{4.6}$$

equation (4.5) also reads

$$\begin{aligned} \Theta^{-2}|\nabla\Theta|^2 &= -\Phi^{-2}|\nabla\Phi|^2 - 2\Phi^{-1}\Theta^{-1}\nabla\Phi\cdot\nabla\Theta + A\Phi^{-1}H\Theta \\ &\quad - H^2\Theta^2 - C\Phi^{-1}\Theta - \frac{D}{4}\Phi^{-2}. \end{aligned} \tag{4.7}$$

The derivative of (4.5) with respect to x and y yields

$$\begin{aligned} 2\Theta^{-1}L_x\left((\Theta^{-1}L_x)_x + \Theta^{-2}L_y\Theta_y - \Theta(M+A)h_1 + C\Theta\right) &= 0, \\ 2\Theta^{-1}L_y\left((\Theta^{-1}L_y)_y + \Theta^{-2}L_x\Theta_x - \Theta(M+A)h_2 + C\Theta\right) &= 0. \end{aligned}$$

When $L_xL_y \neq 0$ the last two equations imply

$$(\Theta^{-1}L_x)_x + (\Theta^{-1}L_y)_y + \Theta^{-2}(L_x\Theta_x + L_y\Theta_y) - 2\Theta H(M+A) + 2C\Theta = 0.$$

This equation combined with (4.6) and (4.7) gives the result. A similar argument applies in the other cases. □

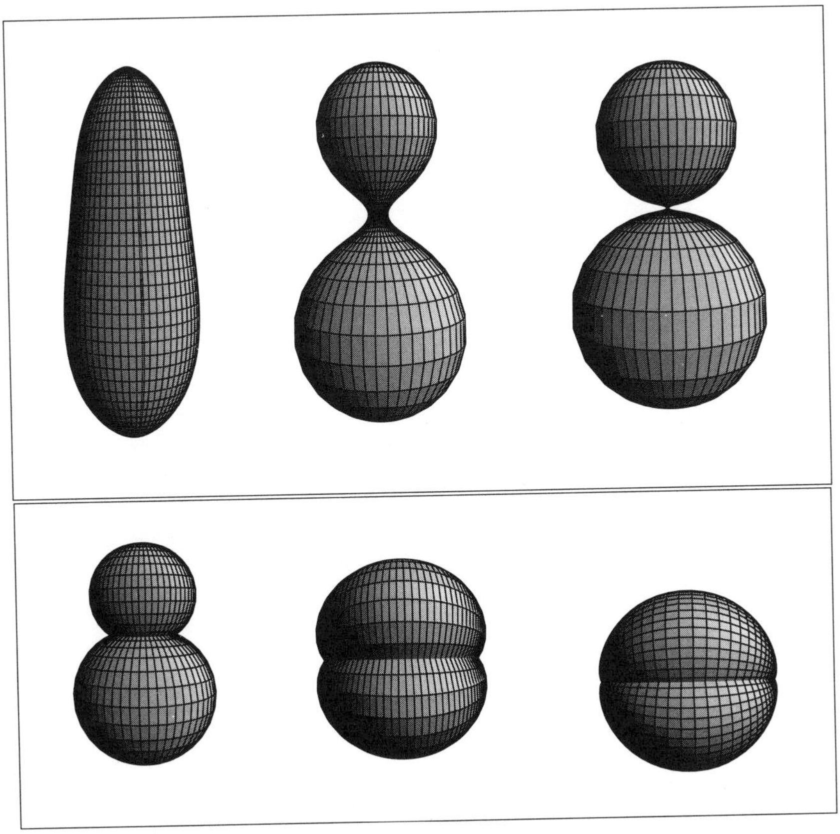

FIGURE 2. T-transforms of the catenoid; special isothermic surfaces of the wave potential $\sqrt{2}\,\mathrm{Sech}(\sqrt{2}y)$ ($m=-3$, $m=-1/4$, $m=-1/40$, $m=1/8$, $m=1$, $m=5/4$ and $s=0$).

REMARK 4.4. In the setting of the previous proposition, notice that when $L_x = L_y = 0$ the immersion f has constant mean curvature. When instead $L_x = 0$ and $L_y \neq 0$ the Calapso potential $\Phi = \Phi(y)$ is a function of the variable y alone and f is an isothermic canal surface. By a classical result of Darboux [**16**], f is then conformally equivalent to either a cone, a cylinder, or a surface of revolution. If f is a surface of revolution, one can prove that the profile curve lies in the hyperbolic half-plane, that its arclength is proportional to y, and that its curvature is parametrized by the Calapso potential. Further, equation (1.2) reduces to the Euler-Lagrange equation of the total square curvature functional $\alpha \mapsto \frac{1}{2}\int \kappa_\alpha^2$ on smooth curves α with fixed length. Therefore, the surface defined by f is obtained by revolving an elastic curve (possibly a free elastic one) in the hyperbolic half-plane about the boundary at infinity. This description indicates, in particular, how to construct special isothermic surfaces with umbilic lines (see also Babich–Bobenko [**1**]) Recall that rotational Willmore isothermic surfaces arise from free elastic curves in the hyperbolic 2-plane [**8**]. See also Langer–Singer [**22**] and Pinkall [**28**]. For the case of Willmore canal surfaces we refer to [**25**].

As for the converse, consider first the following:

DEFINITION 4.5. For any $X \in \dot{\mathcal{L}}$, let $\pi_X : \mathcal{L} \to \mathcal{M}$ be the conformal map defined by $\pi_X = j^{-1} \circ \pi \circ g^+(X)^{-1}$, where $\pi : \mathcal{L} \to \mathbb{P}[\mathcal{L}]$ denote the canonical projection.

THEOREM 4.6. *Let $B(m) : U \to G$ be a normal frame field with wave potential Φ and character s, and let $X \in \dot{\mathcal{L}}$ be a constant vector. Then, $f_m = \pi_X \circ B(m)_0 : U \to \mathcal{M}$ is a special isothermic immersion of type $(-2m, 0, 2\langle b(m), X\rangle, 4s)$ with Calapso potential Φ.*

PROOF. Let V be the solution of the D_m-system with initial condition X, that is $X = B(m)V$. Then, the fundamental forms of $f_m = \pi_X \circ B(m)_0$ can be read off the Maurer–Cartan form of the Euclidean frame $B(m)g^+(V)$. These are computed to be

$$(4.8)\quad I = \frac{\Phi^2}{(v^4)^2}\left((dx)^2 + (dy)^2\right), \quad II = \frac{\Phi^2}{(v^4)^2}\left((v^4 - v^3)(dx)^2 - (v^4 + v^3)(dy)^2\right).$$

This implies that f_m is an isothermic immersion and that

$$H = -v^3, \quad L = \frac{2}{v^4}\Phi^2, \quad M = 2\frac{v^3}{v^4}\Phi^2.$$

Next, by using the constraint $\langle V, V\rangle = 0$ and the conservation of the linear momentum $C_2(V) = -\langle b(m), X\rangle$, a direct computation shows that

$$(4.9)\qquad 4\Theta^2|\nabla H|^2 + M^2 - 4mM + 4\langle b(m), X\rangle L + 4s = 0.$$

Thus f is special isothermic of type $(-2m, 0, 2\langle b(m), X\rangle, 4s)$. □

REMARK 4.7. The solution V in the proof of the theorem can be expressed in terms of the Euclidean invariants of f by the following formulae:

$$v^0 = \frac{1}{h_1 - h_2}\left(\frac{4(H_x)^2}{(h_1 - h_2)^2\Theta^2} + \frac{4(H_y)^2}{(h_1 - h_2)^2\Theta^2} + H^2\right),$$

$$v^1 = \frac{2H_x}{(h_1 - h_2)\Theta}, \quad v^2 = -\frac{2H_y}{(h_1 - h_2)\Theta}, \quad v^3 = -H, \quad v^4 = \frac{1}{2}(h_1 - h_2).$$

REMARK 4.8. Given an umbilic free immersion $f : U \to \mathbb{R}^3$, there exist a canonical frame field $B_f : U \to G$ along f: the *central frame* field of the immersion. The construction of such a frame is due to Bryant [**7**]. If f is special isothermic of type $(A, 0, C, D)$ with Calapso potential Φ, then B_f is exactly the normal frame field with wave potential Φ and spectral parameter $m = -A/2$.

5. The Bäcklund transformation and 1-soliton solutions

DEFINITION 5.1. Let $\Phi : U \to \mathbb{R}$ be a wave potential with character s. A solution V of the linear system (3.2) satisfying $C_1(V) = C_2(V) = 0$ is said to be an *m-system of transforming functions* [8] for the potential Φ. The corresponding *Bäcklund transform* is defined by

$$E(\Phi, V) = \frac{v^3}{v^4}\Phi \tag{5.1}$$

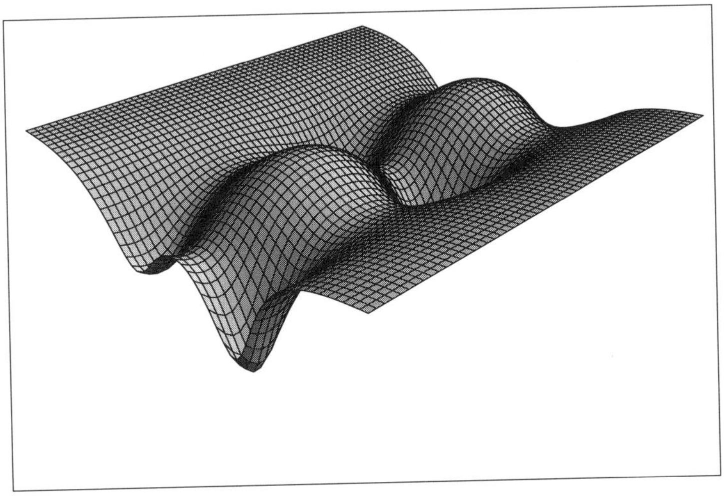

FIGURE 3. A Bäcklund transform $\Psi(x, y)$ of the wave potential $\sqrt{2}\,\mathrm{Sech}(\sqrt{2}y)$.

THEOREM 5.2. *The Bäcklund transform $E(\Phi, V)$ is a wave potential with character s.*

PROOF. Let $\tilde{\Phi} = E(\Phi, V)$. The first derivatives of $\tilde{\Phi}$ are given by

$$\tilde{\Phi}_x = \frac{v^3}{v^4}\left[-\Phi_x + v^1\left(\frac{1}{v^4} - \frac{1}{v^3}\right)\Phi^2\right], \quad \tilde{\Phi}_y = \frac{v^3}{v^4}\left[-\Phi_y + v^2\left(\frac{1}{v^4} + \frac{1}{v^3}\right)\Phi^2\right],$$

[8] Note that m-transforming functions do exist only if $m^2 - s \geq 0$. If $m^2 - s > 0$, then the set of m-transforming functions with potential Φ is a 3-dimensional cone. If $m^2 - s = 0$, the transforming functions are of the form

$$V = r\left(\frac{s}{2}\Phi^{-2}, 0, 0, m, \Phi^2\right)^t$$

for a constant $r \neq 0$. Observe that the last component v^4 of a system of transforming functions never vanishes. If, in addition, the character s is positive, then also the third component v^3 never vanishes.

and then

$$\begin{aligned}\tilde{\Phi}_{xx} &= \frac{v^3}{v^4}\Big\{-\Phi_{xx}+2\Phi_x^2\Phi^{-1}+\Phi^2\left(\frac{1}{v^4}-\frac{1}{v^3}\right)\big[-\Phi v^0-\Phi_x\Phi^{-1}v^1+\Phi_y\Phi^{-1}v^2\\ &\quad +\Phi v^3+\left(\frac{s}{2}\Phi^{-3}-m\Phi^{-1}\right)v^4\big]-2\Phi\Phi_x\frac{v^1}{v^4}-2\Phi^3\frac{(v^1)^2}{v^3v^4}+2\Phi^3\left(\frac{v^1}{v^4}\right)^2\Big\},\\ \tilde{\Phi}_{yy} &= \frac{v^3}{v^4}\Big\{-\Phi_{yy}+2\Phi_y^2\Phi^{-1}+\Phi^2\left(\frac{1}{v^4}+\frac{1}{v^3}\right)\big[-\Phi v^0+\Phi_x\Phi^{-1}v^1-\Phi_y\Phi^{-1}v^2\\ &\quad +\Phi v^3+\left(\frac{s}{2}\Phi^{-3}-m\Phi^{-1}\right)v^4\big]-2\Phi\Phi_y\frac{v^2}{v^4}-2\Phi^3\frac{(v^2)^2}{v^3v^4}+2\Phi^3\left(\frac{v^2}{v^4}\right)^2\Big\}.\end{aligned}$$

We now use the constraint $\langle V,V\rangle=0$ to obtain

$$\tilde{\Phi}^{-1}\Delta\tilde{\Phi}-\tilde{\Phi}^{-2}|\nabla\tilde{\Phi}|^2=-\tilde{\Phi}^2+\frac{v^4}{(v^3)^2}\left(2mv^3-2\Phi^2v^0\right).$$

From this equation and the constraint

$$\Phi^2v^0-mv^3+\frac{s}{2}\Phi^{-2}v^4=0$$

the result follows. □

REMARK 5.3. Let Φ be a potential with character $s\neq 0$, then the *complementary potential* Φ^* is defined by $\sqrt{|s|}\Phi^{-1}$. This is a new potential with the same character of Φ. Note that Φ^* can be obtained as the Bäcklund transform of Φ with respect to the system of transforming functions

$$V=\left(\frac{s}{2}\Phi^{-2},0,0,\sqrt{|s|},\Phi^2\right)^t.$$

EXAMPLE 5.4 (**One-soliton solutions**). Consider the trivial solution $\Phi=1$ of (1.2) with $s=1$. In this case the 1-form $\beta(m)$ corresponding to the Lax pair is given by

$$\beta(m)=\begin{pmatrix}0&(m-\frac{1}{2})dx&-(m+\frac{1}{2})dy&0&0\\ dx&0&0&-dx&(m-\frac{1}{2})dx\\ dy&0&0&dy&-(m+\frac{1}{2})dy\\ 0&dx&-dy&0&0\\ 0&dx&dy&0&0\end{pmatrix}. \tag{5.2}$$

Now set

$$\zeta=\sqrt{2|m-1|},\quad \eta=\sqrt{2|m+1|}.$$

By solving a system of first order linear differential equations with constant coefficients, the normal framing $B(m)$ is computed to be

$$\begin{pmatrix}\frac{\eta\cosh(\zeta x)+2}{\sqrt{2}\zeta\eta}&\frac{\sinh(\zeta x)}{\sqrt{2}}&0&-\frac{\eta\cosh(\zeta x)+2m}{\sqrt{2}\eta\zeta}&\frac{(2m-1)\eta\cosh(\zeta x)+2}{2\sqrt{2}\eta\zeta}\\ \frac{\sinh(\zeta x)}{\zeta}&\cosh(\zeta x)&0&-\frac{\sinh(\zeta x)}{\zeta}&\frac{(2m-1)\sinh(\zeta x)}{2\zeta}\\ \frac{\cos(\eta y)}{\eta}&0&-\sin(\eta y)&\frac{\cos(\eta y)}{\eta}&-\frac{(2m+1)\cos(\eta y)}{2\eta}\\ \frac{\sin(\eta y)}{\eta}&0&\cos(\eta y)&\frac{\sin(\eta y)}{\eta}&-\frac{(2m+1)\sin(\eta y)}{2\eta}\\ \frac{\eta\cosh(\zeta x)-2}{\sqrt{2}\zeta\eta}&\frac{\sinh(\zeta x)}{\sqrt{2}}&0&-\frac{\eta\cosh(\zeta x)-2m}{\sqrt{2}\eta\zeta}&\frac{(2m-1)\eta\cosh(\zeta x)-2}{2\sqrt{2}\eta\zeta}\end{pmatrix},$$

In order to find the solutions of (3.2) satisfying the constraint $C_1=C_2=0$, we may assume that $m>1$. Note that the pointing vector of the normal framing

$B(m)$ is the space-like vector $(1,0,0,0,-1)^t$. Thus, the transforming functions of the D_m-system are given by

$$V(m,a^1,a^2,a^3) = G\cdot B(m)^t \cdot GX(a^1,a^2,a^3),$$

where $X(a^1,a^2,a^3)\in\mathcal{L}$ is defined by

$$X(a^1,a^2,a^3) = \left(\sqrt{\frac{(a^1)^2+(a^2)^2+(a^3)^2}{2}}, a^1, a^2, a^3, \sqrt{\frac{(a^1)^2+(a^2)^2+(a^3)^2}{2}}\right)^t$$

and where $G=(g_{ij})$ is as in (2.1). It follows that

$$\begin{aligned} v^0 &= \frac{2m+1}{2\eta}\left(a^2\cos(\eta y)+a^3\sin(\eta y)\right)+\frac{2m-1}{\sqrt{2}\eta\zeta}(\|a\|\cosh(\zeta x)+ \\ &\quad -\frac{a^1\eta}{\sqrt{2}}\sinh(\zeta x)) \\ v^1 &= a^1\cosh(\zeta x)-\|a\|\sinh(\zeta x) \\ v^2 &= a^1\cos(\eta y)-a^2\sin(\eta y) \\ v^3 &= \frac{a^2}{\eta}\cos(\eta y)+\frac{a^3}{\eta}\sin(\eta y)+\frac{\|a\|}{\zeta}\cosh(\zeta x)-\frac{a^1}{\zeta}\sinh(\zeta x) \\ v^4 &= \frac{\|a\|}{\zeta}\cosh(\zeta x)-\frac{a^1}{\zeta}\sinh(\zeta x)-\frac{a^2}{\eta}\cos(\eta y)-\frac{a^3}{\eta}\sin(\eta y), \end{aligned} \tag{5.3}$$

where

$$\| a \| = \sqrt{(a^1)^2+(a^2)^2+(a^3)^2}.$$

From this we obtain the following formula for the one-soliton solutions

$$\Phi(m,a^1,a^2,a^3) = \frac{\zeta(a^2\cos(\eta y)+a^3\eta\sin(\eta y))+\eta(\|a\|\cosh(\zeta x)-a^1\sinh(\zeta x))}{-\zeta(a^2\cos(\eta y)+a^3\eta\sin(\eta y))+\eta(\|a\|\cosh(\zeta x)-a^1\sinh(\zeta x))}. \tag{5.4}$$

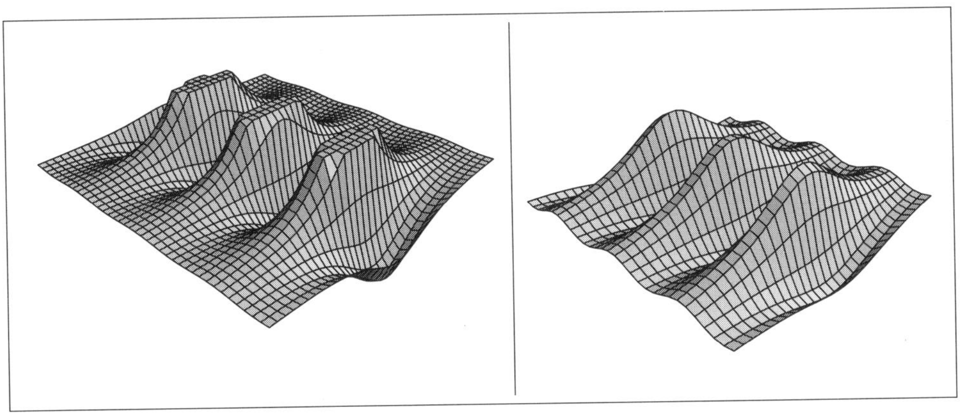

FIGURE 4. One-solitons $\Phi(4/3,0,1/100,1/300)$ and $\Phi(10/9,0,1/500,0)$.

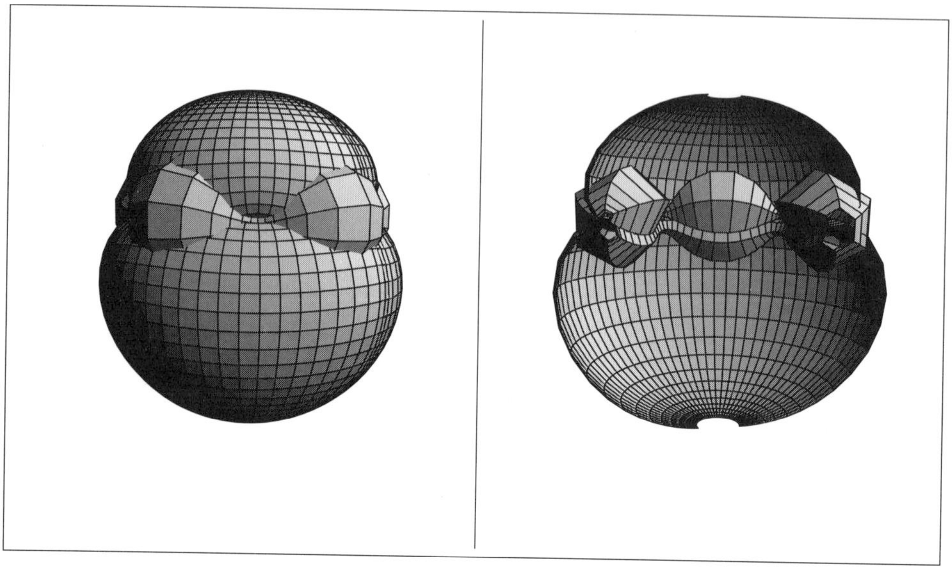

FIGURE 5. Special isothermic surface with wave potential $\Psi(x, y)$ (see Figure 3) and spectral parameter $m = 11/12$.

6. The geometry of the Bäcklund transformation

In this section we briefly describe the geometric transformation of special isothermic surfaces corresponding to the Bäcklund transformation.

DEFINITION 6.1. A *curved flat* [9] *framing* is a smooth map $A : U \to G$ such that $A^{-1}dA$ takes the form

$$\begin{pmatrix} 0 & -r\Theta^{-1}dx & r\Theta^{-1}dy & 0 & 0 \\ \Theta dx & 0 & \Theta^{-1}(\Theta_y dx - \Theta_x dy) & -h_1\Theta dx & -r\Theta^{-1}dx \\ \Theta dy & \Theta^{-1}(\Theta_x dy - \Theta_y dx) & 0 & -h_2\Theta dy & r\Theta^{-1}dy \\ 0 & h_1\Theta dx & h_2\Theta dy & 0 & 0 \\ 0 & \Theta dx & \Theta dy & 0 & 0 \end{pmatrix},$$

where Θ, h_1, h_2 are smooth functions with $\Theta(p) \neq 0$, for each $p \in U$, and r is a constant referred to as the *spectral parameter* of the curved flat. For short, the connection form $A^{-1}dA$ will be denoted by $\alpha_r(\Theta, h_1, h_2)$.

The functions Θ, h_1 and h_2 satisfy the *isothermic Gauss-Codazzi system*

$$(6.1) \qquad \begin{cases} \Theta\Delta(\Theta) = |\nabla\Theta|^2 - h_1h_2\Theta^4, \\ \Theta(h_1)_y = \Theta_y(h_2 - h_1), \\ \Theta(h_2)_x = \Theta_x(h_1 - h_2). \end{cases}$$

REMARK 6.2. Note that if $A : U \to G$ is a curved flat framing, then $f = [A_0] : U \to \mathcal{M}$ and $\bar{f} = [A_4] : U \to \mathcal{M}$ are isothermic immersions with Calapso potentials

$$\Phi = \frac{1}{2}(h_1 - h_2)\Theta, \quad \bar{\Phi} = -\frac{1}{2}(h_1 + h_2)\Theta,$$

[9] For the general notion of a curved flat we refer to Ferus–Pedit [**18**].

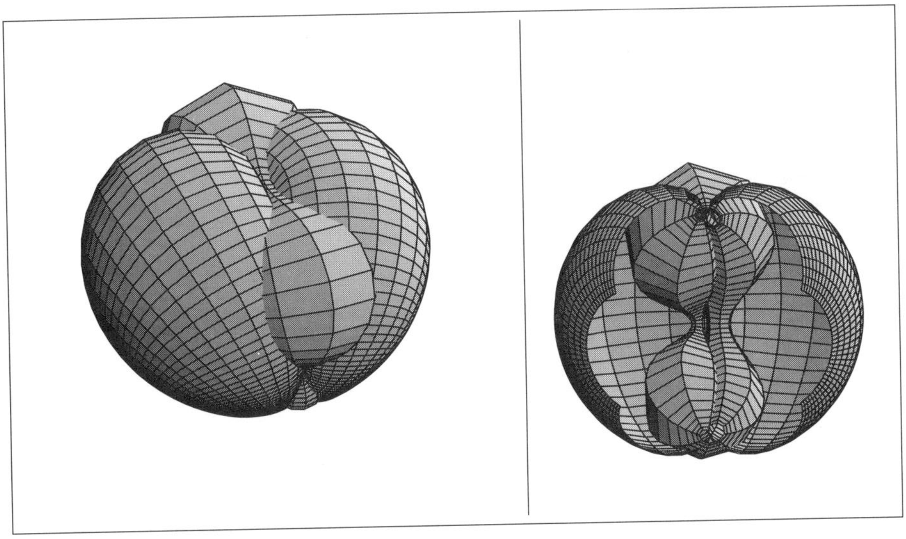

FIGURE 6. Special isothermic surface with wave potential $\Psi(x, y)$ (see Figure 3) and spectral parameter $m = 1$.

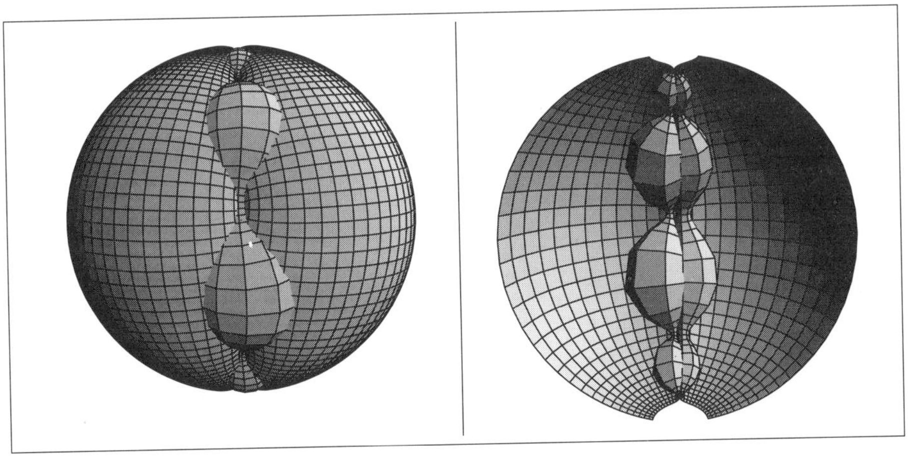

FIGURE 7. Special isothermic surface with wave potential $\Psi(x, y)$ (see Figure 3) and spectral parameter $m = 13/12$.

respectively. We say that A is a curved flat framing along f and call Φ the Calapso potential of A. The *dual frame field* is defined to be $\bar{A} = AJ$, where $J \in G$ is given by

$$\begin{pmatrix} 0 & 0 & 0 & 0 & 1 \\ 0 & -1 & 0 & 0 & 0 \\ 0 & 0 & 1 & 0 & 0 \\ 0 & 0 & 0 & 1 & 0 \\ 1 & 0 & 0 & 0 & 0 \end{pmatrix}. \tag{6.2}$$

In particular, $\bar{A}$ is a curved flat framing along $\bar{f}$. The mapping $\sigma := A_3 : U \to S^4_1 \subset \mathbb{R}^5_1$ defines a sphere congruence whose envelopes are f and $\bar{f}$. Moreover, the correspondence induced by σ preserves the curvature lines and is conformal. This amounts to saying that σ is a Darboux congruence. Accordingly, $\bar{f}$ is called a *Darboux transform* of f. Any Darboux transform of f arises in this way [10].

REMARK 6.3. The first envelope $f : U \to \mathcal{M}$ of a curved flat framing A is a special isothermic immersion if and only if, in addition to (6.1), the triple (Θ, h_1, h_2) satisfies an equation of the form

$$4\Theta^2|\nabla H|^2 + M^2 + 2AM + 2CL + D = 0, \tag{6.3}$$

where A, C, D are real constants and $M = -HL$ being $L = \Theta^2(h_1 - h_2)$ and $2H = h_1 + h_2$. We then say that A is a *special curved flat framing* of type $(A, 0, C, D)$. Notice that both the envelopes of the curved flat are special isothermic immersions if and only if $C = 0$. In this case we shall say that $\bar{f}$ is a *special Darboux transform* of f.

We now have:

PROPOSITION 6.4. *Let $f : U \to \mathcal{M}$ be a special isothermic immersion of type $(A, 0, C, D)$ with deformation parameter m and Calapso potential Φ, and let $B(m) : U \to G$ be the Bryant's central frame field along f. Next, let $V : U \to \mathcal{L}$ be any solution of the D_h-linear system*

$$dV = -\beta(h)V, \quad h \neq m,$$

satisfying the constraint $C_2(V) = 0$. Then

$$\bar{f} := \left[v^0 B(m)_0 + v^1 B(m)_1 + v^2 B(m)_2 + v^3 B(m)_3 + v^4 B(m)_4\right]$$

defines a special isothermic immersion which is a special Darboux transform of f such that

$$\bar{\Phi} = E(\Phi, V).$$

PROOF. The connection form α of the framing $A = B(m)g^+(V)$

$$\alpha = g^+(V)^{-1}d[g^+(V)] + g^+(V)^{-1}\beta(m)g^+(V)$$

takes the form $\alpha = \alpha_r(\Theta, h_1, h_2)$, with

$$\Theta = \frac{\Phi}{v^4}, \quad h_1 = v^4 - v^3, \quad h_2 = -(v^4 + v^3), \quad r = h - m. \tag{6.4}$$

It then follows that A is a special curved flat framing of type $(-2h, 0, -2C_2(V), 4s)$, where s denotes the character of Φ. The first envelope $[A_0]$ represents the original immersion f and the second envelope $[A_4]$ represents the Darboux transform $\bar{f}$. If $C_2(V) = 0$, then $\bar{f}$ is a special isothermic immersion and $\bar{\Phi} = E(\Phi, V)$. □

REMARK 6.5. $\bar{f}$ has deformation parameter m and the normal frame field $\bar{B}(m) : U \to G$ along $\bar{f}$ is given by

$$\bar{B}(m) = B(m)g^+(V)Jg^+(T),$$

[10] The result that an isothermic surface together with a Darboux transform form a curved flat in the pseudo-Riemannian symmetric space of pairs of distinct points in $\mathcal{M}$ has been proved in **[10]**. For the interpretation of the Darboux transforms as dressing transformations of loop groups we refer to Burstall **[9]**

where $T : U \to \mathcal{L}$ is the smooth map defined by

$$T = \left(\frac{\Phi^2 v^0 v^4}{(h-m)(v^3)^3}, \frac{v^1 v^4}{(v^3)^2}, -\frac{v^2 v^4}{(v^3)^2}, -\frac{v^4}{v^3}, \frac{(h-m)\Phi^{-2}(v^4)^2}{v^3} \right)^t . \tag{6.5}$$

EXAMPLE 6.6 (**Special Darboux transforms of Dupin cyclides**). Special isothermic maps with Calapso potential $\Phi = 1$ are given (up to the action of the conformal group) by the following formulae (where m is the deformation parameter of the family)

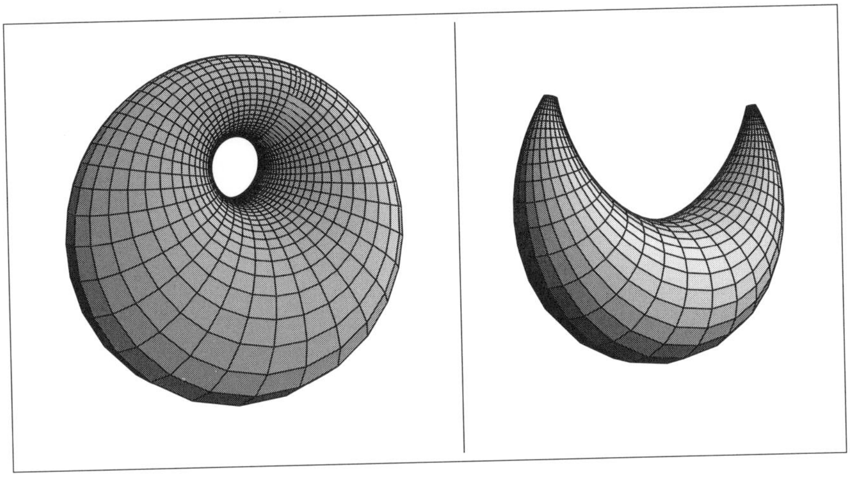

FIGURE 8. Dupin cyclides with spectral parameters $m = 1/8$ and $m = 5/4$, respectively.

- if $m > 1$:

$$f_m(x,y) = \left(\frac{\sqrt{2}\eta \sinh(\zeta x)}{2 + \eta \cosh(\zeta x)}, \frac{\sqrt{2}\zeta \cos(\eta y)}{2 + \eta \cosh(\zeta x)}, \frac{\sqrt{2}\zeta \sin(\eta y)}{2 + \eta \cosh(\zeta x)} \right), \tag{6.6}$$

- if $m = 1$:

$$f_1(x,y) = \left(\frac{8x}{4x^2+1}, \frac{4\cos(2y)}{4x^2+1}, \frac{4\sin(2y)}{4x^2+1} \right), \tag{6.7}$$

- if $0 \leq m < 1$:

$$f_m(x,y) = \left(\frac{\sqrt{2}\eta \cos(\zeta x)}{2 + \eta \sin(\zeta x)}, \frac{\sqrt{2}\zeta \cos(\eta y)}{2 + \eta \sin(\zeta x)}, \frac{\sqrt{2}\zeta \sin(\eta y)}{2 + \eta \sin(\zeta x)} \right). \tag{6.8}$$

The surfaces $S_m \subset \mathbb{R}^3$ parametrized by the maps $f_m : \mathbb{R}^2 \to \mathbb{R}^3$ are the Dupin cyclides. It is a classical result that Dupin cyclides are conformally equivalent to either a circular cone, a circular cylinder, or a torus of revolution [**29, 5**].
The normal frame field along f_m is computed to be:

- if $m > 1$:

$$\begin{pmatrix} \frac{\eta\cosh(\zeta x)+2}{\sqrt{2}\zeta\eta} & \frac{\sinh(\zeta x)}{\sqrt{2}} & 0 & -\frac{\eta\cosh(\zeta x)+2m}{\sqrt{2}\eta\zeta} & \frac{(2m-1)\eta\cosh(\zeta x)+2}{2\sqrt{2}\eta\zeta} \\ \frac{\sinh(\zeta x)}{\zeta} & \cosh(\zeta x) & 0 & -\frac{\sinh(\zeta x)}{\zeta} & \frac{(2m-1)\sinh(\zeta x)}{2\zeta} \\ \frac{\cos(\eta y)}{\eta} & 0 & -\sin(\eta y) & \frac{\cos(\eta y)}{\eta} & -\frac{(2m+1)\cos(\eta y)}{2\eta} \\ \frac{\sin(\eta y)}{\eta} & 0 & \cos(\eta y) & \frac{\sin(\eta y)}{\eta} & -\frac{(2m+1)\sin(\eta y)}{2\eta} \\ \frac{\eta\cosh(\zeta x)-2}{\sqrt{2}\zeta\eta} & \frac{\sinh(\zeta x)}{\sqrt{2}} & 0 & -\frac{\eta\cosh(\zeta x)-2m}{\sqrt{2}\eta\zeta} & \frac{(2m-1)\eta\cosh(\zeta x)-2}{2\sqrt{2}\eta\zeta} \end{pmatrix};$$

- if $m = 1$:

$$\begin{pmatrix} \frac{1}{8}(4x^2+1) & x & 0 & -\frac{1}{8}(4x^2-3) & \frac{1}{16}(4x^2+9) \\ x & 1 & 0 & -x & \frac{x}{2} \\ \frac{1}{2}\cos(2y) & 0 & -\sin(2y) & \frac{1}{2}\cos(2y) & -\frac{3}{4}\cos(2y) \\ \frac{1}{2}\sin(2y) & 0 & \cos(2y) & \frac{1}{2}\sin(2y) & -\frac{3}{4}\sin(2y) \\ 1 & 0 & 0 & -1 & \frac{1}{2} \end{pmatrix};$$

- if $0 \leq m < 1$:

$$\begin{pmatrix} \frac{2+\eta\sin(\zeta x)}{\sqrt{2}\zeta\eta} & \frac{\cos(\zeta x)}{\sqrt{2}} & 0 & -\frac{2m+\eta\sin(\zeta x)}{\sqrt{2}\zeta\eta} & \frac{2+(2m-1)\eta\sin(\zeta x)}{2\sqrt{2}\zeta\eta} \\ \frac{\cos(\zeta x)}{\zeta} & -\sin(\zeta x) & 0 & -\frac{\cos(\zeta x)}{\zeta} & \frac{(2m-1)\cos(\zeta x)}{2\zeta} \\ \frac{\cos(\eta x)}{\eta} & 0 & -\sin(\eta y) & \cos(\eta y) & -\frac{(2m+1)\cos(\eta y)}{2\eta} \\ \frac{\sin(\eta y)}{\eta} & 0 & \cos(\eta y) & \sin(\eta y) & \frac{(2m+1)\sin(\eta y)}{2\eta} \\ \frac{2-\eta\sin(\zeta x)}{\sqrt{2}\zeta\eta} & \frac{\cos(\zeta x)}{\sqrt{2}} & 0 & \frac{-2m+\eta\sin(\zeta x)}{\sqrt{2}\zeta\eta} & \frac{2-(2m-1)\eta\sin(\zeta x)}{2\sqrt{2}\zeta\eta} \end{pmatrix}.$$

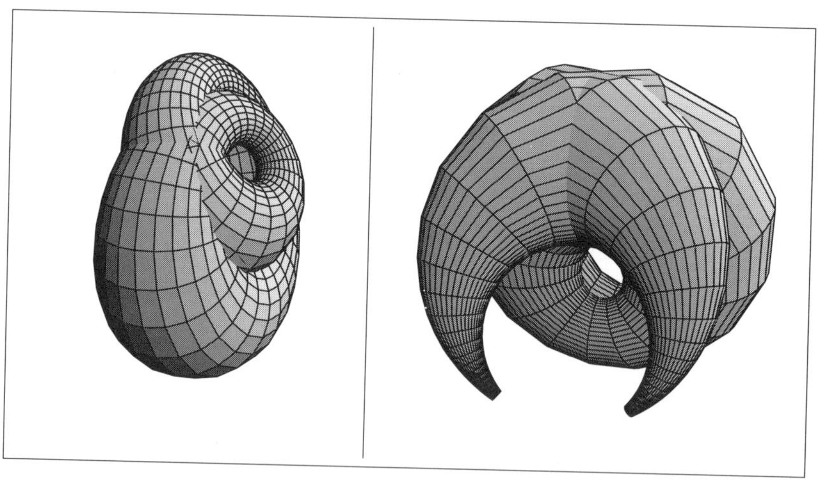

FIGURE 9. Darboux transforms of Dupin cyclides; special isothermic surfaces with wave potentials $\Phi(0,4/3,1/40,0)$, $\Phi(4/3,0,1/100,1/300)$ and spectral parameters $m=0$, $m=1$, respectively.

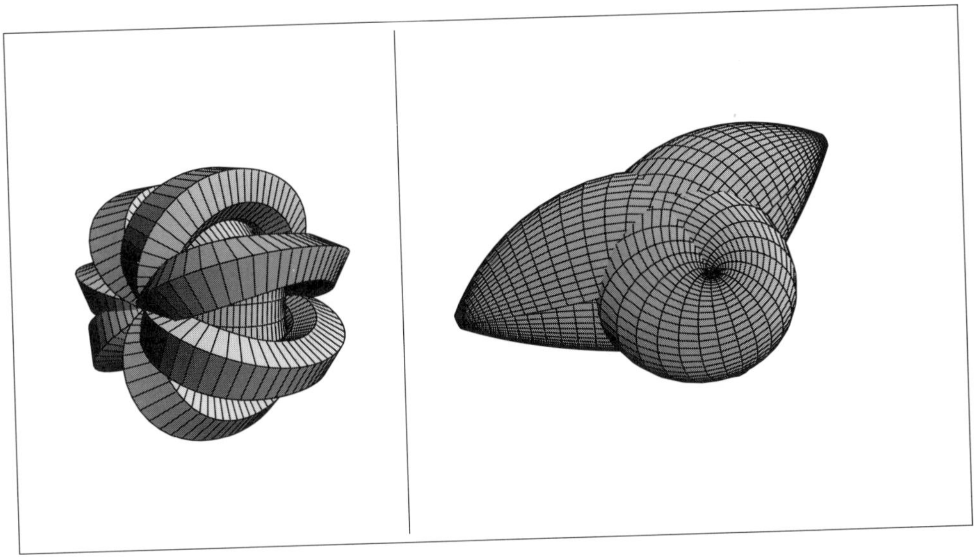

FIGURE 10. Darboux transforms of Dupin cyclides; special isothermic surfaces with wave potentials $\Phi(4/3, 0, 1/100, 1/300)$, $\Phi(4, 0, 1/12, 0)$ and spectral parameters $m = 2$, $m = 3$, respectively.

According to the above discussion, the special Darboux transforms of f_m are given by

$$D(h, a^1, a^2, a^3)(f_m) = \left[B(m)V(h, a^1, a^2, a^3)\right], \tag{6.9}$$

where $V(h, a^1, a^2, a^3) : \mathbb{R}^2 \to \mathcal{L}$ is the solution of the D_h-linear system for $\Phi = 1$ with initial condition $X(a^1, a^2, a^3)$ given by (5.4). Thus, $D(h, a^1, a^2, a^3)(f_m)$ is a special isothermic immersion with Calapso potential $\Phi(h, a^1, a^2, a^3)$.

7. The superposition formula and two-soliton solutions

THEOREM 7.1. *Let $\Phi : U \to \mathbb{R}$ be a wave potential with character s, and let $V = (v^0, ..., v^4)^t$ and $W = (w^0, ..., w^4)^t$ be two systems of h and k-transforming functions, respectively, $h \neq k$. Let $\Phi_1 = E(\Phi, V)$ and $\Phi_2 = E(\Phi, W)$ be the corresponding Bäcklund transforms. Then,*

$$\Phi_3 = \Phi - (h - k)\frac{v^3 w^3}{\langle V, W\rangle}\left[\frac{\Phi_2 - \Phi_1}{\Phi_1 \Phi_2}\right]. \tag{7.1}$$

is a wave potential with character s.

PROOF. Let $B(m) : U \to G$ be the normal frame with potential Φ and spectral parameter m, $m \neq h$. Consider the curved flat framing $A := B(m)g^+(V)$ and its dual framing $\bar{A} := AJ$; then $\bar{A}$ is a curved flat framing and the corresponding normal frame $\bar{B}(m)$ is computed to be

$$\bar{B}(m) = \bar{A}g^+(T) = B(m)g^+(V)Jg^+(T),$$

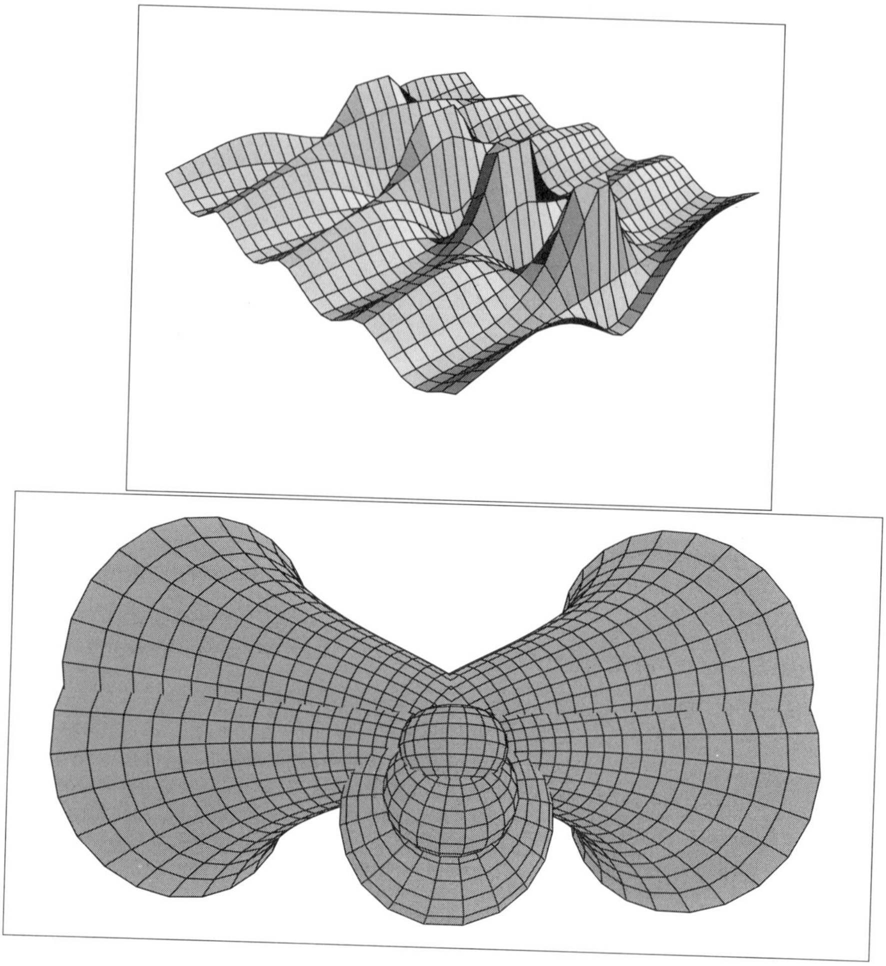

FIGURE 11. The superposition of the wave potentials $\Phi(4/3, 0, 1/100, 1/300)$ and $\Phi(10/9, 0, 1/500, 0)$ and the corresponding special isothermic surface with $m = 1$.

where $T : U \to \mathcal{L}$ is the smooth map defined as in (6.5). Next, consider the map $Y : U \to \mathcal{L}$ given by

(7.2)

$$\left(\frac{(h-k)w^4}{(h-m)v^4}, \frac{v^1w^4 - v^4w^1}{v^4}, \frac{v^4w^2 - v^2w^4}{v^4}, \frac{v^4w^3 - v^3w^4}{v^4}, -\frac{(h-m)}{(h-k)}\langle V, W\rangle\right)^t.$$

Then $\tilde{A} := \bar{A}g^+(Y)$ is a curved flat framing with spectral parameter $k - m$ and Calapso potential Φ_1. Further, let $L : U \to \mathcal{L}$ be defined by

$$L = T^{-1} \star Y = \left(-\langle T, Y\rangle, \frac{y^1t^4 - y^4t^1}{t^4}, \frac{y^2t^4 - y^4t^2}{t^4}, \frac{y^3t^4 - y^4t^3}{t^4}, \frac{y^4}{t^4}\right)^t, \tag{7.3}$$

so that

$$\tilde{A} = [\bar{A}g^+(T)][g^+(T^{-1})g^+(Y)] = \bar{B}(m)g^+(L).$$

Since $\bar{B}(m)g^+(L)$ is a curved flat framing with Calapso potential Φ_1 and spectral parameter $k-m$, then L is a solution of the D_k-system with potential Φ_1. Combining (6.5), (7.2), (7.3) and using the constraints $C_2(V) = C_2(W) = 0$, it is a computational matter to check that also L satisfies the constraint

$$C_2(L) = \Phi_1^2 L^0 - kL^3 + \frac{s}{2}\Phi_1^{-2} L^4 = 0.$$

This implies that L is a system of D_k-transforming functions for the potential Φ_1. In particular, the Bäcklund transform $E(\Phi_1, L)$ is a new solution of the differential equation (1.2). On the other hand, $E(\Phi_1, L)$ is computed to be

$$E(\Phi_1, L) = \frac{\left(v^1w^1 + v^2w^2 + v^3w^3 - v^0w^4 - v^4w^0\right)\Phi^2 - (h-k)\left(w^3v^4 - w^4v^3\right)}{\left(v^1w^1 + v^2w^2 + v^3w^3 - v^0w^4 - v^4w^0\right)\Phi},$$

from which follows that

$$E(\Phi_1, L) = \Phi - v^3w^3\frac{(h-k)}{\langle V, W\rangle}\left[\frac{\Phi_2 - \Phi_1}{\Phi_1\Phi_2}\right].$$

□

REMARK 7.2. The wave potential Φ_3 can be realized as the Calapso potential of the special isothermic immersion

$$[(B(m)g^+(V)Jg^+(Y))_4], \tag{7.4}$$

which can be explicitly computed from the normal frame $B(m)$ and the two sets V and W of transforming functions.

EXAMPLE 7.3 (**Two-solitons and related isothermic surfaces**). Two soliton solutions can be computed by means of (5.4) and (7.1). We then obtain a six-parameter family of wave potentials given by

$$\Phi = 1 - \frac{(h-k)(w^3v^4 - w^4v^3)}{\langle V, W\rangle},$$

where V and W are the h and k-transforming functions corresponding to the initial conditions $X(a^1, a^2, a^3)$ and $X(b^1, b^2, b^3)$. Special isothermic immersions with spectral parameter m and Calapso potentials Φ can be constructed by using (7.4), (5.4) and the explicit formulae for the central frame fields $B(m)$ of Dupin cyclides.

References

[1] M. Babich, A. Bobenko, *Willmore tori with umbilic lines and minimal surfaces in hyperbolic space*, Duke Math. J. **72**(1993), pp. 151–185.

[2] H. E. Bernstein, *Non-special, non-canal, isothermic tori with spherical lines of curvature*, Trans. Amer. Math. Soc. **353**(2001), pp. 2245–2274.

[3] L. Bianchi, *Ricerche sulle superficie isoterme e sulla deformazione delle quadriche*, Ann. Mat. Pura Appl. **11**(1905), pp. 93–157.

[4] L. Bianchi, *Complementi alle ricerche sulle superficie isoterme*, Ann. Mat. Pura Appl. **12**(1905), pp. 20–54.

[5] W. Blaschke, Vorlesungen über Differentialgeometrie und geometrische Grundlagen von Einsteins Relativitätstheorie B. 3, bearbeitet von G. Thomsen, J. Springer, Berlin, 1929.

[6] M. Brück, X. Du, J. Park, C.-L. Terng, *The submanifold geometries associated to Grassmannian systems*, e-print math.DG/0006216.

[7] R. L. Bryant, *A duality theorem for Willmore surfaces*, J. Differential Geom. **20**(1984), pp. 23–53.

[8] R. L. Bryant, P. A. Griffiths, *Reduction for constrained variational problems and* $\int \frac{k^2}{2}$, Amer. J. Math. **108**(1986), pp. 525–570.

[9] F. Burstall, *Isothermic surfaces: conformal geometry, Clifford algebras and integrable systems*, e-print math.DG/0003096.

[10] F. Burstall, U. Hertrich-Jeromin, F. Pedit, U. Pinkall, *Curved flats and isothermic surfaces*, Math. Z. **225**(1997), pp. 199–209.

[11] P. Calapso, *Sulle superficie a linee di curvatura isoterme*, Rend. Circ. Mat. Palermo **17**(1903), pp. 273–286.

[12] P. Calapso, *Sulle trasformazioni delle superficie isoterme*, Ann. Mat. Pura Appl. **24**(1915), pp. 11–48.

[13] D. Carfì, E. Musso, *T-transformations of Willmore isothermic surfaces*, Rend. Sem. Mat. Messina Ser.II, suppl. (2000), 257–266.

[14] J. Cieśliński, P. Goldstein, A. Sym, *Isothermic surfaces in* $\mathbf{E}^3$ *as soliton surfaces*, Physics Letters **A** **205**(1995), pp. 37–43.

[15] J. Cieśliński, *The Darboux–Bianchi transformation for isothermic surfaces. Classical versus the soliton approach.* Differential Geom. Appl. **2**(1997), pp. 1–28.

[16] G. Darboux, *Leçons sur la théorie générale des surfaces*, Chelsea, Bronx, New York (1972).

[17] G. Darboux, *Sur une classe des surfaces isothermiques liées à la deformations dea surfaces du second degré*, C. R. Acad. Sci. Paris **128**(1899), pp. 1483–1487.

[18] D. Ferus, F. Pedit, *Curved flats in symmetric spaces*, Manuscripta Math. **91**(1996), pp. 445–454

[19] U. Hertrich-Jeromin, *Supplement on curved flats in the space of point pairs and isothermic surfaces: a quaternionic calculus*, Documenta Math. J. DMV **2**(1997), pp. 335–350.

[20] U. Hertrich-Jeromin, F. Pedit, *Remarks on the Darboux transform of isothermic surfaces*, Documenta Math. J. DMV **2**(1997), pp. 313–333.

[21] U. Hertrich-Jeromin, E. Musso, L. Nicolodi, *Möbius geometry of surfaces of constant mean curvature 1 in hyperbolic space*, Ann. Global Anal. Geom. **19**(2001), pp. 185–205.

[22] J. Langer, D. Singer, *Curves in the hyperbolic plane and mean curvature of tori in 3-space*, Bull. London Math. Soc. **16**(1984), pp. 531–534.

[23] E. Musso, *Deformazione di superfici nello spazio di Möbius*, Rend. Istit. Mat. Univ. Trieste **27**(1995), pp. 25–45.

[24] E. Musso, Isothermic surfaces in Euclidean space, in *Proc. Workshop on Recent Topics in Differential Geometry*, L.A. Cordero & E. García-Río, edd., Public. Depto. Geometría y Topología, Univ. Santiago de Compostela, #89, Santiago de Compostela, (1998), pp. 237–253.

[25] E. Musso, L. Nicolodi, *Willmore canal surfaces in Euclidean space*, Rend. Istit. Mat. Univ. Trieste **31**(1999), pp. 177–202.

[26] E. Musso, L. Nicolodi, *The Bianchi–Darboux transform of L-isothermic surfaces*, Internat. J. Math. **11**(2000), pp. 911–924.

[27] E. Musso, L. Nicolodi, *Darboux transforms of Dupin surfaces*, Banach Center Publ. (to appear).

[28] U. Pinkall, *Hopf tori in* S^3, Invent. Math. **81**(1985), pp. 379–386.

[29] M.E. Vessiot, *Contribution à la géométrie conforme. Thèorie des surfaces*, Bull. Soc. Math. France **54**(1926), 139–179, **55**(1927), 39–79.

Dipartimento di Matematica Pura ed Applicata, Universitá degli Studi di L'Aquila, Via Vetoio, I-67100 L'Aquila, Italy

E-mail address: `musso@univaq.it`

Dipartimento di Matematica, Universitá degli Studi di Parma, Via M. D'Azeglio 85/A, I-43100 Parma, Italy

E-mail address: `nicolodi@mat.uniroma1.it`

Contemporary Mathematics
Volume **288**, 2001

The Isoperimetric and Willmore Problems

Antonio Ros

ABSTRACT. We introduce some methods to study the isoperimetric problem in 3-dimensional Riemannian manifolds and we show that in the positive curvature case we can control the topology of the isoperimetric regions. We consider the case of the projective space, which was first solved by Ritoré and Ros, and we apply it to the Willmore problem.

1. Introduction

The isoperimetric problem is a classical topic in geometry but at the same time many basic questions about it remain unsolved. In this paper we first introduce some of the methods used in the study of that problem, although we will not try to be exhaustive at all. We will explain some relatively flexible ideas, like symmetrization or stability, which can be adapted to a certain number of situations. As an example we will study the problem for radial metrics on the 3-sphere.

Then we will show that in 3-manifolds with positive Ricci curvature the topology of the isoperimetric regions can be controled. In particular we will prove that, when the volume of the ambient space is large, any isoperimetric surface must be either an sphere or a torus. As consequence, we will solve the isoperimetric problem in the real projective space or, equivalently, the isoperimetric problem for antipodal invariant regions in the 3-sphere. That result was first obtained by Ritoré and Ros [**35**], but here we will give a somewhat different proof.

Finally, as application of the above results, we will solve the Willmore conjecture for tori in euclidean space which are symmetric with respect to a point.

2. The isoperimetric problem

In this paper we will only consider the three dimensional case. Let M be a Riemaniann 3-dimensional manifold with or without boundary and volume $\mathrm{V}(M) \in]0, \infty]$. Given a positive number $v < \mathrm{V}(M)$, we want to study the compact surfaces $\Sigma \subset M$ such that

(1) Σ encloses a region of volume v, and
(2) Σ minimizes area under the constraint (1),

1991 *Mathematics Subject Classification.* 53A10, 53C42.
Key words and phrases. isoperimetric problem, projective space, Willmore conjecture.
Research partially supported by a DGYCYT Grant No. PB97-0785.

see Figure 1. If $\partial M \neq \emptyset$ we do not count the area coming from the boundary of M. Existence and regularity for the isoperimetric problem has been considered by several authors. From the works of Almgren [**5**], Grüter [**19**] and Gonzalez, Massari, Tamanini [**16**], see also [**32**], we have the following satisfactory result.

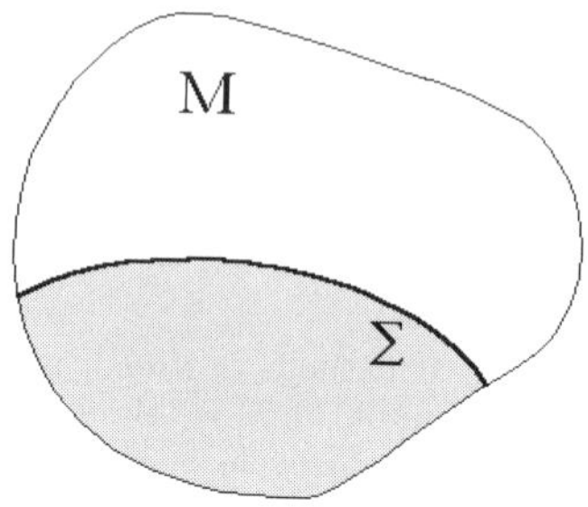

FIGURE 1

THEOREM 1. *If M is either compact or homogeneous, then for any given v there exists a surface which solves the isoperimetric problem above. Any solution Σ is a smooth embedded surface with constant mean curvature and, if $\partial M \cup \Sigma \neq \emptyset$, then ∂M and Σ meet orthogonally.*

The surfaces which solve the problem and the regions enclosed by them will be called *isoperimetric surfaces* and *isoperimetric regions* respectively.

Isoperimetric inequalities, that is, estimates of the area of the isoperimetric surfaces have been also studied by many authors. Later on we will give a couple of examples of this type. This kind of results have a wide number of applications, see for instance [**14**].

However, the study of the geometry and the topology of solutions and the explicit description of these surfaces when the ambient space is simple enough, is an almost unexplored field. In this paper we will put the emphasis just in this kind of properties.

If M is compact, the *isoperimetric profile* of M is defined by

$$I = I_M : [0, \mathrm{V}(M)] \to \mathbb{R}, \qquad I(v) = \min\{\mathrm{A}(\partial\Omega) \,|\, \Omega \subset M \,\text{region},\ \mathrm{V}(\Omega) = v\}.$$

This function is continuous, vanishes at 0 and $\mathrm{V}(M)$, and is positive elsewhere. Moreover, as the complement of an isoperimetric region is an isoperimetric region too, I is always symmetric $I(\mathrm{V}(M) - 2v) = I(v)$.

3. Symmetrization

The method which is more clearly connected with our problem is the symmetrization one. Classically, symmetrization was used in several ways, see Schwarz [**42**] and Steiner [**44**], to show that the round sphere solves the isoperimetric problem in $\mathbb{R}^3$. Since we have existence and regularity results, we can use another version of the idea which was first introduced by Hsiang [**24, 25**]. Roughly speaking, we can say that if M has enough isometries then any isoperimetric surface Σ is connected and if M has many symmetries then Σ is symmetric. Instead of stating general results we will show how the method works in the Euclidean case.

THEOREM 2. *Isoperimetric surfaces in $\mathbb{R}^3$ are round spheres.*

PROOF. Let Ω be an isoperimetric region and $\Sigma = \partial\Omega$. If Ω is not connected, then we move one of its connected components until we touch for the first time the remaining part of Ω. At that moment, we have a new isoperimetric region with a singular point at its boundary, which is impossible by the regularity results above.

The same argument applies if Ω has nonconnected boundary. By moving one of the boundary components we get a new isoperimetric region with a singular point. So we conclude that Σ is connected. Consider now a plane $P \subset \mathbb{R}^3$ cutting Ω in

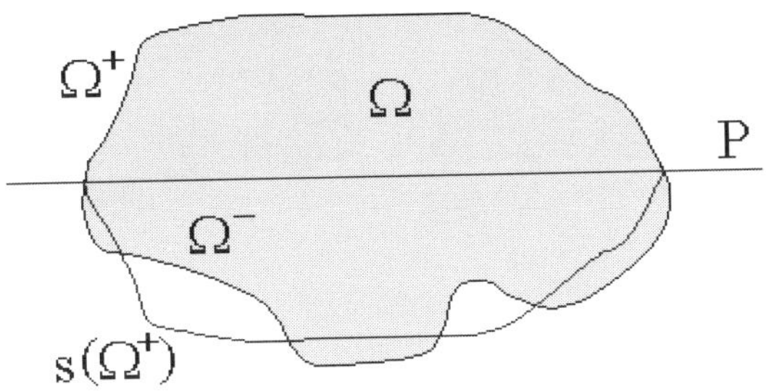

FIGURE 2

two pieces with the same volume, Ω^+ and Ω^-, see Figure 2. Assume that the area of the pieces induced on Σ verifies $A(\Sigma \cap \Omega^+) \leq A(\Sigma \cap \Omega^-)$. The new domain $\Omega^* = \Omega^+ \cup s(\Omega^+)$, where s denotes the symmetry with respect to P, is symmetric and satisfies $V(\Omega^*) = V(\Omega)$ and $A(\partial\Omega^*) \leq A(\Sigma)$. So $A(\partial\Omega^*) = A(\Sigma)$ and Ω^* is also an isoperimetric region. Therefore, using again the regularity of minimizers, we have that $\partial\Omega^*$ and Σ are two closed surfaces in $\mathbb{R}^3$ with constant mean curvature which coincide in a nonempty open set. As surfaces of this kind are analytic, we get finally that $\partial\Omega^* = \Sigma$ or, in other words, Σ is symmetric with respect to P (instead of analitycity we can invoke the unique continuation property [**7**] which works also when the ambient space is not analytic but only C^∞).

As each family of parallel planes in $\mathbb{R}^3$ contains one plane which divides Ω in two equal volumes, we conclude easily that Σ is a round sphere. □

Of course, Theorem 2 follows also from the well-known Alexandrov uniqueness theorem [**3**] for closed constant mean curvature surfaces in $\mathbb{R}^3$. However, when we try to apply these methods to other related situations, symmetrization gives stronger results than Alexandrov reflection technique.

The above symmetrization argument works without any change in the hyperbolic 3-space and in the unit 3-sphere $\mathbf{S}^3$, and gives that isoperimetric surfaces in these spaces are round spheres too. The method works also in higher dimension.

If M is either a halfspace or a slab bounded by parallel planes in $\mathbb{R}^3$ then, using the same idea again, we get that any isoperimetric surface Σ is a connected surface of revolution which meets the boundary of M orthogonally. Note that Σ must meet ∂M: otherwise by moving Σ until we touch the first time the boundary of M, we get an isoperimetric surface which meets ∂M tangentially, a contradiction. If M is a halfspace with $P = \partial M$ and s denotes the symmetry with respect to P, we have that $\Sigma \cup s(\Sigma)$ is an embedded closed surface with constant mean curvature, which must be a round sphere by [**3**]. So, Σ is a halfsphere. When M is a slab, using symmetrization with respect to planes orthogonal to P, we get that Σ is a surface of revolution and, then, it must be a piece of a sphere, an unduloid or a cylinder, see Figure 3. It is known, by a different argument, that unduloids are not isoperimetric surfaces, thus only halfspheres and cylinders appear as minimizers, see [**2**] and [**35**]. However if M is an slab in $\mathbb{R}^n$, $n \geq 10$, Pedrosa and Ritoré [**33**] have shown that unduloids are solutions for some prescribed volumes.

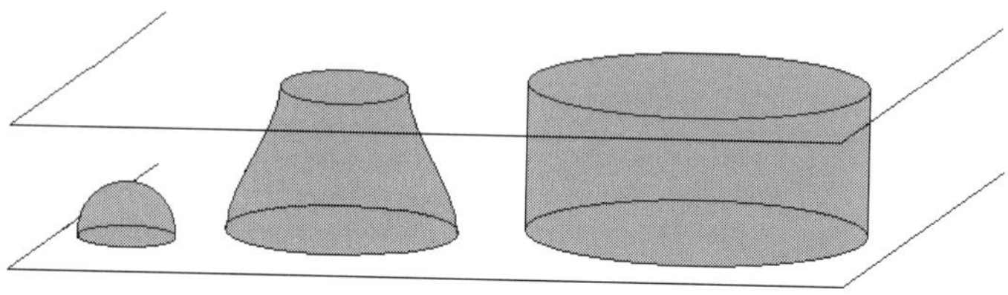

FIGURE 3

4. Stability

Isoperimetric surfaces are *stable*, that is, they minimize area up to the second order (under volume constraint). This means that for any function u in the Sobolev space $H^1(\Sigma)$ with $\int_\Sigma u = 0$ we have $Q(u,u) \geq 0$, where Q is defined by the second variation formula of the area functional. The quadratic form Q is given by

$$(1) \qquad Q(u,u) = \int_\Sigma |\nabla u|^2 - (Ric(N) + |\sigma|^2)u^2 - \int_{\partial\Sigma} \kappa\, u^2,$$

where $Ric(N)$ denotes Ricci curvature of M along the unit normal direction N of Σ, σ is the second fundamental form of Σ and κ is the normal curvature of ∂M along N, see for instance [**8**] and [**39**]. Stability plays an important role in the theory of minimal surfaces. In the context of the isoperimetric problem, the interest of this notion was first pointed out by Barbosa and do Carmo, who gave a proof of the minimizing property of the sphere in $\mathbb{R}^3$ by looking at the stability of the solutions.

THEOREM 3. (Barbosa & do Carmo [**8**]) *Let Σ be a compact orientable surface immersed in $\mathbb{R}^3$ with constant mean curvature. If Σ is stable then it must be a round sphere.*

If M has positive Ricci curvature and ∂M is convex then it follows, using locally constant test functions in (1), that any isoperimetric surface of M must be connected.

The symmetries of M have been used in the symmetrization process and can be used also in the following natural and useful stability argument, see for instance Alías, López, Palmer [**4**], Hutchings, Morgan, Ritoré, Ros [**26**], Ros, Souam [**38**] and Ros, Vergasta [**39**], for concrete applications to the study of constant mean curvature surfaces.

Consider a 1-parameter family $\{f_\theta : |\theta| < \varepsilon\}$ of isometries of M. Then the associated infinitesimal rotation $u = \langle \frac{df}{d\theta}, N\rangle$ is a Jacobi function, that is, $Q(u, v) = 0$ for any $v \in H^1(\Sigma)$ or, equivalently,

$$\Delta u + (Ric(N) + |\sigma|^2)u = 0 \text{ on } \Sigma \qquad \text{and} \qquad \frac{\partial u}{\partial \nu} = \kappa u \text{ along } \partial\Sigma.$$

The well-known Courant nodal argument, [**14**], gives the following restriction on u.

PROPOSITION 1. *Let Σ be an isoperimetric surface in M and u an infinitesimal rotation of Σ. Then either $u \equiv 0$ or $\{u \neq 0\}$ has at most two connected components.*

PROOF. Assume $\{u \neq 0\}$ has at least 3 nodal domains Σ_i, $i = 1, 2, 3$ and define the functions $u_i \in H^1(\Sigma)$ by $u_i = u$ in Σ_i, and $u_i = 0$ in $\Sigma - \Sigma_i$.

So, we have $Q(u_i, u_j) = 0$ for $i \neq j$ and, using that u is a Jacobi function, $Q(u_i, u_i) = Q(u_i, u) = 0$. Therefore there exist nonzero constants a_1 and a_2 such that $v = a_1u_1 + a_2u_2$ verifies $\int_M v = 0$ and $Q(v, v) = 0$. Stability then implies that v is Jacobi function too. As v vanishes on Σ_3, the unique continuation property [**7**] says that $v \equiv 0$. This contradiction proves the result. □

By joining the two ways we dispose now to use the symmetries of the problem, we can give a new proof of the following result. A similar approach has been used in the solution of the double bubble conjecture in $\mathbb{R}^3$, see [**26**].

THEOREM 4. (Bokowski & Sperner [**13**], Almgren [**6**]) *Isoperimetric surfaces in a Euclidean ball are either spherical caps or planar discs.*

PROOF. First observe that, as in the case in which M is $\mathbb{R}^3$ or a halfspace, any isoperimetric surface Σ is connected and meets the boundary of the ball.

Given a line through the center of the ball we have, by continuity, a plane which contains the line and cut the region enclosed by Σ in two equal volumes. Symmetrizacion gives that this plane is a plane of symmetry of Σ. As this implies that Σ as infinitely many symmetries, we conclude that it must be a surface of revolution. In particular, Σ is either a disc or an annulus. In the first case it follows from properties of Delaunay surfaces that Σ is an umbilical surface.

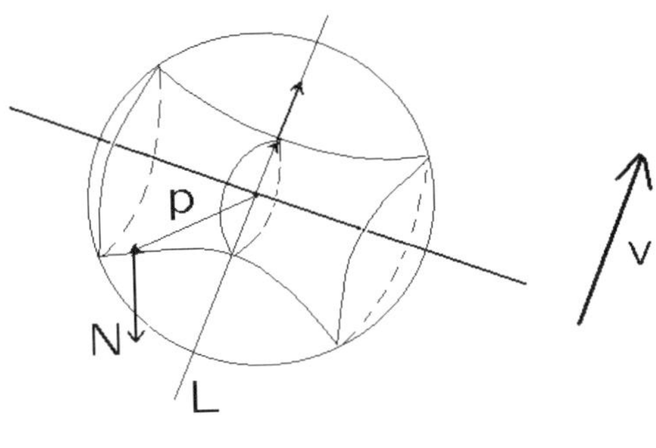

FIGURE 4

Now we prove that the second case is impossible. Take L a line through the origin meeting Σ orthogonally (take for instance the line passing through the point

$q \in \Sigma$ closest to the origin) and consider the family $\{f_\theta\}$ of rotations around L, see Figure 4. Note that, as $\Sigma \cap L \neq \emptyset$, L cannot be the axis of revolution of Σ. The associated infinitesimal rotation on Σ is given by $u = det(p, N, v)$, where p, N and v are the position vector, the unit normal vector of Σ and the unit direction of L respectively. By looking at Figure 4, we see that u vanishes along the curves $\Sigma \cap P$, P being the plane which contains the axis of revolution and the line L, and along the circle in Σ through q. Therefore we see that either $u \equiv 0$ or u has at least 4 nodal domains. From Proposition 1 we conclude that both options are impossible and this contradiction proves the theorem. □

The proof of the above theorem gives without any change the following more general facts.

THEOREM 5. (i) *If the ambient space is the sphere $M = \mathbf{S}^3$ with a $O(3)$-invariant metric, then each connected component of an isoperimetric surface is a (topological) sphere of revolution.*

(ii) *If M is a Euclidean ball with a radial metric, i. e. a $O(3)$-invariant metric, then the components of any isoperimetric surface are either spheres or discs of revolution.*

In the first case $O(3)$ is viewed as the linear subgroup $1 \times O(3) \subset O(4)$. By a surface of revolution we mean a surface which is $O(2)$-invariant.

5. Nonconstant curvature

In this section M will be a complete 3-dimensional Riemannian manifold without boundary. We will denote its Ricci curvature by Ric. When the ambient space M has nonconstant curvature, we have the following basic results.

THEOREM 6. (Kleiner [**27**]) *Assume M is complete, simply-connected and has nonpositive sectional curvature. Let $\Omega \subset M$ be a domain and $B \subset \mathbb{R}^3$ an Euclidean ball. If $\mathrm{V}(\Omega) = \mathrm{V}(B)$ then $\mathrm{A}(\partial\Omega) \geq \mathrm{A}(\partial B)$. The equality holds if and only if $\Omega = B$.*

THEOREM 7. (Lévy [**28**], Gromov [**18**]) *Assume M is compact and has Ricci curvature $Ric \geq 2$. Let $\Omega \subset M$ be a domain and $B \subset \mathbf{S}^3$ a metric ball. Then*

$$\frac{\mathrm{V}(\Omega)}{\mathrm{V}(B)} = \frac{\mathrm{V}(M)}{\mathrm{V}(\mathbf{S}^3)} \Rightarrow \frac{\mathrm{A}(\partial\Omega)}{\mathrm{A}(\partial B)} \geq \frac{\mathrm{V}(M)}{\mathrm{V}(\mathbf{S}^3)},$$

and the equality implies that $M = \mathbf{S}^3$ and $\Omega = B$.

Kleiner's theorem says that the Euclidean isoperimetric inequality extends to Cartan-Hadamard manifolds. The result does not hold for nonsimply-connected ambient spaces. This is a general fact in our setting: most results are sensitive to changes on the fundamental group. In Theorem 7, if the fundamental group of M were larger, then $\mathrm{V}(M)$ would be smaller and so the isoperimetric inequality would be weaker. An improved version of Theorem 7 can be found in [**12**]. Theorem 7 extends to higher dimension but the corresponding version of Theorem 6 remains open.

In the positive curvature case we can control the topology of the isoperimetric surfaces. The result below is based on an idea of Hersch [**22**] which has been extended and used in several related contexts first by Yang and Yau [**48**] and then by many people, see [**15, 29, 30, 31, 38, 39, 41, 47**] and in particular [**35**].

THEOREM 8. *Let Σ be an isoperimetric surface in a compact orientable 3-manifold M with $Ric \geq 2$. Then Σ is connected and genus$(\Sigma) \leq 3$. Moreover, in the cases genus$(\Sigma) = 2$ or 3 we have that $(1 + H^2)A(\Sigma) \leq 2\pi$.*

If M is closed and has positive Ricci curvature, then its isoperimetric profile I_M must be concave, see [**11**]. Therefore, from the symmetry of I_M, we get that the maximum value of the isoperimetric profile is given by $I_M(\mathrm{V}(M)/2)$. Now we show that this maximum gives a lower bound for certain total curvature on closed surfaces Σ bounding a domain in M.

PROPOSITION 2. *If $Ric \geq 2$ on M, then any closed surface $\Sigma \subset M$ which separates the ambient space verifies*

$$\int_\Sigma (1 + H^2)dA \geq \max I_M.$$

If the equality holds, then either Σ is a minimal surface which separates M in two pieces with the same volume or Σ is a round sphere embedded umbilically in M.

PROOF. Let Ω be one of the regions enclosed by Ω and assume $\mathrm{V}(\Omega) \leq \mathrm{V}(M)/2$. Consider $t \geq 0$ such that $\mathrm{V}(\Omega_t) = \mathrm{V}(M)/2$, where $\Omega_t = \{p \in M \mid \mathrm{dist}(p, \Omega) \leq t\}$. If we parametrize M by normal geodesics leaving Σ, then the Riemannian measure on M can be writen as $dV = g(p,t)dAdt$, where dA is the canonical measure on Σ and $g : \Sigma \times \mathbb{R}^+ \longrightarrow \mathbb{R}$ is the Jacobian of the normal exponential map. Define $C(t) = \{p \in \Sigma \mid \text{cut value of } p \ \geq t\}$, that is $p \in C(t)$ iff the geodesic γ leaving p in the direction $N(p)$ minimizes length between Σ and $\gamma(t)$. Given $p \in C(t)$, the Jacobian can be estimated as $g(p,t) \leq (\cos t - H \sin t)^2 \leq 1 + H^2$, where the first inequality follows from Heintze and Karcher [**21**] and the second one follows from Schwarz inequality. Therefore

$$\max I_M = I_M(\mathrm{V}(M)/2) \leq \mathrm{A}(\partial\Omega_t) \leq \int_{C(t)} g(p,t) \leq \int_\Sigma (1 + H^2).$$

Suppose the equality holds. If $t = 0$, it follows directly that $\mathrm{V}(\Omega) = \mathrm{V}(M)/2$ and $H \equiv 0$ (note that $g(p, 0) = 1$). If $t > 0$, then the conclusion follows from [**21**]. □

THEOREM 9. *Let M be a compact 3-manifold with Ricci curvature $Ric \geq 2$ and volume $\mathrm{V}(M) \geq \mathrm{V}(\mathbf{S}^3)/2$. Then, any isoperimetric domain of M is bounded by either a sphere or a torus.*

PROOF. The hypothesis on the volume of M implies that M is a topological 3-sphere or a 3-dimensional projective space. Thus M is orientable. Assume that M is not a round sphere (if $M = \mathbf{S}^3$ the statement is clear). Let Σ be an isoperimetric surface of M and H its mean curvature. From Theorem 8 we know that Σ is a connected orientable surface of genus at most 3. Proposition 2 gives

$$(1 + H^2)\mathrm{A}(\Sigma) \geq I_M(\mathrm{V}(M)/2).$$

Using the volume assumption plus the fact that M is not a metric sphere, we get from Theorem 7 that

$$I_M(\frac{\mathrm{V}(M)}{2}) > 4\pi\frac{\mathrm{V}(M)}{\mathrm{V}(\mathbf{S}^3)} \geq 2\pi.$$

Therefore $(1 + H^2)\mathrm{A}(\Sigma) > 2\pi$ and Theorem 8 implies that genus$(\Sigma) \leq 1$. □

If the ambient space M has nonnegative curvature and the simplest topology, then it is natural to expect that the solutions of our problem will be also as simple as possible. The following conjecture gives two precise versions of this idea.

Conjecture. *Let Σ be an isoperimetric surface in M.*

(i) *If $M \subset \mathbb{R}^3$ is a strictly convex domain, then Σ is a topological disk.*

(ii) *If M is a 3-sphere with a metric of positive Ricci curvature, then Σ is a topological sphere.*

6. Space-forms

Assume now that the ambient space is an orientable 3-dimensional space form $\overline{M}(c)$, with constant sectional curvature c. Only a few results are known about the isoperimetric problem in a nonsimply-connected 3-space-form. The correct idea at this point is that when the fundamental group becomes larger, then the isoperimetric problem gets harder. The first restriction we have for general space-forms is the following.

THEOREM 10. *Let Σ be an isoperimetric surface of $\overline{M}(c)$.*

(i) *If Σ has genus zero, then Σ is an umbilical sphere.*

(ii) *If* $\text{genus}(\Sigma) = 1$, *then Σ is flat.*

In the first case, Σ lifts to the simply-connected covering of $\overline{M}(c)$ and the result follows from Hopf uniqueness theorem for constant mean curvature spheres, [**23**]. Assertion (ii) is proved in [**35**] by using a stability argument.

Now we give the complete solution of the isoperimetric problem on the real projective space $\mathbf{P}^3 = \mathbf{S}^3/\{\pm\}$ obtained as a metric quotient of the unit 3-sphere under the antipodal map. The result was first proved by Ritoré and Ros [**35**]. The proof here uses Theorem 9 and differs somewhat from [**35**]. At the present, $\mathbf{S}^3$ and $\mathbf{P}^3$ are the only elliptic 3-space forms where the problem has been solved. The problem is also open for higher dimensional projective spaces. We remark that in the projective space symmetrization does not work. We must solve the problem by different arguments.

THEOREM 11. ([**35**]). *Isoperimetric surfaces in $\mathbf{P}^3$ are either geodesic spheres or tubes around a line.*

PROOF. Let $\Sigma \subset \mathbf{P}^3$ be an isoperimetric surface. As $\mathrm{V}(\mathbf{P}^3) = \mathrm{V}(\mathbf{S}^3)/2$, Theorem 9 gives that $\text{genus}(\Sigma) = 0$ or 1. Then Theorem 10 implies that Σ is an umbilical sphere or a flat torus. The explicit classification of this kind of surfaces in $\mathbf{P}^3$ is an elementary and well-known fact and this concludes the proof. □

Direct computations give that for small (resp. large) prescribed volume, the isoperimetric domain in $\mathbf{P}^3$ is a geodesic ball (resp. the complementary of a geodesic ball). If the prescribed volume is near $\mathrm{V}(\mathbf{P}^3)/2$, then the isoperimetric solution is a tubular neighborhood of a geodesic line. In particular, if we consider just one half of the volume of $\mathbf{P}^3$, the solution is the domain enclosed by the quotient of the Clifford torus (i.e. a tube of radius $\pi/4$).

COROLLARY 1. *If $\Sigma \subset \mathbf{S}^3$ is a compact surface which divides the 3-sphere in two antipodally invariant pieces of equal volume, then $A(\Sigma) \geq 2\pi^2$. The equality holds if and only if Σ is the minimal Clifford torus.*

If M is a complete flat space with cyclic infinite fundamental group, then Ritoré and Ros [**35**] proved that any isoperimetric surface in M is either an umbilical sphere or a flat cylinder. The problem remains open for other quotients of $\mathbb{R}^3$. If $M = \mathbf{S}^1(r_1) \times \mathbf{S}^1(r_2) \times \mathbf{S}^1(r_3)$ is a rectangular flat 3-torus, then a symmetrization argument gives that the isoperimetric problem in M is equivalent to the isoperimetric problem in a box of $\mathbb{R}^3$. In that case we know the following.

THEOREM 12. *Let $M = [0,1] \times [0,a] \times [0,b] \subset \mathbb{R}^3$ be a box with $1 \leq a \leq b$. Then the rectangle $M \cap \{z = b/2\}$ has least area among surfaces which divide the box in two equal volumes.*

This result was proved by Hadwiger [**20**] for the case of the cube and extended to boxes by Schnell [**43**]. A related general property for the isoperimetric problem in a product space has been obtained recently by Barthe and Maurey [**9**], [**10**]. The natural conjecture is that isoperimetric surfaces in a box must be pieces of spheres, cylinders or planes. This conjecture is known to be true for boxes of the type $M = [0,1] \times [0,a] \times [0,\infty[$, with $a \geq 1$ large enough, see [**36**]. Ross [**41**] has shown that the classical Schwarz P minimal surface is stable in the cubic 3-torus of volume 1. So in this space we have a local minimum of the isoperimetric problem which is not a global one: Schwarz surface divides the 3-torus in two equal pieces and, thus, theorem 12 implies that its area is bigger than 2.

7. The Willmore conjecture

Now we want to study the functional $\int_\Sigma H^2 dA$ on closed surfaces Σ immersed in $\mathbb{R}^3$, where H denotes the mean curvature of the immersion. This functional is the simplest global invariant under conformal transformations of the space. In this sense, it can be viewed as the *area* in the conformal geometry setting. The functional takes its minimum at the round sphere and a natural problem is to try to understand its minima when we restrict to surfaces with fixed topology or which belong to a given isotopy class or under other kind of topological restrictions. From this point of view even the simple next situation is still open: what is the minimum of the functional on tori? This problem was first proposed by Willmore [**46**], who conjectured that the minimum is given by certain anchor ring Σ_0, and today we have a lot of partial results about it, see [**37**] and references therein. I will only mention that the conjecture is known to be true for non embedded tori and for embedded tori outside the standard isotopy class.

As the functional is compatible with conformal geometry, the problem can be stated as a problem in the 3-sphere: if Σ is viewed, via the stereographic projection, as a surface in the unit sphere $\mathbf{S}^3$, then the functional above transforms into

$$\int_\Sigma (1 + H^2) dA,$$

where now H denotes the mean curvature of Σ in $\mathbf{S}^3$. Thus Proposition 2 gives us a relation between the isoperimetric problem and the Willmore problem. For instance, in the 3-sphere we know that the maximum of the isoperimetric profile is 4π, which gives a well-known universal bound for the Willmore functional, see [**46**].

In the same way, Corollary 1 implies the following result.

THEOREM 13. (Ros [**37**]) *For any compact surface $\Sigma \subset \mathbf{S}^3$ of odd genus which is antipodal invariant, we have $\int_\Sigma (1 + H^2) dA \geq 2\pi^2$ and the equality holds if and only if Σ is the minimal Clifford torus.*

PROOF. As the genus of Σ is odd, it follows from [**37**] that the antipodal map preserves the components of $\mathbf{S}^3 - \Sigma$ and, therefore, the quotient surface $\Sigma' = \Sigma/\pm$ separates the projective space $\mathbf{P}^3$. Using Proposition 2 and Corollary 1 we obtain

$$\int_\Sigma (1+H^2)dA = 2\int_{\Sigma'}(1+H^2)dA \geq 2\pi^2$$

and the equality holds just for the Clifford torus. □

Recently Topping [**45**] has given another proof of Theorem 13 by using an integral geometric approach. For another recent result see Ammann [**1**]. Finally I will give a proof of the Willmore conjecture in another case which is more natural from the Euclidean point of view. We show that the result is true for any torus in $\mathbb{R}^3$ with a center of symmetry.

THEOREM 14. *Let $\Sigma \subset \mathbb{R}^3$ be a compact surface of odd genus which is symmetric with respect to a point. Then*

$$\int_\Sigma H^2 dA \geq 2\pi^2$$

and the equality holds if and only if Σ is homothetic to the anchor ring Σ_0.

PROOF. Let s be the symmetry $s(x_1,x_2,x_3,x_4) = (-x_1,-x_2,-x_3,x_4)$ on $\mathbf{S}^3$. The fix points of s are $\pm p$, where $p=(0,0,0,1)$. Using the conformal nature of the Willmore functional, the result we want is equivalent to the following one.

Claim. *If $\Sigma \subset \mathbf{S}^3$ is a compact surface with odd genus and $s(\Sigma)=\Sigma$, then*

$$\int_\Sigma (1+H^2)dA \geq 2\pi^2$$

and the equality holds if and only if Σ is ambiently conformal to the minimal Clifford torus.

Now we prove the claim. Lemma 1 below gives that the points $\pm p$ lie in the same components of $\mathbf{S}^3 - \Sigma$ (in particular s preserves these components). Denote by Ω and Ω' the two domains enclosed by Σ and suppose that $\pm p \in \Omega'$.

Given a smooth domain $W \subset \mathbf{S}^3$ and $0 \leq t$, we consider the enlarged domain $W_t = \{x \in \mathbf{S}^3 : \mathrm{dist}(x,W) \leq t\}$. For $t<0$, we define W_t as the points in W at distance at least t from the boundary, $W_t = \{x \in W : \mathrm{dist}(x,\partial W) \geq |t|\}$. In particular, there exist constants a and b, $a<0<b$, such that $\mathrm{V}(W_a)=0$, $\mathrm{V}(W_b) = \mathrm{V}(\mathbf{S}^3)$ and $\mathrm{V}(W_t)$ is a strictly increasing continuous function, for $a \leq t \leq b$.

Let $\{f_\lambda : 0<\lambda<\infty\}$ be the 1-parameter group of conformal transformations of $\mathbf{S}^3$ which corresponds, via stereographic projection centered at p, to the homotheties centered at the origin of $\mathbb{R}^3$. Each f_λ fixes the points $\pm p$ and verifies $f_\lambda \circ s = s \circ f_\lambda$. Moreover f_1 is the identity map. For any λ, $0<\lambda<\infty$, we can construct a region $\Omega(\lambda) = (f_\lambda(\Omega))_t$, where t is uniquely determined by the condition $\mathrm{V}(\Omega(\lambda)) = \mathrm{V}(\mathbf{S}^3)/2$. Note that $s(\Omega(\lambda)) = \Omega(\lambda)$.

When λ is very small, the domain $f_\lambda(\Omega)$ is close to the point $-p$ and so the enlarged domain $\Omega(\lambda)$ almost coincides with the halfsphere centered at that point, see Figure 5. The same holds when λ is very big: in this case $\Omega(\lambda)$ is close to the halfsphere centered at the point p. Clearly our construction is continuous enough to conclude that, for certain intermediate value λ_0, the hyperplane $x_4=0$ cuts the region $\Omega(\lambda_0)$ in 2 pieces with the same volume.

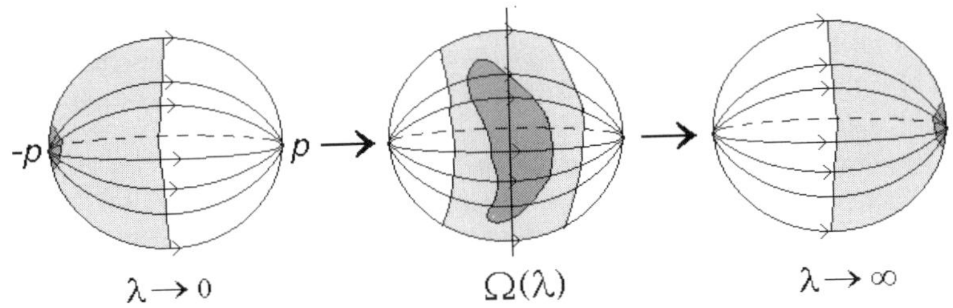

FIGURE 5

Consider the domain Ω^* obtained from $\Omega(\lambda_0)$ by symmetrization with respect to $x_4 = 0$: if the area of $\partial\Omega(\lambda_0)$ in $x_4 > 0$ is not greather than the one in $x_4 < 0$, take $\Omega^* = W \cup s'(W)$, where $W = \Omega(\lambda_0) \cap \{x_4 \geq 0\}$ and $s' = -s$ is the symmetry with respect to $x_4 = 0$. Note that Ω^* is invariant under s and s'. In particular $-\Omega^* = \Omega^*$, $\mathrm{V}(\Omega^*) = \mathrm{V}(\mathbf{S}^3)/2$ and $\mathrm{A}(\partial\Omega(\lambda_0)) \geq \mathrm{A}(\partial\Omega^*)$. As the Willmore functional is preserved by the transformations f_λ, using the argument in the proof of Proposition 2 and the isoperimetric inequality in Corollary 1, we get

$$\int_\Sigma (1 + H^2) = \int_{f_{\lambda_0}(\Sigma)} (1 + H_{\lambda_0}^2) \geq \mathrm{A}(\partial\Omega(\lambda_0)) \geq \mathrm{A}(\partial\Omega^*) \geq 2\pi^2,$$

where H_{λ_0} denotes the mean curvature of $f_{\lambda_0}(\Sigma)$. If the equality holds, we conclude easily, using again Proposition 2 and Corollary 1, that $f_{\lambda_0}(\Sigma)$ is the minimal Clifford torus. □

LEMMA 1. *Let $\Sigma \subset \mathbf{S}^3$ be a compact surface of odd genus and such that $s(\Sigma) = \Sigma$. Then the points $\pm p$ lie in the same connected component of $\mathbf{S}^3 - \Sigma$.*

PROOF. Denote by Ω and Ω' the components of $\mathbf{S}^3 - \Sigma$. If Σ contains some of the points $\pm p$, we conclude by looking at a neighborhood of this point that $s(\Omega) = \Omega'$ and therefore, both points must lie on Σ. As s reverses both the orientation on $\mathbf{S}^3$ and the normal vector on Σ, we have that the quotient surface $\Sigma' = \Sigma/\{1, s\}$ is orientable. Denote by g and g' the genus of the surfaces Σ and Σ' respectively. The quotient map $\Sigma \to \Sigma'$ has degree 2 and just 2 single branch points. So, Riemann-Hurwitz formula gives that $g = 2g'$ and this contradiction proves that $\Sigma \cap \{\pm p\} = \emptyset$.

Consider now a 2-dimensional equator S of $\mathbf{S}^3$ passing through the points $\pm p$ and meeting Σ transversally (after a small perturbation of Σ if necessary), and denote by S^+ and S^- the components of $\mathbf{S}^3 - S$. Note that $s(S) = S$ and $s(S^+) = S^-$. As Σ is obtained by gluing along their boundaries two copies of $\Sigma^+ = \Sigma \cap S^+$, we get that the Euler characteristic of these surfaces are related by the formula $\chi(\Sigma) = 2\chi(\Sigma^+)$ and thus $2 - 2g = 2(2m - r)$, where m is a certain integer number and r is the number of components of $\Gamma = \partial\Sigma^+ = \Sigma \cap S$. By assumption g is odd and so, the relation above implies that r is even.

Finally we study $\Gamma \subset S$, which consists of a disjoint union of r Jordan curves and verifies $s(\Gamma) = \Gamma$. Consider $\Gamma' \subset \Gamma$ (resp. $\Gamma'' \subset \Gamma$) the union of the Jordan curves of Γ which separate (resp. do not separate) the points $\pm p$. Clearly s preserves Γ' and Γ'' and transforms any Jordan curve $C \subset \Gamma''$ into a curve $s(C) \subset \Gamma''$ with $s(C) \neq C$. Hence Γ'' consists of an even number of closed curves and this implies that the number of components of Γ' is even too. Then we conclude that any arc in S transversal to Γ and joining the points $\pm p$ cuts Γ at an even number of times. Therefore, the points $\pm p$ lie in the same component of $\mathbf{S}^3 - \Sigma$, as we claimed.

□

References

[1] B. Ammann, *The Willmore Conjecture for immersed tori with small curvature integral*, Manusc. Math. **101** (2000) 1-22.

[2] M. Athanassenas, *A variational problem for constant mean curvature surfaces with free boundary*, J. Reine Angew. Math. **377** (1987) 97-107.

[3] A.D. Alexandrov, *Uniqueness theorems for surfaces in the large I*, Vestnik Leningrad Univ. Math. **11** (1956) 5-17.

[4] L. Alías, R. López, B. Palmer, *Stable constant mean curvature surfaces with circular boundary*, Proc. Amer. Math. Soc. **127** (1999), 1195-1200.

[5] F. J. Almgren, *Existence and regularity almost everywhere of solutions to elliptic variational problems with constraints*, Mem. AMS **165** (1976).

[6] F. J. Almgren, Spherical symmetrization, Proc. International workshop on integral functions in the calculus of variations, Trieste 1985, Red. Circ. Mat. Palermo (2) Supple. (1987) 11-25.

[7] N. Aronszajn, *A unique continuation theorem for solution of elliptic partial differential equations or inequalities of second order*, J. Math. Pures Apll. **36** (1957) 235-249.

[8] J. L. Barbosa & M. do Carmo, *Stability of hypersurfaces with constant mean curvature*, Math. Z. **185** (1984) 339-353.

[9] F. Barthe, *Extremal properties of central half-spaces for product measures*, preprint.

[10] F. Barthe & Maurey, *Some remarks on isoperimetry of Gaussian type*, Ann. Inst. H. Poincar Probab. Statist. **36** (2000) 419-434.

[11] C. Bavard & P. Pansu, *Sur le volume minimal de* $\mathbb{R}^2$, Ann. Sci. École Norm. Sup. (4) **19** (1986) 479-490.

[12] P. Berard, G. Besson & S. Gallot, *Sur une inegalité isopérimetrique qui généralise celle de Paul Lévy-Gromov*, Invent. Math. **80** (1985) 295-308.

[13] J. Bokowski & E. Sperner, *Zerlegung konvexer Körper durch minimale Trennflchen*, J. Reine Angew. Math. **311/312** (1979) 80–100.

[14] I. Chavel, *Eigenvalues in Riemannian geometry*, Academic Press, 1984.

[15] A. El Soufi & S. Ilias, *Majoration de la seconde valeur propre d'un operateur de Schrödinger sur une varieté compacte et applications*, J. Funct. Anal. **103** (1992) 294-316.

[16] E. Gonzalez, U. Massari & I. Tamanini, *On the regularity of boundaries of sets minimizing perimeter with a volume constraint*, Indiana Univ. Math. J. **32** (1983), 25-37.

[17] P. Griffiths & J. Harris, *Principles of algebraic geometry*, Wiley, New York, (1978).

[18] M. Gromov, *Paul Levy's isoperimetric inequality*, prepublications I.H.E.S. (1980).

[19] M. Grüter, *Boundary regularity for solutions of a partioning problem*, Arch. Rat. Mech. Anal. **97** (1987) 261-270.

[20] H. Hadwiger, *Gitterperiodische Punktmengen und Isoperimetrie*, Monatsh. Math. **76** (1972) 410-418.

[21] E. Heintze & H. Karcher, *A general comparison theorem with applications to volume estimates for submanifolds*, Ann. Sci. Ec. Norm. Super. **11** (1978), 451-470.

[22] J. Hersch, *Quatre propertiés isopérimétriques de membranes sphériques homogenes*, C. R. Acad. Sci. Paris Ser. A-B **270** (1970) A1645-A1648.

[23] H. Hopf, *Differential geometry in the large*, Lecture Notes in Mathematics, **1000** (1989), Springer-Verlag.

[24] W. Y. Hsiang, *A symmetry theorem on isoperimetric regions*, PAM-409 (1988), UC Berkeley.

[25] W. Y. Hsiang, *Isoperimetric regions and soap bubbles*, Proceedings conference in honor to Manfredo do Carmo, Pitman survey in pure and. appl. math . **52** (1991) 229-240.

[26] M. Hutchings, F. Morgan, M. Ritoré & A. Ros, *Proof of the double bubble conjecture* (preprint).

[27] B. Kleiner, *An isoperimetric comparison theorem*, Invent. Math. **108** (1992) 37–47.

[28] P. Lévy, *Le problème des isopérimètriques et les polygones articulés*, Bull. Soc. Math. France **90** (1966) 103-112.

[29] P. Li & S. T. Yau, *A new conformal invariant and its applications to the Willmore conjecture and first eigenvalue of compact surfaces*, Invent. Math. **69**(1982) 269-291.
[30] F. J. López & A. Ros, *Complete minimal surfaces with index one and stable constant mean curvature surfaces*, Comment. Math. Helvet. **64** (1989) 34-43.
[31] S. Montiel & A. Ros, *Schrodinger operators associated with a holomorphic map*, Proceedings conference on global anal. and global diff. geom. at Berlin, 1990. Lecture notes in Math, **1481**, Springer Verlag (1991) 147-174.
[32] F. Morgan, *Geometric measure theory: a beginner's guide*, 2nd ed., Academic Press, 1998.
[33] R. Pedrosa & M. Ritoré, *Isoperimetric domains in the Riemannian product of a circle with a simply connected space form and applications to free boundary problems*, Indiana Univ. Math. J. **48** (1999) 1357-1394.
[34] M. Ritoré, *Index one minimal surfaces in flat three-space forms*, Indiana Univ. Math. J. **46** (1997) 1137-1153.
[35] M. Ritoré & A. Ros, *Stable Constant Mean Curvature Tori and the Isoperimetric Problem in Three Space Forms*, Comment. Math. Helvet. **67** (1992) 293-305.
[36] M. Ritoré & A. Ros, *The spaces of index one minimal surfaces and stable constant mean curvature surfaces embedded in flat three manifolds*, Trans. Amer. Math. Soc. **348** (1996) 391-410.
[37] A. Ros, *The Willmore conjecture in the real projective space*, Math. Research Letters **6** (1999) 487-494.
[38] R. Souam & A. Ros, *On stability of capillary surfaces in a ball*, Pacific Math. J. **178** (1997) 345-361.
[39] A. Ros & E. Vergasta, *Stability for hypersurfaces of constant mean curvature with free boundary*, Geom. Dedicata **56** (1995) 19-33.
[40] M. Ross, *Schwarz' P and D surfaces are stable*, Differential Geom. Appl. **2** (1992) 179-195.
[41] M. Ross, *Complete nonorientable minimal surfaces in* $\mathbf{R}^3$, Comment. Math. Helv. **67** (1992) 64-76.
[42] H. A. Schwarz, *Gesammelte Mathematische Abhandlungen*, Springer Verlag, Berlin (1890).
[43] U. Schnell, *Lattice inequalities for convex bodies and arbitrary lattices*, Monatsh. Math. **116** (1993) 331-337.
[44] J. Steiner, *Sur le maximum et le minimum des figures dans le plan, sur la sphére et dans l'espace en géneral* J. Reine Angew. Math. 24 (1842) 93-152.
[45] P. Topping, *Towards the Willmore conjecture*, Calc. Var. and PDE (to appear).
[46] T. J. Willmore, *Note on embedded surfaces*, An. Sti. Univ. "Al. I. Cuza" Iasi Sect. I a Mat. **11** (1965) 493-496.
[47] S. T. Yau, *Nonlinear analysis in geometry*, L'enseignement Math. **33** (1987) 109-158.
[48] P. Yang & S. T. Yau, *Eigenvalues of the Laplacian of compact Riemann surfaces and minimal submanifolds*, Ann. Scuola Norm. Sup. Pisa Cl. Sci. (4) **7** (1980) 55-63.

Departamento de Geometría y Topología, Universidad de Granada, E–18071 Granada, Spain
E-mail address: aros@ugr.es

Contemporary Mathematics
Volume **288**, 2001

Almost Parallel Structures

Simon Salamon

ABSTRACT. A discussion of torsion of Riemannian G-structures leads to a survey of contributions of Alfred Gray and others on almost Hermitian manifolds, G_2-manifolds, curvature identities, volume expansions, plotting geodesics, and the geometry of nilmanifolds. The paper concludes with a new example of a compact 8-manifold with a quaternionic 4-form that is closed but not parallel.

Introduction

This article is closely based on the lecture entitled 'A Mathematical Tribute' that I gave at the conference in memory of Alfred Gray in Bilbao in September 2000. In preparing the lecture, I was well aware that a large number of his theorems had directly influenced my own work. The aim therefore was to present not just a personal selection of his results, but also an indication of progress that has been made in many of the areas that he pioneered. In this written version, I have given more emphasis to subsequent work, and have added a final section that suits the new title. It is evident from a comparison with the bibliography at the beginning of this volume that the topics below cover a fraction of Gray's overall output. Nonetheless they demonstrate the importance his work has had for the development of differential geometry.

I first met Alfred Gray at the IHES on Friday 13 April 1979, whilst visiting Paris with fellow graduate student Martin Guest. I remember looking forward with anticipation to a meeting on a combination of day and date that in the past had seemed to bring me good fortune, and this occasion was to live up to expectation. It led to my arrival in Maryland that summer, and regular contact with Alfred that lasted until his untimely death in 1998. Although we never formally collaborated, mathematical ideas were exchanged at dozens of memorable occasions and locations.

Whilst finishing my thesis on quaternionic geometry, I had been working on holonomy groups in general and studying properties of various hypothetical geometries. At that time, there was a fair amount of pessimism about the existence of the more exotic structures that Berger's list had highlighted. Alekseevsky and Brown–Gray had supplied independent proofs that any metric with holonomy *Spin* 9 is necessarily locally symmetric, and there appeared to be great difficulty in constructing jets of metrics with holonomy equal to G_2 or *Spin* 7. Equally disheartening was the

1991 *Mathematics Subject Classification.* Primary 53C25; Secondary 53C22, 53C26, 53C30.

belief that there were no compact irreducible hyperkähler manifolds in dimension greater than 4.

Against this background, Gray had been systematically examining G-structures associated with potential holonomy groups, and paying close attention to conditions that generalize the full holonomy reduction. I myself understood his approach in terms explained in §1, a discussion of which probably formed the basis of our first meeting. Although some of the comments below now seem naïve in the light of a greater understanding of exceptional geometry, everything was less obvious twenty years ago.

1. Tensors and representations

Let M be an oriented Riemannian manifold of dimension d, with metric tensor g and Levi Civita connection ∇. Let ϕ be a tensor with associated stabilizer

$$G = \{a \in SO(n) : a \cdot \phi = \phi\},$$

relative to a fixed orthonormal frame at a fixed point $x \in M$. The restricted holonomy group of M is a subgroup of G if and only if $\nabla\phi \equiv 0$.

The space $\bigwedge^2 T^* = \bigwedge^2 T_x^* M$ of 2-forms at $x \in M$ is isomorphic to the Lie algebra $\mathfrak{so}(d)$ of skew-symmetric matrices. This contains the Lie algebra $\mathfrak{g}$ of G and, relative to our fixed frame, there is a consistent splitting

$$\bigwedge^2 T^* = \mathfrak{g} \oplus \mathfrak{g}^\perp. \tag{1.1}$$

This can be used to provide a unified description of the tensor $\nabla\phi$ that measures the failure of the holonomy algebra to lie in $\mathfrak{g}$.

There is no obstruction to the existence of a connection ∇' that preserves the G-structure. If we choose one, then $\nabla_X - \nabla'_X$ is a tensor with values in (1.1), and $\nabla_X\phi = (\nabla_X - \nabla'_X)\phi$ belongs to $\mathfrak{g}^\perp$ at x. Thus,

Lemma 1.2 *$\nabla\phi$ can be identified with an element of the space $T^* \otimes \mathfrak{g}^\perp$.*

It is easy to check that $\nabla\phi$ carries the same information as that component of the torsion of ∇' that is independent of the choice of G-connection. This component was described as the space of 'intrinsic torsion' in a similar context (Bryant, 1987).

For a given G, there will be a decomposition

$$T^* \otimes \mathfrak{g}^\perp = \bigoplus_{i=1}^{N} \mathcal{W}_i, \tag{1.3}$$

in which each $\mathcal{W}_i$ is an irreducible G-module. Determining this and similar decompositions is a relatively straightforward matter that nowadays can be performed quickly by computer. Given his later admiration for computational methods, it is perhaps ironic that Alfred carried out all the decompositions by hand. But in this way, he was able to write down formulae for the various projections and tensors in meticulous detail, and develop properties of the corresponding classes of manifolds that others had overlooked.

Examples (i) $d = 2n$, $\phi = J$ is an orthogonal almost complex structure, and $G = U(n)$. In this case, $N = 4$ provided $n \geqslant 3$. The same theory results if $\phi = \omega$ is a non-degenerate 2-form compatible with g. Indeed, 3 of the 4 real components of $T^* \otimes \mathfrak{g}^\perp$ can be identified with those of the exterior derivative

$$d\omega \in \bigwedge^3 T^* = [\![\Lambda^{3,0}]\!] \oplus [\![\Lambda_0^{2,1}]\!] \oplus T^*,$$

notation as in (Salamon, 1989). The fourth component forms part of the Nijenhuis tensor of J. Given an even-dimensional Riemannian manifold (M, g), one may investigate the existence of compatible (i) complex structures, and (ii) symplectic structures, and there are constraints arising from the curvature tensor of a rather different nature in each of the two cases.

(ii) $d = 7$, ϕ is a 'positive generic 3-form', meaning that there exists an orthonormal basis of 1-forms at each point relative to which

$$\phi = e^{125} - e^{345} + e^{136} - e^{426} + e^{147} - e^{237} + e^{567}, \tag{1.4}$$

and G is the exceptional Lie group G_2. As was well understood from the study of vector cross products (Brown and Gray, 1967), the 3-form ϕ can also be defined in terms of Cayley multiplication on $\operatorname{Im}\mathbb{O} = \mathbb{R}^7$. Once again, $N = 4$. The four components of $T^* \otimes \mathfrak{g}^\perp$ can be recovered from those of $d\phi$ and $d * \phi$ as explained in §3, and the metric has holonomy contained in G_2 if and only if both these forms vanish. The larger class of 'calibrated' G_2-manifolds consists of ones for which $d\phi = 0$; by contrast $d * \phi$ is more akin to the Nijenhuis tensor in case (i).

(iii) $d = 4n$, ϕ is a 'quaternionic' 4-form, meaning that

$$\phi = \sum_1^3 \omega^i \wedge \omega^i,$$

where $\omega_1, \omega_2, \omega_3$ is a triple of 2-forms corresponding to almost complex structures I_1, I_2, I_3 modelled on the Lie algebra $\mathfrak{su}(2)$. The stabilizer of ϕ is $Sp(n)Sp(1)$, and such structures were considered in (Gray, 1969). Whilst he did not study the full decomposition of $T^* \otimes \mathfrak{g}^\perp$, it is known that $N = 6$ provided that $n \geqslant 3$ (Salamon, 1989). The vanishing of exactly 3 of these 6 components characterizes the class of quaternionic manifolds which possess a complex twistor space.

We shall have something to say about each of these structures in §2, §3, §7 respectively. Other examples studied by others include *Spin* 7 structures (for which $N = 2$), almost product structures ($N = 6$), and almost contact structures. See also the analysis by Friedrich of the *Spin* 9 case in this volume. We shall confine ourselves to two additional remarks before leaving the general set-up.

Firstly, the decomposition (1.1) has been exploited to define a type of Dolbeault complex for G-structures, important for the study of instantons associated to special geometries (Reyes-Carrion, 1998). Let $N(G)$ denote the normalizer of G in $SO(d)$; if $N(G)$ is strictly larger than G, it suffices to start with an $N(G)$-structure in order to define (1.1). On can then set $A^0 = \bigwedge^0 T^*$, $A^1 = T^*$, $A^2 = \mathfrak{g}^\perp$, and more generally

$$A^k = (\mathfrak{g} \wedge \bigwedge^{k-2} T^*)^\perp, \quad k \geqslant 3.$$

Let d_k denote the composition of exterior differentiation acting on A^k with projection $\bigwedge^{k+1} T^* \to A^{k+1}$. An 'extendability' condition guarantees that

$$A^0 \to A^1 \to A^2 \to A^3 \to \cdots$$

is a complex, and can be expressed in terms of the vanishing of certain components of $\mathcal{W}$. The case $G = G_2$ was developed by (Fernández and Ugarte, 1998), based on a thesis co-examined by Gray and myself.

Secondly, relative to an orthonormal frame compatible with the holonomy group G, the Riemann curvature tensor has components R_{ijkl} whose symmetries determine an element in the space

$$\mathcal{R} = \ker\left(S^2\mathfrak{g} \hookrightarrow S^2(\textstyle\bigwedge^2 T^*) \to \bigwedge^4 T^*\right), \tag{1.5}$$

in which the second mapping corresponds to the first Bianchi identity. Thus, any G-invariant element B (such as the Killing form) in $S^2\mathfrak{g}$ will have non-zero image in $\bigwedge^4 T^*$ unless it satisfies the same symmetries as a Riemann curvature tensor. It follows from the theory of E. Cartan that any G-structure with $G \subset SO(n)$ possesses a canonical 4-form, provided the structure possesses metrics that are not locally symmetric. Example (iii) above is actually defined by such a 4-form, as is *Spin* 7 geometry, though invariant 4-forms are also present in (i) (namely $\omega \wedge \omega$), and in (ii) (namely $*\phi$).

2. The sixteen classes of almost Hermitian manifolds

Let M be an almost Hermitian manifold of dimension $d = 2n$. It is therefore equipped with a positive-definite metric tensor g, an orthogonal almost complex structure J, and associated 2-form ω. The subgroup of $SO(2n)$ determined by these structures is isomorphic to the unitary group $U(n)$.

The structure is Kähler if and only if $\nabla J \equiv 0$, and the decomposition (1.3) in this case is given by

Proposition 2.1 (Gray and Hervella, 1980) *For $n \geqslant 3$, the tensor ∇J belongs to the direct sum*

$$\mathcal{W}_1 \oplus \mathcal{W}_2 \oplus \mathcal{W}_3 \oplus \mathcal{W}_4 \tag{2.2}$$

of four irreducible $U(n)$-modules of respective dimensions

$$\tfrac{1}{3}n(n-1)(n-2), \quad \tfrac{2}{3}n(n-1)(n+1), \quad n(n+1)(n-2), \quad 2n.$$

The list of dimensions reflects the following facts:

(i) When $n = 2$, $\mathcal{W}_1$ and $\mathcal{W}_3$ reduce to zero, leaving the sum $\mathcal{W}_2 \oplus \mathcal{W}_4$ of two spaces each of dimension 4. The 2-form is closed if and only if $\nabla J \in \mathcal{W}_2$, and J is integrable if and only if $\nabla J \in \mathcal{W}_4$. Thus, in 4 dimensions the symplectic and complex conditions are exactly complementary. A fuller description of this case is included in the paper by Donaldson in these Proceedings.

(ii) The space $\mathcal{W}_4$ is isomorphic to the cotangent space T^*. Such a summand is always present in (1.3), as it reflects the change in $\nabla\phi$ that occurs when the metric is altered by a conformal factor. For example, if M is locally conformally Kähler then necessarily $\nabla J \in \mathcal{W}_4$. Conversely, if $\nabla J \in \mathcal{W}_4$ then ∇J can be identifed with a 1-form θ that satisfies $d\omega = \theta \wedge \omega$. If $n \geqslant 3$ then $d\theta = 0$ and M is locally conformally Kähler, but the 4-dimensional case is more complicated. Of particular interest are those Hermitian 4-manifolds for which $\nabla\theta = 0$.

(iii) The space $\mathcal{W}_1$ underlies the third exterior power of the space of (1, 0)-forms. When $n = 3$, $\mathcal{W}_1$ has dimension 2, and a non-zero element of it determines a reduction from $U(3)$ to $SU(3)$. This leads to some special features of so-called nearly-Kähler manifolds of dimension 6 (see below).

Further to (i), the two halves $\mathcal{W}_1 \oplus \mathcal{W}_2$ and $\mathcal{W}_3 \oplus \mathcal{W}_4$ of equal dimension $n^2(n-1)$ lead to a sort of duality. They are characterized by the conditions

$$\nabla J \in \mathcal{W}_1 \oplus \mathcal{W}_2 \Leftrightarrow (d\omega)^{1,2} = 0,$$

$$\nabla J \in \mathcal{W}_3 \oplus \mathcal{W}_4 \Leftrightarrow M \text{ is Hermitian.}$$

The significance of the first condition was already recognized by Gray in his thesis, and the manifolds satisfying $(d\omega)^{1,2} = 0$ are often called *quasi-Kähler.*

Proposition 2.3 (Gray, 1965) *If M is quasi-Kähler, and $M' \subset M$ is a J-holomorphic submanifold, then M' is minimal.*

This generalizes the well-known fact that a complex submanifold of a Kähler manifold is minimal. A similar result, but expressed in terms of harmonic mappings was presented by (Lichnerowicz, 1980).

The following case has rightly merited special attention.

Definition 2.4 *M is nearly-Kähler if $\nabla J \in \mathcal{W}_1$, equivalently $(\nabla_X J)X = 0$ for all X. M is said to be strictly nearly-Kähler if in addition $\nabla J \neq 0$.*

The basic model of a strictly nearly-Kähler manifold is the sphere $S^6 = G_2/SU(3)$ endowed with a non-integrable almost-Hermitian structure induced from the cross product on $\operatorname{Im}\mathbb{O} = \mathbb{R}^7$. In fact, S^6 belongs to a large class of homogeneous examples that we describe next.

Let G be a real semisimple Lie group. A homogeneous manifold $M = G/H$ is called a *3-symmetric space* if H is the fixed point set of an automorphism θ of G with $\theta^3 = 1$. The complexified Lie algebra of G has the form

$$\mathfrak{g} = \mathfrak{h} \oplus \mathfrak{m}, \quad \mathfrak{m}_c = \mathfrak{m}^{1,0} \oplus \mathfrak{m}^{0,1}, \tag{2.5}$$

and the resulting 3 summands of $\mathfrak{g}_c$ are the eigenspaces of θ. Defining

$$J = \frac{1}{\sqrt{3}}(2\theta + 1) \tag{2.6}$$

determines a canonical almost complex structure on the tangent space $T_x M \cong \mathfrak{m}$. This is non-integrable unless M is Hermitian symmetric, a possibility we exclude below.

Theorem 2.7 (Wolf and Gray, 1968) *Any 3-symmetric space has a nearly-Kähler metric compatible with (2.6).*

Classification. The collection of 3-symmetric spaces can be divided into the following categories, properties of which appear in (Gray, 1972).

(i) generalizations of S^6 with irreducible isotropy, to cite one example $E_8/SU(9)$. Such spaces are in a sense rarer than their complex homogeneous counterparts.

(ii) $G \times G$, where G is a compact Lie group. Just as such a group can be viewed as an ordinary symmetric space by considering the space of cosets $(G \times G)/G$, so $G \times G$ becomes a 3-symetric space when viewed as $(G \times G \times G)/G$.

(iii) the total space of a fibration $\pi : Z \to M$ where M is a symmetric space defined by an inner involution. In this case, $\mathfrak{m}^{1,0}$ is a reducible representation of H, reflecting the horizontal/vertical decomposition of TZ.

The 'twistor spaces' Z in (iii) have been extensively used in the study of minimal surfaces and harmonic maps, in essence because of results (Bryant, 1985; Salamon, 1985) that extend Proposition 2.3. For example,

Proposition 2.8 *If $f : \Sigma \to Z$ is a J-holomorphic curve in a 3-symmetric space then $\pi \circ f$ is a harmonic map into M.*

Example. A 'classical' instance of this construction is provided by the diagram

$$\begin{array}{ccc} & & \dfrac{U(p+q+1)}{U(p)\times U(q)\times U(1)} = Z \\ & \nearrow & \big\downarrow \pi \\ \Sigma & \to & \dfrac{U(n+1)}{U(n)\times U(1)} = \mathbb{CP}^n \end{array} \tag{2.9}$$

in which $p + q = n$. Any twistor space Z has an integrable complex structure J_1 in addition to J_2, and the two structures coincide on the horizontal subspace of each tangent space T_xZ. The most important class of maps into Z are those that are horizontal and (unambibuously) holomorphic. If $\Sigma \cong \mathbb{CP}^1$ has genus 0, any harmonic map into $\mathbb{CP}^n$ is the projection of a horizontal holomorphic one in a flag manifold such as Z above (Eells and Wood, 1983).

Related methors have been successfully used to classify orthogonal complex structures and to extend the description of harmonic maps to symmetric spaces (Burstall and Rawnsley, 1990; Burstall and Guest, 1997).

3. Weak and exceptional holonomy

The importance of nearly-Kähler manifolds is highlighted by

Theorem 3.1 (Gray, 1976b) *A 6-dimensional strictly nearly-Kähler manifold is Einstein.*

Example. It is well known that there are two Einstein metrics on $\mathbb{CP}^3$ for which the Penrose twistor fibration $\pi : \mathbb{CP}^3 \to S^4$ is a Riemannian submersion. The non-standard one is nearly-Kähler relative the non-integrable almost complex structure J_2 associated to π.

Further curvature identities amount to the statement that the curvature of a strictly nearly-Kähler 6-manifold has the form

$$R = sR_{S^6} + R_{CY},$$

where s is the scalar curvature, R_{S^6} is the (suitably normalized) constant curvature tensor of S^6, and R_{CY} stands for the curvature tensor of a 6-manifold with a Calabi-Yau metric, so holonomy equal to $SU(3)$. The only known compact 6-dimensional strictly nearly-Kähler manifolds are

$$S^6, \quad S^3 \times S^3, \quad \mathbb{CP}^3, \quad \mathbb{F}^3.$$

The second arises from (ii) in §2 above with $G = SU(2)$, and the last one denotes the flag manifold Z in (2.9) with $p = q = 1$. Any 6-dimensional strictly nearly-Kähler manifold has a canonical connection with holonomy $SU(3)$, and this reduces to $U(2)$ if and only if M fibres over a self-dual Einstein 4-manifold (Belgun and Morioanu, 2000). This construction is important because it can be used to show that there is an abundance of locally-defined strictly nearly-Kähler metrics.

Nearly-Kähler manifolds play a central role in the theory of weak holonomy developed in (Gray, 1971). A full holonomy reduction to G requires a tensor with stabilizer G to remain constant under parallel translation, whereas G is a weak holonomy group if a certain family of subspaces is mapped into itself by parallel translation. An obvious candidate would be the family of quaternionic lines (which are real 4-dimensional subspaces) for $G = Sp(n)Sp(1)$, but it turns out that in this case a weak holonomy reduction is equivalent to a full holonomy reduction. Instead, the theory is most fruitful in 6 and 7 dimensions.

Manifolds with weak holonomy G_2 can be characterized algebraically, as in the nearly-Kähler case which corresponds to weak holonomy $U(3)$. If a manifold M has a 3-form of type (1.4), then there are G_2-equivariant decompositions

$$\begin{aligned} \textstyle\bigwedge^2 T^* &= \mathfrak{g} \oplus \mathfrak{g}^\perp \cong \mathfrak{g}_2 \oplus T^* \\ \textstyle\bigwedge^3 T^* &\cong \mathbb{R} \oplus T^* \oplus S^2_0 T^*. \end{aligned} \tag{3.2}$$

The presence of a summand T^* in $\bigwedge^2 T^*$ corresponds to the cross product operation, and the presence of T^* in $\bigwedge^3 T^*$ reflects the failure of this product to satisfy the Jacobi identity. The decomposition (1.3) becomes

$$\begin{aligned} T^* \otimes \mathfrak{g}_2^\perp \cong T^* \otimes T^* &\cong S^2 T^* \oplus \textstyle\bigwedge^2 T^* \\ &\cong \mathbb{R} \oplus S^2_0 T^* \oplus T^* \oplus \mathfrak{g}_2, \end{aligned} \tag{3.3}$$

and a comparison of (3.2) and (3.3) reveals

Corollary 3.4 (Fernández and Gray, 1982) *M has holonomy in G_2 iff $d * \phi = 0$ and $d\phi = 0$.*

The manifold has weak holonomy G_2 if and only if $\nabla\phi$ lies in the 1-dimensional component of (3.3). This means that

$$d\phi = c * \phi \tag{3.5}$$

for some c (a universal constant times the scalar curvature that we assume is non-zero), and consequently $d * \phi = 0$. In this case, associative subspaces are ones invariant by the cross product, and are preserved by parallel transport.

The local existence of metrics with holonomy equal to G_2 was proved by (Bryant, 1987). The first space found to exhibit an explicit metric with exceptional holonomy was the cone $Z = Y \times \mathbb{R}^+$, in which

$$Y = \frac{SO(5)}{SO(3)}$$

is an isotropy irreducible space arising from the representation $S^6\mathbb{C}^2$ of $SO(3) = SU(2)/\mathbb{Z}_2$, with a history of providing counterexamples (Berger, 1961). The fact that there is a 1-dimensional space of invariant 3-forms in $\bigwedge^3(S^6\mathbb{C}^2)$ led the author to observe that Y has a metric with weak holonomy G_2 satisfying (3.5). The realization that Z has a closed 4-form defining a metric with holonomy equal to *Spin* 7 was a step away. There is no doubt that the discovery of explicit metrics with exceptional holonomy would have taken longer without Gray's theory of weak holonomy. The subsequent theory of compact manifolds with exceptional holonomy is the subject of work culminating in the book (Joyce, 2000).

With the hindsight provided by the first examples of metrics with exceptional holonomy, it became clear that the most effective way of understanding the relationship between weak and exceptional holonomy is via the theory of Killing spinors, developed by (Baum et al., 1990). Indeed,

Theorem 3.6 (Bär, 1993) *(i) A Riemannian metric on X^6 is strictly nearly-Kähler iff the corresponding conical metric on $X \times \mathbb{R}^+$ is Ricci-flat with holonomy contained in G_2.*

(ii) A metric on Y^7 has weak holonomy G_2 iff the conical metric on $Y \times \mathbb{R}^+$ is Ricci-flat metric with holonomy contained in $Spin\,7$.

Possible holonomy groups of $Y \times \mathbb{R}^+$ in (ii) are $Spin\,7$, $SU(4)$ and $Sp(2)$, and these correspond to 7-manifolds with a space of Killing spinors of dimension 1,2,3 respectively. This is fully discussed in (Friedrich et al., 1998), and examples appear in (Galicki and Salamon, 1996).

4. Curvature and volume

Gray's published papers incorporate extensive studies of the Riemann curvature tensor. As I realized during the Bilbao conference (responding to a query from Peter Gilkey), given any natural condition on the curvature, the chances are that Alfred had already given it a name. His classification of manifolds in terms of the covariant derivative of the Ricci tensor (Gray, 1978) is particularly appealing, and was the subject of talks at the conference. However, in this section, we have chosen to comment on the curvature of Hermitian manifolds, and unrelated volume-type expansions.

A striking result concerning the curvature of Kähler manifolds is

Theorem 4.1 (Gray, 1977) *A compact Kähler manifold with nonnegative sectional curvature and constant scalar curvature is locally symmetric.*

This result is a prototype for many subsequent theorems in the literature, characterizing the curvature of Hermitian symmetric spaces and complex projective space.

The curvature tensor K of a Kähler manifold of real dimension $2n$ satisfies the equation

$$K(W,X,Y,Z) = K(W,X,JY,JZ) \tag{4.2}$$

that reflects the fact that K may be regarded as a 2-form with values in the holonomy algebra $\mathfrak{u}(n)$. The curvature tensor R of an arbitary almost Hermitian manifold can therefore be decomposed as

$$R = K + K^\perp,$$

where K represents the component of R in the subspace of tensors satisfying (4.2). The next result (see for example (Falcitelli et al., 1994)) is a starting point for the analysis of metrics with constant holomorphic sectional curvature.

Proposition 4.3 *The tensor $K^\perp$ is a linear contraction of $\nabla\nabla J$ (where ∇ denotes the Levi Civita connection) and has zero holomorphic sectional curvature.*

There are further decompositions of K and $K^\perp$ under the action of $GL(n,\mathbb{C})$, and decompositions under $U(n)$ are given in (Tricerri and Vanhecke, 1981). A significant fact concerning curvature of a Hermitian manifold is encapsulated in

Proposition 4.4 (Gray, 1976a) *If M is Hermitian then*

$$\begin{aligned}R(W,X,Y,Z)&+R(JW,JX,JY,JZ)\\&=R(JW,JX,Y,Z)+R(JW,X,JY,Z)+R(JW,X,Y,JZ)\\&\quad+R(W,JX,JY,Z)+R(W,JX,Y,JZ)+R(W,X,JY,JZ).\end{aligned}$$

A complex structure that is orthogonal relative to a metric g is also orthogonal relative to any conformally related metric. It follows that the above equation constrains only the Weyl tensor Wey of M^{2n}. In terms of type decomposition it amounts to saying that R has no component in the real subspace underlying $S^2(\Lambda^{2,0})\ominus\Lambda^{4,0}$, and imposes $k=\frac{1}{6}n^2(n^2-1)$ equations on Wey, which itself has dimension asymptotic to $8k$ as $n\to\infty$.

The Weyl tensor of an oriented Riemannian 4-manifold M decomposes as

$$Wey = Wey_+ + Wey_-,\qquad Wey_\pm\in\textstyle\bigwedge^2_\pm T^*M. \tag{4.5}$$

If $Wey_+\neq 0$ then M has at most two distinct pairs $\pm I_1$, $\pm I_2$ of orthogonal complex structures compatible with the metric and orientation. In higher dimensions, less is known about metrics with a 'large' number of isolated orthogonal complex structures, though work of (Pontecorvo, 1997; Apostolov et al., 1999) suggests that the theory will have both a local and global flavour.

The curvature of a Riemannian manifold provides a quantifiable means of comparing the volume of submanifolds with those of Euclidean space, and established theorems relate the volume of balls and spheres in Riemannian manifolds with those of spaces of constant curvature (Günther, 1960; Bishop and Crittenden, 1964). Gray extended this theory by developing asymptotic expansions to study the volume of balls and subsequently tubes around curves and submanifolds.

Let $V(r)$ denote the volume of a ball $\mathbb{B}(r)$ of radius r centred at the origin in $\mathbb{R}^d$. It is convenient to set $d=2n$, whether or not d is even. Then

$$V(r)=V(1)r^d=\frac{(\pi r^2)^n}{n!},$$

where $n!$ stands for $\Gamma(n+1)$ if n is half-integral. The theory of volume in higher dimensions incorporates a number of counter-intuitive features that lend themselves naturally to investigation in curved space:

(i) The table below shows that $V(1)$ reaches a peak for $d=5$ and then decreases to 0.

d	1	2	3	4	5	6	7	8	9	10	15	20
$V(1)$	2	3.14	4.19	4.93	5.26	5.17	4.72	4.06	3.30	2.55	0.38	0.03

To express this in another way, let r_d be the radius for which $V(r_d) = 1$. For instance, $r_{1000} \sim 7.68$, and Stirling's formula implies that

$$r_d \sim \sqrt{\frac{d}{2\pi e}}, \qquad d \to \infty.$$

(ii) A study of the distribution of volume within balls implies that (for example) the slice

$$\mathbb{B}(r_d) \cap (\mathbb{R}^{d-1} \times [-0.4, 0.4])$$

has volume at least 0.8 irrespective of d (as the author discovered whilst browsing the conference bookstall). This is a simple result concerning a 'tube' around a planar hypersurface of $\mathbb{B}(r_d)$.

Now suppose that M is a Riemannian manifold. Fix a point $x \in M$, and let $V(r)$ now denote the volume of a ball $\mathbb{B}(r)$ of radius r centered at x. We assume that r is less than the injectivity radius of M, so that $\mathbb{B}(r)$ is formed of geodesics of length r emanating from x. Early results on the volume of $\mathbb{B}(r)$ include those of (Hotelling, 1939; Weyl, 1939). There is an expansion

$$V(r) = \frac{(\pi r^2)^n}{n!}\left(1 + c_2 r^2 + c_4 r^4 + c_6 r^6 + \cdots\right),$$

in which the coefficient c_2 equals $-\frac{1}{3}s/(n+1)$ where s is the scalar curvature ($n = \frac{1}{2}\dim M$). Gray took up the challenge of determining further coefficients. Let R denote the full curvature tensor (with components R_{ijkl} relative to an orthonormal frame), and *Ric* the Ricci tensor (components $R_{jl} = \sum_i R^i_{jil}$).

Theorem 4.6 (Gray, 1973) $c_4 = \dfrac{8\|Ric\|^2 - 3\|R\|^2 + 5s^2 - 18\Delta s}{1440(n+1)(n+2)}$.

A natural conjecture is that the vanishing of all c_{2k} implies that M is flat, but a proof of this became elusive by an array of examples of $2n$-manifolds for which

$$V(r) = \frac{(\pi r^2)^n}{n!}(1 + O(r^{2k})). \tag{4.7}$$

Results from (Gray and Vanhecke, 1979) showed that in dimension 4, there exist metrics with $s = 0$ and $c_4 = 0$ so that (4.7) holds with $k = 3$. The same paper has an example with $n = 367$ and $k = 4$. These results were drammatically extended by the use of an additive functor to prove

Theorem 4.8 (Kowalski, 1980) *There exists a product of homogeneous spaces such that (4.7) holds with $k = 8$.*

Similar techniques can be applied to other asymptotic expansions involving curvature. An analogous conjecture concerning the class of so-called harmonic manifolds motivated the paper (Carpenter et al., 1982), but was settled in the negative by the discovery of the so-called Damek-Ricci spaces. On a different topic, the article by Pinsky in the Proceedings refers to work on 'mean exit times' for Brownian motion on Riemannian manifolds (Gray and Pinsky, 1983).

The volume of tubes is the subject of many of Alfred Gray's papers that provided the basis of his first book (Gray, 1990). It investigates a theory that has developed from work of H. Weyl, whose starting point is the fact that the volume of a tube of radius r about a space curve of arclength ℓ remarkably depends only on r and ℓ.

Corresponding invariance properties in the context of Kähler manifolds and characteristic classes lead to a number of elegant formulae such as

Theorem 4.9 (Gray, 1985) *A tube of (sufficiently small) radius r surrounding a hypersurface of degree k in $\mathbb{CP}^n$ has volume* $\dfrac{\pi^n}{n!}\Big(1-(1-k\sin^2 r)^n\Big)$.

5. Plotting geodesics on surfaces

We have been talking about geodesic balls on manifolds. Such objects can be graphically illustrated in the 2-dimensional case. The first edition of Gray's book on curves and surfaces contained some of the first programs for constructing geodesics on surfaces.

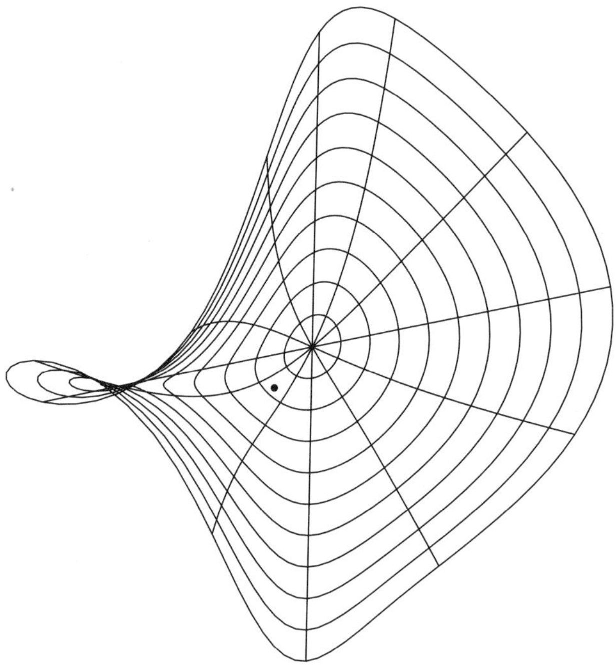

Figure 1

By way of a diversion at the conference, I displayed a geodesic clockface on the paraboloid $z = xy$, plotted using the MATHEMATICA program `solvegeoeqs` and its cousins (Gray, 1998). Another is shown in Figure 1, in which the saddle point is indicated by a dot and one can easily spot two straight line rulings. Analogous curves on a torus are displayed in Figures 2 and 3, and are not entirely inconsistent with the title of this article. The aim of this section is to present a self-explanatory account of how the plotting is quickly achieved, to serve as an introduction to the more extensive programs in (Gray, 1998). This powerful construction kit can be downloaded from the website `math.cl.uh.edu/ gray` maintained by M. Mezzino.

Figure 2 consists of geodesics emanating from a point P on a torus of revolution with conjugate points visible, and Figure 3 displays corresponding 'circles' centered at P up to the point where they become singular. They were plotted with the program below, a condensed version of Gray's that can be copied by hand onto a keyboard without too much effort. It mixes local and global variables in a rather amateur way, but is designed to make the various steps transparent. The instructions can easily be modified so as to plot individual geodesics and related objects on an arbitrary surface.

Let $\mathbf{x}(u, v)$ be a parameterized surface. The aim is to construct geodesics emanating from a fixed point $\mathbf{x}(a, b)$, and to this end we divide the task into three blocks of code that correspond to notebook cells. The first consists of standard definitions that are independent of the choice of the function $\mathbf{x}$. The latter is entered at the second stage, and plotting takes place after loading the required parameters.

(i) First, one defines the coefficients of the first fundamental form and the Christoffel symbols. These basic differential geometric formulae translate readily into computer code, in readiness for all manner of applications.

```
Unprotect[E];
E:= D[x[u,v],u].D[x[u,v],u]
F:= D[x[u,v],u].D[x[u,v],v]
G:= D[x[u,v],v].D[x[u,v],v]
ga[1,1,1]:= D[E,v]F-2D[F,u]F+D[E,u]G
ga[2,2,2]:= D[G,v]E-2D[F,v]F+D[G,u]F
ga[2,1,1]:=-D[E,v]E+2D[F,u]E-D[E,u]F
ga[1,2,2]:=-D[G,v]F+2D[F,v]G-D[G,u]G
ga[1,1,2]:= D[E,v]G-D[G,u]F
ga[2,1,2]:=-D[E,v]F+D[G,u]E
Ga[i_,j_,k_]:= Simplify[ga[i,j,k]/(E G-F^2)/2]
```

(ii) It is convenient to enter the required parameterization at this point. The geodesic equations are then entered, and are then in a form ready to use for the surface in question. Definitions are included of appropriate initial conditions and the command for solving the differential equations.

```
x[u_,v_]:= {(2+Cos[v])Cos[u],(2+Cos[v])Sin[u],Sin[v]}
su:= {u->u[s],v->v[s],p->u'[s],q->v'[s]}
e[j_]:=e[j]= Ga[j,1,1]p^2+2Ga[j,1,2]pq+Ga[j,2,2]q^2 /.su
eqic:= {u''[s]+e[1]==0,v''[s]+e[2]==0,
        u[0]==a,v[0]==b,u'[0]==Cos[th],v'[0]==Sin[th]}
so:= NDSolve[eqic,{u,v},{s,0,r}]
sd[m_]:= Flatten[Table[so,{th,0,2Pi,2Pi/m}],1]
```

(iii) Points along geodesics equidistant from (a, b) are joined and plotted, together with the geodesics themselves. It suffices to re-enter the lines below to re-draw the plot with different parameters.

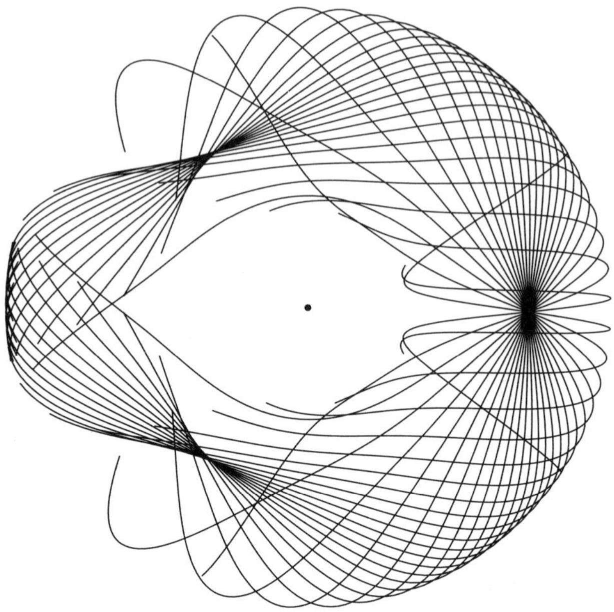

Figure 2

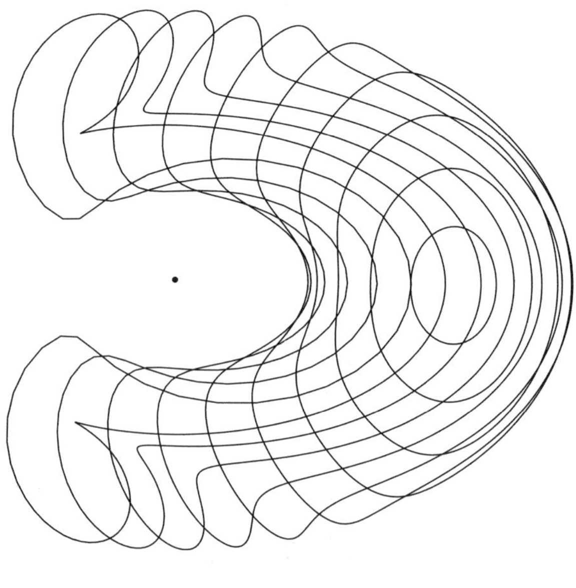

Figure 3

```
a:=0; b:=Pi/2; r:=4; m1:=100; m2:=50; k:=.4
xs:= x@@Sequence[{u[t],v[t]}]
g0:= Graphics3D[Point[{0,0,0}]]
g1:= Graphics3D[Table[Line[xs/.sd[m1]],{t,0,r,k}]]
g2:= ParametricPlot3D[Evaluate[xs/.sd[m2]],{t,0,r},
                          DisplayFunction->Identity]
Show[g0,g1,g2,Boxed->False,BoxRatios->{1,1,1},ViewPoint->{0,0,4}]
```

The graphics object g0 enables the origin to be added to the diagram as a point of reference. Other commands in (iii) that enable the various definitions to be combined are of course fully explained in the Mathematica Book (Wolfram, 1999).

6. Invariant geometry on nilmanifolds

Any compact simple Lie group G of even dimension $d = 2n$ admits left-invariant complex structures, that can easily be described in terms of the root decomposition of the Lie algebra $\mathfrak{g}$. On the other hand, G admits no symplectic structure as $b_2 = 0$. By contrast, a given nilpotent Lie group N may or may not admit left-invariant complex and/or symplectic structures.

Example. There are three simply-connected nilpotent Lie groups in dimension $d = 4$, namely N_1, N_2, N_3 in which the Lie-algebra of N_k is k-step (so N_1 is abelian). All three admit left-invariant symplectic structures, but it is easy to see that only the first two admit invariant complex structures.

If N is a nilpotent Lie groups with rational structure constants, it possesses a discrete subgroup Γ for which $M = N/\Gamma$ is compact (Malcev, 1962). Such a compact nilmanifold N/Γ can only admit a Kähler metric if N is abelian, in which case the quotient is a torus (see for example (Benson and Gordon, 1988)). A celebrated theorem (Nomizu, 1954) implies that the space of left-invariant forms provides a minimal model for the nilmanifold's de Rham cohomology, which is therefore readily computed. Gray pioneered the use of Massey products to detect the non-existence of a Kähler metric, and applied this technique in a number of different situations. The power of this approach is apparent from (Cordero et al., 1986; Hasegawa, 1989).

A Kodaira (complex) surface is a compact quotient of N_2 endowed with an integrable left-invariant complex structure, and actually admits a holomorphic symplectic structure. It corresponds to $g = 1$ in

Theorem 6.1 (Fernández et al., 1991) *A compact real surface U of genus $g \geqslant 1$ with a nowhere zero 2-form and symplectomorphism φ fixing a non-zero class in $H^1(U, \mathbb{Z})$ defines a circle bundle $E \to (U \times [0,1])/\varphi$ that is symplectic and generally non-Kähler.*

A further generalization of this construction is shown to account for all symplectic manifolds with a free S^1 action.

Some tools are available to compute the Dolbeault cohomology of an invariant complex structure on N/Γ (Cordero et al., 2000; Console and Fino, 2001; Pedersen and Poon, 2001), but not in the most general situation. It is therefore natural to study the convergence of the Frölicher spectral sequence $E_r^{p,q}$, that relates the

Dolbeault and de Rham groups. Whereas any Kähler manifold (and complex surface) satisfies $E_2^{p,q} = E_\infty^{p,q}$, few explicit examples of higher order degeneration were known prior to

Proposition 6.2 (Cordero et al., 1991) *There exist invariant complex structures on nilmanifolds of real dimension d such that (i) $d = 8$ and $E_2 \neq E_\infty$, and (ii) $d = 12$ and $E_3 \neq E_\infty$.*

Semisimple examples, but in somewhat higher dimensions, were provided by (Pittie, 1989).

Gray himself introduced me to the classification problem for invariant complex structures on nilmanifolds, and I tackled the 6-dimensional case (Salamon, 2001) building on results of (Cordero et al., 1997). Invariant symplectic structures are classified in (Goze and Khakimdjanov, 1996), again from a Lie algebra perspective. Combining the various approaches yields

Theorem 6.3 *There are 34 isomorphism classes of real 6-dimensional nilpotent Lie algebras of which*

15 admit both complex and symplectic structures,
3 admit complex but not symplectic structures,
11 admit symplectic but not complex structures,
5 admit neither complex nor symplectic structures.

These examples provide a forum for the investigation of the Gray-Hervella classes discussed in §2. An inner product on a 6-dimensional nilpotent Lie algebra gives rise to a metric g on an associated nilmanifold $M = N/\Gamma$. The corresponding set of almost-Hermitian structures on M is isomorphic to

$$\frac{SO(6)}{U(3)} \cong \mathbb{CP}^3, \tag{6.4}$$

a point of which defines an almost complex structure J. For each $S \subseteq \{1,2,3,4\}$ there is a corresponding subset

$$Z_S = \{J \in \mathbb{CP}^3 : \nabla J \in \bigoplus_{i \in S} \mathcal{W}_i\}.$$

Recall that the component of ∇J in $\mathcal{W}_4$ is affected by a conformal change in the metric. In the lecture (available on my homepage), I coloured points of a tetrahedron representing (6.4) red, green, blue to measure the respective components in $\mathcal{W}_1, \mathcal{W}_2, \mathcal{W}_3$, and displayed the resulting spectra for the example below. Whilst this was partly light-hearted, it emphasized that the resulting lattice of 16 subsets of (6.4) is subject to non-trivial constraints arising from the non-existence of a Kähler metric on M.

Example. A full description of the classses for a standard metric on the Iwasawa manifold H/Γ has the following key features. The complex Heisenberg group H possesses (almost by definition) a bi-invariant complex structure J_0 that defines an 'origin' in $\mathbb{CP}^3$. The torsion ∇J has zero $\mathcal{W}_4$-component if and only if J lies in the union of distinct hyperplanes F, F' in $\mathbb{CP}^3$, so $Z_{\{1,2,3\}} = F \cup F'$. Let $L = F \cap F' \cong \mathbb{CP}^1$. Then

(i) $F \cup F'$ contains all 15 proper classes, and $J_0 \in F \setminus F'$.
(ii) The set $Z_{\{3,4\}} = Z_{\{3\}}$ of Hermitian structures equals $\{J_0\} \sqcup L$.
(iii) The set $Z_{\{2\}}$ of 'almost Kähler' or compatible symplectic structures is a 3-sphere in F separating J_0 and L.
More details are contained in (Abbena et al., 2001). Further examples of nilmanifolds of dimension 6 with $b_1 = 4$ exhibit isolated Hermitian structures (see the remark after (4.5)), and illustrate the way the homotopy type of a subset Z_S can depend on the choice of metric.

7. Manifolds whose holonomy is not a subgroup of $Sp(2)Sp(1)$

The note (Gray, 1969) helped to enhance interest in a subject that had already been founded by Berger, Bonan, Ishihara and others. It contains the important result that a quaternionic submanifold of a quaternion-Kähler (or hyperkähler) manifold is totally geodesic. Whilst this put paid to hopes of developing a theory of quaternionic submanifolds, it took many years to realize that the most appropriate way of generating new quaternionic manifolds is instead a quotient construction (Hitchin et al., 1987).

A quaternion-Kähler manifold of dimension $4n$ can be characterized by the existence of a parallel 4-form with stabilizer $Sp(n)Sp(1)$. It is known that closure of the 4-form Ω implies covariant constancy of Ω in dimension $4n \geqslant 12$ (Swann, 1989), but this has left an intriguing open question in dimension 8. Below, we establish the existence of a compact 8-manifold with a closed non-parallel 4-form with stabilizer $Sp(2)Sp(1)$. This is a quaternionic analogue of the symplectic manifolds mentioned in the previous section that do not admit a Kähler metric.

Motivation comes from the (possibly) more familiar theory of *Spin* 7 holonomy. Consider the 2-forms

$$\begin{aligned} \omega_1 &= e^{13} + e^{57} + e^{24} + e^{68}, \\ \omega_2 &= e^{15} - e^{37} + e^{26} - e^{48}, \\ \omega_3 &= e^{17} + e^{35} + e^{28} + e^{46}. \end{aligned} \tag{7.1}$$

on $\mathbb{R}^8$, whose quaternionic structure is defined by first dividing the coordinates into 'odd' and 'even'. If we set

$$\Omega_\pm = \omega_1 \wedge \omega_1 + \omega_2 \wedge \omega_2 \pm \omega_3 \wedge \omega_3, \tag{7.2}$$

then Ω_+ has stabilizer $Sp(2)Sp(1)$, and Ω_- has stabilizer *Spin* 7 (Bryant and Harvey, 1989).

It is instructive to carry out 'dimensional reduction' by regarding the 1-forms e^7, e^8 as constant. Let

$$\sigma = -e^{12} + e^{34} + e^{56}. \tag{7.3}$$

Then

$$\tfrac{1}{2}\Omega_- = \tfrac{1}{2}\sigma^2 + \phi \wedge e^7 + \psi \wedge e^8 - \sigma \wedge e^{78},$$

where

$$\phi + i\psi = (e^1 + ie^2) \wedge (e^3 + ie^4) \wedge (e^5 + ie^6).$$

The equation $d\Omega_- = 0$ is therefore satisfied on $\mathbb{R}^6 \times \mathbb{R}^2$ or a compact manifold of the form $M^6 \times T^2$ if M^6 has a symplectic form σ and certain compatible closed 3-forms ϕ, ψ. The stabilizer of both ϕ and ψ is the same subgroup $SL(3,\mathbb{C})$ of $GL(6,\mathbb{R})$;

such structures have been investigated in (Hitchin, 2001). Since $\sigma\wedge\phi=0=\sigma\wedge\psi$, the 3-form $\phi+i\psi$ determines a complex structure on M for which σ has type $(1,1)$. It follows that M is necessarily Kähler and so Calabi-Yau (or flat). In this case, $M\times T^2$ has holonomy in $SU(3)\times\{e\}\subset Spin\,7$.

The 'plus' case in (7.2) yields more flexibility:

Theorem 7.4 *There exists a closed 4-form with stabilizer $Sp(2)Sp(1)$ on a compact nilmanifold of the form $M^6\times T^2$. The associated Riemannian metric g is reducible and is not therefore quaternion-Kähler.*

Proof. By analogy to (7.3), let

$$\tau=e^{12}+e^{34}+e^{56}.$$

Then

$$\tfrac{1}{2}\Omega_+=-\tfrac{1}{2}\tau^2+\alpha\wedge e^7+\beta\wedge e^8-\tau\wedge e^{78},$$

where

$$\begin{aligned}\alpha&=3e^{135}+e^{146}+e^{236}+e^{245},\\ \beta&=3e^{246}+e^{235}+e^{136}+e^{145}.\end{aligned}\tag{7.5}$$

As before, we seek a 6-manifold M with a symplectic form τ and closed 3-forms α,β, which satisfy $\tau\wedge\alpha=0=\tau\wedge\beta$. This time, the stabilizer of each of α,β is isomorphic to $SL(3,\mathbb{R})\times SL(3,\mathbb{R})$ but (because of the factor 3) these stabilizers do not coincide. Indeed, the overall structure group $Sp(2)Sp(1)\cap SO(6)$ of M is $SO(3)$ acting diagonally on $\mathbb{R}^3\oplus\mathbb{R}^3$.

Take M to be a 6-dimensional nilmanifold associated to the Lie algebra $\mathfrak{g}=\langle e^2\rangle\oplus\langle e^1,e^3,e^4,e^5,e^6\rangle$ determined by the relations

$$\begin{cases} de^i=0, & i=1,2,3,5,\\ de^4=e^{15}, \\ de^6=e^{13}.\end{cases}\tag{7.6}$$

The fact that the structure constants are rational guarantees the existence of a lattice Γ in the associated Lie group G for which $M=G/\Gamma$ is compact.

Equations (7.6) imply that $d\tau=0$. Moreover, of the 8 simple 3-forms in (7.5), all are closed except e^{246}. Let $\widehat{\Omega}$ be the 4-form obtained from Ω by substituting

$$\begin{aligned}e^1&\leftarrow e^1+\sqrt{3}e^2,\\ e^3&\leftarrow e^3-\sqrt{3}e^4.\end{aligned}\tag{7.7}$$

The stabilizer of $\widehat{\Omega}$ is still isomorphic to $Sp(2)Sp(1)$, though the 1-form e^2 will no longer be covariant constant relative to the new metric, leaving only $\langle e^7,e^8\rangle$ as a direct summand. Observe that (7.7) leaves τ unchanged. Its effect on (1) is given by

$$\begin{aligned}\alpha&\leftarrow\alpha-9e^{245}-3\sqrt{3}e^{145}+3\sqrt{3}e^{235},\\ \beta&\leftarrow\beta-3e^{246}-\sqrt{3}e^{146}+\sqrt{3}e^{236}.\end{aligned}$$

The offending term $3e^{246}$ has been eliminated from α at the expense of adding only closed 3-forms. It follows that $\widehat{\Omega}$ is closed.

QED

Note that the tensor $\widehat{\Omega}$ is irrational relative to a basis for which the structure constants of $\mathfrak{g}$ are rational. The fact that $\widehat{\Omega}$ is not parallel also follows from verifying that the ideal generated by (7.1) is not closed under exterior differentiation. A similar reduction to 7 (rather than 6) dimensions will allow such structures to be built up from 7-manifolds with a certain type of G_2 structure arising from the subgroup $Sp(2)Sp(1) \cap SO(7)$ isomorphic to $SO(4)$.

Acknowledgments. Untold thanks are due to the Organizing Committee who worked so hard to make the conference in Bilbao a success. I am also grateful to Mary Gray and Wolfram Research for their generosity in providing me with versions of Mathematica to use in my lecture and beyond.

References

Abbena, E., Garbiero, S., and Salamon, S. (2001). Almost Hermitian geometry of 6-dimensional nilmanifolds. *Ann. Sc. Norm. Sup.*, 30:147–170.

Apostolov, V., Gauduchon, P., and Grantcharov, G. (1999). Bihermitian structures on complex surfaces. *Proc. London. Math. Soc.*, 79:414–428.

Bär, C. (1993). Killing spinors and holonomy. *Commun. Math. Phys.*, 154:509–521.

Baum, H., Friedrich, T., Grunmewald, R., and Kath, I. (1990). *Twistor and Killing Spinors on Riemannian manifolds.* Seminarbericht 108. Humboldt-Universität, Berlin.

Belgun, F. and Morioanu, A. (2000). Nearly-Kähler manifolds with reduced holonomy. preprint.

Benson, C. and Gordon, C. (1988). Kähler and symplectic structures on nilmanifolds. *Topology*, 27:513–518.

Berger, M. (1961). Les variétés riemanniennes homogène normales simplement conexes à courbure strictemente positive. *Ann. Sc. Norm. Sup. Pisa*, 15:179–246.

Bishop, R. and Crittenden, R. (1964). *Geometry of Manifolds.* Academic Press.

Brown, R. and Gray, A. (1967). Vector cross products. *Comment. Math. Helv.*, 42:222–236.

Bryant, R. (1985). Lie groups and twistor spaces. *Duke Math. J.*, 52:223–261.

Bryant, R. (1987). Metrics with exceptional holonomy. *Annals of Math.*, 126:525–576.

Bryant, R. and Harvey, R. (1989). Submanifolds in hyper-Kähler geometry. *J. Amer. Math. Soc.*, 2:1–31.

Burstall, F. and Guest, M. (1997). Harmonic 2-spheres in compact symmetric spaces, revisited. *Math. Ann.*, 309:541–572.

Burstall, F. and Rawnsley, J. (1990). *Twistor theory for Riemannian Symmetric Spaces*, volume 1424 of *Lect. Notes Math.* Springer.

Carpenter, P., Gray, A., and Willmore, T. (1982). The curvature of Einstein symmetric spaces. *Quart. J. Math.*, 33:45–64.

Console, S. and Fino, A. (2001). Dolbeault cohomology of compact nilmanifolds. To appear in Transformation Groups.

Cordero, L., Fernández, M., and Gray, A. (1986). Symplectic manifolds with no Kähler structure. *Topology*, 25:375–380.

Cordero, L., Fernández, M., and Gray, A. (1991). The Frölicher spectral sequence for compact nilmanifolds. *Illinois Math. J.*, 35:56–67.

Cordero, L., Fernández, M., Gray, A., and Ugarte, L. (1997). Nilpotent complex structures on compact nilmanifolds. *Rend. Circolo Mat. Palermo*, 49 suppl.:83–100.

Cordero, L., Fernández, M., Gray, A., and Ugarte, L. (2000). Compact nilmanifolds with nilpotent complex structures: Dolbeault cohomology. *Trans. Amer. Math. Soc.*, 352:5405–5433.

Eells, J. and Wood, J. (1983). Harmonic maps from surfaces to complex projective spaces. *Adv. in Math.*, 49:217–263.

Falcitelli, L., Farinola, A., and Salamon, S. (1994). Almost-Hermitian geometry. *Differ. Geom. Appl.*, 4:259–282.

Fernández, M. and Gray, A. (1982). Riemannian manifolds with structure group G_2. *Ann. Mat. Pura Appl.*, 32:19–45.

Fernández, M., Gray, A., and Morgan, J. (1991). Compact symplectic manifolds with free circle actions and Massey products. *Michigan Math. J.*, 38:271–283.

Fernández, M. and Ugarte, L. (1998). Dolbeault cohomology for G_2-structures. *Geom. Dedicata*, 70:57–86.

Friedrich, T., Kath, I., Morioanu, A., and Semmelmann, U. (1998). On nearly parallel G_2 structures. *J. Geom. Phys.*, 27:155–177.

Galicki, K. and Salamon, S. (1996). Betti numbers of 3-sasakian manifolds. *Geom. Dedicata*, 63:45–68.

Goze, M. and Khakimdjanov, Y. (1996). *Nilpotent Lie Algebras, Mathematics and its Applications 361.* Kluwer.

Gray, A. (1965). Minimal varieties and almost Hermitian submanifolds. *Michigan Math. J.*, 12:273–287.

Gray, A. (1969). A note on manifolds whose holonomy group is a subgroup of $Sp(n)\cdot Sp(1)$. *Michigan J. Math.*, 16:125–128.

Gray, A. (1971). Weak holonomy groups. *Math. Z.*, 123:290–300.

Gray, A. (1972). Riemannian manifolds with geodesic symmetries of order 3. *J. Differ. Geom.*, 7:343–369.

Gray, A. (1973). The volume of a small geodesic ball in a Riemannian manifold. *Michigan Math. J.*, 20:329–344.

Gray, A. (1976a). Curvature identities for Hermitian and almost Hermitian manifolds. *Tôhoku Math. J.*, 28:601–612.

Gray, A. (1976b). The structure of nearly Kähler manifolds. *Math. Ann.*, 223:233–248.

Gray, A. (1977). Compact Kähler manifolds with nonnegative sectional curvature. *Invent. Math.*, 41:33–43.

Gray, A. (1978). Einstein-like manifolds which are not Einstein. *Geom. Dedicata*, 7:259–280.

Gray, A. (1985). Volumes of tubes about complex submanifolds of complex projective space. *Trans. Amer. Math. Soc.*, 291:437–449.

Gray, A. (1990). *Tubes.* The Advanced Book Program. Addison-Wesley.

Gray, A. (1998). *Modern Differential Geometry of Curves and Surfaces with Mathematica.* CRC Press, second edition.

Gray, A. and Hervella, L. (1980). The sixteen classes of almost Hermitian manifolds. *Ann. Mat. Pura e Appl.*, 282:1–21.

Gray, A. and Pinsky, M. (1983). The mean exit time from a small geodesic ball in a Riemannian manifold. *Bull. des Sciences Math.*, 107:345–370.

Gray, A. and Vanhecke, L. (1979). Riemannian geometry as determined by the volumes of small geodesic balls. *Acta Math.*, 142:157–198.

Günther, P. (1960). Einige sätze über das volumenelement eines riemannschen raumes. *Math. Deberecen*, 7:78–93.

Hasegawa, K. (1989). Minimal models of nilmanifolds. *Proc. Amer. Math. Soc.*, 106:65–71.

Hitchin, N. (2001). Geometry of three-forms in six and seven dimensions. To appear in J. Differ. Geom.

Hitchin, N., Karlhede, A., Lindström, U., and Roček, M. (1987). Hyperkähler metrics and supersymmetry. *Commun. Math. Phys.*, 108:535–589.

Hotelling, H. (1939). Tubes and spheres in n-spaces and a class of statistical problems. *Amer. J. Math.*, 61:440–460.

Joyce, D. D. (2000). *Compact manifolds with special holonomy.* Oxford University Press.

Kowalski, O. (1980). Additive volume invarinats of riemannian manifolds. *Acta Math.*, 145:205–225.

Lichnerowicz, A. (1980). Applications harmoniques et variétés kählériennes. *Symposia Math.*, 3:341–402.

Malcev, A. (1962). On a class of homogeneous spaces. *Amer. Math. Soc. Translations, Series 1*, 9:276–307.

Nomizu, K. (1954). On the cohomology of compact homogeneous spaces of nilpotent Lie groups. *Annals of Math.*, 59:531–538.

Pedersen, H. and Poon, Y.-S. (2001). Lie groups, Kodaira manifolds and complex deformations. Preprint.

Pittie, H. (1989). The nondegeneration of the Hodge-de Rham spectral sequence. *Bull. Amer. Math. Soc.*, 20:19–22.

Pontecorvo, M. (1997). Complex structures on Riemannian 4-manifolds. *Math. Ann.*, 309:159–177.

Reyes-Carrion, R. (1998). A generalization of the notion of instanton. *Differ. Geom. Appl.*, 8.

Salamon, S. (1985). *Harmonic and holomorphic maps*, volume 1164 of *Lect Notes Math.*, pages 161–224. Springer.

Salamon, S. (1989). *Riemannian geometry and holonomy groups.* Pitman Research Notes in Mathematics 201. Longman.

Salamon, S. (2001). Complex structures on nilpotent Lie algebras. *J. Pure Applied Algebra*, 157:311–333.

Swann, A. (1989). *Quaternionic Kähler geometry and the fundamental 4-form*, pages 165–174. ULDM Publ.

Tricerri, F. and Vanhecke, L. (1981). Curvature tensors on almost Hermitian manifolds. *Trans. Amer. Math. Soc.*, 267:365–398.

Weyl, H. (1939). On the volume of tubes. *Amer. J. Math.*, 61:461–472.

Wolf, J. and Gray, A. (1968). Homogeneous spaces defined by Lie group automorphisms I. *J. Differ. Geom.*, 2:77–114.

Wolfram, S. (1999). *The Mathematica Book.* Wolfram Media and Cambridge University Press, fourth edition.

Department of Mathematics, Politecnico di Torino, Corso Duca degli Abruzzi 24, 10129 Torino, Italy

E-mail address: salamon@calvino.polito.it

Contemporary Mathematics
Volume **288**, 2001

Technical Computing with *Mathematica*

Paul Wellin

ABSTRACT. It is often said that the release of *Mathematica* marked the beginning of modern technical computing. Ever since the 1960s individual packages had existed for specific numerical, algebraic, graphical and other tasks. But the visionary concept of *Mathematica* was to create once and for all a single system that could handle all the various aspects of technical computing in a coherent and unified way. The key intellectual advance that made this possible was the invention of a new kind of symbolic computer language that could for the first time manipulate the very wide range of objects involved in technical computing using only a fairly small number of basic primitives.

Using *Mathematica* as a Computational Tool

Simple computations

Here are some simple computations much like you would do with a calculator. But with *Mathematica*, you are not limited in the size of the computation.

2^{100}

```
1267650600228229401496703205376
```

2^{10000}

```
1995063116880758384883742162683585083823496831886192454852008949852943883
  0221946631919961684036194597899331129423209124271556491349413781117593785
  9320963239578557300467937945267652465512660598955205500869181933115425086
  0846061810468550907486608962488809048989483800925394163325785062156830947
  3902556912388065225096643874441046759871626985453222868538161694315775629
  6407628368807607322285350916414761839563814589694638994108409605362678210
  6462142733339403652556564953060314268023496940033593431665145929777327966
  5775606172582031407994198179607378245683762280037302885487251900834464581
  4546505579296014148339216157345881392570953797691192778008269577356744441
  2306201875783632550272832378927071037380286639303142813324140162419567169
  0574061419654342324638801248856147305207431992259611796250130992860241708
  3408076059323201612684922884962558413128440615367389514871142563
```

2000 *Mathematics Subject Classification.* Primary 00–01.

Thanks to Luis Carlos de Andres for help in preparation of the manuscript for this paper.

```
315111089745514203313820202931640957596464756010405845841566072044962867
```

This gives a numerical approximation of the previous result. We see that this number has over 3,000 digits.

```
N[%]
```

1.99506×10^{3010}

When given exact input, *Mathematica* will try to return exact output.

Sin$\left[\frac{\pi}{3}\right]$

$\frac{\sqrt{3}}{2}$

This computes an indefinite integral.

$$\int \frac{1}{1-x^3}\,dx$$

$$\frac{\text{Tan}^{-1}\left(\frac{2x+1}{\sqrt{3}}\right)}{\sqrt{3}} - \frac{1}{3}\text{Log}(x-1) + \frac{1}{6}\text{Log}\left(x^2+x+1\right)$$

Here is a polynomial factoring problem.

```
Factor[x^99 + y^99]
```

$$(x+y)\,(x^2 - y\,x + y^2)\,(x^6 - y^3 x^3 + y^6)\,(x^{10} - y\,x^9 + y^2 x^8 - y^3 x^7 + y^4 x^6 - y^5 x^5 + y^6 x^4 - y^7 x^3 + y^8 x^2 - y^9 x + y^{10})\,(x^{20} + y\,x^{19} - y^3 x^{17} - y^4 x^{16} + y^6 x^{14} + y^7 x^{13} - y^9 x^{11} - y^{10} x^{10} - y^{11} x^9 + y^{13} x^7 + y^{14} x^6 - y^{16} x^4 - y^{17} x^3 + y^{19} x + y^{20})\,(x^{60} + y^3 x^{57} - y^9 x^{51} - y^{12} x^{48} + y^{18} x^{42} + y^{21} x^{39} - y^{27} x^{33} - y^{30} x^{30} - y^{33} x^{27} + y^{39} x^{21} + y^{42} x^{18} - y^{48} x^{12} - y^{51} x^9 + y^{57} x^3 + y^{60})$$

You can factor very large integers directly. Pierre de Fermat thought that all numbers of this form were prime.

```
FactorInteger[2^500 + 1]
{{17, 1}, {401, 1}, {4001, 1}, {61681, 1},
  {340801, 1}, {1074001, 1}, {2020001, 1}, {2787601, 1},
  {22624001, 1}, {3173389601, 1}, {1481124532001, 1},
  {8877945148742945001146041439025147034098690503591013177336356694416517527310181938001, 1}}
```

You can solve equations symbolically.

```
Solve[x^2 + 2a == x, x]
```

$$\left\{\left\{x \to \frac{1}{2}\left(1 - \sqrt{1 - 8a}\right)\right\}, \left\{x \to \frac{1}{2}\left(\sqrt{1 - 8a} + 1\right)\right\}\right\}$$

Some equations do not admit simple closed form solutions.

```
Solve[Sin[x^2 + Cos[√x]] == e^(-0.2x), x]

Solve :: tdep: The equations appear to involve the variables to be solved for in
an essentially non-algebraic way.
Solve[Sin(x^2 + Cos(√x)) == e^(-0.2 x), x]
```

We can of course solve such equations numerically. First, let's find where the roots occur.

```
Plot[Sin[x^2 + Cos[√x]] - e^(-0.2x), {x, -2, 2}];
```

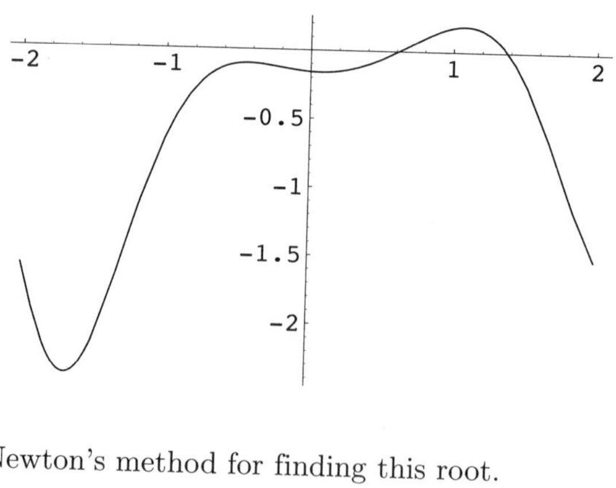

Then we can use Newton's method for finding this root.

```
FindRoot[Sin[x^2 + Cos[√x]] == e^(-0.2x), {x, 0.5}]
```

```
{x → 0.61392}
```

Here is a simple computation on a matrix with some symbolic elements.

$$\mathrm{Det}\left[\begin{pmatrix} \lambda & 0 & -1 \\ 2 & 1-\lambda & 1 \\ 0 & 4 & 1+\lambda \end{pmatrix}\right]$$

$$-\lambda^3 - 3\lambda - 8$$

This computes the inverse of this matrix.

$$\mathrm{Inverse}\left[\begin{pmatrix} \lambda & 0 & -1 \\ 2 & 1-\lambda & 1 \\ 0 & 4 & 1+\lambda \end{pmatrix}\right]$$

$$\begin{pmatrix} \frac{-\lambda^2-3}{-\lambda^3-3\lambda-8} & -\frac{4}{-\lambda^3-3\lambda-8} & \frac{1-\lambda}{-\lambda^3-3\lambda-8} \\ \frac{-2\lambda-2}{-\lambda^3-3\lambda-8} & \frac{\lambda^2+\lambda}{-\lambda^3-3\lambda-8} & \frac{-\lambda-2}{-\lambda^3-3\lambda-8} \\ \frac{8}{-\lambda^3-3\lambda-8} & -\frac{4\lambda}{-\lambda^3-3\lambda-8} & \frac{\lambda-\lambda^2}{-\lambda^3-3\lambda-8} \end{pmatrix}$$

This generates a 100 × 100 matrix consisting of random real numbers uniformly distributed on the interval [0, 1].

```
data = Table[Random[], {100}, {100}];
```

Here is a plot at the eigenvalues of this matrix. We first need to take the absolute value of the (complex) eigenvalues to plot them in the real plane.

```
ListPlot[Abs[Eigenvalues[data]]];
```

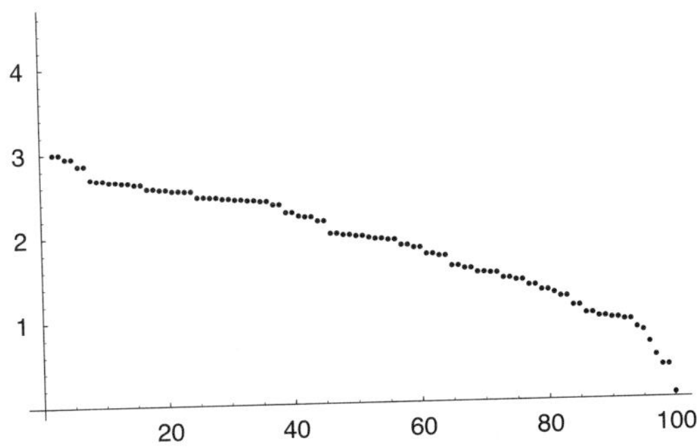

More matrix manipulations

This creates a symbolic 3×3 matrix using the `Table` function.

```
mat1 = Table[a_{i,j}, {i, 3}, {j, 3}]
```

$$\begin{pmatrix} a_{1,1} & a_{1,2} & a_{1,3} \\ a_{2,1} & a_{2,2} & a_{2,3} \\ a_{3,1} & a_{3,2} & a_{3,3} \end{pmatrix}$$

To display the matrix in a standard form using lists, use `StandardForm`.

```
StandardForm[mat1]
```

$\{\{a_{1,1}, a_{1,2}, a_{1,3}\}, \{a_{2,1}, a_{2,2}, a_{2,3}\}, \{a_{3,1}, a_{3,2}, a_{3,3}\}\}$

This selects the first row.

```
mat1[[1]]
```

$\{a_{1,1}, a_{1,2}, a_{1,3}\}$

This selects the first column.

```
mat1[[All, 1]]//MatrixForm
```

$$\begin{pmatrix} a_{1,1} \\ a_{2,1} \\ a_{3,1} \end{pmatrix}$$

Subarrays can be extracted using ranges as selection criteria.

```
mat1[[{1, 2, 3}, {1, 2}]]//MatrixForm
```

$$\begin{pmatrix} a_{1,1} & a_{1,2} \\ a_{2,1} & a_{2,2} \\ a_{3,1} & a_{3,2} \end{pmatrix}$$

For larger matrices, the `Range` function simplifies setting up the selection.

```
mat1[[Range[1, 3], {1, 2}]]//MatrixForm
```

$$\begin{pmatrix} a_{1,1} & a_{1,2} \\ a_{2,1} & a_{2,2} \\ a_{3,1} & a_{3,2} \end{pmatrix}$$

Here is the singular value decomposition for a non-singular 3×3 matrix.

```
mat = Table[Random[Integer, {-10, 10}], {3}, {3}]
```

$$\begin{pmatrix} 2 & 5 & -1 \\ 8 & -6 & 3 \\ 10 & 6 & -9 \end{pmatrix}$$

```
SingularValues[N[mat]]
```

```
({-0.268064, -0.0931002, -0.958892}   {0.223977, -0.97407, 0.0319595}
                                      {-0.937004, -0.206203, 0.281966}

              15.3344                                  10.5996
                                                  2.91623

 {-0.708853, -0.42617, 0.562055}      {-0.662763, 0.675127, -0.323958}
                                      {-0.241397, -0.602148, -0.761016})
```

Here is a convenient way to pick out the pieces of the result.

```
u, md, v = %
```

```
({-0.268064, -0.0931002, -0.958892}   {0.223977, -0.97407, 0.0319595}
                                      {-0.937004, -0.206203, 0.281966}

              15.3344                                  10.5996
                                                  2.91623

 {-0.708853, -0.42617, 0.562055}      {-0.662763, 0.675127, -0.323958}
                                      {-0.241397, -0.602148, -0.761016})
```

This gives something close to the original matrix.

```
Transpose[u].DiagonalMatrix[md].v
```

$$\begin{pmatrix} 2. & 5. & -1. \\ 8. & -6. & 3. \\ 10. & 6. & -9. \end{pmatrix}$$

For additional matrix manipulation functions, see the package documentation for `LinearAlgebra‘MatrixManipulation‘`.

Importing and analyzing data

You can import data from almost any application and use *Mathematica*'s visualization or data analysis tools right away. In this example, we export data from an Excel spreadsheet, import it into *Mathematica*, plot it, and then do a convolution on the data.

```
data = ReadList["bessel.data.prn"]
{0, 0.020019, 0.175637, 0.104957, 0.198025, 0.234222, 0.247778, 0.296866,
  0.43652, 0.45331, 0.50889, 0.636129, 0.551111, 0.605936, 0.543539, 0.729859,
  0.748445, 0.65928, 0.661857, 0.671335, 0.618094, 0.722142, 0.651315,
  0.581076, 0.617591, 0.645583, 0.654417, 0.490475, 0.469613, 0.496912,
  0.527241, 0.348551, 0.421408, 0.264078, 0.269537, 0.192639, 0.177656,
  0.190553, 0.0811921, 0.0383231, 0.0225262, -0.0603398, 0.0399936,
  -0.0029631, -0.159063, -0.142489, -0.170817, -0.145982, -0.143965,
  -0.163611, -0.211774, -0.236665, -0.229993, -0.295537, -0.194854,
  -0.263923, -0.154821, -0.289381, -0.221914, -0.24302, -0.148804,
  -0.12682, -0.18099, -0.19041, -0.111519, -0.0374771, -0.022233, -0.0181579,
  -0.0116075, 0.018655, 0.0364624, 0.191569, 0.173772, 0.203312,
  0.251014, 0.214963, 0.176763, 0.260512, 0.19865, 0.225273, 0.274076,
  0.290983, 0.362711, 0.413907, 0.374298, 0.377269, 0.31335, 0.386157,
  0.281709, 0.426396, 0.364523, 0.251891, 0.360686, 0.295857, 0.23767,
  0.19196, 0.340901, 0.268989, 0.18545, 0.19783, 0.206492, 0.0716381}

ListPlot[data];
```

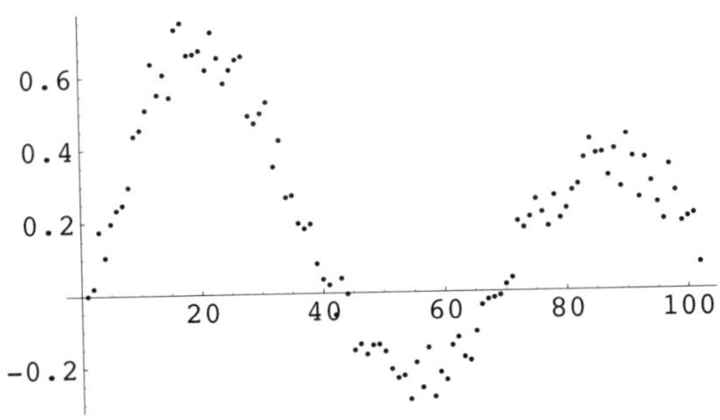

Here is a Gaussian kernel:

```
kern = Table[Exp[-n^2/100]/Sqrt[2.π], {n, -10, 10}];
```

This convolves the kernel with the data.

```
ListConvolve[kern, data, {-1, 1}];
```

The result is a smoothed version of the data.

```
ListPlot[%];
```

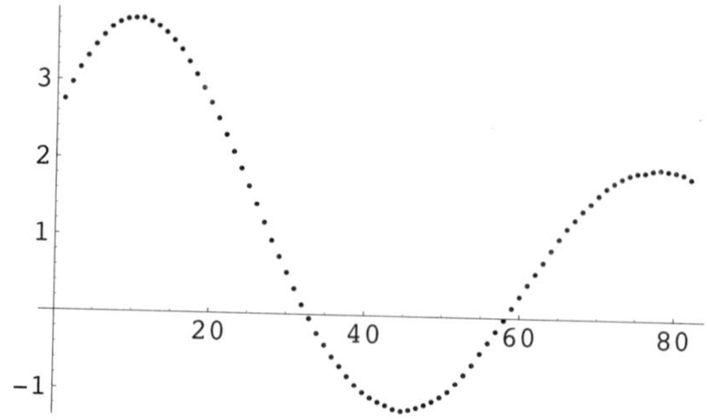

Differential equations

After entering the equations, NDSolve automatically selects the optimal algorithm to use or lets you select it, if you know in advance. *Mathematica*'s unique concept of using interpolation functions instead of simple lists of numbers to represent numerical solutions to differential equations makes it easy to reuse the results of *Mathematica*'s NDSolve. All results can be used like any other function, namely integrated, differentiated, or included in more complex calculations.

Here are some implementation notes about NDSolve:

• By default or with Method→Automatic, NDSolve switches between a non-stiff Adams method and a stiff Gear method.
• Based on LSODE.
• Adams: implicit Adams method with order between 1 and 12.
• Gear: backward difference formula method with order between 1 and 5.
• RungeKutta: Fehlberg order 4–5 Runge-Kutta method for non-stiff equations.
• For linear boundary value problems the Gel'fand-Lokutsiyevskii chasing method is used.
• For 1+1-dimensional PDEs the method of lines is used.
• The differential equations in NDSolve can involve complex numbers.

Mathematica's adaptive algorithms and arbitrary precision control guarantee accurate results, without roundoff errors or spurious solutions due to numerical errors. Re-checking the results against the differential equation itself is as easy as substituting the resulting interpolation function in the equation itself.

Example: Solving the VanDerPol equation with *Mathematica*

This solves the differential equation $x''(t)-0.2\big(1-x^2(t)\big)x'(t)+x(t)=0$ with the given initial conditions.

```
soln = NDSolve[{x''[t] - 0.2 (1 - x[t]^2) x'[t] + x[t] == 0,
     x[0] == 1, x'[0] == 0}, x, {t, 0, 30}]
{{x → InterpolatingFunction[(0.  30.  ), <>]}}
```

The resulting InterpolationFunction can be used like any other Mathematica function, namely integrated, differentiated, plotted, transformed...

```
Plot[x[t]/.soln[[1]], {t, 0, 30}];
```

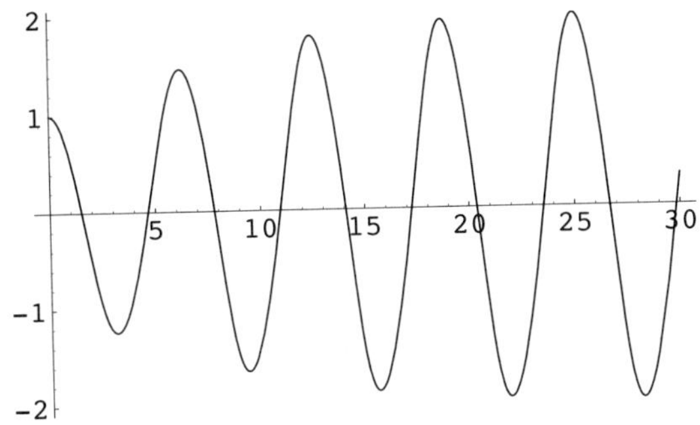

You can also integrate the solution over a domain.

```
NIntegrate[Evaluate[x[t]/.soln], {t, 11, 14}]
```

```
{3.57385}
```

Or evaluate the derivative of the solution,

```
D[Evaluate[x[t]/.soln], t]
```

```
{InterpolatingFunction[(0.   30. ), <>][t]}
```

...and and then plot the derivative.

```
Plot[Evaluate[%], {t, 0, 30}];
```

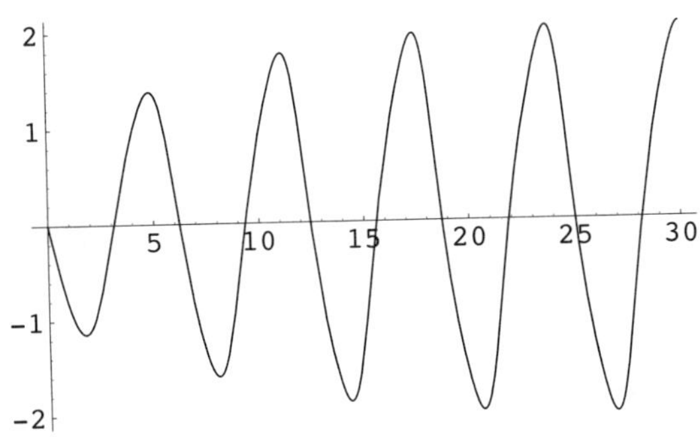

Here is a plot of the original function evaluated at the solution. This gives a good picture of the numerical error in the solution.

```
Plot[N[Evaluate[x''[t] - 0.2 (1 - x[t]^2) x'[t] + x[t]/.soln[[1]]], 3],
  {t, 0, 30}, PlotRange → {-0.02, 0.02}];
```

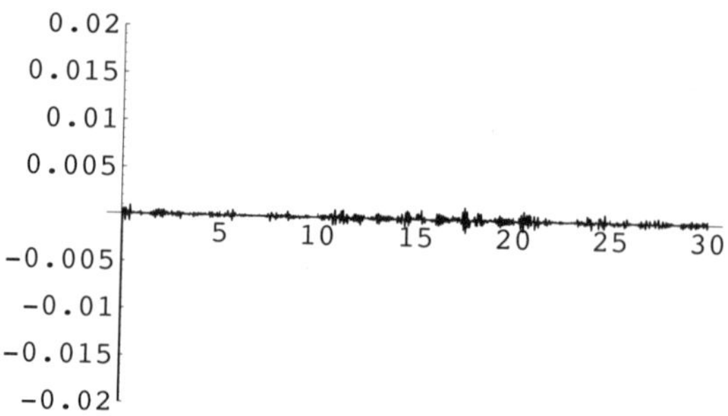

If needed, you could ramp up the number of digits used in internal computations to get higher precision.

```
soln24 = NDSolve[{(x')'[t] - 0.2 (1 - x[t]^2) x'[t] + x[t] == 0, x[0] == 1,
    x'[0] == 0}, x, {t, 0, 30}, WorkingPrecision → 24, MaxSteps → 10000]
{{x → InterpolatingFunction[(0.   30.0000000000000000000000 ), <>]}}

Plot[N[Evaluate[x''[t] - 0.2 (1 - x[t]^2) x'[t] + x[t] /.soln24[[1]]], 3],
    {t, 0, 30}, PlotRange → {- 10^-8, 10^-8}];
```

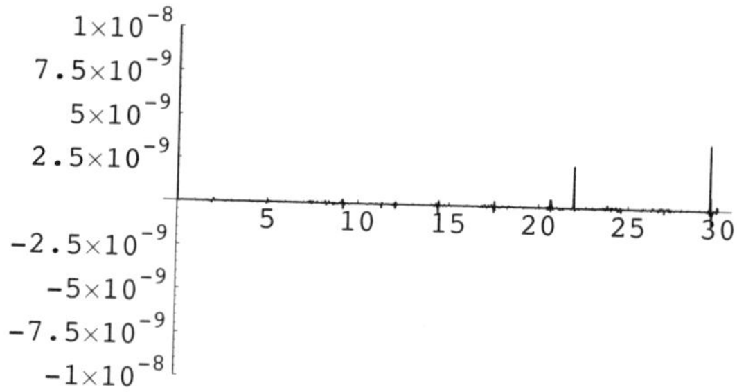

The following commands compare a known exact solution of a simple differential equation with solutions computed with different values of `WorkingPrecision`.

```
s16 = y/.
    First[NDSolve[{y''[x] + y[x] == 0, y[0] == 1, y'[0] == 0}, y, {x, 0, 2 π}]];

s24 = y/.First[NDSolve[{y''[x] + y[x] == 0, y[0] == 1, y'[0] == 0},
        y, {x, 0, 2 π}, WorkingPrecision → 24]];

s32 = y/.
  First[NDSolve[{y''[x] + y[x] == 0, y[0] == 1, y'[0] == 0},
   y, {x, 0, 2 π}, WorkingPrecision → 32]];

s40 = y/.
  First[NDSolve[{y''[x] + y[x] == 0, y[0] == 1, y'[0] == 0},
  y, {x, 0, 2 π}, WorkingPrecision → 40, MaxSteps → 2000]];
```

The values of x at which the steps are taken is kept in the third part of the `InterpolatingFunction` object. This is why the `Length` command gives the number of steps.

```
<< Mathematics`Utilities`

ShowTable[{{16, Abs[1 - s16[2π]], Length[s16[[3, 1]]]},
  {24, Abs[1 - s24[2π]], Length[s24[[3, 1]]]},
   {32, Abs[1 - s32[2π]], Length[s32[[3, 1]]]},
   {40, Abs[1 - s40[2π]], Length[s40[[3, 1]]]}},
   ColumnHeadings → {"Working Precision", "Error", "Number of steps"}]
```

Working Precision	Error	Number of steps
16	0.0000212397	47
24	6.87776×10^{-14}	155
32	3.07024×10^{-22}	346
40	1.81835×10^{-30}	1100

Here is a simple phase plot of this differential equation.

```
ParametricPlot[Evaluate[{x[t], x'[t]}/.soln[[1]]],
  {t, 0, 30}, PlotStyle → {Thickness[0.005], RGBColor[1, 0, 0]}];
```

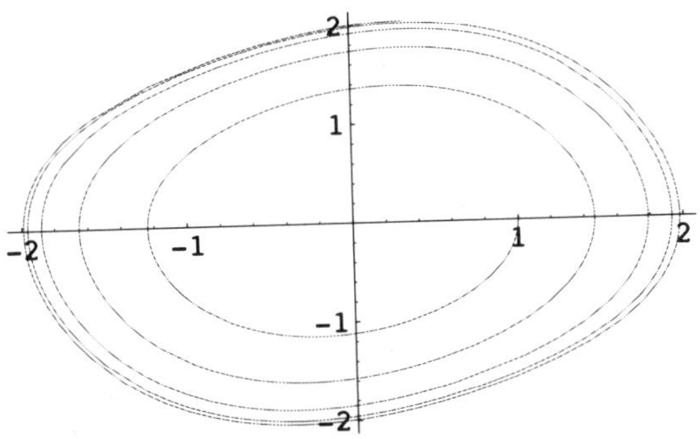

You can calculate and plot the integral over the solution in one step

```
Plot[Evaluate[∫ (x[t]/.soln[[1]]) dt], {t, 0, 30}];
```

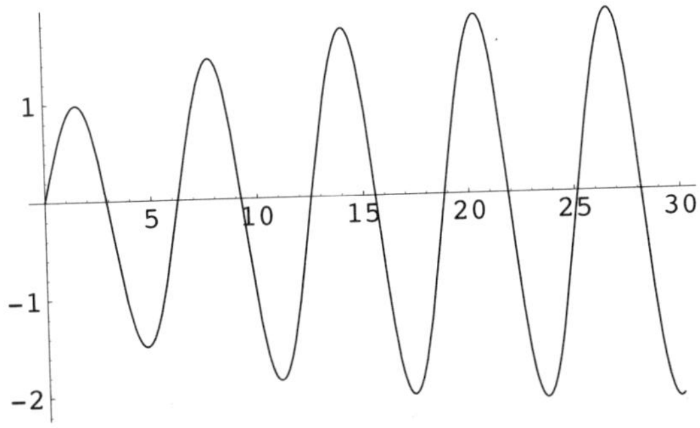

Partial differential equations

This finds the general solution to a simple partial differential equation with two independent variables.

```
DSolve[∂x1 y[x1, x2] + ∂x2 y[x1, x2] == 1/(x1 x2), y[x1, x2], {x1, x2}]
```

$$\left\{\left\{y(x1, x2) \to \frac{-\text{Log}(x1) + \text{Log}(x2) + x1\, c_1[x2 - x1] - x2\, c_1[x2 - x1]}{x1 - x2}\right\}\right\}$$

Numerical solutions work exactly the same, of course. Here, for example we solve the wave equation numerically.

```
res = NDSolve[{∂t,t y[x, t] == ∂x,x y[x, t], y[x, 0] == e^(-x^2),
     y^(0,1)[x, 0] == 0, y[-5, t] == y[5, t]}, y, {x, -5, 5}, {t, 0, 5}]
```

```
{{y → InterpolatingFunction[{{..., -5., 5., ...}, {0., 5.}}, <>]}}
```

Here is a plot of the solution.

```
Plot3D[Evaluate[D[y[x, t], t]/.First[res]],
     x, -5, 5, t, 0, 5, PlotPoints → 30];
```

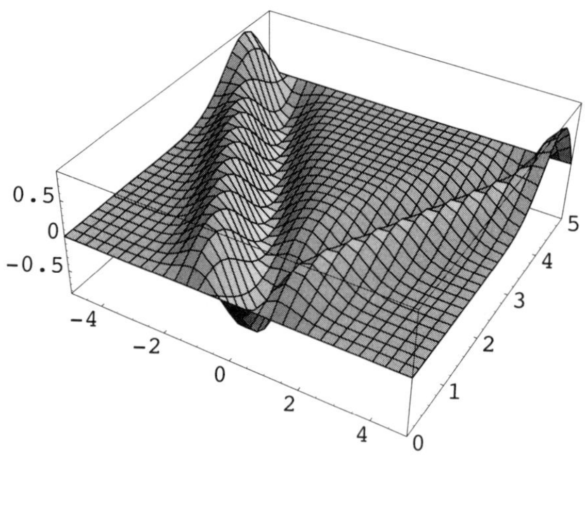

Finally, we make a contour chart of its derivative with respect to t.

```
ContourPlot[Evaluate[∂t y[x, t]/.First[res]],
  {x, -5, 5}, {t, 0, 5}, PlotPoints → 30, Contours → 30,
  ContourLines → False, ColorFunction → Hue];
```

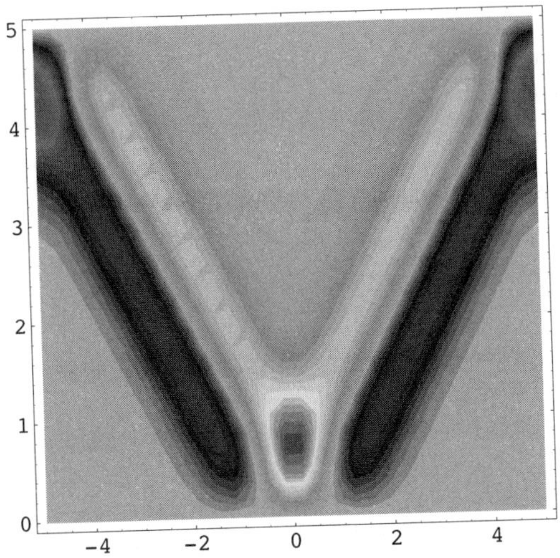

Discrete Fourier transform

```
data = Table[Sin[1/200 30. (2 π n)] + Random[Real, {-1/2, 1/2}], {n, 200}];
```

```
ListPlot[data, PlotJoined → True];
```

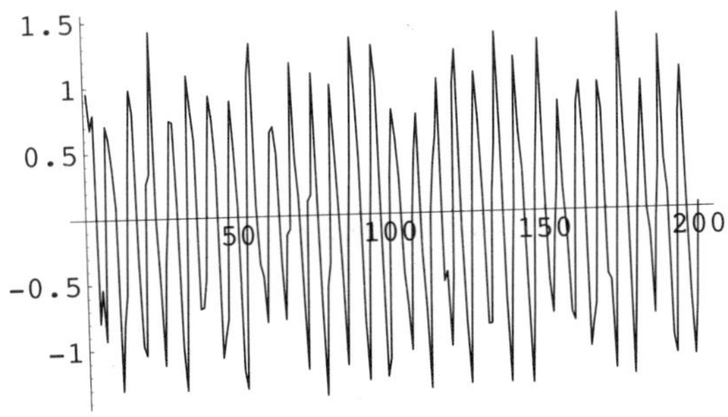


```
ListPlot[Abs[Fourier[data]], PlotJoined → True, PlotRange → All];
```

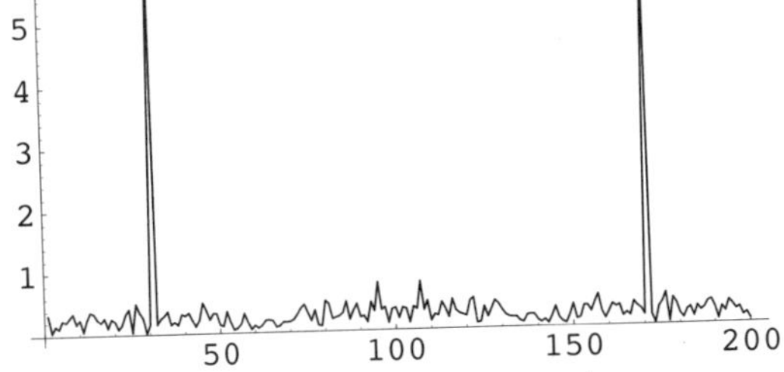

The implementation of the discrete Fourier transform in *Mathematica* works for any dimension and also covers all the different conventions for Fourier transforms. For instance the following computes the discrete Fourier transform of a 500×500 matrix.

```
m = Table[Random[], {500}, {500}];

Fourier[m]; //Timing
{0.99 Second, Null}
```

There are no restrictions such as having to have a size that is a power of two (as in some earlier algorithms in this field), or a tensor of a particular rank.

The *Mathematica* Programming Language

Procedural programming

A variety of programming paradigms are available in the *Mathematica* language. Here are several approaches to one computation—adding up the first 10,000 integers.

First, two procedural approaches:

```
sum = 0;

Do[sum+ = i, {i, 10000}];

sum
50005000

sum = 0;

i = 1;

While[i ≤ 10000, sum+ = i + +];

sum
50005000
```

A functional approach:

```
Apply[Plus, Range[10000]]
50005000
```

The natural way to do this...

$$\sum_{n=1}^{10000} n$$

```
50005000
```

But of course, Gauss knew best!

```
Binomial[10001, 2]
50005000
```

Rule-based programming

With rule-based programming, you can define your own rules for performing transformations.

The following example demonstrates the ease with which you can build quite complicated graphics from the basic building blocks, using some fairly sophisticated programming constructs.

Here are the coordinates for a set of 7 random points in 3-space.

```
coords = Table[Random[Integer, {0, 10}], {7}, {3}]

{{1, 9, 6}, {6, 6, 4}, {1, 8, 1}, {7, 5, 9}, {9, 5, 2}, {5, 4, 3}, {1, 5, 0}}
```

This turns the coordinates into Point graphics objects.

```
points = Map[Point, coords]

{Point[{1, 9, 6}], Point[{6, 6, 4}], Point[{1, 8, 1}],
  Point[{7, 5, 9}], Point[{9, 5, 2}], Point[{5, 4, 3}], Point[{1, 5, 0}]}
```

To connect all the points with lines, we need to generate all the 2-element subsets of this set. We load a combinatorics package that contains a function to make arbitrary subsets.

```
Needs["DiscreteMath`Combinatorica`"]
```

```
?KSubsets

KSubsets[l, k] gives all subsets of set l containing exactly k elements,
ordered lexicographically.  More...
```

Here are the lines connecting each point to every other point.

```
lines = Map[Line, KSubsets[coords, 2]]

{Line[{{1, 9, 6}, {6, 6, 4}}], Line[{{1, 9, 6}, {1, 8, 1}}],
 Line[{{1, 9, 6}, {7, 5, 9}}], Line[{{1, 9, 6}, {9, 5, 2}}],
 Line[{{1, 9, 6}, {5, 4, 3}}], Line[{{1, 9, 6}, {1, 5, 0}}],
 Line[{{6, 6, 4}, {1, 8, 1}}], Line[{{6, 6, 4}, {7, 5, 9}}],
 Line[{{6, 6, 4}, {9, 5, 2}}], Line[{{6, 6, 4}, {5, 4, 3}}],
 Line[{{6, 6, 4}, {1, 5, 0}}], Line[{{1, 8, 1}, {7, 5, 9}}],
 Line[{{1, 8, 1}, {9, 5, 2}}], Line[{{1, 8, 1}, {5, 4, 3}}],
 Line[{{1, 8, 1}, {1, 5, 0}}], Line[{{7, 5, 9}, {9, 5, 2}}],
 Line[{{7, 5, 9}, {5, 4, 3}}], Line[{{7, 5, 9}, {1, 5, 0}}],
 Line[{{9, 5, 2}, {5, 4, 3}}], Line[{{9, 5, 2}, {1, 5, 0}}],
 Line[{{5, 4, 3}, {1, 5, 0}}]}
```

And here is the graphic.

```
network = Show[Graphics3D[{points, lines}]];
```

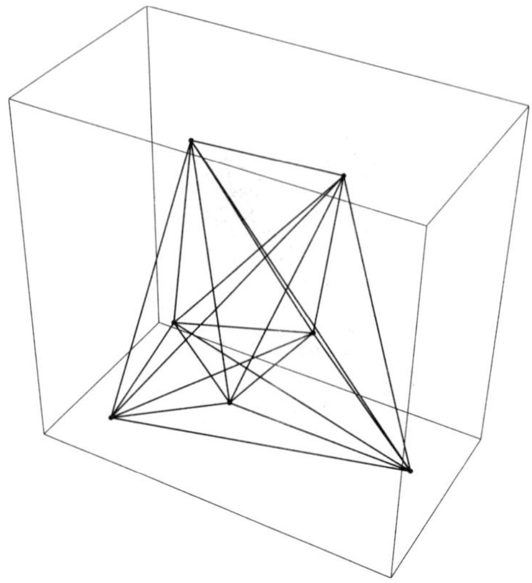

Using a rule-based approach, we can easily make every point red and larger.

```
Show[network/.Point[x_] :> {Hue[0], PointSize[0.05], Point[x]}];
```

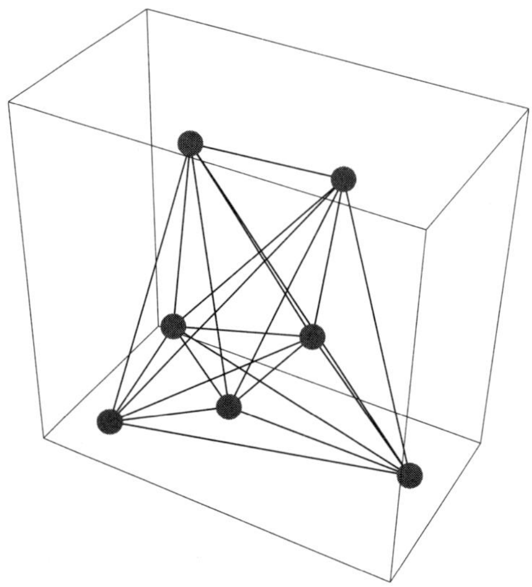

To get a more complicated shape at each point, we first load the Polyhedra package,

```
Needs["Graphics`Polyhedra`"];
```

and then convert each point to an icosahedron and each line to a larger blue line.

```
Show[network/.{Line[x_] :> {Hue[0.7], Thickness[0.007], Line[x]},
        Point[x_] :> {First[Polyhedron[Icosahedron, x, 0.8]]}}];
```

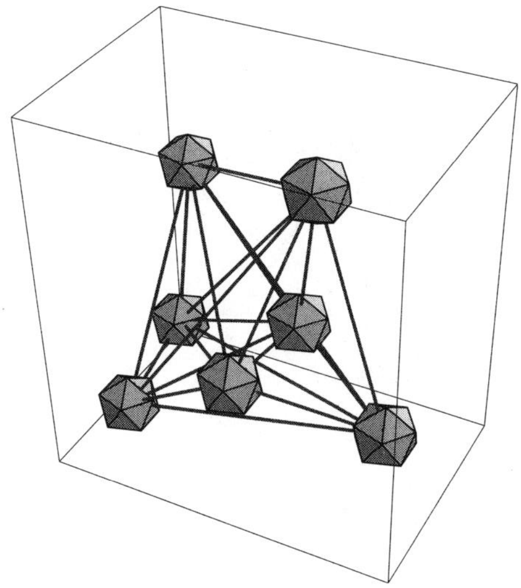

Mathematica as a Technical Environment

Import/Export of graphics and sound

Mathematica 4 introduces the Import and Export commands, which provide a simple mechanism for importing and exporting data in many frequently used formats.

```
$ImportFormats
```

```
{AIFF, AU, BMP, CSV, Dump, DXF, EPS, EPSI, EPSTIFF, Expression, GIF, HDF, JPEG,
  Lines, List, MAT, MathML, MGF, MPS, PBM, PGM, PNM, PSImage, PPM, RawBitmap,
  SND, STL, Table, Text, TIFF, UnicodeText, WAV, Words, XBitmap}
```

```
$ExportFormats
```

```
{AI, AIFF, AU, BMP, CSV, Dump, DXF, EPSI, EPSTIFF, Expression, EPS, GIF, HDF, HTML,
  HTML + MathML, JPEG, Lines, List, MAT, MathML, MGF, MPS, PBM, PCL, PDF, PGM, PICT,
  PNM, PPM, PSImage, RawBitmap, SND, STL, Table, TeX, Text, TIFF, UnicodeText,
  WAV, WMF, Words, XBitmap}
```

These functions also have counterparts named ImportString and ExportString which read and write to string buffers in memory rather than to files. Import and Export can determine the file formats needed by looking at the file name extension (although you can also denote them explicitly).

```
Import["Images\image1.jpg", "JPEG"]
```

```
-Graphics-
```

```
Show[%];
```

```
Plot3D[Sin[x y], {x, 0, π}, {y, 0, π}, PlotPoints → 45, Mesh → False];
```

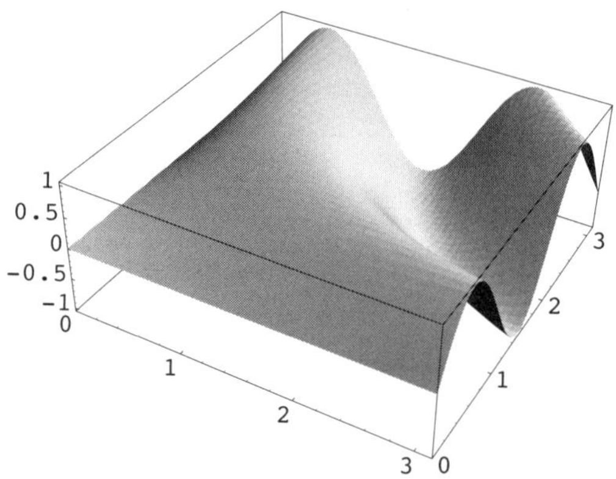


```
Export["plot17.tif", %, "TIFF"]
plot17.tif
```

Web Publishing: Save to HTML and MathML

More and more universities and corporations are using the web as a simple, efficient way to distribute up-to-the-minute information, both inside the organization and to the world. *Mathematica* 4 lets users easily distribute their results over the web through its built-in ability to generate HTML directly from its notebooks. A single menu command is all that's needed to create a full HTML version of your notebook, including graphics and mathematical expressions, ready for uploading to your web server.

Mathematica 4 actually offers two different methods for HTML export. The World Wide Web Consortium, the international organization which sets communications standards for the web, has recently introduced an extension to HTML specifically designed for describing mathematical expressions. This extension, called MathML, was developed by members of Wolfram Research, using the same conceptual framework developed for *Mathematica*'s own formula typesetting algorithms.

HTML, as it stood before the introduction of MathML, could handle $e = m\,c^2$ but not $e = \frac{m\,v^2}{2}$; a web author who wanted to display the latter could only save the formula as a graphic image. One of the conveniences of the web is the ability to cut and paste text from a web site into a word processor, but the same thing couldn't be done with formulas; you couldn't capture a formula from the web, paste it into a *Mathematica* notebook, and perform calculations with it without any retyping. The MathML standard changes that; users can cut expressions from MathML-enabled browsers and paste them directly into *Mathematica*—a real time-saver when the expressions look like this...

$$\left(\frac{z}{2}\right)^{\nu} \sum_{n=0}^{\infty} \frac{(-1)^n \left(\frac{z}{2}\right)^{2n}}{\Gamma(n+\nu+1)\,n!}$$

Documents

The main document formats into which notebooks can be converted are LaTeX and HTML, both of which have been greatly enhanced in Version 4. These formats are designed to preserve much of the look and feel of the original notebook document, but there are, of course, obvious limits to what can be achieved for these formats in that regard. In both cases these conversions are extensible and there is additional information available on how this works in the corresponding reference guide entries.

In the case of LaTeX there is now a version of the *Mathematica* fonts that fits with modern LaTeX distributions and allows these documents to preserve much of the look and feel of a typical notebook document.

Below is an example where HTML + MathML is being exported instead of the usual HTML + GIF. The portion containing the MathML markup has been extracted from resulting string.

```
ExportString[ (x + y)^2, "HTML",
    ConversionOptions →
      {"MathOutput" → "MathML"} ]//
  StringTake[#, {304, 468} ]&
```

```
< math >
< msup >
 < mrow >
  < mo > (< /mo >
  < mrow >
   < mi > x < /mi >
   < mo > + < /mo >
   < mi > y < /mi >
  < /mrow >
  < mo >) < /mo >
 < /mrow >
```

```
< mn > 2 < /mn >
< /msup >
< /math >
```

Compare this output to the box representation used by *Mathematica*. Apart from additional markup in MathML such as <mi> for math italic variables and <mo> for math operators these expressions are structurally identical.

```
ToBoxes[ (x + y)^2]
SuperscriptBox[|
        RowBox[|
            List[(, |
                RowBox[|
                    List[x, +, y]
                ]
            , )]
        ]
    , 2]
```

This is, of course, no accident since Wolfram Research has been closely involved in the MathML design from the very start. And in addition we are hosting a conference on MathML in the fall of 2000.

TEX Publishing

Some scientific and mathematical journals require authors to submit their manuscripts as a TEX file. TEX is a technical typesetting system and macro language developed by Donald E. Knuth, notable for its power but hindered its lack of interactivity. To produce a typeset page in TEX, the author must create an ASCII file containing text and describing, using the TEX language, each mathematical expression. This ASCII file is then used by the TEX system to generate typeset output.

Describing mathematical expressions in TEX is not easy. Take the simple mathematical expression $x = \frac{-b \pm \sqrt{b^2 - 4 a c}}{2 a}$.

The TEX encoding for the same expression is:

$$x={{-b\pm \sqrt{b^2-4 a c}}\over{2a}}$$

Even in this short expression, there's plenty of opportunity to mismatch braces or make a similar syntactic error, which means there's also plenty of opportunity for unexpected results in the typeset output. Because the TEX system separates the creation of the input file from the typesetting phase, often the author must go through the edit-typeset-inspect cycle repeatedly to produce the intended results.

Mathematica 4 offers robust TEX output generation. Documents can be created and edited interactively in the *Mathematica* 4 front end—offering "What-You-See-Is-What-You-Get" equation editing and spell-checking—and then automatically translated into a TEX input file for setting. Just as was the case with the HTML conversion, the translation only takes a single menu command to perform. The mechanics of creating clean, accurate TEX output are taken care of by *Mathematica*, leaving the author free to concentrate on the substance of the document itself.

J/Link and *Mathematica*

With J/Link, Java programs can now access the vast array of mathematical and technical knowledge that is available in *Mathematica*. For example, *Mathematica* can provide high-level numeric and symbolic solvers to applets, servlets, and applications. These can also use the powerful tools in *Mathematica* to generate and render graphical and mathematical objects. For instance, *Mathematica* provides world-class typesetting of mathematical formulas, including full support for input, output, and rendering of MathML. See the J/Link web site (www.wolfram.com/solutions/mathlink/jlink/) for more details.

Examples

Spectral analysis of astronomical time-series data
The Polyhedron Explorer
Periodic Table of the Elements
Surface Explorer (contact the author Paul Wellin and Mike Mezzino)

References

[1] Michael Trott, Visualization of Riemann Surfaces IIa, *The Mathematica Journal*, **7**:4, 465–496, 2000.
[2] Adam Stzebonski, Solving Algebraic Inequalities, *The Mathematica Journal*, **7**:4, 525–541, 2000.

PAUL R. WELLIN
DIRECTOR OF CORPORATE AND ACADEMIC AFFAIRS
WOLFRAM RESEARCH, INC.

E-mail address: `wellin@wolfram.com`

Contemporary Mathematics
Volume **288**, 2001

Complex Geometry and Representations of Lie Groups

Joseph A. Wolf

Dedicated to the memory of my friend and colleague Alfred Gray

ABSTRACT. Let $Z = G/Q$ be a complex flag manifold and let G_0 be a real form of G. Then the representation theory of the real reductive Lie group G_0 is intimately connected with the geometry of G_0–orbits on Z. The open orbits correspond to the discrete series representations and their analytic continuations, closed orbits correspond to the principal series, and certain other orbits give the other series of tempered representations. Here I try to indicate some of that interplay between geometry and analysis, concentrating on the complex geometric aspects of the open orbits and the related representations.

Contents:

§0. Introduction
Part I. **Geometry of Flag Manifolds**
§1. Parabolic Subalgebras and Complex Flags
§2. Real Group Orbits on Complex Flags
§3. The Closed Orbit
§4. Open Orbits
§5. Cycle Spaces
§6. The Double Fibration Transform
Part II. **Representations of Reductive Lie Groups**
§7. The Principal Series
§8. The Discrete Series
§9. The Various Tempered Series
§10. The Plancherel Formula
Part III. **Geometric Realizations of Representations**
§11. L^2 Realizations of the Principal Series
§12. L^2 Realizations of the Discrete Series
§13. L^2 Realizations of the Various Tempered Series
§14. Approaches to Non–Tempered Representations

2000 *Mathematics Subject Classification.* Primary 22E15, 22E46, 43A80; secondary 22E10, 32E10, 32M10, 32T15.

Key words and phrases. reductive Lie group, reductive Lie algebra, semisimple Lie group, semisimple Lie algebra, representation, flag manifold, flag domain.

Research partially supported by NSF Grant DMS 99-88643.

0. Introduction.

There is an mysterious intimate correlation between the theory of homogeneous complex manifolds and the theory of unitary representations of real Lie groups. This relation is becoming clear in the case of a real reductive Lie group G_0. Let G be the complexification of G_0 and suppose that G_0 has a compact Cartan subgroup. Then the open G_0–orbits on a complex flag manifold $Z = G/Q$ (where Q is a parabolic subgroup of G) can be used to construct the discrete series representations of G_0 as the action of G_0 on L^2 cohomologies of various sorts. If G_0 is compact, this is the famous Bott–Borel–Weil Theorem. If G_0 is not compact, one use the same idea, even though noncompactness introduces a number of nontrivial technical problems, but finally one does get good geometric realizations for discrete series representations. Whether G_0 has a compact Cartan subgroup or not, the other series of representations of G_0 that are involved in the Plancherel formula, all come out of discrete series representations of certain subgroups $M_0 \subset G_0$ by a straightforward construction, and those representations occur as cohomologies over certain partially complex G_0–orbits. Here I'll sketch some of the main points in this beautiful liaison of geometry with analysis.

This paper updates [**141**], with fewer details on the older material but with much more material on singular representations, cycle spaces, and double fibration transforms.

Part I addresses geometric aspects of complex flag manifolds $Z = G/Q$ and G_0–orbits on Z. In Sections 1 through 4, we review the basic facts on complex flags, real group orbits and convexity of real group orbits. This material is found in much more detail in [**133**], [**116**] and [**139**]. Then in Section 5 we indicate the current state of information on cycle spaces of open G_0–orbits, from [**128**], [**139**], [**149**], [**150**], [**88**], [**89**] and [**90**]. Finally, in Section 6, we indicate the current state of affairs on the double fibration transforms that come out of these cycle spaces; see [**97**], [**99**] and [**150**] for details.

Part II is concerned with the representation theory of real reductive (or semisimple) Lie groups G_0. It concentrates on the series of unitary representations that are involved in the Plancherel Theorem and Fourier Inversion Formula of G_0. As we will see in Part III, those are the representations that appear in a straightforward geometric setting on G_0–orbits in complex flag manifolds $Z = G/Q$. In Section 7 we illustrate the construction with the "principal series", where a certain subgroup $M_0 \subset G_0$ is compact and thus presents no technical complications. Then in Section 8 we indicate the main results from Harish–Chandra's theory of the "discrete series", which we need both for itself and for the cases where M_0 is noncompact. In Section 9 we fit this together for the construction of all the relevant series of unitary representations of G_0. In Section 10 we give a brief indication of just how these "tempered series" of representations give the Plancherel Theorem and Fourier Inversion Formula of G_0.

Part III indicates the geometric realization of tempered representations as square integrable cohomologies of certain partially holomorphic vector bundles over G_0–orbits in complex flag manifolds $Z = G/Q$. Much as in Part II, we start in Section 11 illustrating the construction for principal series representations where there essentially are no technical problems. In Section 12 we indicate the realization of discrete series representations on certain Hilbert spaces of L^2 harmonic $(0,q)$–forms with values in a homogeneous holomorphic vector bundle over an open G_0–orbit. Then in Section 13 we fit the two together to realize tempered representations in general on certain Hilbert spaces of L^2 partially harmonic partially $(0,q)$–forms with values in a homogeneous partially holomorphic vector bundle over an appropriate G_0–orbit. Finally, in Section 14, we look at some of the many approaches to geometric realization of non–tempered representations.

Part I. Geometry of Flag Manifolds.

Complex flag manifolds are of considerable interest in algebraic geometry, in differential geometry, and in harmonic analysis. The most familiar ones are the complex Grassmann manifolds and the other compact hermitian symmetric spaces. In this Part we discuss the theory of complex flag manifolds and of real group orbits on those flag manifolds. Our emphasis is on the geometry of open orbits, which are instrumental for the realization of discrete series representations for semisimple Lie groups.

1. Parabolic Subalgebras and Complex Flags.

We start with the definitions of Borel subgroups and subalgebras, parabolic subgroups and subalgebras, and complex flag manifolds. Parabolic subgroups and flag manifolds were invented by J. Tits [**121**], and independently Borel subgroups were invented by A. Borel [**19**]. Also see [**20**], [**21**], and [**22**].

Fix a complex reductive Lie algebra $\mathfrak{g}$ and a Cartan subalgebra $\mathfrak{h} \subset \mathfrak{g}$. Thus $\mathfrak{g} = \mathfrak{g}' \oplus \mathfrak{z}$ where $\mathfrak{g}' = [\mathfrak{g}, \mathfrak{g}]$ is semisimple and $\mathfrak{z}$ is the center of $\mathfrak{g}$, and $\mathfrak{h} \cap \mathfrak{g}'$ is a Cartan subalgebra of $\mathfrak{g}'$, and $\mathfrak{h} = (\mathfrak{h} \cap \mathfrak{g}') \oplus \mathfrak{z}$. Let $\Sigma = \Sigma(\mathfrak{g}, \mathfrak{h})$ denote the corresponding root system, and fix a positive subsystem $\Sigma^+ = \Sigma^+(\mathfrak{g}, \mathfrak{h})$. The corresponding *Borel subalgebra*

$$\mathfrak{b} = \mathfrak{h} + \sum_{\alpha \in \Sigma^+} \mathfrak{g}_{-\alpha} \subset \mathfrak{g} \tag{1.1}$$

has its nilradical[1] $\mathfrak{b}^{-n} = \sum \mathfrak{g}_{-\alpha}$ and a Levi complement $\mathfrak{h}$. In general a subalgebra $\mathfrak{s} \subset \mathfrak{g}$ is called a *Borel subalgebra* if it is conjugate by an inner automorphism of $\mathfrak{g}$ to a subalgebra of the form (1.1), in other words if there exist choices of $\mathfrak{h}$ and $\Sigma^+(\mathfrak{g}, \mathfrak{h})$ such that $\mathfrak{s}$ is given by (1.1).

Let G be a connected Lie group with Lie algebra $\mathfrak{g}$. The Cartan subgroup of G corresponding to $\mathfrak{h}$ is $H = Z_G(\mathfrak{h})$, the centralizer of $\mathfrak{h}$ in G. It has Lie algebra $\mathfrak{h}$, and it is connected because G is connected, complex and reductive. The *Borel*

[1] Here we describe the nilradical as a sum of negative root spaces, rather than positive, so that, in applications, positive functionals on $\mathfrak{h}$ will correspond to positive bundles (instead of negative bundles), and holomorphic discrete series representations will be highest weight (instead of lowest weight) representations.

subgroup $B \subset G$ corresponding to a Borel subalgebra $\mathfrak{b} \subset \mathfrak{g}$ is defined to be the G–normalizer of $\mathfrak{b}$, that is,

$$B = \{g \in G \mid \mathrm{Ad}\,(g)\mathfrak{b} = \mathfrak{b}\}\ . \tag{1.2}$$

The basic facts on Borel subgroups are:

LEMMA 1.3. *B has Lie algebra $\mathfrak{b}$, B is a closed connected subgroup of G, and B is its own normalizer in G.*

LEMMA 1.4. (See [**122**], [**144**] and [**145**]) *Let $G_u \subset G$ be a compact real form. Then G_u is transitive on $X = G/B$, and X has a G_u–invariant Kähler metric. In particular X has the structure of compact Kähler manifold.*

LEMMA 1.5. *There is a finite dimensional irreducible representation π of G with the property: let $[v]$ be the image of a lowest weight vector in the projective space $\mathbb{P}(V_\pi)$ corresponding to the representation space of π. Then the action of G on V_π induces a holomorphic action of G on $\mathbb{P}(V_\pi)$, and B is the G–stabilizer of $[v]$. In particular $X = G/B$ is a projective algebraic variety.*

LEMMA 1.6. *B is a maximal solvable subgroup of G.*

A subalgebra $\mathfrak{q} \subset \mathfrak{g}$ is called *parabolic* if it contains a Borel subalgebra. For example, let Ψ be the simple root system corresponding to Σ^+ and let Φ be an arbitrary subset of Ψ. Every root $\alpha \in \Sigma$ has unique expression

$$\alpha = \sum_{\psi \in \Psi} n_\psi(\alpha)\psi \tag{1.7}$$

where the $n_\psi(\alpha)$ are integers, all $\geqq 0$ if $\alpha \in \Sigma^+$ and all $\leqq 0$ if $\alpha \in \Sigma^- = -\Sigma^+$. Set

$$\Phi^r = \{\alpha \in \Sigma \mid n_\psi(\alpha) = 0 \text{ whenever } \psi \notin \Phi\} \tag{1.8}$$

and

$$\Phi^n = \{\alpha \in \Sigma^+ \mid \alpha \notin \Phi^r\} = \{\alpha \in \Sigma \mid n_\psi(\alpha) > 0 \text{ for some } \psi \notin \Phi\}. \tag{1.9}$$

Now define

$$\mathfrak{q}_\Phi = \mathfrak{q}^r + \mathfrak{q}^{-n} \text{ with } \mathfrak{q}^r = \mathfrak{h} + \sum_{\alpha \in \Phi^r} \mathfrak{g}_\alpha \text{ and } \mathfrak{q}^{-n} = \sum_{\alpha \in \Phi^n} \mathfrak{g}_{-\alpha}\ . \tag{1.10}$$

Then $\mathfrak{q}_\Phi$ is a subalgebra of $\mathfrak{g}$ that contains the Borel subalgebra (1.1), so it is a parabolic subalgebra of $\mathfrak{g}$.

PROPOSITION 1.11. *Let $\mathfrak{q} \subset \mathfrak{g}$ be a subalgebra that contains the Borel subalgebra $\mathfrak{b} = \mathfrak{h} + \sum_{\alpha \in \Sigma^+} \mathfrak{g}_{-\alpha}$ of $\mathfrak{g}$. Then there is a set Φ of simple roots such that $\mathfrak{q} = \mathfrak{q}_\Phi$.*

The *parabolic subgroup* $Q \subset G$ corresponding to a parabolic subalgebra $\mathfrak{q} \subset \mathfrak{g}$ is defined to be the G–normalizer of $\mathfrak{q}$, that is,

$$Q = \{g \in G \mid \mathrm{Ad}\,(g)\mathfrak{q} = \mathfrak{q}\}\ . \tag{1.12}$$

The basic facts on parabolic subgroups are most easily derived from the corresponding results for Borel subgroups. However, the two notions were developed separately, and from different viewpoints, in the 1950's.

LEMMA 1.13. *The parabolic subgroup $Q \subset G$ defined by (1.12) has Lie algebra $\mathfrak{q}$. That group Q is a closed connected complex subgroup of G, and Q is its own normalizer in G. In particular, a Lie subgroup of G is parabolic if and only if it contains a Borel subgroup.*

LEMMA 1.14. *Fix a standard parabolic subgroup $Q = Q_\Phi$ in G. Then there is a finite dimensional irreducible representation π of G with the property: let $[v]$ be the image of a lowest weight vector in the projective space $\mathbb{P}(V_\pi)$ corresponding to the representation space of π. Then the action of G on V_π induces a holomorphic action of G on $\mathbb{P}(V_\pi)$, and Q is the G–stabilizer of $[v]$. In particular $Z = G/Q$ is a projective algebraic variety.*

Since Q is its own normalizer in G,

LEMMA 1.15. *Let $Z = G/Q$ where Q is a parabolic subgroup of G. Then we can view Z as the space of all G–conjugates of $\mathfrak{q}$, by the correspondence $gQ \leftrightarrow Ad(g)\mathfrak{q}$. Write $\mathfrak{q}_z$ for the parabolic subalgebra of $\mathfrak{g}$ corresponding to $z \in Z$, and write Q_z for the corresponding parabolic subgroup of G. Then the usual action $g : g'Q \mapsto gg'Q$ of G on Z carries over to $g : \mathfrak{q}_z \mapsto \mathrm{Ad}(g)\mathfrak{q}_z = \mathfrak{q}_{g(z)}$.*

Let $B \subset Q \subset G$ consist of a Borel subgroup contained in a parabolic subgroup. Then we have complex homogeneous quotient spaces $X = G/B$ and $Z = G/Q$, and a G–equivariant holomorphic projection $X \to Z$ given by $gB \mapsto gQ$. In particular, transitivity of G_u on X gives

LEMMA 1.16. *Let $G_u \subset G$ be a compact real form. Then G_u is transitive on $Z = G/Q$.*

Combining Lemmas 1.14 and 1.16 we have

LEMMA 1.17. *Let $G_u \subset G$ be a compact real form. Then the isotropy subgroup $G_u \cap Q$ is the centralizer of a torus subgroup of G_u, and Z has a G_u–invariant Kähler metric. Thus Z has the structure of G_u–homogeneous compact simply connected Kähler manifold.*

At this point we summarize:

PROPOSITION 1.18. *Let Q be a complex Lie subgroup of G. Then the following conditions are equivalent. (1) G/Q is a compact complex manifold. (2) G/Q is a projective algebraic variety. (3) If G_u denotes a compact real form of G then G/Q is a G_u–homogeneous compact Kähler manifold. (4) G/Q is the projective space orbit of an extremal weight vector in an irreducible finite dimensional representation of G. (5) G/Q is a G–equivariant quotient manifold of G/B, for some Borel subgroup $B \subset G$. (6) Q is a parabolic subgroup of G.*

We will simply refer to these spaces $Z = G/Q$ as *complex flag manifolds.*

2. Real Group Orbits on Complex Flags.

We now consider the action of a real group on a complex flag manifold, developed in [**133**] and [**134**]. Also see [**146**], [**134**] and [**135**]. The most familiar case is the action of $SL(2;\mathbb{R})$ on the Riemann sphere $P^1(\mathbb{C})$ by linear fractional transformation.

Let G_0 be a real form of G. In other words, G_0 is a Lie group whose Lie algebra $\mathfrak{g}_0$ is a real form of $\mathfrak{g}$. Although G is connected, G_0 does not have to be connected, but we do need some control over the components. For this reason, and for some technical reasons that will come out in the representation theory, we assume that G_0 is of *Harish–Chandra class* developed in [**69**], [**70**] and [**71**]:

Definition 2.1. *A real Lie group G_0 belongs to the* Harish–Chandra class *of reductive Lie groups if* (i) *the Lie algebra $\mathfrak{g}_0$ is reductive,* (ii) *the component group G_0/G_0^0 is finite,* (iii) *the derived group $[G_0^0, G_0^0]$ is closed in G_0, and* (iv) *if $g \in G_0$ then* $\mathrm{Ad}(g)$ *is an inner automorphism of the complex Lie algebra $\mathfrak{g}$.*

The results we describe in this report hold for a somewhat larger class of reductive Lie groups developed a bit earlier [**135**].

The group G_0 acts on the complex flag manifold $Z = G/Q$ by $g : \mathfrak{q}_z \mapsto \mathrm{Ad}(g)\mathfrak{q}_z$ as in Lemma 1.15, because of condition (iv) in Definition 2.1.

We write τ both for the complex conjugation of $\mathfrak{g}$ over $\mathfrak{g}_0$ and for the corresponding conjugation of G over its analytic subgroup $\exp(\mathfrak{g}_0)$.

The basic results on the G_0–orbit structure of $Z = G/Q$ depends on the consequence

Lemma 2.2. *The intersection of two Borel subalgebras of $\mathfrak{g}$ contains a Cartan subalgebra of $\mathfrak{g}$.*

of the Bruhat decomposition ([**26**], [**62**]) of Z. From this,

Lemma 2.3. *The intersection $\mathfrak{q} \cap \tau\mathfrak{q}$ contains a τ–stable Cartan subalgebra $\mathfrak{h}$ of $\mathfrak{g}$.*

Theorem 2.4. *Consider an orbit $G_0(z)$ on $Z = G/Q$. Then there exist a τ–stable Cartan subalgebra $\mathfrak{h} \subset \mathfrak{q}_z$ of $\mathfrak{g}$, a positive root system $\Sigma^+ = \Sigma^+(\mathfrak{g}, \mathfrak{h})$, and a set Φ of simple roots, such that $\mathfrak{q}_z = \mathfrak{q}_\Phi$ and $Q_z = Q_\Phi$.*

There are only finitely many G_0–conjugacy classes of Cartan subalgebras $\mathfrak{h}_0$ in $\mathfrak{g}_0$, for each $\mathfrak{h}_0$ there are only finitely many positive root systems $\Sigma^+ = \Sigma^+(\mathfrak{g}, \mathfrak{h})$, and for each such Σ^+ there are only finitely many subsets Φ of the simple root system. Thus

Corollary 2.5. *There are only finitely many G_0–orbits on Z. The maximal–dimensional orbits are open and the minimal–dimensional orbits are closed.*

Corollary 2.6. *In the notation of* Theorem 2.4, *$\mathfrak{q}_z \cap \tau\mathfrak{q}_z$ is the semidirect sum of its nilpotent radical $(\mathfrak{q}_\Phi^{-n} \cap \tau\mathfrak{q}_\Phi^{-n}) + (\mathfrak{q}_\Phi^r \cap \tau\mathfrak{q}_\Phi^{-n}) + (\mathfrak{q}_\Phi^{-n} \cap \tau\mathfrak{q}_\Phi^r)$ with the Levi complement $\mathfrak{q}_\Phi^r \cap \tau\mathfrak{q}_\Phi^r = \mathfrak{h} + \sum_{\Phi^r \cap \tau\Phi^r} \mathfrak{g}_\alpha$. In particular,* $\dim_{\mathbb{R}} \mathfrak{g}_0 \cap \mathfrak{q}_z = \dim_{\mathbb{C}} \mathfrak{q}_\Phi^r + |\Phi^n \cap \tau\Phi^n|$.

Corollary 2.7. *In that notation,* $\mathrm{codim}_{\mathbb{R}}(G_0(z) \subset Z) = |\Phi^n \cap \tau\Phi^n|$. *In particular, $G_0(z)$ is open in Z if and only if $\Phi^n \cap \tau\Phi^n$ is empty.*

3. The Closed Orbit.

We now look at the closed real group orbit. This material was developed in [**133**].

There is least one closed G_0–orbit on Z, for minimal dimensional orbits are closed. Consider the case where $G = SL(2; \mathbb{C}), G_0 = SU(1,1)$, and X is the Riemann sphere. G acts as usual by linear fractional transformations. Let $D =$

$G_0(0)$, the unit disk. There are three G_0–orbits:

$$G_0(0), \text{ interior of } D: \; Q_0 = \left\{ \begin{pmatrix} a & 0 \\ c & a^{-1} \end{pmatrix} \right\} \text{ and } H_0 = \left\{ \begin{pmatrix} e^{i\theta} & 0 \\ 0 & e^{-i\theta} \end{pmatrix} \Big| \; \theta \text{ real} \right\},$$

$$G_0(\infty), \text{ exterior of } D: \; Q_\infty = \left\{ \begin{pmatrix} a & b \\ 0 & a^{-1} \end{pmatrix} \right\} \text{ and } H_0 = \left\{ \begin{pmatrix} e^{i\theta} & 0 \\ 0 & e^{-i\theta} \end{pmatrix} \Big| \; \theta \text{ real} \right\},$$

$$G_0(1), \text{ boundary of } D: \; Q_1 = \left\{ \begin{pmatrix} a & b \\ c & d \end{pmatrix} \Big| \; ad - bc = 1, a + b = c + d \right\} \text{ and}$$

$$H_0 = \left\{ \pm \begin{pmatrix} \cosh(t) & \sinh(t) \\ \sinh(t) & \cosh(t) \end{pmatrix} \Big| \; t \text{ real} \right\}.$$

The first two give the open orbits, with H_0 compact, and the third gives the closed orbit, where H_0 is the $T_0 A_0$ of an Iwasawa decomposition of G_0. That mirrors the general case for closed orbits:

THEOREM 3.1. *Let $X = G/Q$ be a complex flag manifold and let G_0 be a real form of G. Then there is a unique closed orbit $G_0(z) \subset Z$. Further, there is an Iwasawa decomposition $G_0 = K_0 A_0 N_0$ such that $G_0 \cap Q_z$ contains $H_0 N_0$ whenever H_0 is a Cartan subgroup of G_0 that contains A_0.*

THEOREM 3.2. *Let $Z = G/Q$ be a complex flag manifold and let G_0 be a real form of G. Let $G_0(z)$ be the unique closed G_0–orbit on Z. Then* $\dim_{\mathbb{R}} G_0(z) \geqq \dim_{\mathbb{C}} Z$, *and the following conditions are equivalent.*

1. $\dim_{\mathbb{R}} G_0(z) = \dim_{\mathbb{C}} Z$.
2. $\tau\Phi^n = \Phi^n$.
3. *View G as the group of complex points, and G_0 as an open subgroup in the group of real points, of a linear algebraic group defined over $\mathbb{R}$. Then Q_z is the group of complex points in an algebraic subgroup defined over $\mathbb{R}$.*
4. *Z is the set of complex points in a projective variety defined over $\mathbb{R}$, and $G_0(z)$ is the set of real points.*

4. Open Orbits.

At the other extreme, in general we have several open real group orbits. The material here is from [**133**], except that the material on the exhaustion function is from [**116**].

Fix a Cartan involution θ of $\mathfrak{g}_0$ and G_0. θ is an automorphism of square 1 and the fixed point set $K_0 = G_0^\theta$ is a maximal compact subgroup of G_0. Thus $\mathfrak{g}_0 = \mathfrak{k}_0 + \mathfrak{s}_0$ where $\mathfrak{k}_0$ is the Lie algebra of K_0 and is the $(+1)$–eigenspace of θ on $\mathfrak{g}_0$, and $\mathfrak{s}_0$ is the (-1)–eigenspace. The Killing form of $\mathfrak{g}_0$ is negative definite on $\mathfrak{k}_0$ and positive definite on $\mathfrak{s}_0$, and $\mathfrak{k}_0 \perp \mathfrak{s}_0$ under the Killing form.

Every Cartan subalgebra of $\mathfrak{g}_0$ is $\mathrm{Ad}(G_0)$–conjugate to a θ–stable Cartan subalgebra. A θ–stable Cartan subalgebra $\mathfrak{h}_0 \subset \mathfrak{g}_0$ is called *fundamental* if it maximizes $\dim(\mathfrak{h}_0 \cap \mathfrak{k}_0)$, *compact* if it is contained in $\mathfrak{k}_0$ (a more stringent condition). More generally, a Cartan subalgebra of $\mathfrak{g}_0$ is called *fundamental* if it is conjugate to a θ–stable fundamental Cartan subalgebra, *compact* if it is conjugate to a θ–stable compact Cartan subalgebra.

THEOREM 4.1. *Let $Z = G/Q$ be a complex flag manifold, G a connected reductive complex Lie group, and let G_0 be a real form of G. The orbit $G_0(z)$ is open in Z if and only if $\mathfrak{q}_z = \mathfrak{q}_\Phi$ where*

(i) *$\mathfrak{q}_z \cap \mathfrak{g}_0$ contains a fundamental Cartan subalgebra $\mathfrak{h}_0$ of $\mathfrak{g}_0$ and*
(ii) *Φ is a set of simple roots for a positive root system $\Sigma^+(\mathfrak{g}, \mathfrak{h})$ such that $\tau\Sigma^+ = \Sigma^-$.*

Fix $\mathfrak{h}_0 = \theta\mathfrak{h}_0$, $\Sigma^+(\mathfrak{g}, \mathfrak{h})$ and Φ as above. Let $W(\mathfrak{g}, \mathfrak{h})^{\mathfrak{h}_0}$ and $W(\mathfrak{q}^r_\Phi, \mathfrak{h})^{\mathfrak{h}_0}$ denote the respective subgroups of Weyl groups that stabilize $\mathfrak{h}_0$. Then the topological components of the open G_0–orbits on Z are parameterized by the double coset space $W(\mathfrak{k}, \mathfrak{h} \cap \mathfrak{k})\backslash W(\mathfrak{g}, \mathfrak{h})^{\mathfrak{h}_0}/W(\mathfrak{q}_z, \mathfrak{h})^{\mathfrak{h}_0}$.

COROLLARY 4.2. *Suppose that G_0 has a compact Cartan subgroup, i.e. that $\mathfrak{k}_0$ contains a Cartan subalgebra of $\mathfrak{g}_0$. Then an orbit $G_0(z)$ is open in Z if and only if $\mathfrak{g}_0 \cap \mathfrak{q}_z$ contains a compact Cartan subalgebra $\mathfrak{h}_0$ of $\mathfrak{g}_0$, and then, in the notation of* Theorem 4.1, *the topological components of the open G_0–orbits on Z are parameterized by $W(\mathfrak{k}, \mathfrak{h})\backslash W(\mathfrak{g}, \mathfrak{h})/W(\mathfrak{q}^r_z, \mathfrak{h})$.*

A careful examination of the way $\mathfrak{k}_0$ sits in both $\mathfrak{k}$ and $\mathfrak{g}_0$, relative to the integrability condition [**45**] for homogeneous complex structures, gives us

THEOREM 4.3. *Let $Z = G/Q$ be a complex flag manifold, G a connected reductive complex Lie group, and let G_0 be a real form of G. Let $z \in Z$ such that $G_0(z)$ is open in Z, and let $\mathfrak{h}_0 \subset \mathfrak{g}_0 \cap \mathfrak{q}_z$ be a θ–stable fundamental Cartan subalgebra of $\mathfrak{g}_0$. Then $K_0(z)$ is a compact complex submanifold of $G_0(z)$. Let K be the complexification of K_0, analytic subgroup of G with Lie algebra $\mathfrak{k} = \mathfrak{k}_0 \otimes \mathbb{C}$. If G_0 (and thus K_0) is connected, then $K_0(z) = K(z) \cong K/(K \cap Q_z)$, complex flag manifold of K.*

If G_0 is not connected then of course one has essentially the same result, but one must be careful about topological components.

COROLLARY 4.4. *The compact subvariety $K_0(z)$ is a deformation retract of $G_0(z)$. In particular, $G_0(z)$ is simply connected and has connected isotropy subgroup $(Q_z \cap \tau Q_z)_0$ at z.*

Fix a complex flag manifold $Z = G/Q$. An open orbit $G_0(z) \subset Z$ is called *measurable* if it carries a G_0–invariant volume element. If that is the case, then the invariant volume element is the volume element of a G_0–invariant, possibly indefinite, Kähler metric on the orbit, and the isotropy subgroup $G_0 \cap Q_z$ is the centralizer in G_0 of a (compact) torus subgroup of G_0. In more detail, measurable open orbits are characterized by

PROPOSITION 4.5. *Let $D = G_0(z)$ be an open G_0–orbit on the complex flag manifold $Z = G/Q$. Then the following conditions are equivalent.*

1. *The orbit $G_0(z)$ is measurable.*
2. *$G_0 \cap Q_z$ is the G_0–centralizer of a (compact) torus subgroup of G_0.*
3. *D has a G_0–invariant possibly–indefinite Kähler metric, thus a G_0–invariant measure obtained from the volume form of that metric.*
4. *$\tau\Phi^r = \Phi^r$, and $\tau\Phi^n = -\Phi^n$ where $\mathfrak{q}_z = \mathfrak{q}_\Phi$.*

5. $\mathfrak{q}_z \cap \tau\mathfrak{q}_z$ *is reductive, i.e.* $\mathfrak{q}_z \cap \tau\mathfrak{q}_z = \mathfrak{q}_z^r \cap \tau\mathfrak{q}_z^r$.

6. $\mathfrak{q}_z \cap \tau\mathfrak{q}_z = \mathfrak{q}_z^r$.

7. $\tau\mathfrak{q}$ *is* Ad (G)*–conjugate to the parabolic subalgebra* $\mathfrak{q}^- = \mathfrak{q}^r + \mathfrak{q}^n$ *opposite to* $\mathfrak{q}$.

In particular, if one open G_0*–orbit on* Z *is measurable, then they all are measurable.*

Note that condition 4 of Proposition 4.5 is automatic if the Cartan subalgebra $\mathfrak{h}_0$, relative to which $\mathfrak{q}_z = \mathfrak{q}_\Phi$, is the Lie algebra of a compact Cartan subgroup of G_0, for in that case $\tau\alpha = -\alpha$ for every $\alpha \in \Sigma(\mathfrak{g}, \mathfrak{h})$. In particular, if G_0 has discrete series representations, so that by a result of Harish–Chandra it has a compact Cartan subgroup, then every open G_0–orbit on Z is measurable.

Condition 4 is also automatic if P is a Borel subgroup of G, and more generally Condition 7 provides a quick test for measurability.

Bounded symmetric domains $D \subset \mathbb{C}^n$ are convex, and thus Stein, so cohomologies $H^k(D; \mathcal{F}) = 0$ for $k > 0$ whenever $\mathcal{F} \to D$ is a coherent analytic sheaf. This is a key point in dealing with holomorphic discrete series representations. More generally, for general discrete series representations and their analytic continuations, one has

THEOREM 4.6. *Let* $Z = G/P$ *be a complex flag manifold,* G *a connected reductive complex Lie group, and let* G_0 *be a real form of* G. *Let* $D = G_0(z) \subset Z = G/P$ *be a measurable open orbit. Let* $Y = K_0(z)$, *maximal compact subvariety of* D, *and let* $s = \dim_{\mathbb{C}} Y$. *Then* D *is* $(s+1)$*–complete in the sense of Andreotti–Grauert* [**2**]. *In particular, if* $\mathcal{F} \to D$ *is a coherent analytic sheaf then* $H^k(D; \mathcal{F}) = 0$ *for* $k > s$.

Indication of Proof. Let $\mathbb{K}_Z \to Z$ and $\mathbb{K}_D = K_Z|_D \to D$ denote the canonical line bundles. Their dual bundles $\mathbb{L}_Z = \mathbb{K}_Z^* \to Z$ and $\mathbb{L}_D = \mathbb{K}_D^* \to D$ are the homogeneous holomorphic line bundles over Z associated to the character

$$e^\lambda : Q_z \to \mathbb{C} \text{ defined by } e^\lambda(q) = \text{ trace Ad}(q)|_{\mathfrak{q}_z^n} . \tag{4.7}$$

Write $D = G_0/V_0$ where V_0 is the real form $G_0 \cap Q_z$ of Q_z^r. Write V for the complexification Q_z^r of V_0, $\rho_{G/V}$ for half the sum of the roots that occur in $\mathfrak{q}_z^n$, and $\lambda = 2\rho_{G/V}$. If $\alpha \in \Sigma(\mathfrak{g}, \mathfrak{h})$ then (i) $\langle\alpha, \lambda\rangle = 0$ and $\alpha \in \Phi^r$, or (ii) $\langle\alpha, \lambda\rangle > 0$ and $\alpha \in \Phi^n$, or (iii) $\langle\alpha, \lambda\rangle < 0$ and $\alpha \in \Phi^{-n}$. Now $\tau\lambda = -\lambda$. Decompose $\mathfrak{g}_0 = \mathfrak{k}_0 + \mathfrak{s}_0$ under the Cartan involution with fixed point set $\mathfrak{k}_0$, thus decomposing the Cartan subalgebra $\mathfrak{h}_0 \subset \mathfrak{g}_0 \cap \mathfrak{q}_z$ as $\mathfrak{h}_0 = \mathfrak{t}_0 + \mathfrak{a}_0$ with $\mathfrak{t}_0 = \mathfrak{h}_0 \cap \mathfrak{k}_0$ and $\mathfrak{a}_0 = \mathfrak{h}_0 \cap \mathfrak{s}_0$. Then $\lambda(\mathfrak{a}_0) = 0$.

View $D = G_0/V_0$ and $Z = G_u/V_u$ where G_u is the analytic subgroup of G for the compact real form $\mathfrak{g}_u = \mathfrak{k}_0 + \sqrt{-1}\,\mathfrak{s}_0$, and where V_u is its analytic subgroup for $\mathfrak{q}_u \cap \mathfrak{g}_z$, compact real form of Q_z^r. Then e^λ is a unitary character both on V_0 and on V_u, so

$$\begin{aligned} &\mathbb{L}_Z \to Z = G_u/V_0 \text{ has a } G_u\text{–invariant hermitian metric } h_u, \\ &\mathbb{L}_D \to D = G_0/V_0 \text{ has a } G_0\text{–invariant hermitian metric } h_0. \end{aligned} \tag{4.8}$$

With this information we obtain

LEMMA 4.9. *The hermitian form* $\sqrt{-1}\,\partial\bar\partial h_u$ *on the holomorphic tangent bundle of* Z *is negative definite. The hermitian form* $\sqrt{-1}\,\partial\bar\partial h_0$ *on the holomorphic tangent bundle of* D *has signature* $n - 2s$ *where* $n = \dim_{\mathbb{C}} D$.

COROLLARY 4.10. *Define* $\phi : D \to \mathbb{R}$ *by* $\phi = \log(h_0/h_u)$. *Then the Levi form* $\mathcal{L}(\phi)$ *has at least* $n - s$ *positive eigenvalues at every point of* D.

The next point is to show that ϕ is an exhaustion function for D, in other words that $\{z \in D \mid \phi(z) \leqq c\}$ is compact for every $c \in \mathbb{R}$. It suffices to show that $e^{-\phi}$ has a continuous extension from D to the compact manifold Z that vanishes on the topological boundary $\mathrm{bd}(D)$ of D in Z. For that, choose a G_u–invariant metric h_u^* on $\mathbb{L}_Z^* = \mathbb{K}_Z$ normalized by $h_u h_u^* = 1$ on Z, and a G_0–invariant metric h_0^* on $\mathbb{L}_D^* = \mathbb{K}_D$ normalized by $h_0 h_0^* = 1$ on D. Then $e^{-\phi} = h_0^*/h_u^*$. So it suffices to show that h_0^*/h_u^* has a continuous extension from D to Z that vanishes on $\mathrm{bd}(D)$.

The holomorphic cotangent bundle $\mathbb{T}_Z^* \to Z$ has fiber $\mathrm{Ad}(g)(\mathfrak{q}_z^n)^* = \mathrm{Ad}(g)(\mathfrak{q}_z^{-n})$ at $g(z)$, so its G_u–invariant hermitian metric is given on the fiber $\mathrm{Ad}(g)(\mathfrak{q}_z^{-n})$ at $g(z)$ by $F_u(\xi, \eta) = -\langle \xi, \tau\theta\eta \rangle$ where $\langle , \rangle$ is the Killing form. Similarly the G_0–invariant indefinite–hermitian metric on $\mathbb{T}_D^* \to D$ is given on the fiber $\mathrm{Ad}(g)(\mathfrak{q}_z^{-n})$ at $g(z)$ by $F_0(\xi, \eta) = -\langle \xi, \tau\eta \rangle$. But $\mathbb{K}_Z = \det \mathbb{T}_Z^*$ and $\mathbb{K}_D = \det \mathbb{T}_D^*$, so

$$h_0^*/h_u^* = c \cdot (\text{determinant of } F_0 \text{ with respect to } F_u) \tag{4.11}$$

for some nonzero real constant c. This extends from D to a C^∞ function on Z given by

$$f(g(z)) = c \cdot \det F_0|_{\mathrm{Ad}(g)(\mathfrak{q}_z^{-n})} . \tag{4.12}$$

It remains only to show that the function f of (4.12) vanishes on $\mathrm{bd}(D)$. If $g(z) \in \mathrm{bd}(D)$ then $G_0(g(z))$ is not open in Z, so

$$\mathrm{Ad}(g)(\mathfrak{q}_z) + \tau\mathrm{Ad}(g)(\mathfrak{q}_z) \neq \mathfrak{g}.$$

Thus $\mathfrak{g}_\alpha \subset \mathrm{Ad}(g)(\mathfrak{q}_z^{-n})$ while there exists an $\alpha \in \Sigma(\mathfrak{g}, \mathrm{Ad}(g)\mathfrak{h})$ such that

$$\mathfrak{g}_{-\alpha} \not\subset \mathrm{Ad}(g)(\mathfrak{q}_z) + \tau\mathrm{Ad}(g)(\mathfrak{q}_z) .$$

If $\beta \in \Sigma(\mathfrak{g}, \mathrm{Ad}(g)\mathfrak{h})$ with $\mathfrak{g}_\beta \subset \mathrm{Ad}(g)(\mathfrak{q}_z^{-n})$ then $F_0(\mathfrak{g}_\alpha, \mathfrak{g}_\beta) = 0$, so $f(g(z)) = 0$. Thus ϕ is an exhaustion function for D in Z. In view of Corollary 4.10 now D is $(s+1)$–complete, and the Theorem is proved.

In the non–measurable case, $(s+1)$–completeness is not completely settled.

5. Cycle Spaces

We now look at the linear cycle space associated to an open real group orbit. It is contained in a component of the Barlet cycle space ([**14**], [**15**], and also see [**30**]) for that open orbit. The general material here was developed in [**133**] and [**139**], and the explicit information comes from [**149**], [**88**], [**89**] and [**90**].

As above, $Z = G/Q$ is a complex flag manifold, G_0 is a real form of G, and $D = G_0(z) \subset Z$ is an open orbit. Theorem 4.3 says that

$$Y = K_0(z) \cong K_0/(K_0 \cap P_z) \cong K/(K \cap P_z)$$

is a complex submanifold of D. Y is not contained in any compact complex submanifold of D of greater dimension, so it is a maximal compact subvariety of D.

Let $L = \{g \in G \mid gY = Y\}$. Then L is a closed complex subgroup of G, so $M_Z = \{gY \mid g \in G\} \cong G/L$ has a natural structure of G–homogeneous complex manifold. Since Y is compact and D is open in Z, $\widetilde{M_D} = \{gY \mid g \in G \text{ and } gY \subset D\}$

is open in M_Z, and thus has a natural structure of complex manifold. The *linear cycle space* of D is

(5.1) $\quad M_D$: topological component of Y in $\widetilde{M_D} = \{gY \mid g \in G \text{ and } gY \subset D\}$.

Thus M_D has a natural structure of complex manifold. Its structure is given by

THEOREM 5.2. *If D is measurable then M_D is a Stein manifold.*

This is proved by taking the exhaustion function ϕ of Corollary 4.10 and "pushing it down" from D to M_D. In general, G, Q, Z, D, K and Y break up as direct products according to any decomposition of $\mathfrak{g}_0$ as a direct sum of ideals, equivalently any decomposition of G_0 as a local direct product. Here we may assume that G is connected and simply connected. So, for purposes of determining L we may, and do, assume that G_0 is noncompact and simple, in other words that G_0/K_0 is an irreducible riemannian symmetric space of noncompact type.

As usual we say that G_0 is of *hermitian type* if the irreducible riemannian symmetric space G_0/K_0 is an hermitian symmetric space.

Let θ be the Cartan involution of G_0 with fixed point set K_0 and $\mathfrak{g} = \mathfrak{k} + \mathfrak{s}$ under θ, as usual. By irreducibility of G_0/K_0, the adjoint action of K_0 on $\mathfrak{s}_0 = \mathfrak{g}_0 \cap \mathfrak{s}$ is irreducible. G_0 is of hermitian type if and only if this action fails to be absolutely irreducible. Then there is a positive root system $'\Sigma^+ = {}'\Sigma^+(\mathfrak{g}, \mathfrak{h})$ such that $\mathfrak{s} = \mathfrak{s}_+ + \mathfrak{s}_-$ where $\mathfrak{s}_+$ is a sum of $'\Sigma^+$–positive root spaces and represents the holomorphic tangent space of G_0/K_0, and $\mathfrak{s}_- = \overline{\mathfrak{s}_+}$ is a sum of $'\Sigma^+$–negative root spaces and represents the antiholomorphic tangent space. Write $S_\pm = \exp(\mathfrak{s}_\pm) \subset G$. Then G_0/K_0 is an open G_0–orbit on G/KS_-.

Now we have three cases, based on the possibilities for $L = \{g \in G \mid gY = Y\}$:

1. (Trivial Case) $K = G$, i.e. $D = Z$, so M_Z (and thus M_D) is reduced to a single point. These cases have been classified [**142**].

2. (Hermitian Case) $\mathfrak{l} = \mathfrak{k} + \mathfrak{s}_\pm$, i.e., $\mathfrak{k} \subsetneq \mathfrak{l} \subsetneq \mathfrak{g}$, so $M_Z = G/KS_-$ is a projective variety and $B = G_0/K_0$ is a bounded symmetric domain. This is the situation where the two maps of the double fibration

$$\begin{array}{ccc} & G_0/(K_0 \cap Q_z) & \\ \swarrow & & \searrow \\ D & & B \end{array}$$

are simultaneously holomorphic for some choice between B and $\overline{B}$ and some choice of invariant complex structure on $G_0/(L_0 \cap K_0)$. Then $M_D = B$.

3. (Generic Case) $\mathfrak{l} = \mathfrak{k}$, so K is the identity component of L and M_Z is a nontrivial affine variety. Here a number of cases are known:

- $D \cong SU(2p, q)/(SO(2p) \times U(q))$. Then M_D has an explicit description, and that description shows that it is a Stein manifold [**127**].
- $B = G_0/K_0$ is a bounded symmetric domain and G_0 is a classical group. Then $M_D = B \times \overline{B}$ [**149**]. Some special cases had been worked out in [**103**], in [**34**], in [**97**] and in [**98**].

- $B = G_0/K_0$ is a bounded symmetric domain and D is a pseudo–riemannian symmetric space. Then $M_D = B \times \overline{B}$ [**150**].
- $G_0 = SL(n;\mathbb{R})$, real special linear group. Then M_D has an explicit description, and that description shows that it is a Stein manifold [**88**].
- $G_0 = SL(n;\mathbb{H})$, quaternion special linear group. Then M_D has an explicit description, and that description shows that it is a Stein manifold [**90**].

The proof [**139**] of Theorem 5.2 is rather easy in Cases 1 and 3 above, but in Case 2 it involves detailed information [**146**] on the G_0–orbit structure of G/KS_-.

The case [**127**] where $D \cong SU(2p,q)/(SO(2p) \times U(q))$ was the first nontrivial case worked out. It was done in view of then–recent work on moduli spaces for compact Kähler manifolds in dimension > 1 ([**52**] and [**53**]). In [**128**] it was first shown that M_D is Stein in some degree of generality, but the emphasis was on automorphic cohomology in connection with [**52**] and [**53**]. This theory of automorphic cohomology has been extended in [**137**], [**125**], [**129**], [**130**], and [**131**].

The proof of Theorem 5.2 does not quite work in the non–measurable case, but there we have explicit descriptions (see just above), which in particular prove Stein, except when G_0 is a complex simple group or is of type E_6 with maximal compact subgroup of type F_4 or C_4. Theorem 5.2 would be shown in general, and probably with less effort, if we could show that M_D is closed in the Barlet cycle space of D. There is some progress in that direction ([**16**], [**143**]).

For other approaches, viewing M_D as a Stein neighborhood of G_0/K_0 in G/K, see [**1**], [**11**], [**24**], [**25**], [**45**], [**43**], [**44**], [**47**], [**87**], [**101**] and [**147**]. Also see [**28**], [**29**] and [**92**] for an approach based on Grauert tubes and the Monge–Ampère equations.

6. The Double Fibration Transform

Double fibrations are an old topic, starting with the Crofton formulae and Chern's intersection theory of two homogeneous spaces of the same group, the classical Radon and X–ray transforms, the Gelfand–Graev horocycle transforms [**48**] and Helgason's group–theoretic reformulation of the horocycle transform. Real–analytic double fibration transforms come up in [**12**], [**117**] and [**151**], with the Identity Theorem [**112**] as a degenerate early case, in the study of cohomology representations. Holomorphic double fibration transforms were first used in the study of automorphic cohomology [**128**]. Penrose' twistor theory (cf. [**3**], [**104**], [**105**], [**106**], [**107**] and [**86**]) took advantage of those transforms, and the special case $G_0 = SU(2,2)$ was reworked, with physical applications, in [**41**], as the *Penrose Transform.* The material here is taken from [**150**].

In general let $D = G_0(z)$ be an open orbit in the complex flag manifold $Z = G/P$, let Y be the maximal compact linear subvariety $K_0(z)$, and consider the linear cycle space M_D: component of Y in $\{gY \mid g \in G \text{ and } gY \subset D\}$. Then we have a double fibration

$$\begin{array}{ccc} & W_D & \\ {}^{\mu}\swarrow & & \searrow^{\nu} \\ D & & M_D \end{array} \tag{6.1}$$

where $W_D = \{(Y', y') \mid y' \in Y' \in M_D\}$ is the incidence space. The projections are given by $\mu(Y', y') = y'$ and $\nu(Y', y') = Y'$.

Let $n = \dim_{\mathbb{C}} D$ and $s = \dim_{\mathbb{C}} Y$ as before. Consider a *negative* homogeneous holomorphic vector bundle $\mathbb{E} \to D$. Then we can expect nonzero cohomology only in degree s. For many purposes, for example for making estimates of one sort or another, it is better to have representations of G_0 occur on spaces of functions rather than on cohomology spaces. Here we indicate a double fibration transform that carries $H^s(D; \mathcal{O}(\mathbb{E}))$ to a space of functions on M_D. Given a coherent analytic sheaf $\mathcal{E} \to D$ we construct a coherent sheaf $\mathcal{E}' \to M$ and a transform

$$P : H^s(D; \mathcal{E}) \to H^0(M; \mathcal{E}'). \tag{6.2}$$

We refer to the transform (6.2) as a *double fibration transform.* The Penrose transform is the case where $G_0 = SU(2,2)$ and Z is complex projective space $P^3(\mathbb{C})$.

One wants two things in (6.2): that P be injective, and that there be an explicit description of its image. Assuming (6.5) below, injectivity of P is equivalent to injectivity of $j^{(p)}$ in (6.4) below. There are several ways to ensure this. The most general is the collection of vanishing conditions in Theorem 6.6 below. Another, more specific to our situation, is that in many cases we know that the fibers of μ are Stein manifolds. Finally, in some cases one knows that $H^p(D; \mathcal{E})$ is an irreducible representation space for a group under which all our constructions are equivariant, so P is an intertwining operator, thus zero or injective.

The double fibration transform is constructed in several steps.

We first pull cohomology back from D to W_D in (6.1). Let $\mu^{-1}(\mathcal{E}) \to W_D$ denote the inverse image sheaf. As μ is open, it is the sheaf defined by the presheaf whose value at an open set $\widetilde{U} \subset W_D$ is $\Gamma(U, \mathcal{E})$ where $U = \mu(\widetilde{U})$. Here, as usual, Γ denotes the space of sections. For every integer $r \geqq 0$ there is a natural map $\mu^{(r)} : H^r(D; \mathcal{E}) \to H^r(W_D; \mu^{-1}(\mathcal{E}))$ given on the Cech cocycle level by $\mu^{(r)}(c)(\sigma) = c(\mu(\sigma))$ where $c \in Z^r(D; \mathcal{E})$ and where $\sigma = (w_0, \ldots, w_r)$ is a simplex.

PROPOSITION 6.3. **[27]** *Suppose that the fiber F of $\mu : W_D \to D$ is connected and that $H^r(F; \mathbb{C}) = 0$ for $1 \leqq r \leqq p-1$. Then the map $\mu^{(r)}$ is an isomorphism for $r \leqq p-1$ and is injective for $r = p$. In particular, if the fibers of μ are contactable then $\mu^{(r)}$ is an isomorphism for all r.*

To complete the pull–back, we change the inverse image sheaf $\mu^{-1}(\mathcal{E}) \to W_D$ into a coherent analytic sheaf over W_D. This is necessary for the push–down step.

As usual, if X is a complex manifold then $\mathcal{O}_X \to X$ denotes its structure sheaf, the sheaf of germs of holomorphic $\mathbb{C}$–valued functions on X. If $\mathbb{E} \to X$ is a holomorphic vector bundle then $\mathcal{O}(\mathbb{E}) \to X$ is its sheaf of germs of holomorphic sections.

Denote $\mu^*(\mathcal{E}) = \mu^{-1}(\mathcal{E}) \otimes_{\mu^{-1}(\mathcal{O}_D)} \mathcal{O}_{W_D}$. It is a sheaf of $\mathcal{O}_{W_D}$–modules. If it happens that $\mathcal{E} = \mathcal{O}(\mathbb{E})$ for some holomorphic vector bundle $\mathbb{E} \to D$, then $\mu^*(\mathcal{E}) = \mathcal{O}(\mu^*(\mathbb{E}))$, where $\mu^*(\mathbb{E})$ is the pull–back bundle. In any case, $[\sigma] \mapsto [\sigma] \otimes 1$ defines a map $i : \mu^{-1}(\mathcal{E}) \to \mu^*(\mathcal{E})$ which in turn specifies the coefficient morphisms $i_p : H^p(W_D; \mu^{-1}(\mathcal{E})) \to H^p(W_D; \mu^*(\mathcal{E}))$ for $p \geqq 0$. Our natural pull–back maps are the compositions

$$j^{(p)} : H^p(D; \mathcal{E}) \to H^p(W_D; \mu^*(\mathcal{E})) \quad \text{for } p \geqq 0, \tag{6.4}$$

that is, $j^{(p)} = i_p \cdot \mu^{(p)}$.

Consider the case $\mathcal{E} = \mathcal{O}(\mathbb{E})$ for some holomorphic vector bundle $\mathbb{E} \to D$. Then $\mu^*(\mathcal{E}) = \mathcal{O}(\mu^*(\mathbb{E}))$, we realize these sheaf cohomologies as Dolbeault cohomologies, and the pull–back maps are given by pulling back $[\omega] \mapsto [\mu^*(\omega)]$ on the level of differential forms.

Proposition 6.3 is not quite enough to show that the maps $j^{(p)}$ of (6.4) are isomorphism. That involves some study of relative Dolbeault cohomology and the vanishing hypothesis of Theorem 6.6 below.

The second step is to push cohomology down from W_D to M_D. This requires

(6.5) $\nu : W_D \to M$ is a proper map and M is a Stein manifold.

THEOREM 6.6. *Suppose that the fiber F of $\mu : W_D \to D$ is connected and, for some fixed integer $s \geqq 0$, that $H^r(F; \mathbb{C}) = 0$ for $1 \leqq r < s$. Assume* (6.5) *that $\nu : W_D \to M$ is a proper map and that M_D is a Stein manifold. Suppose further that $H^p(\nu^{-1}(Y'); \Omega^q_\mu(\mathbb{E})|_{\nu^{-1}(Y')}) = 0$ for all $Y' \in M_D$, all $p < s$, and $1 \leqq q \leqq m$. Then $P : H^s(D; \mathcal{E}) \to H^0(M_D; \mathcal{R}^s(\mu^*(\mathcal{E})))$ is injective.*

The argument is a little bit technical. Write $\mathcal{R}^p$ gives the pth Leray derived sheaf in the The Leray spectral sequence for $\mu : W_D \to M_D$. Write Ω^q_μ for relative holomorphic q–forms on W_D. The assumption on F ensures, that $\mu^{(s)} : H^s(D; \mathcal{E}) \to H^s(W_D; \mu^{-1}(\mathcal{E}))$ is injective. The Leray spectral sequence for $\mu : W_D \to M_D$ and $\Omega^q_\mu(\mathcal{E}) \to W_D$, and the Stein condition on M_D, give $H^p(W_D; \Omega^q_\mu(\mathcal{E})) \cong H^0(M_D; \mathcal{R}^p(\Omega^q_\mu(\mathcal{E})))$. The vanishing assumption for certain $H^p(\nu^{-1}(Y'); \Omega^q_\mu(\mathcal{E})|_{\nu^{-1}(Y')})$ says

$$\mathcal{R}^p(\Omega^q_\mu(\mathcal{E})) = 0 \text{ for } p < s \text{ and } 1 \leqq q \leqq m.$$

So

$$H^p(W_D; \Omega^q_\mu(\mathcal{E})) = 0 \text{ for } p < s \text{ and } 1 \leqq q \leqq m,$$

and

$$H^s(W_D; \Omega^q_\mu(\mathcal{E})) \cong H^0(M_D; \mathcal{R}^s(\Omega^q_\mu(\mathcal{E}))) \text{ for } 1 \leqq q \leqq m.$$

Now $i_s : H^s(W_D; \mu^{-1}\mathcal{E}) \to H^s(W_D; \mu^*\mathcal{E})$ is injective, and Proposition 6.3 shows that $j^{(s)} : H^s(D; \mathcal{E}) \to H^s(W_D; \mu^*(\mathcal{E}))$ is injective. We conclude that the double fibration transform $P : H^s(D; \mathcal{E}) \to H^0(M_D; \mathcal{R}^s(\mu^*(\mathcal{E})))$ is injective.

In the cases of interest to us, $\mathbb{E} = \mathcal{O}(\mathbb{E})$ for some holomorphic vector bundle $\mathbb{E} \to D$, and P has an explicit formula. Let ω be an $\mathbb{E}$–valued $(0, s)$–form on D representing a Dolbeault cohomology class $[\omega] \in H^s_{\bar\partial}(D, \mathbb{E})$. Note $\mathcal{R}^s(\mu^*(\mathcal{E})) = \mathcal{O}(\mathbb{H}^s(\mu^*(\mathbb{E})|_{\mu^{-1}(Y)})$ where the latter bundle has fiber $H^s(Y'; \mu^*(\mathbb{E})|_{\mu^{-1}(Y')})$ over

$Y' \in M_D$. Thus

$$(6.7)\qquad \begin{aligned} &P([\omega])(Y') \text{ is the section of } \mathcal{R}^s(\mu^*(\mathcal{E})) \to M_D \\ &\qquad \text{whose value at } Y' \in M_D \text{ is } \mu^*(\omega)|_{\nu^{-1}(Y')}. \end{aligned}$$

In other words,

$$(6.8)\qquad P([\omega])(Y') = [\mu^*(\omega)|_{\nu^{-1}(Y')}] \in H^0_{\overline{\partial}}(M_D; \mathbb{H}^s(\mu^*(\mathbb{E})|_{\nu^{-1}(Y)})).$$

This is most conveniently interpreted by viewing $P([\omega])(Y')$ as the Dolbeault class of $\omega|_{Y'}$, and by viewing $Y' \mapsto [\omega|_{Y'}]$ as a holomorphic section of the holomorphic vector bundle over M_D whose fiber at Y' is $H^s(Y'; \mu^*(\mathbb{E})|_{\nu^{-1}(Y')})$.

In order to make the double fibration transform explicit one needs to know

- the exact structure of M_D and
- the differential equations that pick out image of P

The structure of M_D was addressed at the end of the previous section.

There has been a lot of work on the image of double fibration transforms. Almost all of this has been on the image of the Radon transform [**108**], the X–ray transform (see [**5**] and [**38**]), the Funk transform [**46**], and the Penrose transform ([**104**], [**105**], [**106**], [**107**], [**37**], [**39**],). The Funk and Radon transforms are essentially the same, and they can be viewed as Penrose transforms [**6**]. For the image of the Penrose transform in twistor theory see, for example, [**3**], [**4**], [**7**], [**17**], [**35**], [**36**], [**41**], [**86**], [**126**], and some of the references in [**49**] and in [**37**]. See [**73**], [**74**], [**75**], [**76**], [**77**], [**78**], [**79**], [**80**] and [**81**] for connections between Radon transforms and analysis on Riemannian symmetric spaces of noncompact type, with consequences for principal series representations. There also is a $\mathcal{D}$–module approach ([**31**]and [**32**]) which, I expect, will soon have consequences in semisimple representation theory through connections with [**72**]. See [**112**], [**128**], [**109**] and [**151**] for the image of the first double fibration transforms related to semisimple representation theory (rather than particular examples). See [**37**] and [**40**] for a more geometric viewpoint, and, finally, see [**118**], [**99**] (and eventually, [**100**]) for the image of the double fibration transform in our setting.

Part II. Representations of Reductive Lie Groups.

In this Part we describe those unitary representations of the reductive Lie group that come into the Plancherel formula, and we indicate the Plancherel Plancherel formula based on their characters. For the most part, these are the representations that have the cleanest geometric realization in real group orbits on complex flag manifolds. See [**69**], [**70**] and [**71**] for Harish–Chandra's treatment of the Plancherel formula, [**135**] and then [**84**] and [**85**] for another approach.

7. The Principal Series.

The "principal series" of unitary representations of a semisimple or reductive Lie group G_0 was the first series to be constructed in some degree of generality. We consider it separately from the other tempered series in order to demonstrate the general construction without the technical considerations of the discrete series.

A subalgebra $\mathfrak{p}_0 \subset \mathfrak{g}_0$ is a (*real*) *parabolic subalgebra* of $\mathfrak{g}_0$ if it is a real form of a parabolic subalgebra $\mathfrak{p} \subset \mathfrak{g}$. A subgroup $P_0 \subset G_0$ is a *parabolic subgroup* of G_0 if it is the G_0–normalizer of a parabolic subalgebra, say $P_0 = \{g \in G_0 \mid \operatorname{Ad}(g)\mathfrak{p}_0 = \mathfrak{p}_0\}$ where $\mathfrak{p}_0$ is a parabolic subalgebra of $\mathfrak{g}_0$. In that case $\mathfrak{p}_0$ is the Lie algebra of P_0.

For example, let $G_0 = K_0A_0N_0$ is an Iwasawa decomposition. Let M_0 be the centralizer of A_0 in K_0. Then $M_0A_0N_0$ is a minimal parabolic subgroup of G_0. Any two minimal parabolic subgroups of G_0 are conjugate. Now fix a minimal parabolic subgroup $P_0 = M_0A_0N_0$.

Whenever E is a topological group we write $\widehat{E}$ for its unitary dual. Thus $\widehat{E}$ consists of the unitary equivalence classes $[\eta]$ of (strongly continuous) topologically irreducible unitary representations η of E. Now $[\eta] \in \widehat{M_0}$ and $\sigma \in \mathfrak{a}_0^*$ determine $[\alpha_{\eta,\sigma}] \in \widehat{P_0}$ by

$$\alpha_{\eta,\sigma}(man) = \eta(m)e^{i\sigma(\log a)}. \tag{7.1}$$

The corresponding *principal series representation* of G_0 is

$$\pi_{\eta,\sigma} = \operatorname{Ind}_{P_0}^{G_0}(\alpha_{\eta,\sigma}), \quad \text{unitarily induced representation.} \tag{7.2}$$

The *principal series* of G_0 consists of the unitary equivalence classes of these representations. A famous result of Bruhat says that if σ satisfies a certain nonsingularity condition then $\pi_{\eta,\sigma}$ is irreducible.

8. The Discrete Series.

The representations of the "discrete series" are the basic building blocks for the representations involved in the Plancherel formula. We recall the definition and Harish–Chandra parameterization of the discrete series for reductive Lie groups. For application we have to make G_0 more general so that the description applies to certain subgroups of G_0 as well as to G_0. Later we will show how discrete series representations can be realized over certain open orbits.

The *discrete series* of a unimodular locally compact group G_0 is the subset $\widehat{G_{0,d}} \subset \widehat{G_0}$ consisting of all classes $[\pi]$ for which π is equivalent to a subrepresentation of the left regular representation of G_0. These are equivalent: (i) π is a discrete series representation of G_0, (ii) every coefficient $f_{u,v}(x) = \langle u, \pi(x)v\rangle$ belongs to $L^2(G_0)$, (iii) for some nonzero u, v in the representation space H_π, the coefficient $f_{u,v} \in L^2(G_0)$. Then one has orthogonality relations much as in the case of finite groups: there is a real number $\deg(\pi) > 0$ such that the $L^2(G_0)$-inner product of coefficients of π is given by

$$\langle f_{u,v}, f_{s,t}\rangle = \frac{1}{\deg(\pi)}\langle u, s\rangle\overline{\langle v, t\rangle} \text{ for } s, t, u, v \in H_\pi. \tag{8.1}$$

Furthermore, if π' is a discrete series representation not equivalent to π, then

$$\langle f_{u,v}, f_{u',v'}\rangle = 0 \text{ for } u, v \in H_\pi \text{ and } u', v' \in H_{\pi'}. \tag{8.2}$$

These orthogonality relations come out of convolution formulae. With the usual $f * h(x) = [L(f)h](x) = \int_G f(y)h(y^{-1}x)dy$ we have $f_{u,v} * f_{s,t} = \frac{1}{\deg(\pi)}\langle u, t\rangle f_{s,v}$ for

$s, t, u, v \in H_\pi$. Also $f_{u,v} * f_{u',v'} = 0$ for $u, v \in H_\pi$ and $u', v' \in H_{\pi'}$ whenever π and π' are inequivalent discrete series representations of G_0.

If G_0 is compact, then every class in $\widehat{G_0}$ belongs to the discrete series, and if Haar measure is normalized as usual to total volume 1 then $\deg(\pi)$ has the usual meaning, the dimension of H_π. The orthogonality relations for irreducible unitary representations of compact groups are more or less equivalent to the Peter–Weyl Theorem. More generally, if G_0 is a unimodular locally compact group then $L^2(G_0) = {}^0L^2(G_0) \oplus {}'L^2(G_0)$, orthogonal direct sum, where ${}^0L^2(G_0) = \sum_{[\pi]\in\widehat{G_{0,d}}} H_\pi \otimes H_\pi^*$, the "discrete" part, and ${}'L^2(G_0) = {}^0L^2(G_0)^\perp$, the "continuous" part. If, further, G_0 is a group of type I then ${}'L^2(G_0)$ is a continuous direct sum (direct integral) over $\widehat{G_0} \setminus \widehat{G_{0,d}}$ of the Hilbert spaces $H_\pi \otimes H_\pi^*$.

Recall our assumption that G_0 belongs to the Harish–Chandra class of reductive Lie groups (2.1): the Lie algebra $\mathfrak{g}_0$ is reductive, the component group G_0/G_0^0 is finite, the derived group $[G_0^0, G_0^0]$ is closed in G_0, and if $g \in G_0$ then $\mathrm{Ad}(g)$ is an inner automorphism of the complex Lie algebra $\mathfrak{g}$. The results we describe hold somewhat more generally, but that is not the point here.

If $[\pi] \in \widehat{G_0}$ and $f \in \mathbb{C}_c^\infty(G_0)$ then $\pi(f) = \int_G f(x)\pi(x)dx$ is a trace class operator on H_π, and the map

(8.3) $$\Theta_\pi : C_c^\infty(G_0) \to \mathbb{C} \text{ defined by } \Theta_\pi(f) = \text{ trace } \pi(f)$$

is a distribution on G_0. Θ_π is called the *character*, the *distribution character* or the *global character* of π. The equivalence class $[\pi]$ determines Θ_π and, conversely, Θ_π determines $[\pi]$.

Let $\mathcal{Z}(\mathfrak{g})$ denote the center of the universal enveloping algebra $\mathcal{U}(\mathfrak{g})$. If we interpret $\mathcal{U}(\mathfrak{g})$ as the algebra of all left-invariant differential operators on G_0 then $\mathcal{Z}(\mathfrak{g})$ is the subalgebra of those which are also invariant under right translations. If π is irreducible then $d\pi|_{\mathcal{Z}(\mathfrak{g})}$ is an associative algebra homomorphism $\chi_\pi : \mathcal{Z}(\mathfrak{g}) \to \mathbb{C}$ called the *infinitesimal character* of π. We say that π is *quasi-simple* if it has an infinitesimal character, i.e. if it is a direct sum of irreducible representations that have the same infinitesimal character.

Let π be quasi-simple. Then the distribution character Θ_π satisfies a system of differential equations

(8.4) $$z \cdot \Theta_\pi = \chi_\pi(z)\Theta_\pi \text{ for all } z \in \mathcal{Z}(\mathfrak{g}).$$

Now let π be irreducible. A serious study of these equations shows that Θ_π is integration against a locally L^1 function T_π that is real analytic on a dense open subset G_0' of G_0,

(8.5) $$\Theta_\pi(f) = \int_{G_0} f(x)T_\pi(x)dx \text{ for all } f \in C_c^\infty(G_0).$$

So we identify Θ_π with the function T_π, and it makes sense to talk about *a priori* estimates on characters and coefficients as well as explicit character formulae.

Fix a Cartan involution θ of G_0: $\theta \in \mathrm{Aut}(G_0)$, $\theta^2 = 1$, and the fixed point set $K_0 = G_0^\theta$ is a maximal compact subgroup of G_0. The choice is essentially unique because any two are conjugate in $\mathrm{Aut}(G_0)$. If $G_0 = U(p,q)$ then $\theta(x) = {}^t x^{-1}$ and $K_0 = U(p) \times U(q)$.

Every Cartan subgroup of G_0 is $\mathrm{Ad}(G_0^0)$-conjugate to a θ-stable Cartan subgroup. In particular, G_0 has compact Cartan subgroups if and only if K_0 contains a Cartan subgroup of G_0.

Harish–Chandra proved that G_0 has discrete series representations if and only if it has a compact Cartan subgroup. Suppose that this is the case and fix a compact Cartan subgroup $T_0 \subset K_0$ of G_0. Let $\Sigma = \Sigma(\mathfrak{g}, \mathfrak{t})$ be the root system, $\Sigma^+ = \Sigma^+(\mathfrak{g}, \mathfrak{t})$ a choice of positive root system, and let $\rho = \frac{1}{2}\sum_{\alpha \in \Sigma^+} \alpha$. If $\xi \in \mathfrak{t}$ then $\rho(\xi)$ is half the trace of $ad(\xi)$ on $\sum_{\alpha \in \Sigma^+} \mathfrak{g}_\alpha$.

If π is a discrete series representation of G_0 and Θ_π is its distribution character, then the equivalence class of π is determined by the restriction of Θ_π to $T_0 \cap G_0'$. Harish–Chandra parameterized the discrete series of G_0 by parameterizing those restrictions.

Let $G_0^\dagger$ denote the finite index subgroup $T_0 G_0^0 = Z_{G_0}(G_0^0) G_0^0$ of G_0. Here $T_0 = Z_{G_0}(G_0^0) T_0^0$, so $T_0 = T_0^\dagger$. Lemma 11.2 says that the group M_0 of a minimal parabolic subgroup of G_0 satisfies $M_0 = M_0^\dagger$, and similarly in that context we have $U_{\Phi,0} = U_{\Phi,0}^\dagger$. In general, where M_0 may be noncompact, this need not hold. In any case, the Weyl group $W^\dagger = W(G_0^\dagger, T_0)$ coincides with $W^0 = W(G_0^0, T_0^0)$ and is a normal subgroup of $W = W(G_0, T_0)$.

$\widehat{T_0} = (Z_{G_0}(G_0^0) T_0^0)^\frown$ consists of the $\chi \otimes e^{i(\lambda - \rho)}$ where $\lambda \in i\mathfrak{t}_0^*$ and $\lambda - \rho$ satisfies an integrality condition, where $\chi \in Z_{G_0}(G_0^0)^\frown$, and where χ and $e^{i(\lambda-\rho)}$ restrict to (multiples of) the same unitary character on the center of G_0^0.

Given $\chi \otimes e^{i(\lambda-\rho)} \in \widehat{T_0}$ as above, with λ is regular, i.e., there are unique discrete series representations π_λ^0 of G_0^0 and $\pi_{\chi,\lambda}^\dagger = \chi \otimes \pi_\lambda^0$ of $G_0^\dagger$, whose distribution characters satisfy

(8.6)

$$\Theta_{\pi_\lambda^0}(x) = (-1)^{q(\lambda)} \frac{\sum_{w \in W^0} \mathrm{sign}(w) e^{w(\lambda)}}{\prod_{\alpha \in \Sigma^+} (e^{\alpha/2} - e^{-\alpha/2})} \text{ and } \Theta_{\pi_{\chi,\lambda}^\dagger}(zx) = \mathrm{trace}\, \chi(z) \Theta_{\pi_\lambda^0}(x)$$

for $z \in Z_{G_0}(G_0^0)$ and $x \in T_0^0 \cap G_0'$, where $q(\lambda)$ is the cardinality

$$q(\lambda) = |\{\alpha \in \Sigma^+(\mathfrak{k}, \mathfrak{t}) \mid \langle \alpha, \lambda \rangle < 0\}| + |\{\beta \in \Sigma^+(\mathfrak{g}, \mathfrak{t}) \setminus \Sigma^+(\mathfrak{k}, \mathfrak{t}) \mid \langle \beta, \lambda \rangle > 0\}|.$$

The same datum (χ, λ) specifies a discrete series representation $\pi_{\chi,\lambda}$ of G_0, by the formula $\pi_{\chi,\lambda} = \mathrm{Ind}_{G_0^\dagger}^{G_0}(\pi_{\chi,\lambda}^\dagger)$. This induced representation is irreducible because its conjugates by elements of $G_0/G_0^\dagger$ are mutually inequivalent, consequence of regularity of λ. $\pi_{\chi,\lambda}$ is characterized by the fact that its distribution character is supported in $G_0^\dagger$ and is given on $G_0^\dagger$ by

$$\Theta_{\pi_{\chi,\lambda}} = \sum_{1 \leqq i \leqq r} \Theta_{\pi_{\chi,\lambda}^\dagger} \cdot \gamma_i^{-1}, \tag{8.7}$$

with $\gamma_i = \mathrm{Ad}(g_i)|_{G_0^\dagger}$ where $\{g_1, \ldots, g_r\}$ is any system of coset representatives of G_0 modulo $G_0^\dagger$. To combine these into a single formula one chooses the g_i so that they normalize T_0, i.e. chooses the γ_i to be a system of coset representatives of W modulo $W^\dagger$.

Every discrete series representation of G_0 is equivalent to a representation $\pi_{\chi,\lambda}$ as just described. Discrete series representations $\pi_{\chi,\lambda}$ and $\pi_{\chi',\lambda'}$ are equivalent if and only if $\chi'\otimes e^{i\lambda'} = (\chi\otimes e^{i\lambda})\cdot w^{-1}$ for some $w \in W$. And λ is both the infinitesimal character and the Harish–Chandra parameter for the discrete series representation $\pi_{\chi,\lambda}$.

9. The Various Tempered Series.

A representation of G_0 is called *tempered* if its character is a tempered distribution in a suitable sense. This means that it is weakly contained in the left regular representation or, roughly, that it is involved in the Plancherel formula for G_0. Tempered representations are constructed from a class of real parabolic subgroups of G_0 called *cuspidal parabolic subgroups*, constructed using minimal parabolic subgroups, but using a discrete series representation on the part of the parabolic corresponding to M_0.

A (real) parabolic subgroup $P_0 \subset G_0$ is called *cuspidal* if the commutator subgroup of the Levi component (reductive part) has a compact Cartan subgroup.

Let H_0 be a Cartan subgroup of G_0. Fix a Cartan involution θ of G_0 such that $\theta(H_0) = H_0$. Write K_0 for the fixed point set G_0^θ, maximal compact subgroup of G_0. Decompose

(9.1)
$$\mathfrak{h}_0 = \mathfrak{t}_0 \oplus \mathfrak{a}_0 \text{ and } H_0 = T_0 \times A_0 \text{ where}$$
$$T_0 = H_0 \cap K_0, \theta = -1 \text{ on } \mathfrak{a}_0, \text{ and } A_0 = \exp_G(\mathfrak{a}_0).$$

Then the centralizer $Z_{G_0}(A_0)$ of A_0 in G_0 has form $M_0 \times A_0$ where $\theta(M_0) = M_0$. The group M_0 is a reductive Lie group of Harish–Chandra class. T_0 is a compact Cartan subgroup of M_0, so M_0 has discrete series representations. Now suppose that $\Sigma^+ = \Sigma^+(\mathfrak{g}, \mathfrak{h})$ is defined by positive root systems $\Sigma^+(\mathfrak{m}, \mathfrak{t})$ and $\Sigma^+(\mathfrak{g}_0, \mathfrak{a}_0)$ as in (11.1).

The Cartan subgroup $H_0 \subset G_0$ and the positive system $\Sigma^+ = \Sigma^+(\mathfrak{g}, \mathfrak{h})$ define a cuspidal parabolic subgroup $P_0 = M_0A_0N_0$ of G_0 as follows. The Lie algebra of N_0 is $\mathfrak{n}_0 = \sum_{\alpha\in\Sigma^+(\mathfrak{g}_0,\mathfrak{a}_0)}(\mathfrak{g}_0)_{-\alpha}$, M_0 and A_0 are as above, and $M_0A_0 = M_0 \times A_0$ is the Levi component of P_0. One extreme is the case where $\dim \mathfrak{a}_0$ is maximal; then P_0 is a minimal parabolic subgroup of G_0. The other extreme is where $\dim \mathfrak{a}_0$ is minimal; if $\mathfrak{a}_0 = 0$ then $P_0 = G_0$.

Every cuspidal parabolic subgroup of G_0 is produced by the construction just described, as H_0 varies. Two cuspidal parabolic subgroups of G_0 are *associated* if they are constructed as above from G_0–conjugate Cartan subgroups; then we say that the G_0–conjugacy class of Cartan subgroups is *associated* to the G_0–association class of cuspidal parabolic subgroups.

In the same manner as the construction of the principal series representations,

(9.2) $[\eta] \in \widehat{M_0}$ and $\sigma \in \mathfrak{a}_0^*$ determine $[\alpha_{\eta,\sigma}] \in \widehat{Q_0}$ by $\alpha_{\eta,\sigma}(man) = \eta(m)e^{i\sigma(\log a)}$.

Then we have

(9.3)
$$\pi_{\eta,\sigma} = \operatorname{Ind}_{P_0}^{G_0}(\alpha_{\eta,\sigma}), \quad \text{unitarily induced representation.}$$

The H_0*–series* or *principal* H_0*–series* of G_0 consists of the unitary equivalence classes of the representations (9.3) for which η is a discrete series representation of M_0.

The character formula for the $\pi_{\eta,\sigma}$ is a bit complicated, and we just refer to [**135**] and [**84**]. We mention that the formula is independent of choice of $\Sigma^+(\mathfrak{g}_0, \mathfrak{a}_0)$ so, as the terminology indicates, $\pi_{\eta,\sigma} = \operatorname{Ind}_{P_0}^{G_0}(\alpha_{\eta,\sigma})$ is independent of choice of $\Sigma^+(\mathfrak{g}_0, \mathfrak{a}_0)$. In fact this is the case even if η does not belong to the discrete series of M_0.

10. The Plancherel Formula.

There are two approaches to the Plancherel formula. Harish–Chandra starts with an analysis of his Schwartz space $\mathcal{C}(G_0)$ and construction of functions on G_0 as wave packets of Eisenstein integrals ([**69**], [**70**], [**71**]). He knew that it is better to step through the conjugacy classes of Cartan subgroups from maximally compact to minimally compact, using explicit character formulae [**54**]. This was first carried out in real rank 1 [**111**], and then carried out in general in [**84**] and [**85**]. That is the argument we indicate here.

We start with Kostant's "cascade construction" for the conjugacy classes of Cartan subgroups of G_0. Suppose first that G_0 has a compact Cartan subgroup T_0. Fix a Cartan involution θ of G_0 such that $\theta(T_0) = T_0$ and the corresponding ± 1 eigenspace decomposition $\mathfrak{g}_0 = \mathfrak{k}_0 + \mathfrak{s}_0$ where $\mathfrak{k}_0$ is the Lie algebra of the maximal compact subgroup $K_0 = \{g \in G_0 \mid \theta(g) = g\}$. If $\alpha \in \Sigma(\mathfrak{g}, \mathfrak{t})$ then either $\mathfrak{g}_\alpha \subset \mathfrak{k}$ and we say that α is *compact*, or $\mathfrak{g}_\alpha \subset \mathfrak{s}$ and we say that α is *noncompact.*

Let $\alpha \in \Sigma(\mathfrak{g}, \mathfrak{t})$ be noncompact. Let $\mathfrak{g}[\alpha] = \mathfrak{g}_\alpha + \mathfrak{g}_{-\alpha} + [\mathfrak{g}_\alpha, \mathfrak{g}_{-\alpha}] \cong \mathfrak{sl}(2;\mathbb{C})$, let $G[\alpha]$ denote the corresponding analytic subgroup of G, and consider the corresponding real forms $\mathfrak{g}_0[\alpha] = \mathfrak{g}_0 \cap \mathfrak{g}[\alpha] \cong \mathfrak{sl}(2;\mathbb{R})$ and $G_0[\alpha] = G_0 \cap G[\alpha]$. Then $G_0[\alpha] \cap T_0$ is a compact Cartan subgroup, and we can simply replace it by the noncompact Cartan subgroup of $G_0[\alpha]$. Let $\mathfrak{a}_0[\alpha]$ denote the Lie algebra of that noncompact Cartan subgroup. Then we have a new Cartan subalgebra and a new Cartan subgroup

$$\mathfrak{h}_0\{\alpha\} = (\mathfrak{t}_0 \cap (\mathfrak{g}_0[\alpha] \cap \mathfrak{t}_0)^\perp) + \mathfrak{g}_0[\alpha] \cap \mathfrak{a}_0 \text{ and } H_0\{\alpha\} = Z_{G_0}(\mathfrak{h}_0\{\alpha\}),$$

where Z_{G_0} denotes the centralizer in G_0. The point is that $H_0\{\alpha\}$ has one compact dimension less than that of T_0 and one noncompact dimension more.

Let $\alpha, \beta \in \Sigma(\mathfrak{g}, \mathfrak{t})$ be noncompact. We can carry out the above construction for α and β independently, one after the other, if α and β are *strongly orthogonal* in the sense that neither of $\alpha \pm \beta$ are roots. (We write this relation as $\alpha \underline{\perp} \beta$. It implies $\alpha \perp \beta$.) If $\alpha \underline{\perp} \beta$ then we have the new Cartan subalgebra and a new Cartan subgroup given by $\mathfrak{h}_0\{\alpha, \beta\} = (\mathfrak{t}_0 \cap ((\mathfrak{g}_0[\alpha] \oplus \mathfrak{g}_0[\beta]) \cap \mathfrak{t}_0)^\perp) + (\mathfrak{a}_0[\alpha] \oplus \mathfrak{a}_0[\beta])$ and $H_0\{\alpha, \beta\} = Z_{G_0}(\mathfrak{h}_0\{\alpha, \beta\})$. Here $H_0\{\alpha, \beta\}$ has two compact dimensions less than that of T_0 and two noncompact dimensions more.

We say that a set S of noncompact roots is *strongly orthogonal* if its elements are mutually strongly orthogonal. Then as above we have a Cartan subalgebra by

$$\mathfrak{h}_0\{S\} = \left(\mathfrak{t}_0 \cap \left(\left(\textstyle\sum_{\alpha \in S} \mathfrak{g}_0[\alpha]\right) \cap \mathfrak{t}_0\right)^\perp\right) + \left(\textstyle\sum_{\alpha \in S} \mathfrak{a}_0[\alpha]\right), \tag{10.1}$$

and the corresponding Cartan subgroup is $H_0\{S\} = Z_{G_0}(\mathfrak{h}_0\{S\})$. Here $H_0\{S\}$ has $|S|$ compact dimensions fewer than T_0 has, and $H_0\{S\}$ has $|S|$ noncompact dimensions more than T_0 has.

Cartan subgroups $H_0\{S_1\}$ and $H_0\{S_2\}$ are G_0–conjugate just when some element $w \in W(G_0, T_0)$ sends S_1 to S_2. Every Cartan subgroup of G_0 is conjugate to $\mathfrak{h}_0\{S\}$ for some set S of strongly orthogonal noncompact roots. This sets up a hierarchy among the conjugacy classes of Cartan subgroups of $G_0 : H_0\{S_1\} \leqq H_0\{S_2\}$ if and only if $w(S_2) \subset S_1$ for some $w \in W(G_0, T_0)$. That in turn sets up a hierarchy among parts of the regular set G'_0. If H_0 is any Cartan subgroup of G_0 we denote $G'_{H_0} = G'_0 \cap \mathrm{Ad}(G)H_0$, the set of all regular elements G'_0 that are conjugate to an element of H_0. Now $G'_{H_0\{S_1\}} \leqq G'_{H_0\{S_2\}}$ if and only if some Weyl group element $w \in W(G_0, T_0)$ sends S_2 to a subset of S_1. Here G'_{T_0} sits at the top, the $G'_{H_0\{\alpha\}}$ sit just below, the $G'_{H_0\{\alpha,\beta\}}$ are on the next level down, and finally the part of G'_0 corresponding to the Cartan subgroup of the minimal parabolic subgroups sit at the bottom.

If G_0 does not have a compact Cartan subgroup, let $H_0 = T_0 \times A_0$ be a fundamental (maximally compact) Cartan subgroup, so T_0 is a Cartan subgroup of a maximal compact subgroup $K_0 \subset G_0$. Let $P_0 = M_0 A_0 N_0$ be an associated cuspidal parabolic subgroup. Then just do the cascade construction for M_0, obtaining a family of Cartan subgroups $H_{M,0}\{S\} \subset M_0$ as S runs over the $W(M_0, T_0)$–conjugacy classes of strongly orthogonal sets $S \subset \Sigma(\mathfrak{m}, \mathfrak{t})$ of noncompact roots of $\mathfrak{m}$. Then the $H_0\{S\} = H_{M,0}\{S\} \times A_0$ give the conjugacy classes of Cartan subgroups of G_0.

The character formulae for the various tempered series exhaust enough of $\widehat{G_0}$ for a decomposition of $L_2(G_0)$ essentially as

$$\sum_{[H_0] \in Car(G_0)} \sum_{\chi \otimes e^{\nu - \rho_{\mathfrak{m}}} \in \widehat{T_0}} \int_{\mathfrak{a}_0^*} H_{\pi_{\chi,\nu,\sigma}} \otimes H^*_{\pi_{\chi,\nu,\sigma}} m(H_0 : \chi : \nu : \sigma) d\sigma. \tag{10.2}$$

Here $Car(G_0)$ denotes the set of G_0–conjugacy classes $[H_0]$ of Cartan subgroups H_0 and the Borel measure $m(H_0 : \chi : \nu : \sigma)d\sigma$ is the *Plancherel measure* on $\widehat{G_0}$. In general the Plancherel density $m(H_0 : \chi : \nu : \sigma)$ has a formula that varies with the component of the regular set. This was worked out by Harish–Chandra for groups of Harish–Chandra class, and somewhat more generally by Herb and myself. Harish–Chandra's approach is based on an analysis of the structure of the Schwartz space, while Herb and I use explicit character formulae. These explicit formulae allow us to prove (10.2), as follows.

Start with G'_{H_0} where H_0 represents the conjugacy class of Cartan subgroups of G_0 that are as compact as possible. The H_0–series representations suffice to expand functions $f \in C_0^\infty(G'_{H_0})$. That expansion formula gives us a map $C_0^\infty(G_0) \to C^\infty(G_0 \setminus G'_{H_0})$ by $f \mapsto f_1$ where r_x denotes right translation by $x \in G_0$ and $f_1(x) = f(x) - \sum_{\chi \otimes e^{\nu - \rho_{\mathfrak{m}}} \in \widehat{T_0}} \int_{\mathfrak{a}_0^*} \Theta_{\pi_{\chi,\nu,\sigma}}(r_x f) m(H_0 : \chi : \nu : \sigma) d\sigma$. This map requires an exact knowledge of the characters of the H_0–series. Now let $\{H_0\{\alpha_1\}, \ldots, H_0\{\alpha_{m_1}\}$ be a set of representatives of the conjugacy classes of Cartan subgroups just below H_0. The $H_0\{\alpha_i\}$–series representations suffice to expand functions $f \in C_0^\infty(G'_{H_0\{\alpha_i\}})$. Those expansions do not interact at this level, nor do they introduce nonzero values in G'_{H_0}, so they give us a map

$$C^\infty\left(G_0 \setminus G'_{H_0}\right) \to C^\infty\left(G_0 \setminus \left(G'_{H_0} \cup \bigcup G'_{H_0\{\alpha_i\}}\right)\right) \text{ by } f_1 \mapsto f_2 ,$$

where

$$f_2(x)-f_1(x) = \sum_{1\leqq i\leqq m_1} \sum_{\chi\otimes e^{\nu-\rho_{\mathfrak{m}}}\in\widehat{T_0\{\alpha_i\}}} \int_{\mathfrak{a}_0^*\{\alpha_i\}} \Theta_{\pi_{\chi,\nu,\sigma}}(r_x f)m(H_0\{\alpha_i\} : \chi : \nu : \sigma)d\sigma.$$

Now simply proceed down one level at a time. The tricky point here is to know the character formulae completely, so that one knows f_j well enough to compute f_{j+1}. Thus one obtains exact information on the Plancherel densities $m(H_0 : \chi : \nu : \sigma)$ and the final form

$$f(x) = \sum_{H_0\in Car(G_0)} \sum_{\chi\otimes e^{\nu-\rho_{\mathfrak{m}}}\in\widehat{T_0}} \int_{\mathfrak{a}_0^*} \Theta_{\pi_{\chi,\nu,\sigma}}(r_x f)m(H_0 : \chi : \nu : \sigma)d\sigma \tag{10.3}$$

of the Plancherel formula.

Part III. Geometric Realizations of Representations.

Now we show how the tempered representations $\pi_{\eta,\sigma}$ are realized on appropriate real G_0–orbits on complex flags $Z = G/Q$. The material on L^2 realizations is taken from [**135**].

11. L^2 Realizations of the Principal Series.

In order to realize the principal series of G_0 on closed orbits, we need the Bott–Borel–Weil Theorem for M_0, where $P_0 = M_0A_0N_0$ is a minimal parabolic subgroup of G_0. We have to be careful here because the compact group M_0 need not be connected. We will first decompose M_0 as the product $Z_{M_0}(M_0^0)M_0^0$ where M_0^0 is its identity component, then indicate the analog of the Cartan highest weight description for $\widehat{M_0}$. That done, the standard Bott–Borel–Weil Theorem for M_0^0 will carry over to M_0.

A Cartan subgroup $T_0 \subset M_0$ specifies a Cartan subgroup $H_0 = T_0A_0 \cong T_0 \times A_0$ in G_0. Our choice of P_0 specifies a choice of positive restricted root system $\Sigma^+(\mathfrak{g}_0, \mathfrak{a}_0)$: the Lie algebra of N_0 is given by $\mathfrak{n}_0 = \sum_{\alpha\in\Sigma^+(\mathfrak{g}_0,\mathfrak{a}_0)}(\mathfrak{g}_0)_{-\alpha}$. Now any positive root system $\Sigma^+(\mathfrak{m}, \mathfrak{t})$ specifies a positive system $\Sigma^+(\mathfrak{g}, \mathfrak{h})$ by

$$\begin{aligned} \alpha \in \Sigma^+(\mathfrak{g}, \mathfrak{h}) \text{ if and only if either } & \alpha|_{\mathfrak{a}_0} = 0 \text{ and } \alpha|_{\mathfrak{t}} \in \Sigma^+(\mathfrak{m}, \mathfrak{t}) \\ \text{or } & \alpha|_{\mathfrak{a}_0} \neq 0 \text{ and } \alpha|_{ga_0} \in \Sigma^+(\mathfrak{g}_0, \mathfrak{a}_0). \end{aligned} \tag{11.1}$$

Lemma 11.2. $M_0 = Z_{M_0}(M_0^0)M_0^0$. *Given a representation class* $[\eta] \in \widehat{M_0}$, *there exist unique classes* $[\chi] \in \widehat{Z_{M_0}(M_0^0)}$ *and* $[\eta^0] \in \widehat{M_0^0}$ *such that* $[\eta] = [\chi \otimes \eta^0]$, *and* $[\chi]$ *and* $[\eta^0]$ *restrict to multiples of the same unitary character on the center of* M_0^0.

Let $\Psi_{\mathfrak{m}}$ denote the set of simple roots in $\Sigma^+(\mathfrak{m},\mathfrak{t})$. Every subset $\Phi \subset \Psi_{\mathfrak{m}}$ defines

$$(11.3)\quad \begin{cases} \mathfrak{z}_\Phi = \{\xi \in \mathfrak{t} \mid \phi(\xi) = 0 \text{ for all } \phi \in \Phi\} \text{ with real form } \mathfrak{z}_{\Phi,0} = \mathfrak{m}_0 \cap \mathfrak{x}_\Phi\ , \\ U_\Phi = Z_M(\mathfrak{z}_\Phi), U_{\Phi,0} = M_0 \cap U_\Phi, \text{ and their Lie algebras } \mathfrak{u}_\Phi \text{ and } \mathfrak{u}_{\Phi,0}\ , \\ \mathfrak{r}_\Phi = \mathfrak{u}_\Phi + \sum_{\gamma \in \Sigma^+(\mathfrak{m},\mathfrak{t})} \mathfrak{m}_{-\gamma}, \text{ parabolic subalgebra of } \mathfrak{m}\ , \\ R_\Phi = N_M(\mathfrak{r}_\Phi), \text{ corresponding parabolic subgroup of } M\ , \text{ and} \\ S_\Phi = M/R_\Phi, \text{ associated complex flag manifold.} \end{cases}$$

Lemma 11.2 holds for $U_{\Phi,0}$, M_0 acts transitively on S_Φ , and $M_0 \cap R_\Phi = U_{\Phi,0}$, so

LEMMA 11.4. *S_Φ is a compact homogeneous Kähler manifold under the action of M_0, and $S_\Phi = M_0/U_{\Phi,0}$ as coset space. Furthermore $U_{\Phi,0} = Z_{M_0}(M_0^0)U_{\Phi,0}^0$, so $\widehat{U_{\Phi,0}}$ decomposes as does $\widehat{M_0}$ in* Lemma 11.2.

An irreducible unitary representation μ of $U_{\Phi,0}$, say with representation space V_μ , gives us

$$(11.5)\quad \begin{cases} \mathbb{V}_\mu \to S_\Phi : U_{\Phi,0}\text{–homogeneous, hermitian, holomorphic vector bundle,} \\ A^{p,q}(S_\Phi;\mathbb{V}_\mu): \text{ space of } C^\infty\ \mathbb{V}_\mu\text{–valued } (p,q)\text{–forms on } S_\Phi\ , \\ \mathcal{O}(\mathbb{V}_\mu): \text{ sheaf of germs of holomorphic sections of } \mathbb{V}_\mu \to S_\Phi\ . \end{cases}$$

$A^{p,q}(S_\Phi;\mathbb{V}_\mu)$ is the space of C^∞ sections of $\mathbb{V}_\mu^{p,q} = \mathbb{V}_\mu \otimes \Lambda^p(\mathbb{T}^*) \otimes \Lambda^q(\overline{\mathbb{T}}^*) \to S_\Phi$. $\mathbb{V}_\mu^{p,q}$ has an M_0–invariant hermitian metric, so we also have the Hodge–Kodaira orthocomplementation operators

$$(11.6)\quad \begin{aligned} &\sharp : A^{p,q}(S_\Phi;\mathbb{V}_\mu) \to A^{n-p,n-q}(S_\Phi;\mathbb{V}_\mu^*) \\ &\qquad\qquad \text{and } \tilde{\sharp} : A^{n-p,n-q}(S_\Phi;\mathbb{V}_\mu^*) \to A^{p,q}(S_\Phi;\mathbb{V}_\mu), \end{aligned}$$

where $n = \dim_{\mathbb{C}} S_\Phi$. The global M_0–invariant inner product on $A^{p,q}(S_\Phi;\mathbb{V}_\mu)$ is given by taking the inner product in each fiber of $\mathbb{V}_\mu^{p,q}$ and integrating over S_Φ ,

$$(11.7)\quad \langle F_1, F_2\rangle_{S_\Phi} = \int_{M_0} \langle F_1, F_2\rangle_{mU_{\Phi,0}} d(mU_{\Phi,0}) = \int_{S_\Phi} F_1 \bar{\wedge} \sharp F_2\ .$$

In the usual way that gives us the Kodaira–Hodge–Laplace operator $\square = \bar\partial\bar\partial^* + \bar\partial^*\bar\partial$ on $A^{p,q}(S_\Phi;\mathbb{V}_\mu)$, where $\bar\wedge$ means exterior product followed by contraction of V_μ against V_μ^* . So we arrive at the Hilbert space

$$(11.8)\quad \mathcal{H}_2^{p,q}(S_\Phi;\mathbb{V}_\mu): \mathbb{V}_\mu\text{–valued square integrable harmonic}(p,q)\text{–forms on } S_\Phi$$

Everything is invariant under the action of M_0, and the natural action of the group M_0 on $\mathcal{H}_2^{p,q}(S_\Phi;\mathbb{V}_\mu)$ is a unitary representation. Denote $\mathcal{H}_2^q(S_\Phi;\mathbb{V}_\mu) = \mathcal{H}_2^{0,q}(S_\Phi;\mathbb{V}_\mu)$.

Let $\rho_{\mathfrak{m}}$ denote half the sum of the roots in $\Sigma^+(\mathfrak{m},\mathfrak{t})$, and let η_ν^0 denote the irreducible representation of M_0^0 of highest weight $\nu - \rho_{\mathfrak{m}}$ (corresponding to infinitesimal character ν). With these conventions, the Bott–Borel–Weil Theorem for M_0 is

THEOREM 11.9. *Let* $[\mu] = [\chi \otimes \mu_\beta^0] \in \widehat{U_{\Phi,0}}$ *and fix an integer* $q \geqq 0$.

1. *If* $\langle \beta - \rho_{\mathfrak{u}_\Phi} + \rho_{\mathfrak{m}}, \alpha \rangle = 0$ *for some* $\alpha \in \Sigma(\mathfrak{m}, \mathfrak{t})$ *then* $\mathcal{H}_2^q(S_\Phi; \mathbb{V}_\mu) = 0$.

2. *If* $\langle \beta - \rho_{\mathfrak{u}_\Phi} + \rho_{\mathfrak{m}}, \alpha \rangle \neq 0$ *for all* $\alpha \in \Sigma(\mathfrak{m}, \mathfrak{t})$, *let* w *be the unique element in* $W(\mathfrak{m}, \mathfrak{t})$ *such that* $\langle w(\beta - \rho_{\mathfrak{u}_\Phi} + \rho_{\mathfrak{m}}), \alpha \rangle > 0$ *for all* $\alpha \in \Sigma^+(\mathfrak{m}, \mathfrak{t})$. *Then* $\mathcal{H}_2^q(S_\Phi; \mathbb{V}_\mu) = 0$ *for* $q \neq q_0$, *and* M_0 *acts irreducibly on* $\mathcal{H}_2^{q_0}(S_\Phi; \mathbb{V}_\mu)$ *by* $[\chi \otimes \eta_\nu^0]$.

Fix $[\mu] = [\chi \otimes \mu_\beta^0] \in \widehat{U_{\Phi,0}}$ and $\sigma \in \mathfrak{a}_0^*$. We will use the Bott–Borel–Weil Theorem 11.9 to find the principal series representation $\pi_{\chi\otimes\eta_\nu^0,\sigma}$ of G_0 on a cohomology space related to the closed orbit in the complex flag manifold $Z_\Phi = G/Q_\Phi$. Here the simple root system $\Psi_{\mathfrak{m}} \subset \Psi$ by the coherence in our choice of $\Sigma^+(\mathfrak{g}, \mathfrak{h})$, so $\Phi \subset \Psi$ and Φ defines a parabolic subgroup $Q_\Phi \subset G$.

Let $z_\Phi = 1Q_\Phi \in G/Q_\Phi = Z_\Phi$. As $A_0N_0 \subset G_0 \cap Q_\Phi$ we have $G_0 \cap Q_\Phi$ (in the sense of the action on Z_Φ) equal to $U_{\Phi,0}A_0N_0$. Thus $Y_\Phi = G_0(z_\Phi)$ is the closed G_0–orbit on Z_Φ, and S_Φ sits in Y_Φ as the orbit $M_0(z_\Phi)$. Here $P_0 = M_0A_0N_0 = \{g \in G_0 \mid gS_\Phi = S_\Phi\}$.

LEMMA 11.10. *The map* $Y_\Phi \to G_0/P_0$, *given by* $g(z_\Phi) \mapsto gP_0$, *defines a* G_0–*equivariant fiber bundle with structure group* M_0 *and whose fibers* gS_Φ *are the maximal complex analytic submanifolds of* Y_Φ.

The data (μ, σ) defines a representation $\gamma_{\mu,\sigma}(uan) = e^{(\rho_{\mathfrak{g}}+i\sigma)(\log a)}\mu(u)$ of $U_{\Phi,0}A_0N_0$ where $\rho_{\mathfrak{g}} = \frac{1}{2}\sum_{\alpha\in\Sigma^+}\alpha$. That defines a G_0–homogeneous complex vector bundle

$$\mathbb{V}_{\mu,\sigma} \to G_0/U_{\Phi,0}A_0N_0 = Y_\Phi \text{ such that } \mathbb{V}_{\mu,\sigma}|_{S_\Phi} = \mathbb{V}_\mu . \tag{11.11}$$

Each $\mathbb{V}_{\mu,\sigma}|_{gS_\Phi}$ is an $\mathrm{Ad}(g)P_0$–homogeneous holomorphic vector bundle. Also, as $[\mu]$ is unitary we have a K_0–invariant hermitian metric on $\mathbb{V}_{\mu,\sigma}$.

The complexified tangent bundle of Y_Φ has a subbundle $\mathbb{T} \to Y_\Phi$ defined by

$$\mathbb{T}|_{gS_\Phi} \to gS_\Phi \text{ is the holomorphic tangent bundle of } gS_\Phi .$$

It defines

$$\begin{cases} \mathbb{V}_{\mu,\sigma}^{p,q} = \mathbb{V}_{\mu,\sigma} \otimes \Lambda^p(\mathbb{T}^*) \otimes \Lambda^q(\overline{\mathbb{T}}^*) \to Y_\Phi , \\ A^{p,q}(Y_\Phi; \mathbb{V}_{\mu,\sigma}) : C^\infty \text{ sections of } \mathbb{V}_{\mu,\sigma}^{p,q} \to Y_\Phi , \text{ and} \\ \mathcal{O}(\mathbb{V}_{\mu,\sigma}) : \text{ sheaf of } C^\infty \text{ sections of } \mathbb{V}_{\mu,\sigma} \text{ holomorphic over every } gS_\Phi . \end{cases} \tag{11.12}$$

$A^{p,q}(Y_\Phi; \mathbb{V}_{\mu,\sigma})$ is the space of $\mathbb{V}_{\mu,\sigma}$–valued partially (p,q)–forms on Y_Φ.

We take the positive definite $U_{\Phi,0}$–invariant hermitian inner product on fiber V_μ of $\mathbb{V}_\mu \to S_\Phi$, and translate it around by K_0 to obtain a K_0–invariant hermitian structure on $\mathbb{V}_{\mu,\sigma}^{p,q} \to Y_\Phi$. In the same way we have a K_0–invariant hermitian metric on $\mathbb{T} \to Y_\Phi$. So we have K_0–invariant Hodge–Kodaira orthocomplementation operators $\sharp : A^{p,q}(Y_\Phi; \mathbb{V}_{\mu,\sigma}) \to A^{n-p,n-q}(Y_\Phi; \mathbb{V}_{\mu,\sigma}^*)$ and $\tilde{\sharp} : A^{n-p,n-q}(Y_\Phi; \mathbb{V}_{\mu,\sigma}^*) \to A^{p,q}(Y_\Phi; \mathbb{V}_{\mu,\sigma})$. where $n = \dim_{\mathbb{C}} S_\Phi$.

We obtain the global G_0–invariant hermitian inner product on $A^{p,q}(Y_\Phi; \mathbb{V}_{\mu,\sigma})$ from the inner product along each fiber of $Y_\Phi \to G_0/P_0$ by integrating over G_0/P_0,

$$\langle F_1, F_2 \rangle_{Y_\Phi} = \int_{K_0/M_0} \left(\int_{kS_\Phi} F_1 \bar{\wedge} \sharp F_2 \right) d(kM_0). \tag{11.13}$$

where $\bar{\wedge}$ means exterior product followed by contraction of V_μ against V_μ^*.

The $\bar\partial$ operator of Z_Φ induces the $\bar\partial$ operators on each of the gS_Φ, so they fit together to give us an operator $\bar\partial : A^{p,q}(Y_\Phi;\mathbb{V}_{\mu,\sigma}) \to A^{p,q+1}(Y_\Phi;\mathbb{V}_{\mu,\sigma})$. It has formal adjoint $\bar\partial^* : A^{p,q+1}(Y_\Phi;\mathbb{V}_{\mu,\sigma}) \to A^{p,q}(Y_\Phi;\mathbb{V}_{\mu,\sigma})$ given by $\bar\partial^* = -\tilde\sharp\bar\partial\sharp$. That defines the "partial Kodaira–Hodge–Laplace operator"

(11.14) $$\square = \bar\partial\bar\partial^* + \bar\partial^*\bar\partial : A^{p,q}(Y_\Phi;\mathbb{V}_{\mu,\sigma}) \to A^{p,q}(Y_\Phi;\mathbb{V}_{\mu,\sigma}).$$

$A^{p,q}(Y_\Phi;\mathbb{V}_{\mu,\sigma})$ is a pre Hilbert space with the global inner product (11.13). Denote

(11.15) $$L_2^{p,q}(Y_\Phi;\mathbb{V}_{\mu,\sigma}) : \text{ Hilbert space completion of } A^{p,q}(Y_\Phi;\mathbb{V}_{\mu,\sigma}).$$

Essentially as in the case of the Bott–Borel–Weil Theorem, we apply Weyl's Lemma along each gS_Φ to see that the closure of $\widetilde\square$ of $\square$, as a densely defined operator on $L_2^{p,q}(Y_\Phi;\mathbb{V}_{\mu,\sigma})$ from the domain $A^{p,q}(Y_\Phi;\mathbb{V}_{\mu,\sigma})$, is essentially self–adjoint. Its kernel

(11.16) $$\mathcal{H}_2^{p,q}(Y_\Phi;\mathbb{V}_{\mu,\sigma}) = \{\omega \in \text{Domain}(\widetilde\square) \mid \widetilde\square\omega = 0\}$$

is the space of $\mathbb{V}_{\mu,\sigma}$*–valued square integrable partially harmonic* (p,q)–forms on Y_Φ.

The factor $e^{\rho_\mathfrak{a}}$ in the representation $\gamma_{\mu,\sigma}$ that defines $\mathbb{V}_{\mu,\sigma}$ insures that the global inner product on $A^{p,q}(Y_\Phi;\mathbb{V}_{\mu,\sigma})$ is invariant under the action of G_0. It corresponds to the $\Delta(G_0/P_0)$ in the definition (7.2) of (unitarily induced) principal series representation.

The other ingredients in the construction of $\mathcal{H}_2^{p,q}(Y_\Phi;\mathbb{V}_{\mu,\sigma})$ are invariant as well, so G_0 acts naturally on $\mathcal{H}_2^{p,q}(Y_\Phi;\mathbb{V}_{\mu,\sigma})$ by isometries. This action is a unitary representation of G_0.

Write $\mathcal{H}_2^{q}(Y_\Phi;\mathbb{V}_{\mu,\sigma})$ for $\mathcal{H}_2^{0,q}(Y_\Phi;\mathbb{V}_{\mu,\sigma})$. We combine the Bott–Borel–Weil Theorem 11.9 with the definition (7.2) of the principal series, obtaining a geometric realization of the principal series of G_0 based on the closed G_0–orbit in the complex flag $Z_\Phi = G/Q_\Phi$, as follows.

THEOREM 11.17. *Let* $[\mu] = [\chi \otimes \mu_\beta^0] \in \widehat{U_{\Phi,0}}$ *and* $\sigma \in \mathfrak{a}_0^*$, *and fix* $q \geqq 0$.

1. *If* $\langle \beta - \rho_{\mathfrak{u}_\Phi} + \rho_\mathfrak{m}, \alpha\rangle = 0$ *for some* $\alpha \in \Sigma(\mathfrak{m},\mathfrak{t})$ *then every* $\mathcal{H}_2^q(Y_\Phi;\mathbb{V}_{\mu,\sigma}) = 0$.
2. *If* $\langle \beta - \rho_{\mathfrak{u}_\Phi} + \rho_\mathfrak{m}, \alpha\rangle \neq 0$ *for all* $\alpha \in \Sigma(\mathfrak{m},\mathfrak{t})$, *let* w *be the unique element in* $W(\mathfrak{m},\mathfrak{t})$ *such that* $\langle w(\beta - \rho_{\mathfrak{u}_\Phi} + \rho_\mathfrak{m}), \alpha\rangle > 0$ *for all* $\alpha \in \Sigma(\mathfrak{m},\mathfrak{t})$. *Then* $\mathcal{H}_2^q(Y_\Phi;\mathbb{V}_{\mu,\sigma}) = 0$ *for* $q \neq q_0$, *and the natural action of* G_0 *on* $\mathcal{H}_2^{q_0}(Y_\Phi;\mathbb{V}_{\mu,\sigma})$ *is the principal series representation* $\pi_{\chi\otimes\eta_\nu^0,\sigma}$.

12. L^2 Realizations of the Discrete Series.

Suppose that G_0 has a compact Cartan subgroup $T_0 \subset K_0$. Let $Z = G/Q$ be a complex flag manifold, let $z \in Z$, set $D = G_0(z)$, and suppose that

(12.1) D is open in Z and G_0 has compact isotropy subgroup U_0 at z.

Passing to a conjugate, equivalently moving z within D, we may suppose $T_0 \subset U_0$.

Let $\mu \in \widehat{U_0}$, let E_μ denote the representation space, and let $\mathbb{E}_\mu \to D \cong G_0/U_0$ denote the associated holomorphic homogeneous vector bundle. Then μ is finite dimensional and is constructed as follows. First, $U_0 \cap G_0^0$ is the identity component

U_0^0, and $U_0 = Z_{G_0}(G_0^0)U_0^0$. Second there are irreducible unitary representations $[\chi] \in \widehat{Z_{G_0}(G_0^0)}$ and $[\mu^0] \in \widehat{U_0^0}$ that agree on Z_{G_0} such that $[\mu] = [\chi \otimes \mu^0]$.

Let $\beta - \rho_{\mathfrak{u}}$ denote the highest weight of μ^0, corresponding to infinitesimal character β, and suppose that

$$\lambda = \beta - \rho_{\mathfrak{u}} + \rho_{\mathfrak{g}} \text{ is regular.} \tag{12.2}$$

Then G_0 has a discrete series representation $\pi_{\chi,\lambda}$, whose infinitesimal character has Harish–Chandra parameter λ.

Since μ is unitary, the bundle $\mathbb{E}_\mu \to D$ has a G_0–invariant hermitian metric. Essentially as in the compact case, we have the spaces

$$A_0^{(p,q)}(D;\mathbb{E}_\mu) : C^\infty \text{ compactly supported } \mathbb{E}_\mu\text{–valued } (p,q)\text{–forms on } D, \tag{12.3}$$

the Kodaira–Hodge orthocomplementation operators

$$\begin{aligned} \sharp : &A_0^{(p,q)}(D;\mathbb{E}_\mu) \to A_0^{(n-p,n-q)}(D;\mathbb{E}_\mu^*) \\ \text{and } \tilde{\sharp} : &A_0^{(n-p,n-q)}(D;\mathbb{E}_\mu^*) \to A_0^{(p,q)}(D;\mathbb{E}_\mu) \end{aligned} \tag{12.4}$$

where $n = \dim_{\mathbb{C}} D$. Thus we have a positive definite inner product on $A_0^{(p,q)}(D;\mathbb{E}_\mu)$ give by

$$\langle F_1, F_2\rangle_D = \int_{G_0} \langle F_1, F_2\rangle_{gU_0} d(gU_0) = \int_D F_1 \overline{\wedge} \sharp F_2 \tag{12.5}$$

and thus

$$L_2^{(p,q)}(D;\mathbb{E}_\mu) : \text{ Hilbert space completion of } \big(A_0^{(p,q)}(D;\mathbb{E}_\mu), \langle\cdot,\cdot\rangle_D\big). \tag{12.6}$$

Let $\square$ denote the Kodaira–Hodge–Laplace operator $\overline{\partial}\,\overline{\partial}^* + \overline{\partial}^*\overline{\partial}$ of $\mathbb{E}_\mu$. Then $\square$ is a hermitian–symmetric elliptic operator on $L_2^{(0,q)}(D;\mathbb{E}_\mu)$ with domain $A_0^{(p,q)}(D;\mathbb{E}_\mu)$, and a result of Andreotti and Vesentini allows one to conclude that its closure $\widetilde{\square}$ is self–adjoint. Accordingly, we have the Hilbert spaces

$$\mathcal{H}_2^{(p,q)}(D;\mathbb{E}_\mu) = \{\omega \in \text{ Domain}\,(\widetilde{\square}) \mid \widetilde{\square}(\omega) = 0\} \tag{12.7}$$

of L^2 harmonic $\mathbb{E}_\mu$–valued $(0,q)$–forms on D. G_0 acts on $\mathcal{H}_2^{(p,q)}(D;\mathbb{E}_\mu)$ by a unitary representation.

We write $\mathcal{H}_2^q(D;\mathbb{E}_\mu)$ for $\mathcal{H}_2^{(0,q)}(D;\mathbb{E}_\mu)$ and we write π_μ^q for the unitary representation of G_0 on $\mathcal{H}_2^q(D;\mathbb{E}_\mu)$.

THEOREM 12.8. *Let $[\mu] = [\chi \otimes \mu^0] \in \widehat{U_0}$ where μ^0 has highest weight $\beta - \rho_{\mathfrak{u}}$ and thus has infinitesimal character β. If $\lambda + \rho$ (as in (12.2)) is $\Sigma(\mathfrak{g},\mathfrak{t})$–singular then every $\mathcal{H}_2^q(D;\mathbb{E}_\mu) = 0$. Now suppose that $\lambda = \beta - \rho_{\mathfrak{u}} + \rho_{\mathfrak{g}}$ is $\Sigma(\mathfrak{g},\mathfrak{t})$–regular and define*

$$\begin{aligned} q_{\mathfrak{u}}(\lambda) = &|\{\alpha \in \Sigma^+(\mathfrak{k},\mathfrak{t}) \setminus \Sigma^+(\mathfrak{u},\mathfrak{t}) \mid \langle\lambda,\alpha\rangle < 0\}| \\ &+ |\{\beta \in \Sigma^+(\mathfrak{g},\mathfrak{t}) \setminus \Sigma^+(\mathfrak{k},\mathfrak{t}) \mid \langle\lambda,\beta\rangle > 0\}|. \end{aligned} \tag{12.9}$$

Then $\mathcal{H}_2^q(D;\mathbb{E}_\mu) = 0$ for $q \neq q_{\mathfrak{u}}(\lambda)$, and G_0 acts irreducibly on $\mathcal{H}_2^{q_{\mathfrak{u}}(\lambda)}(D;\mathbb{E}_\mu)$ by the discrete series representation $\pi_{\chi,\lambda}$ of infinitesimal character λ.

One can also realize the discrete series on spaces of L_2 bundle–valued harmonic spinors [**136**].

13. L^2 Realizations of the Various Tempered Series.

Choose a Cartan subgroup $H_0 \subset G_0$. We are going to realize the H_0–series representations of G_0 in a way analogous to the way we realized the principal series in §11, with Theorem 12.8 in place of the Bott–Borel–Weil Theorem 11.9.

Let θ be the Cartan involution of G_0 that stabilizes H_0, split $H_0 = T_0 \times A_0$ and let $Z_{G_0}(A_0) = M_0 \times A_0$ as before. Fix a positive root system $\Sigma^+ = \Sigma^+(\mathfrak{g}, \mathfrak{h})$ defined by positive root systems $\Sigma^+(\mathfrak{m}, \mathfrak{t})$ and $\Sigma^+(\mathfrak{g}_0, \mathfrak{a}_0)$ as in (11.1). Let $P_0 = M_0 A_0 N_0$ be the corresponding cuspidal parabolic subgroup of G_0 associated to H_0.

Following the idea of the geometric realization of the principal series, we fix a set $\Phi \subset \Psi_\mathfrak{m}$ where $\Psi_\mathfrak{m}$ is the simple root system for $\Sigma^+(\mathfrak{m}, \mathfrak{t})$. Then as in (11.2) we have

$$(13.1)\quad \begin{cases} \mathfrak{z}_\Phi = \{\xi \in \mathfrak{t} \mid \phi(\xi) = 0 \ \forall \phi \in \Phi\} \text{ and its real form } \mathfrak{z}_{\Phi,0} = \mathfrak{m}_0 \cap \mathfrak{x}_\Phi \mathfrak{z}_\Phi , \\ U_\Phi = Z_M(\mathfrak{z}_\Phi), U_{\Phi,0} = M_0 \cap U_\Phi, \text{ and Lie algebras } \mathfrak{u}_\Phi \text{ and } \mathfrak{u}_{\Phi,0} , \\ \mathfrak{r}_\Phi = \mathfrak{u}_\Phi + \sum_{\gamma \in \Sigma^+(\mathfrak{m},\mathfrak{t})} \mathfrak{m}_{-\gamma}, \text{ parabolic subalgebra of } \mathfrak{m} , \\ R_\Phi = N_M(\mathfrak{r}_\Phi), \text{ corresponding parabolic subgroup of } M , \text{ and} \\ S_\Phi = M/R_\Phi, \text{ associated complex flag manifold.} \end{cases}$$

Let r_Φ denote the base point, $r_\Phi = 1R_\Phi \in R_\Phi$. As T_0 is a compact Cartan subgroup of M_0 contained in $U_{\Phi,0}$, $D_\Phi = M_0(r_\Phi) \subset S_\Phi$ is a measurable open M_0–orbit on R_Φ. We now assume that $U_{\Phi,0}$ is compact, so the considerations of §11 apply to $D_\Phi \subset S_\Phi$.

Fix $[\mu] = [\chi \otimes \mu_\beta^0] \in \widehat{U_{\Phi,0}}$ as before. Given $\sigma \in \mathfrak{a}_0^*$ we will use the Theorem 12.8 to find the H_0–series representation $\pi_{\chi \otimes \eta_\nu^0, \sigma}$ on a cohomology space related to a particular orbit in the complex flag manifold $Z_\Phi = G/Q_\Phi$. Here as before, the simple root system $\Psi_\mathfrak{m} \subset \Psi$ by the coherence in our choice of $\Sigma^+(\mathfrak{g}, \mathfrak{h})$, so $\Phi \subset \Psi$ and Φ defines a parabolic subgroup $Q_\Phi \subset G$.

Let $z_\Phi = 1Q_\Phi \in G/Q_\Phi = Z_\Phi$. As $A_0 N_0 \subset G_0 \cap Q_\Phi$ we have $G_0 \cap Q_\Phi = U_{\Phi,0} A_0 N_0$. Thus $Y_\Phi = G_0(z_\Phi)$ is a G_0–orbit on Z_Φ, and D_Φ sits in Y_Φ as the orbit $M_0(z_\Phi)$. Here note that $P_0 = M_0 A_0 N_0 = \{g \in G_0 \mid gD_\Phi = D_\Phi\}$.

LEMMA 13.2. *The map $Y_\Phi \to G_0/P_0$, given by $g(z_\Phi) \mapsto gP_0$, defines a G_0–equivariant fiber bundle with structure group M_0 and whose fibers gD_Φ are the maximal complex analytic submanifolds of Y_Φ.*

The data (μ, σ) defines a representation $\gamma_{\mu,\sigma}(uan) = e^{(\rho_\mathfrak{g} + i\sigma)(\log a)} \mu(u)$ of $U_{\Phi,0} A_0 N_0$ where $\rho_\mathfrak{g} = \frac{1}{2} \sum_{\alpha \in \Sigma^+} \mathfrak{g}_\alpha$. That defines a G_0–homogeneous vector bundle

$$(13.3)\qquad \mathbb{E}_{\mu,\sigma} \to G_0/U_{\Phi,0} A_0 N_0 = Y_\Phi \text{ such that } \mathbb{E}_{\mu,\sigma}|_{D_\Phi} = \mathbb{E}_\mu .$$

Each $\mathbb{E}_{\mu,\sigma}|_{gD_\Phi}$ is an $\mathrm{Ad}(g)P_0$–homogeneous holomorphic vector bundle.

Since $[\mu]$ is unitary and K_0 acts transitively on G_0/P_0 we have a K_0–invariant hermitian metric on $\mathbb{E}_{\mu,\sigma}$. We will use it without explicit reference.

Consider the subbundle $\mathbb{T} \to Y_\Phi$ of the complexified tangent bundle to Y_Φ defined by

$$\text{(13.4)} \qquad \mathbb{T}|_{gD_\Phi} \to gD_\Phi \text{ is the holomorphic tangent bundle of } gD_\Phi \,.$$

It defines

$$\text{(13.5)} \quad \begin{cases} \mathbb{E}^{p,q}_{\mu,\sigma} = \mathbb{E}_{\mu,\sigma} \otimes \Lambda^p(\mathbb{T}^*) \otimes \Lambda^q(\overline{\mathbb{T}}^*) \to D_\Phi \,, \\ A^{p,q}_0(Y_\Phi; \mathbb{E}_{\mu,\sigma}) : C^\infty \text{ compactly supported sections of } \mathbb{E}_{\mu,\sigma} \to Y_\Phi \,, \\ \mathcal{O}(\mathbb{E}_{\mu,\sigma}) : \text{ sheaf of } C^\infty \text{ sections of } \mathbb{E}_{\mu,\sigma} \text{ holomorphic over every } gD_\Phi \,. \end{cases}$$

$A^{p,q}(Y_\Phi; \mathbb{E}_{\mu,\sigma})$ is the space consisting of all $\mathbb{E}_{\mu,\sigma}$–valued partially (p,q)–forms on Y_Φ, and $A^{p,q}_0(Y_\Phi; \mathbb{E}_{\mu,\sigma})$ is the subspace of compactly supported forms.

The fiber E_μ of $\mathbb{E}_\mu \to D_\Phi$ has a positive definite $U_{\Phi,0}$–invariant hermitian inner product because μ is unitary; we translate this around by K_0 to obtain a K_0–invariant hermitian structure on the vector bundle $\mathbb{E}^{p,q}_{\mu,\sigma} \to Y_\Phi$. Similarly $\mathbb{T} \to Y_\Phi$ carries a K_0–invariant hermitian metric. Using these hermitian metrics we have K_0–invariant Hodge–Kodaira orthocomplementation operators

$$\text{(13.6)} \quad \begin{aligned} &\sharp : A^{p,q}_0(Y_\Phi; \mathbb{E}_{\mu,\sigma}) \to A^{n-p,n-q}_0(Y_\Phi; \mathbb{E}^*_{\mu,\sigma}) \\ &\qquad\qquad \text{and } \tilde{\sharp} : A^{n-p,n-q}_0(Y_\Phi; \mathbb{E}^*_{\mu,\sigma}) \to A^{p,q}_0(Y_\Phi; \mathbb{E}_{\mu,\sigma}), \end{aligned}$$

where $n = \dim_{\mathbb{C}} D_\Phi$. The global G_0–invariant inner product on $A^{p,q}_0(Y_\Phi; \mathbb{E}_{\mu,\sigma})$ is given by taking the M_0–invariant inner product along each fiber of $Y_\Phi \to G_0/P_0$ and integrating over G_0/P_0,

$$\text{(13.7)} \qquad \langle F_1, F_2 \rangle_{Y_\Phi} = \int_{K_0/(K_0 \cap M_0)} \left(\int_{kD_\Phi} F_1 \bar{\wedge} \sharp F_2 \right) d(k(K_0 \cap M_0)),$$

where $\bar{\wedge}$ means exterior product followed by contraction of E_μ against E^*_μ.

The $\overline{\partial}$ operator of Z_Φ induces the $\overline{\partial}$ operators on each of the gD_Φ, so they fit together to give us an operator

$$\text{(13.8)} \qquad \overline{\partial} : A^{p,q}_0(Y_\Phi; \mathbb{E}_{\mu,\sigma}) \to A^{p,q+1}_0(Y_\Phi; \mathbb{E}_{\mu,\sigma})$$

that has formal adjoint

$$\text{(13.9)} \qquad \overline{\partial}^* : A^{p,q+1}_0(Y_\Phi; \mathbb{E}_{\mu,\sigma}) \to A^{p,q}_0(Y_\Phi; \mathbb{E}_{\mu,\sigma}) \text{ given by } \overline{\partial}^* = -\tilde{\sharp}\overline{\partial}\sharp \,.$$

That in turn defines an elliptic operator, the "partial Kodaira–Hodge–Laplace operator"

$$\text{(13.10)} \qquad \square = \overline{\partial}\,\overline{\partial}^* + \overline{\partial}^*\overline{\partial} : A^{p,q}_0(Y_\Phi; \mathbb{E}_{\mu,\sigma}) \to A^{p,q}_0(Y_\Phi; \mathbb{E}_{\mu,\sigma}).$$

$A^{p,q}_0(Y_\Phi; \mathbb{E}_{\mu,\sigma})$ is a pre Hilbert space with the global inner product (13.10). Denote

$$\text{(13.11)} \qquad L^{p,q}_2(Y_\Phi; \mathbb{E}_{\mu,\sigma}) : \text{Hilbert space completion of } A^{p,q}_0(Y_\Phi; \mathbb{E}_{\mu,\sigma}).$$

Apply Andreotti–Vesentini along each gD_Φ to see that the closure of $\widetilde{\square}$ of $\square$, as a densely defined operator on $L^{p,q}_2(Y_\Phi; \mathbb{E}_{\mu,\sigma})$ from the domain $A^{p,q}_0(Y_\Phi; \mathbb{E}_{\mu,\sigma})$, is essentially self–adjoint. Its kernel

$$\text{(13.12)} \qquad \mathcal{H}^{p,q}_2(Y_\Phi; \mathbb{E}_{\mu,\sigma}) = \{\omega \in \text{ Domain}(\widetilde{\square}) \mid \widetilde{\square}\omega = 0\}$$

is the space of *square integrable partially harmonic* (p,q)–forms on Y_Φ with values in $\mathbb{E}_{\mu,\sigma}$.

The factor $e^{\rho_{\mathfrak{g}}}$ in the representation $\gamma_{\mu,\sigma}$ that defines $\mathbb{E}_{\mu,\sigma}$ insures that the global inner product on $A_0^{p,q}(Y_\Phi;\mathbb{E}_{\mu,\sigma})$ is invariant under the action of G_0. The other ingredients in the construction of $\mathcal{H}_2^{p,q}(Y_\Phi;\mathbb{E}_{\mu,\sigma})$ are invariant as well, so G_0 acts naturally on $\mathcal{H}_2^{p,q}(Y_\Phi;\mathbb{E}_{\mu,\sigma})$ by isometries. This action is a unitary representation of G_0.

As before we write $\mathcal{H}_2^{q}(Y_\Phi;\mathbb{E}_{\mu,\sigma})$ for $\mathcal{H}_2^{0,q}(Y_\Phi;\mathbb{E}_{\mu,\sigma})$.

We can now combine Theorem 12.8 with the definition (9.3) of the H_0–series, obtaining

THEOREM 13.13. *Let $[\mu]=[\chi\otimes\mu^0_\beta]\in\widehat{U_{\Phi,0}}$ where μ^0 has highest weight $\beta-\rho_{\mathfrak{u}}$ and thus has infinitesimal character β. Let*

$$\nu=\beta-\rho_{\mathfrak{u}_\Phi}+\rho_{\mathfrak{m}}, \tag{13.14}$$

suppose $\sigma\in\mathfrak{a}_0^*$, and fix an integer $q\geqq 0$.

1. *If $\langle\nu,\alpha\rangle=0$ for some $\alpha\in\Sigma(\mathfrak{m},\mathfrak{t})$ then $\mathcal{H}_2^{q}(Y_\Phi;\mathbb{E}_{\mu,\sigma})=0$.*
2. *If $\langle\nu,\alpha\rangle\neq 0$ for all $\alpha\in\Sigma(\mathfrak{m},\mathfrak{t})$, define*

$$\begin{aligned} q_{\mathfrak{u}_\Phi}(\nu)=&\,|\{\alpha\in\Sigma^+((\mathfrak{k}\cap\mathfrak{m}),\mathfrak{t})\setminus\Sigma^+(\mathfrak{u},\mathfrak{t})\mid\langle\nu,\alpha\rangle<0\}|\\ &+|\{\beta\in\Sigma^+(\mathfrak{m},\mathfrak{t})\setminus\Sigma^+((\mathfrak{k}\cap\mathfrak{m}),\mathfrak{t})\mid\langle\nu,\beta\rangle>0\}|.\end{aligned} \tag{13.15}$$

Then $\mathcal{H}^q(D;\mathbb{E}_{\mu,\sigma})=0$ for $q\neq q_{\mathfrak{u}_\Phi}(\nu)$, and the action of G_0 on $\mathcal{H}^{q_{\mathfrak{u}_\Phi}(\nu)}(D;\mathbb{E}_{\mu,\sigma})$ is the H_0–series representation $\pi_{\chi,\nu,\sigma}$ of infinitesimal character $\nu+i\sigma$.

A variation on this theorem [**136**] realizes the tempered series on spaces of L_2 bundle-valued partially harmonic spinors.

14. Approaches to Non–Tempered Representations.

There are a number of approaches to the problem of finding good geometric realizations of representations that need not be tempered — or even need not be unitary. There is no general agreement here on which approaches are "best", so I'll just give thumbnail descriptions of the four that I find most interesting. I apologize in advance to the many researchers in this area whose approach is not mentioned or is only mentioned in passing, and whose papers are not referenced.

Geometric quantization. In the Kostant–Souriau theory of geometric quantization, one considers a linear functional $\lambda\in\mathfrak{g}_0^*$ on the Lie algebra $\mathfrak{g}_0$ of G_0 and the coadjoint orbit $\mathcal{O}=\mathrm{Ad}^*(G_0)(\lambda)$. Let L_0 denote the G_0–stabilizer of λ, so $\mathcal{O}\cong G_0/L_0$ as homogeneous space, and let $\mathfrak{l}_0$ be the Lie algebra of L_0. For complexification we drop the subscript 0, as before. A *polarization* for $\mathcal{O}$ is a (complex) subalgebra $\mathfrak{q}\subset\mathfrak{g}$ that (i) contains $\mathfrak{l}$ with $\dim\mathfrak{g}-\dim\mathfrak{q}=\dim\mathfrak{q}-\dim\mathfrak{l}$ and (ii) is $\mathrm{Ad}^*(L_0)$–invariant. The coadjoint orbit $\mathcal{O}$ is *integral* if $\exp(2\pi\sqrt{-1}\lambda):L_0^0\to\mathbb{C}$ is well defined. Then every extension χ of $\exp(2\pi\sqrt{-1}\lambda)$ to an irreducible representation of L_0, say on a vector space E_χ, leads to a G_0–homogeneous vector bundle $\mathbb{E}_\chi\to\mathcal{O}$ with typical fiber E_χ.

By $\mathrm{Ad}(L_0)$–invariance, $\mathfrak{q}$ acts on E_χ in a manner consistent with the action of L_0. This can happen in several different ways; choose one. Denote this action just by $d\chi$. Also, $\mathfrak{q}$ acts by differential operators on local sections, $s(x;\xi)=$

$\frac{d}{dt}|_{t=0}(s(x\exp(t\xi_1))) + \sqrt{-1}\frac{d}{dt}|_{t=0}(s(x\exp(t\xi_2)))$ for $\xi = \xi_1 + \sqrt{-1}\xi_2 \in \mathfrak{q}$ with $\xi_1, \xi_2 \in \mathfrak{g}_0$. Let $\mathcal{E}_\chi$ denote the sheaf of local sections s of $\mathbb{E}_\chi$ such that $s(x;\xi) + d\chi(\xi)s(x) = 0$ for all $\xi \in \mathfrak{q}$. Then one studies the representations of G_0 on the cohomologies $H^q(\mathcal{O};\mathcal{E}_\chi)$.

The most familiar example is where $\mathcal{O}$ has an invariant complex structure, $\mathfrak{q} = \mathfrak{l} + \mathfrak{n}$ in such a way that $\mathfrak{n}$ represents the antiholomorphic tangent space, $d\chi(\mathfrak{n}) = 0$ so that $s(x;\xi) + d\chi(\xi)s(x) = 0$ simplifies to $s(x;\xi) = 0$, and $s(x;\xi) = 0$ is the Cauchy–Riemann equation, so $\mathcal{E}_\chi$ is the sheaf of germs of holomorphic sections of $\mathbb{E}_\chi \to \mathcal{O}$. Our L^2 realizations of tempered representations were done by an L^2 partially harmonic form variation on this pattern. Essentially this works because we are dealing with semisimple coadjoint orbits there, and their polarizations are parabolic subalgebras.

Several things can go wrong here. It can happen that there is no invariant polarization. A method of moving polarizations was developed to deal with the case where the polarization is invariant only under a certain subgroup of L_0. Another approach to this problem is given by the theory of metalinear structures. It can happen that one is looking for representations in some topological category, so one looks for the $H^q(\mathcal{O};\mathcal{E}_\chi)$ as cohomologies of some complex of topological vector spaces, and one needs a closed range theorem for the differentials of the complex in order that the cohomologies inherit a topology. This can be very delicate, but it has been settled in a few cases by using hyperfunction (instead of C^∞) coefficients; see **[115]**, **[117]**, **[138]** and **[151]**. It can happen that one looks to unitarize these representations, but the positive definite hermitian metrics of the previous three sections are not available. This has been carried out in a few cases; see the subsection "Indefinite metric harmonic theory" below. It can happen that one tries to realize representations that cannot occur on semisimple coadjoint orbits; there one goes in principle to the nascent technical theory of unipotent representations. And finally there is the problem (which the theory of unipotent representations tries to address) that, for most real semisimple Lie groups, the unitary dual is not completely known.

Double fibration transforms. This is a tool for investigating growth properties of Fréchet space representations on Dolbeault cohomology, with an eye toward questions of unitarity and Lebesgue class, and as such is essentially a tool for use in geometric quantization. It hasn't really been applied yet, but I think that it will be extremely useful for analysis of non–tempered representations. Historically this sort of thing was done the other way around, starting with differential equations and then getting a representation on Dolbeault (or some other) cohomology whose double fibration transform was the set of solutions to the differential equations.

Let $D = G_0(z) \subset Z = G/Q$ be a measurable open orbit and let $\mathbb{E} \to D$ be a *negative* G_0–homogeneous holomorphic vector bundle. Let s denote the complex dimension of the maximal compact subvariety $Y_0 = K_0(z)$ of D. Theorem 4.6 says that $H^q(D;\mathcal{O}(\mathbb{E})) = 0$ for $q > s$ and the negativity ensures that $H^q(D;\mathcal{O}(\mathbb{E})) = 0$ for $q < s$. We have (6.5). In many cases we have verified that the fiber F of $\mu : W_D \to D$ is contractible, so Theorem 6.6 says that the double fibration transform map $P : H^s(D;\mathcal{O}(\mathbb{E})) \to H^0(M_D;\mathcal{R}^s(\mu^*(\mathcal{O}(\mathbb{E}))))$ is injective. Now the point is to characterize its image by a set of PDE which will allow us to make *a priori* growth estimates on the coefficients of the representation of G_0 on $H^s(D;\mathcal{O}(\mathbb{E}))$.

Dual reductive pairs and the ϑ correspondence. A pair of subgroups $G_0, G_0' \subset Sp(n;\mathbb{R})$ is a *dual reductive pair* if they are reductive in the real symplectic group $Sp(n;\mathbb{R})$ and each is the centralizer of the other in $Sp(n;\mathbb{R})$. There are just a few such pairs:

$$\begin{aligned}&(O(p,q), Sp(k;\mathbb{R})) \text{ in } Sp((p+q)k;\mathbb{R})),\\ &(U(p,q), U(k,\ell)) \text{ in } Sp((p+q)(k+\ell);\mathbb{R}), \text{ and}\\ &(Sp(p,q), SO^*(2k)) \text{ in } Sp((p+q)k;\mathbb{R}),\end{aligned}$$

which Howe calls Type I, and the Type II pairs

$$(GL(u,\mathbb{F}), GL(v,\mathbb{F})) \text{ in } Sp(uvd;\mathbb{R}), d = \dim_{\mathbb{R}} \mathbb{F}, \text{ for } \mathbb{F} = \mathbb{R}, \mathbb{C}, \text{ and } \mathbb{H}.$$

The point, for us, is Howe's theorem that the metaplectic representation μ of (the double cover of) $Sp(n;\mathbb{R})$ has restriction to G_0G_0' of the form $\int \alpha_i \otimes \alpha_i'\, dm(i)$ with $\alpha_i \in \widehat{G_0}$ and $\alpha_i' \in \widehat{G_0'}$, and such that α_i determines α_i' and α_i' determines α_i , a.e. $m(i)$. The correspondence $\widehat{G_0} \ni \alpha_i \leftrightarrow \alpha_i' \in \widehat{G_0'}$ is the *ϑ correspondence.* If one knows one of $\widehat{G_0}, \widehat{G_0'}$ it gives a lot of useful information about the other. This is especially useful when one of G_0, G_0' is compact. Also, representations that occur on dual reductive pairs are automatically unitary, and they are sufficiently concrete so that one can often check questions of growth and irreducibility.

Indefinite metric harmonic theory. Consider a variation on the L^2 realizations of the discrete series in Section 12, in which the isotropy subgroup U_0 of G_0 may be noncompact. Thus the open orbit $D = G_0(z) \subset Z = G/Q$ carries an invariant pseudo–kähler metric from the Killing form of $\mathfrak{g}$, but perhaps does not carry an invariant positive definite hermitian metric. In the language of geometric quantization, D is an elliptic coadjoint orbit.

Let $[\mu] \in \widehat{U_0}$ and let E_μ be the representation space. The associated vector bundle $\mathbb{E}_\mu \to D \cong G_0/U_0$ is a G_0–homogeneous hermitian holomorphic vector bundle. As in (12.3) one has the space $A_0^{(p,q)}(D;\mathbb{E}_\mu)$ of compactly supported $\mathbb{E}_\mu$–valued C^∞ (p,q)–forms on D, and, from the invariant pseudo–kähler metric, one has the Kodaira–Hodge orthocomplementation operators (12.4). The problem is that the inner product $\langle F_1, F_2\rangle_D = \int_D F_1 \overline{\wedge} \sharp F_2$ on $A_0^{(p,q)}(D;\mathbb{E}_\mu)$ will in general be indefinite rather than positive definite, so the definition of $L_2^{(p,q)}(D;\mathbb{E}_\mu)$ is problematical. The approach of [**109**] is to use an auxiliary hermitian metric which is not G_0–invariant, but for which $L_2^{(p,q)}(D;\mathbb{E}_\mu)$ is defined and is a space on which G_0 acts by a bounded representation. Then, relative to the invariant pseudo–kähler metric one still has $\overline{\partial}^*$ and the indefinite metric analog $\square = \overline{\partial}\,\overline{\partial}^* + \overline{\partial}^*\overline{\partial}$ of the Kodaira–Hodge–Laplace operator. We say that a form $\omega \in L_2^{(p,q)}(D;\mathbb{E}_\mu)$ is *harmonic* if it is annihilated by the closure $\widetilde{\square}$ of $\square$, *strongly harmonic* if $\overline{\partial}\omega = 0 = \overline{\partial}^*\omega$.

Let $\widetilde{\mathcal{H}}_2^q(D;\mathbb{E}_\mu)_{K_0}$ denote the space of K_0–finite L^2 strongly harmonic $\mathbb{E}_\mu$–valued $(0,q)$-forms on D. It has a natural map $\omega \mapsto [\omega]$ to K_0–finite Dolbeault cohomology $H^q(D;\mathbb{E}_\mu)_{K_0}$ of $\mathbb{E}_\mu \to D$. Let $\mathcal{H}_2^q(D;\mathbb{E}_\mu)_{K_0}$ denote the image. Under certain circumstances, in [**109**] it is shown that every Dolbeault class $[\omega] \in \widetilde{\mathcal{H}}_2^q(D;\mathbb{E}_\mu)_{K_0}$

has a "best" representative ω, that $\langle\omega,\omega\rangle_D = \int_D \omega\overline{\wedge}\sharp\omega > 0$ for $[\omega] \neq 0$, and that $\langle\cdot,\cdot\rangle_D$ is null on the kernel of $\widetilde{\mathcal{H}}^q_2(D;\mathbb{E}_\mu)_{K_0} \to H^q(D;\mathbb{E}_\mu)_{K_0}$. That unitarizes the action of G_0 on certain Dolbeault cohomologies as the Hilbert space completions of the $\mathcal{H}^q_2(D;\mathbb{E}_\mu)_{K_0}$. Although it moves rather slowly, this program remains quite active and has since been carried much further. See [**8**], [**9**], [**10**], [**13**], [**18**], [**138**], [**152**], [**153**], [**154**] and [**155**].

References

[1] D. N. Akhiezer & S. G. Gindikin, On the Stein extensions of real symmetric spaces. Math. Ann. **286** (1990), 1–12.

[2] A. Andreotti & H. Grauert, Théorèmes de finitude pour la cohomologie des espaces complèxes. Bull. Soc. Math. France **90** (1962), 193–259.

[3] M. F. Atiyah, N. J. Hitchin & I. M. Singer, Self duality in four–dimensional Riemannian geometry. Proc. Roy. Soc. London **A 362** (1978), 425–461.

[4] T. N. Bailey & L. David, The Penrose transform for compactly supported cohomology. University of Edinburgh preprint, 9/2000.

[5] T. N. Bailey & M. G. Eastwood, Zero-energy fields on real projective space. Geom. Dedicata **67** (1997) 245–258.

[6] T. N. Bailey, M. G. Eastwood, A. R. Gover & L. J. Mason, The Funk transform as a Penrose transform. Math. Proc. Cambridge Philos. Soc. **125** (1999), 67–81.

[7] T. N. Bailey & M. A. Singer, Twistors, massless fields and the Penrose transform. In: "Twistors in Mathematics and Physics", London Math. Soc. Lecture Notes **156** (1990), 299–338.

[8] L. Barchini, Strongly harmonic differential forms on elliptic orbits. Contemporary Math. **154** (1992), 77–88.

[9] L. Barchini, Szegö mappings, harmonic forms, and Dolbeault cohomology. J. Funct. Anal. **118** (1993), 351–406.

[10] L. Barchini, Strongly harmonic forms for representations in the discrete series. J. Funct. Anal. **161** (1999), 111–131.

[11] L. Barchini, S. G. Gindikin & H. W. Wong, Geometry of flag manifolds and holomorphic extensions of Szegö kernels for $SU(p,q)$. Pacific J. Math. **179** (1997), 201–220.

[12] L. Barchini, A. W. Knapp & R. Zierau, Intertwining operators into Dolbeault cohomology representations. J. Funct. Anal. **107** (1992), 302–341.

[13] L. Barchini & R. Zierau, Square integrable harmonic forms and representation theory. Duke Math. J. **92** (1998), 645–664.

[14] D. Barlet, Espace analytique réduit des cycles analytiques complexes compacts d'un espace analytique complexe de dimension finie. In: "Séminaire F. Norguet: Fonctions de Plusiers Variables Complexes, 1974/75", Lecture Notes in Mathematics **482**, Springer-Verlag, 1975, 1–158.

[15] D. Barlet, Convexité de l'espace des cycles. Bull. Soc. Math. de France **106** (1978), 373–397.

[16] D. Barlet & V. Koziarz, Fonctions holomorphes sur l'espace des cycles: La méthode d'intersection. Math. Research Letters **7** (2000), 537–549.

[17] R. J. Baston & M. G. Eastwood, "The Penrose Transform: Its Interaction with Representation Theory", Clarendon Press, Oxford, 1989.

[18] B. Binegar & R. Zierau, Unitarization of a singular representation of SO(p,q). Comm. Math. Phys. **138** (1991), 245–258.

[19] A. Borel, Groupes linéaires algébriques, Ann. of Math **64** (1956), 20–82.

[20] A. Borel, "Linear algebraic groups", Benjamin, 1961.

[21] A. Borel, "Linear algebraic groups", 2nd Ed., GTM **126**, Springer–Verlag, 1991.

[22] A. Borel & J. Tits, "Groupes réductifs", Publ. Math. I.H.E.S. **27** (1965), 55–150.

[23] R. Bott, Homogeneous vector bundles. Annals of Math. **66** (1957), 203–248.

[24] R. J. Bremigan, Invariant analytic domains in complex semisimple groups. Transform. Groups **1** (1996), 279–305.

[25] R. J. Bremigan & G. Fells, Bi-invariant domains in semisimple Lie groups: Non-principal orbits and boundaries. Geometriae Dedicata **82** (2000), 225–283.

[26] F. Bruhat, Sur les représentations induites des groupes de Lie. Bull. Soc. Math. France **84** (1956), 97–205.

[27] N. Buchdahl, On the relative DeRham sequence. Proc. Amer. Math. Soc. **87** (1983), 363–366.

[28] D. Burns, On the uniqueness and characterization of Grauert tubes. In: "Complex analysis and geometry (Trento, 1993)", Lecture Notes in Pure and Appl. Math. **173**, Dekker, New York, 1996, 119–133.

[29] D. Burns & R. Hind, Geometry of Grauert tubes and complexifications of symmetric spaces, in preparation.

[30] F. Campana & Th. Peternell, Cycle spaces. In: "Several Complex Variables VII", Encyclopaedia of Mathematical Sciences, **74** (1994), 319–349.

[31] A. D'Agnolo & P. Schapira, Radon–Penrose transform for $\mathcal{D}$–modules. J. Funct. Anal. **139** (1996), 349–382.

[32] A. D'Agnolo & P. Schapira, The Radon–Penrose correspondence, II: Line bundles and simple $\mathcal{D}$–modules. J. Funct. Anal. **153** (1998), 343–356.

[33] J. Dixmier, "Les C*-algèbres et leurs représentations", Gauthier–Villars, Paris, 1964.

[34] E. G. Dunne & R. Zierau, Twistor theory for indefinite Kähler symmetric spaces. Contemporary Math. **154** (1993), 117–132.

[35] M. G. Eastwood, Twistor Theory (the Penrose transform), Springer Lecture Notes in Math **926** (1982), 1–11.

[36] M. G. Eastwood, The Penrose transform. In: "Twistors in Mathematical Physics", London Math. Soc. Lecture Notes **156** (1990), 87–103.

[37] M. G. Eastwood, Introduction to the Penrose transform. In: "Analytic Cohomology in Representation Theory", Contemporary Math. **154** (1993), 71–75.

[38] M. G. Eastwood, Complex methods in real integral geometry. In: Proc. 16th Winter School on Geometry and Physics, Srní, Suppl. Rend. Circ. Mat. Palermo **46** (1997), 55–71.

[39] M. G. Eastwood, Some examples of the Penrose transform, Kyoto Univ. R. I. M. S. Kokyuroku **1058** (1998), 22–28.

[40] M. G. Eastwood, Variations on the de Rham complex. Notices Amer. Math. Soc. **46** (1999), 1368–1376.

[41] M. G. Eastwood, R. Penrose & R. O. Wells, Cohomology and massless fields. Communic. Math. Physics **78** (1981), 305–351.

[42] G. Fels, A differential geometric characterization of invariant domains of holomorphy. Ann. Inst. Fourier (Grenoble) **45** (1995), 1329–1351.

[43] G. Fells & L. Geatti, Invariant domains in complex symmetric spaces. J. Reine Angew. Math. **454** (1994), 97–118.

[44] G. Fells & L. Geatti, Geometry of bi-invariant subsets of complex semisimple Lie groups. Ann. Scuola Norm. Sup. Pisa **26** (1998), 329–356.

[45] A. Frölicher, Zur Differentialgeometrie der komplexen Strukturen. Math. Ann. **129** (1955) 50–95.

[46] P. Funk, Über Flächen mit lauter geschossenen geodätischen Linien. Math. Ann **74** (1913), 278–300.

[47] I. M. Gelfand & S. G. Gindikin, Complex manifolds, the skeletons of which are semisimple real Lie groups, and analytic discrete series of representations. Funkt. Anal. Appl. **11** (1977) 19–27.

[48] I. M. Gelfand, M. I. Graev & N. Ya. Vilenkin, "Generalized Functions (Volume 5): Integral Geometry and Representation Theory", Academic Press, 1966.

[49] S. G. Gindikin, Between integral geometry and twistors. In: "Twistors in Mathematics and Physics", London Math. Soc. Lecture Notes **156** (1990), 30–35.

[50] S. G. Gindikin, Some notes on the Radon transform and integral geometry. Monat. Math. **113** (1992), 23–32.

[51] R. Godement, Sur les rélations d'orthogonalité de V. Bargmann. C. R. Acad. Sci. Paris **225**, (1947), 521–523 and 657–659.

[52] P. A. Griffiths, Periods of integrals on algebraic manifolds, I. Amer. J. Math. **90** (1968), 568–626.

[53] P. A. Griffiths, Periods of integrals on algebraic manifolds: Summary of main results and discussion of open problems, Bull. Amer. Math. Soc. **76** (1970), 228–296.

[54] Harish–Chandra, Plancherel formula for the 2×2 real unimodular group, Proc. Nat. Acad. Sci. USA **38** (1952), 337–342.

[55] Harish–Chandra, Representations of a semisimple Lie group on a Banach space, I. Trans. Amer. Math. Soc. **75** (1953), 185–243.
[56] Harish–Chandra, Representations of a semisimple Lie group on a Banach space, II. Trans. Amer. Math. Soc. **76** (1954), 26–65.
[57] Harish–Chandra, Representations of a semisimple Lie group on a Banach space, III. Trans. Amer. Math. Soc. **76** (1954), 234–253.
[58] Harish–Chandra, The Plancherel formula for complex semisimple Lie groups. Trans. Amer. Math. Soc. **76** (1954), 485–528.
[59] Harish–Chandra, Representations of semisimple Lie groups, IV. Amer. J. Math. **77** (1955), 743–777.
[60] Harish–Chandra, Representations of semisimple Lie groups, V. Amer. J. Math. **78** (1956), 1–41.
[61] Harish–Chandra, Representations of semisimple Lie groups, VI. Amer. J. Math. **78** (1956), 564–628.
[62] Harish–Chandra, On a lemma of F. Bruhat. J. Maths. Pures Appl. **35** (1956), 203–210.
[63] Harish–Chandra, Characters of semisimple Lie groups. Trans. Amer. Math. Soc. **83** (1956), 98–163.
[64] Harish–Chandra, Automorphic forms on a semisimple Lie group. Proc. Nat. Acad. Sci. USA **45** (1959), 570–573.
[65] Harish–Chandra, Invariant eigendistributions on a semisimple Lie group. Trans. Amer. Math. Soc. **119** (1965), 457–508.
[66] Harish–Chandra, Discrete series for semisimple Lie groups, I. Acta Math .**113** (1965), 241–317.
[67] Harish–Chandra, Discrete series for semisimple Lie groups, II. Acta Math .**116** (1966), 1–111
[68] Harish–Chandra, "Automorphic Forms on Semisimple Lie Groups", Lecture Notes in Math. **62**, Springer–Verlag, 1968.
[69] Harish–Chandra, Harmonic analysis on real reductive groups, I. J. Funct. Anal. **19** (1975), 104–204.
[70] Harish–Chandra, Harmonic analysis on real reductive groups, II. Inventiones Math. **36** (1976), 1–55.
[71] Harish–Chandra, Harmonic analysis on real reductive Lie groups, III: The Maass-Selberg relations and the Plancherel formula. Annals of Math. **104** (1976), 117–201.
[72] H. Hecht, D. Miličić , W. Schmid & J. A. Wolf, Localization and standard modules for real semisimple Lie groups, I: The duality theorem. Inventiones Math., **90** (1987), 297–332.
[73] S. Helgason, Duality and Radon transform for symmetric spaces. Amer. J. Math. **85** (1963), 667–692.
[74] S. Helgason, The Radon transform on Euclidean spaces, compact two-point homogeneous spaces and Grassmann manifolds. Acta Math. **113** (1965), 153–180.
[75] S. Helgason, A duality for symmetric spaces with applications to group representations. Advances in Math. **5** (1970), 1–154.
[76] S. Helgason, The X-ray transform on a symmetric space. Lecture Notes in Math. **838** (1981), 145–148.
[77] S. Helgason, Some results on Radon transforms, Huygens' principle and X-ray transforms. Contemp. Math. **63** (1987), 151–177.
[78] S. Helgason, Geometric and analytic features of the Radon transform. Mitt. Math. Ges. Hamburg **12** (1992), 977–988.
[79] S. Helgason, Radon transforms for double fibrations. Examples and viewpoints. Internat. Press Lecture Notes Math. Phys.**IV** (1994), 175–188
[80] S. Helgason, Radon transforms and wave equations. Springer Lecture Notes in Math. **1684** (1998), 99–121,
[81] S. Helgason, "The Radon transform, Second edition". Progress in Mathematics **5**, Birkhuser, 1999.
[82] R. A. Herb, Fourier inversion and the Plancherel formula for semisimple Lie groups. Amer. J. Math. **104** (1982), 9–58.
[83] R. A. Herb, Discrete series characters and Fourier inversion on semisimple real Lie groups. Trans. Amer. Math. Soc. **277** (1983), 241–261.
[84] R. A. Herb and J. A. Wolf, The Plancherel theorem for general semisimple Lie groups. Compositio Math. **57** (1986), 271–355.

[85] R. A. Herb and J. A. Wolf, Rapidly decreasing functions on general semisimple Lie groups. Compositio Math. **58** (1986), 73–110.

[86] N. J. Hitchin, Linear field equations on self–dual spaces. Proc. Roy. Soc. London **A370** (1980),

[87] A. T. Huckleberry & G. Fels, A characterization of K-invariant Stein domains in symmetric embeddings. In: "Complex analysis and geometry", Univ. Ser. Math., Plenum, New York, 1993, 223–234.

[88] A. T. Huckleberry & A. Simon, On cycle spaces of flag domains of $SL_n(\mathbb{R})$, to appear in Crelle's Journal in 2001.

[89] A. T. Huckleberry & J. A. Wolf, Flag duality. Annals of Global Analysis and Geometry **18** (2000), 331-340.

[90] A. T. Huckleberry & J. A. Wolf, Cycle Spaces of Real Forms of $SL_n(\mathbb{C})$. Submitted to the Göttingen Academy of Sciences for a special volume in honor of Hans Grauert's 70th birthday.

[91] A. Korányi & J. A. Wolf, Realization of hermitian symmetric spaces as generalized half–planes. Annals of Math. **81** (1965), 265–288.

[92] S.-J. Kan & D. Ma, On rigidity of Grauert tubes over locally symmetric spaces. J. Reine Angew. Math. **524** (2000), 205–225.

[93] B. Kostant, Lie algebra cohomology and the generalized Borel–Weil theorem. Annals of Math. **74** (1961), 329–387.

[94] Y. I. Manin, "Gauge Field Theory and Complex Geometry", Springer-Verlag, 1988.

[95] C. C. Moore, Compactifications of symmetric spaces II (The Cartan domains). Amer. J. Math. **86** (1964), 358–378.

[96] H. Moscovici, L^2–index of elliptic operators on locally symmetric spaces of finite volume. Contemporary Math. **10** (1982), 129–137.

[97] J. D. Novak, Parameterizing Maximal Compact Subvarieties. Proc. Amer. Math. Soc. **124** (1996), 969–975.

[98] J. D. Novak, "Explicit Realizations of Certain Representations of $Sp(n;\mathbb{R})$ via the Penrose Transform, Ph.D. thesis, Oklahoma State University, 1996.

[99] J. D. Novak, Explicit realizations of certain representations of $Sp(n,\mathbb{R})$ via the double fibration transform. Pacific J. Math. **195** (2000) 417–451.

[100] J. D. Novak, J. A. Wolf & R. Zierau, The image of the holomorphic double fibration transform, paper planned.

[101] G. I. Ol'shansky, Complex Lie groups, Hardy spaces and the program of Gelfand and Gindikin. In: "Problems of Group Theory and Homological Algebra", Yaroslav, 1982, 85–98.

[102] C. M. Patton & H. Rossi, Unitary structures on cohomology. Trans. Amer. Math. Soc. **290** (1985), 235–258.

[103] C. M. Patton & H. Rossi, Cohomology on complex homogeneous manifolds with compact subvarieties. Contemporary Math. **58** (1986), 199–211.

[104] R. Penrose, Twistor algebra. J. Math. Phys. **8** (1967), 345–366.

[105] R. Penrose, Solutions of the zero rest–mass equations. J. Math. Phys. **10** (1969), 38–39.

[106] R. Penrose, Non–linear gravitons and curved twistor theory. Gen. Rel. & Grav. **7** (1976), 31–52.

[107] R. Penrose, On the twistor description of massless fields. In: "Complex Manifold Techniques in Theoretical Physics", Pitman Res. Notes in Math **32** (1979), 55–91.

[108] J. Radon, Über die Bestimmung von Funktionen durch ihre Integralwerte längs gewisser Mannigfaltigkeiten. Sächs. Akad. Wiss. Leipzig, Math.–Nat. Kl. **69** (1917), 262–277.

[109] J. Rawnsley, W. Schmid & J. A. Wolf, Singular unitary representations and indefinite harmonic theory. J. Funct. Anal. **51** (1983), 1–114.

[110] M. A. Rieffel, Square integrable representations of Hilbert algebras. J. Funct. Anal. **3** (1969), 265–300.

[111] P. Sally & G. Warner, The Fourier transform on semisimple Lie groups of real rank one. Acta Math. **131** (1973), 11–26.

[112] W. Schmid, "Homogeneous complex manifolds and representations of semisimple Lie groups", Ph.D. thesis, University of California at Berkeley, 1967. Reprinted in "Representation Theory and Harmonic Analysis on Semisimple Lie Groups", Mathematical Surveys and Monographs **31**, 223–286, eds. P. Sally & D. Vogan, Amer. Math. Soc., 1988.

[113] W. Schmid, On a conjecture of Langlands. Annals of Math. **93** (1971), 1–42.

[114] W. Schmid, L^2–cohomology and the discrete series. Annals of Math. **103** (1976), 375–394.
[115] W. Schmid, Boundary value problems for group invariant differential equations. In: "Proceedings, Cartan Symposium, Lyon, 1984," Astérisque, 1985, 311–322.
[116] W. Schmid & J. A. Wolf, A vanishing theorem for open orbits on complex flag manifolds. Proc. Amer. Math. Soc. **92** (1984, 461-464.
[117] W. Schmid & J. A. Wolf, Geometric quantization and derived functor modules for semisimple Lie groups. J. Funct. Anal. **90** (1990), 48–112.
[118] H. Sekiguchi, The Penrose transform for certain non-compact homogeneous manifolds of $U(n,n)$. J. Math. Sci. Univ. Tokyo **3** (1996), 655–697.
[119] R. J. Stanton, Analytic extension of the holomorphic discrete series. Amer. J. Math. **108** (1986), 1411–1424.
[120] J. A. Tirao & J. A. Wolf, Homogeneous holomorphic vector bundles. Indiana Univ. Math. J. **20** (1970), 15–31.
[121] J. Tits, "Sur certains classes d'espaces homogènes de groupes de Lie", Memoir, Belgian Academy of Sciences, 1955.
[122] J. Tits, Espaces homogènes complexes compacts. Comment. Math. Helv. **37** (1962), 111–120.
[123] N. Wallach, "Real reductive groups", I, Academic Press, 1988.
[124] N. Wallach, "Real reductive groups", II, Academic Press, 1992.
[125] N. R. Wallach & J. A. Wolf, Completeness of Poincaré series for automorphic forms associated to the integrable discrete series. In: "Representation Theory of Reductive Groups", Progress in Mathematics **40**, Birkhäuser, 1983, 265–281.
[126] R.S. Ward and R.O. Wells, Jr, "Twistor Geometry and Field Theory", Cambridge University Press, 1989.
[127] R. O. Wells, Parameterizing the compact submanifolds of a period matrix domain by a Stein manifold. In: "Symposium on Several Complex Variables", Lecture Notes in Mathematics **104** (1971), 121–150.
[128] R. O. Wells & J. A. Wolf Poincaré series and automorphic cohomology on flag domains. Annals of Math. **105** (1977, 397–448.
[129] F. L. Williams, On the finiteness of the L^2 automorphic cohomology of a flag domain. J. Funct. Anal. **72** (1987) 33–43.
[130] F. L. Williams, On the dimension of spaces of automorphic cohomology. In: "Algebraic Groups and Related Topics", Advanced Studies in Pure Mathematics **6**, Kinokuniya & North–Holland, 1985, 1–15.
[131] F. L. Williams, Finite spaces of non–classical Poincaré theta series. Contemporary Mathematics **53** (1986), 543–554.
[132] J. A. Wolf, Locally symmetric homogeneous spaces. Commentarii Mathematici Helvetici **37** (1962), 65–101.
[133] J. A. Wolf, The action of a real semisimple Lie group on a complex manifold, I: Orbit structure and holomorphic arc components. Bull. Amer. Math. Soc. **75** (1969), 1121–1237.
[134] J. A. Wolf, Fine structure of hermitian symmetric spaces. In: "Symmetric Spaces", ed. W. M. Boothby & G. L. Weiss, Dekker, 1972, 271–357.
[135] J. A. Wolf, "The action of a real semisimple Lie group on a complex flag manifold, II: Unitary representations on partially holomorphic cohomology spaces", Memoirs. Amer. Math. Soc. **138**, 1974.
[136] J. A. Wolf, Partially harmonic spinors and representations of reductive Lie groups. J. Funct. Analysis 15 (1974), 117–154.
[137] J. A. Wolf, Completeness of Poincaré series for automorphic cohomology. Annals of Math. **109** (1979), 545–567.
[138] J. A. Wolf, Geometric realizations of discrete series representations in a nonconvex holomorphic setting. Bull. Soc. Math. de Belgique **42** (1990), 797–812.
[139] J. A. Wolf, The Stein condition for cycle spaces of open orbits on complex flag manifolds. Annals of Math. **136** (1992), 541–555.
[140] J. A. Wolf, Exhaustion functions and cohomology vanishing theorems for open orbits on complex flag manifolds Mathematical Research Letters **2** (1995), 179–191.
[141] J. A. Wolf, Flag Manifolds and Representation Theory. In: "Geometry and Representation Theory of Real and p-adic Lie Groups", Progress in Mathematics **158**, Birkhäuser, 1997, 273–323.

[142] J. A. Wolf, Real groups transitive on complex flag manifolds. Proc. Amer. Math. Soc. **129** (2001), 2483–2487.

[143] J. A. Wolf, Hermitian Symmetric Spaces, Cycle Spaces, and the Barlet–Koziarz Intersection Method for Construction of Holomorphic Functions. Math. Research Letters **7** (2000), 551–564.

[144] J. A. Wolf & A. Gray, Homogeneous spaces defined by Lie group automorphisms, I. J. Diff. Geometry **2** (1968), 77–114.

[145] J. A. Wolf & A. Gray, Homogeneous spaces defined by Lie group automorphisms, II. J. Diff. Geometry **2** (1968), 115–159.

[146] J. A. Wolf & A. Korányi , Generalized Cayley transformations of bounded symmetric domains. Amer. J. Math. **87** (1965), 899–939.

[147] J. A. Wolf & R. Zierau, Riemannian exponential maps and decompositions of reductive Lie groups. In: "Topics in Geometry: In Memory of Joseph D'Atri", Progress in Nonlinear Differential Equations **20**, Birkhäuser, 1996, 349–353.

[148] J. A. Wolf & R. Zierau, Cayley transforms and orbit structure in complex flag manifolds. Transformation Groups **2** (1997), 391-405.

[149] J. A. Wolf & R. Zierau, Linear cycle spaces in flag domains Math. Annalen, **316** (2000), 529–545.

[150] J. A. Wolf & R. Zierau, Holomorphic double fibration transforms. In "The Mathematical Legacy of Harish–Chandra", PSPM **68**, AMS, 2000, 527–551.

[151] H.–W. Wong, Dolbeault cohomologies and Zuckerman modules associated with finite rank representations. J. Funct. Anal. **127** (1995), 428–454.

[152] R. Zierau, Geometric construction of certain highest weight modules. Proc. Amer. Math. Soc. **95** (1985), 631–635.

[153] R. Zierau, A construction of harmonic forms on $U(p+1,q)/U(p,q) \times U(1)$. Pacific J. Math. **139** (1989), 377–399.

[154] R. Zierau, Unitarity of certain Dolbeault cohomology representations. Contemporary Math. **154** (1992), 239–259,

[155] R. Zierau, Harmonic forms and unitary representations of $SO(2n,1)$. Nova J. Algebra Geom. **1** (1992), 359–369.

Department of Mathematics, University of California, Berkeley, California 94720–3840, U.S.A.

E-mail address: `jawolf@math.berkeley.edu`

Contemporary Mathematics
Volume **288**, 2001

On the Volume of Compact Spacelike Hypersurfaces with Hyperbolic Boundary in Minkowski Space

Juan A. Aledo and José A. Pastor

Abstract. We give an integral formula for the volume of the domain enclosed by the hyperbolic boundary of a compact spacelike hypersurface in Minkowski space. Moreover, we deduce some inequalities involving the volume of the hypersurface.

1. Introduction.

In this paper we will deal with compact spacelike hypersurfaces (necessarily with nonempty boundary) in Minkowski space $\mathbf{L}^{n+1}$. There exists some existence and uniqueness results when the boundary is prescribed and the hypersurface verifies a geometric condition (see, for example, [**AP2**], [**AM**] and [**BS**]). On the other hand, in the paper [**AP**] the second author jointly with Alias gave an integral formula for the volume of the domain enclosed by the hyperplanar boundary of a compact spacelike hypersurface as well as some inequalities involving the volume of the hypersurface.

In this work we consider a spacelike hypersurface such that its boundary is contained in a hyperbolic space. When the domain determined by the boundary in the hyperbolic space is simply connected, we are able to prove (see lemma 3.1) that the hypersurface is diffeomorphic precisely to the hyperbolic domain enclosed by the boundary and, moreover, we can state an integral formula for the volume of this domain (theorem 4.1). Finally, we derive an inequality from this integral formula and we present some examples.

2. Preliminaries.

Let us denote by $\mathbf{L}^{n+1}$ the $(n+1)$-dimensional Minkowski space, that is, the real vector space $\mathbf{R}^{n+1}$ endowed with the Lorentzian metric

$$\langle , \rangle = \mathrm{d}x_1^2 + \cdots + \mathrm{d}x_n^2 - \mathrm{d}x_{n+1}^2$$

2000 *Mathematics Subject Classification.* Primary 53C42; Secondary 53B30, 53C50.
Key words and phrases. Minkowski space, spacelike hypersurfaces, volume.
The authors were supported by DGICYT Grant. No. PB97-0784-C03-02 and Consejeria de Educación y Cultura CARM Grant No. PB/5/FS/97, Programa Séneca (PRIDTYC).

where $(x_1, x_2, ..., x_{n+1})$ are the canonical coordinates in $\mathbf{R}^{n+1}$. We will say that a smooth immersion $\psi : M \longrightarrow \mathbf{L}^{n+1}$ of a n-dimensional connected manifold M (possibly with nonempty boundary) is a *spacelike hypersurface* if the induced metric via ψ is a Riemannian metric on M which, as usual, is also denote by $\langle , \rangle$. Now, observe that $(0, \ldots, 0, 1)$ is a unitary timelike vector field globally defined on $\mathbf{L}^{n+1}$ which determines a time-orientation on $\mathbf{L}^{n+1}$. Thus, in every point of M we can choose the normal vector N in the same time orientation as $(0, \ldots, 0, 1)$ and we will assume that M is oriented by N. We will refer to N as the *future-directed Gauss map* of the hypersurface M. Throughout this paper we are going to consider only *compact* spacelike hypersurfaces immersed in $\mathbf{L}^{n+1}$. In this sense, let us remark that there exists no closed (compact without boundary) spacelike hypersurfaces in $\mathbf{L}^{n+1}$. To prove this fact, let $a \in \mathbf{L}^{n+1}$ be a fixed arbitrary vector, and consider the height function $\langle a, x \rangle$ defined on the spacelike hypersurface M. The gradient on M of $\langle a, x \rangle$ satisfies

$$|\nabla \langle a, x \rangle|^2 = \langle a, a \rangle + \langle a, N \rangle^2 \geq \langle a, a \rangle.$$

In particular, when a is spacelike the height function has no critical points in M, so that M cannot be closed. Therefore, every compact spacelike hypersurface M necessarily has non-empty boundary ∂M. As usual, if Σ is an $(n-1)$-dimensional closed submanifold in $\mathbf{L}^{n+1}$, a spacelike hypersurface $\psi : M \to \mathbf{L}^{n+1}$ is said to be a hypersurface *with boundary* Σ if the restriction of the immersion ψ to the boundary ∂M is a diffeomorphism onto Σ.

3. Hypersurfaces with hyperbolic boundary

Let us consider now the n-dimensional hyperbolic space of radius 1 centered at the origin, that is, the hypersurface given by

$$\mathbf{H}^n = \left\{ x \in \mathbf{L}^{n+1} : \langle x, x \rangle = -1, x_{n+1} > 0 \right\}.$$

Since $\mathbf{H}^n$ is a complete spacelike hypersurface, it can be easily checked that $\mathbf{H}^n$ divides $\mathbf{L}^{n+1}$ into two connected components and we will denote by $\mathbf{L}^{n+1}_+$ the one towards the future-directed Gauss map of $\mathbf{H}^n$ (that is, the position vector field x) is pointing, that is,

$$\mathbf{L}^{n+1}_+ = \left\{ x \in \mathbf{L}^{n+1} : \langle x, x \rangle \leq -1, x_{n+1} > 0 \right\}.$$

We define the *central projection* onto $\mathbf{H}^n$ as the map $F : \mathbf{L}^{n+1}_+ \longrightarrow \mathbf{H}^n$ given by

$$F(x) = \frac{x}{\sqrt{-\langle x, x \rangle}},$$

which is a surjective smooth function that maps *radially* $\mathbf{L}^{n+1}_+$ onto $\mathbf{H}^n$. Given $\psi : M \to \mathbf{L}^{n+1}$ a compact spacelike hypersurface with boundary $\Sigma = \psi(\partial M)$, we will say that M has *hyperbolic boundary* if the boundary Σ is contained in the hyperbolic space $\mathbf{H}^n$. If we denote by Ω the n-dimensional (open) domain bounded by Σ in $\mathbf{H}^n$, we can state the following result.

LEMMA 3.1. *Let $\psi : M \longrightarrow \mathbf{L}^{n+1}_+$ be a compact spacelike immersion with hyperbolic boundary Σ such that the domain Ω bounded by Σ in the hyperbolic space is simply connected. Then, the central projection of M onto the hyperbolic space $\mathbf{H}^n$ is a diffeomorphism between M and $\overline{\Omega}$.*

PROOF. We will denote the central projection of M onto $\mathbf{H}^n$ by $\widetilde{F}$, that is,

$$\widetilde{F} = F \circ \psi : M \longrightarrow \mathbf{H}^n.$$

Our first step is to show that $\widetilde{F}$:int$(M) \longrightarrow \mathbf{H}^n$ is a local diffeomorphism and, therefore, is an open map. In order to do that, take a point $x \in$int(M) and $v \in T_xM$, $v \neq 0$, a tangent vector. Let $\alpha : (-\delta, \delta) \longrightarrow M$ be a curve with initial conditions x, v and let us identify, as usual, x, v and α with $\psi(x), \mathrm{d}\psi_x(v)$ and $\psi \circ \alpha$, the corresponding elements via the immersion ψ. Then,

$$\mathrm{d}F_x v = \mathrm{d}\widetilde{F}_x(v) = \frac{\mathrm{d}}{\mathrm{d}s}\bigg|_{s=0} F(\alpha(s)) = \frac{\mathrm{d}}{\mathrm{d}s}\bigg|_{s=0} \frac{\alpha(s)}{\sqrt{-\langle \alpha(s), \alpha(s)\rangle}} = \frac{|x|^2 v + \langle x, v\rangle x}{|x|^3}$$

where $|x| = \sqrt{-\langle x, x\rangle}$. Now, if v is a vector such that $\mathrm{d}\widetilde{F}_x(v) = 0$, then the last formula implies $\langle v, v\rangle < 0$ in contradiction with the fact that v is a spacelike vector. Thus, $\mathrm{d}\widetilde{F}_x$ has maximum rank and $\widetilde{F}$ is a local diffeomorphism defined on int(M). Immediately, $\widetilde{F}$ is an open map from int(M) to $\mathbf{H}^n$. Our aim now is to show that $\widetilde{F}\,(\text{int}(M)) = \Omega$. Let us first see that

$$\partial\widetilde{F}\,(\text{int}(M)) = \partial\widetilde{F}\,(M) \subset \Sigma. \tag{3.1}$$

Observe that since M is compact, $\widetilde{F}(M)$ is closed and $\partial\widetilde{F}(M) \subset \widetilde{F}(M)$. Thus, for any $y \in \partial\widetilde{F}\,(M)$ there exists $x \in M$ such that $\widetilde{F}(x) = y$. We would like to show that $x \in \partial M$. If $x \in$int(M), then there is an open neighbourhood $\mathcal{U}(x)$ of x in int(M) and an open neighbourhood $\mathcal{V}(y)$ of y in $\widetilde{F}\,(\text{int}(M))$ such that $\widetilde{F} : \mathcal{U}(x) \longrightarrow \mathcal{V}(y)$ is a diffeomorphism. This implies that y is an interior point of $\widetilde{F}\,(\text{int}(M))$, in contradiction to the fact that y is a boundary point of $\widetilde{F}\,(\text{int}(M))$. Therefore, we have the inclusion (3.1). Let us see now that $\widetilde{F}\,(\text{int}(M)) = \Omega$. If there exists a point in $\widetilde{F}\,(\text{int}(M))$ which is not in Ω, since $\widetilde{F}\,(\text{int}(M))$ is bounded, there are points in $\partial\widetilde{F}\,(\text{int}(M))$ outside Ω which is not possible (it contradicts (3.1)). Analogously, if there is a point in Ω which is not in $\widetilde{F}\,(\text{int}(M))$, there are points in $\partial\widetilde{F}\,(\text{int}(M))$ inside Ω which again is not possible. Thus we have the equality

$$\widetilde{F}\,(\text{int}(M)) = \overline{\Omega}$$

and, consequently, $\widetilde{F} : M \longrightarrow \overline{\Omega} = \Omega \cup \Sigma$ is a local diffeomorphism. The compactness of M implies that $\widetilde{F}$ is a covering map and since $\overline{\Omega}$ is simply connected, $\widetilde{F}$ must be a global diffeomorphism from M to $\overline{\Omega}$. □

REMARK 3.2. In order to get the preceding result, once we know that the central projection $\widetilde{F}$ is a covering map, it suffices to assume that $\widetilde{F} : M \longrightarrow \overline{\Omega}$ is a 1-fold covering of $\overline{\Omega}$.

4. A volume formula.

In this section we are going to derive a volume formula for the hyperbolic domain Ω in terms of an integral on the hypersurface M. Our approach is based on the *change of variables* theorem and the diffeomorphism defined on the previous section. Let us denote by $\mathrm{d}A$ the n-dimensional volume element of M with respect to the induced metric and the orientation given by N, the future-directed Gauss map. Also, let us denote by $\mathrm{d}A_\Omega$ the n-dimensional volume element of Ω with respect to the metric induced on $\mathbf{H}^n$ by the Lorentzian metric of $\mathbf{L}^{n+1}$ and the orientation given by the future-directed Gauss map x. Given $y \in \Omega$ a point and $v_1, v_2, \ldots, v_n$ tangent vectors to Ω at y, it's easy to see that

$$\mathrm{d}A_\Omega(y)\,(v_1, v_2, \ldots, v_n) = \det\,(v_1, v_2, \ldots, v_n, y)\,,$$

where det stands for the usual determinant on $\mathbf{R}^{n+1}$. A well known theorem for manifolds states the following equation

$$\mathrm{vol}(\Omega) = \int_\Omega 1\ \mathrm{d}A_\Omega = \int_{\mathrm{int}M} F^*\,(\mathrm{d}A_\Omega)\,, \tag{4.1}$$

where $F^*\,(\mathrm{d}A_\Omega)$ stands for the *pull-back* by F of the n-dimensional volume element $\mathrm{d}A_\Omega$. For each $x \in \mathrm{int}M$ and $v_1, v_2, \ldots, v_n$ tangent vectors at x we have

$$\begin{aligned} F^*\,(\mathrm{d}A_\Omega)\,(x)\,(v_1, v_2, \ldots, v_n,) &= \mathrm{d}A_\Omega\,(F\,(x))\,(\mathrm{d}F_x(v_1), \mathrm{d}F_x(v_2), \ldots, \mathrm{d}F_x(v_n)) \\ &= \det\,(\mathrm{d}F_x(v_1), \mathrm{d}F_x(v_2), \ldots, \mathrm{d}F_x(v_n), F(x))\,. \end{aligned}$$

Let $\{e_1, e_2, \ldots, e_n\}$ be an orthonormal basis positively oriented in T_xM. For each $i = 1, 2, \ldots, n$ we know that

$$\mathrm{d}F_x(e_i) = \frac{|x|^2 e_i + \langle x, e_i\rangle x}{|x|^3},$$

and then

$$\begin{aligned} F^*\,(\mathrm{d}A_\Omega)\,(x)\,(e_1, e_2, \ldots, e_n,) &= \det\left(\mathrm{d}F_x(e_1), \mathrm{d}F_x(e_2), \ldots, \mathrm{d}F_x(e_n), \frac{x}{|x|}\right) \\ &= \frac{1}{|x|^{n+1}} \det\,(e_1, e_2, \ldots, e_n, x) = \frac{-\langle x, N\rangle}{|x|^{n+1}}. \end{aligned} \tag{4.2}$$

Since $F^*\,(\mathrm{d}A_\Omega)$ and $\mathrm{d}A$ are differential n-forms over the n-dimensional manifold M, there exists a function $\lambda \in \mathcal{C}^\infty(M)$ such that $\lambda \mathrm{d}A = F^*\,(\mathrm{d}A_\Omega)$ where λ is precisely given by (4.2) and then

$$\frac{-\langle x, N\rangle}{|x|^{n+1}} \mathrm{d}A(x) = F^*\,(\mathrm{d}A_\Omega)\,(x).$$

Finally, using (4.1) we get the following result concerning the volume of the domain Ω.

THEOREM 4.1. *Let $\psi : M \longrightarrow \mathbf{L}_+^{n+1}$ be a compact spacelike immersion with hyperbolic boundary Σ. If the domain bounded by the boundary Σ in $\mathbf{H}^n$ is simply connected , then the following equation holds*

$$\mathrm{vol}(\Omega) = \int_{\mathrm{int}M} \frac{-\langle x, N\rangle}{|x|^{n+1}}\ \mathrm{d}A$$

where N is the future-directed Gauss map of the hypersurface M, Ω is the domain bounded by Σ and $\mathrm{d}A$ is the n-dimensional volume element with respect to the induced metric and the chosen orientation.

As a first consequence of (4.1), let us observe that x and N are timelike vectors with the same time-orientation. Thus $-\langle x, N\rangle$ is positive and

$$-\langle x, N\rangle \geq |x||N| = |x| \geq 1$$

which implies

$$\mathrm{vol}(\Omega) = \int_{\mathrm{int}M} \frac{-\langle x, N\rangle}{|x|^{n+1}}\ \mathrm{d}A \geq \int_{\mathrm{int}M} \frac{1}{|x|^n}\ \mathrm{d}A \geq \frac{1}{k_0^n}\ \mathrm{vol}(M), \tag{4.3}$$

where $k_0 = \max\{|x| : x \in M\} \geq 1$, with equality if and only if $\overline{\Omega} = M$.

REMARK 4.2. Let us consider

$$M_\infty = \left\{ \left(x_1, x_2, \ldots, x_n, \sqrt{2} \right) : x_1^2 + x_2^2 + \cdots + x_n^2 \leq 1 \right\}.$$

Then, M_∞ is a compact spacelike hypersurface with boundary $\partial M_\infty \equiv \mathbf{S}^{n-1} \subset \mathbf{H}^n$, which determines a domain $\Omega \subset \mathbf{H}^n$. Since M_∞ is maximal (its mean curvature vanishes) and has a common boundary with Ω, it is well known that $\mathrm{vol}(M_\infty) \geq \mathrm{vol}(\Omega)$. Thus, we cannot expect to get the inequality $\mathrm{vol}(\Omega) \geq \mathrm{vol}(M)$ without and additional factor like, in this case, $1/k_0^n$.

It would be interesting to get a reverse inequality involving the same terms. Again, we can predict the necessity of include an additional factor as we observe in the next example.

EXAMPLE 4.3. Let r be a positive real number and let M_r be the *lower hyperbolic cap* of radius r, that is, the hypersurface given by

$$M_r = \left\{ x \in \mathbf{L}^{n+1} : \langle x, x \rangle = -r^2, -\sqrt{1+r^2} \leq x_{n+1} \leq -r \right\}.$$

It's clear that M_r is a compact spacelike hypersurface with boundary

$$\partial M_r = \left\{ x \in M_r : x_{n+1} = -\sqrt{1+r^2} \right\} \equiv \mathbf{S}^{n-1}$$

and, after an appropiate translation on the (x_{n+1})-axis (which is, obviously, an isometry of the Minkowski space) we have a compact hypersurface whose boundary is contained in the hyperbolic space $\mathbf{H}^n$ determining the same domain Ω for every $r > 0$. A computation for the volume of this hypersurface give us

$$\mathrm{vol}(M_r) = \mathrm{vol}\left(\mathbf{S}^{n-1}\right) r^n \int_0^{\arctan(1/r)} \frac{\sin^{n-1}\theta}{\cos^n\theta} \, \mathrm{d}\theta$$

and let us observe that

$$\lim_{r \to 0} \mathrm{vol}(M_r) = 0.$$

Thus, we cannot expect to get the reverse inequality $\mathrm{vol}(M) \geq \mathrm{vol}(\Omega)$ since the second term is a positive constant.

REMARK 4.4. In [**AP**] the second author jointly with Alias obtained similar formulae for the volume of the hyperplanar domain enclosed by the hyperplanar boundary of a spacelike hypersurface. In that case, the authors were able to get some inequalities involving the size of the hyperbolic image of the hypersurface M.

References

[AM] L.J. Alias and J.M. Malacarne, *Spacelike hypersurfaces with constant higher order mean curvature in the Lorentz-Minkowski space*, preprint (2000).

[AP] L.J. Alias and J.A. Pastor, *Spacelike hypersurfaces with constant scalar curvature in the Lorentz-Minkowski space.* Ann. Global Anal. Geom. **18** (2000), 75–83.

[AP2] L.J. Alias and J.A. Pastor, *Constant mean curvature spacelike hypersurfaces with spherical boundary in the Lorentz-Minkowski space.* J. Geom. Phys. **28** (1998), 85–93.

[BS] R. Bartnik and L. Simon, *Spacelike hypersurfaces with prescribed boundary values and mean curvature.* Comm. Math. Phys. **87** (1982), 131–152.

DEPARTAMENTO DE MATEMÁTICAS, UNIVERSIDAD DE CASTILLA-LA MANCHA, ESCUELA POLITÉCNICA SUPERIOR DE ALBACETE, E-02071 ALBACETE, SPAIN

E-mail address: `jaledo@pol-ab.uclm.es`

DEPARTAMENTO DE MATEMÁTICAS, UNIVERSIDAD DE MURCIA, CAMPUS DE ESPINARDO, E-30100 ESPINARDO, MURCIA, SPAIN

E-mail address: `jpastor@um.es`

Contemporary Mathematics
Volume **288**, 2001

Stability of Zero Mean Curvature Surfaces in Flat Lorentzian 4-manifolds

Luis J. Alías and Bennett Palmer

ABSTRACT. Similarly to minimal surfaces in Riemannian manifolds, spacelike zero mean curvature surfaces in Lorentzian manifolds are also characterized as critical points for the area functional. However, the stability aspects of both variational problems are completely different. In this paper we study the stability of spacelike zero mean curvature surfaces in Lorentzian 4-spaces.

1. Introduction

In this paper we study the stability of spacelike zero mean curvature surfaces in Lorentzian 4-spaces. As is well-known, spacelike zero mean curvature surfaces in Lorentzian manifolds arise naturally as critical points for the area functional. Although such a variational characterization is similar to the corresponding characterization of minimal surfaces in Riemannian manifolds, when one looks at the second variation for the variational problem, the situation is completely different. When the dimension of the Lorentzian ambient space is greater than three, neither the terms maximal nor minimal is appropriate to describe these surfaces, since they are neither local maxima nor local minima, even when the support of the variations are taken to be sufficiently small. This motivates our interest in the subject and justifies the generic term *stationary* for refering to spacelike zero mean curvature surfaces.

2. Mean isotropic stability of stationary surfaces

In this section, we will formulate the second variation for stationary surfaces in such a certain way so that they will locally extremize. In particular, we will see that under fairly weak conditions on the ambient space, stationary surfaces locally minimize area among nearby surfaces having the same boundary and having null mean curvature.

A spacelike surface Σ immersed into a Lorentzian 4-space $\mathbb{M}_1^4$ is said to have *isotropic mean curvature* if its mean curvature vector $\vec{H}$ is isotropic (null), $\langle \vec{H}, \vec{H} \rangle \equiv$

2000 *Mathematics Subject Classification.* Primary 53C42; Secondary 53C50.
Key words and phrases. stationary surfaces, mean isotropic stability.
L.J. Alías is partially supported by DGICYT Grant No PB97-0784-C03-02 and Fundación Séneca, Centro de Coordinación de la Investigación, CARM.

0. Let us remark that spacelike surfaces with isotropic mean curvature play an important role in several diverse contexts. In relativity theory, they are called *marginally-trapped* surfaces and are used in the study of spacetime singularities. They also play an important part in *conformal geometry* since conformal Gauss maps turn out to be surfaces with isotropic mean curvature.

DEFINITION 1. A *mean isotropic variation* of a stationary surface $\psi : \Sigma \longrightarrow \mathbb{M}_1^4$ will be a smooth variation of ψ, $\psi_t : \Sigma \longrightarrow \mathbb{M}_1^4$, $t \in (-\varepsilon, \varepsilon)$, such that for each $t \in (-\varepsilon, \varepsilon)$ the spacelike surface ψ_t has isotropic mean curvature, $\langle \vec{H}_t, \vec{H}_t \rangle \equiv 0$.

Before stating our stability result, let us recall that $\mathbb{M}_1^4$ is said to satisfy the *null convergence condition* if its Ricci curvature tensor is non-negative on null vectors, that is, $\mathrm{Ric}(X, X) \geq 0$ for all null vectors X in $\mathbb{M}_1^4$. This condition is more general than the *timelike convergence condition*, which means that $\mathrm{Ric}(X, X) \geq 0$ for all timelike vectors X in $\mathbb{M}_1^4$. The timelike convergence condition is of interest in general relativity and it means that, on average, gravity attracts. Trivial examples of manifolds satisfying null convergence condition are the Einstein manifolds and, in particular, manifolds with constant sectional curvature. Now, we are ready to state the following stability result [**1**].

THEOREM 2. *Let $\psi : \Sigma \longrightarrow \mathbb{M}_1^4$ be a stationary surface into a four-dimensional Lorentzian manifold which satisfies the null convergence condition. Then each point $p \in \Sigma$ has a neighborhood $\mathcal{U}$ in Σ such that $\psi|_{\mathcal{U}}$ is a stable local minimum for the area functional restricted to mean isotropic variations of ψ with compact support in $\mathcal{U}$ and agreeing with ψ on some neighborhood of $\partial\mathcal{U}$.*

3. Global mean isotropic stability in flat spacetime

Let Λ be a co-compact lattice in $\mathbb{R}^3$. In this section we will consider the global mean isotropic stability of stationary surfaces in the case where the ambient manifold is a flat spacetime $\mathbb{M}_1^4$ given by $\mathbb{M}_1^4 = \mathbb{R} \times (\mathbb{R}^3/\Lambda)$ or $\mathbb{M}_1^4 = \mathbb{S}^1 \times (\mathbb{R}^3/\Lambda)$ endowed with the Lorentzian metric

$$\langle , \rangle = -dt^2 + d\sigma^2,$$

where $d\sigma^2$ is the standard flat metric on $\mathbb{R}^3/\Lambda$. Observe that if we identify $\mathbb{R}^3/\Lambda$ with a slice $t =$ constant, then every minimal surface $\chi : \Sigma \longrightarrow \mathbb{R}^3/\Lambda$ can be seen as a stationary surface in $\mathbb{M}_1^4$ via the immersion $\psi : \Sigma \longrightarrow \mathbb{M}_1^4$ given by $\psi(p) = (\mathrm{constant}, \chi(p))$. Throughout this section we will investigate the stability of these stationary surfaces. In particular we will prove the following result.

THEOREM 3. *Let $\chi : \Sigma \longrightarrow \mathbb{R}^3/\Lambda$ be a minimal immersion of a closed, oriented surface in $\mathbb{R}^3/\Lambda$ and let $\psi : \Sigma \longrightarrow \mathbb{M}_1^4$ be as above. Then ψ is mean isotropically stable if and only if* $\mathrm{index}(\chi) \leq 1$.

Let $\chi : \Sigma \longrightarrow \mathbb{R}^3/\Lambda$ be a smooth minimal immersion of a closed, oriented surface in $\mathbb{R}^3/\Lambda$. Let ν be a unit normal field along χ and let K denote the Gaussian curvature of the induced metric ds^2. Let us recall that the second variation of area of χ in the normal direction $f\nu$, for a smooth function f on Σ, is given by

$$\delta_f^2 \mathrm{Area}(\chi) = -\int_\Sigma f L[f] dA,$$

where $L = \Delta - 2K$ is the Jacobi operator of χ. The immersion χ is called *stable* (as a minimal surface in $\mathbb{R}^3/\Lambda$) if and only if the second variation is non-negative for all normal variations. Let $\lambda_1 \leq \lambda_2 \leq \cdots$ denote the eigenvalues of L. Then clearly

χ is stable if and only if $\lambda_1 \geq 0$ holds. The number of negative eigenvalues of L is called the *(Morse) index* of χ. The following lemma is well known.

LEMMA 4. *Let $\chi : \Sigma \longrightarrow \mathbb{R}^3/\Lambda$ be a minimal immersion of a closed, oriented surface. Then χ is stable if and only if χ is flat and hence totally geodesic.*

REMARK 5. Minimal surfaces in $\mathbb{R}^3/\Lambda$ were studied classicaly by Schwartz and more recently by A. H. Schoen and Karcher. Closed minimal surfaces with Morse index one in three dimensional Riemannian manifolds arise naturally as stable critical points for the isoperimetric problem of minimizing area with a constraint on the enclosed 3-volume. One can obtain a closed minimal surface in $\mathbb{R}^3/\Lambda$ by solving the free boundary problem for a minimal surface with boundary on the sides of a rectangular box B and then by considering the interior of the box as a fundamental domain for a 3-torus. The minimal surface descends to a smooth surface since the critical points for the free boundary problem meet the box sides in an right angle. The isoperimetric problem of minimizing the area of a compact surface with boundary constrained to lie on the sides of the box has critica which in general have constant mean curvature and meet the box sides in a right angle. There are many examples of this type with zero mean curvature [**3, 4**]. For these surfaces the condition that the Morse index of the surface in $\mathbb{R}^3/\Lambda$ is one is exactly the condition that the surface in B is a *stable* critical point for the isoperimetric problem of minimizing the area with the volume constraint. In [**6**], Ross showed that a particular classical example, the genus three Schwartz P-surface, has index one. Surfaces with larger index can be obtained by considering coverings of any nonflat example of sufficiently high degree. An upper bound for the Morse index of a compact, oriented minimal surface in $\mathbb{R}^3/\Lambda$ was given by Tysk in [**7**]. In [**5**], Ritoré and Ros proved a compactness theorem for index one, minimal embedded surfaces in quotients of $\mathbb{R}^3$.

The immersion χ lifts to a complete minimal immersion $\tilde{\chi}$ of the universal cover $\tilde{\Sigma}$ into $\mathbb{R}^3$. Let $\tilde{\nu} : \tilde{\Sigma} \longrightarrow \mathbb{S}^2$ denote its Gauss map. Although $\tilde{\nu}$ may not descend to Σ, the quantity $|d\nu|^2 = -2K$ is well defined there. Since the map ν is holomorphic, if $K \not\equiv 0$ then the branch points of ν, that is, the zeros of K, are isolated and we can consider the singular Riemannian metric $d\tilde{s}^2 = \nu^* dS^2 = -K ds^2$, where dS^2 denotes the usual metric on $\mathbb{S}^2$. It is well known that although $d\tilde{s}^2$ is in general singular, the spectrum $0 = \mu_0 < \mu_1 \leq \cdots$ of its Laplacian $\tilde{\Delta}$ is still well-defined. The proof of our Theorem 3 is based on the following result.

LEMMA 6. *Let $\chi : \Sigma \longrightarrow \mathbb{R}^3/\Lambda$ be a minimal, non-totally geodesic immersion of a closed, oriented surface. Then*

$$\text{index}(\chi) = 1 \quad \text{if and only if} \quad \text{spec}(\tilde{\Delta}) \cap (0, 2) = \emptyset.$$

PROOF. If $\mu \in (0, 2) \cap \text{spec}(\tilde{\Delta})$, let f be an eigenfunction belonging to μ. Note that almost everywhere on Σ

$$\Delta f - \mu K f = 0 \tag{1}$$

holds. Let g solve $L[g] + \lambda_1 g = 0$ and recall that g does not change sign [**2**], so we may assume that $g > 0$ holds. Define $\tilde{f} = \alpha + \beta f$ where α, β are real constants, not both zero such that

$$\int_\Sigma \tilde{f} g dA = 0. \tag{2}$$

A computation gives

$$-\int_\Sigma \tilde{f}L[\tilde{f}]dA = 2\alpha^2\int_\Sigma KdA + (2-\mu)\beta^2\int_\Sigma Kf^2dA + 4\alpha\beta\int_\Sigma KfdA.$$

Note that by (1) and Stoke's theorem, the last integral above is zero. Also note that $\beta \neq 0$ since g is positive and since (2) holds. Therefore, since K is non-positive we have

$$-\int_\Sigma \tilde{f}L[\tilde{f}]dA < 0$$

and so $\lambda_2 < 0$. Since χ is not totally geodesic this implies index$(\chi) > 1$.

Now assume index$(\chi) > 1$. Let $f_j, j = 1,2$ solve $L[f_j] + \lambda_j f_j = 0$. Choose constants α, β not both zero such that

$$\int_\Sigma (\alpha f_1 + \beta f_2)d\tilde{A} = \int_\Sigma (\alpha f_1 + \beta f_2)(-K)dA = 0.$$

Then if $\tilde{f} = \alpha f_1 + \beta f_2$ then

$$-\int_\Sigma \tilde{f}(\tilde{\Delta}+2)\tilde{f}d\tilde{A} = -\int_\Sigma \tilde{f}L[\tilde{f}]dA = \lambda_1\alpha^2\int_\Sigma f_1^2dA + \lambda_2\beta^2\int_\Sigma f_2^2dA < 0.$$

This shows that $\mu_1 < 2$ holds. □

For stationary surfaces in a flat Lorentzian 4-space, the second variation formula is given by

$$\delta_\xi^2 \text{ Area}(\psi) = \frac{d^2}{dt^2}|_{t=0} \text{ Area}(\psi) = -\int_\Sigma \langle \xi, J[\xi]\rangle dA, \tag{3}$$

where $J : \mathcal{X}^\perp(\Sigma) \longrightarrow \mathcal{X}^\perp(\Sigma)$ is the *Jacobi operator*, which is given here by $J[\xi] = \tilde{\Delta}^\perp\xi + \mathcal{B}(\xi)$. Here $\tilde{\Delta}^\perp$ is the *rough Laplacian* in the normal bundle, and $\mathcal{B}$ is the endomorphism of the normal bundle defined by $\langle \mathcal{B}(\xi), \eta\rangle = \langle D^T\xi, D^T\eta\rangle$.

PROOF OF THEOREM 3. Let ν denote a unit normal to Σ in $\mathbb{R}^3/\Lambda$ and let ∂_t denote a unit timelike normal to $\mathbb{R}^3/\Lambda$ in $\mathbb{M}_1^4$. Observe that

$$\xi_1 = \frac{\nu + \partial_t}{\sqrt{2}}, \qquad \xi_2 = \frac{\nu - \partial_t}{\sqrt{2}},$$

is a global null framing of the normal bundle. Let ξ be a normal field. Then the condition that to first order the normal section ξ is the normal component of the variation field of a mean isotropic variation of ψ is that at each point either

$$\langle J[\xi], \xi_1\rangle = 0 \qquad \text{or} \qquad \langle J[\xi], \xi_2\rangle = 0 \tag{4}$$

holds. Since minimal surfaces in a flat space are real analtic, one of these two conditions must hold globally.

The second variation of the area in the direction ξ is given by (3). If ξ is expressed as $\xi = f_1\xi_1 + f_2\xi_2$, where f_1, f_2 are smooth functions on Σ, then a calculation gives

$$J[\xi] = (\Delta f_1 - Kf_1 - Kf_2)\xi_1 + (\Delta f_2 - Kf_2 - Kf_1)\xi_2.$$

If χ has index zero then by Lemma 4 $K \equiv 0$ and the two conditions in (4) lead to respectively either $f_2 =$ constant or $f_1 =$ constant. If f_2 is constant then $J[\xi] = (\Delta f_1)\xi_1$ and hence

$$\delta_\xi^2\text{Area}(\psi) = -f_2\int_\Sigma \Delta f_1 dA = 0.$$

Similarly if f_1 is constant then $\delta^2_\xi \text{Area}(\psi) = 0$ holds.

We now assume that $K \neq 0$ on Σ. If $\langle J[\xi], \xi_1 \rangle \equiv 0$ holds then

$$\Delta f_2 - K f_2 - K f_1 = -K(\tilde{\Delta} f_2 + f_2 + f_1) = 0 \tag{5}$$

almost everywhere on Σ. Let $\{u_\alpha\}_{\alpha=1,2,\ldots}$ be an orthonormal (with respect to the measure of $d\tilde{s}^2$) set of eigenfunctions for $\tilde{\Delta}$. For $f \in C^\infty(\Sigma)$, let

$$\hat{f}(\alpha) = \int_\Sigma u_\alpha f dA$$

be the Fourier coeficients of f. Then by (5),

$$(1 - \mu_\alpha)\hat{f}_2(\alpha) + \hat{f}_1(\alpha) = 0, \quad \alpha = 1, 2, \ldots \tag{6}$$

Therefore

$$\begin{aligned} -\int_\Sigma \langle \xi, J[\xi] \rangle dA &= -\int_\Sigma f_2(\Delta f_1 - K f_1 - K f_2) dA = -\int_\Sigma f_2(\tilde{\Delta} f_1 + f_1 + f_2) d\tilde{A} \\ &= \sum_{\alpha=1}^{\infty} ((\mu_\alpha - 1)\hat{f}_1(\alpha)\hat{f}_2(\alpha) - \hat{f}_2(\alpha)^2) = \sum_{\alpha=1}^{\infty} \mu_\alpha(\mu_\alpha - 2)\hat{f}_2(\alpha)^2 \end{aligned}$$

using (6). If χ has index one then by Lemma 6 $\text{spec}(\tilde{\Delta}) \cap (0, 2) = \emptyset$ and each term in the last sum is non-negative. Reversing the roles of f_1 and f_2 shows that if $\text{index}(\chi) = 1$ then $\delta^2_\xi \text{Area}(\psi) \geq 0$ holds for all variations with $\langle J[\xi], \xi_2 \rangle = 0$.

On the other hand if $\text{index}(\chi) > 1$ then $\text{spec}(\tilde{\Delta}) \cap (0, 2) \neq \emptyset$. Let f_2 solve $(\tilde{\Delta} + \mu) f_2 = 0$ with $0 < \mu < 2$ and define f_1 by $f_1 = (\mu - 1) f_2$ so that $\xi = f_1 \xi_1 + f_2 \xi_2$ satisfies $J[\xi] = \mu(\mu - 2) K f_2 \xi_1$ and $\langle J[\xi], \xi_1 \rangle = 0$. Then by the above

$$\delta^2_\xi \text{Area}(\psi) = \mu(\mu - 2) \int_\Sigma f_2^2 < 0.$$

□

References

[1] L.J. Alías and B. Palmer, *Deformations of stationary surfaces*, Class. Quantum Grav. **14** (1997), 2107–2111.

[2] R. Courant and D. Hilbert, *Methods of mathematical physics, vol. II*, John Wiley and Sons, 1962.

[3] H. Karcher, *The triply periodic minimal surfaces of Alan Schoen and their constant mean curvature companions*, Manuscripta Math. **64** (1989), 291–357.

[4] H. Karcher and K. Polthier, *Construction of triply periodic minimal surfaces*, Philos. Trans. Roy. Soc. London A **354** (1996), 2077–2104.

[5] M. Ritoré and A. Ros, *The spaces of index one minimal surfaces and stable constant mean curvature surfaces embedded in flat three manifolds*, Trans. Amer. Math. Soc. **348** (1996), 391–410.

[6] M. Ross, *Schwarz' P and D surfaces are stable*, Differential Geom. Appl. **2** (1992), 179–195.

[7] J. Tysk, *Eigenvalue estimates with applications to minimal surfaces*, Pacific J. Math. **128** (1987), 361–366.

Departamento de Matemáticas, Universidad de Murcia, E-30100 Espinardo, Murcia, Spain

E-mail address: ljalias@um.es

Department of Mathematical Sciences, University of Durham, Durham DH1 3LE, England

E-mail address: bennett.palmer@durham.ac.uk

Contemporary Mathematics
Volume **288**, 2001

Hitchin–Kobayashi correspondence for equivariant bundles on $X \times \mathbb{P}^1$

Luis Álvarez–Cónsul and Oscar García–Prada

Abstract. This paper is concerned with the study of gauge theory, stability and dimensional reduction on $\mathrm{SL}(2,\mathbb{C})$–equivariant bundles over $X \times \mathbb{P}^1$, where X is a compact Kähler manifold, $\mathbb{P}^1$ is the complex projective line, and the action of $\mathrm{SL}(2,\mathbb{C})$ is trivial on X and standard on $\mathbb{P}^1$. We show that there is a Hitchin–Kobayashi correspondence relating the existence of solutions to a natural deformation of the Hermitian–Einstein equation and an appropriate notion of stability for an equivariant bundle on $X \times \mathbb{P}^1$. The invariance of the solutions and the stability condition lead, via dimensional reduction, to a Hitchin–Kobayashi correspondence for vortex type equations and a stability condition for a chain of holomorphic bundles on X linked by homomorphisms.

1. Introduction

Let M be a compact Kähler manifold and let $\mathcal{F}$ be a holomorphic vector bundle over M. It is well-known that a natural differential equation to consider for a Hermitian metric h on $\mathcal{F}$ is the *Hermitian–Einstein* equation, also referred sometimes as the *Hermitian–Yang–Mills* equation. This says that F_h, the curvature of the Chern connection of h, must satisfy

$$\sqrt{-1}\,\Lambda F_h = \lambda I,$$

were Λ is the contraction with the Kähler form of M, λ is a real number determined by the topology and I is the identity endomorphism of $\mathcal{F}$. A theorem of Donaldson, Uhlenbeck and Yau [**D2, D3, UY**], also known as the *Hitchin–Kobayashi* correspondence, establishes that the existence of such metric is equivalent to the stability of $\mathcal{F}$ in the sense of Mumford–Takemoto. Suppose now that a compact Lie group K acts on M by isometries so that M/K is a smooth Kähler manifold and the action on M can be lifted to an action on $\mathcal{F}$. One can then apply *dimensional reduction* techniques to study K–invariant solutions to the Hermitian–Einstein equation on $\mathcal{F}$ and the corresponding stability condition to obtain a theory expressed entirely

---

2000 *Mathematics Subject Classification.* Primary 53C07; Secondary 53C30, 32M05, 35Q40.

Key words and phrases. Equivariant bundle, stability, Hermitian–Einstein equations, vortex equations, holomorphic chains, Hitchin–Kobayashi correspondence, dimensional reduction.

The authors are members of VBAC (Vector Bundles on Algebraic Curves), which is partially supported by EAGER (EC FP5 Contract no. HPRN-CT-2000-00099) and by EDGE (EC FP5 Contract no. HPRN-CT-2000-00101).

in terms of the orbit space M/K. Many important equations in gauge theory arise in this way [**H**].

In this paper we give a brief account of this programme for the manifold $M = X \times \mathbb{P}^1$, where X is a compact Kähler manifold, and $K = \mathrm{SU}(2)$ acts trivially on X and in the standard way on $\mathbb{P}^1$, with the Kähler structure of $X \times \mathbb{P}^1$ defined by the product of that of X and the Fubini-Study metric on $\mathbb{P}^1$. More details, references, and some applications can be found in [**AG**]. In this paper, however, we take the oportunity to explain in a more thorough way (than has been done in [**AG**]) the classification of $\mathrm{SL}(2,\mathbb{C})$–equivariant holomorphic bundles and coherent sheaves over $X \times \mathbb{P}^1$ —where the action of $\mathrm{SU}(2)$ has been extended to its complexification, $\mathrm{SL}(2,\mathbb{C})$. We will indeed need to understand $\mathrm{SL}(2,\mathbb{C})$–equivariant coherent sheaves in order to study Hitchin–Kobayashi correspondences. It turns out that every $\mathrm{SL}(2,\mathbb{C})$–equivariant holomorphic vector bundle on $X \times \mathbb{P}^1$ admits an equivariant holomorphic filtration by equivariant subbundles, which, in turn, is in one-to-one correspondence with a *chain* consisting of holomorphic vector bundles $\mathcal{E}_i$ on X and morphisms $\phi_i : \mathcal{E}_i \to \mathcal{E}_{i-1}$. The filtered structure of the equivariant bundle on $X \times \mathbb{P}^1$ dictates (by symplectic arguments) that the right equation to consider is a certain deformation of the Hermitian–Einstein equation which involves as many paremeters as terms are in the filtration (notice that the Hermitian–Einstein equation involves just one parameter, λ, which is actually determined by the topology). We show that there is a Hitchin–Kobayashi correspondence relating this equation to a stability criterion for the filtration, and this gives, via dimensional reduction, a Hitchin–Kobayashi correspondence for the holomorphic chains on X, generalising the theory of triples studied in [**G, BG**].

The second author wishes to thank Marisa Fernández and the other organizers for giving him the opportunity to participate in a most stimulating conference to honour the memory of Alfred Gray.

2. Equivariant bundles, sheaves and holomorphic chains

Given a complex Lie group G, a complex G–manifold is a complex manifold M with a holomorphic G–action, i.e. a left G–action by biholomorphisms on M; a G–equivariant holomophic vector bundle on M is a holomophic vector bundle $\pi : \mathcal{F} \to M$, together with holomorphic G–action $\sigma : G \times \mathcal{F} \to \mathcal{F}$ on its total space which commutes with π, and such that for all $(g,p) \in G \times M$, the map $\sigma_{g,p} : \mathcal{E}_p \to \mathcal{E}_{g\cdot p}$, from $\mathcal{E}_p = \pi^{-1}(p)$ into $\mathcal{E}_{g\cdot p} = \pi^{-1}(g \cdot p)$, induced by this action, is an isomorphism of vector spaces.

Let us now consider $M = X \times \mathbb{P}^1$, and $G = \mathrm{SL}(2,\mathbb{C})$ acting trivially on X and with action on $\mathbb{P}^1$ given by the identification $\mathbb{P}^1 = \mathrm{SL}(2,\mathbb{C})/P$, where P is the subgroup of lower triangular matrices

$$P = \begin{pmatrix} * & 0 \\ * & * \end{pmatrix} \subset \mathrm{SL}(2,\mathbb{C}).$$

An $\mathrm{SL}(2,\mathbb{C})$–equivariant holomophic vector bundle $\mathcal{F}$ on $X \times \mathbb{P}^1$ defines a P–equivariant holomorphic vector bundle $i^*\mathcal{F}$ on X by restriction to the slice $i : X \cong X \times P/P \hookrightarrow X \times \mathbb{P}^1$. Conversely, a P–equivariant holomorphic vector bundle $\mathcal{E}$ on X defines by induction an $\mathrm{SL}(2,\mathbb{C})$–equivariant holomophic vector bundle $\mathcal{F} = \mathrm{SL}(2,\mathbb{C}) \times_P \mathcal{E}$ on $X \times \mathbb{P}^1$, which by definition is the quotient of $\mathrm{SL}(2,\mathbb{C}) \times \mathcal{E}$ by the P–action, given by $p \cdot (g,e) = (gp^{-1}, p \cdot e)$, for $p \in P$, $(g,e) \in \mathrm{SL}(2,\mathbb{C}) \times \mathcal{E}$.

The $\mathrm{SL}(2,\mathbb{C})$–equivariant action on $\mathcal{F}$ is $g'\cdot[g,e]=[g'g,e]$, for $g'\in\mathrm{SL}(2,\mathbb{C})$ and $[g,e]\in\mathcal{F}$.

When X is a single point, this induction process supplies an equivalence between the category of holomorphic homogenous vector bundles on $\mathbb{P}^1$ (i.e. $\mathrm{SL}(2,\mathbb{C})$–equivariant vector bundles on $\mathbb{P}^1$) and the category of holomorphic representations of P. When X is not a point, we can still consider $X\times\mathbb{P}^1$ as a family of projective lines $\{x\}\times\mathbb{P}^1$, parametrised by the points $x\in X$, and a holomorphic $\mathrm{SL}(2,\mathbb{C})$–equivariant vector bundle $\mathcal{F}$ on $X\times\mathbb{P}^1$ as a family of holomorphic homogeneous vector bundles $i_x^*\mathcal{F}$ on the projective lines $i_x:\{x\}\times\mathbb{P}^1\hookrightarrow X\times\mathbb{P}^1$. Thus, a detailed study of the holomorphic representations of P is important in classifying the $\mathrm{SL}(2,\mathbb{C})$–equivariant holomophic vector bundles on $X\times\mathbb{P}^1$. We shall see that P–modules admit two equivalent descriptions, both of them with a counterpart in the case of equivariant bundles on $X\times\mathbb{P}^1$. We begin with some notation. The Lie algebra $\mathfrak{p}$ of P is generated by the matrices

$$H=\begin{pmatrix}1&0\\0&-1\end{pmatrix},X^-=\begin{pmatrix}0&0\\1&0\end{pmatrix},\quad\text{with } [H,X^-]=-2X^-.$$

Let W be a representation of P. Since P is not semisimple, W is not generally semisimple. The action of the complex torus $\mathbb{C}^*\subset P$, is given by a weight decomposition

$$W=\oplus_{\lambda\in\Delta(W)}W_\lambda\otimes M_\lambda,\ \text{ with } W_\lambda=\mathrm{Hom}_{\mathbb{C}^*}(M_\lambda,W)\text{ for }\lambda\in\mathbb{Z},\tag{2.1}$$

where M_λ is an irreducible $\mathbb{C}^*$–module with weight λ (i.e. $z\cdot v=z^\lambda v$ for $z\in\mathbb{C}^*,v\in M_\lambda$), for $\lambda\in\mathbb{Z}$, W_λ are the multiplicity spaces, and $\Delta(W)=\{\lambda\in\mathbb{Z}|W_\lambda\neq 0\}$ is the set of weights of W. The rest of the P–module structure of W is determined by the action of $X^-\in\mathfrak{p}$. Since $[H,X^-]=-2X^-$, the action of X^- on $W_\lambda\otimes M_\lambda$ is defined by a linear map $\varphi_\lambda:W_\lambda\to W_{\lambda-2}$, and the trivial X^-–action on M_λ. Thus, a P–module is exactly specified by a diagram of linear maps

$$\begin{array}{ccccccccc}\cdots&\xrightarrow{\varphi_4}&W_2&\xrightarrow{\varphi_2}&W_0&\xrightarrow{\varphi_0}&W_{-2}&\xrightarrow{\varphi_{-2}}&\cdots\\ \cdots&\xrightarrow{\varphi_5}&W_3&\xrightarrow{\varphi_3}&W_1&\xrightarrow{\varphi_1}&W_{-1}&\xrightarrow{\varphi_{-1}}&\cdots\end{array}\tag{2.2}$$

Any P–module is the direct sum of an even part ($W_\lambda=0$ if λ is odd) and an odd part ($W_\lambda=0$ if λ is even), so the set of weights of an *indecomposable* P–module form a sequence $m_-,m_-+2,\dots,m_+-2,m_+$ of consecutive odd or even integers. Such an indecomposable P–module is given by a sequence $W_{m_-},W_{m_-+2},\dots,W_{m_+}$ of non-zero finite dimensional vector spaces, and non-zero linear maps $\varphi_\lambda:W_\lambda\to W_{\lambda-2}$ between consecutive vector spaces. Twisting by M_{-m_-}, we get a representation $V=W\otimes M_{-m_-}$ whose $\mathbb{C}^*$–action is given by

$$V=\oplus_{i=0}^m V_i\otimes M_{2i},\tag{2.3}$$

where $V_i:=W_{m_-+2i}$, $\Delta(V)=\{0,2,\dots,2m\}$, and $m:=(m_+-m_-)/2$, and whose X^-–action is given by maps ϕ_i, which can be represented by the diagram

$$V_m\xrightarrow{\phi_m}V_{m-1}\xrightarrow{\phi_{m-1}}\cdots\xrightarrow{\phi_1}V_0.\tag{2.4}$$

This is our first description of a P–module. The second description is given in terms of flags of P–modules. Actually, any indecomposable P–module V which has the structure given by (2.3) and (2.4) (after twisting by some M_{-m_-} if necessary), admits a flag

$$\begin{aligned} \boldsymbol{V}_{(\le\bullet)} : 0 \subset V_{(\le 0)} \subset V_{(\le 1)} \subset \ldots \subset V_{(\le m)} = V, \\ V_{(\le i)}/V_{(\le i-1)} \cong V_i \otimes M_{2i}, \quad 0 \le i \le m, \end{aligned} \tag{2.5}$$

where $V_{(\le i)} := \oplus_{j\le i} V_j \otimes M_{2j}$ are nested P–submodules of V.

We need now some definitions to generalize this to equivariant bundles and sheaves on $X \times \mathbb{P}^1$.

A *sheaf chain* on X is a pair $\boldsymbol{C} = (\boldsymbol{\mathcal{E}}, \boldsymbol{\phi})$, where $\boldsymbol{\mathcal{E}} = (\mathcal{E}_0, \mathcal{E}_1, \ldots, \mathcal{E}_m)$ is an $(m+1)$–tuple of coherent sheaves on X, and $\boldsymbol{\phi} = (\phi_1, \ldots, \phi_m)$ is an m–tuple of homomorphisms $\phi_i : \mathcal{E}_i \to \mathcal{E}_{i-1}$ $(1 \le i \le m)$. This is represented by the diagram

$$\boldsymbol{C} : \mathcal{E}_m \xrightarrow{\phi_m} \mathcal{E}_{m-1} \xrightarrow{\phi_{m-1}} \cdots \xrightarrow{\phi_1} \mathcal{E}_0. \tag{2.6}$$

We say that $\boldsymbol{C}$ is a *holomorphic chain* if the sheaves $\mathcal{E}_i$ are locally free.

A *morphism* $\boldsymbol{C}' \to \boldsymbol{C}$ between a sheaf chain

$$\boldsymbol{C}' : \mathcal{E}'_m \xrightarrow{\phi'_m} \mathcal{E}'_{m-1} \xrightarrow{\phi'_{m-1}} \cdots \xrightarrow{\phi'_1} \mathcal{E}'_0, \tag{2.7}$$

and (2.6), is given morphisms $f_i : \mathcal{E}'_i \to \mathcal{E}_i$ for each i, such that $\phi_i \circ f_i = f_{i-1} \circ \phi'_i$ for $1 \le i \le m$. It is immediate that sheaf chains form an abelian category.

Let M be a complex G–manifold. By a *coherent G–equivariant sheaf* on M we will mean a coherent sheaf $\mathcal{F}$ together with a *holomorphic* G–equivariant action on $\mathcal{F}$ (cf. e.g. [**Ak**]). The coherent G–equivariant sheaves, with the G–equivariant morphisms, form an abelian category. A *G–equivariant sheaf filtration* $\boldsymbol{\mathcal{F}}$ on M is a finite sequence of G–invariant coherent subsheaves of a coherent G–equivariant sheaf $\mathcal{F}$ on M,

$$\boldsymbol{\mathcal{F}} : 0 \hookrightarrow \mathcal{F}_0 \hookrightarrow \mathcal{F}_1 \hookrightarrow \cdots \hookrightarrow \mathcal{F}_m = \mathcal{F}. \tag{2.8}$$

We say that $\boldsymbol{\mathcal{F}}$ is a *holomorphic filtration* if the sheaves $\mathcal{F}_i$ are locally free.

A *G–equivariant morphism* from a G–equivariant sheaf filtration

$$\boldsymbol{\mathcal{F}}' : 0 \hookrightarrow \mathcal{F}'_0 \hookrightarrow \mathcal{F}'_1 \hookrightarrow \cdots \hookrightarrow \mathcal{F}'_m = \mathcal{F}', \tag{2.9}$$

to another one $\boldsymbol{\mathcal{F}}$ given by (2.8) is a G–equivariant morphism $f : \mathcal{F}' \to \mathcal{F}$, with $f(\mathcal{F}'_i) = \mathcal{F}_i \cap \operatorname{Im}(f)$ for each i.

Let $p : X \times \mathbb{P}^1 \to X$ and $q : X \times \mathbb{P}^1 \to \mathbb{P}^1$ be the canonical projections. Unless otherwise stated, by an $\mathrm{SL}(2,\mathbb{C})$–equivariant sheaf filtration on $X \times \mathbb{P}^1$ we shall mean an $\mathrm{SL}(2,\mathbb{C})$–equivariant sheaf filtration on $X \times \mathbb{P}^1$, such that there are isomorphisms of $\mathrm{SL}(2,\mathbb{C})$–sheaves

$$\mathcal{F}_i/\mathcal{F}_{i-1} \cong p^*\mathcal{E}_i \otimes q^*\mathcal{O}(2i), \qquad \text{for } 1 \le i \le m. \tag{2.10}$$

THEOREM 2.1. *There is a one-to-one correspondence between the indecomposable coherent* $\mathrm{SL}(2,\mathbb{C})$*–equivariant sheaves, up to twisting by powers of* $q^*\mathcal{O}(1)$*, on* $X \times \mathbb{P}^1$*, and the indecomposable objects in any of the following equivalent categories:*
(i) *The category of* $\mathrm{SL}(2,\mathbb{C})$*–equivariant sheaf filtrations on* $X \times \mathbb{P}^1$*.*
(ii) *The category of sheaf chains on* X*.*

We prove it first for holomorphic $\mathrm{SL}(2,\mathbb{C})$–equivariant vector bundles on $X \times \mathbb{P}^1$. As we explained before, there is an equivalence between $\mathrm{SL}(2,\mathbb{C})$–equivariant holomorphic vector bundles $\mathcal{H}$ on $X \times \mathbb{P}^1$ and P–equivariant holomorphic vector bundles $\mathcal{G}$ on X. The $\mathbb{C}^*$–action on any such $\mathcal{G}$ is given by a weight decomposition

(2.11)
$$\mathcal{G} = \oplus_{\lambda\in\Delta(\mathcal{G})}\mathcal{G}_\lambda \otimes M_\lambda,\ \mathcal{G}_\lambda = \mathrm{Hom}_{\mathbb{C}^*}(M_\lambda,\mathcal{G}) \text{ for } \lambda\in\mathbb{Z},\ \Delta(\mathcal{G}) = \{\lambda\in\mathbb{Z}|\mathcal{G}_\lambda\neq 0\},$$

where M_λ is now the irreducible $\mathbb{C}^*$–equivariant vector bundle over X associated to the irreducible representation M_λ (we use the same letter for both), for $\lambda\in\mathbb{Z}$, and the set of weights $\Delta(\mathcal{G})$ is finite because $\mathcal{G}$ has finite rank. The key ingredient to prove the previous decomposition is the existence of a (unique) invariant holomorphic projection operator $\Pi : \mathcal{G}\to\mathcal{G}$ onto the $\mathbb{C}^*$–invariant part of $\mathcal{E}$. This is proved e.g. in [**AB**]. Using the familiar notation $(-)^{\mathbb{C}^*} := \Pi(-)$, the multiplicity bundles are then $\mathcal{G}_\lambda = \mathrm{Hom}_{\mathbb{C}^*}(M_\lambda,\mathcal{G}) = \mathrm{Hom}(M_\lambda,\mathcal{G})^{\mathbb{C}^*}$, where the bundle $\mathrm{Hom}(M_\lambda,\mathcal{G})$ has the canonical $\mathbb{C}^*$–action. As we saw before, the rest of the P–action on $\mathcal{G}$, given by the action of X^-, is specified by a collection of morphisms $\varphi_\lambda : \mathcal{G}_\lambda\to\mathcal{G}_{\lambda-2}$, and if $\mathcal{G}$ is indecomposable, then $\Delta(\mathcal{G})$ is a finite sequence $m_-, m_-+2,\dots, m_+-2, m_+$ of consecutive odd or even integers. Twisting $\mathcal{E} := \mathcal{G}\otimes M_{-m_-}$, we translate the weights into

$$\mathcal{E} = \oplus_{i=0}^m \mathcal{E}_i\otimes M_{2i},\quad \mathcal{E}_i := \mathcal{G}_{m_-+2i},\ \Delta(\mathcal{E}) = \{0,2,\dots,2m\}, \tag{2.12}$$

where $m := (m_+ - m_-)/2$. The X^-–action, given by morphisms $\phi_i : \mathcal{E}_i\to\mathcal{E}_{i-1}$, defines the diagram (2.6). As in (2.5), this holomorphic chain defines a filtration of P–invariant holomorphic vector subbundles

$$\begin{aligned} &\mathcal{E}_{(\leq\bullet)} : 0\subset \mathcal{E}_{(\leq 0)}\subset\mathcal{E}_{(\leq 1)}\subset\dots\subset\mathcal{E}_{(\leq m)},\\ &\mathcal{E}_{(\leq i)}/\mathcal{E}_{(\leq i-1)}\cong \mathcal{E}_i\otimes M_{2i},\quad 0\leq i\leq m.\end{aligned} \tag{2.13}$$

Defining $\mathcal{F}_i = \mathrm{SL}(2,\mathbb{C})\times_P\mathcal{E}_{(\leq i)}$, and noting that the homogeneous bundle induced by M_λ on $\mathbb{P}^1$ is $\mathcal{O}(\lambda) = \mathrm{SL}(2,\mathbb{C})\times_P M_\lambda$, we obtain the filtration

$$\begin{aligned} &\boldsymbol{\mathcal{F}} : 0\hookrightarrow\mathcal{F}_0\hookrightarrow\mathcal{F}_1\hookrightarrow\cdots\hookrightarrow\mathcal{F}_m = \mathcal{F},\\ &\mathcal{F}_i/\mathcal{F}_{i-1}\cong p^*\mathcal{E}_i\otimes q^*\mathcal{O}(2i),\quad 0\leq i\leq m.\end{aligned} \tag{2.14}$$

Finally, if $\mathcal{H} := \mathrm{SL}(2,\mathbb{C})\times_P\mathcal{G}$ then $\mathcal{F} := \mathrm{SL}(2,\mathbb{C})\times_P\mathcal{E}$ is given by $\mathcal{F} = \mathcal{H}\otimes q^*\mathcal{O}(-m_-)$. This proves the theorem for equivariant bundles.

To prove the theorem for general coherent $\mathrm{SL}(2,\mathbb{C})$–equivariant sheaves on $X\times\mathbb{P}^1$, we combine the arguments given for holomorphic $\mathrm{SL}(2,\mathbb{C})$–equivariant vector bundles on $X\times\mathbb{P}^1$ with the lemmas below.

LEMMA 2.2. *There is an equivalence between the category of coherent* $\mathrm{SL}(2,\mathbb{C})$*–equivariant sheaves on $X\times\mathbb{P}^1$, and the category of coherent P–equivariant sheaves on X.*

LEMMA 2.3. *A coherent* $\mathrm{SL}(2,\mathbb{C})$*–equivariant sheaf $\mathcal{G}$ on X admits a weight decomposition*

$$\mathcal{E}\cong\oplus_{\lambda\in\Delta(\mathcal{E})}\mathcal{E}_\lambda\otimes M_\lambda, \tag{2.15}$$

where $\Delta(\mathcal{G})\subset\mathbb{Z}$ is finite, and $\mathcal{E}_\lambda$ is a coherent sheaf with trivial $\mathbb{C}^*$–action, for each $\lambda\in\Delta(\mathcal{G})$.

To prove these lemmas, we need the following definition. Let G be a complex Lie group and M a complex G–manifold such that the quotient M/G is a complex manifold. We define a functor $\mathcal{F}\mapsto\mathcal{F}^G$ from the category of coherent G–equivariant sheaves on M to the category of (not necessarily coherent) sheaves on M/G. Given a coherent G–equivariant sheaf $\mathcal{F}$ on M, and $U\subset M/G$ open, by definition $\mathcal{F}^G(U) := \mathcal{F}(\pi^{-1}(U))^G$, where $\pi : M\to M/G$ is projection.

PROOF OF LEMMA 2.2. Let $\mathcal{E}$ be a coherent P–equivariant sheaf on X. The P–action on $\mathrm{SL}(2,\mathbb{C})$, given by $p \cdot g = gp^{-1}$, for $p \in P$, $g \in \mathrm{SL}(2,\mathbb{C})$, induces a P–action on the structure sheaf $\mathcal{O}_{\mathrm{SL}}$ of $\mathrm{SL}(2,\mathbb{C})$, hence on $\mathcal{O}_M$, where $M = X \times \mathrm{SL}(2,\mathbb{C})$. Let $\pi_X : M \to X$ and $\pi_{\mathrm{SL}} : M \to \mathrm{SL}(2,\mathbb{C})$ be projections. The coherent sheaf $\mathcal{H} = \pi_X^*\mathcal{E} \otimes \pi_{\mathrm{SL}}^*\mathcal{O}_{\mathrm{SL}}$ on M has a holomorphic P–equivariant action. By definition, the $\mathrm{SL}(2,\mathbb{C})$–equivariant sheaf induced by $\mathcal{E}$ on $X \times \mathbb{P}^1$ is its P–invariant part $\mathcal{F} = \mathcal{H}^P$. To prove that it is coherent, one sees that the set of points $y \in X \times \mathbb{P}^1$ where $\mathcal{F}$ does not admit a presentation by locally free sheaves, is $S(\mathcal{F}) = S(\mathcal{E}) \times \mathbb{P}^1$, where $S(\mathcal{E})$ is the set of points $x \in X$ where $\mathcal{E}$ does not admit a presentation by locally free sheaves. But $S(\mathcal{E})$ is empty by hypothesis, so $S(\mathcal{F})$ is also empty. □

PROOF OF LEMMA 2.3. Firstly, for any coherent $\mathbb{C}^*$–equivariant sheaf on X (the $\mathbb{C}^*$–action on X is trivial), and for each $x \in X$, there is a neighbourhood U of x, finite dimensional complex representations V, W of $\mathbb{C}^*$, and a $\mathbb{C}^*$–equivariant exact sequence

$$\mathcal{O}_U \otimes W \xrightarrow{\quad g \quad} \mathcal{O}_U \otimes V \xrightarrow{\quad f \quad} \mathcal{E}|_U \longrightarrow 0, \tag{2.16}$$

such that, for any $U' \subset U$ open, each $s \in \mathcal{E}(U')$ is $\mathbb{C}^*$–finite, i.e. the $\mathbb{C}$–linear span of the orbit $L \cdot s$ is finite-dimensional. This result is an easy consequence of [**R**, Proposition 2.1]. Secondly, we show that for each coherent $\mathbb{C}^*$–equivariant sheaf $\mathcal{E}$ on X, there is a unique $\mathbb{C}^*$–invariant projection operator $\Pi : \mathcal{E} \to \mathcal{E}$ onto $\mathcal{E}^{\mathbb{C}^*}$, and that it commutes with restrictions of sections of the sheaf $\mathcal{E}$, as well as with ($\mathbb{C}^*$–equivariant) morphisms of coherent $\mathbb{C}^*$–equivariant sheaves on X. This is an immediate consequence of [**R**, Proposition 2.2]. The last result that we need is that the functor which takes a coherent $\mathbb{C}^*$–equivariant sheaf $\mathcal{E}$ on X into its invariant part $\mathcal{E}^{\mathbb{C}^*}$ is exact (this is standard in representation theory of Lie groups; in our case it follows from the existence of Π), and that $\mathcal{E}^{\mathbb{C}^*}$ is coherent. The coherence of $\mathcal{E}^{\mathbb{C}^*}$ follows from the exactness of the previous functor and from the existence of $\mathbb{C}^*$–equivariant exact sequences (2.16). The isotopical decomposition now follows from the fact that each section in $\mathcal{E}(U')$ is $\mathbb{C}^*$–finite, and from the existence of the invariant projection Π. □

3. Hitchin–Kobayashi correspondence and dimensional reduction

Let $\mathcal{F}$ be a holomorphic vector bundle over a compact n–dimensional Kähler manifold (M, ω), whose volume is normalised to 2π. A hermitian metric h on $\mathcal{F}$ defines a unique unitary connection A_h —the so-called *Chern connection*— on $(\mathcal{F}, h)$ whose associated $\bar\partial$–operator is compatible with the holomorphic structure of $\mathcal{F}$; let F_h be its curvature. We say that h is *Hermite–Einstein* if it satisfies

$$\sqrt{-1}\,\Lambda F_h = \mu(\mathcal{F}) I, \tag{3.1}$$

where Λ is the contraction with the Kähler form of M, and $\mu(\mathcal{F})$, is the *slope* of $\mathcal{F}$ which, for any torsion free coherent sheaf $\mathcal{F}'$, is defined as

$$\mu(\mathcal{F}') = \frac{\deg \mathcal{F}'}{\operatorname{rk} \mathcal{F}'}.$$

Here

$$\deg \mathcal{F}' = \int_M c_1(\mathcal{F}') \wedge \frac{\omega^{n-1}}{(n-1)!}$$

is the *degree* of $\mathcal{F}'$, where $c_1(\mathcal{F}')$ is the first Chern class of $\mathcal{F}'$, and $\operatorname{rk}\mathcal{F}'$ is the *rank* of $\mathcal{F}'$.

The existence of a Hermitian–Einstein metric is related to the algebro-geometric condition of stability by the so-called *Hitchin–Kobayashi correspondence*. Recall that a holomorphic vector bundle $\mathcal{F}$ is said to be *stable* if for every proper coherent subsheaf $\mathcal{F}' \hookrightarrow \mathcal{F}$, $\mu(\mathcal{F}') < \mu(\mathcal{F})$, where proper means that $0 < \operatorname{rk}\mathcal{F}' < \operatorname{rk}\mathcal{F}$. A direct sum of stable holomorphic vector bundles, all of them with the same slope, is called *polystable*. The Hitchin–Kobayashi correspondence states that the existence of a Hermite–Einstein metric on $\mathcal{F}$ is equivalent to the polystability of $\mathcal{F}$. For Riemann surfaces, this is equivalent to the theorem of Narasimhan and Seshadri [**NS**] (also proved using gauge-theoretic methods in [**D1**]), and for higher dimensions is the Theorem of Donaldson, Uhlenbeck and Yau [**D2, D3, UY**] (see also [**K**]).

Let $\boldsymbol{\mathcal{F}}$ be a holomorphic filtration over a compact Kähler manifold (M, ω) given by

$$\boldsymbol{\mathcal{F}} : 0 \hookrightarrow \mathcal{F}_0 \hookrightarrow \mathcal{F}_1 \hookrightarrow \cdots \hookrightarrow \mathcal{F}_m = \mathcal{F}. \tag{3.2}$$

The *deformed Hermite–Einstein equation* for a hermitian metric on $\mathcal{F}$ involves as many parameters as steps are in the filtration, and has the form

$$\sqrt{-1}\,\Lambda F_h = \begin{pmatrix} \tau_0 I_0 & & & \\ & \tau_1 I_1 & & \\ & & \ddots & \\ & & & \tau_m I_m \end{pmatrix}, \tag{3.3}$$

where F_h is the curvature of the Chern connection of h, and the RHS is a diagonal matrix, with constants $\tau_0, \tau_1, \dots, \tau_m \in \mathbb{R}$, written in blocks corresponding to the splitting which h defines in the filtration $\boldsymbol{\mathcal{F}}$. Taking traces in (3.3) and integrating over M, we see that the parameters are constrained by

$$\sum_{i=0}^{m} \tau_i \operatorname{rk}(\mathcal{F}_i/\mathcal{F}_{i-1}) = \deg \mathcal{F}, \tag{3.4}$$

which means that there are only m independent parameters, as one would expect. If $\tau_0 = \cdots = \tau_m$, (3.3) reduces of course to the Hermite–Einstein equation (3.1). A metric h on $\mathcal{F}$ satisfying (3.3) is called a $\boldsymbol{\tau}$–Hermite–Einstein.

As in the ordinary Hermite–Einstein equation, the existence of solutions to the deformed equation is related to a stability condition for the holomorphic filtration.

Let $\boldsymbol{\mathcal{F}}$, as in (3.2), be a holomorphic filtration on M. A *sheaf subfiltration* of $\boldsymbol{\mathcal{F}}$ is a sheaf filtration

$$\boldsymbol{\mathcal{F}}' : 0 \hookrightarrow \mathcal{F}_0' \hookrightarrow \mathcal{F}_1' \hookrightarrow \cdots \hookrightarrow \mathcal{F}_m' = \mathcal{F}', \tag{3.5}$$

where $\mathcal{F}'$ is a subsheaf of $\mathcal{F}$, such that $\mathcal{F}_i' = \mathcal{F}_i \cap \mathcal{F}'$ for $0 \le i \le m$. The sheaf subfiltration $\boldsymbol{\mathcal{F}}' \hookrightarrow \boldsymbol{\mathcal{F}}$ is called *proper* if $0 < \operatorname{rk}\mathcal{F}' < \operatorname{rk}\mathcal{F}$. Let $\boldsymbol{\alpha} = (\alpha_0, \dots, \alpha_{m-1}) \in \mathbb{R}^m$, and let $\boldsymbol{\mathcal{F}}'$ be a sheaf filtration of $\boldsymbol{\mathcal{F}}$. We define its $\boldsymbol{\alpha}$*–degree* and $\boldsymbol{\alpha}$*–slope* respectively by

$$\deg_{\boldsymbol{\alpha}} \boldsymbol{\mathcal{F}}' = \deg \mathcal{F}' + \sum_{i=0}^{m-1} \alpha_i \operatorname{rk}(\mathcal{F}_i'), \quad \mu_{\boldsymbol{\alpha}}(\boldsymbol{\mathcal{F}}') = \frac{\deg_{\boldsymbol{\alpha}} \boldsymbol{\mathcal{F}}'}{\operatorname{rk}\mathcal{F}'}.$$

We say that the sheaf filtration $\mathcal{F}$ is $\boldsymbol{\alpha}$–*stable* if for all proper sheaf subfiltrations $\mathcal{F}' \hookrightarrow \mathcal{F}$, we have $\mu_{\boldsymbol{\alpha}}(\mathcal{F}') < \mu_{\boldsymbol{\alpha}}(\mathcal{F})$. A $\boldsymbol{\alpha}$–*polystable* sheaf filtration is a direct sum of $\boldsymbol{\alpha}$–stable sheaf filtrations, all of them with the same $\boldsymbol{\alpha}$–slope.

One has the following Hitchin–Kobayashi correspondence ([**AG**, Theorem 2.1]).

THEOREM 3.1. *Let $\mathcal{F}$ be a holomorphic filtration on M. Let $\boldsymbol{\tau} = (\tau_0, \dots, \tau_m) \in \mathbb{R}^{m+1}$ satisfy (3.4), and let $\boldsymbol{\alpha} = (\alpha_0, \dots, \alpha_{m-1}) \in \mathbb{R}^m$ be an m–tuple of* positive *real numbers defined by*

$$\alpha_i = \tau_{i+1} - \tau_i \quad (0 \leq i \leq m-1). \tag{3.6}$$

Then $\mathcal{F}$ admits a $\boldsymbol{\tau}$–Hermite–Einstein metric if and only if it is $\boldsymbol{\alpha}$–polystable.

One has an equivariant version of this theorem for the $\mathrm{SL}(2,\mathbb{C})$–manifold $M = X \times \mathbb{P}^1$, by considering $\mathrm{SL}(2,\mathbb{C})$–invariant sheaf subfiltrations of $\mathcal{F}$ and $\mathrm{SU}(2)$–invariant hermitian metrics on $\mathcal{F}$ ([**AG**, Theorem 2.3]).

THEOREM 3.2. *Let $\mathcal{F}$ be an* $\mathrm{SL}(2,\mathbb{C})$*–equivariant holomorphic filtration on $X \times \mathbb{P}^1$. Let $\boldsymbol{\tau} = (\tau_0, \dots, \tau_m) \in \mathbb{R}^{m+1}$ satisfy (3.4), and let $\boldsymbol{\alpha} = (\alpha_0, \dots, \alpha_{m-1}) \in \mathbb{R}^m$ be an m–tuple of* positive *real numbers given by (3.6). Then $\mathcal{F}$ is* $\mathrm{SL}(2,\mathbb{C})$*–invariantly $\boldsymbol{\alpha}$–polystable if and only if it admits an* $\mathrm{SU}(2)$*–invariant $\boldsymbol{\tau}$–Hermite–Einstein metric.*

By means of gauge-theoretic dimensional reduction methods we can analyse $\mathrm{SU}(2)$–invariant solutions to the deformed Hermitian-Einstein equation for a filtration on $X \times \mathbb{P}^1$ to obtain some equations on X. To be more precise, let $\mathcal{F}$ be an $\mathrm{SL}(2,\mathbb{C})$–equivariant holomorphic filtration on $X \times \mathbb{P}^1$ and let $\mathcal{C} = (\mathcal{E}, \phi)$ be its corresponding holomorphic chain over X (given by (2.6)). Let $\boldsymbol{\tau} = (\tau_0, \dots, \tau_m), \boldsymbol{\tau}' = (\tau_0', \dots, \tau_m') \in \mathbb{R}^{m+1}$ be related by

$$\tau_i' = \tau_i + 4\pi i \quad (0 \leq i \leq m).$$

Then one can prove [**AG**, Theorem 4.1] that $\mathcal{F}$ admits an $\mathrm{SU}(2)$–invariant $\boldsymbol{\tau}'$–Hermite–Einstein metric if and only if $\mathcal{C}$ admits an $(m+1)$–tuple $\boldsymbol{h} = (h_0, h_1, \dots, h_m)$ of hermitian metrics where h_i is a metric on $\mathcal{E}_i$ satisfying

$$\begin{aligned} \sqrt{-1}\Lambda F_{h_0} + \tfrac{1}{2}\phi_1 \circ \phi_1^* &= \tau_0 \,\mathrm{id}_{\mathcal{E}_0}, \\ \sqrt{-1}\Lambda F_{h_i} - \tfrac{1}{2}(\phi_i^* \circ \phi_i - \phi_{i+1} \circ \phi_{i+1}^*) &= \tau_i \,\mathrm{id}_{\mathcal{E}_i}, \qquad (1 \leq i \leq m-1), \\ \sqrt{-1}\Lambda F_{h_m} - \tfrac{1}{2}\phi_m^* \circ \phi_m &= \tau_m \,\mathrm{id}_{\mathcal{E}_m}. \end{aligned} \tag{3.7}$$

These are called the *chain $\boldsymbol{\tau}$–vortex equations*, and generalize the vortex equations studied in [**G, BG**].

Let $\mathcal{C} = (\mathcal{E}, \phi)$ a holomorphic chain, as in (2.6). A sheaf subchain of $\mathcal{C}$ is a sheaf chain

$$\mathcal{C}' : \mathcal{E}_m' \xrightarrow{\phi_m'} \mathcal{E}_{m-1}' \xrightarrow{\phi_{m-1}'} \cdots \xrightarrow{\phi_1'} \mathcal{E}_0', \tag{3.8}$$

such that $\mathcal{E}_i'$ is a subsheaf of $\mathcal{E}_i$ for $0 \leq i \leq m$, and $\phi_i \circ f_i = f_{i-1} \circ \phi_i'$ for $1 \leq i \leq m$, where $f_i : \mathcal{E}_i' \hookrightarrow \mathcal{E}_i$ are the inclusion morphisms. The sheaf subchain $\mathcal{C}' \hookrightarrow \mathcal{C}$ is called *proper* if $0 < \sum_{i=0}^m \operatorname{rk} \mathcal{E}_i' < \sum_{i=0}^m \operatorname{rk} \mathcal{E}_i$. Let $\boldsymbol{\alpha} = (\alpha_0, \alpha_1, \dots, \alpha_m) \in \mathbb{R}^m$. The $\boldsymbol{\alpha}$–*degree* and $\boldsymbol{\alpha}$–*slope* of a coherent sheaf chain $\mathcal{C}'$, are defined by

$$\deg_{\boldsymbol{\alpha}} \mathcal{C}' = \sum_{i=0}^m \deg \mathcal{E}_i' - \sum_{i=0}^m \alpha_i \operatorname{rk} \mathcal{E}_i', \quad \mu_{\boldsymbol{\alpha}}(\mathcal{C}') = \frac{\deg_{\boldsymbol{\alpha}} \mathcal{C}'}{\sum_{i=0}^m \operatorname{rk} \mathcal{E}_i'},$$

respectively. We say that the coherent sheaf chain $\boldsymbol{C}$ is $\boldsymbol{\alpha}$*–stable* if for all proper sheaf subchains $\boldsymbol{C}' \hookrightarrow \boldsymbol{C}$, $\mu_{\boldsymbol{\alpha}}(\boldsymbol{C}') < \mu_{\boldsymbol{\alpha}}(\boldsymbol{C})$. A direct sum of $\boldsymbol{\alpha}$–stable coherent sheaf chains, all of them with the same $\boldsymbol{\alpha}$–slope, is called $\boldsymbol{\alpha}$*–polystable*.

Using the correspondence proved in the previous section, one can show a dimensional reduction reduction result relating filtrations to chains. More precisely, let $\boldsymbol{\mathcal{F}}$ be an $\mathrm{SL}(2,\mathbb{C})$–equivariant holomorphic filtration on $X \times \mathbb{P}^1$, and let $\boldsymbol{C}$ be the corresponding holomorphic chain on X. Let $\boldsymbol{\alpha} = (\alpha_0, \dots, \alpha_{m-1}) \in \mathbb{R}^m$, $\boldsymbol{\alpha}' = (\alpha'_0, \dots, \alpha'_m) \in \mathbb{R}^{m+1}$ be related by

$$\alpha_0 = 0, \qquad \alpha_i = \sum_{j=0}^{i-1} \alpha'_j - 4\pi i \quad (1 \le i \le m).$$

Then $\boldsymbol{\mathcal{F}}$ is $\mathrm{SL}(2,\mathbb{C})$–invariantly $\boldsymbol{\alpha}'$–stable if and only if $\boldsymbol{C}$ is $\boldsymbol{\alpha}$–stable.

Combining all these results we have the following Hitchin–Kobayashi correspondence [**AG**, Theorem 3.1] generalizing that obtained for triples in [**BG**].

THEOREM 3.3. *Let $\boldsymbol{C}$ be a holomorphic chain. Let $\boldsymbol{\tau} = (\tau_0, \tau_1, \dots, \tau_m) \in \mathbb{R}^{m+1}$ be such that $\deg_{\boldsymbol{\tau}}(\boldsymbol{C}) = 0$. The holomorphic chain $\boldsymbol{C}$ admits an $(m+1)$–tuple $\boldsymbol{h} = (h_0, h_1, \dots, h_m)$ of hermitian metrics satisfying the chain $\boldsymbol{\tau}$–vortex equation if and only if it is $\boldsymbol{\tau}$–polystable.*

References

[Ak] D.N. Akhiezer, *Lie group actions in complex analysis*, Aspects of Mathematics, **E27**, Vieweg, 1995.

[AB] M.F. Atiyah and R. Bott, *On the periodicity theorem for complex vector bundles*, Acta Mathematica **112** (1964), 229-247.

[AG] L. Álvarez–Cónsul and O. García–Prada, *Dimensional reduction*, SL(2, ℂ)*–equivariant bundles and stable holomorphic chains*, Int. J. Math. **12** (2001), 159–201.

[BG] S.B. Bradlow and O. García–Prada, *Stable triples, equivariant bundles and dimensional reduction*, Math. Ann. **304** (1996), 225–252.

[D1] S.K. Donaldson, *A new proof of a theorem of Narasimhan and Seshadri*, J. Diff. Geom. **18** (1983), 269–278.

[D2] S.K. Donaldson, *Anti-self-dual Yang–Mills connections on a complex algebraic surface and stable vector bundles*, Proc. Lond. Math. Soc. **3** (1985), 1–26.

[D3] S.K. Donaldson, *Infinite determinants, stable bundles and curvature*, Duke Math. J. **54** (1987), 231–247.

[G] O. García–Prada, *Dimensional reduction of stable bundles, vortices and stable pairs*, Int. J. Math. **5** (1994), 1–52.

[H] N.J. Hitchin, *The self-duality equations on a Riemann surface*, Proc. Lond. Math. Soc. **55** (1987), 59–126.

[K] S. Kobayashi, *Differential Geometry of Complex Vector Bundles*, Princeton University Press, New Jersey, 1987.

[NS] M.S. Narasimhan, C.S. Seshadri, *Stable and unitary bundles on a compact Riemann surfaces*, Ann. Math. **82** (1965), 540–567.

[R] M. Roberts, *A note on coherent G–sheaves*, Math. Ann. **275** (1986), 573–582.

[UY] K.K. Uhlenbeck and S.T. Yau, *On the existence of Hermitian–Yang–Mills connections on stable bundles over compact Kähler manifolds*, Comm. Pure and Appl. Math. **39–S** (1986), 257–293.

DEPARTMENT OF MATHEMATICS, UNIVERSITY OF ILLINOIS AT URBANA-CHAMPAIGN, URBANA, IL 61801,USA

E-mail address: lalvarez@math.uiuc.edu

DEPARTAMENTO DE MATEMÁTICAS, UNIVERSIDAD AUTÓNOMA DE MADRID, 28049 MADRID, SPAIN

E-mail address: oscar.garcia-prada@uam.es

Contemporary Mathematics
Volume **288**, 2001

Hopf Vesicles in $S^3(1)$

J. Arroyo and O.J. Garay

Abstract. We study the existence and stability of closed elastic curves in $\mathbf{S}^2(\mathbf{1})$ which are circular at rest. Then, by pulling them back via the Hopf map and using the symmetric criticality Principle of Palais, we obtain examples of closed membranes in $\mathbf{S}^3(\mathbf{1})$.

It was around 1810 that Sophie Germain in her study of the elastic plates, set a model for the *elastic energy* $\mathbf{E}(\mathbf{S})$ of a surface $\mathbf{S}$ (see §2). Moreover, a model of Canhan-Helfrich **[2]**, **[4]**, states that in order to obtain the mechanical equilibrium of a biological *membrane* $\mathbf{S}$ (called *vesicle,* if $\mathbf{S}$ is closed) we have to compute the critical points of the *shape energy,* $\mathbf{E}(\mathbf{S})$. Observe that for suitable choices of the parameters in (2.1), particular examples of membranes are *minimal surfaces, surfaces of constant mean curvature and Willmore surfaces.* Several authors have studied the existence of vesicles of different topological configuration in $\mathbf{R}^3$ and confront it with experimental observations.

On the other hand, D. Bernoulli proposed around 1740 a simple geometric model for a *elastic curve* in $\mathbf{R}^3$, which has been considered recently in more general Riemannian manifolds (see for instance **[6]**). In this connection, a curve $\gamma : I \to M^n$, which is a critical point of $\int_\gamma (\kappa+\lambda)^2$, κ being the curvature of γ, would correspond with the elastic curves whose position at rest is not "straight" but circular instead. We call them *circular λ-elastic curves.*

It has been pointed out by Hsu-Kusner-Sullivan **[5]**, that it would be interesting to study the existence of vesicles in 3-dimensional space forms other than $\mathbf{R}^3$. We want to study the existence of vesicles in $\mathbf{S}^3(\mathbf{1})$. By using the Hopf map and the symmetric criticality principle of Palais (an idea used by Pinkall in connection with Willmore tori,**[8]**) this problem is reduced in section §2 to find closed $\lambda-$elastic curves in $\mathbf{S}^2(\mathbf{1})$. The latter problem is analyzed in §1.

1. $\lambda-$elastic curves in $S^2(1)$

Let $\mathbf{S}^2(\mathbf{1})$ be the 2-dimensional unit sphere. We consider the functional

$$\mathcal{F}^\lambda(\gamma) = \int_\gamma (\kappa+\lambda)^2 ds, \tag{1.1}$$

2000 *Mathematics Subject Classification.* Primary 53C40, 49Q10; Secondary 74K10, 74K15.
Key words and phrases. Elastic curves, membranes, Hopf map, submanifolds.
Supported by a DGICYT grant PB97-0784-C03-03 and by UPV grant 127.310-EA-4505/98.

where $\kappa(s)$ is the oriented curvature of $\gamma(s)$, acting on a suitable space of immersed closed curves $\gamma(s)$ in $\mathbf{S}^2(\mathbf{1})$. Then we consider a variation of $\gamma(s)$ within the specified space of curves and use standard arguments which involve some integrations by parts, the Frenet equations of γ and a technical proposition similar to proposition 1.1 of ([**6**]), to obtain the first variation formula of this functional. In particular, γ is a critical point of $\mathcal{F}^\lambda$, if and only if the following Euler-Lagrange equation is satisfied

$$2\kappa_{ss} + \kappa^3 + (2-\lambda^2)\kappa + 2\lambda = 0. \tag{1.2}$$

Circles of curvature $-\lambda$ are trivial solutions of (1.2) and absolute minima of (1.1). For $\lambda^2 < 8$, there are no other circular solutions. For $\lambda^2 > 8$, circles of curvature $\frac{\lambda-\sqrt{\lambda^2-8}}{2}$ and $\frac{\lambda+\sqrt{\lambda^2-8}}{2}$ are also critical points. Assume that κ is not constant. Multiplying by κ_s we obtain a first integral of (1.2)

$$\kappa_s^2 = \frac{1}{4}\left(d - (\kappa+\lambda)^2\left(((\kappa-\lambda)^2)+4\right)\right), \tag{1.3}$$

where $d \in \mathbf{R}^+$. Denote by $Q(x) = \frac{1}{4}\left(d - (x+\lambda)^2\left(((x-\lambda)^2)\right)+4\right)$. By using the method of phase portraits, for instance, one can see that the non-constant solutions of (1.3) are periodic functions. Actually, it can be proved that they are given by elliptic functions. One can discuss the solutions of (1.3) in terms of the roots of the polynomial $Q(x)$, being its multiple roots those which affect the integrability. $Q(x)$ can only have simple or double roots. Avoiding the cases of double roots, $Q(x)$ may have either 2 or 4 simple roots depending on the values of the parameters λ and d. Explicit solutions and a much more detailed study for all possible values of λ and d can be found in [**1**]. For brevity, we restrict ourselves here to the case $\lambda^2 > 8$ and $d \in (\delta_1, \delta_2)$, where $\delta_1 = \frac{1}{2}\lambda^4 - \frac{1}{2}\lambda(\lambda^2-8)^{\frac{3}{2}} + 10\lambda^2 - 4$; $\delta_2 = \frac{1}{2}\lambda^4 + \frac{1}{2}\lambda(\lambda^2-8)^{\frac{3}{2}} + 10\lambda^2 - 4$. In this case, for every d, we have four simple roots $\alpha_1^d < \alpha_2^d < \alpha_3^d < \alpha_4^d$, and there are two solutions of (1.3), $\kappa_d^\wedge(s)$ and $\kappa_d^\vee(s)$, which can be written in terms of the Jacobi elliptic functions. $\kappa_d^\wedge(s)$ oscillates between $\alpha_1^d = \kappa_d^\wedge(0)$ and $\kappa_d^\wedge\left(\frac{K(M)}{r}\right) = \alpha_2^d$, $K(M)$ being the complete elliptic integral of the first kind of modulus $M = \sqrt{\frac{(\alpha_4-\alpha_3)(\alpha_2-\alpha_1)}{(\alpha_4-\alpha_2)(\alpha_3-\alpha_1)}}$ and $r = \frac{\sqrt{(\alpha_4-\alpha_2)(\alpha_3-\alpha_1)}}{4}$. Analogously, $\kappa_d^\vee(s)$ oscillates between $\alpha_3^d = \kappa_d^\vee(0)$ and $\kappa_d^\vee\left(\frac{K(M)}{r}\right) = \alpha_4^d$. If we denote by $\gamma_d^\wedge, \gamma_d^\vee$ the associated $\lambda-$elastic curves in $\mathbf{S}^2(\mathbf{1})$, then they are both convex if $\lambda^2 > 5+3\sqrt{3}$ and $d < \lambda^4 + 4\lambda^2$. One is convex and the other not, otherwise. In order to determine the closed ones among then, we use the following argument of Langer and Singer [**6**]. By using (1.3) and Lemma 1 of [**6**], one can check by direct computation that

PROPOSITION 1.1. *Let γ be a $\lambda-$elastic curve in $\mathbf{S}^2(\mathbf{1})$ and $\{T, N\}$ its Frenet frame. Then the vector field $J = (\kappa^2-\lambda^2)T + 2\kappa_s N$ is the restriction to γ of a Killing field on $\mathbf{S}^2(\mathbf{1})$ (which we also denote by J).*

Now, choose spherical coordinates $x(\theta,\psi) = (\cos\theta\sin\psi, \sin\theta\sin\psi, \cos\psi)$, so that its equator gives the only integral geodesic of $J : x_\theta = bJ$. Using this and the above proposition, we have $\theta'(s) = \frac{<T, x_\theta>}{\sin^2\psi} = \frac{\kappa^2-\lambda^2}{b\left(d-4(\kappa-\lambda)^2\right)}$. Therefore

PROPOSITION 1.2. *Let γ be a $\lambda-$elastic curve in $\mathbf{S}^2(\mathbf{1})$ with periodic curvature κ of period ρ. Then γ is closed, if and only if, its total angle in one period*

$$\Lambda = \int_0^\rho \sqrt{d}\frac{\left(\kappa^2-\lambda^2\right)}{d-4\left(\kappa+\lambda\right)^2}ds \tag{1.4}$$

is a rational multiple of π.

Hence, in order to find closed $\lambda-$elastic curves, we must check the closedness condition (1.4) for the periodic solutions of (1.3). In the next proposition take δ_1, δ_2 as given just after (1.3)

PROPOSITION 1.3. *Given a real number $\lambda^2 > 8$, then for every pair of natural numbers $m, n \in \mathbf{N}$, such that $(m,n) = 1$ and $\frac{m}{n} > \sqrt{\frac{2\lambda^2+2\lambda(\lambda^2-8)^{\frac{1}{2}}-4}{\lambda^2+3\lambda(\lambda^2-8)^{\frac{1}{2}}-8}}$, there exist a real number $d_{mn} \in (\delta_1, \delta_2)$ and, associated to it, a couple of closed $\lambda-$elastic curves in $\mathbf{S}^2(\mathbf{1})$, $\gamma^\wedge_{mn}, \gamma^\vee_{mn}$, which close up after n periods and m trips around the equator. Moreover, the respective angle variations in one period, $\Lambda^\wedge(d_{mn})$ and $\Lambda^\vee(d_{mn})$, are related by*

$$\Lambda^\wedge(d_{mn}) - \Lambda^\vee(d_{mn}) = \begin{cases} 2\pi \ \textit{if} \ d_{mn} < 16\lambda^2 \\ 0 \ \ \textit{if} \ d_{mn} > 16\lambda^2. \end{cases}$$

PROOF. Fix a real number $\lambda^2 > 8$ and let d be in (δ_1, δ_2) . We use the notation introduced before Proposition 1. Let us denote $\Lambda^\wedge(d), \Lambda^\vee(d)$ the total angles associated to the $\lambda-$elastic curves $\gamma^\wedge_d, \gamma^\vee_d$. To compute (1.4), we see that as $\alpha_1^d \le \kappa_d^\wedge(s) \le \alpha_2^d$, then $Q(\kappa) > 0$; $(\kappa_d^\wedge(s))_s > 0$ when $0 < x < \frac{K(M)}{r}$ and $\kappa^\wedge(s)$ is symmetric with respect to $y = \frac{K(M)}{r}$. Then, using (1.3) and (1.4) we have that

$$\Lambda^\wedge(d) = 2\int_{\alpha_1^d}^{\alpha_2^d} \sqrt{d}\frac{\kappa^2-\lambda^2}{d-4\left(\kappa-\lambda\right)^2}\frac{d\kappa}{Q(\kappa)^{\frac{1}{2}}}, \tag{1.5}$$

is finite and it moves continuously as d varies in (δ_1, δ_2). Thus, in order to check the condition of Proposition 2, we must analyze the range of variation of $\Lambda^\wedge(d)$. We assume that $d \neq 16\lambda^2$, then using formulas 3.147 and 3.151 of ([**3**]) one can obtain

$$\begin{aligned}
\Lambda^\wedge(d) &= \frac{2\sqrt{d}}{\sqrt{(\alpha_4-\alpha_2)(\alpha_3-\alpha_1)}}\left\{\left(\frac{\left(\lambda+\frac{\sqrt{d}}{4}\right)}{\left(\frac{\sqrt{d}}{2}+\lambda+\alpha_2\right)}+\frac{\left(\lambda-\frac{\sqrt{d}}{4}\right)}{\left(-\frac{\sqrt{d}}{2}+\lambda+\alpha_2\right)}-1\right)K\left(M\right)\right.\\
&+\left(\frac{-\left(\lambda+\frac{\sqrt{d}}{4}\right)(\alpha_3-\alpha_2)}{\left(\alpha_3+\frac{\sqrt{d}}{2}+\lambda\right)\left(\frac{\sqrt{d}}{2}+\lambda+\alpha_2\right)}\right)\Pi\left(\frac{(\alpha_4-\alpha_3)\left(\frac{\sqrt{d}}{2}+\lambda+\alpha_2\right)}{(\alpha_4-\alpha_2)\left(\frac{\sqrt{d}}{2}+\lambda+\alpha_3\right)},M\right)\\
&+\left.\left(\frac{-\left(\lambda-\frac{\sqrt{d}}{4}\right)(\alpha_3-\alpha_2)}{\left(\alpha_3-\frac{\sqrt{d}}{2}+\lambda\right)\left(\lambda-\frac{\sqrt{d}}{2}+\alpha_2\right)}\right)\Pi\left(\frac{(\alpha_4-\alpha_3)\left(\lambda-\frac{\sqrt{d}}{2}+\alpha_2\right)}{(\alpha_4-\alpha_2)\left(\lambda-\frac{\sqrt{d}}{2}+\alpha_3\right)},M\right)\right\},
\end{aligned}$$

$K\left(M\right)$ and $\Pi\left(\nu, M\right)$ being the complete elliptic integral of the first and third kind of modulus M, respectively. Now, using that $\lim_{d\to\delta_1^+} M = \lim_{d\to\delta_1^+} \nu = 0$ and $K(0) = \Pi\left(0,0\right) = \frac{\pi}{2}$ we have $\Lambda^\wedge_{\delta_1} = \lim_{d\to\delta_1^+}\Lambda^\wedge(d) = -2\pi\sqrt{\frac{2\lambda^2+2\lambda(\lambda^2-8)^{\frac{1}{2}}-4}{\lambda^2+3\lambda(\lambda^2-8)^{\frac{1}{2}}-8}}$. Analogously one can check that $\lim_{d\to\delta_2^-}\Lambda^\wedge(d) = -\infty$. This proves the first part. Now, interchanging $1 \longleftrightarrow 4$ and $2 \longleftrightarrow 3$ in the above formula, one can obtain the corresponding one for $\Lambda^\vee(d)$. On the other hand, given real numbers $0 < \nu \leq M \leq \mu < 1$, satisfying $\nu\mu = M^2$, then $\Pi(\nu, M) + \Pi(\mu, M) = \frac{1}{\sqrt{1-\nu}}\frac{1}{\sqrt{1-\mu}}\frac{\pi}{2} + K(M)$. Thus

$$\Lambda^\wedge(d) - \Lambda^\vee(d) = \pi\left\{\frac{\left(\lambda - \frac{\sqrt{d}}{4}\right)\sqrt{d}}{-4Q(\left(-\lambda - \frac{\sqrt{d}}{2}\right))} + \frac{\left(\lambda + \frac{\sqrt{d}}{4}\right)\sqrt{d}}{-4Q(\left(-\lambda - \frac{\sqrt{d}}{2}\right))}\right\},$$

which finishes the proof. □

Now, let us denote by ε_α^m the m-fold circle of curvature α, where α is one of the following : $-\lambda$; $\frac{\lambda+\sqrt{\lambda^2-8}}{2}$; or $\frac{\lambda-\sqrt{\lambda^2-8}}{2}$. These circles are critical points as we know. We also denote by $\varpi = \frac{P(\alpha)}{2(\alpha^2+1)^2}$, where $P(x) = (x+\lambda)^2$. As for the stability, we have

PROPOSITION 1.4. *Assume $\lambda^2 > 8$. If $\alpha = -\lambda$, then ε_α^m is stable for every $m \in \mathbf{N}$. If $\alpha = \frac{\lambda-\sqrt{\lambda^2-8}}{2}$, ε_α^m is unstable for every $m \in \mathbf{N}$. If $\alpha = \frac{\lambda+\sqrt{\lambda^2-8}}{2}$, then ε_α^m is stable, if and only if, $m(1-\sqrt{1-\varpi}) < 1$, and therefore the circle ε_α^1 is stable for every λ in this case.*

PROOF. We consider a critical point γ of $\mathcal{F}^\lambda$ and a normal variation of γ with variation normal field $W = \phi N$. Computing the second variation formula at a critical point give

$$\frac{d^2\mathcal{F}^\lambda}{dt^2}(\gamma) = \int_\gamma (2\phi_{ss}^2 - (5\kappa^2 + 4 - \lambda^2)\phi_s^2 + \eta\phi^2)\, ds, \tag{1.6}$$

where $\eta = 3\kappa^4 + (5-\lambda^2)\kappa^2 + 6\kappa_s^2 + 8\kappa\kappa_{ss} + 2 - \lambda^2$. Assume that $\gamma = \varepsilon_\alpha^m$ is one of the above m-fold circles and take $r = \frac{1}{\sqrt{1+\alpha^2}}$. Then ϕ must have a Fourier series of the form $\frac{a_o}{2} + \sum_{n=1}^\infty a_n \cos(\frac{ns}{mr}) + b_n \sin\left(\frac{ns}{mr}\right)$. Suppose first that $\alpha = -\lambda$, then substitution in (1.6) and integration by parts give $\frac{d^2\mathcal{F}^\lambda}{dt^2}(\epsilon_\alpha^m) = 2\int_{\varepsilon_\alpha^m}(\phi_{ss} + (\lambda^2 + 1)\phi)^2 \geq 0$, as it should be. For the two other values of α, we have

$$\frac{d^2\mathcal{F}^\lambda}{dt^2}(\varepsilon_\alpha^m) = A_o\left\{(1-\varpi)\frac{a_o^2}{2} + \sum_{n=1}^\infty (a_n^2 + b_n^2)\left(\frac{m^2}{n^2} - 1\right)\left(\frac{m^2}{n^2} - 1 + \varpi\right)\right\}, \tag{1.7}$$

where $A_o = 2\pi m\left(\alpha^2+1\right)^{\frac{3}{2}}$. If $\alpha = \frac{\lambda-\sqrt{\lambda^2-8}}{2}$, then $1-\varpi < 0$ so that any choice of $W = hN$ with constant h, makes (1.7) negative and then ε_α^m is unstable. On the other hand, if $\alpha = \frac{\lambda+\sqrt{\lambda^2-8}}{2}$ then $1-\varpi > 0$ and the second variation is positive semidefinite whenever $1-(1-\varpi)^{\frac{1}{2}} > \frac{1}{m}$. Equality occurs when $a_n = b_n = 0$, $n \neq m$ and $\phi = a_m \cos(\frac{s}{r}) + b_m \sin\left(\frac{s}{r}\right)$, which corresponds to the space of Killing fields along γ. □

2. Hopf vesicles

According to Germain's model and assuming that the surface $\mathbf{S}$ is immersed in 3-space form, that the two sides of the elastic surface are distinguished (as in polymer interface) and that we may have some constraints on the area and enclosed volume, then the *elastic energy* is of the form

$$\mathbf{E}(\mathbf{S}) = \int_S (b(H-c_o)^2 + c\, G)\cdot dA + a\int_S dA + d\int_\Omega dV, \tag{2.1}$$

where $a, b, c, d \in \mathbf{R}$, c_o is a constant (the *spontaneous curvature* of S), dA is the area element of $\mathbf{S}$, dV is the volume element of the domain bounded by $\mathbf{S}$, and H and G are the mean and Gaussian curvature of the surface. Critical points of (2.1) are called *elastic surfaces or membranes* (*vesicles* if $\mathbf{S}$ is closed). For simplicity we consider the functional obtained taking $a = 0 = d$ in (2.1)

$$\int_S (H+\lambda)^2 dA + c\int_S G\, dA. \tag{2.2}$$

The Hopf fibration $\pi : \mathbf{S}^3(\mathbf{1}) \to \mathbf{S}^2(\frac{1}{2})$ is a Riemannian submersion with geodesic fibers. If $\gamma : [0, L] \to \mathbf{S}^2(\frac{1}{2})$ is a unit parametrized closed curve, its complete lift to $\mathbf{S}^3(\mathbf{1})$, $M_\gamma = \pi^{-1}(\gamma)$, is an immersed flat torus called the Hopf torus on γ. Using the O'Neill formulas, the mean curvature of M_γ is given by $H(t,s) = \frac{\kappa(s)}{2}$, where $\kappa(s)$ is the curvature of $\gamma(s)$ and t is the fiber parameter. Thus we can relate the Hopf tori which are critical for (2.2), that is, those which are a type of vesicles in $\mathbf{S}^3(\mathbf{1})$ of spontaneous curvature λ , with the $2\lambda-$elastic curves in $\mathbf{S}^2(\frac{1}{2})$:

PROPOSITION 2.1. *A Hopf torus M_γ is a critical point of (2.2) in $\mathbf{S}^3(\mathbf{1})$, if and only if, γ is a closed critical point of $\int_\gamma(\kappa - 2\lambda)^2$ in $\mathbf{S}^2(\frac{1}{2})$.*

PROOF. Let M_γ be a Hopf torus in $\mathbf{S}^3(\mathbf{1})$, by using the symmetric criticality principle of Palais [**7**], we see that for M_γ to be a critical point of (2.2) is enough to consider variations of M_γ by Hopf tori. But then, we use the above formula for the mean curvature of M_γ to get

$$\int_{M_\gamma} (H+\lambda)^2 dA = \iint (\frac{\kappa(s)}{2} + \lambda)^2 ds\cdot dt = \frac{\pi}{2}\int_0^L (\kappa(s)+2\lambda)^2 ds,$$

which proves the statement. □

Combining Propositions 3 and 5, one obtains plenty of examples of Hopf vesicles in $\mathbf{S}^3(\mathbf{1})$.

References

[1] J. Arroyo. *Presión calibrada total. Estudio Variacional y aplicaciones al problema de Chen-Willmore.* Ph. D. thesis. Univ. del País Vasco. Januay 2001.
[2] P.B. Canham. J. Theor. Biol. 26 (1970), 61.
[3] I.S. Gradsteyn, I.M. Ryzhik. *Table of Integrals, series and products.* Academic Press. 1980.
[4] W. Helfrich. Z. Naturforsch. 28c (1973), 693
[5] L. Hsu, R. Kusner and J. Sullivan. Minimizing the squared mean curvature integral for surfaces in space forms. Experimental Mathematics 1 (1992), 191-207.
[6] J. Langer and D.A. Singer. The total squared curvature of closed curves. J. Diff. Geom. 20 (1984), 1-22.
[7] R.S. Palais, Critical point theory and the minimax principle. Global Analysis, Proc. Sympos. Pure Math. 15 (1970), 185-212
[8] U. Pinkall. Hopf tori in S^3. Inventiones Math. 81 (1985), 379-386.

DEPARTAMENTO DE MATEMÁTICAS. FACULTAD DE CIENCIAS, U.P.V./E.H.U., APTDO 644, 48080 BILBAO, SPAIN.
E-mail address: mtparolj@lg.ehu.es

E-mail address: mtpgabeo@lg.ehu.es

Contemporary Mathematics
Volume **288**, 2001

Invariant Nearly Kähler f–structures on Homogeneous Spaces

Vitaly V. Balashchenko

ABSTRACT. We introduce a concept of nearly Kahler f-structures naturally generalizing well-known nearly Kähler structures. A remarkable collection of invariant structures of such a kind is presented. It is based on the canonical f-structures on homogeneous Φ-spaces, i.e. homogeneous spaces defined by Lie group automorphisms. As a result, the analogy with homogeneous nearly Kähler manifolds is obtained. Some general and particular examples including the 6-dimensional generalized Heisenberg group are considered.

1. Introduction

Homogeneous spaces of Lie groups are of great importance in differential geometry. In particular, homogeneous spaces defined by Lie group automorphisms Φ provided the remarkable examples for main classes in Hermitian geometry [WG]. The most important role belongs to Riemannian 3-symmetric spaces (i.e. $\Phi^3 = id$). The key point is that homogeneous 3-symmetric spaces possess the canonical almost complex structure constructed by means of Φ (see [S2], [WG], [G2]).

It turns out that the situation can be effectively extended. On the one hand, an almost complex structure is the particular case of *f-structures* by K.Yano, which are smooth tensor fields f of type $(1,1)$ satisfying the condition $f^3 + f = 0$. On the other hand, homogeneous Φ-spaces possess a remarkable collection of canonical structures including f-structures [BS1,2]. It gives the opportunity to introduce a class of f-structures such that it both generalizes classical nearly Kähler structures and contains many homogeneous examples on the base of canonical f-structures.

Let (M,g) be a (pseudo-)Riemannian manifold, ∇ the Levi-Civita connection of g, J an almost complex structure on M. Recall that (g,J) is *an almost Hermitian structure* if $g(JX,JY) = g(X,Y)$ for any $X, Y \in \mathfrak{X}(M)$. If, in addition, $\nabla_X(J)X = 0$, then J is called *a nearly Kähler structure* (breafly, *NK-structure*) [G1].

Let us now turn to f-structures on (M,g). f is *a metric f-structure* if $g(fX,Y) + g(X,fY) = 0$ (see [Ki2]). Further, a metric f-structure on (M,g) is said to be *a Killing f-structure* if $\nabla_X(f)X = 0$ (see [Gr]). Now we introduce the following

---

2000 *Mathematics Subject Classification.* Primary 53C15, 53C30; Secondary 53C10, 53C35.

Key words and phrases. Nearly Kähler structure, naturally reductive space, homogeneous Φ–space, k–symmetric space, f–structure.

Partially supported by the Education Ministry of Belarus and SFB 288.

more general concept. A metric f-structure on (M, g) is called *a nearly Kähler f-structure* (breafly, *NKf-structure*) if $\nabla_{fX}(f)fX = 0$ for all $X \in \mathfrak{X}(M)$ [B1].

It should be mentioned that invariant NK-structures were intensively investigated in [WG], [G2], [Ki1] and other papers. In particular, A.Gray proved (see [G2]) that a (pseudo-)Riemannian homogeneous Φ-space $(G/H, g)$ of order 3 with the canonical structure J is nearly Kähler iff the metric g is naturally reductive.

In the paper we concentrate on classes and remarkable examples of invariant NKf-structures. In particular, it is proved that the canonical f-structure on any naturally reductive homogeneous Φ-space of order 4 is NKf-structure.

2. Invariant NKf-structures on naturally reductive spaces

Let G be a connected Lie group, H its closed subgroup, g an invariant (pseudo-)Riemannian metric on the homogeneous space G/H. Denote by $\mathfrak{g}$ and $\mathfrak{h}$ the Lie algebras corresponding to G and H respectively. Suppose that G/H is a reductive homogeneous space, $\mathfrak{g} = \mathfrak{h} \oplus \mathfrak{m}$ the reductive decomposition of the Lie algebra $\mathfrak{g}$. As usual, we identify $\mathfrak{m}$ with the tangent space $T_o(G/H)$ at the point $o = H$. For convenience we denote by the same manner both any invariant structure on G/H and its value at o throughout the rest of the paper.

Let λ and μ be linear endomorphisms on $\mathfrak{m}$. Denote by $\mathfrak{m}^{(\lambda,\mu)}$ the linear span of all vectors $[\lambda X, \mu X]$, where $X \in \mathfrak{m}$ (see [B1]).

Recall that $(G/H, g)$ is *naturally reductive* with respect to a reductive decomposition $\mathfrak{g} = \mathfrak{h} \oplus \mathfrak{m}$ if $g([X, Y]_{\mathfrak{m}}, Z) = g(X, [Y, Z]_{\mathfrak{m}})$ for all $X, Y, Z \in \mathfrak{m}$.

THEOREM 2.1. *Let $(G/H, g)$ be a naturally reductive homogeneous space, f an invariant metric f-structure, $\mathfrak{g} = \mathfrak{h} \oplus \mathfrak{m}$ the corresponding reductive decomposition.*

(i) *f is a Killing f-structure iff $\mathfrak{m}^{(1,f)} \subset \mathfrak{h}$.*

(ii) *f is a nearly Kähler f-structure iff $\mathfrak{m}^{(f,f^2)} \subset \mathfrak{h}$.*

PROOF. For instance, we prove (ii). The condition $\nabla_{fX}(f)fX = 0$ means that $\nabla_{fX}f^2X = f\nabla_{fX}fX$ for all $X \in \mathfrak{X}(G/H)$. Denote by α the Nomizu function of an invariant affine connection∇ [N]. Then, using special vector fields in a neighborhood of the point o [N] it can be shown that the previous condition is equivalent to that of $\alpha(fX, f^2X) = f\ \alpha(fX, fX)$, $X \in \mathfrak{m}$. Since the metric g is naturally reductive, $\alpha(X, Y) = 1/2\ [X, Y]_{\mathfrak{m}}$, $X, Y \in \mathfrak{m}$. Hence we obtain $[fX, f^2X] \in \mathfrak{h}$ for any $X \in \mathfrak{m}$. This is the condition $\mathfrak{m}^{(f,f^2)} \subset \mathfrak{h}$. Statement (i) can be proved in the same way. □

Any invariant metric f-structure on G/H determines the orthogonal decomposition $\mathfrak{m} = \mathfrak{m}_1 \oplus \mathfrak{m}_2$ such that $\mathfrak{m}_1 = Im\, f$, $\mathfrak{m}_2 = Ker\, f$. Applying Theorem 2.1, it is not difficult to prove the following result.

THEOREM 2.2. *Let $(G/H, g, f)$ be a naturally reductive homogeneous space with a metric f-structure, $\mathfrak{m} = \mathfrak{m}_1 \oplus \mathfrak{m}_2$ the decomposition with respect to f. The following conditions are equivalent:*
(i) *f is a Killing f-structure;* (ii) *f is NKf-structure and $[\mathfrak{m}_1, \mathfrak{m}_2] \subset \mathfrak{h}$.*

3. Canonical structures on regular Φ-spaces

Let G be a connected Lie group, Φ its (analytic) automorphism. Denote by G^Φ the subgroup of all fixed points of Φ and G^Φ_o the identity component of G^Φ. Suppose a closed subgroup H of G satisfies the condition $G^\Phi_o \subset H \subset G^\Phi$. Then G/H is called *a homogeneous Φ-space*. Further, let $\mathfrak{g}$ and $\mathfrak{h}$ be the Lie algebras for

G and H, $\varphi = d\Phi_e$ the automorphism of $\mathfrak{g}$. Consider the linear operator $A = \varphi - id$ and the Fitting decomposition $\mathfrak{g} = \mathfrak{g}_0 \oplus \mathfrak{g}_1$ with respect to A, where $\mathfrak{g}_0$ and $\mathfrak{g}_1$ denote 0– and 1–component respectively. It is clear that $\mathfrak{h} = Ker\, A$ and $\mathfrak{h} \subset \mathfrak{g}_0$. A homogeneous Φ-space G/H is called *a regular Φ-space* if $\mathfrak{h} = \mathfrak{g}_0$ ([S1], [BS2], [F]).

We recall two basic facts due to N.A. Stepanov (see [S1]):

- *Any homogeneous Φ-space of order k ($\Phi^k = id$) is a regular Φ-space.*
- *Any regular Φ-space is reductive. More exactly, the Fitting decomposition $\mathfrak{g} = \mathfrak{h} \oplus \mathfrak{m}$, $\mathfrak{m} = A\mathfrak{g}$ is a reductive one. It is called* the canonical reductive decomposition *corresponding to G/H.*

Denote by θ the restriction of φ to $\mathfrak{m}$. Now we turn to *affinor structures*, i.e. tensor fields of type $(1,1)$. An invariant affinor structure F on a regular Φ-space G/H is called *canonical* [BS1,2] if its value at the point o is a polynomial in θ.

No doubt, almost product structures P ($P^2 = 1$), almost complex structures J ($J^2 = -1$), and, more generally, f-structures ($f^3 + f = 0$) belong to the most important affinor structures. We shall say that P, J, f are structures of *classical type*. All the canonical structures of classical type on regular Φ-spaces were completely described [BS1,2]. In particular, for homogeneous Φ-spaces of arbitrary finite order k, precise computational formulae were indicated. As an example, for $k = 3$ there exists a unique (up to sign) canonical structure of classical type. This is the well-known structure $J = \frac{1}{\sqrt{3}}(\theta - \theta^2)$ (see [S2], [WG], [G2]). As to $k = 4$, there exist only (up to sign) the following canonical structures: $P = \theta^2$, $f = \frac{1}{2}(\theta - \theta^3)$.

4. Canonical NKf-structures on regular Φ-spaces

THEOREM 4.1. *Let G/H be a regular Φ-space, g a naturally reductive metric on G/H with respect to the canonical reductive decomposition $\mathfrak{g} = \mathfrak{h} \oplus \mathfrak{m}$, f a metric canonical f-structure on G/H. Suppose the f-structure satisfies the condition $f^2 = \pm\theta f$. Then $(G/H, g, f)$ is a nearly Kähler f-manifold.*

PROOF. Using the assumption, for any $X \in \mathfrak{m}$ we have

$$\varphi[fX, f^2X] = [\theta fX, \theta f^2X] = [f^2X, f^3X] = [f^2X, -fX] = [fX, f^2X].$$

It follows that $[fX, f^2X] \in \mathfrak{h}$. By Theorem 2.1(ii), it proves the required result. □

COROLLARY 4.2. *Let $(G/H, g)$ be a naturally reductive homogeneous Φ-space of order $k = 4n$, $n \geq 1$. If $\{i, -i\} \subset$ spec θ, then there exists a non-trivial canonical NKf-structure on G/H.*

PROOF. The procedure of describing canonical structures on homogeneous Φ-spaces of finite order is constructive (see [BS2], §4 and §5). We use it to present a canonical f-structure required. Consider the decomposition $\mathfrak{m} = \mathfrak{m}_1 \oplus \mathfrak{m}_2$ such that $\mathfrak{m}_1$ and $\mathfrak{m}_2$ are determined by the eigenvalues $\{i, -i\}$ and all the others of θ respectively. Now we put f to be the complex structure on $\mathfrak{m}_1$ and vanish on $\mathfrak{m}_2$. In other words, consider the collection $(\zeta_1, ..., \zeta_l, ..., \zeta_u)$, $\zeta_j \in \{-1, 0, 1\}$ for constructing canonical f-structures (see [BS2], Theorem 5.3), where ζ_l corresponds to the eigenvalue i of θ. We put $\zeta_l = 1$, $\zeta_m = 0$, $m \in \{1, 2, ..., u\}$, $m \neq l$. It is not difficult to see that we obtain $f^2 = \theta f$. It remains to apply Theorem 4.1. □

COROLLARY 4.3. *Any naturally reductive homogeneous Φ-space $(G/H, g)$ of order 4 is NKf-manifold with respect to the canonical f-structure $f = \frac{1}{2}(\theta - \theta^3)$.*

5. Examples

5.1. Riemannian 4-symmetric spaces. Riemannian homogeneous Φ-spaces of order 4 of classical compact Lie groups were classified in [J]. By Corollary 4.3, it presents a collection of homogeneous NKf-manifolds.

5.2. The sphere S^5. It is well known (see [Sa],[Ko]) that the sphere $S^5 \cong SU(3)/SU(2)$ admits the structure of a Riemannian homogeneous Φ-space of order 4. Its canonical structure $f = \frac{1}{2}(\theta - \theta^3)$ was indicated in [BS2]. Clearly, this is NKf-structure. It can be easily shown that f is not a Killing f-structure.

5.3. Homogeneous Φ-spaces of order 8. The canonical NKf-structure described in Corollary 4.2 can be effectively constructed for homogeneous Φ-spaces of any order $k = 4n$. For example, consider the case $k = 8$. As a result, we have

THEOREM 5.1. *Let $(G/H, g)$ be a naturally reductive homogeneous Φ-space of order 8. If $\{i, -i\} \subset spec\,\theta$, then the canonical f-structure $f = \frac{1}{4}(\theta - \theta^3 + \theta^5 - \theta^7)$ is a non-trivial NKf-structure on G/H.*

5.4. The 6-dimensional generalized Heisenberg group. All necessary notions and results related to *generalized Heisenberg groups* can be found in [Ka1,2], [TV]. The 6*–dimensional generalized Heisenberg group* (N, g) is of especial interest (see [Ka2], [TV]). The brackets for the Lie algebra $\mathfrak{n} = L(x_1, x_2, x_3, x_4) \oplus L(a_1, a_2) = V \oplus Z$ were explicitly indicated in ([TV], p.111): $[x_1, x_2] = a_1$, $[x_1, x_3] = a_2$, $[x_2, x_4] = -a_2$, $[x_3, x_4] = a_1$, all the other brackets being zero.

It is known (see [TV], p.112) that (N, g) is a Riemannian homogeneous Φ-space of order 4. The corresponding isometric automorphism φ of the Lie algebra $\mathfrak{n}$ can be written in the form $\varphi : (x_1, x_2, x_3, x_4, a_1, a_2) \to (-x_4, -x_3, x_2, x_1, -a_1, -a_2)$. Since $\theta = \varphi$, for the canonical f-structure $f = \frac{1}{2}(\theta - \theta^3)$ we obtain

$$f : (x_1, x_2, x_3, x_4, a_1, a_2) \to (-x_4, -x_3, x_2, x_1, 0, 0).$$

It is easy to show that f is a metric f-structure on (N, g). Note that (N, g) is not a naturally reductive homogeneous space (see [Ka2], [TV],p.96). The Levi-Civita connection ∇ of the metric g was indicated in [Ka1]. Using this result, we can get $\nabla_{fX}(f)fX = 0$. However this f-structure is not Killing. Besides, using the results of [BD] we can show that f is non-integrable. To sum up, we have

THEOREM 5.2. *The 6-dimensional generalized Heisenberg group (N, g) is a nearly Kähler f-manifold with respect to the canonical f-structure $f = \frac{1}{2}(\theta - \theta^3)$ of a homogeneous Φ-space of order 4. This f-structure is not integrable on (N, g).*

Remark. We already mentioned in the Introduction the result of A. Gray concerning homogeneous Φ-spaces of order 3. As to homogeneous Φ-spaces of order 4 with the canonical f-structure, the situation is not completely similar. More precisely, by Corollary 4.3 any naturally reductive space of such a kind is a nearly Kähler f-manifold. However the above example illustrates that the inverse is not true.

5.5. The group of hyperbolic motions of the plane $\mathbb{R}^2$. Finally, we consider the well-known example of a 3-dimensional Riemannian homogeneous Φ-space of order 4 (see [Ko],p.18). Let

$$G = \left\{ \begin{pmatrix} e^{-c} & 0 & a \\ 0 & e^{c} & b \\ 0 & 0 & 1 \end{pmatrix} \middle| \, a, b, c \in \mathbb{R} \right\}$$

be the Lie group of hyperbolic motions of the plane $\mathbb{R}^2$. This is a solvable Lie group diffeomorphic to $\mathbb{R}^3$. The Riemannian metric g on G determined by the formula $ds^2 = e^{2c}da^2 + e^{-2c}db^2 + \lambda^2 dc^2$, $\lambda > 0$ is invariant with respect to G for any λ indicated. The automorphism Φ of the Lie group G given by the formula $\Phi : (a, b, c) \to (-b, a, -c)$ is an isometry of order 4 with the only fixed point. Hence $(G = \mathbb{R}^3(a, b, c), g)$ is a Riemannian Φ-space of order 4.

Now using the results of [Ko] and calculating as above, we obtain

THEOREM 5.3. *The canonical f-structure $f = \frac{1}{2}(\theta - \theta^3)$ on the Riemannian homogeneous Φ-space $(\mathbb{R}^3(a, b, c), g)$ of order 4 is a metric f-structure but f is neither integrable nor nearly Kähler.*

References

[B1] Balashchenko V.V., *Riemannian geometry of canonical structures on regular Φ-spaces*, Preprint No.174/1994, Fakultät für Mathematik der Ruhr-Universität Bochum, 1994, 1-19.

[BD] Balashchenko V.V., Dashevich O.V., *Geometry of canonical structures on homogeneous Φ-spaces of order* 4, Russian Math. Surveys, 1994, vol. 49, no. 4, pp. 149-150.

[BS1] Balashchenko V.V., Stepanov N.A., *Canonical affinor structures on regular Φ-spaces*, Russian Math. Surveys, 1991, vol. 46, no. 1, pp. 247-248.

[BS2] Balashchenko V.V., Stepanov N.A., *Canonical affinor structures of classical type on regular Φ-spaces*, Sbornik: Mathematics, 1995, vol. 186, no. 11, pp. 1551–1580.

[F] Fedenko A.S., *Spaces with symmetries*, Belarusian State University, Minsk, 1977 (in Russian).

[G1] Gray A., *Nearly Kähler manifolds*, J. Diff. Geom., 1970, vol. 4, no. 3, pp. 283-309.

[G2] Gray A., *Riemannian manifolds with geodesic symmetries of order* 3, J. Diff. Geom., 1972, vol. 7, no. 3–4, pp. 343–369.

[Gr] Gritsans A.S., *On construction of Killing f–manifolds*, Izv. Vyssh. Uchebn. Zaved. Mat., 1992, no. 6, pp. 49–57; English transl. in Soviet Math. (Iz. VUZ), 1992, vol. 36, no. 6.

[J] Jimenez J.A., *Riemannian 4-symmetric spaces*, Trans. Amer. Math. Soc., 1988, vol. 306, no. 2, pp. 715–734.

[Ka1] Kaplan A., *Riemannian nilmanifolds attached to Clifford modules*, Geom. Dedicata, 1981, vol. 11, pp. 127–136.

[Ka2] Kaplan A., *On the geometry of groups of Heisenberg type*, Bull. London Math. Soc., 1983, vol. 15, pp. 35–42.

[Ki1] Kirichenko V.F., *On geometry of homogeneous K-spaces*, Mat. Zametki, 1981, vol. 30, no. 4, pp. 569–582; English transl. in Math. Notes, 1981, vol. 30.

[Ki2] Kirichenko V.F., *Methods of generalized Hermitian geometry in the theory of almost contact manifolds*, Itogi Nauki i Tekhniki: Probl. Geom. vol. 18, VINITI, Moscow, 1986, pp. 25–71; English transl. in J. Soviet Math. 1988, vol. 42, no. 5.

[Ko] Kowalski O., *Generalized symmetric spaces*, LN in Math., vol.805, Berlin, Heidelberg, New York: Springer-Verlag, 1980.

[N] Nomizu K., *Invariant affine connections on homogeneous spaces*, Amer. J. Math., 1954, vol. 76, no. 1, pp. 33–65.

[Sa] Sanchez C.U., *Regular s-structure on spheres*, Indiana Univ. Math. J., 1988, vol. 37, no. 1, pp. 165–180.

[S1] Stepanov N.A., *Basic facts of the theory of φ-spaces*, Izv. Vyssh. Uchebn. Zaved. Mat., 1967, no. 3, pp. 88-95; English transl. in Soviet Math. (Iz. VUZ), 1967, vol. 11, no. 3.

[S2] Stepanov N.A., *Homogeneous 3-cyclic spaces*, Izv. Vyssh. Uchebn. Zaved. Mat., 1967, no.12, pp. 65–74; English transl. in Soviet Math. (Iz. VUZ), 1967, vol. 11, no. 12.

[TV] Tricerri F., Vanhecke L., *Homogeneous structures on Riemannian manifolds*, London Math. Soc., LN Ser., no. 83, 1983.

[WG] Wolf J.A., Gray A., *Homogeneous spaces defined by Lie group automorphisms*, J. Diff. Geom., 1968, vol. 2, no. 1–2, pp. 77–159.

FACULTY OF MATHEMATICS AND MECHANICS, BELARUSIAN STATE UNIVERSITY, F.SCORINA AV. 4, MINSK 220080, BELARUS

E-mail address: `vitaly@mmf.bsu.unibel.by`

Contemporary Mathematics
Volume **288**, 2001

An Extension of A. Gray's Investigations on Small Geodesic Balls

N. Bokan, M. Djorić, and U. Simon

Dedicated to Alfred Gray.

ABSTRACT. We modify investigations of A. Gray and other authors on small geodesic balls and consider an analytic structure (M, D, g) defined by a torsion-free, Ricci-symmetric connection D and a Riemannian metric g on an oriented and connected manifold M; D is not necessarily the Levi-Civita connection of g. D induces an exponential map $\exp^D$, while g induces a Euclidean structure on each tangent space. Under $\exp^D$, we map small balls from the tangent space to M and compute the Taylor expansion for the volume of their images. We list examples from different fields (hypersurface theory, statistical manifolds, transformations of PDEs, decomposition of curvature tensors) where such a structure (M, D, g) naturally appears and where one could apply the Taylor expansion; exemplary we sketch one application.

Introduction

The history of power series expansions of area and volume functions on Riemannian manifolds was sketched by A. Gray [**G-1**] and A. Gray and L. Vanhecke [**G-V**]; both authors and others like O. Kowalski [**KOW**] contributed to the theory, [**G-V**] contains many references. In 1982, V. Miquel [**M-1**] computed the first two non-zero terms in the expansion for the volume of small geodesic balls for a more general analytic structure (M, D, g) where the connection D has torsion and the metric g is D-parallel: Dg=0. In [**M-2**] he computed the first two non-zero terms of the Taylor expansion for the volumes of geodesic spheres.
In our paper we consider an analytic structure (M, D, g), where D is a torsion-free and Ricci-symmetric (that means its Ricci tensor is symmetric) connection which not necessarily is the Levi-Civita connection of the metric g; we do not assume any further compatibility condition for the pair (D, g). In a heuristic description, all metric notions are defined via the metric g, while – and this is the difference

2000 *Mathematics Subject Classification.* 53B05, 53B20, 53A15.
Key words and phrases. Small geodesic balls, conjugate connections, exponential map of connections, affine hypersurfaces.
We thank A. Gray, O. Kowalski, L. Vanhecke and L. Vrancken for discussions, and we thank the referee for important hints.

from the Riemannian approach – the exponential map is induced by the connection D. Before we sketch this in section 2 we would like to list the following examples (A) - (D) of such a structure (M, D, g); in the examples we admit g to be semi-Riemannian.

(A) Conjugate triples. A triple (D, g, D^*) of two symmetric connections D and D^* and a semi-Riemannian metric g is called a conjugate triple if the compatibility condition $Xg(Y,Z) = g(D_X Y, Z) + g(Y, D^*_X Z)$ is satisfied for all vector fields X, Y, Z. The compatibility condition can be equivalently reformulated in terms of Codazzi equations $(D_X g)(Y,Z) = (D_Y g)(X,Z)$ for the pair (D, g). For conjugate triples in relative hypersurface theory see e.g. [**S-S-V**], section 4. Conjugate triples also appear on hypersurfaces in space forms with Weingarten operator of maximal rank; see e.g. [**B-L-O-S-W**].

(B) Statistical manifolds. Amari introduced the notion of a statistical manifold which is equivalent to a structure given by a pair (D, g) satisfying Codazzi equations; this notion is of interest in stochastics; see e.g. [**KUR**].

(C) Codazzi transformations for PDEs. Consider a pair (D, g) satisfying Codazzi equations and with symmetric Ricci tensor of D. The metric g generates a conformal class, while D generates a projective class of Ricci-symmetric connections. Simultaneous transformation of PDEs in these classes are of interest, see [**SI**].

(D) The curvature of a torsion-free connection. This topic has been studied from the viewpoint of representation theory; for more details see e.g. [**BOK**].

Our examples motivate a further study of a structure (M, D, g). In section 1 we state elementary relations between geometric invariants of the structure (M, D, g). In section 2 we consider a metric ball $B_m(r)$ in the tangent space M_m of an arbitrary point m of M and its image (called a D-geodesic ball) under the exponential map $\exp^D$ of D. We compute the Taylor expansion of the volume of $\exp^D(B_m(r))$ in terms of local invariants of the geometry of (D, g), in particular in terms of the invariants from section 1 and curvature invariants of D. As our space is limited we give the basic ideas and state our computation of the Taylor series expansion only up to the term of second order. In section 3 we show how the volume of small D-geodesic balls determines invariants of (M, D, g). We give exemplary applications for affine hypersurfaces. Further terms and applications will appear in a forthcoming paper.

1. Elementary relations for the geometry of (M, D, g)

For the structure (M, D, g) we introduce the following notations: $\nabla(g)$ denotes the Levi-Civita connection of g, $\omega(g)$ its oriented Riemannian volume form, and ω an oriented volume form parallel to D, that means $D\omega = 0$; the existence of such a form (which then is unique up to a non-zero constant factor) is equivalent to the symmetry of the Ricci tensor Ric of D. Assume that $\omega(g)$ and ω induce the same orientation; then there exists a positive function μ such that $\omega(g) = \mu\omega$. The torsion-free connections define a $(1,2)$-tensor field $C := D - \nabla(g)$; we denote its trace by $nT(X) := tr\{Y \longrightarrow C(X,Y)\}$. The proof of the following Lemma is straightforward.

LEMMA 1.1. *(i)* $nT = -dln\mu$;
(ii) $D\omega(g) = nT \otimes \omega(g)$;
(iii) $DD\omega(g) = n(DT + nT \otimes T) \otimes \omega(g)$.

The foregoing lemma and the introduction of the difference tensor C admit it to replace the covariant differentiation in terms of D by the covariant differentiation in terms of Levi-Civita and pointwise algebraic operations using C. From this our calculations up to the second order term of the power series expansion in section 2 below are similar to the ones in the Riemannian case.

2. The power series expansion for the volume of a D-geodesic ball

Consider the analytic structure (M, D, g) as above. To avoid any possibility of a misunderstanding we introduce notations s. t. the dependence of invariants induced by D and g, resp., should be clear. The sign of the D-curvature operator R is given by $R_{XY} := D_{[X,Y]} - [D_X, D_Y]$. For $v \in M_m$, let ${}^D\xi_v$ denote the unique D-geodesic in M with ${}^D\xi_v(0) = m$ and ${}^D\xi'_v(0) = v$. We write $\exp_m^D(v) = {}^D\xi_v(1)$, provided that ${}^D\xi_v(t)$ can be defined for $t = 1$. The map $\exp_m^D$ may be defined only on a neighborhood of $0 \in M_m$, where it is a diffeomorphism. To compute power series expansions for volumes in the structure (M, D, g) we shall modify a method described in [**G-1**], [**G-2**], [**G-V**] and extend the notion of "normal coordinates" to the structure (M, D, g). Let $\{e_1, \ldots, e_n\}$ be a g-orthonormal basis of M_m. For $1 \leq j \leq n$ define a real-valued function x_j on a neighborhood of m by $x_j\left(\exp_m^D\left(\sum t_i e_i\right)\right) = t_j$. Then $(x_1, \ldots, x_n)$ is called the system of *D-normal coordinates* corresponding to $\{e_1, \ldots, e_n\}$. Consider the local Gauss basis $\partial_1, \ldots, \partial_n$ associated to a D-normal coordinate system $(x_1, \ldots, x_n)$. A *D-normal coordinate vector field* at m is a local vector field X of the form $X = \sum a_i\, \partial_i$, where the a_i's are constants. For many purposes it turns out to be far easier to work with D-normal coordinate vector fields instead of D-normal coordinates. For example, we can prove that the notion of a D-normal coordinate vector field at m does not depend on the choice of the D-normal coordinate system at m.
In case that D is not the Levi-Civita connection of g, it follows that the length $\|\left(\exp_m^D(p)\right)^{-1}\|$ is not necessarily constant if p varies along a D-geodesic σ through $m \in M$. Thus define a D-geodesic ball as follows: Let r be small enough so that the map $\exp_m^D$ is defined on a ball of radius r in the tangent space M_m. Now, let $G_m^D(r) = \exp_m^D(B_0(r))$, where $B_0(r) = \{x \in M_m |\, \|x\| \leq r\}$. We shall call $G_m^D(r)$ a D-geodesic ball with center m and radius r. Let $\mathcal{V}_m^D(r)$ denote the volume of $G_m^D(r)$ with respect to the metric g. Then

$$\mathcal{V}_m^D(r) = \int_0^r t^{n-1}\left(\int_{S^{n-1}(1)} \omega(g)(\exp_m^D(tu))du\right) dt,$$

where u varies on the unit sphere $S^{n-1}(1)$ in M_m. Let $(x_1, \ldots, x_n)$ be a D-normal coordinate system at m, and write $x_i = a_i\, r$ for $i = 1, \ldots, n$. Following Gray [**G-1**], we can expand $\omega(g)(\exp_m^D(ru))$ in a power series in r, where the coefficients of this series are homogeneous polynomials in the terms a_i, while the coefficients of the polynomials are local invariants of the structure (M, D, g). We use an obvious local notation for these invariants, the Ricci tensor of D and its D-covariant derivative, while the norm we use is defined in terms of the metric g. We compute:

$$\mathcal{V}_m^D(r) = \pi^{\frac{n}{2}}\Gamma\left(\frac{n+2}{2}\right)^{-1} r^n\Big\{1 + \frac{1}{2(n+2)}\Big[n(g^{ij}D_iT_j + n\|T\|^2) - \frac{1}{3}tr_g Ric\Big]r^2 + O(r^4)\Big\}(m).$$

3. Affine hypersurfaces as determined by the volumes of D-geodesic balls

Consider an $(n+1)$-dimensional affine space A^{n+1} with associated vector space V, and a determinant form Det fixing an oriented volume. Let $x : M^n \longrightarrow A^{n+1}$ be an analytic, locally strongly convex, embedded hypersurface with so called Blaschke structure induced by the unimodular affine normal, see e.g. [**S-S-V**], chapters 4 and 6. The convexity condition implies that the Blaschke metric g on M is Riemannian; we denote the associated Levi-Civita connection by $\nabla(g)$. The metric g, together with the connection D, induced from the affine normal, and the conormal connection D^*, define a conjugate triple, see example (A), and again we have an analytic structure (M, D, g). As in section 1, we define the invariants C and T (here called the Tchebychev form). Within the relative theories the unimodular theory of hypersurfaces (so called Blaschke theory) can be characterized by the vanishing of T. Recall that the fundamental theorem of affine hypersurface theory states that one can choose the pair (D, g) or (C, g), resp., as system of fundamental geometric invariants which determines the hypersurface modulo unimodular affine transformations of the ambient space. In terms of these fundamental invariants of the hypersurface and from further relations (in particular curvature relations) one can compute the invariants which appear in the power series expansion of $\mathcal{V}_m^D(r)$ in the foregoing section. As space is limited we omit details and just state the result.

THEOREM 3.1. *Consider an embedded, locally strongly convex, analytic hypersurface in affine space with fundamental structure (M, D, g) with induced connection D and Blaschke metric g. Denote by H the affine mean curvature. Then*

$$\mathcal{V}_m^D(r) = \pi^{\frac{n}{2}}\Gamma\left(\frac{n+2}{2}\right)^{-1} r^n\{1 + \frac{n(n-1)}{6(n+2)}Hr^2 + O(r^4)\}(m).$$

We want to demonstrate how one can use this result for geometric conclusions and give an example.

THEOREM 3.2. *Let $x_1, x_2 : M^n \longrightarrow A^{n+1}$ be two embedded, locally strongly convex, analytic hypersurfaces. For $m \in M$, consider D-geodesic balls with the same radius r around $x_1(m)$ and $x_2(m)$, denoted by $G_{m_1}^D(r)$ and $G_{m_2}^D(r)$, resp. Assume that, for all m and all sufficiently small r, the volumes of the D-geodesic balls $G_{m_1}^D(r)$ and $G_{m_2}^D(r)$ are equal. Then the affine mean curvature functions of both hypersurfaces coincide on M.*

COROLLARY 3.3. *Let $x_1, x_2 : M^n \longrightarrow A^{n+1}$ be two analytic hyperovaloids and let x_1 be an ellipsoid. As above, if the volumes of the D-geodesic balls are equal for all $m \in M$ and all sufficiently small r, then the hyperovaloids are unimodularly equivalent.*

References

[BOK] N. Bokan, *On the complete decomposition of curvature tensors of Riemannian manifolds with symmetric connection*, Rend. Circ. Mat. Palermoi, **XXXIX**, 1990, pp. 331–380.

[B-L-O-S-W] F. Brito, H.L. Liu, V.I. Oliker, U. Simon, C.P. Wang, *Polar hypersurfaces in spheres*, Geometry and topology of submanifolds, vol. IX, (F. Defever et al., eds.), World Scientific, Singapore, 1999 pp. 33–47.

[G-1] A. Gray, *The volume of a small geodesic ball of a Riemannian manifold*, Michigan Math. J., vol. 20, 1973, pp. 329–344.

[G-2] A. Gray, *Tubes*, Addison-Wesley Publ. Co., Reading, 1990.

[G-V] A. Gray and L. Vanhecke, *Riemannian geometry as determined by the volumes of small geodesic balls*, Acta Math., vol. 142, 1979, pp. 157–198.

[KOW] O. Kowalski, *Additive volume invariants of Riemannian manifolds*, Acta mathematica, vol. 145, 1980, pp. 205-225.

[KUR] T. Kurose, *Affine differential geometry and statistics*, Geometry and topology of submanifolds, vol. VIII (F. Dillen et al., eds.), 1996, pp. 12–13.

[M-1] V. Miquel, *Volumes of small geodesic balls for a metric connection*, Compositio Math., vol. 46, 1982, pp 121–132.

[M-2] V. Miquel, *Volumes of geodesic balls and spheres associated to a metric connection with torsion*, preprint.

[SI] U. Simon, *Affine Differential Geometry*, Handbook of Differential Geometry (F. Dillen, L. Verstraelen, eds.), Elsevier Science, 2000, pp. 905–961.

[S-S-V] U. Simon, A. Schwenk-Schellschmidt, H. Viesel, *Introduction to the affine differential geometry of hypersurfaces*, Lecture Notes, Science University Tokyo, 1991, ISBN 3 7983 1529 9.

[V] L. Vanhecke, *Geometry in normal and tubular neighborhoods*, Rend. Sem. Fac. Sci. Univ. Cagliari Supplemento al, vol. 58, 1988, pp. 73–176.

Department of Mathematics, University of Belgrade, PP550, 11001 Beograd, YU.
E-mail address: neda@matf.bg.ac.yu

Department of Mathematics, University of Belgrade, PP550, 11001 Beograd, YU.
E-mail address: mdjoric@matf.bg.ac.yu

FB Mathematik MA 8-3, TU Berlin, D-10623-Berlin.
E-mail address: simon@math.tu-berlin.de

Contemporary Mathematics
Volume **288**, 2001

An Energy Minimizing Family of Unit Vector Fields on Odd-dimensional Spheres

Vincent Borrelli, Fabiano Brito, and Olga Gil-Medrano

This paper is dedicated to the memory of Alfred Gray

ABSTRACT. On the sphere $\mathbb{S}^{2k+1}$ with $k \geq 2$, we construct a one-parameter family of globally defined unit smooth vector fields with energy converging to the energy of the unit radial vector field defined on the complementary of two antipodal points.

1. Introduction

Let M be a closed, n-dimensional Riemannian manifold and T^1M be the unit tangent bundle of M considered as a closed Riemannian manifold with the Sasaki metric. The energy of a unit smooth vector field $v \in \chi^1(M)$ is :

$$E(v) = \frac{1}{2}\int_M (n + \|\nabla v\|^2)\, dM,$$

where dM is the density defined by the metric and ∇ is its Levi-Civita connection.

Since for parallel v the energy attains its minimum value $E(v) = \frac{n}{2}$ vol (M), the problem of minimizing the energy E among unit vector fields becomes interesting when M does not admit a global parallel unit vector field, as it is the case for an odd-dimensional sphere. In [**1**], it was established that Hopf vector fields are the unique absolute minima in the case of $\mathbb{S}^3$. For the sphere $\mathbb{S}^{2k+1}$ $(k \geq 2)$ it was proven in [**2**] that every unit smooth vector field v (possibly with isolated singularities) satisfies $E(v) \geq \frac{1}{2}(2k+1+\frac{2k}{2k-1})$ vol $(\mathbb{S}^{2k+1}) = c_k$. Moreover the equality holds if and only if v is geodesic and $v^\perp$ is integrable and with umbilical leaves, from where the authors concluded that c_k is not attained by any globally defined smooth vector field but it is the value of the energy of the unit radial vector field

1991 *Mathematics Subject Classification.* Primary 53C20; Secondary 58E15.

The first and the third authors are grateful to the second for his invitation at the University of São Paulo. During the preparation of this paper, the first author was supported by FAPESP, proj. tem. no. 1999/02684-5. The second author was supported by CNPq Brazil. The third was supported partially by FAPESP, proc. no. 2000/07143-1 and partially by DGES Grant No. PB97-1425.

defined on the complementary of two antipodal points.

In this paper, we will outline a proof of the following theorem.

Theorem. – *There exists a family* $\{T_R \mid R \in]0,1]\}$ *of smooth globally defined unit vector fields on* $\mathbb{S}^{2k+1}$ *such that :* $\lim_{R\to 0} E(T_R) = c_k$.

Consequently, we can conclude that c_k is the infimum of the energy and that the family T_R is energy minimizing.

Variational aspects of the energy of unit vector fields have been widely studied, as can be seen in [**6**], [**7**], [**3**],[**4**] and [**5**].

2. Sketch of the proof of the Theorem

Let p be a point of $\mathbb{S}^{2k+1} \subset \mathbb{R}^{2k+2}$ and let T be the unit vector field tangent to radial geodesics from p. The field T is defined on $\mathbb{S}^{2k+1} - \{-p,p\}$. Our purpose is to construct, for each $R \in]0,1]$, a globally defined field T_R by modifying T in a neighbourhood of one geodesic joining p and $-p$.

Let us assume for simplicity that $p = (1,0,...,0)$. We represent by $(x,y) \in \mathbb{R} \times \mathbb{R}^{2k}$ the coordinates corresponding to the stereographic projection from the north pole $(0,...,0,1)$ to the plane $\{x_n = 1\}$. In this chart p has coordinates $(2,0)$ and the geodesic segment joining p and $-p$ is just $I = [-2,2] \times \{0\}$. Let E_0, E_y be two orthogonal vector fields that are unit with respect to the metric of the sphere and such that E_0 is on the direction of the x-axis and E_y is a radial field on the plane $\{x = 0\}$. Of course, since T is invariant by rotation around the x-axis it can be written as :

$$T(x,y) = a(x,\rho)E_0(x,y) + b(x,\rho)E_y(x,y),$$

where $\rho(y) = \sum_{i=1}^{2k} y_i^2$.

For each $R > 0$ we consider the open subset G_R bounded by the ellipsoide $\frac{x^2}{(2+R)^2} + \frac{\rho(y)}{R^2} = 1$. That is :

$$G_R = \{(x,y) \in \mathbb{R} \times \mathbb{R}^{2k} | d_R(x,y) < 1\}$$

where :

$$d_R(x,y) = \frac{\sqrt{R^2x^2 + \rho(y)(2+R)^2}}{R(2+R)}.$$

The field T_R will be defined such that :
- it is invariant by rotation,
- it is equal to T on the complementary of G_R.

Let $f : \mathbb{R} \to \mathbb{R}$ be a smooth function such that $0 \le f \le 1$, $f_{|]-\infty,\frac{1}{10}]} \equiv 0$ and $f_{|[\frac{9}{10},\infty[} \equiv 1$. The function $h_R = f \circ d_R$ vanishes on $G_{\frac{R}{10}}$ and is equal to 1 on the complementary of $G_{\frac{9R}{10}}$. We are going to use this function to define a smooth vector field U on the complementary of $A = \{0\} \times (B_{\frac{9R}{10}} - B_{\frac{R}{10}})$ (see fig. 2). Here $B_r = \{(0,y) | \ |y| < r\}$ where $|y|$ is the Euclidean norm (we reserve the notation $\|.\|$

for the norm of the usual metric of the sphere). Let φ be the angle of T with the horizontal axis so that $a = \cos\varphi$ and $b = \sin\varphi$. Let $G_R^+ = \{(x,y) \in G_R | x > 0\}$. On $G_R^+ - \{p\}$ the angle φ is between 0 and π. We define $U_{|G_R^+}$ to be :

$$U = \cos(h_R\varphi)E_0 + \sin(h_R\varphi)E_y.$$

According to the definition of h_R, the angle of U with the horizontal axis is 0 near the axis and is φ near the boundary of G_R. On $G_R^- = \{(x,y) \in G_R | x < 0\}$ we define U with the same formula but now the angle φ is taken to be between $-\pi$ and 0. Notice that on the set $\{x = 0\}$ our choices do not match together except for $\{0\} \times (B_{\frac{R}{10}} \cup (B_R - B_{\frac{9R}{10}}))$. Outside G_R the vector field U is taken to be equal to T. So, U is a smooth unit vector field, invariant by rotation, defined on the connected set A^c. In order to overcome this difficulty, we introduce the disc D_R (resp. $\overline{D}_R$) on the plane $\{y_2 = ... = y_{2k} = 0\}$ with center $(0, R/2)$ and radius $R/2$ (resp. $\frac{2}{5}R$) .

We are going to modify U on $C_R = (D_R - \overline{D}_R) \times \mathbb{S}^{2k-1}$ such that the resulting vector field V has the following properties :

1) V restricted to $\partial D_R \times \mathbb{S}^{2k-1}$ is U,
2) V restricted to $\partial \overline{D}_R \times \mathbb{S}^{2k-1}$ is the outward normal field.

Since V is going to be also invariant by rotation, it is enough to define it on $D_R - \overline{D}_R$. To do this, let us use polar coordinates on D_R. On these coordinates, U has the form :

$$U(r,\theta) = \cos\psi(r,\theta)E_r + \sin\psi(r,\theta)E_\theta.$$

Note that $\psi(r,\theta) \in] - \pi, \pi[$. We take V to be :

$$V = \cos(g_R\psi)E_r + \sin(g_R\psi)E_\theta$$

where $g_R(r,\theta) = f(\frac{10r}{R} - 4)$. By construction, V satisfies conditions 1) and 2).

It remains to define the vector field on $\overline{D}_R \times \mathbb{S}^{2k-1}$. Let $X \in \chi^1(\mathbb{S}^{2k-1}(1))$ and $\alpha(r) = \frac{\pi}{2} f(\frac{10r}{9R})$. On $\{(r,\theta,s) \in \mathbb{R}^{2k+1} | r < \frac{9R}{10}$ and $|s| = \frac{R}{2}\}$ we put :

$$W_R(r,\theta,s) = \cos\alpha(r)X_R(s) + \sin\alpha(r)E_r(r,\theta,s)$$

with $X_R(s) = \frac{1}{\lambda}X(\frac{2s}{R})$ where λ is the conformal factor of the metric on the sphere, that is $\lambda(x,y) = 4/(4 + x^2 + \rho)$. In particular $X_R(s) = \frac{16+R^2}{16}X(\frac{2s}{R})$.

Now T_R is defined to be W_R on $\overline{D}_R \times \mathbb{S}^{2k-1}$, V on C_R, U on $G_R - (D_R \times \mathbb{S}^{2k-1})$ and T otherwise.

To show that in fact the energy of T_R goes to c_k as R goes to zero is now a rather technical question. The details will appear in a subsequent article.

References

[1] F. Brito, *Total Bending of flows with mean curvature correction*, Diff. Geom. and its Appl. 12 (2000), 157-163.

[2] F. Brito and P. Walczak, *On the Energy of Unit vector Fields with Isolated Singularities*, Ann. Polon. Math. 73 (2000), no. 3, 269–274.

[3] O. Gil-Medrano, *Relationship between volume and energy of vector fields*, Diff. Geom. and its Appl. to appear.

[4] O. Gil-Medrano and E. Llinares-Fuster, *Second variation of Volume and Energy of vector fields. Stability of Hopf vector fields*, Math. Ann. to appear.
[5] D.-S. Han and J.-W. Yim, *Unit vector fields on spheres which are harmonic maps*, Math Z. 227 (1998), 83-92.
[6] T. Ishihara, *Harmonic sections of tangent bundles*, J. Math. Tokushima Univ. 13 (1979), 23-27.
[7] G. Wiegmink, *Total bending of vector fields on Riemannian manifolds*, Math. Ann. 303 (1995), 325-344.

V. Borrelli, Institut Girard Desargues, Université Claude Bernard, Lyon 1, 43, boulevard du 11 Novembre 1918, 69622 Villeurbanne Cedex, France,
E-mail address: `borrelli@desargues.univ-lyon1.fr`

F. G. B. Brito, Departamento de Matemática - IME - USP, Caixa Postal 66281 - CEP 05315-970, São Paulo - SP, Brazil,
E-mail address: `fabiano@ime.usp.br`

O. Gil-Medrano, Departamento de Geometría y Topología, Facultad de Matemáticas, Universidad de Valencia, 46100 Burjassot, Valencia, España,
E-mail address: `Olga.Gil@uv.es`

Contemporary Mathematics
Volume **288**, 2001

Is it Admissible to Linearize the Einstein Equation in the Presence of Matter?

Lluís Bruna and Joan Girbau

ABSTRACT. Under what mathematical conditions is it admissible to linearize the Einstein equations in Robertson-Walker cosmological models? The aim of this article is to explain and discuss some recent results on this subject, some of them contained in [**1**], [**2**], and some others not yet published.

In general relativity the equation governing gravitation is the Einstein's one: $G(g) = \chi T$, where G is the Einstein tensor of g, $G(g) = \mathrm{Ric}(g) - (1/2)Rg$, T is the stress-energy tensor and χ is a constant. The Minkowski metric η in $\mathbf{R}^4$ fulfills this equation in the vacuum (when $T = 0$). In case T is small, Einstein wrote the metric g in the form $g = \eta + h$ with h small. Then he realized that the linear terms in h in the equation $G(\eta + h) = \chi T$ were strongly related with the classical wave equation (gravitational waves). In the seventies a lot of papers appeared trying to answer the following question: under what mathematical conditions is it admissible to linearize the Einstein equation? Most of them (Y. Choquet-Bruhat and S. Deser [**3**], Y. Choquet-Bruhat, A. Fisher and J.E. Marsden [**4**], V. Moncrief [**5**], [**6**]) concerned the vacuum. But little work was done in case of non-empty spaces. Recently we have studied the linearization stability of the Einstein equation in Robertson-Walker cosmological models ([**1**], [**2**]) and proved that such a model is stable in case of null curvature and unstable in case of positive curvature. The aim of this article is to explain and discuss these results.

1. The concept of linearization stability of Einstein equation

Let's begin with the concept of linearization stability of the Einstein equation in *empty space*, introduced in the seventies. Let $V = M \times I$ be a product manifold, where M is a 3-manifold and I an $\mathbb{R}$-interval. Let g be a Lorentz initial metric on V that satisfies the Einstein equation in the vacuum $G(g) = 0$. Then one is interested in the Lorentz metrics $g' = g + h$ close to g that satisfy $G(g') = 0$. In order to solve $G(g') = 0$, it is common to linearize this equation. This means to consider the equation $(D_gG)(h) = 0$ consisting of the linear terms in h of the true

---

2000 *Mathematics Subject Classification.* Primary 83C05; Secondary 83C15 83C40.
Key words and phrases. linearization stability, Robertson-Walker, Einstein equations.

equation $G(g') = 0$,

$$G(g+h) = G(g) + (D_gG)(h) + \cdots$$

This simplification, however, only makes sense if each solution h of the linearized equation $(D_gG)(h) = 0$ is tangent to a curve $\lambda \to g'(\lambda)$ of exact solutions of the true equation $G(g') = 0$. That is, for any solution h of $(D_gG)(h) = 0$ there exists a curve $g'(\lambda)$ of true solutions (curve parametrized by a real parameter λ) such that

$$g'(\lambda) = g + h\lambda + \cdots$$

and $G(g'(\lambda)) = 0$. If this condition happens it is said (by many authors of the seventies) that the Einstein *empty* equation is linearization stable at g. In an easier context, the equation $x^2 - y^2 = 0$ in $\mathbb{R}^2$ is not linearization stable at 0 because its linearized equation is $0 = 0$ which is fulfilled by any vector of $\mathbb{R}^2$. Using this definition of *linearization stability* Y. Choquet-Bruhat and S. Deser [**3**], and Y. Choquet-Bruhat, A. Fisher and J.E. Marsden [**4**] proved that the Minwoswki metric η in $\mathbb{R}^4$ is linearization stable.

Nevertheless this definition of stability is not suitable for the Einstein equation *with matter*. In fact Einstein himself considered the Minkowski metric η (which satisfies $G(\eta) = 0$) and tried to study the equation $G(\eta + h) = \chi T$ for small h. Let ψ be the 2-covariant tensor defined in terms of h by

$$\psi_{\alpha\beta} = h_{\alpha\beta} - \frac{1}{2}(\mathrm{tr}_\eta h)\eta_{\alpha\beta}\ .$$

Einstein realized that if h satisfies the condition $\mathrm{div}_\eta \psi = 0$ then the linearized equation in h of $G(\eta + h) = \chi T$ can be written with respect to ψ in the following form:

$$-\frac{1}{2}\left(\frac{\partial^2}{\partial x_1^2} + \frac{\partial^2}{\partial x_2^2} + \frac{\partial^2}{\partial x_3^2} - \frac{\partial^2}{\partial x_4^2}\right)\psi_{\alpha\beta} = \chi T_{\alpha\beta}\ . \tag{1.1}$$

Einstein also realized that for any h there exists an appropriate *gauge* in which the condition $\mathrm{div}_\eta \psi = 0$ is fulfilled, and therefore this supplementary condition is physically irrelevant. Then Einstein gave the solutions ψ of (1.1) vanishing at infinity in terms of T. Briefly, given T close to zero, he solved the linear part $(D_\eta G)(h) = \chi T$ of the equation $G(\eta + h) = \chi T$. Remark that Einstein solved the equation $(D_\eta G)(h) = \chi T$ and not the equation $(D_\eta G)(h) = 0$ as the authors of the seventies considered. We have given a new definition of *linearization stability* for Einstein equation *with matter* which, shortly speaking, can be formulated in the following way. Let $\mathcal{F}$ be a subspace of the space of stress-energy tensors. The Einstein equation $G(g) = \chi T$ is said to be linearization stable at the initial metric g and the initial tensor T in the direction of $\mathcal{F}$, if or any $H \in \mathcal{F}$ close to zero and for any small solution h of the linearized equation $(D_gG)(h) = \chi H$, $g + h$ is *close* to a solution g' of the true equation $G(g') = \chi(T + H)$. Note that if $\mathcal{F} = 0$ and $T = 0$ one obtains the old definition. We say that the Einstein equation $G(g) = \chi T$ is linearized stable if it is stable in any direction $\mathcal{F}$ of the space of stress-energy tensors.

2. Our results for Robertson-Walker models

Let (S, g) be a 3-Riemannian manifold with constant curvature K. Let $V = S \times I$, where I is an $\mathbb{R}$-interval. Let $\tilde{g}$ be a Lorentz metric in V of the form $\tilde{g} = -dt^2 + \zeta(t)^2 g$, where t is the coordinate of I. In V we consider a perfect fluid with $u = \partial/\partial t$ as a velocity field, so that the stress-energy tensor of the fluid has the form $\tilde{T} = (\rho + p)u^* \otimes u^* + p\tilde{g}$, where u^* is the 1-form associated to u by $\tilde{g}$. Obviously, the metric $\tilde{g}$ and the tensor $\tilde{T}$ are related by the Einstein equation $G(\tilde{g}) = \chi\tilde{T}$, which gives well-known relations between ρ, p, ζ and the curvatura K of g. The Lorentz manifold $(V, \tilde{g})$ with the stress-energy tensor $\tilde{T}$ is called a Robertson-Walker model. At present our main results concerning the linearization stability or the Einstein equations for Robertson-Walker models are the following:

THEOREM 2.1. *Let $(V = S \times I\,, \tilde{g}\,, \tilde{T})$ be a Robertson-Walker model. Suppose that S is connected and simply connected. Let K be the curvature of (S, g) (constant by definition of Robertson-Walker model).*

a) *If $K = 0$ then the Einstein equation is stable under linearization at the initial metric $\tilde{g}$ and the initial stress-energy tensor $\tilde{T}$.*
b) *If $K > 0$ then there is no subspace $\mathcal{F}$ in the space of stress-energy tensors such that the Einstein equation (at the initial metric $\tilde{g}$ and the initial stress-energy tensor $\tilde{T}$) is stable in the direction of $\mathcal{F}$.*
c) *If $K < 0$ then the Einstein equation is stable under linearization (at the initial metric $\tilde{g}$ and the initial stress-energy tensor $\tilde{T}$).*

Of course this is a simplification of our true statement. Precise definitions of the spaces of metrics and the spaces of stress-energy tensors should be given in order to define accurately the concept of linearization stability and to state the theorem in a precise way. The proof of a) is contained in [**1**]. The proof of b) is contained in [**2**]. We have not yet published the proof of c). It is based on an unpublished result on functional analysis due to Joaquim Bruna concerning the Laplace-Beltrami operator associated to the Poincaré metric in the hyperbolic space of dimension 3.

REMARK 2.2. (Remark concerning Physics) Imagine that you replace the real universe by a particular Robertson-Walker model. Gravitation due to any particular star is not furnished by the model. If one wishes to study a gravitational wave produced by a distant supernova explosion, for example, one is led to consider a small perturbation $\delta\tilde{T}$ (produced by the explosion) of the stress-energy tensor $\tilde{T}$ of the Robertson-Walker model. This perturbation $\delta\tilde{T}$ gives rise to another perturbation $\delta\tilde{g}$ of the metric $\tilde{g}$. Such perturbations are related by the Einstein equation $G(\tilde{g}+\delta\tilde{g}) = \chi(\tilde{T}+\delta\tilde{T})$. To study the perturbation $\delta\tilde{g}$ of $\tilde{g}$, physicists might usually be tempted to deal with the linearized Einstein equation instead of the true equation. Our theorem says that if the manifold (S, g) of the Robertson-Walker model $V = S \times I$ is connected and simply connected this procedure is admissible when the curvature of S is ≤ 0, but it is not admissible when the curvature is positive.

REMARK 2.3. (Remark concerning Mathematics) The above theorem concerns models $V = S \times I$ with S connected and simply connected. When S is not simply connected the behaviour of the model with regard to the linearization stability of the Einstein equation may be very different and it is not due to the sign of the curvature as one might suppose, but to the existence of many or few Killing vector fields in the 3-manifold (S, g), in a similar way that the results obtained by A. Fisher, J.E. Marsden and V. Moncrief in the seventies for the vacuum spacetime. To make this behaviour clear we have also proved the following result:

THEOREM 2.4. *Let* $(V = S \times I\,, \tilde{g}\,, \tilde{T})$ *be a Robertson-Walker model.*

(1) *Suppose that* (S, g) *is an euclidean torus of dimension 3. Then the Einstein equation is not stable under linearization for any subspace* $\mathcal{F}$ *of the space of the stress-energy tensors.*

(2) *Suppose that* (S, g) *is a double torus of dimension 3 endowed with a metric* g *of curvature* $K = -1$. *Then the Einstein equation is stable under linearization (at the initial metric* $\tilde{g}$ *and the initial stress-energy tensor* $\tilde{T}$*).*

References

[1] **L. Bruna and J. Girbau**, Linearization stability of the Einstein equation for Robertson-Walker models (I). Journal of Math. Phys. 40, 5117-5130 (1999).

[2] **L. Bruna and J. Girbau**, Linearization stability of the Einstein equation for Robertson-Walker models (II). Journal of Math. Phys. 40, 5131-5137 (1999).

[3] **Y. Choquet-Bruhat and S. Deser**, On the stability of flat space. Ann. Phys. 81, 165-168 (1973)

[4] **Y. Choquet-Bruhat, A. Fischer and J.E. Marsden**, Équations des contraintes sur une variété non compacte. Comptes Rendus Acad. Sci. Paris, Ser I: Math. 284, A975-A978 (1977)

[5] **V. Moncrief**, Space-time symmetries and linearization stability of the Einstein equations (I), J. Math. Phys. 16, 493-498 (1975).

[6] **V. Moncrief**, Space-time symmetries and linearization stability of the Einstein equations (II), J. Math. Phys. 17, 1893-1902 (1976).

DEPARTAMENT DE FÍSICA APLICADA, ETS D'ENGINYERS DE TELECOMUNICACIONS,, UPC, C/JORDI GIRONA S/N, 08034-BARCELONA, SPAIN

E-mail address: lbruna@pie.xtec.es

DEPARTAMENT DE MATEMÀTIQUES, UAB, 08193 BELLATERRA, BARCELONA, SPAIN

E-mail address: girbau@mat.uab.es

Contemporary Mathematics
Volume **288**, 2001

Solutions of some conformally invariant equations of the first order

Jarolím Bureš

This paper is dedicated to the memory of A.Gray.

ABSTRACT. First order conformally invariant elliptic equations are related to some irreducible representation spaces of the group $Spin(m)$ with weight λ on an oriented riemannian spin manifold M. On the flat space R^m the structure of special (polynomial type) solutions of equations for spinor-valued differential forms and symmetric spinor-valued tensor fields is described.

1. Introduction

Let M be an oriented conformal spin m-manifold. There is a set of conformally invariant differential operators naturally defined on M. The simplest first order invariant operator is the Dirac operator, other first order invariant operators can be constructed from twisted Dirac operators. Harmonic spinors (solutions of the Dirac equation) are studied intensively, only few facts are known about the properties of solutions of the other equations.

A special system $\mathcal{E}l$ of the first order invariant operators consists of elliptic operators $D_\lambda : \Gamma(M, V_\lambda) \to \Gamma(M, V_\lambda)$ which have source and target in the same space of sections. The system $\mathcal{E}l$ can be indexed by a special set Λ^{el} of weights of $spin(m)$, dominant weights for m odd, and sums of the related dominant weights for m even (for details see[**3,4**]).

The set Λ^{el} is ordered by the order defined on the weight space. For any $\lambda \in \Lambda^{el}$ the set $B(\lambda) := \{\lambda' \in \Lambda^{el}; \lambda' < \lambda\}$ is finite. The study of properties of the space of solutions of equations $D_\lambda\psi = 0$ (called spin λ-fields) is an interesting and open problem. The main question of the paper is to find a relation between solutions of D_λ and $D'_\lambda, \lambda' \in B(\lambda)$ on M using other (twistor type) invariant operators ([**3,6**]) $D_{\lambda',\lambda} : \Gamma(M, V_{\lambda'}) \to \Gamma(M, V_\lambda)$ and properties of Bernstein-Gelfand-Gelfand sequences.

A conformal spin structure on a (real, smooth) manifold M is usually given by a choice of Riemannian metric g from the conformal class, together with the spin

1991 *Mathematics Subject Classification.* Primary 58G25, 53C25; Secondary 43A85,.
Key words and phrases. Conformallly invariant operators, polynomial solutions.
The author was supported by grant GAUK No. 247/2000 and by VZ MSM 113200007

structure we have the principal fibre bundle $\tilde{P}$ with structural group $Spin(n)$ on M. Consider the Levi-Civita connection for (M, g) and its covariant derivative ∇_λ on the associated vector bundle $V_\lambda = \tilde{P} \times_{\tilde{H}} \mathbb{V}_\lambda$, and projection $\pi_{\lambda\mu} : \Gamma(V_\lambda \otimes T^*(M) \to V_\mu$ into irreducible component, we can define $D_{\lambda\mu} := \pi_{\lambda\mu} \circ \nabla_\lambda$.

It can be shown ([**8**]) that if the conformal weight is chosen properly, the operator just constructed does not depend on the choice of the metric in the conformal class.

1.1. Elliptic operators of the first order.

A list of elliptic first order operators in even, resp. odd dimensions is presented in [**3**], by means of a suitable twisted Dirac operators as follows.

Let $\mathbb{S}$ (for $m = 2k+1$), resp. $\mathbb{S} = \mathbb{S}^+ \oplus \mathbb{S}^-$ (for $m = 2k$), denote the basic spinor representations with highest weights $\sigma = (\frac{1}{2}, \dots, \frac{1}{2}, \frac{1}{2})$, resp. $\sigma^\pm = (\frac{1}{2}, \dots, \frac{1}{2}, \pm\frac{1}{2})$.

Let $\lambda \in \Lambda^+$, (for $m = 2k+1$), resp. $\lambda^\pm \in \Lambda^+$ (for $m = 2k$) be dominant weights with $\lambda_k = (\lambda_1, \dots, \lambda_{k-1}, \frac{1}{2})$, resp. $\lambda^\pm = (\lambda_1, \dots, \lambda_{k-1}, \pm\frac{1}{2})$. Denote further $\lambda' = \lambda - \sigma \in \Lambda^+$, resp. $\lambda' = \lambda^+ - \sigma^+ \in \Lambda^+$. In even dimensions, we shall use the notation $\mathbb{V}_\lambda = \mathbb{V}_{\lambda^+} \oplus \mathbb{V}_{\lambda^-}$.

The representation $\mathbb{V}_\lambda$ appears with multiplicity one in the decomposition of the tensor product $\mathbb{S} \otimes \mathbb{V}_{\lambda'}$ (it is the highest weight part of the product, i.e. the Cartan product of both representations). Hence we can write the product as $\mathbb{S} \otimes \mathbb{V}_{\lambda'} = \mathbb{V}_\lambda \oplus \mathbb{W}$, where $\mathbb{W}$ is the sum of all other irreducible components in the decomposition. Let $D_T(\lambda')$ be the twisted Dirac operator on $S \otimes V_{\lambda'}$. We have a decomposition :

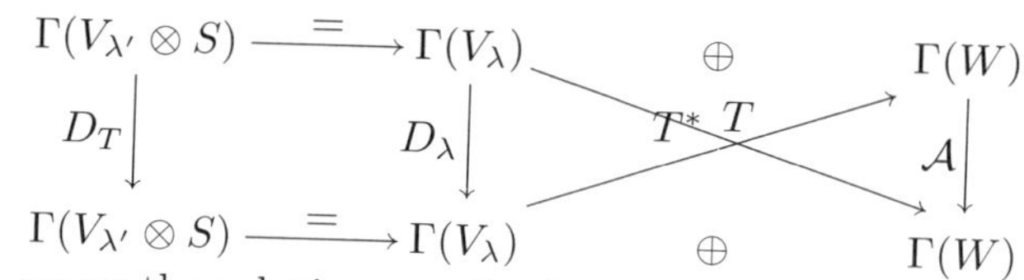

In the present paper the solutions on the flat space will be studied. On a general riemannian spin manifold the curvature of a connection and its components (Weyl, Ricci and scalar curvatures) is a natural obstruction for (anti-)commutativity of diagrams of operators, as well as for the existence of solutions of $D_\lambda\psi = 0$ equations. Some results for curved cases (mainly for Einstein metrics and self-dual 4-dimensional manifolds) are also available, they will appear in a future paper.

2 The flat case

Let us consider as a basic example the flat space $M = R^{m+1}$ with the standard euclidean metric and flat conformal structure. In the rest of the paper the standard euclidean metric will be used as representative metric in the conformal class and conformal behavior (invariance) of the solutions follows from the general theory. One of main related analytic problems is to define and study analogy of Taylor (Fourier) series for the corresponding fields with center in the origin $O \in R^{m+1}$, which are given by polynomial solutions of the corresponding equations. Polynomial solutions (of the fixed degree) of the equations on flat space (called polynomial λ-fields) are finite-dimensional representation spaces of the group $Spin(m)$ which acts on R^{m+1} in the standard way fixing 0, and can be decomposed into irreducible pieces. There is an idea of existence of an inductive procedure how to get all polynomial λ-fields mainly from the polynomial λ'-fields, with $\lambda' \le \lambda$ and the rest from the eigenspaces of the induced operators on the unit sphere $S^{m-1} \subset R^m$. This problem is solved for the Rarita-Schwinger fields ($\lambda = (3/2, 1/2, ..., 1/2)$ in the paper [**2**] and for the case of the solutions of symmetric analogues of Rarita-Schwinger equations ($\lambda = (k + 1/2, 1/2..., 1/2)$ by induction on k in two related papers [**4,5**], and also by (finite) induction procedure for the case of monogenic differential s-forms ($\lambda = (3/2, ..., 3/2, 1/2...1/2)$, s-times 3/2) with $s < m/2$ in [**6**]. In all those cases there is a possibility to use Clifford analysis as an alternative tool for description of polynomial solutions of the corresponding equations [**5**]. In some cases thus explicit analytic expressions for solutions are available.

2.1. **The symmetric analogues of the Rarita-Schwinger equations.** Let us consider a integral dominant weight $\lambda'_j = (j, 0, \dots, 0)$. Let us consider the tensor product $\mathbb{V}_{\lambda'_j} \otimes \mathbb{S}$, where $\mathbb{S}$ is the basic spinor representation (in even dimension, $\mathbb{S} = \mathbb{S}^+ \oplus \mathbb{S}^-$). The twisted Dirac operator $D_T(j)$ is defined on sections of $V_{\lambda'_j} \otimes S$. The twisting bundle is $V_{\lambda'_j} = \tilde{P} \times_{\tilde{H}} \mathbb{V}_{\lambda'_j}$ with its associated covariant derivative. The splitting $\mathbb{V}_{\lambda'_j} \otimes \mathbb{S} \simeq \mathbb{V}_{\lambda_j} \oplus \mathbb{V}'_{\lambda_{j-1}}$ then induces the splitting of the associated vector bundles and it leads to the following block diagram.

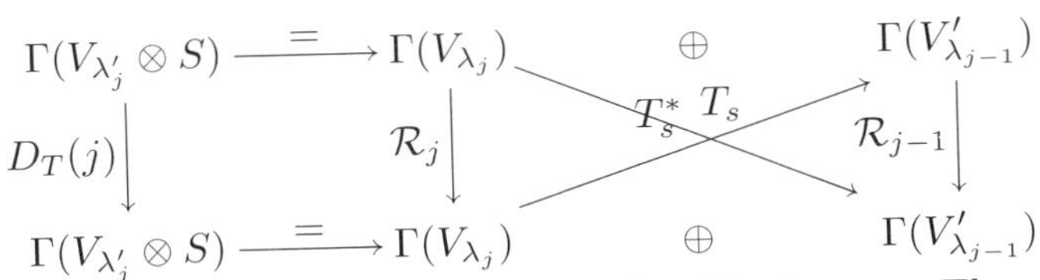

The four operators $R_j, T_j, T_j^*, R'_{j-1}$ are defined by this diagram. The modules $\mathbb{V}'_{\lambda_{j-1}}$ are canonically isomorphic to $\mathbb{V}_{\lambda_{j-1}}$.

The structure of all polynomial solutions of the symmetric Rarita-Schwinger equation $R_j(\psi) = 0$ is described in the following theorem.

THEOREM 1. ([**4**]) *Suppose that $k \geq j$. The space $\mathcal{M}_j(k)$ of all polynomial solutions of degree k of the equation $\mathcal{R}_j(\psi) = 0$ can be written as a direct sum of $\binom{j+2}{2}$ subspaces $\mathcal{M}_j^{ps}(k)$; $p, s = 0, \dots, j$; $p + s \leq j$. Each of these subspaces is an irreducible Spin(m)-module with the highest weight $(\frac{2k+1+2s-2p}{2}, \frac{2j+1-2p-2s}{2}, \frac{1}{2}, ..., \frac{1}{2})$.*

For any $p > 1$, the Laplace operator Δ maps the space $\mathcal{M}_j^{ps}(k+2)$ onto the space $\mathcal{M}_j^{p-1,s}(k)$ (this is an isomorphism of Spin(n)-modules).

Let us remark, that in terms of irreducible components the theorem is saying that $\{\psi \in \mathcal{M}_j(k) | \Delta^l(\psi) = 0\} = \sum_{p \leq l, s=0}^{j-p} \mathcal{M}_j^{ps}(k)$.

We have a decomposition of the space of all solutions into three disjoint types.

$$\mathcal{M}_j(k) = \mathcal{M}_j^A(k) \oplus \mathcal{M}_j^{B1}(k) \oplus \mathcal{M}_k^{B2}(k).$$

Solutions of different types will be constructed by three different procedures. Solutions ψ of $R_j(\psi) = 0$ of type A and $B1$ will be obtained by induction from solutions of $R_{j-1}(\phi) = 0$. In the first case, by solving the equation $T_{j-1}^*(\psi) = \iota(\phi)$; in the second case by using the image of the twistor operator $\psi = T_j(\phi)$. Note that in the first case, $\psi \in \mathcal{M}_j(k)$ implies that $\phi \in \mathcal{M}_{j-1}(k-1)$, while in the second case $\psi \in \mathcal{M}_j(k)$ implies that $\phi \in \mathcal{M}_{j-1}(k+1)$.

The space $\mathcal{M}_j^A(k)$ of solutions of type A is the space of all pieces with exception of all solutions of the Laplace equation, i.e.

$$\mathcal{M}_j^A(k) = \oplus_{p,s,p \neq 0} \mathcal{M}_j^{ps}(k).$$

The space $\mathcal{M}_j^{B1}(k)$ of solutions of type $B1$ is the space of all solutions of the Laplace equation with exception of the bottom piece, i.e.

$$\mathcal{M}_j^{B1}(k) = \oplus_{s>0}\mathcal{M}_j^{0s}(k).$$

Finally, the space of solutions of type B2 is just one piece

$$\mathcal{M}_j^{B2}(k) = \mathcal{M}_j^{00}(k)$$

in the decomposition. Its construction has to use different tools (its restriction to the sphere [**4**]).

2.2 **The monogeneity operators for spinor valued forms.**
Let us consider $\lambda = (3/2, ...3/2, 1/2...1/2)$, s-times $3/2$, the case of special subspace of spinor-valued forms. There is the following diagram:

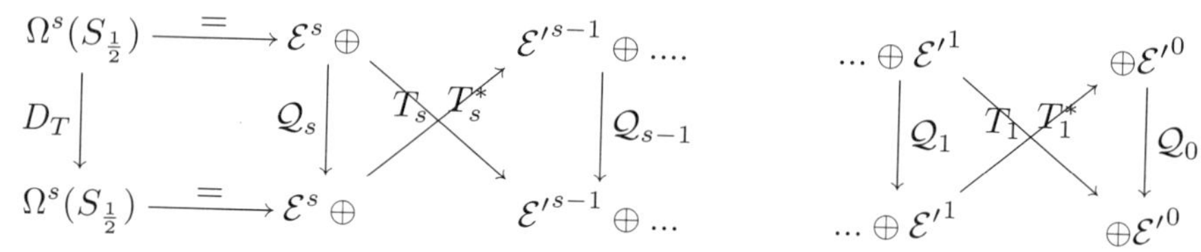

Let us consider the case for general $s < m/2$. We would like to find a procedure how to find a structure of homogeneous solutions of $\mathcal{Q}_s = 0$. As above it can be possible to get some types of polynomial solutions of $\mathcal{Q}_s = 0$ from the polynomial solutions of the equation $\mathcal{Q}_{s-1} = 0$ and one type of solutions from special eigenspace of induced operator on the sphere [**6**]. There is canonical invariant isomorphism $\iota_p : \mathcal{E}^p \to \mathcal{E}'^p$.

Let us introduce the following short notation for the representation spaces of $Spin(m)$: $[k, q] = (\frac{2k-1}{2}, \frac{3}{2}, ..\frac{3}{2}, ..., \frac{1}{2})$ where $\frac{3}{2}$ appears q-times starting from the second place, and $\frac{1}{2}$ on the remaining places. For $q < 0$ we set $[k, q] = 0$.

Let us remark that there exists an invariant scalar product on the space of spinor-valued differential forms, denote $U^\perp$ the ortogonal complement to U with respect to this product. Moreover let us define the operator $\mathcal{L}(\sum_{|I|=k} \psi_I \otimes dx^I) = \sum_I x_i\psi_I \otimes \iota(\frac{\partial}{\partial x_i})dx^I$.

THEOREM 2. *Suppose in the following $k \geq s$. The space of polynomial solutions of the equation $Q_s\psi = 0$ degree k has a decomposition as representation space of $Spin(m)$ of the form*

$$\mathcal{M}_k(s) \simeq [k+1, s-2] \oplus [k+2, s-1] \oplus [k, s-1] \oplus [k+1, s].$$

Remark : If we need to consider $[p, q]$ as subspace of $\mathcal{M}_k^s$ we shall use the more precise notation $[p, q]_s^k$. The construction is done by induction, it follows the method used in [**2**]. There are again three types of the homogeneous solutions of degree k of $Q_s\psi = 0$.

Type A : The space $\mathcal{M}_k^A(s)$ consists of solutions of the equation $D_T\psi = \iota_{s-1}(\phi)$ with $\phi \neq 0$. Then $Q_{s-1}(\phi) = 0$ and the map $\mathcal{T}^*$ is an isomorphism from the space $\mathcal{M}_k^A(s) := (Ker\mathcal{T}_s^*)^\perp$ onto the space $[k+1, s-2]_{s-1}^{k-1} \oplus [k, s-1]_{s-1}^{k-1}$. Let us denote $\mathcal{T}^{*-1}$ the inverse mapping defined on the orthogonal complement.

Type B. This type is characterized by the condition $\phi = 0$ and consists of solutions of the whole twisted Dirac equation.

There are still two subcases available :

Type B1: This type is characterized by conditions $\mathcal{L}(\psi) = 0$ and the corresponding space of k-homogeneous solutions $\mathcal{M}_k^{B1}(s)$ corresponds to the eigenspace of induced operator Q_s^S on the unit sphere and induces solutions of $Q_s\psi =$) of the type $[k+1, s]$. The correspondence is done by restriction of the solution ψ to the unit sphere.

Type B2: There is $\mathcal{L}(\psi) \neq 0$, $Q_{s-1}\mathcal{L}(\psi) = 0$, The corresponding space $\mathcal{M}_k^{B1}(s)$ is a part of the image of the twistor operator $\mathcal{T}_s$. Namely the map $\mathcal{T}_s$ maps the space $[k+2, s-1]_{s-1}^{k+1}$ isomorphically onto $\mathcal{M}_k^{B2}(s)$ and the space $[k+1, s-2]_{s-1}^{k+1}$ isomorphically onto $[k+1, s-2]_s^k$.

References

[1] T.Branson, *Stein-Weiss operators and ellipticity*, Journal of Funct. Analysis **151** (1997), 334-383.

[2] J.Bureš, *The Rarita-Schwinger operator and spherical monogenic forms*, Complex Variables **43** (2000), 77-108.

[3] J.Bureš , V.Souček, *Eigenvalues of conformally invariant operators on spheres*, Proceedings of 18th Winter School "Geometry and Physics", vol. 59, Suppl. Rend. Circ. Mat. Palermo, 1999, pp. 109-122.

[4] J.Bureš, P.Van Lancker, F.Sommen, V.Souček, *Symmetric analogies of Rarita-Schwinger equations*, to appear in Journal of Global Analysis and Geometry.

[5] J.Bureš, P.Van Lancker, F.Sommen, V.Souček, *Rarita-Schwinger type operators in Clifford Analysis*, to appear in Journal of Funct. Analysis.

[6] J.Bureš, *Monogenic forms of the polynomial type*, to appear in Proc. of the Int. Conference on Clifford analysis, 2000.

[7] A. Čap, J.Slovák , V.Souček, *Invariant operators for almost hermitian symmetric structures*III, Diff. Geom. Appl. **12** (2000), 51-84.

[8] H.D.Fegan, *Conformally invariant first order operators*, Quat. J. of Math. **27** (1976), 371-378.

[9] V.Severa, *Invariant Differential operators on spinor-valued forms*, PhD thesis (1998).

MATHEMATICAL INSTITUTE, CHARLES UNIVERSITY, SOKOLOVSKA 83 18675 PRAGUE, CZECH REPUBLIC

E-mail address: jbures@karlin.mff.cuni.cz

Contemporary Mathematics
Volume **288**, 2001

On Biharmonic Maps

Renzo Caddeo, Stefano Montaldo, and Paola Piu

In memory of Alfred Gray

ABSTRACT. Biharmonic maps are critical points of the bienergy functional, and generalize the harmonic maps. In this note we consider some general aspects of the theory of biharmonic maps and we give new examples of non-geodesic biharmonic curves.

1. Introduction

Since their first work on harmonic maps, J. Eells and J.H. Sampson in [**ES**] suggested the idea of studying k-harmonic maps. For $k = 2$, the idea is the following. First define harmonic maps $\phi : (M, g) \to (N, h)$ between two Riemannian manifolds as critical points of the energy $E(\phi) = \frac{1}{2}\int_M |d\phi|^2\, dv$. The corresponding Euler-Lagrange equation for the energy is given by the vanishing of the tension field $\tau = \operatorname{trace} \nabla d\phi$. Then define the *bienergy* of a map ϕ by $E_2(\phi) = \frac{1}{2}\int_M |\tau|^2\, dv$, and say that ϕ is *biharmonic* if it is a critical point of the bienergy.

In [**Ji**$_1$, **Ji**$_2$] G.Y. Jiang derived the first variation formula of the bienergy showing that the Euler-Lagrange equation for E_2 is

$$\tau_2 = J(\tau) = 0, \tag{1.1}$$

where J is the Jacobi operator of ϕ. The equation $\tau_2 = 0$ will be called the *biharmonic equation.*

In a different setting, in [**Ch**], B.Y.Chen defined biharmonic submanifolds of the Euclidean space as those with harmonic mean curvature vector, and stated the following

Conjecture: *Any biharmonic submanifold of the Euclidean space is harmonic, i.e. minimal.*

If we consider the biharmonic equation $\tau_2 = 0$ for isometric immersions into the Euclidean space we recover Chen's notion of biharmonic submanifolds, so the two definitions agree.

2000 *Mathematics Subject Classification.* 58E20.
Key words and phrases. Harmonic maps, Biharmonic.
The first author was supported by 60% Italian M.U.R.S.T. funds.
The second author was supported by "Contratto giovani ricercatori", University of Cagliari.
The third author was supported by G.N.S.A.G.A., Italy.

The advantage in using the Eells-Sampson definition is that it works for all Riemannian manifolds, and if the target is not the Euclidean space Chen's conjecture is not true, that is, there exist examples of nonharmonic biharmonic submanifolds. For example, it was observed in Jiang paper that the generalized Clifford torus

$$\mathbb{S}^p(\tfrac{1}{\sqrt{2}}) \times \mathbb{S}^q(\tfrac{1}{\sqrt{2}}) \hookrightarrow \mathbb{S}^{m+1}$$

with $p+q=m$ and $p \neq q$ is a biharmonic nonminimal submanifold of $\mathbb{S}^{m+1}$. In **[CMO]** C. Oniciuc and the first two authors have produced many other examples of biharmonic nonharmonic submanifolds of the round sphere and some partial classifications. In this last case, therefore, the biharmonic theory seems to be more interesting.

In this note we restrict our attention to curves in a surface giving examples of nongeodesic biharmonic curves.

Notation. We shall place ourselves in the C^∞ category, i.e. manifolds, metrics, connections, maps will be assumed to be smooth. By (M^m, g) we shall mean a connected manifold, of dimension m, without boundary, endowed with a Riemannian metric g. We shall denote by ∇ the Levi-Civita connection on (M,g). For vector fields X, Y, Z on M we define the Riemann curvature operator by $R(X,Y)Z = \nabla_{[X,Y]}Z - [\nabla_X, \nabla_Y]Z$.

Acknowledgements. The authors wish to thank the organizers of the "International Congress on Differential Geometry in memory of Alfred Gray" for their exquisite hospitality and the opportunity of presenting this lecture.

2. Biharmonic maps

Let M be a submanifold of $\mathbb{R}^n$ and denote by $\mathbf{i} : M \hookrightarrow \mathbb{R}^n$ the canonical inclusion and by H the mean curvature vector. Regarding H as a map $H \in C^\infty(M, \mathbb{R}^n)$, $H = (H_1, \cdots, H_n)$ we have

DEFINITION 2.1 ([**Ch**]). A submanifold $M \subset \mathbb{R}^n$ is *biharmonic* if

$$\Delta H = (\Delta H_1, \cdots, \Delta H_n) = 0$$

where Δ is the Beltrami-Laplace operator of the metric induced on M by $\mathbf{i}$.

The terminology is justified by the following identities:

$$\Delta H = \Delta(\frac{-1}{m} \Delta \mathbf{i}) = \frac{-1}{m} \Delta^2 \mathbf{i};$$

so M is biharmonic if and only if $\Delta^2 \mathbf{i} = 0$.

It would be of course interesting to find examples of biharmonic nonharmonic submanifolds. Unfortunately such submanifolds could not exist, as the Chen conjecture stated in the introduction suggests. As yet the conjecture has not been either proved or disproved, although some positive answers are known, namely for

- *curves in* $\mathbb{R}^n$ [**Di**];
- *hypersurfaces in* $\mathbb{R}^n$ *with at most two different principal curvatures* [**Di**];
- *conformally flat hypersurfaces in* $\mathbb{R}^4$ [**Ga**] - *hypersurfaces in* $\mathbb{R}^4$ [**HV**];
- *submanifolds of finite type in* $\mathbb{R}^n$ [**Di**].

Let's now consider the Eells–Sampson definition. As we have stated in the introduction, for a map $\phi : M \to N$, Jiang derived the first variation formula for the bienergy

$$E_2(\phi) = \frac{1}{2} \int_M |\tau|^2 \, dv,$$

proving that a map ϕ is biharmonic (in the sense of Eells–Sampson) if and only if

$$\tau_2 = J^{\phi}(\tau) = \Delta^{\phi}\tau - \operatorname{trace} R^{N}(d\phi, \tau)d\phi = 0,$$

where $\Delta^{\phi} = -\operatorname{trace} \nabla\nabla^{\phi}$ is the Laplacian on sections of $\phi^{-1}TN$.
We have the following properties of biharmonic maps:

- **(A)** ϕ is biharmonic if its tension field is in the kernel of the Jacobi operator;
- **(B)** a harmonic map ($\tau = 0$) is obviously a biharmonic map;
- **(C)** a harmonic map is an absolute minimum of the bienergy;
- **(D)** if M is compact and $Riem^{N} \leq 0$, then $\phi : M \to N$ is biharmonic if and only if it is harmonic [**Ji**$_2$];
- **(E)** if $\phi : M \to N$ is an isometric immersion with $|\tau| = constant$ and $Riem^{N} \leq 0$, then ϕ is biharmonic if and only if it is harmonic [**On**].

If now we consider the inclusion $\mathbf{i} : M \hookrightarrow \mathbb{R}^n$, the tension field is given by $\tau = mH$ and we deduce from (**E**):

1. if $|H| = constant$, then $\mathbf{i}$ is biharmonic if and only if it is minimal;
2. if $|H| \neq constant$, then $\mathbf{i}$ is biharmonic if and only if $\Delta H = 0$.

The two properties (**D**) and (**E**) suggest the following
Generalized Chen's Conjecture: *The only biharmonic submanifolds of a manifold N with $Riem^{N} \leq 0$ are the minimal ones.*

It is now clear that, in order to find examples of nonharmonic biharmonic submanifolds, a fair candidate for the ambient space is a manifold with **positive** sectional curvature.

3. Biharmonic curves

Let (M^2, g) be a 2-dimensional Riemannian manifold and let $\gamma : I \to (M^2, g)$ be a differentiable curve parametrized by arc length. Let $\{T, N\}$ be an orthonormal frame field tangent to M^2 along γ, where $T = \dot{\gamma}$ is the unit tangent vector field of γ. Then we have the following Frenet equations

$$\nabla_T T = k_g N\,; \qquad \nabla_T N = -k_g T\,,$$

where $k_g = |\tau_\gamma| = |\nabla_{\dot{\gamma}}\dot{\gamma}|$ is the geodesic curvature of γ. Using these Frenet equations the Euler-Lagrange equation of the bienergy can be written

$$\begin{aligned}\tau_2(\gamma) &= -\nabla^3_T T - R(T, k_g N)T \\ &= 3k_g \dot{k}_g T - (\ddot{k}_g - k_g^3 + k_g G)N \\ &= 0\,,\end{aligned}$$

where $G = R(T, N, T, N)$ is the Gauss curvature of the surface. It follows that γ is a biharmonic curve if and only if

$$\begin{cases} k_g \dot{k}_g = 0 \\ \ddot{k}_g - k_g^3 + k_g G = 0 \end{cases}$$

If we look for nongeodesic solutions, that is for curves with $k_g \neq 0$, we obtain

$$\begin{cases} k_g = constant \neq 0 \\ k_g^2 = G \end{cases}$$

Thus we have:

PROPOSITION 3.1. *Let $\gamma : I \to (M^2, g)$ be a nongeodesic biharmonic curve in a surface M^2. Then, along γ the Gauss curvature is constant, positive and equal to the square of the geodesic curvature of γ. Therefore, if M^2 has nonpositive Gaussian curvature, any biharmonic curve is a geodesic of M^2.*

This gives a positive answer to the generalized Chen's conjecture, that is, *any biharmonic curve in a surface with nonpositive Gauss curvature is a geodesic.*

REMARK 3.2. By means of a similar argument one can see that the generalized conjecture is also true for a curve in a 3-dimensional manifold with nonpositive **constant** sectional curvature (for more details see [**CMO**]).

4. Biharmonic curves on a surface of revolution

Let $\alpha(u) = (f(u), 0, g(u))$ be a curve in the xz-plane and consider the surface of revolution obtained by rotating this curve around the z-axis with the standard parametrization

$$X(u,v) = (f(u)\cos(v), f(u)\sin(v), g(u)),$$

where v is the rotation angle around the z-axis. If α is parametrized by arc length, so that $\dot{f}^2 + \dot{g}^2 = 1$, the Gauss curvature of the surface of revolution is given by

$$G(u,v) = -\frac{\ddot{f}(u)}{f(u)}.$$

REMARK 4.1. The Gauss curvature G depends only on u, that is, G is constant along a parallel. This implies that if the Gauss curvature is constant along a curve, then either the curve is a parallel or the curve lies in a part of the surface with constant Gauss curvature. In particular a nongeodesic biharmonic curve is either a parallel or lies on the surface with constant Gauss curvature.

Directly from Proposition 3.1 we have

PROPOSITION 4.2. *A parallel $u = u_0 =$ constant is biharmonic if and only if u_0 satisfies the equation*

$$\dot{f}^2(u_0) + \ddot{f}(u_0)f(u_0) = 0.$$

For example, in a torus of revolution with its standard parametrization $X(u,v) = ((a + r\cos u)\cos v, (a + r\cos u)\sin v, r\sin u)$, $a > r$, the biharmonic parallels are

$$u = \arccos\left(\frac{-a \pm \sqrt{a^2 + 8r^2}}{4r}\right)$$

For more examples see [**CMP**].

EXAMPLE 4.3 (Sphere). There is a geometric way of understanding the behaviour of biharmonic curves in a sphere. In fact, for a curve α in $\mathbb{S}^2$ the following equation holds

$$\tau k^2 = \frac{1}{r}\dot{k}_g, \tag{4.1}$$

where τ and k are the torsion and curvature of α as seen from the ambient space $\mathbb{R}^3$. Equation (4.1) says that if α is a biharmonic curve, then $\tau k^2 = 0$, and since $k^2 = k_n^2 + k_g^2 = \frac{2}{r^2} = const. \neq 0$ we must have that $\tau = 0$. Thus, a biharmonic curve on a sphere of radius r is the intersection of the sphere with a plane such that the resulting circle has radius $\frac{r}{\sqrt{2}}$. Now, using the local isometries between the sphere and the other two types of surfaces of revolution with constant positive Gauss curvature, we get the following picture

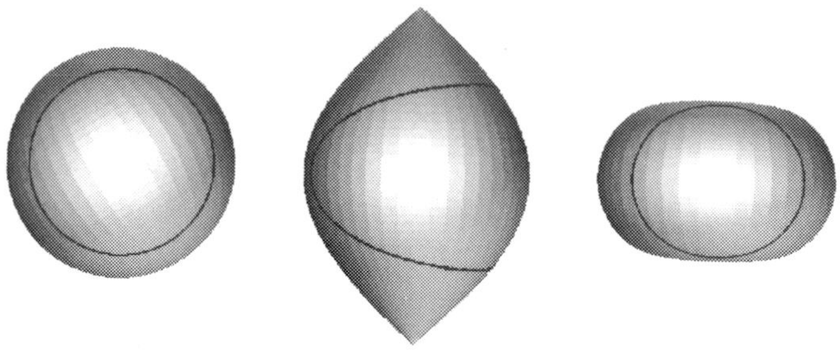

FIGURE 1. Biharmonic curves in the three types of constant positive Gauss curvature surfaces of revolution.

References

[CMP] R. Caddeo, S. Montaldo and P. Piu, *Biharmonic curves on a surface*, Rendiconti di Matematica e delle sue Applicazioni, 21 (2001).

[CMO] R. Caddeo, S. Montaldo and C. Oniciuc, *Biharmonic submanifolds of* $\mathbb{S}^3$, Internat. J. Math, to appear.

[Ch] B.Y. Chen, *Some open problems and conjectures on submanifolds of finite type*, Soochow J. math. 17 (1991), 169-188.

[Di] I. Dimitric, *Submanifolds of* $\mathbb{E}^m$ *with harmonic mean curvature vector*, Bull. Inst. Math. Acad. Sinica 20 (1992), 53-65.

[ES] J. Eells and J.H. Sampson, *Harmonic mappings of Riemannian manifolds*, Amer. J. Math. 86 (1964), 109–160.

[Ga] O.J. Garay, *A classification of certain 3-dimensional conformally flat Euclidean hypersurfaces*, Pacific J. Math.. 162 (1994), 13–25.

[Ji_1] G.Y. Jiang, *2-harmonic isometric immersions between Riemannian manifolds*, Chinese Ann. Math. Ser. A 7 (1986), 130–144.

[Ji_2] G.Y. Jiang, *2-harmonic maps and their first and second variation formulas*, Chinese Ann. Math. Ser. A 7 (1986), 389–402.

[HV] T. Hasanis and T.Vlachos, Hypersurfaces in $\mathbb{E}^4$ with harmonic mean curvature vector field, *Math. Nachr.* 172 (1995), 145–169.

[On] C. Oniciuc, *Biharmonic maps between Riemannian manifolds*, An. Stiint. Univ. Al. I. Cuza Iasi. Mat. (N.S.), to appear.

UNIVERSITÀ DEGLI STUDI DI CAGLIARI, DIPARTIMENTO DI MATEMATICA, VIA OSPEDALE 72, 09124 CAGLIARI, ITALY.

E-mail address: caddeo@unica.it, montaldo@unica.it, piu@unica.it

Contemporary Mathematics
Volume **288**, 2001

Characterizations of Pfaffian Systems. Consequences in Dimension Five

María A. Cañadas-Pinedo and Ceferino Ruiz

ABSTRACT. We present some classification results for Pfaffian systems. These results have been obtained by means of an intrinsic technique based in the study of the incidence relations between a Pfaffian system and the characteristic system of the first derived system. By using these results we obtain the complete classification of Pfaffian systems up to dimension four as well as the first classification in dimension five.

1. Introduction

The main goal of this work is to present some results leading to the classification of Pfaffian systems. These are used to obtain what Élie Cartan calls *Première classification des systèmes de deux ou trois équations de Pfaff à cinq variables* (see [**C**]). We want to point out that these classification results are obtained in an intrinsic way, and not by using coordinates.

The technique we use is based on the study of the incidence relations between a Pfaffian system and the characteristic of its first derived system (see [**CP-R;98**]) together with a structure tensor defined by Kumpera-Ruiz ([**K-R**]). This tensor is an invariant of the system which contains all the information about the class of the system and allows us to reformulate Frobenius' theorem as well as to determine the characteristic and derived systems of a given Pfaffian system.

By using this technique, that we develop in [**CP-R;97**], we obtain several classification results for Pfaffian systems. In particular we characterize Pfaffian systems with derived length one, that is, Pfaffian systems whose first derived system is a completely integrable system, and we also give a characterization for flag systems in terms of the class.

These results are used to obtain the complete classification of Pfaffian systems up to dimension four as well as the first classification in dimension five in an intrinsic way (see [**CP-R;00**]). Related results concerning dimension five can be found in [**CP-R;93**].

1991 *Mathematics Subject Classification.* Primary 58A17; Secondary 53C10.

Key words and phrases. Pfaffian system, structure tensor, derived system, characteristic system, flag system, covariant system.

The first author was supported in part by DGICYT Grant #PB97-1095.

The second author was supported in part by DGICYT Grant #PB97-0785.

2. Preliminary results

Let M be a connected differentiable manifold of finite dimension n and let us denote by T and T^* the tangent and cotangent bundles of M respectively. Let us consider a Pfaffian system S on M, that is, a locally trivial vector sub-bundle of T^*. Then, if we suppose S to have rank p, S is locally generated by p 1-forms $\omega^1, ..., \omega^p \in \Gamma_\ell(S)$. We shall represent $S = span\{\omega^1, ..., \omega^p\}$ omitting both the point and the neighborhood unless necessary to avoid misunderstandings. To a Pfaffian system S, a differentiable distribution is associated, namely $\Sigma = S^\perp$. We will identify $(T/\Sigma)^* \equiv S$ and $(T^*/S) \equiv \Sigma^*$. $C(S)$ and Δ shall denote, respectively, the characteristic system and the characteristic distribution of S.

The *structure tensor* of a Pfaffian system S ([**K-R**]) is the vector bundle morphism

$$\mathbf{k} : \Sigma \otimes S \longrightarrow \Sigma^* \equiv T^*/S$$

given by

$$\mathbf{k}(\xi \otimes \omega) = (\mathbf{q} \circ L_\xi)(\omega) = \mathbf{q}(i_\xi(d\omega))$$

for all $\xi \in \Gamma(\Sigma)$ and $\omega \in \Gamma(S)$, where $\mathbf{q} : T^* \longrightarrow T^*/S$ is the quotient morphism.

In the next theorem, we summarize some properties of the tensor $\mathbf{k}$:

THEOREM 2.1. *Let S be a Pfaffian system and let $\mathbf{k}$ be the structure tensor.*

(1) *The rank of $\mathbf{k}$ is constant if and only if the dimension of Δ_x is constant. In this case $C(S) = \mathbf{q}^{-1}(\mathbf{k}(\Sigma \otimes S))$ and $rank(\mathbf{k}) = class(S) - rank(S)$.*
(2) *S is completely integrable if and only if $\mathbf{k} \equiv 0$.*
(3) *The rank of $\mathbf{k}$ is never equal to one.*

For each $x \in M$ we define

$$(S_1)_x = \{\omega \in S_x \,/\, \mathbf{k}(\xi \otimes \omega) = 0, \forall \xi \in \Sigma_x\}$$

Whenever the dimension of these spaces is constant, the Pfaffian system S_1 obtained is called, following Élie Cartan, the first derived system of S. We shall denote $\Delta(S_1) = \Delta_1$, $C(S_1) = C_1$ and $\mathbf{k}_1$ the structure tensor of S_1.

The successive derived systems are defined inductively by:

$$S_{\mathrm{r}+1} = (S_{\mathrm{r}})_1$$

The Pfaffian system is called totally regular if all its derived systems exist. In this case we obtain a decreasing sequence of Pfaffian systems,

$$S = S_0 \supseteq S_1 \supseteq ... \supseteq S_{\mathrm{r}} \supseteq S_{\mathrm{r}+1} \supseteq ...$$

whose last term is a completely integrable system (which might be trivial). The derived length of a totally regular Pfaffian system S is the smallest nonnegative integer N such that $S_N = S_{N+1}$. Whenever $dim(S_{\mathrm{r}-1}/S_{\mathrm{r}}) = 1$, for $1 \leq \mathrm{r} \leq N$, the totally regular Pfaffian system S is called a flag system. In case $S_N = \{0\}$ the flag is said to be complete. Otherwise we call S a non-complete flag system.

THEOREM 2.2. *If $dim(S/S_1) \leq 1$, then, for every $p \in M$, $rank(\mathbf{k}_p)$ is even.*

The proofs of the next two results can be found in [**CP-R;97**].

THEOREM 2.3. *If $rank(\mathbf{k}) = 2$ then S_1 has codimension one in S.*

THEOREM 2.4. *One has $dim(S/S_1) = dim(S_1/S_2) = 1$ if and only if $rank(\mathbf{k}) = 2$ and $rank(\mathbf{k}_1) = 2$. Thus, in this case, S and S_1 have constant class equal to $rank(S) + 2$ and $rank(S_1) + 2$ respectively.*

3. First classification in dimension five

By using the results of the previous section we can obtain the complete classification in dimension less than or equal to four and the first classification in dimension five. The proofs can be found in [**CP-R;00**]. For the local models see [**K-R**]. All Pfaffian systems are supposed to have constant class.

For Pfaffian systems of rank one the differential invariant which provides a complete classification is the class. The first case that cannot be obtained directly from Frobenius' and Darboux theorems arises for Pfaffian systems of rank two in dimension four. In this case the differential invariant which provides the classification is the derived length.

THEOREM 3.1. *Let S be a Pfaffian system of rank 2 on a differentiable manifold of dimension 4. Then, either S is completely integrable, or S has class 4 and is a flag system (which might be non-complete). Two Pfaffian systems of rank 2 in dimension 4 are locally equivalent if and only if they have the same derived length.*

Let S be a Pfaffian system of rank 3 on a 5-dimensional differentiable manifold M. Let us consider the structure tensor of S, $\mathbf{k} : \Sigma \otimes S \longrightarrow T^*/S$. If S is not completely integrable, as the rank of $\mathbf{k}$ is always different from 1, we have $rank(\mathbf{k}) = 2$ and we obtain:

THEOREM 3.2. *Let S be a Pfaffian system with $rank(S) = 3$ on a differentiable manifold M with $dim\,(M) = 5$. Then either S is integrable or the first derived system S_1 of S has codimension one in S.*

For the completely integrable case ($rank(\mathbf{k}) = 0$) Frobenius' theorem provides a local model for S. In other cases, $rank(\mathbf{k}) = 2$ and the derived system S_1 is a Pfaffian system of rank two. Since $dim\,(M) = 5$, $rank(\mathbf{k}_1) = 0$ or $rank(\mathbf{k}_1) = 2$ or $rank(\mathbf{k}_1) = 3$. If $rank(\mathbf{k}_1) \leq 2$, the system S is a flag system and, as before, it is possible to obtain a complete classification. Nevertheless, in this case, for systems with derived length 3 there are two non equivalent local models (see [**G-K-R**]).

The case $rank(\mathbf{k}) = 2$, $rank(\mathbf{k}_1) = 3$ corresponds to what Élie Cartan calls *the general case*. In this case, as $rank(\mathbf{k}) = 2$, $dim\ (S/S_1) = 1$. Also if $dim\ (S_1/S_2) = 1$ it would imply $rank(\mathbf{k}_1) = 2$. Therefore, the general case corresponds to those Pfaffian systems of rank three in dimension five which are totally regular but are not flag systems because $S_2 = \{0\}$.

For the case $rank(\mathbf{k}) = rank(\mathbf{k}_1) = 2$, we obtain $dim\ (S/S_1) = dim\ (S_1/S_2) = 1$. Therefore $rank(S_2) = 1$ and S is a flag system. On the other hand, since $rank(S_1) = 2$ and $rank(\mathbf{k}_1) = 2$ we can, by restricting to the characteristic variables of S_1, consider S_1 as defined on a differentiable manifold of dimension 4. Thus, the classification before gives the local models for S_1.

Let us consider the structure tensor $\mathbf{k}$ of a given Pfaffian system S of rank two on a differentiable manifold of dimension five. There are three possibilities: $rank(\mathbf{k}) = 0$ or $rank(\mathbf{k}) = 2$ or $rank(\mathbf{k}) = 3$. If $rank(\mathbf{k}) \leq 2$, either S is completely integrable ($rank(\mathbf{k}) = 0$) or S has class four ($rank(\mathbf{k}) = 2$). Therefore for $rank(\mathbf{k}) = 2$ we can consider S as a Pfaffian system on a manifold of dimension four, and so S is a flag system and the local model is well determined, depending on the derived length of S.

Let us consider the case $rank(\mathbf{k}) = 3$. At each point, $S = S_1$ if and only if $rank(\mathbf{k}) = 0$ and the condition $dim\ (S/S_1) = 1$ implies that the rank of $\mathbf{k}$ is even.

We deduce that if $rank(S) = 2$ and $rank(\mathbf{k}) = 3$, necessarily S_1 has codimension two in S, that is, $S_1 = \{0\}$ (independently of the dimension of M).

THEOREM 3.3. *Let S be a Pfaffian system of rank 2 and let $\mathbf{k}$ be the structure tensor of S. If $rank(\mathbf{k}) = 3$ (i.e., $class\,(S) = 5$), then $S_1 = \{0\}$.*

The converse is not true in general. Nevertheless, in dimension five, since for a rank 2 Pfaffian system S, $rank(\mathbf{k}) \leq 2$ implies that S is completely integrable or S is a flag system, we deduce that, if $S_1 = \{0\}$, then $rank(\mathbf{k}) = 3$. Hence,

THEOREM 3.4. *Let S be a Pfaffian system of rank 2 on a differentiable manifold of dimension 5. Then $S_1 = \{0\}$ iff $rank(\mathbf{k}) = 3$, that is, iff $class\,(S) = 5$.*

The condition $rank(\mathbf{k}) = 3$ allows us to associate to a Pfaffian system S of rank two in dimension five a Pfaffian system $\widetilde{S}$ of rank three, with $S \subset \widetilde{S}$. If $\widetilde{S}$ has constant class, then either $\widetilde{S}$ is completely integrable or $\widetilde{S}_1 = S$. In the second case $\widetilde{S}$ corresponds to the general case and it is the system that Élie Cartan calls the *covariant system* of the system S. The proof is based on the next lemmas:

LEMMA 3.5. *Let S be a Pfaffian system of rank r on a differentiable manifold of finite dimension n, with $1 \leq \mathrm{r} \leq \mathrm{n}-2$. If $S_1 = \{0\}$, then given $\omega \in \Gamma_\ell(S)$, $\omega \neq 0$, there exist $\xi_1, \xi_2 \in \Gamma_\ell(\Sigma)$ such that $\mathbf{q}(i_{\xi_1} d\omega)$, $\mathbf{q}(i_{\xi_2} d\omega)$ are linearly independent in C/S. As a consequence, $dim\,(\mathbf{k}\,(\Sigma \otimes S(\omega))) \geq 2$, with $S(\omega) = span\,\{\omega\}$.*

LEMMA 3.6. *Let S be a Pfaffian system of rank 2 on a differentiable manifold of dimension 5, and let $\omega \in \Gamma_\ell(S)$, $\omega \neq 0$. If $rank(\mathbf{k}) = 3$, then $dim\,(\mathbf{k}\,(\Sigma \otimes S(\omega))) = 2$ with $S(\omega) = span\,\{\omega\}$.*

LEMMA 3.7. *Let S be a Pfaffian system of rank 2 on a differentiable manifold of dimension 5. Let ω^1 and ω^2 be local generators of S and let us denote $S(\omega^{\mathrm{i}}) = span\,\{\omega^{\mathrm{i}}\}$, $\mathrm{i} = 1, 2$. If $rank(\mathbf{k}) = 3$ then*

$$dim\,\left(\mathbf{k}\,\left(\Sigma \otimes S(\omega^1)\right) \cap \mathbf{k}\,\left(\Sigma \otimes S(\omega^2)\right)\right) = 1$$

Moreover $\alpha \in \mathbf{q}^{-1}\left(\mathbf{k}\,\left(\Sigma \otimes S(\omega^1)\right) \cap \mathbf{k}\,\left(\Sigma \otimes S(\omega^2)\right)\right)$ if and only, for all $\omega \in \Gamma_\ell(S)$, $\alpha \wedge d\omega \equiv 0 \quad mod\,S$.

THEOREM 3.8. *On a differentiable manifold of dimension 5, to each Pfaffian system S of rank 2 with $rank(\mathbf{k}) = 3$, a unique Pfaffian system $\widetilde{S}$ of rank 3 with $S \subseteq \widetilde{S}_1$ can be associated. Moreover, if $\widetilde{S}$ has constant class, either $\widetilde{S}$ is completely integrable or $\widetilde{S}_1 = S$*

Following Élie Cartan ([**C**]), we shall call $\widetilde{S}$ the covariant system of S.

From now on, we shall assume $\widetilde{S}$ to have constant class.

As a consequence of this theorem, two cases can be distinguished for Pfaffian systems of rank two in dimension five whose first derived system is trivial, depending on whether the covariant system $\widetilde{S}$ is completely integrable or not. In case $\widetilde{S}$ is completely integrable about every point, it is possible to find a local model well determined for S. More precisely:

THEOREM 3.9. *Let S be a Pfaffian system of rank 2 on a differentiable manifold M of dimension 5 and such that the first derived system S_1 is trivial. If the covariant system of S is completely integrable, then, about every point $p \in M$, there exists a chart centered at p, $(U, (x^1, x^2, x^3, x^4, x^5))$, such that*

$$S|_U = span\,\{dx^1 + x^4 dx^3, dx^2 + x^5 dx^3\}$$

In the case that the covariant system $\widetilde{S}$ is not completely integrable, its first derived system is the system S. Since $\widetilde{S}_2 = \{0\}$, $\widetilde{S}$ is a Pfaffian system of rank three in dimension five corresponding to the *general case*. Therefore

THEOREM 3.10. *On a manifold of dimension 5 there exists a one-to-one correspondence between Pfaffian systems, $\widetilde{S}$, of rank 3 whose second derived system is trivial, and Pfaffian systems S of rank 2 and class 5 whose covariant system is not completely integrable. That is:*

(1) *If $\widetilde{S}$ is a system of rank 3 with $\widetilde{S}_2 = \{0\}$, then $\widetilde{S}_1$ is a system of rank 2 and class 5 whose covariant system is $\widetilde{S}$.*
(2) *If S is a system of rank 2 and class 5 whose covariant system $\widetilde{S}$ is not completely integrable, then $\widetilde{S}_1 = S$ and $\widetilde{S}_2 = \{0\}$.*

So, in dimension five, for systems of constant class, the classification in the *general case* of systems of rank two is equivalent to the one of systems of rank three. In the words of Élie Cartan: *Dans ce cas général, il y a correspondance univoque entre les deux systèmes covariants*

$$\begin{matrix} \omega_1 = 0, \\ \omega_2 = 0 \end{matrix} \quad et \quad \begin{matrix} \omega_1 = 0, \\ \omega_2 = 0, \\ \omega_3 = 0. \end{matrix}$$

We deal with the general case in [**CP-R;93**].

References

[CP-R;93] M.A. Cañadas-Pinedo and C. Ruiz, *Sternberg's structure function of differential systems in dimension five*, Czechoslovak Math. J. **43** (1993), 429–438.

[CP-R;97] M.A. Cañadas-Pinedo and C. Ruiz, *Pfaffian Systems with Derived Length One. The Class of Flag Systems*, Trans. Amer. Math. Soc., **353(5)** (2001), 1755–1766.

[CP-R;98] M.A. Cañadas-Pinedo and C. Ruiz, *Incidence relations between a Pfaffian system and the characteristic of its first derived system*, Proceedings del Encuentro de Otoño de Geometría y Aplicaciones (Valencia), vol I, Publicaciones de la R.S.M.E., 31–40.

[CP-R;00] M.A. Cañadas-Pinedo and C. Ruiz, *First Classification of Pfaffian Systems in Dimension Five*, Preprint (Univ. Málaga, Univ Granada) (2000).

[C] É. Cartan, *Les systèmes de Pfaff à cinq variables et les équations aux dérivées partielles du second ordre*, Ann. Sci. Ec. Normale Sup. Paris, **27** (1910), 109–192.

[G-K-R] A. Giaro, A. Kumpera and C. Ruiz, *Sur la lecture correcte d'un résultat d'Élie Cartan*, C.R. Acad. Sci. Paris, Série A, **287** (1978), 241–244.

[K-R] A. Kumpera and C.Ruiz, *Sur l'équivalence locale des systèmes de Pfaff en drapeau*, Ist. Naz. Alta Mat. Francesco Severi (Rome), 1982.

DEPARTAMENTO DE ÁLGEBRA, GEOMETRÍA Y TOPOLOGÍA, FACULTAD DE CIENCIAS, UNIVERSIDAD DE MÁLAGA, APDO. 59, 29080-MÁLAGA, SPAIN.

E-mail address: pinedo@uma.es

DEPARTAMENTO DE GEOMETRÍA Y TOPOLOGÍA, UNIVERSIDAD DE GRANADA, 18071- GRANADA, SPAIN.

E-mail address: ruiz@ugr.es

Contemporary Mathematics
Volume **288**, 2001

On a New Class of Contact Riemannian Manifolds

Jong Taek Cho

Dedicated to the memory of Alfred Gray

ABSTRACT. N.Tanaka ([17]) defined the canonical affine connection on a non-degenerate integrable CR-manifold. In [8] we have introduced a new class of contact Riemannian manifolds satisfying (C) $(\hat{\nabla}_{\dot{\gamma}}R)(\cdot,\dot{\gamma})\dot{\gamma} = 0$ for any unit $\hat{\nabla}$-geodesic $\gamma(\hat{\nabla}_{\dot{\gamma}}\dot{\gamma} = 0)$, where $\hat{\nabla}$ is the generalized Tanaka connection. In this article, we give the brief survey about this new class, including the 3-dimensional classification and the determination of this class restricted on the unit tangent sphere bundle.

There are two typical examples of contact manifolds ; one is the odd dimensional spheres and the other is the unit tangent sphere bundles of Riemannian manifolds. These spaces have the standard Riemannian structures and their associated CR-structures. The contact structure η is a global differentiable 1-form on a smooth manifold M^{2n+1} such that $\eta \wedge (d\eta)^n \neq 0$ everywhere on M. It is well-known that there exists an associated Riemannian structure (metric) g and a (1,1)-type tensor ϕ, where g and ϕ are canonically related. We call the pair (η, g) a contact Riemannian structure and $M = (M;\eta, g)$ a contact Riemannian manifold. S.Sasaki and Y.Hatakeyama [14] defined the normality of the contact Riemannian structure. A normal contact Riemannian manifold is said to be a Sasakian manifold. For a more detailed treatment about contact (Riemannian) geometry we refer to [2]. In [11] it was proved that a Sasakian manifold which is locally symmetric ($\nabla R = 0$) must have constant curvature +1, where ∇ is the Levi-Civita connection and R is the curvature tensor. This fact means that local symmetry is a very strong condition for a Sasakian manifold. For this reason, T. Takahashi([16]) introduced the notion of Sasakian locally ϕ-symmetric spaces which may be considered as the analogues of locally Hermitian symmetric spaces. A contact Riemannian locally ϕ-symmetric space is defined as a generalization of a Sasakian locally ϕ-symmetric space and investigated in [4].

1991 *Mathematics Subject Classification.* 53C15, 53C25.
Key words and phrases. contact Riemannian manifolds, the generalized Tanaka connection, Jacobi operators, unit tangent sphere bundles.

One the other hand, the associated CR-structure of a given contact Riemannian manifold $M = (M; \eta, g)$ is given by the holomorphic subbundle

$$\mathcal{H} = \{X - i\bar{\phi}X : X \in D\}$$

of the complexification $TM^{\mathbb{C}}$ of the tangent bundle TM, where D is the subbundle of TM defined by the kernel of η and $\bar{\phi} = \phi|D$, the restriction of ϕ to D. Then we see that each fiber $\mathcal{H}_x$ $(x \in M)$ is of complex dimension n and $\mathcal{H} \cap \bar{\mathcal{H}} = \{0\}$. Furthermore, we have $\mathbb{C}D = \mathcal{H} \oplus \bar{\mathcal{H}}$. We say that *the associated CR-structure is integrable* if $[\mathcal{H}, \mathcal{H}] \subset \mathcal{H}$. For $\mathcal{H}$ we define the Levi form by

$$L : D \times D \to \mathcal{F}(M), \quad L(X, Y) = -d\eta(X, \phi Y)$$

where $\mathcal{F}(M)$ denotes the algebra of differential functions on M. Then we see that the Levi form is Hermitian and positive definite, that is, the CR-structure is *strongly pseudo-convex, pseudo-Hermitian CR-structure*. In fact, for a contact manifold $(M; \eta)$, there is a correspondence between the contact Riemannian structure (η, g) and strongly pseudo-convex, pseudo-Hermitian CR-structure (η, L) by the relation $g = L + \eta \otimes \eta$, where we denote by the same letter L the natural extension of the Levi form to a (0,2)-tensor field on M. N.Tanaka [17] defined the canonical affine connection on a nondegenerate integrable CR-manifold. In [18] S.Tanno proved that for a given contact Riemannian manifold M the associated CR-structure is strongly pseudo-convex integrable if and only if M satisfies the integrability condition $Q = 0$, where Q is a (1,2)-tensor field on M defined by

$$Q(X, Y) = (\nabla_X \phi)(Y) + (\nabla_X \eta)(\phi Y)\xi + \eta(Y)\phi\nabla_X \xi$$

for all vector fields X and Y, and further, he defined the generalized Tanaka connection $\hat{\nabla}$ on contact Riemannian manifolds which coincides with the Tanaka connection when the associated CR-structures are integrable. Here we note that the normality of a contact Riemannian structure implies the integrability of the associated CR-structure, but the converse does not always hold. The associated CR-structures of 3-dimensional contact Riemannian manifolds are always integrable (see [18]). Also, we see that their associated CR-structures are integrable for a class of contact Riemannian manifolds whose characteristic vector field ξ belongs to the (k, μ)-nullity distribution for some real numbers k and μ, that is, the curvature tensor R satisfies

$$R(X, Y)\xi = k(\eta(Y)X - \eta(X)Y) + \mu(\eta(Y)hX - \eta(X)hY)$$

for all vector fields X and Y, where $h = \frac{1}{2}\mathfrak{L}_\xi \phi$, $\mathfrak{L}_\xi$ denotes the Lie differentiation in the direction ξ. The unit tangent sphere bundles of spaces of constant curvature c satisfy the condition with $k = c(2 - c)$ and $\mu = -2c$, and further, these are the only ones with this property within the class of unit tangent sphere bundles (cf. [5] or [8]).

Recently, in [1] a locally symmetric space M is characterized by the remarkable property that the Jacobi operator field $R_{\dot\gamma} = R(\cdot, \dot\gamma)\dot\gamma$ is diagonalizable by a ∇-parallel orthonormal frame field along γ and its eigenvalues are constant along γ for any geodesic γ on M. In the preceding papers [8] [9], we have introduced a new class of contact Riemannian manifolds satisfying the property (C) for any unit $\hat{\nabla}$-geodesic γ $(\hat{\nabla}_{\dot\gamma}\dot\gamma = 0)$, i.e., the Jacobi operator field $R_{\dot\gamma}$ is diagonalizable by a

$\hat{\nabla}$-parallel orthonormal frame field along γ and its eigenvalues are constant along γ, or equivalently,

$$(\hat{\nabla}_{\dot{\gamma}}R)(\cdot,\dot{\gamma})\dot{\gamma}=0 \tag{C}$$

for any unit $\hat{\nabla}$-geodesic γ, where $\hat{\nabla}$ is the generalized Tanaka connection. We observe that the geodesics of the Levi-Civita connection and the generalized Tanaka connection do not coincide in general. A Sasakian locally ϕ-symmetric space is characterized by the property (C) for any unit $\hat{\nabla}$-geodesic γ (see [8]). Furthermore, in [8] we have shown that contact Riemannian manifolds with ξ belonging to the $(k,2)$-nullity distribution $(k \neq 1)$, including the standard contact Riemannian structure of the unit tangent sphere bundle T_1M of M with constant curvature -1, are examples that are neither Sasakian nor locally symmetric but satisfy the condition (C) for any $\hat{\nabla}$-geodesic γ. Also, it is remarkable that a contact Riemannian manifold M of dimension ≥ 5 whose characteristic vector field ξ belonging to the (0,0)-nullity distribution, which is a product of $(n+1)$-dimensional flat manifold and n-dimensional space of constant curvature 4, is locally symmetric but M fails to satisfy the condition (C) for any $\hat{\nabla}$-geodesic γ. In particular, we have proved ([10])

THEOREM A. *The standard contact Riemannian structure of the unit tangent sphere bundle T_1M satisfies the condition (C) for any $\hat{\nabla}$-geodesic γ if and only if*
1. *the base manifold M is of constant curvature 1 and T_1M is Sasakian locally ϕ-symmetric;*
2. *the base manifold M is 2-dimensional and T_1M is flat, i.e. $c=0$;*
3. *the base manifold M is of constant curvature -1.*

REMARK. *Recently, it was proved in [3] ([12] and [13], respectively) that the base manifold is of constant curvature $c=-1$, $c=1$ $(c=0, c=1$, respectively) if and only if the standard contact Riemannian structure on the unit tangent sphere bundle is a critical point of some functional on the set of associated Riemannian metrics $\mathcal{M}(\eta)$ of a given contact form η.*

Moreover, we have given a local and a global classification of 3-dimensional contact Riemannian manifolds satisfying the condition (C) for any $\hat{\nabla}$-geodesic γ. More precisely, we have proved ([9])

THEOREM B (LOCAL CLASSIFICATION). *Let M be a 3-dimensional contact Riemannian manifold. Then M satisfies the condition (C) for any $\hat{\nabla}$-geodesic γ if and only if M is locally isometric to one of the following spaces :*
1. *a Sasakian ϕ-symmetric space;*
2. *$SU(2)$ (or $SO(3)$), $SL(2,\mathbb{R})$ (or $O(1,2)$) with a special left-invariant contact metric which is not Sasakian, respectively;*
3. *a flat manifold.*

In [6] the authors gave a classification of Sasakian ϕ-symmetric spaces (complete and simply connected Sasakian locally ϕ-symmetric spaces). Together with this classification we have

THEOREM C (GLOBAL CLASSIFICATION). *Let M be a complete and simply connected 3-dimensional contact Riemannian manifold. Then M satisfies the condition*

(C) for any $\hat{\nabla}$-geodesic γ if and only if M is isometric to one of the following spaces :

1. *the standard unit sphere S^3; $SU(2)$, $\widetilde{SL(2,\mathbb{R})}$ (the universal covering space of $SL(2,\mathbb{R})$) or the Heisenberg group H with a left invariant Sasakian metric, respectively;*
2. *$SU(2)$, $\widetilde{SL(2,\mathbb{R})}$ with a special left-invariant contact metric which is not Sasakian, respectively;*
3. *$\mathbb{R}^3$.*

References

[1] J. Berndt and L. Vanhecke, *Two natural generalizations of locally symmetric spaces*, Diff. Geom. Appl. **2** (1992), 57–80.

[2] D. E. Blair, *Contact manifolds in Riemannian geometry*, Lecture Notes in Math. Springer-Verlag, Berlin-Heidelberg-New-York **509** (1976).

[3] D. E. Blair, *Critical associated metrics on contact manifolds III*, J. Austral. Math. Soc. **50** (1991), 189–196.

[4] D. E. Blair, T. Koufogiorgos and R. Sharma, *A classification of 3-dimensional contact metric manifolds with $Q\phi = \phi Q$*, Kodai Math. J. **13** (1990), 391–401.

[5] D. E. Blair, T. Koufogiorgos and B. J. Papantoniou, *Contact metric manifolds satisfying a nullity condition*, Israel J. Math. **91** (1995), 189–214.

[6] D. E. Blair and L. Vanhecke, *Symmetries and ϕ-symmetric spaces*, Tôhoku Math. J. **39** (1987), 373–383.

[7] E. Cartan, *Leçons sur la géométrie des espaces de Riemann, 2nd éd.*, Gauthier-Villars, Paris (1946).

[8] J. T. Cho, *A new class of contact Riemannian manifolds*, Israel J. Math **109** (1999), 299–318.

[9] J. T. Cho, *A new class of contact Riemannian manifolds II*, preprint.

[10] J. T. Cho and S. H. Chun, *Unit tangent sphere bundle satisfying a certain condition*, preprint.

[11] M. Okumura, *Some remarks on spaces with a certain contact structure*, Tôhoku Math. J. **14** (1962), 135–145.

[12] D. Perrone, *Torsion tensor and critical metrics on contact (2n+1)-manifolds*, Monatsh. Math. **114** (1992), 245–259.

[13] D. Perrone, *Tangent sphere bundles satisfying $\nabla_\xi \tau = 0$*, J. Geom. **49** (1994), 178–188.

[14] S. Sasaki and Y. Hatakeyama, *On differentiable manifolds with certain structures which are close related to almost contact structure II*, Tôhoku Math. J. **13** (1961), 281–294.

[15] S. Sasaki and Y. Hatakeyama, *On differentiable manifolds with contact metric structure*, J. Math. Soc. Japan **14** (1962), 249–271.

[16] T. Takahashi, *Sasakian ϕ-symmetric spaces*, Tôhoku Math. J. **29** (1977), 91–113.

[17] N. Tanaka, *On non-degenerate real hypersurfaces, graded Lie algebras and Cartan connections*, Japan J. Math. **2** (1976), 131–190.

[18] S. Tanno, *Variational problems on contact Riemannian manifolds,*, Trans. Amer. Math. Soc. **314** (1989), 349–379.

Department of Mathematics, Chonnam National University, CNU The Institute of Basic Sciences, Kwangju 500-757, Korea.

E-mail address: `jtcho@chonnam.ac.kr`

Contemporary Mathematics
Volume **288**, 2001

Description of $(1,2)$-Symplectic Metrics on Flag Manifolds

Nir Cohen, Caio J. C. Negreiros, and Luiz A. B. San Martin

This paper is dedicated to the memory of Alfred Gray

ABSTRACT. In this note we announce the result establishing the bijective correspondence between $(1,2)$-symplectic structures on the manifold of flags of subspaces of $\mathbb{C}^n$ and the cone-free tournaments. This result provides an independent proof of a Theorem due to Wolf and Gray. For further developments see, for example, [**3**].

1. Introduction

Let M^2 be a closed Riemann surface and N a Riemannian manifold, with metric ds^2, endowed with an almost-complex structure J. A theorem due to Lichnerowicz [**7**], proved independently by Gray [**5**], states: "If $\phi : M^2 \to N$ is J-holomorphic and ds^2 is $(1,2)$-symplectic with respect to J, then ϕ is harmonic". This theorem drives the attention to the problem of understanding the $(1,2)$-symplectic metrics on manifolds endowed with almost complex structures. Apart from this relation to harmonic maps the $(1,2)$-symplectic metrics appear in twistor theory, as shown, for example, in Eells and Salamon [**4**].

In this note we focus attention on invariant $(1,2)$-symplectic metrics and invariant almost-complex structures on flag manifolds. Denote by $\mathbb{F}(n)$ the usual manifold of full flags of subspaces of $\mathbb{C}^n$. Our approach to this problem is based on a method derived by Burstall and Salamon [**2**], relating in a natural way an invariant almost-complex structure J on $\mathbb{F}(n)$ to a tournament T_J.

Specifically, Mo and Negreiros [**8**] singled out the class of the so-called cone-free tournaments, a property which involves the 4 -subtournaments of T_J.

In this note we discuss the proof that an invariant almost complex structure J admits a $(1,2)$-symplectic metric if and only T_J is cone-free. In addition, given J satisfying the necessary and sufficient condition we exhibit an explicit n-dimensional parametrization for all the possible metrics.

1991 *Mathematics Subject Classification.* Primary 53C55; Secondary 05C20, 58E20.

Key words and phrases. Flag manifold, Tournament, Harmonic maps and Hermitian geometry.

Supported by CNPq grant no. 300019/96-3.

Supported by CNPq grant no. 301060/94-0.

Our approach is based on a concrete characterization of the structure of cone-free tournaments. Specifically, in Theorem 2.3 we state that a tournament T is cone-free if and only if its incidence matrix is permutation-similar to a stair-shaped incidence matrix of the following type

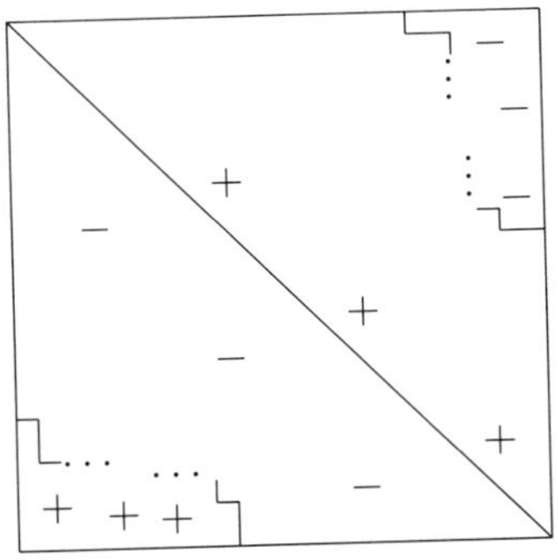

(1)

A more detailed version of this communication which contains the formal definitions of (1) and complete proofs will appear in [**3**]. Similar results for flag manifolds associated to complex semi-simple Lie groups were derived in [**12**].

2. Tournaments

An n-player tournament is a complete directed graph $T = (N, E)$, where N is an ordered set and $|N| = n$. For concreteness we shall always assume that $N = \{1, \ldots, n\}$. Two tournaments are isomorphic if one is obtained from the other via a rearrangement of N.

With each tournament T we assign its incidence matrix $\varepsilon = \varepsilon_T$, which is a real skew-symmetric matrix with all off-diagonal entries ± 1. If $(a, b) \in E$ we say that a wins against b and set $\varepsilon_{ab} = 1$, $\varepsilon_{ba} = -1$. By our conventions, T would then contain an arrow pointing from vertex a to vertex b. Up to isomorphism, there are four distinct 4-player tournaments. The two of them which contain a single (directed) 3-cycle are called *coned 3-cycles*. Each of them contains a cycle and a winner or a loser.

DEFINITION 2.1. An n-player tournament T is called cone-free if its restriction to any four vertices is never a coned 3-cycle.

We call a tournament T *transitive* if it contains no 3-cycle. In this case, T is obviously cone-free, and is isomorphic to the *canonical* transitive tournament T' for which the incidence matrix has a triangular form.

DEFINITION 2.2. A tournament T' is called *stair-shaped* if its incidence matrix has the form displayed in (1).

A special case of a stair-shaped tournament is the canonical transitive one.

THEOREM 2.3. *A tournament T is cone-free if and only if it is isomorphic to a stair-shaped tournament T'.*

3. Flag manifolds

Consider the complex full flag manifold

$$\mathbb{F}(n) = \{(V_1, \ldots, V_n) : V_j \subset V_{j+1}, \dim V_j = j\}.$$

The natural action of the unitary group $\mathrm{U}(n)$ on $\mathbb{F}(n)$ is transitive, turning the flag manifold into the homogeneous space $\mathrm{U}(n)/M$ where M is a maximal torus of $\mathrm{U}(n)$.

Let $\mathfrak{u}(n)$ be the Lie algebra of skew-symmetric matrices. The tangent space of $\mathbb{F}(n)$ at the origin identifies with the subspace $\mathfrak{p} \subset \mathfrak{u}(n)$ of zero-diagonal matrices. We have $\mathfrak{p} = \bigoplus_{j \neq k} \mathfrak{p}_{jk}$, where $\mathfrak{p}_{jk} = (\mathbb{C}E_{jk} + \mathbb{C}E_{kj}) \cap \mathfrak{u}(n)$, and $\{E_{jk}\}$ stands for the canonical basis matrix.

Tensors on $\mathbb{F}(n)$ that are invariant under $\mathrm{U}(n)$ are determined by their values on $\mathfrak{p}$. In particular, an invariant almost complex structure is given by $J : \mathfrak{p} \to \mathfrak{p}$, $J^2 = -1$, which commutes with the adjoint representation of the torus M on $\mathfrak{p}$. This condition implies that $J(\mathfrak{p}_{jk}) = \mathfrak{p}_{jk}$ for all $j \neq k$, which in turn guarantees that $J(A) = A'$ with $A'_{jk} = \varepsilon_{jk}\sqrt{-1}A_{jk}$ such that $\varepsilon_{jk} = \pm 1$ and satisfies $\varepsilon_{kj} = -\varepsilon_{jk}$. Thus an invariant almost complex structure is completely determined by a skew-symmetric matrix (ε_{jk}), with off-diagonal entries in $\{\pm 1\}$. Of course, (ε_{jk}) is the incidence matrix of a tournament, establishing a one-to-one correspondence between the invariant almost complex structures on $\mathbb{F}(n)$ and n-players tournaments.

Analogously, an $\mathrm{U}(n)$-invariant Riemannian metric on $\mathbb{F}(n)$ is given by an inner product in $\mathfrak{p}$ invariant under M. Such an inner product has the form $(X, Y)_\Lambda = \langle \Lambda X, Y \rangle$ where $\langle X, Y \rangle = -\mathrm{tr}(XY)$ is the Cartan-Killing form and $\Lambda : \mathfrak{p} \to \mathfrak{p}$ positive-definite with respect to $\langle \cdot, \cdot \rangle$. Furthermore, $(\cdot, \cdot)_\Lambda$ is M-invariant if and only if the elements of the standard basis $\sqrt{-1}(E_{jk} + E_{kj})$, $E_{jk} - E_{kj}$ are eigenvectors of Λ. Thus $\Lambda(E_{jk}) = \lambda_{jk}E_{jk}$ with $\lambda_{jk} > 0$ and $\lambda_{kj} = \lambda_{jk}$. We denote by ds^2_Λ the invariant metric given by Λ.

It is easily checked that ds^2_Λ is Hermitian with respect to J. Let $\Omega = \Omega_{J,\Lambda}$ stand for the corresponding Kähler form

$$\Omega(X, Y) = ds^2_\Lambda(X, JY) = -\mathrm{tr}(\Lambda(X) J(Y)).$$

This form is also invariant under $\mathrm{U}(n)$. Its exterior differential $d\Omega$ has the decomposition

$$d\Omega = d\Omega^{30} + d\Omega^{21} + d\Omega^{12} + d\Omega^{03}.$$

The structure $(\mathbb{F}(n), J, ds^2)$ is said to be *almost Kähler* if $d\Omega = 0$. It is *Kähler* if furthermore J is integrable.Also, $(\mathbb{F}(n), J, ds^2)$ is (r,s)-*symplectic* if $d\Omega^{rs} = 0$. Of course, the triple is (r,s)-symplectic in case it is almost Kähler. Furthermore, Burstall and Salamon [**1**] proved that in the invariant setup, $(\mathbb{F}(n), J, ds^2_\Lambda)$ is an almost Kähler manifold if and only if it is a Kähler manifold.

4. Main results

We can now state the main results characterizing the invariant almost complex structures on $\mathbb{F}(n)$ admitting $(1,2)$-symplectic metrics.

THEOREM 4.1. *A necessary and sufficient condition for an invariant almost complex structure J on $\mathbb{F}(n)$ to admit an invariant $(1,2)$-symplectic metric is that the associated tournament is cone-free.*

The proof that the condition is necessary was provided in [**8**]. The sufficiency of the condition was studied by Paredes [**9**], [**10**], where affirmative answers was obtained for certain classes of tournaments, including those with $5 \leq n \leq 7$ players. For $n = 3$ and 4 this result was proved in [**8**]. The general proof of the sufficient condition relies on Theorem 2.3. In fact, we have the following n -parameter description for all invariant metrics which are $(1,2)$-symplectic with respect to an invariant almost complex structure associated to a stair-shaped tournament.

THEOREM 4.2. *Let ε be the incidence matrix of the tournament associated to a given complex structure J on $\mathbb{F}(n)$. Assume that ε is stair-shaped. Then Λ corresponds to a $(1,2)$-symplectic metric on $(\mathbb{F}(n), J)$ if and only if there are n positive values λ_1 through λ_n such that for all $i<j$:*

(i): $\lambda_{ij} = \sum_{k=i}^{j-1} \lambda_k$ *if* $\varepsilon_{ij} = 1$, *and*
(ii): $\lambda_{ij} = \lambda_n + \sum_{k=1}^{i} \lambda_k + \sum_{k=j}^{n-1} \lambda_k$, *otherwise. Equivalently,* $\lambda_{ij} = q - \sum_{k=i+1}^{j-1} \lambda_k$, *where* $q = \sum_{k=1}^{n} \lambda_k$.

A special case of this theorem concerns the standard complex structure in $\mathbb{F}(n)$, whose corresponding tournament is the canonical one, having upper triangular incidence matrix. In this case we have only a $(n-1)$-parameter family of $(1,2)$-symplectic metrics, since we must take $\lambda_n = \lambda_1 + \cdots + \lambda_{n-1}$. Furthermore the metrics are Kähler, a result already noted by Borel [**1**].

Our result also provides an independent proof of a theorem by Wolf and Gray [**13**], namely that the Cartan-Killing metric on $\mathbb{F}(n)$ is not $(1,2)$-symplectic for any J if $n \geq 4$. In fact, the possible $(1,2)$-symplectic metrics Λ built above cannot have all entries $\lambda_{ij} = 1$ if $n \geq 4$, as happens to the Cartan-Killing metric.

Finally, we mention that these results are easily extended to partial flag manifolds, by making similar computations with the underlying block structure.

References

[1] A. Borel, Kählerian coset spaces of semi-simple Lie groups. *Proc. Nat. Acad. of Sci.* **40**:1147-1151, 1954.

[2] F.E. Burstall and S. Salamon, Tournaments, flags and harmonic maps. *Math Ann.* **277**:249-265, 1987.

[3] N. Cohen, C.J.C. Negreiros and L.A.B. San Martin, (1,2)-Symplectic metrics on flag manifolds and tournaments. To appear.

[4] J. Eells and S. Salamon, Twistorial constructions of harmonic maps of surfaces into four-manifolds. *Ann. Scuola Norm. Sup. Pisa (4)* **12**:589-640, 1985.

[5] A. Gray, Minimal varieties and almost Hermitian submanifolds, *Michigan Math. J.* **2**:273-287, 1965.

[6] A. Gray and L. M. Hervella, The sixteen classes of almost hermitian manifolds and their linear invariants. *Ann. Mat. Pura Appl* **123**:35-58, 1980.

[7] A. Lichnerowicz, Applications harmoniques et variétés Kähleriennes. *Symp. Math. 3* (Bologna), 341-402, 1970.

[8] X. Mo and C.J.C. Negreiros, (1,2)-Symplectic structure on flag manifolds. *Tohoku Math. J.*, **52**:271-282, 2000.

[9] M. Paredes, *Aspectos da geometria complexa das variedades bandeira.* Ph.D. thesis, State Univ. of Campinas, 2000.

[10] M. Paredes, On the existence of families of (1,2)-symplectic metrics on flag manifolds. International Journal of Mathematics and Mathematical Sciences. To appear.

[11] S. Salamon, *Harmonic and holomorphic maps.* Lec. Notes in Math. **1164**, Springer 1986.

[12] L.A.B. San Martin and C.J.C. Negreiros, Invariant almost complex structures on flag manifolds. Preprint, 2000.

[13] J. A. Wolf and A. Gray, Homogeneous spaces defined by Lie group automorphisms II. *J. Diff. Geom.* **2**:115-159, 1968.

Department of Applied Mathematics. Universidade Estadual de Campinas. Cx. Postal 6065. 13.081-970, Campinas - SP, Brasil

E-mail address: `nir@ime.unicamp.br`

Department of Mathematics. Universidade Estadual de Campinas. Cx. Postal 6065. 13.081-970, Campinas - SP, Brasil

E-mail address: `caione@ime.unicamp.br`

Department of Mathematics. Universidade Estadual de Campinas. Cx. Postal 6065. 13.081-970, Campinas - SP, Brasil

E-mail address: `smartin@ime.unicamp.br`

Contemporary Mathematics
Volume **288**, 2001

Complex Surfaces with Betti Numbers $b_1 = 1$, $b_2 > 0$ and Finite Quotients

Georges Dloussky

ABSTRACT. We show that any quotient of a compact complex surface S with Betti numbers $b_1(S) = 1$, $b_2(S) > 0$ and containing a Global Spherical Shell by a finite group of automorphisms still contains a GSS, contrarily to the case $b_2 = 0$ where there exists non primary Hopf surfaces.

1. Compact surfaces of the VII_0 class with $b_2 > 0$

The VII_0 class of Kodaira may be defined as being the class of (non kắhlerian) minimal compact complex surfaces with first Betti number $b_1(S) = 1$. In the sequel we shall suppose that surfaces are minimal and we shall focus on the case $b_2 > 0$. For the case $b_2 = 0$ see [**LT**]. Following Ma. Kato, we shall say that a compact complex surface X contains a Global Spherical Shell (GSS) if there is an open embedding of a neighbourhood of the sphere $S^3 \subset \mathbb{C}^2 \setminus \{0\}$ into X which does not disconnect X. All known examples of surfaces with $b_1(S) = 1$ and $b_2(S) > 0$ contain GSS. To each contracting germ $F = \Pi\sigma$ composed of n well-chosen blowing-ups and of a germ σ of isomorphism, we may associate a minimal surface $S = S(\Pi, \sigma)$ with $b_2(S) = n$ and containing a GSS. Conversely, to each rational curve C of S we may associate a contracting germ F_C of this type (see [**D1**]). Normal forms of such contracting germs have been investigated in [**D1**], [**D2**], [**DK**], [**DO1**]. Finally, C. Favre [**FA**] achieved this step (see thm 2.3) by giving a complete classification of a wider class of contracting germs called *rigid germs*. Rigid germs appear naturally in 2-dimensional dynamical systems (see [**DO2**], [**FA**]).
All surfaces containing a GSS admit at least one (singular) holomorphic foliation [**DO1**]. Conversely, if S admits a foliation, does S contain a GSS ? The answer is yes when there exists a global vector field [**DOT1**], [**DOT2**] or when there exists two linearly independant twisted vector fields [**OTZ**].

2. Compact surfaces with $b_2 > 0$ containing GSS

2.1. Basic constructions. A surface with GSS contains exactly $n := b_2(S)$ rational curves $D_0, \dots, D_{n-1}$. A rational curve is regular or has a double point.

2000*Mathematics Subject Classification.* Primary 32J15, 32M05, 32Q55.
Key words and phrases. Compact surfaces, Kodaira classification, automorphism group.
With support of LATP UMR 6632 - CNRS.

We introduce the following invariants [**D1**]

$$\sigma_n(S) = -\sum_{i=0}^{n-1} D_i^2 + 2\,Card\{\text{double points}\}, \quad tr(S) := tr(DF_C(0))$$

where $tr(S)$, called the *trace* of S, is the trace of the tangent map of any germ associated to S. The inequalities $2n \leq \sigma_n(S) \leq 3n$ hold. We have the following cases

- $\sigma_n(S) = 2n$: this condition is equivalent to the existence of a divisor D such that $D^2 = 0$ or the condition $tr(S) \neq 0$ (see [**E**], [**D1**], [**DK**], [**N1**]).
- $2n < \sigma_n(S) < 3n$: such surfaces will be called *intermediate surfaces* [**DO1**].
- $\sigma_n(S) = 3n$: these surfaces are Inoue-Hirzebruch surfaces [**I**], [**D3**].

We denote by $(\tilde{S}, \tilde{\omega}, S)$ the universal covering space of S. By [**D1**], there exists a canonical total order on the rational curves of $\tilde{S}$, "the order of creation ". The canonical order explains the notation $C+1$ used later, $C+1$ is the curve obtained just after C blowing up a point of C. We choose the covering automorphism $\tilde{g} : \tilde{S} \to \tilde{S}$ induced by the action of $\pi_1(S) \simeq \mathbb{Z}$ in such a way that for all curve C, $\tilde{g}(C) = C + n$. For every curve $C \subset \tilde{S}$, there exists a surface $\hat{S}_C$ and a mapping $p_C : \tilde{S} \to \hat{S}_C$ which collapses all the curves $C' > C$ onto a point $\hat{O}_C \in \hat{S}_C$ and which is an isomorphism from $\tilde{S} \backslash p_C^{-1}(\hat{O}_C)$ onto $\hat{S}_C \backslash \{\hat{O}_C\}$. By Hartogs theorem, we deduce from this property a non invertible contraction $F_C : \hat{S}_C \to \hat{S}_C$ with a fixed point $\hat{O}_C$ and an isomorphism $\sigma_C^{C+n} : \hat{S}_C \to \hat{S}_{C+n}$ such that $p_C\tilde{g} = F_C p_C$. There is a blowing-up $\Pi_C^{C+1} : \hat{S}_{C+1} \to \hat{S}_C$ of $\hat{S}_C$ at the point $\hat{O}_C$ such that $C + 1 = (\Pi_C^{C+1})^{-1}(\hat{O}_C)$ and $\Pi_C^{C+1} p_{C+1} = p_C$ (see [**D1**] or [**DK**] for a detailed explanation).

2.2. Finite quotients. Let $Aut^0(S) \subset Aut(S)$ be the subgroup of all automorphisms h wich leave each rational curve globally invariant. Let H be a finite subgroup of $Aut^0(S)$ and let $h \in H$. Let $\tilde{h} : \tilde{S} \to \tilde{S}$ be the automorphism on $\tilde{S}$ induced by h which leaves each rational curve of $\tilde{S}$ globally invariant. Of course, $\tilde{g}\tilde{h} = \tilde{h}\tilde{g}$. The proof of the following lemmas is straightforward

LEMMA 2.1. *There exist automorphisms $h_C : \hat{S}_C \to \hat{S}_C$ such that the following diagrams are commutative*

$$\begin{array}{ccc} \tilde{S} & \xrightarrow{\tilde{h}} & \tilde{S} \\ {\scriptstyle p_C}\downarrow & & \downarrow{\scriptstyle p_C} \\ \hat{S}_C & \xrightarrow{h_C} & \hat{S}_C \end{array} \qquad \begin{array}{ccc} \hat{S}_{C+1} & \xrightarrow{h_{C+1}} & \hat{S}_{C+1} \\ {\scriptstyle \Pi_C^{C+1}}\downarrow & & \downarrow{\scriptstyle \Pi_C^{C+1}} \\ \hat{S}_C & \xrightarrow{h_C} & \hat{S}_C \end{array}$$

Let $\tilde{H} \subset Aut(\tilde{S})$ and $H_C \subset Aut(\hat{S}_C)$ be the subgroups of automorphisms induced by H. If $\tilde{T} = \tilde{S}/\tilde{H}$ is the quotient normal analytic space with canonical mapping $\tilde{\alpha} : \tilde{S} \to \tilde{T}$, the image $D = \tilde{\alpha}(C)$ of any rational curve is still rational. Define the normal analytic space $\hat{T}_D := \hat{S}_C/\hat{H}_C$, with canonical mapping $\alpha_D : \hat{S}_C \to \hat{T}_D$.

LEMMA 2.2. *There exist holomorphic mappings $\tilde{u} : \tilde{T} \to \tilde{T}$, $\hat{U}_D : \hat{T}_D \to \hat{T}_D$, $q_D : \tilde{T} \to \hat{T}_D$, $\Psi_D^{D+1} : \hat{T}_{D+1} \to \hat{T}_D$ and $\tau_D^{D+n} : \hat{T}_D \to \hat{T}_{D+n}$ such that the following diagrams are commutative*

$$\begin{array}{ccc} \tilde S & \xrightarrow{\tilde g} & \tilde S \\ {\scriptstyle\tilde\alpha}\downarrow & & \downarrow{\scriptstyle\tilde\alpha} \\ \tilde T & \xrightarrow{\tilde u} & \tilde T \end{array} \qquad \begin{array}{ccc} \hat S_C & \xrightarrow{F_C} & \hat S_C \\ {\scriptstyle\alpha_D}\downarrow & & \downarrow{\scriptstyle\alpha_D} \\ \hat T_D & \xrightarrow{U_D} & \hat T_D \end{array} \qquad \begin{array}{ccc} \tilde S & \xrightarrow{p_C} & \hat S_C \\ {\scriptstyle\tilde\alpha}\downarrow & & \downarrow{\scriptstyle\alpha_D} \\ \tilde T & \xrightarrow{q_D} & \hat T_D \end{array} \qquad \begin{array}{ccc} \hat S_{C+1} & \xrightarrow{\Pi_C^{C+1}} & \hat S_C \\ {\scriptstyle\alpha_{D+1}}\downarrow & & \downarrow{\scriptstyle\alpha_D} \\ \hat T_{D+1} & \xrightarrow{\Psi_D^{D+1}} & \hat T_D \end{array}$$

$$\begin{array}{ccc} \hat S_C & \xrightarrow{\sigma_C^{C+n}} & \hat S_{C+n} \\ {\scriptstyle\alpha_D}\downarrow & & \downarrow{\scriptstyle\alpha_{D+n}} \\ \hat T_D & \xrightarrow{\tau_D^{D+n}} & \hat T_{D+n} \end{array} \qquad \begin{array}{ccc} \tilde T & \xrightarrow{\tilde u} & \tilde T \\ {\scriptstyle q_D}\downarrow & & \downarrow{\scriptstyle q_D} \\ \hat T_D & \xrightarrow{U_D} & \hat T_D \end{array} \qquad \begin{array}{ccc} \hat T_{D+1} & \xrightarrow{U_{D+1}} & \hat T_{D+1} \\ {\scriptstyle\Psi_D^{D+1}}\downarrow & & \downarrow{\scriptstyle\Psi_D^{D+1}} \\ \hat T_D & \xrightarrow{U_D} & \hat T_D \end{array} \qquad \begin{array}{ccc} \tilde T & \xrightarrow{\tilde u} & \tilde T \\ {\scriptstyle q_D}\downarrow & & \downarrow{\scriptstyle q_{D+n}} \\ \hat T_D & \xrightarrow{\tau_D^{D+n}} & \hat T_{D+n} \end{array}$$

For convenience of the reader we gather several results about automorphism groups of surfaces with $b_2(S) > 0$ containing GSS: 1) is proved in [**DK**] thm 1.4 (replacing λ by λ^{-1} with a necessary correction); 2) in [**FA**]; 3) in [**D2**] and [**FA**].

THEOREM 2.3. *Let S be a surface containing a GSS such that $n = b_2(S) > 0$.*
1) Case $\sigma_n(S) = 2n$. After conjugation, F_C has the normal form

$$F_C(z_1, z_2) = (z_1 z_2^n t^n + \sum_{k=0}^{n-1} \alpha_k t^{k+1} z_2^{k+1}, t z_2).$$

Moreover: a) If S is a Inoue surface i.e. all α_k vanish, then $h_C = (a z_1, \lambda z_2)$ where $a \in \mathbb{C}^\star$ and $\lambda^n = 1$. In particular $Aut^0(S) \simeq \mathbb{C}^\star \times \mathbb{U}_n$.
b) If there is exactly one index j such that $\alpha_j \neq 0$ then $h_C = (\lambda^{j+1} z_1, \lambda z_2)$. In particular, $Aut^0(S) \simeq \mathbb{U}_n$.
c) If there are at least two indices $\alpha_j \neq 0$, $\alpha_{j'} \neq 0$, let $m = gcd\{n, j-j' \mid \alpha_j \alpha_{j'} \neq 0\}$. Then $h_C = (\lambda^{j+1} z_1, \lambda z_2)$ where j is any index such that $\alpha_j \neq 0$ and $\lambda^m = 1$. In particular $Aut^0(S) \simeq \mathbb{U}_m$.
2) Case $2n < \sigma_n(S) < 3n$. Let C be a curve such that the fixed point $\hat O_C$ of F_C belongs only to the irreducible curve C, then after conjugation F_C has the normal form

$$F_C(z_1, z_2) = (\mu z_1 z_2^s + P(z_2), z_2^k)$$

where $k = k(S) \geq 2$ and $P \neq 0$ is a polynomial. Put $l = gcd\{k-1, s, i-i' \mid a_i a_{i'} \neq 0\}$ where the complex numbers a_i are the coefficients of P.
a) If $s/(k-1)$ is an integer and $\mu = 1$, then ($k \nmid i$ means that k does not divide i)

$$P(z_2) = a_{ks/(k-1)} z_2^{ks/(k-1)} + \sum_{1 \leq i \leq s,\ k \nmid i} a_i z_2^i, \quad h_C(z_1, z_2) = (\xi z_1 + c z_2^{s/(k-1)}, \zeta z_2),$$

where $\zeta \in \mathbb{U}_{k-1} \cap \mathbb{U}_s$, $a_i(\zeta^i - \xi) = 0$ and $c \in \mathbb{C}$. In particular $Aut^0(S) \simeq \mathbb{C} \ltimes \mathbb{U}_l$.
b) If $s/(k-1)$ is not an integer or $\mu = 1$, then

$$P(z_2) = \sum_{1 \leq i \leq s,\ k \nmid i} a_i z_2^i, \quad h_C(z_1, z_2) = (\xi z_1, \zeta z_2),$$

where $\zeta \in \mathbb{U}_{k-1} \cap \mathbb{U}_s$, $a_i(\zeta^i - \xi) = 0$. In particular $Aut^0(S) \simeq \mathbb{U}_l$.
3) Case $\sigma_n(S) = 3n$. After conjugation F_C has the normal form

$$F_C(z_1, z_2) = (z_1^p z_2^q, z_1^r z_2^s),$$

where $A = \begin{pmatrix} p & q \\ r & s \end{pmatrix}$ is the composition of matrices of type $\begin{pmatrix} 1 & 1 \\ 0 & 1 \end{pmatrix}$ or $\begin{pmatrix} 0 & 1 \\ 1 & 1 \end{pmatrix}$ with at least one of the second type. Moreover $h_C(z_1, z_2) = (\mu_1 z_1, \mu_2 z_2)$ with $\mu_i \in \mathbb{U}_{d_i}$, $i = 1, 2$ and $d_1 d_2 = \det(A - Id)$. In particular $Aut^0(S) \simeq \mathbb{U}_{d_1} \times \mathbb{U}_{d_2}$.

THEOREM 2.4. *Let S be a surface containing a GSS such that $n = b_2(S) > 0$. Let H be a finite subgroup of $Aut^0(S)$. Then*
1) The quotient $T = S/H$ is a normal surface containing a GSS and the singular points are Jung-Hirzebruch singularities.
2) The minimal desingularization (T', π) of T contains a GSS. Moreover

- *If S is a Inoue surface S_t^n with $tr(S) = t$, then there exists integers $p \geq 1$ and a divisor q of n such that $H \simeq \mathbb{U}_p \times \mathbb{U}_q$, T' is the Inoue surface $S_{t^q}^{pn/q}$ with $tr(T') = t^q$ and $b_2(T') = pn/q$.*
  If S is not a Inoue surface but $tr(S) = t \neq 0$, there is a divisor q of n, $1 \leq q \leq n$ such that $H \simeq \mathbb{U}_q$, $tr(T') = t^q$ and $b_2(T') = n/q$.
- *If S is an intermediate (resp. Inoue-Hirzebruch) surface then T' is an intermediate (resp. Inoue-Hirzebruch) surface.*

PROOF. 1) Case $\sigma_n(S) = 2n$: a) If S is a Inoue surface, any automorphism is of the type $h_C(z) = (az_1, \lambda z_2)$ where $a \in \mathbb{C}^\star$ and $\lambda^n = 1$. Since H is finite, H is a subgroup of $S^1 \times S^1$. Let pr_2 be the second projection, $\mathbb{U}_q = pr_2(H)$. The kernel $R_C := \ker pr_2 \simeq \mathbb{U}_p$ is a reflection group (see [**BR**]), therefore the quotient germ $(\hat{S}_C, \hat{O}_C)/R_C$ is regular. Let (a', λ) be a generator of the cyclic group $H_C/R_C \simeq \mathbb{U}_q$, where $\lambda = exp(2\pi i/q)$. It is clear that $a'^q = 1$ hence $a' = \lambda^j$ for some j. For $h_C(z_1, z_2) = (\lambda^j z_1, \lambda z_2)$, a straightforward computation shows that $h_{C+1}(z_1, z_2) = (\lambda^{j-1} z_1, \lambda z_2)$. Commutative diagrams of lemma 2.1, show that H_C and H_{C+1} are isomorphic and that if h_C is a generator of H_C/R_C then h_{C+1} is a generator of H_{C+1}/R_{C+1}. Therefore replacing C by $C+j$, H_C is a reflection group isomorphic to $\mathbb{U}_p \times \mathbb{U}_q$. The quotient $\hat{T}_D$ is regular at the fixed point $\hat{P}_D$ and has regular points or Jung-Hirzebruch singularities at other fixed points. The boundary of any small ball centered at $\hat{P}_D$ defines a GSS in T. Moreover $\alpha_C(z_1, z_2) = (z_1^p, z_2^q)$, hence we obtain easily $U_D(u, v) = (uv^{pn/q}t^{pn}, t^q v)$ and we have the result.
b) If S is not a Inoue surface, $F_C(z_1, z_2) = (z_1 z_2^n t^n + \sum_{k=0}^{n-1} \alpha_k t^{k+1} z_2^{k+1}, tz_2)$ where at least one coefficient α_k is different from 0. Let $j = Max\{k \mid \alpha_k \neq 0\}$ and let m be the integer defined by $m = n$ (resp. $m = gcd\{k - k', n \mid \alpha_k \alpha_{k'} \neq 0\}$) if there is only one coefficient (resp. at least two coefficients) different from 0.
Replacing F_C by F_{C+1} we notice that j is increased by one but m does not change. Therefore we may choose the curve C in such a way that $j = n - 1$. Hence $h_C(z_1, z_2) = (z_1, \lambda z_2)$, with $\lambda^m = 1$. Therefore H_C is a reflection group, $\hat{P}_D$ is a regular point and we conclude as before.
2) Case $2n < \sigma_n(S) < 3n$: The hypothesis on $\sigma_n(S)$ implies that there exists a curve C such that $\hat{O}_C \in \hat{S}_C$ is not at the intersection of two rational curves of $\hat{S}_C$, however C meets two other curves. An automorphism h_C has three fixed points hence C is a set of fixed points. If $s/(k-1)$ is an integer and $\mu = 1$, $h_C(z) = (z_1 + cz_2^{s/(k-1)}, \zeta z_2)$ with $\zeta \in \mathbb{U}_s \cap \mathbb{U}_{k-1}$. Furthermore, if $c \neq 0$ the group generated by h_C is infinite, therefore $c = 0$, H_C is a reflection group and T contains a GSS. The curve $D \subset T$ meets three other curves transversaly and singularities are Jung-Hirzebruch ones; hence by [**BPV**] p80, the curve induced in T' meets three others; it means that T' is an intermediate surface. If $s/(k-1)$ is not an integer the result is clear.
3) Case $\sigma_n(S) = 3n$: Similar arguments as in 1) yield the result. □

Let G be a finite group of automorphisms; then $H = G \cap Aut^0(S)$ is a normal subgroup in G. Let $K := G/H$ be the quotient group. By the previous theorem, $X := S/H$ is a normal analytic surface containing a GSS with Jung-Hirzebruch singularities, such that $K \cap Aut^0(X) = \{Id\}$. Let (X', π, X) be the minimal desingularization of X and K' be the group of automorphisms induced by K. Blowing down, if necessary, the exceptional curves we may suppose that X' is minimal and $K' \cap Aut^0(X') = \{Id\}$. The group K' acts on X' without fixed point by [**D3**], Proposition 3.7., therefore by [**D3**] Corollaire 3.9, K' is cyclic, $(X', \omega', X'/K')$ is a galois covering and X'/K' contains a GSS. The desingularizations (X', π, X) consists in adding chains of non singular rational curves which do not alter the configuration of curves (see [**BPV**], p 80) in particular the number of cycles and the number of trees does not change. Since X'/K' is the desingularization of $Y = S/G = X/K$, we have proved the

COROLLARY 2.5. *Let S be a surface with $b_2(S) > 0$ containing a GSS and G be a finite group of automorphisms. Then :*
1) The quotient $Y = S/G$ is a normal surface containing a GSS and the singular points are Jung-Hirzebruch singularities.
2) The minimal desingularization (Y', π) of Y contains a GSS. Moreover $tr(Y') \neq 0$ if and only if $tr(S) \neq 0$, and the number of cycles (resp. trees) of rational curves in Y' and in S are equal.

References

[BPV] W. Barth, C. Peters, A. Van de Ven A, *Compact Complex Surfaces.* Springer (1984).
[BR] K. Behnke, O. Riemenschneider, *Quotient surface singularities and their deformations.* Singularity Theory I.C.T.P. Ed. D.T. Lê, K. Saito & B. Teissier. World Sc. (1995).
[D1] G. Dloussky, *Structure des surfaces de Kato.* Mémoire de la S.M.F **112**. n°14 (1984).
[D2] G. Dloussky, *Sur la classification des germes d'applications holomorphes contractantes.* Math. Ann. **280**, (1988), 649-661.
[D3] G. Dloussky, *Une construction élémentaire des surfaces d'Inoue-Hirzebruch.* Math. Ann. **280**, (1988), 663-682.
[DK] G. Dloussky, F. Kohler *Classification of singular germs of mappings and deformations of compact surfaces of class VII_0.* Ann. Polonici Math. **LXX** (1998), 49-83.
[DO1] G. Dloussky, K. Oeljeklaus K, *Vector fields and foliations on surfaces of class VII_0.* Ann. Inst. Fourier **49** (1999), 1503-1545.
[DO2] G. Dloussky, K. Oeljeklaus, *Surfaces de la classe VII_0 et automorphismes de Hénon.* Note aux CRAS **328**, Série I (1999), 609-612.
[DOT1] G. Dloussky, K. Oeljeklaus, M. Toma, *Surfaces de la classe VII_0 admettant un champ de vecteurs.* Comment. Math. Helv. **75** (2000) 255-270.
[DOT2] G. Dloussky, K. Oeljeklaus, M. Toma, *Surfaces de la classe VII_0 admettant un champ de vecteurs II.* Preprint LATP. UMR 6632 CNRS Marseille (2000).
[E] I. Enoki I, *Surfaces of class VII_0 with curves.* Tôhoku Math. J. **33**, (1981), 453-492.
[FA] C. Favre, *Classification of 2-dimensional contracting rigid germs and Kato surfaces: I* J. Math. Pures Appl. **79**, 5 (2000), 475-514.
[I] M. Inoue, *New surfaces with no meromorphic functions II.* Complex Analysis and Alg. Geom. W.L. Baily and T. Shioda ed. 91-106. Cambridge Univ. Press and Iwanami Shoten Pb. (1977).
[LT] M. Lűbke and A.D. Teleman, *The Kobayashi-Hitchin correspondance.* World Sc. (1995).
[N1] I. Nakamura, *On surfaces of class* VII_0 *with curves.* Invent. Math. **78** (1984), 393-443.
[N2] I. Nakamura, *On surfaces of class VII_0 with curves II.* Tohoku Math. J. **42** (1990), 475-516.
[OTZ] K. Oeljeklaus, M. Toma, D. Zaffran, *Une caractérisation des surfaces d'Inoue-Hirzebruch.* Preprint LATP Marseille (2000).

CMI, UNIVERSITÉ D'AIX-MARSEILLE 1, 39 RUE F. JOLIOT-CURIE, 13013 MARSEILLE
E-mail address: dloussky@cmi.univ-mrs.fr

Contemporary Mathematics
Volume **288**, 2001

Hypercomplex nilpotent Lie groups

Isabel G. Dotti and Anna Fino

Abstract. We prove that an abelian hypercomplex structure can only occur on solvable Lie groups and every hypercomplex 8-dimensional nilmanifold is quaternionic (i.e. the Obata connection is flat). We provide also an example of a 3-step, 12-dimensional nilpotent Lie group with an hypercomplex structure which is not Obata flat.

1. Introduction

An invariant hypercomplex structure on a Lie group G is a triple of integrable complex structures $\{J_i\}_{i=1,2,3}$ on the corresponding Lie algebra $\mathfrak{g}$ satisfying the quaternion identities $J_i^2 = -1, i = 1,2,3$ and $J_1J_2 = J_3 = -J_2J_1$. Invariant hypercomplex structures on compact Lie groups were studied in [**STVS**] in the context of $N = 4$ supersymmetry. D. Joyce (see [**J1**]) constructed, for any compact Lie group G, an invariant hypercomplex structure on $T^{2n-r} \times G$, where r is the rank of G. This extends a result of Samelson who showed that every compact Lie group G of even dimension has a complex structure such that left translations are holomorphic mappings (see [**Sam**]).

There have been a number of recent results on invariant hypercomplex structures on non-compact Lie groups. The classification, for 4-dimensional real Lie groups was given in [**B1**]. In [**BD**] it was shown that every solvable Lie group associated to a rank one symmetric space of non compact type with dimension divisible by 4, admits such a structure. On the other hand, examples were given of solvable Lie groups G such that no product with $\mathbb{R}^s$ admits a hypercomplex structure, in contrast to the compact case discussed in [**J1**]. The case of nilpotent Lie groups had already been considered in [**J2**], [**BDM**] and [**B2**]. The construction given in [**J2**] allows to obtain as a byproduct, many non homogeneous hypercomplex manifolds.

In [**DF2**] a classification of hypercomplex structures on 8-dimensional nilpotent Lie algebras is given. In the present note, using the description in the main result in [**DF2**], we prove that every hypercomplex 8-dimensional nilmanifold is quaternionic, i.e the Obata connection is flat (see Section 2). In Section 3 we concentrate on *abelian* hypercomplex structures, which have the special property

2000 *Mathematics Subject Classification.* Primary 53C56; Secondary 53C30.
Key words and phrases. Hypercomplex, nilpotent Lie groups.
Research partially supported by Conicet, SecytU.N.C, Argentina.
Research partially supported by MURST and CNR of Italy.

that for one (hence for all) of the complex structures in the family, the associated (1, 0)-vector fields commute. We will prove that such a structure can only occur on solvable Lie groups (Proposition 3.1). For such groups these structures are very common. For instance, in 2-step nilpotent Lie groups many examples have been constructed (see [**DF1**], [**B2**]). In a nilpotent Lie group of dimension 8, the existence of a hypercomplex structure forces $\mathfrak{g}$ to be 2-step (see [**DF2**]). However, in Section 3 we provide an example of an abelian hypercomplex structure on a 3-step, 12-dimensional nilpotent Lie group that is not Obata flat.

2. 8-dimensional hypercomplex nilmanifolds

We note that the only nilpotent Lie group of dimension 4 which admits an hypercomplex structure is abelian. On the other hand, in the case of nilpotent Lie groups of dimension 8, the condition of admitting an invariant hypercomplex structure, imposes strong restrictions on G. In [**DF2**] the following result is proved

THEOREM 2.1. *Let G be an 8-dimensional nilpotent Lie group endowed with a hypercomplex structure. Then G is at most 2-step nilpotent and the first Betti number of the corresponding Lie algebra $\mathfrak{g}$ satisfies $b_1(\mathfrak{g}) \geq 4$.*

It is also proved (in [**DF2**]) that there exists a basis of left invariant 1-forms

$$\{e^1, e^2 = J_1e^1, e^3 = J_2e^1, e^4 = J_3e^1, e^5, e^6 = J_1e^5, e^7 = J_2e^5, e^8 = J_3e^5\}$$

such that, if $b_1(\mathfrak{g}) = 4$ and e^{kl} stands for $e^k \wedge e^l$ then

$$\begin{aligned}
de^i &= 0, i = 1, \dots, 4,\\
de^5 &= a_1e^{12} + a_2e^{13} + a_3e^{14} + (-a_3 + t_3)e^{23} + (a_2 - t_2)e^{24} + (-a_1 + t_1)e^{34},\\
de^6 &= b_1e^{12} + b_2e^{13} + b_3e^{14} + (-b_3 + t_2)e^{23} + (b_2 + t_3)e^{24} + (-b_1 + t_4)e^{34},\\
de^7 &= c_1e^{12} + c_2e^{13} + c_3e^{14} + (-c_3 - t_1)e^{23} + (c_2 - t_4)e^{24} + (-c_1 + t_3)e^{34},\\
de^8 &= d_1e^{12} + d_2e^{13} + d_3e^{14} + (-d_3 + t_4)e^{23} + (d_2 - t_1)e^{24} + (-d_1 - t_2)e^{34},
\end{aligned}$$

with $t_1^2 + t_2^2 + t_3^2 + t_4^2 \neq 0$. If $b_1(\mathfrak{g}) \geq 5$, then

$$\begin{aligned}
de^i &= 0, i = 1, \dots, 5,\\
de^6 &= b_1(e^{12} - e^{34}) + b_2(e^{13} + e^{24}) + b_3(e^{14} - e^{23}) + te^{34},\\
de^7 &= c_1(e^{12} - e^{34}) + c_2(e^{13} + e^{24}) + c_3(e^{14} - e^{23}) - te^{24},\\
de^8 &= d_1(e^{12} - e^{34}) + d_2(e^{13} + e^{24}) + d_3(e^{14} - e^{23}) + te^{23}.
\end{aligned}$$

REMARK 2.2. Note that the center $\mathfrak{z}$ of $\mathfrak{g}$ contains the 4-dimensional subspace $W = span\{e_5, e_6, e_7, e_8\}$, which is invariant with respect to J_i, $i = 1, 2, 3$.

It is known (see [**O**]) that every hypercomplex structure $\{J_i\}_{i=1,2}$ on a $4k$-dimensional differentiable manifold M uniquely determines an affine, torsion free connection (called the Obata connection), with respect to which the complex structures $J_i, i = 1, 2, 3$ are parallel. In other words, the $GL(k, \mathbb{H})$-structure they determine admits a torsion free connection. The above description of the hypercomplex structures on 8-dimensional nilpotent Lie groups allows us to show that its Obata connection is flat. Indeed one has

PROPOSITION 2.1. *Every hypercomplex structure on a 8-dimensional nilpotent Lie group is flat.*

PROOF. The Obata connection associated to an invariant hypercomplex structure $J_1, J_2, J_3 = J_1J_2$ on a nilpotent Lie group G is given by

$$\nabla_X Y = \frac{1}{2}[X,Y] + \frac{1}{12}\sigma([J_1X, J_3Y] + [J_2X, J_3Y]) + \frac{1}{6}\sum_{i=1}^{i=3}[J_iX, Y] + [J_iY, X],$$

where X, Y lie in $\mathfrak{g}$ and σ denotes the cyclic sum with respect to the indices $1, 2, 3$.

Using Remark 2.2 one can prove that

$$\nabla_X Y = 0, \quad \textit{for } X \in W \quad \textit{or} \quad Y \in W,$$

since W is J_i-invariant, for $i = 1, 2, 3$. Moreover, for any $X, Y \in \mathfrak{g}$, we have that $\nabla_X Y \in W$. Hence, the curvature tensor R associated to ∇ vanishes. □

If the connection is flat, then M is quaternionic in the sense of Sommese ([**S**]), that is, M can be covered by coordinate neighborhoods such that the transition functions are quaternionic. In particular, the $GL(k, \mathbb{H})$-structure is integrable, hence a flat affine structure exists on M. Thus, as a consequence of the previous proposition it will follow that any 8-dimensional hypercomplex nilmanifold M is quaternionic in the sense of Sommese.

Furthermore using the previous Proposition it is easy to prove that (G, J_i, ∇) is a special complex manifold in the sense of [**ACD**], since the Obata connection ∇ is a flat torsion-free connection such that $\nabla J_i = 0$. In this case any affine function f (i.e. a function satisfying $\nabla df = 0$) can be extended to a holomorphic function F such that $Re\, F = f$.

3. Abelian hypercomplex structures

Among the left invariant almost complex structures J on a Lie group G one may consider as a special class, those $J's$ satisfying

$$[JX, JY] = [X, Y], \tag{1}$$

for all $X, Y \in \mathfrak{g}$.

One verifies easily that these almost complex structures are integrable, that is,

$$0 = N_J(X, Y) = J[X, Y] - [JX, Y] - [X, JY] - J[JX, JY].$$

Complex structure satisfying condition (1) are called *abelian* since the associated $(1, 0)$ vector fields commute. They have been studied previously in [**BD**], [**DF1**], [**BDM**], [**Sal**], [**B2**]. Abelian complex structures are common on solvable Lie algebras. See [**Sn**], [**Ov**] for families of abelian complex structures on solvable, non nilpotent, 4-dimensional Lie algebras. The following proposition shows that only solvable Lie groups can carry this class of structures.

PROPOSITION 3.1. *If a Lie group G carries an invariant abelian complex structure then G is solvable.*

PROOF. We note first that semisimple Lie groups do not admit abelian complex structures of the above type. To prove the assertion, let $\langle\,,\,\rangle$ be any $\mathrm{Ad}(G)$-invariant non degenerate bilinear form on $\mathfrak{g}$. Then, if J is abelian one has $[JX, Y] = -[X, JY]$ giving

$$\begin{aligned}\langle J^*[X,Y], Z\rangle &= \langle X, [Y, JZ]\rangle = -\langle X, [JY, Z]\rangle = \\ &= -\langle [X, JY], Z\rangle = \langle [JX, Y], Z\rangle,\end{aligned}$$

where J^* is the adjoint with respect to $\langle\, ,\, \rangle$. Hence

$$J^*[X,Y] = [JX,Y] = -[X,JY] = [JY,X] = J^*[Y,X]$$

thus $J^* = 0$ on $\mathfrak{g}^1 = \mathfrak{g}$, which is not possible.

Let $\mathfrak{d}$ be the radical of $\mathfrak{g}$, the Lie algebra of G. Then J abelian implies $\mathfrak{d} + J\mathfrak{d}$ is a solvable ideal and by maximality it must coincide with $\mathfrak{d}$. We claim $\mathfrak{d} = \mathfrak{g}$. Otherwise, J induces on the quotient semisimple Lie algebra $\mathfrak{g}/\mathfrak{d}$ an abelian complex structure contradicting what we have proved above. □

Every hypercomplex structure gives rise to a sphere of complex structures. We show next that if one of the (invariant) complex structures on the sphere is abelian then all of the complex structures in the sphere are abelian.

LEMMA 3.1. *If J, K are anticommuting complex structures on $\mathfrak{g}$ and J satisfies (1) then K also satisfies (1).*

PROOF. Since K is integrable

$$\begin{aligned}[KX,Y] + [X,KY] + K[KX,KY] = K[X,Y] =\\ = K[JX,JY] = [KJX,JY] + [JX,KJY] + K[KJX,KJY] =\\ = -[KX,Y] - [X,KY] + K[KX,KY],\end{aligned}$$

showing that K is abelian. □

The hypercomplex structure is called *abelian* if $[J_iX, J_iY] = [X,Y]$, for any $i = 1,2,3$ and $X, Y \in \mathfrak{g}$. For this, it is sufficient by the above lemma, that one of the complex structures in the associated sphere is abelian.

In [**DF1**] it is proved that an 8-dimensional nilpotent Lie group with an abelian hypercomplex structure is either abelian or a trivial extension of a group of Heisenberg type. In [**B2**], the 2-step nilpotent Lie algebras carrying abelian hypercomplex structures are characterized and it is shown in particular that the result quoted above is not longer true in higher dimensions.

A natural question that comes is whether there exist examples of s-step nilpotent Lie groups (with $s > 2$) in dimension greater than eight endowed with a hypercomplex structure. The following example gives a positive answer to this question.

EXAMPLE 3.2. Consider the 3-step nilpotent Lie algebra $\mathfrak{g}$ of dimension 12 defined by

$$\begin{aligned} de^i &= 0, i = 1, \dots, 9,\\ de^{10} &= e^1 \wedge e^2 - e^5 \wedge e^6,\\ de^{11} &= e^2 \wedge e^5 - e^1 \wedge e^6,\\ de^{12} &= e^1 \wedge e^4 + e^2 \wedge e^{10} + e^5 \wedge e^8 + e^6 \wedge e^{11}.\end{aligned}$$

It is possible to check that $\mathfrak{g}$ admits an abelian hypercomplex structure $\{J_1, J_2\}$, with corresponding basis of $(1,0)$-forms,

$$\begin{cases} \omega_1^1 = e^1 - ie^2, & \omega_2^1 = e^3 - ie^{12}, & \omega_3^1 = e^4 - ie^{10},\\ \omega_4^1 = e^5 - ie^6, & \omega_5^1 = e^7 - ie^9, & \omega_6^1 = e^8 - ie^{11},\end{cases}$$

$$\begin{cases} \omega_1^2 = e^1 - ie^6, & \omega_2^2 = e^2 - ie^5, & \omega_3^2 = e^3 - ie^9,\\ \omega_4^2 = e^4 - ie^{11}, & \omega_5^2 = e^{10} - ie^8, & \omega_6^2 = e^{12} - ie^7.\end{cases}$$

Furthermore we note that the assertion in Proposition 2.3 does not hold in this case. Indeed, the curvature tensor R associated to Obata connection is not zero; for example one can check that $R_{e_1 e_2} e_6 = \frac{1}{3} e_{12}$, where $\{e_i\}$ is the dual basis of $\{e^i\}$.

References

[ACD] D. V. Alekseevsky, V. Cortes, C. Devchand, *Special complex manifolds*, preprint math. DG/9910091.

[B1] M. L. Barberis, *Hypercomplex structures on four-dimensional Lie groups*, Proc. Amer. Math. Soc. **125** (1997), no. 4, 1043-1054.

[B2] M. L. Barberis, *Abelian hypercomplex structures on central extensions of H-type algebras*, Journal of Pure and Applied Algebra, to appear.

[BDM] M. L. Barberis, I. Dotti Miatello, R. Miatello, *On certain locally homogeneous Clifford manifolds*, Ann. Glob. Anal. and Geom. **13** (1995), 289-301.

[BD] M. L. Barberis, I. Dotti Miatello, *Hypercomplex structures on a class of solvable Lie groups*, Quart. J. Math. Oxford (2) **47** (1996), 389-404.

[DF1] I. Dotti, A. Fino, *Abelian hypercomplex 8-dimensional nilmanifolds*, Ann. Glob. Anal. and Geom. **18** (2000), 47-59.

[DF2] I. Dotti, A. Fino, *Hypercomplex 8-dimensional nilpotent Lie groups*, preprint (2000).

[J1] D. Joyce, *Compact hypercomplex and quaternionic manifolds*, J. Diff. Geom. **35** (1992), 743-761.

[J2] D. Joyce, *Manifolds with many complex structures*, Quart. J. Math. Oxford (2) **35** (1992), 743-761.

[O] M. Obata, *Affine connections on manifolds with almost complex, quaternionic or Hermitian structures*, Japan. Jour. Math. **26** (1956), 43-79.

[Ov] G. Ovando, *Invariant complex structures on solvable real Lie groups*, Manuscripta mathematica, to appear.

[Sal] S. Salamon, *Complex structures on nilpotent Lie algebras*, J. Pure Appl. Algebra, to appear.

[Sam] H. Samelson, *A class of complex analytic manifolds*, Portugal. Math. **12** (1953) 129-132.

[Sn] J. E. Snow, *Invariant complex structures on four dimensional solvable real Lie groups*, Manuscripta Math. **66** (1990), 397-412.

[S] A. Sommese, *Quaternionic manifolds*, Math. Ann. **212** (1975), 191-214.

[STVS] A. Sevrin, W. Troost, A. Van Proeyen, Ph. Spindel, *Extended supersymmetric σ-models on group manifolds. II. Current algebras*, Nuclear Phys. B **311** (1988/89), no. 2, 465-492.

FAMAF, CIEM, UNC, CONICET, 5000 Cordoba, Argentina

E-mail address: `idotti@mate.uncor.edu`

Dipartimento di Matematica, Universitá di Torino, via Carlo Alberto 10, 10123 Torino, Italy

E-mail address: `fino@dm.unito.it`

Contemporary Mathematics
Volume **288**, 2001

$\mathfrak{C}$-Spaces of Iwasawa Type and Damek-Ricci Spaces

Maria J. Druetta

ABSTRACT. A Lie group of Iwasawa type is a $\mathfrak{C}$-space if the eigenvalues of the Jacobi operators along geodesics are constant. We show that a Lie group of Iwasawa type and algebraic rank one which is a $\mathfrak{C}$-space is a Damek-Ricci space. Hence, it is a symmetric space of noncompact type and rank one. In particular, irreducible, non-flat homogeneous $\mathfrak{C}$-spaces with nonpositive curvature and algebraic rank one, are rank one symmetric spaces of noncompact type.

1. Condition (C) on Lie algebras of Iwasawa type and rank one

A solvable Lie algebra $\mathfrak{s}$ with inner product $\langle , \rangle$ is a metric Lie algebra of Iwasawa type and algebraic rank one if it satisfies the following conditions:

(i) $\mathfrak{s} = \mathfrak{n} \oplus \mathfrak{a}$ where $\mathfrak{n} = [\mathfrak{s}, \mathfrak{s}]$ and $\mathfrak{a}$, the orthogonal complement of $\mathfrak{n}$, is one-dimensional.

(ii) An element $H \in \mathfrak{a}$, $|H| = 1$, can be chosen so that the adjoint operator $\mathrm{ad}_H|_{\mathfrak{n}}$ is symmetric and its eigenvalues are all positive.

The simply connected Lie group S with Lie algebra $\mathfrak{s}$ and left invariant metric induced by the inner product $\langle , \rangle$ will be called of Iwasawa type. The Levi Civita connection and the curvature tensor associated to the metric can be computed by

$$\begin{aligned} 2\langle \nabla_X Y, Z \rangle &= \langle [X,Y], Z \rangle - \langle [Y,Z], X \rangle + \langle [Z,X], Y \rangle, \\ R(X,Y) &= [\nabla_X, \nabla_Y] - \nabla_{[X,Y]} \end{aligned}$$

for any X, Y, Z in $\mathfrak{s}$. For a unit $X \in \mathfrak{s}$, the Jacobi operator R_X is defined by $R_X Y = R(Y,X)X$. Following [1], we will say that either S is a $\mathfrak{C}$-space or the metric Lie algebra $\mathfrak{s}$ satisfies condition (C) if for every geodesic γ in S the eigenvalues of the associated Jacobi operator $R_{\gamma'(t)}$ are constant functions of t.

If $\mathfrak{s} = \mathfrak{n} \oplus \mathbf{R}H$ is a metric Lie algebra of Iwasawa type and algebraic rank one, let $\mathfrak{z}$ denote the center of $\mathfrak{n} = [\mathfrak{s}, \mathfrak{s}]$ and let $\mathfrak{v}$ be the orthogonal complement of $\mathfrak{z}$ with respect to the metric $\langle , \rangle$ restricted to $\mathfrak{n}$. Thus $\mathfrak{n}$ decomposes as $\mathfrak{n} = \mathfrak{z} \oplus \mathfrak{v}$, and for all $H \in \mathfrak{a}$, $\mathrm{ad}_H : \mathfrak{z} \to \mathfrak{z}$, hence $\mathrm{ad}_H : \mathfrak{v} \to \mathfrak{v}$ since it is symmetric.

For any $Z \in \mathfrak{z}$ the skew-symmetric linear operator $j_Z : \mathfrak{v} \to \mathfrak{v}$ is defined by

$$\langle j_Z X, Y \rangle = \langle [X,Y], Z \rangle \quad \text{for all } X,\ Y \in \mathfrak{v} \text{ and } Z \in \mathfrak{z}.$$

2000 *Mathematics Subject Classification.* Primary 53C30, 53C55.
Key words and phrases. $\mathfrak{C}$-spaces, Iwasawa type, Jacobi nonpositive curvature.
Partially supported by CIEM (Conicet), Conicor and Secyt (UNC).

We recall that $\mathfrak{n}$ has an orthogonal direct sum decomposition into eigenspaces $\mathfrak{n}_\mu$, for all eigenvalues μ of $\mathrm{ad}_H|_{\mathfrak{n}}$, which are invariant by ad_H with the property $[\mathfrak{n}_\mu, \mathfrak{n}_{\mu'}] \subset \mathfrak{n}_{\mu+\mu'}$ (by the Jacobi identity), whenever $\mu+\mu'$ is an eigenvalue of $\mathrm{ad}_H|_{\mathfrak{v}}$. Moreover, since $\mathfrak{z}$ and $\mathfrak{v}$ are ad_H-invariant, they also have decompositions into their eigenspaces $\mathfrak{z} = \sum_\lambda \mathfrak{z}_\lambda$, $\mathfrak{v} = \sum_\mu \mathfrak{v}_\mu$.

We assume that $\mathfrak{n}$ is non-abelian. If λ is an eigenvalue of $\mathrm{ad}_H|_{\mathfrak{z}}$ and $Z \in \mathfrak{z}_\lambda$, then $j_Z|_{\mathfrak{v}_\mu} : \mathfrak{v}_\mu \to \mathfrak{v}_{\lambda-\mu}$ for all eigenvalues μ of $\mathrm{ad}_H|_{\mathfrak{v}}$. In fact, for $X \in \mathfrak{v}_\mu$ and $Y \in \mathfrak{v}_{\mu'}$

$$\langle j_Z X, Y\rangle = \langle [X,Y], Z\rangle \neq 0 \Rightarrow [X,Y] \neq 0,$$

and thus $\mu + \mu' = \lambda$ since $[X,Y] \in \mathfrak{n}_{\mu+\mu'}$ and it has non-zero component in $\mathfrak{z}_\lambda$.

Hence, if j_Z is non-singular then $\lambda > \mu$, $\lambda - \mu$ is also an eigenvalue of $\mathrm{ad}_H|_{\mathfrak{v}}$ and the symmetric operator j_Z^2 preserves $\mathfrak{v}_\mu$, for all eigenvalues μ of $\mathrm{ad}_H|_{\mathfrak{v}}$.

Note that homogeneous $\mathfrak{C}$-spaces represented as Lie group of Iwasawa type and algebraic rank one are Einstein spaces, since they are irreducible and the Ricci curvatures are invariant under the geodesic flow, which means that $\nabla_X \mathrm{Ric}(X,X) = 0$ for all $X \in \mathfrak{s}$. (see Theorem 3.1 of [7]).

In order to describe the eigenvalues of the Jacobi operators along certain geodesics, by applying the connection formulas we compute $\nabla_H = 0$ and for $Z, Z^* \in \mathfrak{z}$ and $X \in \mathfrak{v}$, $\nabla_Z Z^* = \nabla_{Z^*} Z = \langle [H,Z], Z^*\rangle H$, $\nabla_X Z = \nabla_Z X = -\frac{1}{2} j_Z X$. Thus,

(i) $R_H = -\mathrm{ad}_H^2$.

(ii) If either $Z \in \mathfrak{z}_\lambda$, $|Z| = 1$, or $X \in \mathfrak{v}_\mu$, $|X| = 1$,

$$R_Z H = -\lambda^2 H \text{ and } R_X H = -\mu^2 H.$$

(iii) If $Z \in \mathfrak{z}_\lambda$, $|Z| = 1$, for any $Z^* \in \mathfrak{z}$ and $X \in \mathfrak{v}$, we have

$$R_Z Z^* = \lambda\ (\langle Z, \mathrm{ad}_H Z^*\rangle Z - \mathrm{ad}_H Z^*) \text{ and } R_Z X = -\frac{1}{4} j_Z^2(X) - \lambda \mathrm{ad}_H X.$$

The proofs of the following lemma and proposition are given in [3].

LEMMA 1. *If $Z \in \mathfrak{z}$ is an eigenvector of ad_H with associated eigenvalues λ (that is $Z \in \mathfrak{z}_\lambda$), then for any $Z^* \in \mathfrak{z}$ and $Y \in \mathfrak{v}$, we have*

(i) $R_{\gamma'_Z(t)} Z^*(t)$

$$= \frac{1}{\cosh^2 \lambda t}(dL_{\gamma_Z(t)})_e R_Z Z^* - \sinh^2 \lambda t\ ad_H^2 Z^* - \lambda^2 \langle Z, Z^*\rangle \sinh \lambda t H.$$

(ii) $R_{\gamma'_Z(t)} Y(t)$

$$= \frac{1}{\cosh^2 \lambda t}(dL_{\gamma_Z(t)})_e \left(R_Z Y - \sinh^2 \lambda t\ ad_H^2 Y - \sinh \lambda t j_Z \left(\frac{1}{2}\lambda\, Id - ad_H\right) Y\right).$$

PROPOSITION 2. *Let $Z \in \mathfrak{z}_\lambda$ and $X \in \mathfrak{v}_\mu$ be unit vectors such that X is an eigenvector of the symmetric operator j_Z^2 restricted to $\mathfrak{v}_\mu$. Then for all real t, $R_{\gamma'_Z(t)}$ has an eigenvector $U(t) = x(t)X(t) + y(t)Y(t)$, $Y = \frac{j_Z X}{|j_Z X|}$, for some $x(t)$, $y(t)$ with $x(t)^2 + y(t)^2 = 1$, and associated eigenvalue $a_Z(t)$ satisfying*

$$a_Z(t)\cosh^2 \lambda t = \langle R_Z X, X\rangle - \mu^2 \sinh^2 \lambda t - (\frac{1}{2}\lambda - \mu)\,|j_Z X|\,\frac{y(t)}{x(t)} \sinh \lambda t, \quad or$$

$$a_Z(t)\cosh^2 \lambda t = \langle R_Z Y, Y\rangle - (\lambda-\mu)^2 \sinh^2 \lambda t - (\frac{1}{2}\lambda - \mu)\,|j_Z X|\,\frac{x(t)}{y(t)} \sinh \lambda t$$

according to $x(t) \neq 0$ or $y(t) \neq 0$, respectively.

PROOF. Let $T_t|_{\mathfrak{v}_\mu \oplus \mathfrak{v}_{\lambda-\mu}} : \mathfrak{v}_\mu \oplus \mathfrak{v}_{\lambda-\mu} \to \mathfrak{v}_\mu \oplus \mathfrak{v}_{\lambda-\mu}$ be the symmetric operator

$$T_t = \frac{1}{\cosh^2 \lambda t}\left(R_Z - \sinh^2 \lambda t \, \mathrm{ad}_H^2 - (\frac{1}{2}\lambda - \mu)\sinh \lambda t \,(j_Z, -j_Z)\right)$$

where $(j_Z, -j_Z) : X + Y \to -j_Z Y + j_Z X$. We have that $R_{\gamma_Z'(t)}$ and T_t satisfy

$$R_{\gamma_Z'(t)} X(t) = R_{\gamma_Z'(t)} (\mathrm{d}L_{\gamma_Z(t)})_e X = (\mathrm{d}L_{\gamma_Z(t)})_e (T_t X).$$

Consequently, $U(t) = (\mathrm{d}L_{\gamma_Z(t)})_e U_t$ is an eigenvector of $R_{\gamma_Z'(t)}$ with eigenvalue $a_Z(t)$, whenever $U_t \in \mathfrak{v}_\mu \oplus \mathfrak{v}_{\lambda-\mu}$ is an eigenvector of T_t with eigenvalue $a_Z(t)$. □

THEOREM 3. *If S is a Lie group of Iwasawa type of algebraic rank one which is a $\mathfrak{C}$-space, then S is a rank one symmetric space of noncompact type.*

PROOF. Let $\mathfrak{s} = \mathfrak{n} \oplus \mathbf{R}H$, $|H| = 1$, be the associated metric Lie algebra of Iwasawa type of S which satisfies the condition (C). Then:
(i) $\mathrm{ad}_H|_{\mathfrak{z}} = \lambda \mathrm{Id}$.

Let $Z \perp Z^*$ be unit eigenvectors of $\mathrm{ad}_H|_{\mathfrak{z}}$ with associated eigenvalues λ and λ^*, respectively. Applying that $R_Z Z^* = -\lambda\lambda^* Z^*$ in the expression of $R_{\gamma_Z'(t)}$ we get

$$\begin{aligned} R_{\gamma_Z'(t)} Z^*(t) &= \frac{1}{\cosh^2 \lambda t} (\mathrm{d}L_{\gamma_Z(t)})_e \left(-\lambda\lambda^* - (\lambda^*)^2 \sinh^2 \lambda t\right) Z^* \\ &= -\frac{\lambda^*}{\cosh^2 \lambda t}\left(\lambda + \lambda^* \sinh^2 \lambda t\right) Z^*(t). \end{aligned}$$

Hence,

$$-\frac{\lambda^*}{\cosh^2 \lambda t}(\lambda + \lambda^* \sinh^2 \lambda t) = -\lambda\lambda^* \text{ for all real } t,$$

or equivalently $\lambda^*(\lambda - \lambda^*)\sinh^2 \lambda t = 0$, implying that $\lambda^* = \lambda$ and $\mathrm{ad}_H|_{\mathfrak{z}} = \lambda \mathrm{Id}$.

Note that if $\mathfrak{n}$ is non-abelian, it follows that $\lambda > \mu$ for all eigenvalues μ of $\mathrm{ad}_H|_{\mathfrak{v}}$, since for the maximum eigenvalue μ_1 of $\mathrm{ad}_H|_{\mathfrak{v}}$, we have $\mu_1 + \mu^* = \lambda$ for some μ^*.
(ii) j_Z is non-singular for all $Z \neq 0$.

Indeed, if $j_Z X = 0$ for $Z \neq 0$ in $\mathfrak{z}$ and $X \neq 0$ in $\mathfrak{v}$, we can assume that $X \in \mathfrak{n}_\mu$ since $\ker j_Z$ is invariant by ad_H. Using the lemma and the curvature formulas,

$$\begin{aligned} R_{\gamma_Z'(t)} X(t) &= \frac{1}{\cosh^2 \lambda t} (\mathrm{d}L_{\gamma_Z(t)})_e \left(R_Z X - \sinh^2 \lambda t \, \mu^2 X\right) \\ &= -\frac{\mu}{\cosh^2 \lambda t}\left(\lambda + \mu \sinh^2 \lambda t\right) X(t). \end{aligned}$$

It follows for the eigenvalue of $R_{\gamma_Z'(t)}$ attained by $X(t)$ that

$$-\frac{\mu}{\cosh^2 \lambda t}\left(\lambda + \mu \sinh^2 \lambda t\right) = -\mu\lambda \text{ for all real } t$$

since the condition (C) is satisfied. Hence, $\mu \sinh^2 \lambda t \,(\lambda - \mu) = 0$ for all t, which is impossible since $\lambda > \mu$.
(iii) $\mathrm{ad}_H|_{\mathfrak{v}} = \frac{1}{2}\lambda \mathrm{Id}$.

Fix an eigenvalue μ of $\mathrm{ad}_H|_{\mathfrak{v}}$. Let $Z \in \mathfrak{z}$, $|Z| = 1$, and $X \in \mathfrak{v}_\mu$, $|X| = 1$, an eigenvector of j_Z^2. It follows from the proposition that $R_{\gamma_Z'(t)}$ has an eigenvector

$U(t) = x(t)X(t) + y(t)Y(t)$ ($Y = \frac{j_Z X}{|j_Z X|}$) whose eigenvalue $a_Z(t) = a_Z(0)$ satisfies

$$a_Z(0)\cosh^2 \lambda t + \mu^2 \sinh^2 \lambda t = \langle R_Z X, X\rangle - (\frac{1}{2}\lambda - \mu)\,|j_Z X|\,\frac{y(t)}{x(t)} \sinh \lambda t \quad \text{and}$$

$$a_Z(0)\cosh^2 \lambda t + (\lambda - \mu)^2 \sinh^2 \lambda t = \langle R_Z Y, Y\rangle - (\frac{1}{2}\lambda - \mu)\,|j_Z X|\,\frac{x(t)}{y(t)} \sinh \lambda t$$

according toh $x(t) \neq 0$ or $y(t) \neq 0$, respectively. Note that it follows from the expression of T_t given in Proposition 1.2 that

$$\lim_{t\to\infty} T_t|_{\mathfrak{v}_\mu \oplus \mathfrak{v}_{\lambda-\mu}} = -\operatorname{ad}_H^2|_{\mathfrak{v}_\mu \oplus \mathfrak{v}_{\lambda-\mu}} \quad (\lim_{t\to\infty} \frac{\sinh \lambda t}{\cosh \lambda t} = 1)$$

and consequently, under the hypothesis of condition (C),

$$\text{either } a_Z(0) = -\mu^2 \text{ or } a_Z(0) = -(\lambda - \mu)^2,$$

since the eigenvalues of $\operatorname{ad}_H^2|_{\mathfrak{v}_\mu \oplus \mathfrak{v}_{\lambda-\mu}}$ are μ and $\lambda - \mu$, and $a_Z(t) = a_Z(0)$ for all t.

If $x(0) \neq 0$ and $y(0) \neq 0$ we have that $\mu = \frac{1}{2}\lambda$ since $a_Z(0) = \langle R_Z X, X\rangle = \langle R_Z Y, Y\rangle$ gives the equality $\frac{1}{4}|j_Z X|^2 - \lambda\mu = \frac{1}{4}|j_Z X|^2 - \lambda(\lambda - \mu)$.

Assume that $\mu \neq \frac{1}{2}\lambda$. In the case $x(0) = 1$, $y(0) = 0$ ($y(0) = 1$, $x(0) = 0$), we also have $a_Z(0) = \langle R_Z X, X\rangle = \frac{1}{4}|j_Z X|^2 - \lambda\mu$ ($a_Z(0) = \langle R_Z Y, Y\rangle$) since $R_Z X = a_Z(0)X$ ($R_Z Y = a_Z(0)Y$). Therefore, from the first expression of $a_Z(t)$ above,

$$(a_Z(0) + \mu^2)\sinh^2 \lambda t = -(\frac{1}{2}\lambda - \mu)\,|j_Z X|\,\frac{y(t)}{x(t)} \sinh \lambda t$$

and consequently,

$$(a_Z(0) + \mu^2)\sinh \lambda t = -(\frac{1}{2}\lambda - \mu)\,|j_Z X|\,\frac{y(t)}{x(t)} \quad \text{for all real } t \neq 0.$$

If $a_Z(0) = -\mu^2$, we have that

$$(\frac{1}{2}\lambda - \mu)\,|j_Z X|\,\frac{y(t)}{x(t)} = 0 \text{ for all } t \neq 0,$$

giving a contradiction since $x(t) \neq 0$ and $y(t) \neq 0$ for all $t \neq 0$. If $a_Z(0) = -(\lambda-\mu)^2$, it follows from the above expression that

$$(-(\lambda - \mu)^2 + \mu^2)\sinh \lambda t = -(\frac{1}{2}\lambda - \mu)\,|j_Z X|\,\frac{y(t)}{x(t)}$$

or equivalently,

$$(2\mu - \lambda)\left(\lambda \sinh \lambda t - \frac{|j_Z X|}{2}\frac{y(t)}{x(t)}\right) = 0 \text{ for all real } t \neq 0.$$

This implies that

$$\frac{x(t)}{y(t)} \sinh \lambda t = \frac{|j_Z X|}{2\lambda} \quad \text{for all } t \neq 0.$$

By substituting this equality in the second expression of $a_Z(t)$ in the beginning of the proof of (iii), we obtain

$$\frac{1}{4}|j_Z X|^2 - \lambda\mu = \frac{1}{4}|j_Z X|^2 - \lambda(\lambda - \mu) + \frac{1}{4\lambda}(2\mu - \lambda)\,|j_Z X|^2,$$

which implies that $(2\mu - \lambda)\left(\lambda + \frac{1}{4\lambda}|j_Z X|^2\right) = 0$. The asumption $\mu \neq \frac{1}{2}\lambda$ gives a contradiction again since $\lambda > 0$ and $|j_Z X| \neq 0$.

(iv) $j_Z^2 = -|Z|^2 \lambda^2 \mathrm{Id}$ for all $Z \in \mathfrak{z}$.

Let $X \in \mathfrak{v}$ be an eigenvector of j_Z^2, $Z \in \mathfrak{z}$ a unit vector. By Lemma 1.1 (ii),

$$\frac{1}{\cosh^2 \lambda t}\left(\frac{1}{4}|j_Z X|^2 - \frac{1}{2}\lambda^2 - \frac{1}{4}\lambda^2 \sinh^2 \lambda t\right) = -\frac{1}{4}\lambda^2 \text{ for all real } t$$

since $\mathrm{ad}_H|_{\mathfrak{z}} = \lambda \mathrm{Id}$ and $\mathrm{ad}_H|_{\mathfrak{v}} = \frac{1}{2}\lambda \mathrm{Id}$. Hence, $\frac{1}{4}|j_Z X|^2 - \frac{1}{2}\lambda^2 = -\frac{1}{4}\lambda^2$, taking the value at $t = 0$ and limit as $t \to +\infty$ in the above expression, respectively. Thus, we get $|j_Z X|^2 = \lambda^2$. The assertion follows since $j_Z^2 X = -|j_Z X|^2 X$ for all $X \in \mathfrak{v}$.

Then, up to scaling, S is a Damek-Ricci space and applying Theorem 1 of [1, Section 4.3], S is a rank one symmetric space of noncompact type. □

COROLLARY 4. If M is an irreducible, non-flat homogeneous $\mathfrak{C}$-space with nonpositive curvature and algebraic rank one, then M is a symmetric space of noncompact type and rank one.

PROOF. Since M is irreducible and non-flat, by applying [6, Corollary 1] it is well known that it is represented as a simply connected, solvable Lie group with left invariant metric and nonpositive curvature. Moreover, M is Einstein since it is a $\mathfrak{C}$-space (see Corollary 3.8 of [4]), and it follows from [5, Proposition 4.9 and Theorem 4.10] that M is isometric to a Lie group of Iwasawa type and algebraic rank one. Theorem 1.3 implies that M is a rank one symmetric space of noncompact type. □

REMARK 5. Note that Osserman spaces (Jacobi operators in any direction have the same eigenvalues, counted with their multiplicities) are $\mathfrak{C}$-spaces, and in [2] we have proved that Osserman spaces of Iwasawa type are rank one symmetric spaces of noncompact type. Theorem 1.3 here, is not deduced from that result.

2. References

[1] Berndt J., Tricerri F., Vanhecke L., Generalized Heisenberg groups and Damek-Ricci harmonic spaces, Lecture Notes in Math. 1598, Springer, Berlin (1995).
[2] Dotti I, Druetta M.J., Negatively curved homogeneous Osserman spaces, Differential Geometry and Applications 11 (1999), 163-178.
[3]Druetta M.J., $\mathfrak{P}$-spaces of Iwasawa type and algebraic rank one (2000), pre-print.
[4] Heber J., Homogeneous spaces of nonpositive curvature and their geodesic flow, International J. Math. 6 (1995), 279-296.
[5] Heber J., Noncompact homogeneous Einstein spaces, Invent. Math. 133 (1998), 279-352.
[6] Wolf J., Homogeneity and bounded isometries in manifolds of negative curvature, Illinois J. Math. 8 (1964), 14-18.
[7] Wolter T., Einstein metrics on solvable Lie groups, Math. Zeitschrift 206 (1991), 457-471.

FAMAF, CIUDAD UNIVERSITARIA, 5000 CÓRDOBA, ARGENTINA
E-mail address: druetta@mate.uncor.edu

Contemporary Mathematics
Volume **288**, 2001

Deformations of Constant Mean Curvature Surfaces in Half Space Models

M. Elisa E.L. Galvão and Célia C. Góes

ABSTRACT. In this paper we construct deformations of surfaces with constant mean curvature one immersed in the hyperbolic space into euclidean minimal surfaces

Surfaces with mean curvature $c > 0$ in the half space model with curvature $-c^2$ have a local Weierstrass type representation, just as it happens in the case $c = 1$ ([B], [G1]). Using results given in [G2], we succeded in relating the hyperbolic and euclidean geometry of these surfaces and in providing a general representation theory for them. The main pourpose in this paper is to deform a surface with constant mean curvature one, lying in a half space model whose curvature is -1, into a minimal surface in the euclidean space. This deformation is obtained by means of an appropriated sequence of constant mean curvature surfaces, which will be explicitly given for hyperbolic Cathenoids, Enneper and Bonnet type hyperbolic surfaces.

We will denote by $Y_c : M \longrightarrow R^3_{1-1/c}$ a regular immersion of an oriented Riemann surface M in the half space $R^3_{1-1/c} = \{y = (y_1, y_2, y_3) \in R^3 \mid y_3 > 1 - 1/c\}$. In this model the relationship between the *hyperbolic Gauss map* $h_c : M \longrightarrow \hat{C}$ and the complex representation f_c of the euclidean unitary normal vector $\overrightarrow{N}_c$(that is, $f_c(p) = \Pi \circ \overrightarrow{N_c}(p)$) is given by

$$h_c(p) = y_1(p) + iy_2(p) + (y_3 - 1 + 1/c)(p)\, f_c(p).$$

A Representation Theorem. The representation theorem that we have in [G1] has a similar version for surfaces with mean curvature c in the half space model with curvature $-c^2$:

THEOREM. *Let M be an oriented Riemann surface and $Y_c = (y_1, y_2, y_3) : M \longrightarrow R^3_{1-1/c}$, a conformal immersion distinct of the horosphere with hyperbolic*

1991 *Mathematics Subject Classification.* Primary 53A10; Secondary 53C42.
Key words and phrases. Differential geometry, hyperbolic space, constant mean curvature, Gauss map, Weierstrass representation.

Gauss map given by $h_c : M \longrightarrow \hat{C}$. *The functions*

$$\alpha_2^c = \frac{1}{(y_3 - 1 + 1/c)},$$

$$\alpha_3^c = \frac{y_1 + iy_2}{(y_3 - 1 + 1/c)}$$

and

$$\alpha_1^c = \frac{y_1^2 + y_2^2 + (y_3 - 1 + 1/c)^2}{(y_3 - 1 + 1/c)}$$

satisfy

$$\alpha_1^c\,\alpha_2^c = 1 + |\alpha_3^c|^2$$

and

$$\frac{\partial \alpha_1^c}{\partial z} = h_c\,\frac{\partial \bar{\alpha}_3^c}{\partial z}$$
$$\frac{\partial \alpha_2^c}{\partial z} = \frac{1}{h_c}\,\frac{\partial \alpha_3^c}{\partial z}$$

Conversely, if we have three functions of one complex variable, α_1^c, α_2^c *and* α_3^c, *defined in a domain* U, α_1^c *and* α_2^c *real valued, positive, satisfying*

$$\alpha_1^c\,\alpha_2^c = 1 + |\alpha_3^c|^2$$

and the system above for a given map $h_c : U \subset C \longrightarrow C$, *then the map* $Y_c : U \subset C \longrightarrow R^3_{1-1/c}$,

$$Y_c = \frac{1}{\alpha_2^c}\,(\Re\,\alpha_3^c,\ \Im\,\alpha_3^c, 1) + (0, 0, 1 - \frac{1}{c})$$

is either a conformal immersion with hyperbolic Gauss map h_c *or* h_c *is constant in which case it gives a horosphere.*

Construction of examples. To find solutions for the system we first observe that it admits solutions as $\alpha_3^c(z) = F(z)\,\overline{G(z)}$ where F and G are holomorphic functions satisfying $F'(z) = h_c(z)G'(z)$; here the integrability condition $\Im\,\{\bar{h}_c\,\Delta\alpha_3^c\} = 0$ is already verified. We will suit solutions of this kind so as to have the equation $\alpha_1^c\,\alpha_2^c = 1 + |\alpha_3^c|^2$ satisfied. In this case we have F_1, G_1, F_2 and G_2 as above, A_c and B_c real constants such that $A_c\,B_c(\overline{F}_1\overline{G}_2 - \overline{F}_2\overline{G}_1)(F_1G_2 - F_2G_1) = 1$, and

$$\alpha_3 = A_c\,F_1\,\overline{G}_1 + B_c\,F_2\,\overline{G}_2$$

$$\alpha_1^c = A_c|F_1|^2 + B_c|F_2|^2$$
$$\alpha_2^c = A_c|G_1|^2 + B_c|G_2|^2$$

1. The Cathenoid Cousins. We have the called Cathenoid Cousins choosing $h_c(z) = \dfrac{a(a+1)}{(2a+1)}c^{\beta}e^{(2a+1)z}$, $G_1(z) = e^{-(a+1)z}$, $G_2(z) = e^{-az}$ with$a(a+1) = c$ or $a(a+1) = -c$. The first choice will provide embedded solutions and the second, immersed surfaces. When c goes to zero and $\beta = -2$, $A_c = c^{(1-\beta)}$, $B_c = c^{(1-\beta)}$, both of these families converge to the euclidean minimal Cathenoid.

2. The Enneper Cousins. The Enneper cousins with constant mean curvature c will be given by $h_c(z) = c^{\alpha}\tanh(\lambda z)$, $G_1'(z) = \lambda\cosh(\lambda z)$, $G_2'(z) = \lambda z\cosh(\lambda z)$ with A_c and B_c such that $\lambda\,A_c\,c^{\alpha-1} = A$, $\lambda\,B_c\,c^{\alpha-1} = B$, with A, B constants. The one parameter family with $\alpha = -3/2$, $B = A^3$, $\lambda = c^{1/2}A$ has the minimal Enneper surface as the limit surface when c goes to zero.

3. The Bonnet Cousins. Bonnet minimal surface can also be regarded as a limit of hyperbolic Bonnet cousins described explicitly in [GN]. The functions $h_c(z) = c^\alpha[\sinh(\lambda_1 z) + \sinh(\lambda_2 z)]/[\cosh(\lambda_1 z) + \cosh(\lambda_2 z)]$, $G_1'(z) = \cosh(\lambda_1 z) + \cosh(\lambda_2 z)$, $G_2'(z) = \sinh(\lambda_1 z)+\sinh(\lambda_2 z)$ generate the solutions according the calculations described above. The parameters values are such that $A_c = A\,c^{1-\alpha}\,\lambda/(2\lambda_1 + 2\lambda_2)$, $B_c = B\,c^{1-\alpha}/(2\lambda_1+2\lambda_2)$, with $\lambda_1 = (1+\sqrt{1+4\lambda})/2$, $\lambda_2 = (-1+\sqrt{1+4\lambda})/2$, $A = 4B$, $\lambda = B/,c$, B constant.

The pictures in the following pages illustrate the convergence of these deformations.

References

[B] R. L. Bryant, *Surfaces of mean curvature one in hyperbolic space*, Astérisque **155** (1987), 321–347.

[G1] M. E. E. L. Galvão, C. C. Góes, *A Weierstrass type representation for mean curvature one surfaces in the hyperbolic space*, Note di Matematica **18 (1)** (1998), 43–61.

[G2] M. E. E. L. Galvão, C. C. Góes, *Constant mean curvature surfaces in half space models*, preprint.

[GN] M. E. E. L. Galvão, C. C. Góes, B. Nelli, *A Weierstrass type representation for mean curvature one surfaces in the hyperbolic space*, An. Ac. Bras. Ci. **70 (1)** (1998), 1–6.

[UY] M. Umehara, K. Yamada, *A parametrization of the Weierstrass formulae and perturbation of complete minimal surface in R^3 into the hyperbolic 3-space*, J. Reine Angew. Math. **432** (1992), 93–116.

UNIV. DE MOGI DAS CRUZES/UNIV. DE SÃO PAULO, SÃO PAULO, BRASIL

Current address: Centro de Ciências Exatas e de Tecnologia, Av. Dr. Cândido X. de Souza 200, 08780-911/ IME - Departamento de Matemática, Rua do Matão 1010, 05889-970

E-mail address: `lgalvao@umc.br/elisa@ime.usp.br`

UNIVERSIDADE DE SÃO PAULO, SÃO PAULO, BRASIL

Current address: IME - Departamento de Matemática, R. do Matão 1010, 05889-970

E-mail address: `goes@ime.usp.br`

CATHENOID COUSINS

IMMERSED **EMBEDDED**

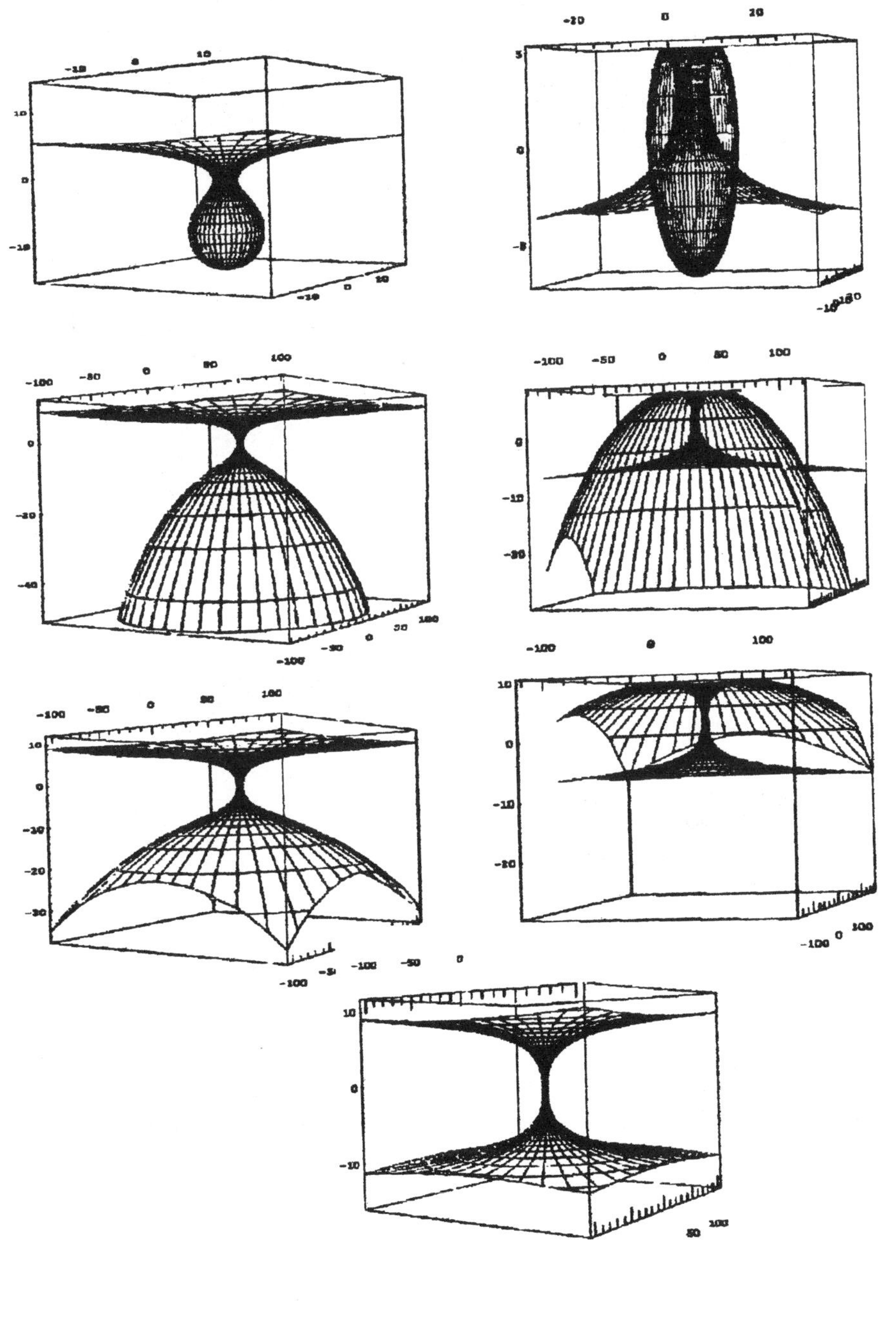

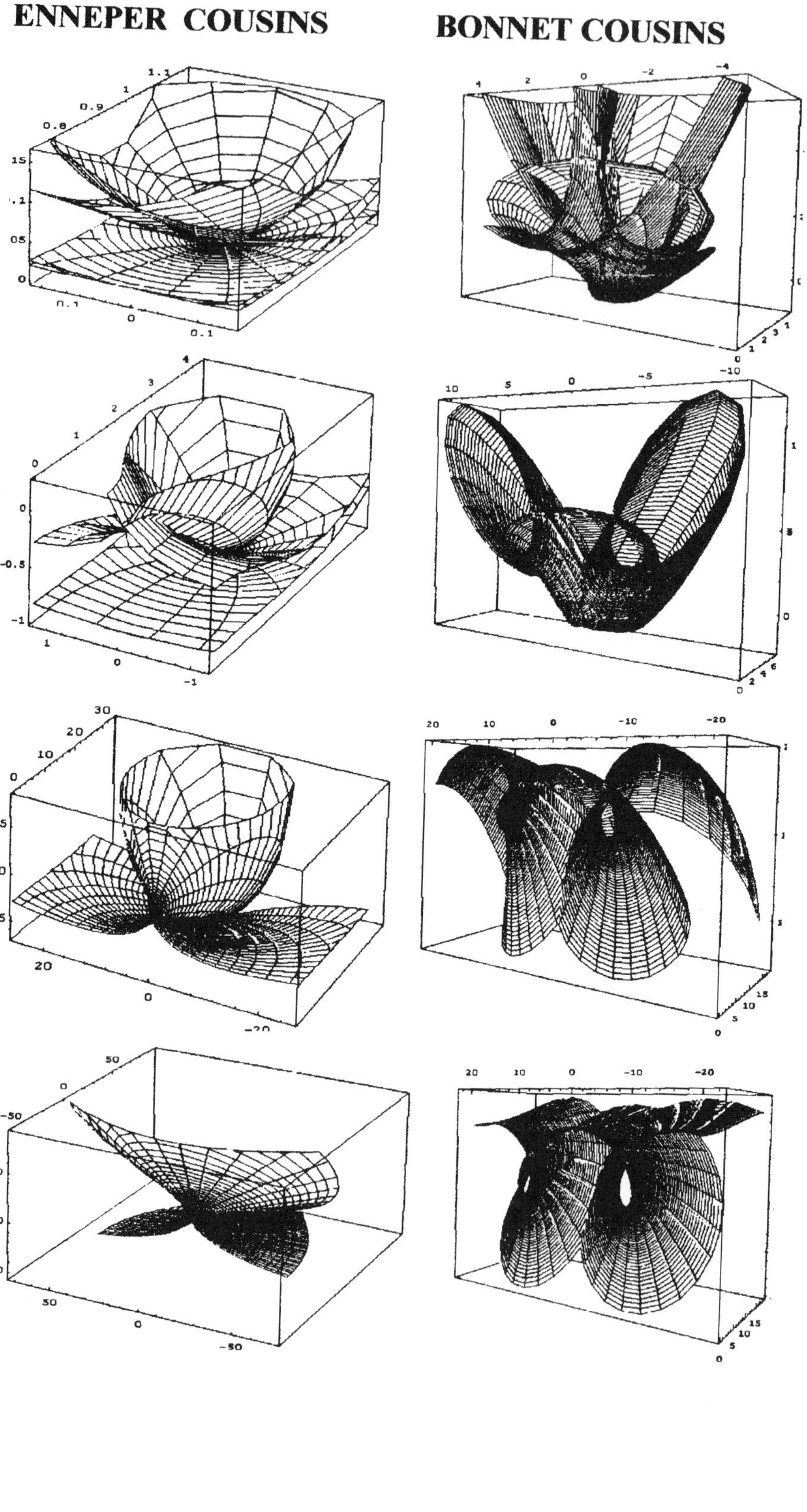
ENNEPER COUSINS
BONNET COUSINS

Contemporary Mathematics
Volume **288**, 2001

The Geometry of the Skew-Symmetric Curvature Operator in the Complex Setting

Peter Gilkey and Raina Ivanova

This paper is dedicated to Alfred Gray

ABSTRACT. We shall construct almost complex algebraic curvature tensors for pseudo Hermitian inner products whose skew-symmetric curvature operator has constant Jordan normal form on the set of non-degenerate complex lines.

Let (M, g) be a connected pseudo-Riemannian manifold of signature (p, q) and dimension $m = p + q$. We are interested in local theory so we take M to be a small ball around a point P of $\mathbb{R}^m$ and take g to be the germ of a metric defined near P. Let ${}^g\nabla$ be the Levi-Civita connection. The associated curvature operator and curvature tensor are then given by:

$$\begin{aligned} &{}^gR(X,Y) := {}^g\nabla_X {}^g\nabla_Y - {}^g\nabla_Y {}^g\nabla_X - {}^g\nabla_{[X,Y]}, \text{ and} \\ &{}^gR(X,Y,Z,W) := g({}^gR(X,Y)Z,W). \end{aligned} \tag{1}$$

We have the following symmetries:

$$ {}^gR(X,Y,Z,W) = -{}^gR(Y,X,Z,W), \tag{2}$$
$$ {}^gR(X,Y,Z,W) = {}^gR(Z,W,X,Y), \text{ and} \tag{3}$$
$$ {}^gR(X,Y,Z,W) + {}^gR(Y,Z,X,W) + {}^gR(Z,X,Y,W) = 0. \tag{4}$$

It is convenient to work in a purely algebraic setting. Let $(\cdot, \cdot)$ be a symmetric bilinear form of signature (p, q) on a finite dimensional real vector space V. A 4 tensor $R \in \otimes^4 V$ is said to be an *algebraic curvature tensor* if R satisfies the symmetries given in equations (2)-(4). A pseudo-Riemannian manifold (M, g) is said to be a *geometrical realization* of an algebraic curvature tensor R at a point P of M if there exists an isometry Ψ from the tangent space T_PM to M at P to the vector space V so that $\Psi^* R = {}^gR_P$ - i.e. we have:

$$\begin{aligned} &(\Psi X, \Psi Y) = (X, Y) \text{ for all } X, Y \in T_PM \text{ and} \\ &R(\Psi X, \Psi Y, \Psi Z, \Psi W) = {}^gR(X,Y,Z,W) \text{ for all } X, Y, Z, W \in T_PM. \end{aligned}$$

2000 *Mathematics Subject Classification.* 53B20.
Key words and phrases. Skew-symmetric curvature operator, algebraic curvature tensor, almost complex curvature tensor, IP tensor, pseudo-Hermitian structure, Gray class $\mathcal{L}_3$.

Since every algebraic curvature tensor is geometrically realizable by the germ of a pseudo-Riemannian metric, algebraic curvature tensors have a central role in differential geometry. To distinguish between the geometric and the algebraic contexts, we shall use capital letters X, Y, Z, W for tangent vectors and lower case letters x, y, z, w for elements of the vector space V.

If $\{x, y\}$ is an oriented basis for a non-degenerate 2 plane π, we define the *skew-symmetric curvature operator:*

$$R(\pi) := |(x,x)(y,y) - (x,y)^2|^{-1/2} R(x,y).$$

Let J be a pseudo-Hermitian almost complex structure on V, i.e. J is an isometry with $J^2 = -\operatorname{Id}$. We use J to give a complex structure to V by defining $\sqrt{-1}x := Jx$. A 2 plane π is said to be a *complex line* if $J\pi = \pi$. An algebraic curvature tensor R is said to be *almost complex* if $JR(\pi) = R(\pi)J$ for every non-degenerate complex line π; this means that $R(\pi)$ is complex linear.

THEOREM 1. *An algebraic curvature tensor R is almost complex if and only if $J^*R = R$ i.e. $R(x,y,z,w) = R(Jx,Jy,Jz,Jw)$ for all (x,y,z,w) in V.*

The algebraic curvature tensors satisfying the symmetry $J^*R = R$ form the class $\mathcal{L}_3$ in the classification of Gray [**10**]; we also refer Falcitelli, Farolina, and Salamon [**1**] for related work.

Since the curvature tensor encodes much of the geometry of the manifold, it can be a very complicated object to study. There are natural operators associated to R which are useful to be examined. One wants to know what are the geometric consequences if the eigenvalues (or more generally the Jordan normal form) of such an operator are constant on the natural domain of definition. We refer to the survey articles [**3, 5**] for a more comprehensive introduction to the subject than we can present here. In this paper, we shall study the geometry of the skew-symmetric curvature operator for an almost complex algebraic curvature tensor.

Let ϕ be a linear transformation of V. We use the inner product $(\cdot,\cdot)$ to define the adjoint map ϕ^* by the identity $(\phi v, w) = (v, \phi^* w)$ for all v and w. We say that ϕ is *self-adjoint* if $\phi = \phi^*$ and that ϕ is *skew-adjoint* if $\phi = -\phi^*$. We shall need two different families of algebraic curvature tensors. If ϕ is self-adjoint, then we define:

$$\begin{aligned} &R_\phi(x,y)z := (\phi(y),z)\phi(x) - (\phi(x),z)\phi(y) \text{ and}\\ &R_\phi(x,y,z,w) := (\phi(y),z)(\phi(x),w) - (\phi(x),z)(\phi(y),w). \end{aligned} \tag{5}$$

If ϕ is skew-adjoint, then we define:

$$\begin{aligned} &R_\phi(x,y)z := (\phi y,z)\phi x - (\phi x,z)\phi y - 2(\phi x,y)\phi z \text{ and}\\ &R_\phi(x,y,z,w) := (\phi y,z)(\phi x,w) - (\phi x,z)(\phi y,w) - 2(\phi x,y)(\phi z,w). \end{aligned} \tag{6}$$

Let $\mathbb{R}^{(p,q)}$ be Euclidean space with the standard metric of signature (p,q). Let S be the shape operator of a non-degenerate hypersurface in $\mathbb{R}^{(p,q)}$. Then ${}^gR = \pm R_S$. Thus the tensor of equation (5) arises in the study of hypersurfaces in flat space. For example the algebraic curvature tensor $\pm R_{\operatorname{Id}}$ has constant sectional curvature ± 1 and can be realized geometrically by the pseudo-spheres $S^\pm := \{P : (P,P) = \pm 1\}$.

In the Riemannian setting, $R_{\operatorname{Id}} + R_J$ is the algebraic curvature tensor of the Fubini-Study metric on $\mathbb{CP}^n$. The following Theorem gives some of the algebraic properties of the curvature tensors R_ϕ which we shall need.

THEOREM 2. *Assume that $\phi^* = \varepsilon\phi$ and that $J\phi = \varrho\phi J$, where $\varepsilon = \pm 1$ and $\varrho = \pm 1$. Then R_ϕ is an almost complex algebraic curvature tensor.*

Gray [**10**] (Corollary 3.2) showed that if an algebraic curvature tensor is geometrically realizable by a holomorphic Hermitian manifold (i.e. by a manifold with an integrable almost complex structure), then R satisfies the additional symmetry:

$$\begin{aligned} &R(x,y,z,w)+R(Jx,Jy,Jz,Jw)\\ &\quad=R(Jx,Jy,z,w)+R(Jx,y,Jz,w)+R(Jx,y,z,Jw)\\ &\quad+R(x,Jy,Jz,w)+R(x,Jy,z,Jw)+R(x,y,Jz,Jw). \end{aligned} \tag{7}$$

THEOREM 3. *If $J\phi=\phi J$ and if $\phi^*=\pm\phi$, then R_ϕ satisfies equation* (7).

Remark: If $\operatorname{rank}(\phi)>2$, if $J\phi=-\phi J$ and if $\phi^*=\pm\phi$, then R_ϕ does not satisfy equation (7); we omit details in the interests of brevity.

Two linear maps T_1 and T_2 of V are said to be *Jordan equivalent* if there exists an invertible linear map ψ so that $T_1=\psi T_2\psi^{-1}$ or equivalently, if there exists bases for V so that the matrix representations of T_1 and T_2 with respect to these two bases are equal. This means that the Jordan normal forms of T_1 and T_2 are equal. Two such maps necessarily have the same eigenvalues but two maps with the same eigenvalues need not be Jordan equivalent. In the positive definite setting ($p=0$), the Jordan normal form of a symmetric or of a skew-symmetric map is determined by the eigenvalue structure. As the eigenvalue structure does not determine the Jordan normal form in the higher signature setting ($p>0$), we work with the Jordan normal form rather than with the eigenvalues.

An algebraic curvature tensor R is said to be *Jordan IP* if the Jordan normal form of the skew-symmetric curvature operator is constant on the Grassmannian of oriented spacelike 2 planes, is constant on the Grassmannian of oriented timelike 2 planes, and is constant on the Grassmannian of oriented mixed 2 planes; the Jordan normal form is allowed to vary with the type of the 2 plane. One can show that if R is Jordan IP, then the rank r is independent of the type of the plane. A pseudo-Riemannian manifold (M,g) is said to be *Jordan IP* if ${}^gR$ is Jordan IP at every point P of M; the Jordan normal forms are allowed to vary with P.

Suppose that $p=0$, i.e. that we are in the Riemannian setting. Ivanova and Stanilov [**15**] began the study of the geometry of the skew-symmetric curvature operator with constant eigenvalues (see also related work by Ivanova [**12, 13, 14**]). Later Ivanov and Petrova [**11**] classified the Riemannian Jordan IP algebraic curvature tensors and metrics in case $m=4$; for this reason the notation "IP" has been used by some authors. Subsequently, the Riemannian Jordan IP algebraic curvature tensors and metrics were classified in dimensions $m=5$, $m=6$, and $m\geq 9$ by Gilkey, Leahy, and Sadofsky [**6**] and in dimension $m=8$ by Gilkey [**2**]; there are some partial results due to Gilkey and Semmelman [**7**] if $m=7$ but the classification is incomplete in this dimension.

The first step in the study of Jordan IP algebraic curvature tensors is to control the structure of the eigenspaces. The following theorem was proved using topological methods [**6, 17**]. It shows that the rank is 2 in many cases. The case $m=4$ is exceptional; there are Jordan IP algebraic curvature tensors of rank 4 in signatures (4,0), (2,2), and (0,4) [**11, 17**].

THEOREM 4 (Gilkey, Leahy, and Sadofsky; Zhang). *Let R be a Jordan IP algebraic curvature tensor of rank $r>0$ for a metric of signature (p,q).*

(1) *Let $p\leq 1$. Let $q=5$, $q=6$, or $q\geq 9$. Then $r=2$.*
(2) *Let $p=2$. Let $q\geq 10$. Let neither q nor $q+2$ be powers of 2. Then $r=2$.*

Theorem 4 focuses attention on the Jordan IP algebraic curvature tensors of rank 2. The following classification theorem follows from work of Gilkey and Zhang [**8**] extending previous work of Gilkey, Leahy, and Sadofsky [**6**] from the Riemannian setting to arbitrary signature:

THEOREM 5 (Gilkey-Zhang). *Let V have signature (p, q), where $q \geq 5$. Let R be an algebraic curvature tensor on V. Then R is a rank 2 Jordan IP algebraic curvature tensor if and only if there exists a non-zero constant C so that $R = CR_\phi$ where ϕ is self-adjoint and $\phi^2 = \pm \operatorname{Id}$.*

There is a corresponding classification of rank 2 Jordan IP pseudo-Riemannian manifolds. We refer to Gilkey and Zhang [**9**] for the proof of the following result:

THEOREM 6 (Gilkey-Zhang). *Let (M, g) be a rank 2 Jordan IP pseudo-Riemannian manifold of signature (p, q) for $q \geq 5$. Then either (M, g) has constant sectional curvature or (M, g) is a warped product of an interval $I \subset \mathbb{R}$ with a pseudo-Riemannian manifold N of constant sectional curvature where*

$$ds^2_M = \varepsilon dt^2 + (\varepsilon K t^2 + At + B) ds^2_N \text{ and } \varepsilon = \pm 1.$$

Let J be a pseudo-Hermitian almost complex structure on V. Let R be an almost complex algebraic curvature tensor. If the Jordan normal form of $R(\cdot)$ is constant on the Grassmanian of non-degenerate complex lines, then R is said to be *almost complex Jordan IP*.

If $p = 0$, then the complex linear map $JR(\pi)$ is self-adjoint and thus is diagonalizable with real eigenvalues. We refer to [**4**] for the proof of the following result which controls the eigenvalue structure; the proof is topological in nature.

THEOREM 7 (Gilkey). *Let V have signature $(0, q)$. Let R be an almost complex Jordan IP algebraic curvature tensor on V. Let $\{\lambda_s, \mu_s\}$ be the eigenvalues and multiplicities of JR, where we order the multiplicities μ_s so that $\mu_0 \geq ... \geq \mu_\ell$. Suppose that $\ell \geq 1$. If $q \equiv 2 \bmod 4$, then $\ell = 1$ and $\mu_1 = 1$. If $q \equiv 0 \bmod 4$, then either $\ell = 1$ and $\mu_1 \leq 2$ or $\ell = 2$ and $\mu_1 = \mu_2 = 1$.*

There are no general classification results for almost complex IP algebraic curvature tensors and metrics. Kath [**16**] has obtained some partial results if $(p, q) = (0, 4)$ or $(p, q) = (2, 2)$. Theorem 8 provides examples of almost complex Jordan IP algebraic curvature tensors. We adopt the following notational conventions. A linear transformation ϕ of V is said to be *admissible* if

(1) We have $\phi = \phi^*$ and $\phi J = \pm J\phi$ or $\phi = -\phi^*$ and $J\phi = -\phi J$.
(2) We have $\phi^2 = \operatorname{Id}$, $\phi^2 = -\operatorname{Id}$, or $\phi^2 = 0$ and $\ker \phi = \operatorname{range} \phi$.

A pair $\{\phi_1, \phi_2\}$ of linear transformations of V is said to be *admissible* if

(1) Both ϕ_1 and ϕ_2 are admissible.
(2) We have $\phi_1 J = J\phi_1$, $\phi_2 J = -J\phi_2$, and $\phi_1^*\phi_2 + \phi_2^*\phi_1 = 0$.
(3) If $\phi_1^2 = \phi_2^2 = 0$ and if π is any non-degenerate complex line, then we have that $\phi_1\pi \cap \phi_2\pi = \{0\}$.

THEOREM 8. *Let ϕ, ϕ_1, and ϕ_2 be linear transformations of V, and let c, c_0, c_1, c_2, c_3 be arbitrary constants.*

(1) *If ϕ is admissible, then cR_ϕ is an almost complex Jordan IP algebraic curvature tensor.*
(2) *If $\{\phi_1, \phi_2\}$ is admissible, then $c_1R_{\phi_1} + c_2R_{\phi_2}$ is an almost complex Jordan IP algebraic curvature tensor.*

(3) *The tensor* $R := c_0R_{\mathrm{Id}} + c_1R_J$ *is an almost complex Jordan IP algebraic curvature tensor.*
(4) *Let* $\{i, j, k\}$ *be a skew-adjoint quaternion structure on* V. *Then the tensor* $c_0R_{\mathrm{Id}} + c_1R_i + c_2R_j + c_3R_k$ *is an almost complex Jordan IP algebraic curvature tensor relative to the almost complex Hermitian structure given by* $J = i$.

Let J be the usual complex structure on $\mathbb{R}^{(0,2s)} = \mathbb{C}^s$. Let $\{i, j, k\}$ be the usual skew-symmetric quaternion structure on $\mathbb{R}^{(0,4s)} = \mathbb{C}^{2s}$. We can use Theorem 8 to construct examples realizing all the possibilities of Theorem 7.

THEOREM 9. *Let* λ_s *be distinct real numbers.*
(1) *Let* $R = c_0R_{\mathrm{Id}} + c_1R_J$. *We can choose* c_0 *and* c_1 *so* $JR(\cdot)$ *has eigenvalues* λ_0 *and* λ_1 *with multiplicities* $\mu_0 = s - 1$ *and* $\mu_1 = 1$ *on* $\mathbb{C}^s$.
(2) *Let* $R = c_0R_{\mathrm{Id}} + c_1R_i + c_2R_j + c_3R_k$. *We can choose* c_0, c_1, c_2, *and* c_3 *so*
  (a) $JR(\cdot)$ *has eigenvalues* λ_0 *and* λ_1 *with multiplicities* $\mu_0 = 2s - 2$ *and* $\mu_1 = 2$ *on* $\mathbb{C}^{2s}$ *or*
  (b) $JR(\cdot)$ *has eigenvalues* λ_0, λ_1, *and* λ_2 *with multiplicities* $\mu_0 = 2s - 2$, $\mu_1 = 1$, *and* $\mu_2 = 1$ *on* $\mathbb{C}^{2s}$.

The remainder of this paper is devoted to the proof of Theorems 1, 2, 3, 8 and 9.

Proof of Theorem 1. Let R be an algebraic curvature tensor. We follow an argument shown to us by Salamon. Note that $JR(\pi) = R(\pi)J$ for all non-degenerate complex lines if and only if $JR(x, Jx) = R(x, Jx)J$ for all x in V. We have the following series of equivalences and implications:

a) $JR(x, Jx) = R(x, Jx)J \ \forall x$,

$\Leftrightarrow$ b) $(JR(x, Jx)z, w) - (R(x, Jx)Jz, w) = 0 \ \forall x, z, w$,

$\Leftrightarrow$ c) $R(x, Jx, z, Jw) + R(x, Jx, Jz, w) = 0 \ \forall x, z, w$.

We polarize c), which is a quadratic identity in x, to obtain a multilinear identity:

$\Leftrightarrow$ d) $R(y, Jx, z, Jw) + R(x, Jy, z, Jw) + R(y, Jx, Jz, w) + R(x, Jy, Jz, w) = 0$

for all x, y, z, w. Replacing (x, w) by (Jx, Jw) and using symmetry (2) we obtain:

$\Leftrightarrow$ e) $-R(x, y, z, w) - R(Jx, Jy, z, w) + R(x, y, Jz, Jw) + R(Jx, Jy, Jz, Jw) = 0$

for all x, y, z, w. Now symmetry (3) implies:

$\Leftrightarrow$ f) $-R(z, w, x, y) - R(z, w, Jx, Jy) + R(Jz, Jw, x, y) + R(Jz, Jw, Jx, Jy) = 0$

for all x, y, z, w. Interchange $(x, y) \leftrightarrow (z, w)$ to see:

$\Leftrightarrow$ g) $-R(x, y, z, w) - R(x, y, Jz, Jw) + R(Jx, Jy, z, w) + R(Jx, Jy, Jz, Jw) = 0$

for all x, y, z, w. We add e) and g) to derive the identity:

$\Rightarrow$ h) $-R(x, y, z, w) + R(Jx, Jy, Jz, Jw) = 0$ for all x, y, z, w

$\Leftrightarrow$ i) $J^*R = R$.

Conversely, we replace (x, y) by (Jx, Jy) in h) and obtain:

$\Leftrightarrow$ j) $-R(Jx, Jy, z, w) + R(x, y, Jz, Jw) = 0$ for all x, y, z, w.

Now adding h) and j), we derive e) which is equivalent to a). □

Proof of Theorem 2. Let $\phi^* = \pm\phi$. To prove that R_ϕ is an algebraic curvature tensor, we must verify that equations (2), (3), and (4) are satisfied. Curvature symmetry (2) is immediate from the defining relations:

$$R_\phi(x,y,z,w) = (\phi(y),z)(\phi(x),w) - (\phi(x),z)(\phi(y),w) \text{ if } \phi = \phi^*, \text{ and}$$
$$R_\phi(x,y,z,w) := (\phi y,z)(\phi x,w) - (\phi x,z)(\phi y,w) - 2(\phi x,y)(\phi z,w) \text{ if } \phi = -\phi^*.$$

We prove that R_ϕ satisfies equation (3) by showing that the pairs (x,y) and (z,w) play symmetric roles. Next, we verify that each of the terms comprising R_ϕ has this symmetry by computing:

$$\begin{aligned}(\phi y,z)(\phi x,w) &= (y,\phi z)(x,\phi w),\\ (\phi x,z)(\phi y,w) &= (x,\phi z)(y,\phi w), \text{ and}\\ (\phi x,y)(\phi z,w) &= (\phi z,w)(\phi x,y).\end{aligned}$$

If $\phi = \phi^*$, then we check that equation (4) is satisfied by computing:

$$\begin{aligned}&R_\phi(x,y,z,w) + R_\phi(y,z,x,w) + R_\phi(z,x,y,w)\\ &\quad = (\phi y,z)(\phi x,w) - (\phi x,z)(\phi y,w)\\ &\quad + (\phi z,x)(\phi y,w) - (\phi y,x)(\phi z,w)\\ &\quad + (\phi x,y)(\phi z,w) - (\phi z,y)(\phi x,w) = 0.\end{aligned}$$

If $\phi^* = -\phi$, then similarly we obtain:

$$\begin{aligned}&R_\phi(x,y)z + R_\phi(y,z)x + R_\phi(z,x)y\\ &\quad = (\phi y,z)\phi x - (\phi x,z)\phi y - 2(\phi x,y)\phi z\\ &\quad + (\phi z,x)\phi y - (\phi y,x)\phi z - 2(\phi y,z)\phi x\\ &\quad + (\phi x,y)\phi z - (\phi z,y)\phi x - 2(\phi z,x)\phi y = 0.\end{aligned}$$

Let $\phi^* = \pm\phi$ and let $\phi J = \pm J\phi$. We have:

$$\begin{aligned}(\phi Jy,Jz)(\phi Jx,Jw) &= (J\phi y,Jz)(J\phi x,Jw) = (\phi y,z)(\phi x,w),\\ (\phi Jx,Jz)(\phi Jy,Jw) &= (J\phi x,Jz)(J\phi y,Jw) = (\phi y,z)(\phi x,w),\\ (\phi Jx,Jy)(\phi Jz,Jw) &= (J\phi x,Jy)(J\phi z,Jw) = (\phi x,y)(\phi z,w).\end{aligned}$$

Since R_ϕ is a linear combination of these terms, we combine these three identities to obtain $R(Jx,Jy,Jz,Jw) = R(x,y,z,w)$. We use Theorem 1 to see that R is almost complex.

□

Proof of Theorem 3. Let $\phi^* = \pm\phi$ and suppose $J\phi = \phi J$. By Theorem 2, R_ϕ is almost complex. Thus by Theorem 1 we see that:

$$\begin{aligned}&R(x,y,z,w) = R(Jx,Jy,Jz,Jw),\ R(Jx,Jy,z,w) = R(x,y,Jz,Jw),\\ &R(Jx,y,Jz,w) = R(x,Jy,z,Jw),\ R(Jx,y,z,Jw) = R(x,Jy,Jz,w).\end{aligned} \tag{8}$$

To prove Theorem 3, it suffices to show that equation (7) holds. Using the relations of display (8), we represent equation (7) in the form:

$$R(x,y,z,w) = R(Jx,Jy,z,w) + R(Jx,y,Jz,w) + R(Jx,y,z,Jw). \tag{9}$$

Equation (9) is an additional symmetry. We complete the proof by showing that it is satisfied by each of the terms comprising R_ϕ:

$$\begin{aligned}(\phi Jy,z)(\phi Jx,w) + (\phi y,Jz)(\phi Jx,w) + (\phi y,z)(\phi Jx,Jw) &= (\phi y,z)(\phi x,w),\\ (\phi Jx,Jz)(\phi y,w) + (\phi Jx,z)(\phi Jy,w) + (\phi Jx,z)(\phi y,Jw) &= (\phi x,z)(\phi y,w),\\ (\phi Jx,Jy)(\phi z,w) + (\phi Jx,y)(\phi Jz,w) + (\phi Jx,y)(\phi z,Jw) &= (\phi x,y)(\phi z,w).\end{aligned}$$

□

Proof of Theorem 8 (1). Let ϕ be admissible. As $\phi^* = \pm\phi$ and $J\phi = \pm\phi J$, R_ϕ is an almost complex algebraic curvature tensor (see Theorem 2). If $\phi^* = -\phi$, then we also assume that $J\phi = -\phi J$. Thus

$$(\phi x, Jx) = (-J\phi x, x) = (\phi Jx, x) = -(Jx, \phi x) = 0$$

so the third term in equation (6) vanishes and we have:

$$R_\phi(x, Jx)z = (\phi Jx, z)\phi x - (\phi x, z)\phi Jx \tag{10}$$

for both self-adjoint and skew-adjoint ϕ.

Let $\pi := \operatorname{span}\{x, Jx\}$ be a non-degenerate complex line where x is either a unit spacelike or a unit timelike vector. Then π is either spacelike or timelike, π is not mixed. Because $\phi J = \pm J\phi$, the set

$$\phi\pi = \operatorname{span}\{\phi x, \phi Jx\} = \operatorname{span}\{\phi x, J\phi x\}$$

is a complex subspace of V. To complete the proof, we must show the the Jordan normal form of $R_\phi(\pi)$ is independent of π. We distinguish two cases.

First, suppose that $\phi^2 = \varepsilon\,\mathrm{Id}$ where $\varepsilon = \pm 1$ and $\phi^* = \varrho\phi$ where $\varrho = \pm 1$. Then $(\phi x, \phi y) = (\phi^*\phi x, y) = \varepsilon\varrho(x, y) = \pm(x, y)$ so either ϕ is an isometry or ϕ is a para isometry. Thus $\phi\pi$ is either spacelike or timelike since π is either spacelike or timelike. We use equation (10) to see that in this case $R_\phi(\pi)$ is a 90 degree rotation in the plane $\phi\pi$ and $R_\phi(\pi)$ vanishes on $(\phi\pi)^\perp$. This implies that the Jordan normal form of $R_\phi(\pi)$ is independent of π.

Second, suppose that $\phi^2 = 0$. We then have

$$(\phi x, \phi y) = (\phi^*\phi x, y) = \pm(\phi^2 x, y) = 0 \text{ for any } x, y$$

so the range of ϕ is totally isotropic. Since $\ker\phi = \operatorname{range}\phi$, the kernel of ϕ contains neither spacelike nor timelike vectors. Because π is either spacelike or timelike this means that $\ker\phi \cap \pi = \{0\}$. Consequently, ϕ is an isomorphism from π to $\phi\pi$. Thus $\dim\phi\pi = 2$ and $\phi\pi = \operatorname{span}\{\phi x, J\phi x\} \subset \operatorname{range}\phi$ is a totally isotropic complex line. We use equation (10) to prove that in this case $R_\phi(\pi)^2 = 0$ and $\operatorname{rank}(R_\phi(\pi)) = 2$. This implies that the Jordan normal form of $R_\phi(\pi)$ is independent of π. $\square$

Proof of Theorem 8 (2). Let $R := c_1 R_{\phi_1} + c_2 R_{\phi_2}$, where $\{\phi_1, \phi_2\}$ is an admissible pair. By assertion (1), R_{ϕ_1} and R_{ϕ_2} are almost complex algebraic curvature tensors and hence R is an almost complex algebraic curvature tensor.

Let $\pi := \operatorname{span}\{x, Jx\}$ be a non-degenerate complex line where x is either a unit spacelike or a unit timelike vector. We show that $\{\phi_1 x, \phi_1 Jx, \phi_2 x, \phi_2 Jx\}$ is an orthogonal set by computing:

$$\begin{aligned}
(\phi_1 x, \phi_1 Jx) &= (\phi_1 x, J\phi_1 x) = -(J\phi_1 x, \phi_1 x) = 0,\\
(\phi_1 x, \phi_2 x) &= (\phi_2^*\phi_1 x, x) = -(\phi_1^*\phi_2 x, x) = -(\phi_2 x, \phi_1 x) = 0,\\
(\phi_1 x, \phi_2 Jx) &= -(J\phi_2^*\phi_1 x, x) = (\phi_2^*\phi_1 Jx, x) = -(\phi_1^*\phi_2 Jx, x)\\
&= -(\phi_2 Jx, \phi_1 x) = 0,\\
(\phi_1 Jx, \phi_2 x) &= (\phi_2^*\phi_1 Jx, x) = -(J\phi_2^*\phi_1 x, x) = (\phi_2^*\phi_1 x, Jx)\\
&= -(\phi_1^*\phi_2 x, Jx) = -(\phi_2 x, \phi_1 Jx) = 0,\\
(\phi_1 Jx, \phi_2 Jx) &= (\phi_2^*\phi_1 Jx, Jx) = -(\phi_1^*\phi_2 Jx, Jx) = -(\phi_2 Jx, \phi_1 Jx) = 0,\\
(\phi_2 x, \phi_2 Jx) &= -(\phi_2 x, J\phi_2 x) = (J\phi_2 x, \phi_2 x) = 0.
\end{aligned}$$

Thus the two complex lines $\phi_1\pi$ and $\phi_2\pi$ are orthogonal complex lines. To complete the proof, we must show that the Jordan normal form of $R(\pi)$ is independent of π. We distinguish three cases:

First, suppose that $\phi_1^2 = \pm\operatorname{Id}$ and that $\phi_2^2 = \pm\operatorname{Id}$. In proving assertion (1), we showed that $R_{\phi_s}(\pi)$ is a 90 degree rotation in the plane $\phi_s\pi$ and $R_{\phi_s}(\pi)$ vanishes on $\phi_s\pi^\perp$ $(s = 1, 2)$. Since $\phi_1\pi$ and $\phi_2\pi$ are orthogonal non-degenerate complex lines, the Jordan normal form of $R(\pi)$ is independent of π.

Second, suppose that $\phi_1^2 = \pm\operatorname{Id}$ and that $\phi_2^2 = 0$ (the argument if $\phi_1^2 = 0$ and $\phi_2^2 = \pm\operatorname{Id}$ is similar and is therefore omitted). In proving assertion (1), we showed that $R_{\phi_1}(\pi)$ is a 90 degree rotation in the plane $\phi_1\pi$, that $\phi_1\pi$ is a timelike or spacelike complex line, and that $R_{\phi_1}(\pi)$ vanishes on $\phi_1\pi^\perp$. We also showed that $\operatorname{range} R_{\phi_2}(\pi)$ is $\phi_2\pi$, that $\phi_2\pi$ is a totally isotropic complex line, and that $R_{\phi_2}^2 = 0$. Consequently, $\phi_1\pi \cap \phi_2\pi = \{0\}$ and the Jordan normal form of $R(\pi)$ is independent of π.

Third, suppose that $\phi_1^2 = \phi_2^2 = 0$. In proving assertion (1), we showed that $\operatorname{range} R_{\phi_s}(\pi) = \phi_s\pi$, that $\phi_s\pi$ is a totally isotropic complex line, and that $R_{\phi_s}^2 = 0$ $(s = 1, 2)$. In this case, we have assumed that $\phi_1\pi \cap \phi_2\pi = \{0\}$. It now follows that $\operatorname{rank}(R_\phi(\pi)) = 4$ and that $R_\phi(\pi)^2 = 0$. These two properties determine the Jordan normal form of $R_\phi(\pi)$ and one sees easily that it does not depend on π. □

Proof of Theorem 8 (3,4). We consider the following two cases in parallel: either let $R := c_0R_{\operatorname{Id}} + c_1R_J$ or let $R := c_0R_{\operatorname{Id}} + c_1R_i + c_2R_j + c_3R_k$, where $\{i, j, k\}$ is a skew-adjoint quaternion structure and $J = i$. The first case is not always contained in the second one as not all almost complex structures can be extended to quaternion structures. However the proof follows the same lines.

The tensor R is an almost complex algebraic curvature tensor (see Theorem 2). Let $\pi := \operatorname{span}\{x, ix\}$ be a non-degenerate complex line where x is a unit spacelike vector; the timelike case is similar. We shall complete the proof by showing that the Jordan normal form of $R(\pi)$ is independent of π. Since $\{i, j, k\}$ is a skew-symmetric quaternion structure, $\{x, ix, jx, kx\}$ is an orthonormal set. It follows from equations (5) and (6) that:

$$\begin{aligned} &R(\pi)x = -(c_0 + 3c_1)ix, \ R(\pi)ix = (c_0 + 3c_1)x, \\ &R(\pi)jx = -(2c_1 - c_2 - c_3)kx, \ R(\pi)kx = (2c_1 - c_2 - c_3)jx, \\ &R(\pi)y = -(2c_1)iy \text{ if } y \perp \operatorname{span}\{x, ix, jx, kx\}. \end{aligned} \tag{11}$$

Let $V_1 := \operatorname{span}\{x, ix\}$, $V_2 := \operatorname{span}\{jx, kx\}$, and $V_3 := (V_1 + V_2)^\perp$. We have an orthogonal direct sum decomposition $V = V_1 \oplus V_2 \oplus V_3$. These subspaces are preserved by the almost complex structure i and by $R(\pi)$. Let R_s be the restriction of $R(\pi)$ to V_s for $1 \le s \le 3$. It follows from equation (11) that $R_1 = -(c_0 + 3c_1)i$, $R_2 = -(2c_1 - c_2 - c_3)i$, and $R_3 = -(2c_1)i$. These are multiples of positive rotations through 90 degrees. If x is timelike, then the rotations are negative. In this way we obtain the Jordan normal form of $R(\pi)$ and verify it does not depend on π. □

Proof of Theorem 9. Let $R = c_0R_0 + c_1R_i + c_2R_j + c_3R_k$ on $\mathbb{C}^{2s}$ and let $J = i$. In proving Theorem 8, we determined the Jordan normal form of $R(\pi)$. It now follows that $(c_0 + 3c_1)$ is an eigenvalue of multiplicity 1, that $(2c_1 - c_2 - c_3)$ is an eigenvalue of multiplicity 1, and that $2c_1$ is an eigenvalue of multiplicity $2s - 2$ for the self-adjoint operator $JR(\pi)$. We choose the constants c_s for $s = 0, 1, 2, 3$ so that $\lambda_1 = c_0 + 3c_1$, $\lambda_2 = 2c_1 - c_2 - c_3$, and $\lambda_0 = 2c_1$ to complete the proof of the second assertion; the first follows similarly. □

Acknowledgements. It is a pleasure for the first author to acknowledge helpful conversations with L. Vanhecke and S. Salamon concerning curvature decompositions. He is also grateful to A. Gray who first introduced him to the subject of almost Hermitian pseudo-Riemannian geometry and curvature decompositions.

The research of the first author was supported in part by the NSF (USA) and the MPI (Leipzig, Germany). The research of the second author was supported in part by the JSPS Post Doctoral Fellowship Program (Japan) and the MPI (Leipzig, Germany).

References

[1] M. Falcitelli, A. Farinola, and S. Salamon, Almost-Hermitian geometry, *Diff. Geo. and its Appl*, **4**, (1994), 259–282.

[2] P. Gilkey, Riemannian manifolds whose skew symmetric curvature operator has constant eigenvalues II, *Differential geometry and applications (ed. Kolar, Kowalski, Krupka, and Slovak) Publ. Masaryk University Brno, Czech Republic ISBN 80-210-2097-0*, (1999), 73–87.

[3] —,Geometric properties of the curvature operator, **Proceedings of Beijing Conference: vol X Geometry and Topology of Submanifolds**, World Scientific, Singapore.

[4] —, Bundles over projective spaces and algebraic curvature tensors, *Journal of Geometry*, to appear.

[5] P. Gilkey and R. Ivanova, Geometric consequences of some algebraic properties of the curvature tensor, *Bull. Mathematiques Soc. Math. Roumanie*, to appear.

[6] P. Gilkey, J. V. Leahy, and H. Sadofsky, Riemannian manifolds whose skew-symmetric curvature operator has constant eigenvalues, *Indiana Journal*, **48**, (1999), 615–634.

[7] P. Gilkey and U. Semmelman, Spinors, self-duality, and IP algebraic curvature tensors of rank 4, (*preprint*).

[8] P. Gilkey and T. Zhang, Algebraic curvature tensors whose skew-symmetric curvature operator has constant rank 2, (*preprint*).

[9] —, Rank 2 spacelike Ivanov Petrova manifolds in the pseudo-Riemannian setting, (*preprint*).

[10] A. Gray, Curvature identities for Hermitian and almost Hermitian manifolds, *Tohoku Math. J.*, **28** (1976), 601–612.

[11] S. Ivanov and I. Petrova, Riemannian manifold in which the skew-symmetric curvature operator has pointwise constant eigenvalues, *Geometriae Dedicata*, **70**, (1998) 269–282.

[12] R. Ivanova, About a classification of some Einstein manifolds, *Mathematics and Education in Mathematics, Proc. XXIII Spring Conference of the Union of Bulgarian Mathematicians, Stara Zagora,* (1994), 175–182.

[13] —, An orthogonal tangential group of transformations of a 4-dimensional Riemannian manifold, *Mathematica Balkanica*, **11**, (1997), 53–63.

[14] —, Point-wise constancy of the skew-symmetric curvature operator's characteristical coefficients, *Tensor, N. S.,* **57, No. 1**, (1996), 97–104.

[15] R. Ivanova and G. Stanilov, A skew-symmetric curvature operator in Riemannian geometry, *Symposia Gaussiana, Conf. A, eds. Behara, Frutsch and Lintz* , (1995), 391–395.

[16] I. Kath, Almost complex algebraic curvature tensors in dimension 4, (*preprint*).

[17] T. Zhang, Manifolds with indefinite metrics whose skew-symmetric curvature operator has constant eigenvalues, *Ph. D. Thesis, University of Oregon*, (2000).

PG: Mathematics Department, University of Oregon, Eugene Or 97403 USA
E-mail address: gilkey@darkwing.uoregon.edu

RI: Dept. of Descriptive Geometry, University of Architecture, Civil Engineering & Geodesy, 1, Christo Smirnenski Blvd., 1421 Sofia, Bulgaria
E-mail address: ivanovar@is.tsukuba.ac.jp and ivanovar@hopf.uoregon.edu

Contemporary Mathematics
Volume **288**, 2001

Representing Minimal Curves in $\mathbb{C}^3$ by Differential Operators

Hubert Gollek

ABSTRACT. For a Riemannian surface Σ we consider the space Min_Σ of parametrized minimal curves, i. e., of meromorphic mappings $\Phi : \Sigma \to \mathbb{C}^3$ with vanishing infinitesimal arc length and define a sequence of nonlinear differential operators $\mathbf{V}^{(n)}$ acting from the space $\mathcal{A}_\Sigma \times \mathcal{X}_\Sigma$ of pairs $(\mathbf{f}, \mathbf{h})$ of meromorphic functions $\mathbf{f}$ and meromorphic vector fields $\mathbf{h}$ to Min_Σ. We show that $\mathbf{V}^{(3)}$ is bijective and equivariant and that it agrees with an algebraic representation formula studied in earlier papers of the author (see [**2**] and [**3**]).

1. Algebraic Representation Formulas.

Let Σ be a Riemannian surface, $\mathcal{A}_\Sigma$ the field of meromorphic functions, $\mathcal{X}_\Sigma$ the space of meromorphic vector fields, $\mathcal{L}_\Sigma$ the space of meromorphic differential 1-forms on Σ and Min_Σ the space of all meromorphic mappings $\Phi : \Sigma \to \mathbb{C}^3$ such that $\mathrm{d}\Phi \cdot \mathrm{d}\Phi = 0$, where the dot $\cdot$ denotes the bilinear extension of the ordinary scalar product of $\mathbb{R}^3$ to $\mathbb{C}^3$. Elements of Min_Σ are called minimal curves. In Min_Σ we will identify mappings that differ by a constant vector.

Denote by $(\mathcal{A}_\Sigma \times \mathcal{L}_\Sigma)^*$ the set of pairs $(\mathbf{f}, \omega) \in \mathcal{A}_\Sigma \times \mathcal{L}_\Sigma$ such that the three 1-forms $\omega, \mathbf{f}\,\omega$ and $\mathbf{f}^2\,\omega$ are exact. The Weierstraß representation formula (see [**1**] or [**4**]) can be considered as an operator $\mathbf{W} : (\mathcal{A}_\Sigma \times \mathcal{L}_\Sigma)^* \to \text{Min}_\Sigma$ assigning to a pair $(\mathbf{f}, \omega) \in (\mathcal{A}_\Sigma \times \mathcal{L}_\Sigma)^*$ the minimal curve

$$\mathbf{W}(\mathbf{f}, \omega) = (\mathbf{g}_0(\mathbf{f}, \omega) - \mathbf{g}_2(\mathbf{f}, \omega), i(\mathbf{g}_0(\mathbf{f}, \omega) + \mathbf{g}_2(\mathbf{f}, \omega)), 2\,\mathbf{g}_1(\mathbf{f}, \omega))\,/2, \tag{1}$$

where $\mathbf{g}_0(\mathbf{f}, \omega), \mathbf{g}_1(\mathbf{f}, \omega), \mathbf{g}_2(\mathbf{f}, \omega) \in \mathcal{A}_\Sigma$ are three functions with $\mathrm{d}\mathbf{g}_i(\mathbf{f}, \omega) = \mathbf{f}^i\,\omega$ for $i = 0, 1, 2$.

We define a sequence $\mathbf{P}_n : \mathcal{A}_\Sigma \times \mathcal{X}_\Sigma \to \mathcal{L}_\Sigma$ of operators such that $(\mathbf{f}, \mathbf{P}_n(\mathbf{f}, \mathbf{h})) \in (\mathcal{A}_\Sigma \times \mathcal{L}_\Sigma)^*$ for $n \geq 3$ and $(\mathbf{f}, \mathbf{h}) \in \mathcal{A}_\Sigma \times \mathcal{X}_\Sigma$ by:

$$\mathbf{P}_0(\mathbf{f}, \mathbf{h}) = \langle \mathrm{d}\mathbf{f}, \mathbf{h} \rangle\ \mathrm{d}\mathbf{f} \quad \text{and} \quad \mathbf{P}_n(\mathbf{f}, \mathbf{h}) = \mathrm{d}\left(\frac{\mathbf{P}_{n-1}(\mathbf{f}, \mathbf{h})}{\mathrm{d}\mathbf{f}}\right), \tag{2}$$

where $\langle .,. \rangle$ denotes the pairing of vector fields with differential 1-forms and the quotient $\mathbf{P}_{n-1}(\mathbf{f}, \mathbf{h})/\mathrm{d}\mathbf{f}$ is the meromorphic function $\widetilde{\mathbf{f}}$ such that $\mathbf{P}_{n-1}(\mathbf{f}, \mathbf{h}) = \widetilde{\mathbf{f}}\,\mathrm{d}\mathbf{f}$.

Key words and phrases. Minimal surfaces, Representation formulas.
AMS classification code: 49Q05, 53A10
The author was partially supported by SFB 288

In a local chart $z : V \subset \Sigma \to U \subset \mathbb{C}$ on Σ, putting $f = z^*\mathbf{f} = \mathbf{f} \circ z^{-1}$ and $h = z_*\mathbf{h}$, the recursion formula (2) becomes $\mathbf{P}_1(f,h) = (f'h)'$ and $\mathbf{P}_n(f,h) = (\mathbf{P}_{n-1}(f,h)/f')'$. The first three members of the sequence $\mathbf{P}_n(f,h)$ are $\mathbf{P}_1 = f'\,h' + h\,f''$, $\mathbf{P}_2 = \left(f'\,h'\,f'' - h\,f''^2 + f'^2\,h'' + h\,f'\,f^{(3)}\right)/f'^2$ and $\mathbf{P}_3 = -\Big(3\,f'\,h'\,f''^2 + 3\,h\,f''^3 + 2\,f'^2\,h'\,f^{(3)} 4\,h\,f'\,f''\,f^{(3)} + f'^3\,h^{(3)} + h\,f'^2\,f^{(4)}\Big)/f'^4$.

PROPOSITION 1. *The 1-forms $\mathbf{f}^k\,\mathbf{P}_n(\mathbf{f},\mathbf{h})$ are exact for all $0 \le k \le n-1$ and all $(\mathbf{f},\mathbf{h}) \in \mathcal{A}_\Sigma \times \mathcal{X}_\Sigma$.*

PROOF. We have to show that there exist functions $\mathbf{g}_{k,n}(\mathbf{f},\mathbf{h}) \in \mathcal{A}_\Sigma$ with $\mathrm{d}\mathbf{g}_{k,n}(\mathbf{f},\mathbf{h}) = \mathbf{f}^k\,\mathbf{P}_n(\mathbf{f},\mathbf{h})$. From (2) we infer that $\mathbf{g}_{0,1}(\mathbf{f},\mathbf{h}) = \langle \mathrm{d}\mathbf{f}, \mathbf{h}\rangle$ and, for $n > 1$, $\mathbf{g}_{0,n}(\mathbf{f},\mathbf{h}) = \mathbf{P}_{n-1}(\mathbf{f},\mathbf{h})/\mathrm{d}\mathbf{f}$ have this property. Moreover, the product rule and (2) show that the operators $\mathbf{g}_{k,n}$ defined for $0 < k < n$ recursively by $\mathbf{g}_{k,n}(\mathbf{f},\mathbf{h}) = \mathbf{f}^k\,\mathbf{P}_{n-1}(\mathbf{f},\mathbf{h})/\mathrm{d}\mathbf{f} - k\,\mathbf{g}_{k-1,n-1}(\mathbf{f},\mathbf{h})$ have the property $\mathrm{d}\mathbf{g}_{k,n}(\mathbf{f},\mathbf{h}) = \mathbf{f}^k\,\mathbf{P}_n(\mathbf{f},\mathbf{h})$. □

Consequently, for $n \ge 3$, $\mathcal{A}_\Sigma \times \mathcal{X}_\Sigma$ is mapped to $(\mathcal{A}_\Sigma \times \mathcal{L}_\Sigma)^*$ under the operator $(\mathbf{f},\mathbf{h}) \to (\mathbf{f},\mathbf{P}_n(\mathbf{f},\mathbf{h}))$. Therefore the composition

$$\mathcal{A}_\Sigma \times \mathcal{X}_\Sigma \xrightarrow{\quad (\mathbf{f},\mathbf{h}) \to (\mathbf{f},\mathbf{P}_n(\mathbf{f},\mathbf{h})) \quad} (\mathcal{A}_\Sigma \times \mathcal{L}_\Sigma)^* \xrightarrow{\quad \mathbf{W} \quad} \mathrm{Min}_\Sigma$$

of this operator with the Weierstraß formula gives algebraic representation formulas for minimal curves on Σ.

Local expressions of $\mathbf{g}_{k,3}(f,h)$ for $k = 0,1,2$ are

$$(3)\quad \begin{cases} \mathbf{g}_{0,3}(f,h) = \dfrac{1}{f'(z)^3}\Big(f'\,h'\,f'' - h\,f''^2 + f'^2\,h'' + h\,f'\,f^{(3)}\Big), \\ \mathbf{g}_{1,3}(f,h) = \dfrac{1}{f'(z)^3}\Big(f\,f'^2\,h'' + f\,h\,f'\,f^{(3)} - f'^3\,h' \\ \qquad -h\,f'^2\,f'' + f\,f'\,h'\,f'' - f\,h\,f''^2\Big), \\ \mathbf{g}_{2,3}(f,h) = \dfrac{1}{f'(z)^3}\Big(2\,h\,f'^4 - 2\,f\,f'^3\,h' - 2\,f\,h\,f'^2\,f'' \\ \qquad +f^2\,f'\,h'\,f'' - f^2\,h\,f''^2 + f^2\,f'^2\,h'' + f^2\,h\,f'\,f^{(3)}\Big). \end{cases}$$

Observe that $\mathbf{g}_{k,n}(f,h)$ is $\mathbb{C}$−linear in h for all f and all $k < n$.

DEFINITION 1. *For $n \ge 3$ define the differential operators $\mathbf{V}^{(n)} : \mathcal{A}_\Sigma \times \mathcal{X}_\Sigma \to \mathrm{Min}_\Sigma$ by $\mathbf{V}^{(n)}(\mathbf{f},\mathbf{h}) = \mathbf{W}(\mathbf{f},\mathbf{P}_n(\mathbf{f},\mathbf{h}))$. The explicit form of $\mathbf{V}^{(n)}$ is*

$$(4)\quad \mathbf{V}^{(n)}(\mathbf{f},\mathbf{h})(z) = \begin{pmatrix} \mathbf{g}_{0,n}(\mathbf{f},\mathbf{h})(z) - \mathbf{g}_{2,n}(\mathbf{f},\mathbf{h})(z) \\ i\,(\mathbf{g}_{0,n}(\mathbf{f},\mathbf{h})(z) + \mathbf{g}_{2,n}(\mathbf{f},\mathbf{h})(z)) \\ 2\,\mathbf{g}_{1,n}(\mathbf{f},\mathbf{h})(z) \end{pmatrix}.$$

PROPOSITION 2. *$\mathbf{V}^{(n)}(\mathbf{f},\mathbf{h})$ is expressed in terms of $\mathbf{P}_{n-1}$, $\mathbf{P}_{n-2}$, $\mathbf{P}_{n-3}$ by*

$$(5)\quad \frac{1}{\mathrm{d}\mathbf{f}}\begin{pmatrix} \left(1-\mathbf{f}^2\right)\mathbf{P}_{n-1}(\mathbf{f},\mathbf{h}) + 2\,\mathbf{f}\,\mathbf{P}_{n-2}(\mathbf{f},\mathbf{h}) - 2\,\mathbf{P}_{n-3}(\mathbf{f},\mathbf{h}) \\ i\,\left(\left(1+\mathbf{f}^2\right)\mathbf{P}_{n-1}(\mathbf{f},\mathbf{h}) - 2\,\mathbf{f}\,\mathbf{P}_{n-2}(\mathbf{f},\mathbf{h}) + 2\,\mathbf{P}_{n-3}(\mathbf{f},\mathbf{h})\right) \\ 2\,\left(\mathbf{f}\,\mathbf{P}_{n-1}(\mathbf{f},\mathbf{h}) - \mathbf{P}_{n-2}(\mathbf{f},\mathbf{h})\right) \end{pmatrix}.$$

$\mathbf{V}^{(3)}$ is the first algebraic one among the operators $\mathbf{V}^{(n)}$. All other operators $\mathbf{V}^{(n)}$ with $n \geq 3$ can be expressed by $\mathbf{V}^{(3)}$:

PROPOSITION 3. *For each $n \geq 3$ the operator $\mathbf{V}^{(n)}$ satisfies the equation $\mathbf{V}^{(n)}(\mathbf{f}, \mathbf{h}) = \mathbf{V}^{(3)}(\mathbf{f}, \mathbf{h}_n)$, where the sequence $\mathbf{h}_n$ of meromorphic vector fields is defined by $\mathbf{h}_3 = \mathbf{h}$, $\mathbf{h}_n = \left\langle \mathrm{d}\left\langle \mathrm{d}\mathbf{f}, \mathbf{h}_{n-1}\right\rangle, \partial\mathbf{f}\right\rangle \partial\mathbf{f}$, and where $\partial\mathbf{f}$ denotes the dual vector field of $\mathrm{d}\mathbf{f}$ defined by $\langle \partial\mathbf{f}, \mathrm{d}\mathbf{f}\rangle = 1$.*

PROOF. It is sufficient to prove in local coordinates that $\mathbf{P}_n(f,h) = \mathbf{P}_3(f, h_n)$ where f, h, h_n are the local expressions of $\mathbf{f}, \mathbf{h}$ and $\mathbf{h}_n$ respectively. The local version of the recursion rule for $\mathbf{h}_n$ is $h_n = (f' h_{n-1})'/f'^2$. We have $\mathbf{P}_3(f,h) = \mathbf{P}_3(f, h_3)$ by assumption. The inductive proof reduces to the proof of the equation $(\mathbf{P}_3(f,h)/f')' = \mathbf{P}_3\left(f, (f'h)'/f'^2\right)$ for arbitrary functions f and h. This can be achieved with (5). □

2. Equivariance and Bijectivity of $\mathbf{V}^{(3)}$.

Next we are going to show that $\mathbf{V}^{(3)}$ is an equivariant and bijective map from $\mathcal{A}_\Sigma \times \mathcal{X}_\Sigma$ onto the set of minimal curves Φ with $\Phi_1' - i\,\Phi_2' \neq 0$. We confine our considerations to minimal curves $\Phi(z)$ in $\mathrm{Min}_\mathbb{C}$ and assume that Φ', Φ'', Φ''' are linearly independent. The *natural parameter* $p(z)$ and the *Study parameter* $\kappa_\Phi(z)$ of $\Phi(z)$ are defined by

$$p'(z) = i\,\sqrt[4]{\Phi''(z)\cdot\Phi''(z)} \text{ and } \kappa_\Phi^2(p) = \frac{\mathrm{d}^3\Phi(p)}{\mathrm{d}p^3}\cdot\frac{\mathrm{d}^3\Phi(p)}{\mathrm{d}p^3}, \tag{6}$$

respectively.

Locally, $\Phi(z)$ can be reparametrized to p. Let us denote by $\Phi', \Phi'', \ldots$ the derivatives with respect to p. Then $\Phi'\cdot\Phi^{(4)} = 0$, $\Phi''\cdot\Phi'' = -1$, $\Phi^{(3)}\cdot\Phi^{(3)} = -\Phi''\cdot\Phi^{(4)} = \kappa^2$, $\Phi^{(3)}\cdot\Phi^{(4)} = \kappa'\,\kappa$ and $\Phi^{(4)}\cdot\Phi^{(4)} = -\kappa^4$

From this we infer:

PROPOSITION 4. *For any minimal curve Φ one has $\Phi^{(4)} = \kappa\,\kappa'\,\Phi' + \kappa^2\,\Phi''$, for any meromorphic function h the new curves*

$$\mathbf{D}(\Phi,h) = \left(h'' - h\,\kappa_\Phi^2\right)\Phi' - h'\Phi'' + h\Phi^{(3)} \text{ and } \Psi(\Phi,h) = \Phi + \mathbf{D}(\Phi,h) \tag{7}$$

are again minimal and the tangent vector $\mathbf{D}'(\Phi,h)$ is parallel to Φ':

$$\mathbf{D}'(\Phi,h) = \left(h''' - h'\,\kappa_\Phi^2 - h\,\kappa_\Phi\,\kappa_\Phi'\right)\Phi'. \tag{8}$$

A version of the Weierstraß formula representing minimal curves that are parametrized by their natural parameter is

$$\mathbf{W}_f^*(z) = \int_{z_0}^{z} \frac{1}{2\,f'(\zeta)}\begin{pmatrix} i\,(1 - f^2(\zeta)) \\ -1 - f^2(\zeta) \\ 2\,f(\zeta)\end{pmatrix}\mathrm{d}\zeta.$$

The Study parameter of $\mathbf{W}_f^*(z)$ is the *Schwarzian derivative*

$$\kappa_{\mathbf{W}_f^*}^2(z) = \left(3\,f''^2(z) - 2\,f'(z)\,f^{(3)}(z)\right)/f'^2(z).$$

Using this expression, a direct computation of the coefficient of Φ' in (8) yields $(h''' - h'\,\kappa_\Phi^2 - h\,\kappa_\Phi\,\kappa'_\Phi) = \mathbf{P}_3(f,h)/f'$. This fact indicates, that there could exist a relation between the operators $\mathbf{D}$ and $\mathbf{V}^{(3)}$.

The combination the operator $\mathbf{D}$ with $\mathbf{W}^*$ defines an algebraic representation formula $\mathbf{V} : \mathcal{A}_\mathbb{C} \times \mathcal{A}_\mathbb{C} \to \mathrm{Min}_\mathbb{C}$ by $\mathbf{V}(f,h) = \mathbf{D}\left(\mathbf{W}_f^*, h\right)$. $\mathbf{V}$ is almost the same as the operator $\mathbf{V}^{(3)}$ on $\mathbb{C}$, namely, one can show that $\mathbf{V}(f,h) = i\,\mathbf{V}^{(3)}(f,h)$ for $f,h \in \mathcal{A}_\mathbb{C}$. From this fact follows that $\mathbf{V}^{(3)}$ and $\mathbf{V}$ have the same property of equivariance as described below for $\mathbf{V}$. Also the property of bijectivity of $\mathbf{V}^{(3)}$ as described in proposition 5 is the same as that of $\mathbf{V}$ proved in [**2**].

$\mathbf{W}_f^*$ is equivariant with respect to the covering homomorphism $\mu : \mathrm{Sl}(2,\mathbb{C}) \to \mathrm{SO}(3,\mathbb{C})$ given by.

$$\mu : \mathrm{g} = \begin{pmatrix} a & b \\ c & d \end{pmatrix} \in \mathrm{Gl}(2,\mathbb{C}) \to \mu(\mathrm{g}) = \frac{1}{2(ad-bc)} *$$

$$\begin{pmatrix} a^2 - b^2 - c^2 + d^2 & i\left(a^2 + b^2 - c^2 - d^2\right) & 2\,(c\,d - a\,b) \\ -i\left(a^2 - b^2 + c^2 - d^2\right) & a^2 + b^2 + c^2 + d^2 & 2\,i\,(a\,b + c\,d) \\ 2\,(b\,d - a\,c) & -2\,i\,(a\,c + b\,d) & 2\,(b\,c + a\,d) \end{pmatrix},$$

where we assume that a matrix $g \in \mathrm{Sl}(2,\mathbb{C})$ acts on a function $f : \mathbb{C} \to \mathbb{C}$ by linear fractional transformations and $\mu(g) \in \mathrm{SO}(3,\mathbb{C})$ by matrix multiplication. Consequently, since $\mathbf{V}(f,h)$ is a linear combination of derivatives of $\mathbf{W}_f^*$, it is also equivariant with respect to μ.

PROPOSITION 5. *(i) For any* $\mathbf{f} \in \mathcal{A}_\Sigma$ *and* $a,b,c \in \mathbb{C}$ *denote by* $\mathbf{h_f}(a,b,c)$ *the vector field* $\mathbf{h_f}(a,b,c) = (a + b\mathbf{f} + c\mathbf{f}^2)\partial\mathbf{f} \in \mathcal{X}_\Sigma$, *where* $\partial\mathbf{f}$ *denotes the vector field defined by* $\langle \partial\mathbf{f}, \mathrm{d}\mathbf{f}\rangle = 1$. *Then we have* $\mathbf{V}(\mathbf{f}, \mathbf{h_f}(a,b,c)) = (i(c-a), -a-c, -ib)$.

(ii)Let $\mathcal{K} \subset \mathcal{A}_\Sigma \times \mathcal{X}_\Sigma$ *be the subset of all pairs* $(\mathbf{f}, \mathbf{h_f}(a,b,c))$. $\mathbf{V}^{(3)}$ *is a bijection of* $\mathcal{A}_\Sigma \times \mathcal{X}_\Sigma \setminus \mathcal{K}$ *onto the subset* $\mathrm{Min}_\Sigma^* \subset \mathrm{Min}_\Sigma$ *of minimal curves* $\Phi = (\Phi_1, \Phi_2, \Phi_3)$ *such that* $\Phi'_1 \neq i\,\Phi'_2$.

The inverse operator of $\mathbf{V}$ is also a differential operator. It is given by $\mathbf{V}^{-1}(\Phi) = (f,h)$, where $f = \Phi'_3/\left(\Phi'_1 - i\,\Phi'_2\right)$ and $h = \Phi \cdot \mathbf{W}'^*(f)(z)$.

3. A Representation Formula for arbitrary Curves.

The formula (7) can be generalized to a representation formula of meromorphic curves with arbitrary arc length element $\mathrm{d}s^2 = -d^2(z)$. Putting

$$\widetilde{\mathbf{D}}(\Phi,h,d) = \left(-h\,\kappa_\Phi^2 + h''\right)\Phi' - h'\Phi'' + h\Phi^{(3)} + d\Phi' \tag{9}$$

we obtain $\widetilde{\mathbf{D}}' = \left(-h'\,\kappa_\Phi^2 - h\,\kappa_\Phi\,\kappa'_\Phi + h''' + d'\right)\Phi' + d\,\Phi''$ and the equations $\Phi' \cdot \Phi' = \Phi' \cdot \Phi'' = 0$, $\Phi'' \cdot \Phi'' = -1$ give $\widetilde{\mathbf{D}}'(\Phi,h,d) \cdot \widetilde{\mathbf{D}}'(\Phi,h,d) = -d^2$.

We replace the curve Φ in $\widetilde{\mathbf{D}}(\Phi,h,d)$ by $\mathbf{W}_f^*$ and define an differential operator $\mathbf{V} : \mathcal{A}_\mathbb{C}^3 \to \mathcal{A}_\mathbb{C}^3$ by $\mathbf{V}(f,h,d) = \widetilde{\mathbf{D}}(\mathbf{W}_f^*, h, d)$. $\mathbf{V}(f,h,d)$ is a bijective, equivariant algebraic representation formula for curves with arbitrary arc length element $\mathrm{d}s^2 = -d^2(z)\mathrm{d}z^2$.

For $\mathbf{V}(f,h,d)$ there exists also a global version $\mathbf{V}_\Sigma(\mathbf{f},\mathbf{h},\omega)$ defined invariantly for arbitrary Riemann surfaces Σ. One observes that the derivative $\mathbf{W}'^*$ of $\mathbf{W}^*$ is the local version of the operator $\mathbf{W}_\Sigma'^* : \mathcal{A}_\Sigma \to \mathcal{X}_\Sigma \times \mathcal{X}_\Sigma \times \mathcal{X}_\Sigma$ given by

$\mathbf{W}'^{*}_{\Sigma}(\mathbf{f}) = \big(1-\mathbf{f}^2, i\,(1+\mathbf{f}^2), 2\,\mathbf{f}\big)\,\partial\mathbf{f}/2$. We define $\mathbf{V}_\Sigma(\mathbf{f},\mathbf{h},\omega)$ in analogy to (9) by $\mathbf{V}(\mathbf{f},\mathbf{h},\omega) = i\,\mathbf{V}^{(3)}(\mathbf{f},\mathbf{h}) + \langle \omega, \mathbf{W}'^{*}_{\Sigma}(\mathbf{f})\rangle$. $\mathbf{V}_\Sigma$ maps triples $(\mathbf{f},\mathbf{h},\omega) \in \mathcal{A}_\Sigma \times \mathcal{X}_\Sigma \times \mathcal{L}_\Sigma$ to meromorphic curves in $\mathbb{C}^3$ with prescribed infinitesimal arc length ω and has properties of injectivity and equivariance similar to those of $\mathbf{V}^{(3)}$ and $\mathbf{V}$.

For $d \neq 0$ there is a 1-1 relation between the pairs $(\widetilde{\kappa},\widetilde{\tau})$ consisting of curvature $\widetilde{\kappa}$ and torsion $\widetilde{\tau}$ of $\mathbf{V}(f,h,d)$ and the pairs $(\pi_{f,h},\kappa_f)$ where $\pi_{f,h}$ is the natural parameter of $\mathbf{V}(f,h)$ and κ_f the Study parameter of $\mathbf{W}^*(f)$:

$$(10)\quad \begin{cases} \widetilde{\kappa}^2 = \dfrac{d^2\,\kappa_f^2 - \pi_{f,h}^4 - 4\,\pi_{f,h}^2\,d' - 3\,d'^2 + 4\,d\,\pi_{f,h}\,\pi'_{f,h} + 2\,d\,d''}{d^4}, \\ \widetilde{\tau} = \dfrac{1}{\widetilde{\kappa}_\Phi^2\,d^6}\Big\{ {\pi_{f,h}}^6 + 6\,{\pi_{f,h}}^4\,d' + 11\,{\pi_{f,h}}^2\,d'^2 + 6\,d'^3 + d^3\,\kappa_f\,\kappa'_f \\ \quad -2\,d\left(3\,{\pi_{f,h}}^3\,\pi'_{f,h} + 5\,\pi_{f,h}\,d'\,\pi_{f,h} + 2\,{\pi_{f,h}}^2\,d'' + 3\,d'\,d''\right) + \\ \quad d^2\left(2\,{\pi'_\Phi}^2 + 2\,\pi_{f,h}\,\pi''_{f,h} + d^{(3)} - \kappa_f^2\left({\pi_{f,h}}^2 + d'\right)\right)\Big\}. \end{cases}$$

The mapping $(\kappa_f, \pi_{f,h}) \to (\widetilde{\kappa}^2, \widetilde{\tau})$ is inverted by an other differential operator, namely by

$$(11)\quad \begin{cases} \kappa_f = d^2\,\widetilde{\kappa}^2 + d^2\,\widetilde{\tau}^2 + \dfrac{3\,d'^2}{d^2} - \dfrac{d\,\widetilde{\tau}\,\widetilde{\kappa}'^2}{\widetilde{\kappa}^2} + \dfrac{d'\,\widetilde{\kappa}'^2}{d\,\widetilde{\kappa}^2} + \dfrac{5\,\widetilde{\kappa}'^4}{4\,\widetilde{\kappa}^4} + \\ \quad 2\,d\,\widetilde{\tau}' - \dfrac{2\,d''}{d} - \dfrac{\widetilde{\kappa}''^2}{\widetilde{\kappa}^2}, \\ \pi_{f,h} = \dfrac{d}{2}\left(-2\,d\,\widetilde{\tau} + \dfrac{\widetilde{\kappa}'^2}{\widetilde{\kappa}^2}\right). \end{cases}$$

A bijective, equivariant representation formula for meromorphic minimal curves in $\mathbb{C}^4$ is obtained by appending to $\mathbf{V}(f,h,d')$ the function $i\,d(z)$.

References

[1] U. Dierkes, S. Hildebrandt, A. Küster, O. Wohlrab, Minimal Surfaces, Vol. 1, Springer-Verlag, Berlin-New York, 1992.

[2] H. Gollek, Deformations of minimal curves in $\mathbb{C}^3$, Proceedings of 1-st NOSONGE Conference, Warsaw, Sept. 1996

[3] H. Gollek, Deformations of minimal Surfaces containing planar Geodesics, Proceedings of 18-th Winter School "Geometry and Physics", Srni (Czechoslovakia), Jan. 1998

[4] A. Gray, Modern Differential Geometry of Curves and Surfaces, CRC Press, 1995, 2-nd edition

Hubert Gollek
Humboldt University, Institute of Mathematics
10099 Berlin (Germany)
Sitz: Berlin-Adlershof, Rudower Chaussee 25
e-mail: gollek@mathematik.hu-berlin.de

Contemporary Mathematics
Volume **288**, 2001

Isometric Immersions without Positive Ricci Curvature

Luis Guijarro

ABSTRACT. In this note we study isometric immersions of Riemannian manifolds with positive Ricci curvature into an Euclidean space.

1. Introduction

When studying isometric immersions, one often realizes that restrictions on the curvature result in the need for extra space. For instance, a classical result of Hilbert affirms that in $\mathbb{R}^3$ there are no complete surfaces whose induced metrics have constant negative curvature [**5**]. In the same spirit, Chern and Kuiper proved that if an n-dimensional compact Riemannian manifold M with nonpositive sectional curvature immerses isometrically in some $\mathbb{R}^N$, then $N \geq 2n$ [**1**], [**2**].

When turning to the positive sectional curvature case, the main observation was due to A. Weinstein [**14**], who showed that for immersions of M^n in $\mathbb{R}^{n+2}$, the curvature operator $R : \Lambda^2 M \to \Lambda^2 M$ had to be positive. As a consequence, if M^n is simply connected, then it has to be homeomorphic to a sphere. Related research appeared in [**8**] and [**11**].

Our aim in this note is to collect a few observations about the positive Ricci curvature case, by taking into account the type of the normal bundle to the immersion as a possible obstruction. Our main tool is Lemma 2.1. We then apply it to several situations, and among them, we get

THEOREM 1.1. *Let $M^4 = \mathbb{CP}^2 \# \dots \# \mathbb{CP}^2$ a connected sum of complex projective spaces ($\mathbb{CP}^2$ itself is allowed). Then no immersion $f : M^4 \to \mathbb{R}^7$ induces a metric with positive Ricci curvature.*

The author would like to thank B. Wilking for a conversation related to the proof of the above theorem.

2. Positive Ricci curvature and normal bundles

Let (M, g) be a Riemannian manifold of dimension n and $f : M \to \mathbb{R}^m$ an isometric immersion. Denote by $\nu(f)$ the normal vector bundle of the immersion;

2000 *Mathematics Subject Classification.* 53C20.

The author was fully supported by a Postdoctoral Fellowship from the Comunidad Autónoma de Madrid.

i.e,

$$\nu(f) = \{ (x, v) : x \in M, v \in T_x\mathbb{R}^n, v \perp f_*(T_xM) \}.$$

There is an isomorphism of vector bundles $TM \oplus \nu(f) \simeq \mathbb{R}^n$, where, without risk of confusion, $\mathbb{R}^n$ denotes the product n–bundle over M.

LEMMA 2.1. *If (M, g) has positive Ricci curvature, then the mean curvature vector H is a non-vanishing section of $\nu(f)$.*

Proof: Denote by $B : TM \times TM \to \nu(f)$ the second fundamental form of the immersion, and $x \in T_pM$ a unit vector field that we complete to an orthonormal basis $\{x = e_1, e_2, \ldots, e_n\}$. Gauss' formula implies that

$$Ric(x, x) = \sum_{i=2}^{n} K(x, e_i) = \sum_{i=2}^{n} \langle B(x, x), B(e_i, e_i)\rangle - \|B(x, e_i)\|^2.$$

Thus

$$\langle B(x, x), H\rangle = Ric(x, x) + \sum_{i=1}^{n} \|B(x, e_i)\|^2 > 0.$$

□

Although we have phrased the Theorem in terms of positive Ricci curvature, the proof shows that it suffices to have a vector x with $Ric(x) > 0$ at each point.

3. Immersions of $\mathbb{CP}^2\# \ldots \#\mathbb{CP}^2$ in $\mathbb{R}^7$

Whitney's immersion theorem asserts that any four-dimensional manifold can be immersed in $\mathbb{R}^7$. We use the lemma to give the application mentioned in the introduction:

THEOREM 3.1. *$M^4 = \mathbb{CP}^2\# \ldots \#\mathbb{CP}^2$ cannot be immersed into $\mathbb{R}^7$ with an induced metric of positive Ricci curvature.*

Proof: If not, then by Lemma 2.1 we have a plane bundle $E \to M$ such that

$$M \times \mathbb{R}^7 \simeq TM \oplus E \oplus \mathbb{R}.$$

Since M is simply connected, E is oriented, and a simple topological argument shows that $H^2(M, \mathbb{Z})$ is torsion–free ([**4**], proposition E.1). Thus, the product formula for Pontryagin classes (see [**9**], theorem 15.3) gives:

$$(1 + p_1(M))(1 + p_1(E)) = 0.$$

Hence $p_1(E) = -p_1(M)$. By Hirzebruch's theorem in dimension 4 ([**9**] Theorem 19.4), $\langle p_1(M), [M]\rangle = 3\sigma(M)$. Since by hypothesis, this is positive, we have

$$\langle p_1(E), [M]\rangle < 0.$$

On the other hand, by Corollary 15.8 in [**9**], $p_1(E) = e(E)^2$. Thus we can use the intersection form Q of M to give an alternate computation of $p_1(E)$ as

$$\langle p_1(E), [M]\rangle = \langle e(E)^2, [M]\rangle = Q(e(E), e(E)).$$

However, M was chosen so that Q was positive definite (see [**3**], for instance), and we obtain a contradiction.

□

The theorem is optimal, since the Veronese embedding

$$\phi : \mathbb{CP}^2 \longrightarrow \mathrm{Herm}_0(\mathbb{C}, 3)$$

where $\mathrm{Herm}_0(3, \mathbb{C})$ are the 3×3 trace–free hermitian matrices, defined by

$$\phi[z_0, z_1, z_2] = I - 2(\overline{z_0}, \overline{z_1}, \overline{z_2})^t (z_0, z_1, z_2)$$

induces the Fubini–Study metric in $\mathbb{CP}^2$, which has positive sectional curvature. In fact, the natural action of $SU(3)$ on $\mathbb{CP}^2$ induces an isometric action on its image by ϕ with $U(2)$ as isotropy, and hence the induced metric agrees with the quotient metric $SU(3)/U(2)$. Observe that the dimension of $\mathrm{Herm}_0(\mathbb{C}, 3)$ is eight.

Remark: There are several extensions of the above proof. For instance, the same argument shows that the complex projective plane $\mathbb{CP}^n$ cannot be immersed with positive curvature in codimension 3. But also any proof of the impossibility of immersing a given manifold M^n into R^N that uses the impossibility of splitting the trivial rank N bundle as a sum $TM \oplus E$ can be adapted to our situation to show that M^n does not immerse in R^{N+1} with positive curvature. As a concrete example, RP^n can be immersed in R^{2n-1} by Whitney's theorem, but when n is a power of two, the induced metric cannot have positive curvature everywhere. Otherwise, the normal bundle would split a line and we would get a contradiction to the product formula for Stiefel-Whitney classes as in [**9**].

4. Immersions of the sphere with positive Ricci curvature

Due to the work of Smale in [**12**], we have a good understanding of the immersions of n-spheres in the euclidean space of dimension $n + k$. Up to isotopy, they are in one-to-one correspondence with elements of $\pi_n(V_{n+k,n})$, where $V_{n+k,n}$ is the Stiefel manifold of n-orthonormal frames in $\mathbb{R}^{n+k}$.

In a development of [**12**], Kervaire gave a characterization of the bundles over S^n that can appear as a normal bundle of an immersion into $\mathbb{R}^{n+k}$. Namely, the isomorphism classes of rank k vector bundles over S^n correspond to elements in $\pi_{n-1}(SO(k))$ through their clutching maps [**13**]. On the other hand, the usual fibration of the Stiefel manifold $SO(k) \to SO(n+k) \to V_{n+k,k}$ induces a homotopy sequence

$$\cdots \to \pi_n(SO(n+k)) \to \pi_n(V_{n+k,k}) \to \pi_{n-1}(SO(k)) \to \cdots$$

where $\partial : \pi_n(V_{n+k,k}) \to \pi_{n-1}(SO(k))$ is the boundary homomorphism. Kervaire proved that there is a bijection between the set of clutching maps for normal bundles of immersions in $\mathbb{R}^{n+k}$ and the image of ∂. He denoted this subgroup by $J_{n,k}$ and proceeded to compute it for some values of k and n.

The results in this section follow from combining some of these facts with the information given by Lemma 2.1.

THEOREM 4.1. *Let n and k positive integers such that $J_{n,k} = 0$. Then any immersion $f : S^n \to \mathbb{R}^{n+k+1}$ with positive Ricci curvature can be isotoped to the map $S^n \xrightarrow{i} \mathbb{R}^{n+1} \to \mathbb{R}^{n+k+1}$, where i is the standard inclusion.*

Proof: The normal bundle $\nu(f)$ to the immersion admits a section without zeros due to lemma 2.1. Thus there is a splitting $\nu(f) = \mathbb{R} \oplus E$, where E has rank k. Theorem 6.1 in [**6**] asserts that f can now be isotoped to an immersion $g : S^n \to \mathbb{R}^{n+k}$, and the bundle E is isomorphic to $\nu(g)$. The hypothesis on $J_{n,k}$ $(= 0)$ implies that E is the trivial bundle, and correspondingly $\nu(f)$ is just $S^n \times \mathbb{R}^{k+1}$. But corollary 6.2

from [**6**] affirms that in this case, f can be isotoped into R^{n+k+1}, and the theorem is proved. □

COROLLARY 4.2. *Any immersion of S^{8s+5} into $\mathbb{R}^{16s+6}$ or $\mathbb{R}^{16s+7}$ with an induced metric of positive Ricci curvature can be isotoped into the standard immersion of S^{8s+5} in $\mathbb{R}^{8s+6}$.*

Proof: In [**7**], it was shown that $\pi_{8s+4}(SO(8s)) = \pi_{8s+4}(SO(8s+1)) = 0$. Thus $J_{8s+5,8s} = J_{8s+5,8s+1} = 0$, and we are in the hypothesis of theorem 4.1. □

It is somewhat unsatisfactory that the theorem does not give any information about the induced metrics at each stage of the isotopy between f and the inclusion $S^n \to \mathbb{R}^{n+1}$. The induced metrics at times $t = 0$ and $t = 1$ of the isotopy have both positive Ricci curvature, and we would like to know whether the deformation could be done entirely by such metrics. However, in the codimension two case, Moore showed that there are not obstructions to such deformations in the class of *positive sectional curvature metrics* [**10**].

References

[1] S.S. Chern and N. Kuiper, *Some theorems on the isometric imbedding of compact Riemannian manifolds in Euclidean space*, Annals of Mathematics, 56 (1952), pp. 422-430.

[2] M. Dajczer, *Submanifolds and Isometric Immersions*, Mathematics Lecture Series, 13 , Publish or Perish Inc. (1990).

[3] S. Donaldson and P. Kronheimer, *The geometry of four-manifolds*, Oxford Science Publications (1990).

[4] D. Freed and K. Uhlenbeck, *Instantons and Four-Manifolds*, MSRI publications 1 (1984).

[5] D. Hilbert, *Über Flächen von konstanter Gausscher Krümmung*, Transactions of the American Mathematical Society, 2 (1901), pp. 87-99.

[6] M. Kervaire, *Sur le fibré normal à une sphère immergée dans un espace euclidien*, Commentarii Mathematici Helvetici, 33 (1959), pp. 121-131.

[7] M. Kervaire, *Some nonstable homotopy groups of Lie groups*, Illinois Journal of Mathematics, 4 (1960), pp. 161-169.

[8] N. Kleinjohann and R. Walter, *Nonnegativity of the Curvature Operator and Isotropy for Isometric Immersions*, Mathematische Zeitschrift, 181 (1982), pp. 129-142.

[9] J. Milnor and J. Stasheff, *Characteristic Classes*, Annals of Mathematical Studies, Princeton University Press, 76 (1974).

[10] D. Moore, *Isometric homotopy in codimension two*, Transactions of the American Mathematical Society, 292 (1985), pp. 653-663.

[11] K. Sarkaria and S. Zoltek, *Positively curved bilinear forms*, Proceedings of the AMS, 97 (1986), pp. 577-584.

[12] S. Smale, *The classification of immersions of spheres in euclidean spaces*, Annals of Mathematics, 69 (1959), pp. 327-344.

[13] N. Steenrod, *The topology of fiber bundles*, Princeton University Press (1951).

[14] A. Weinstein, *Positively curved n-manifolds in $\mathbb{R}^{n+2}$*, Journal of Differential Geometry, 4 (1970), pp. 1-4.

DEPARTMENT OF GEOMETRY AND TOPOLOGY, FACULTAD DE CIENCIAS MATEMÁTICAS, UNIVERSIDAD COMPLUTENSE DE MADRID, 28040 MADRID, SPAIN
E-mail address: `lguijarro@mat.ucm.es`

Contemporary Mathematics
Volume **288**, 2001

Natural Poisson and Jacobi Structures on the Tangent Bundle of a Pseudo-Riemannian Manifold

Josef Janyška

ABSTRACT. Let M be a differentiable manifold with a pseudo-Riemannian metric g and a linear symmetric connection K. We deduce conditions to define Poisson or Jacobi structures on TM for all vector fields and 2-vector fields naturally induced on TM by g and K.

Introduction

In this paper M is a differentiable manifold with a pseudo-Riemannian metric g and $p_M : TM \to M$ is the tangent bundle of M. The natural fibred coordinates on TM are denoted by $(x^i, \dot{x}^i)$ and the canonical local bases of modules of vector fields and forms are denoted by $(\partial_i, \dot{\partial}_i)$ and $(d^i, \dot{d}^i)$.

In general relativistic theories $\dim M = 4$ and g is a Lorentz metric, but this is not relevant for our purposes.

The isomorphism $T^*M \to TM$ given by the metric tensor will be denoted by $\sharp$. The canonical natural function (kinetic energy) on TM will be denoted by $h(u) = \frac{1}{2}g(u,u)$, $u \in TM$.

We also assume a linear symmetric connection (gravitational field) K on M.

We use the term "natural" in the sense of [**4, 5, 8**].

In [**1, 2**] we have classified all natural 2-forms on TM generated by g and K and we have found conditions for such 2-forms to be symplectic. The aim of this paper is to describe conditions under which natural vector fields and 2-vector fields on TM described in [**3**] define Poisson or Jacobi structures. Evidently, the conditions for the natural nondegenerate Poisson structure have to be equivalent with conditions for the natural symplectic structure we have found in [**1, 2**].

In the paper $[\,,\,]$ denotes the Schouten-Nijenhuis bracket (S-N bracket).

All manifolds and mappings are assumed to be smooth.

2000 *Mathematics Subject Classification.* Primary 53D17; Secondary 53C50.

Key words and phrases. Poisson structure, Jacobi structure, natural operator.

This research has been partially supported by the Ministry of Education of the Czech Republic under the project MSM 143100009, the grant of the GA ČR No. 201/99/0296 and the agreement "Collaboration in the Differential Equations and Differential Geometry" between Masaryk University and University of Florence.

1. Natural vector fields on TM

We can define two canonical natural vector fields on TM . The first one is the *vertical lift* u^V at $u \in T_xM$. In coordinates

$$u^V = \dot x^i \dot\partial_i \,. \tag{1.1}$$

A linear connection K on TM is a linear TTM-valued 1-form $K : TM \to T^*M \otimes TTM$ with coordinate expression $K = d^i \otimes (\partial_i + K_i{}^j{}_k \dot x^k \dot\partial_j)$, $K_i{}^j{}_k \in C^\infty(M)$. If we consider K as the horizontal lift $K : TM \times_M TM \to TTM$, we can define the *horizontal lift* at u as the vector field $u^H = K(u,u)$, i.e.

$$u^H = \dot x^i \partial_i + K_r{}^i{}_s \dot x^r \dot x^s \dot\partial_i \,. \tag{1.2}$$

The vertical and the horizontal lifts of u are natural 0-order (with respect to the connection field) vector fields on TM. In [**3**] all natural vector fields on TM induced by the metric field and the torsion-free linear connection were classified. We have

THEOREM 1.1. *Let M be a differentiable manifold with a pseudo-Riemannian metric g and a linear symmetric connection K. Then all natural 0-order vector fields on TM are in the form*

$$E(u) = \alpha(h(u))\, u^H + \beta(h(u))\, u^V \,, \tag{1.3}$$

where α, β are smooth real functions defined on $\mathbb{R}$. □

We can prove the following Lemmas by using (1.1), (1.2) and properties of the S-N bracket.

LEMMA 1.2. *We have*

(1) $[u^V, u^H] = u^H$; (2) $[u^V, u^V \wedge u^H] = u^V \wedge u^H$;

(3) $[u^H, u^V \wedge u^H] = 0$; (4) $[u^V \wedge u^H, u^V \wedge u^H] = 0$. □

LEMMA 1.3. *Let γ be a smooth real function defined on $\mathbb{R}$. Then we have*

$$[u^V, \gamma(h(u))] = 2h(u)\dot\gamma(h(u)) \,,$$

$$[u^H, \gamma(h(u))] = \frac{1}{2}\dot\gamma(h(u))(\nabla g)(u,u)(u) \,,$$

where $\dot\gamma = \frac{d\gamma}{dt}$ and ∇ is the covariant differentiation with respect to K. □

2. Natural 2-vector fields on TM

Let us denote by ϑ the VTM-valued 1-form on M given by the vertical lift, i.e. $\vartheta : M \to T^*M \otimes VTM$, $\vartheta = d^i \otimes \dot\partial_i$.

The metric g and the connection K induce naturally a 2-vector field

$$\Lambda(g,K) = K \bar\wedge \vartheta : TM \to \wedge^2 T(TM) \,, \tag{2.1}$$

where $\bar\wedge$ denotes the wedge product followed by the contraction through the inverse metric $\tilde g$. In coordinates

$$\Lambda(g,K) = g^{ij}(\partial_i + K_i{}^m{}_k \dot x^k \dot\partial_m) \wedge \dot\partial_j \,. \tag{2.2}$$

The metric g can be considered to be a T^*M-valued 1-form on M which will be denoted by $\bar g$ to distinguish it from the metric. We denote by $d_K \bar g$ the *covariant exterior differential* of $\bar g$ with respect to K, [**4**]. Then $d_K^2 \bar g = R \wedge \bar g$, where R is the curvature tensor of K.

THEOREM 2.1. *$\Lambda(g,K)$ defines a Poisson structure, i.e.* $[\Lambda(g,K),\Lambda(g,K)]=0$, *if and only if the condition*

$$d_K\bar{g}=0 \tag{A}$$

is satisfied.

PROOF. We have in coordinates

$$[\Lambda(g,K),\Lambda(g,K)] = g^{ip}g^{rj}g^{mk}(g_{pm,r}+g_{ms}K_p{}^s{}_r)\partial_i\wedge\dot{\partial}_j\wedge\dot{\partial}_k + (R^{ijk}{}_r + R^{jki}{}_r + R^{kij}{}_r)\dot{x}^r\dot{\partial}_i\wedge\dot{\partial}_j\wedge\dot{\partial}_k\,.$$

On the other hand

$$d_K\bar{g} = (g_{ij,k}+g_{js}K_i{}^s{}_k)d^i\otimes d^j\wedge d^k\,.$$

The condition $d_K\bar{g}=0$ imply $d_K^2\bar{g}=0$ which implies Theorem 2.1. □

The following Lemmas can be easily proved in coordinates.

LEMMA 2.2. *Let γ be a smooth real function defined on $\mathbb{R}$. Then we have:*

(1) $[\Lambda(g,K),\gamma(h(u))] = \dot{\gamma}(h(u))\big(-u^H+\frac{1}{2}(((\nabla g)(u,u))^\sharp)^V\big)$,

(2) $[\Lambda(g,K),u^V] = -\Lambda(g,K)$. □

LEMMA 2.3. *We have*

$$(((\nabla g)(u,u))^\sharp)^V\wedge u^H = 0 \Longleftrightarrow \nabla g = 0\,.$$ □

LEMMA 2.4. *The following equalities are equivalent:*

(B) $g(X_1,X_3)(\nabla_{X_2}g)(X_4,X_5) = g(X_2,X_3)(\nabla_{X_1}g)(X_4,X_5)$ *for any vector fields X_i, $i=1,\dots,5$;*

(1) $(((\nabla g)(u,u))^\sharp)^V\wedge\Lambda(g,K)=0$;

(2) $(((\nabla g)(u,u))^\sharp)^V\wedge u^V=0$. □

LEMMA 2.5. *We have*

$$[\Lambda(g,K),u^H]=0\Longleftrightarrow\nabla g=0\,.$$ □

LEMMA 2.6. *The conditions* (A) *and* (B) *imply*

$$(\nabla g)(u,u)(u)u^V\wedge\Lambda(g,K)=0\,,$$
$$[\Lambda(g,K),u^H]\wedge u^V=0\,.$$ □

In [**3**] we have classified all natural 2-vector fields on TM. We have

THEOREM 2.7. *Let (M,g) $(\dim M>3)$ be an oriented pseudo-Riemannian manifold endowed with a symmetric linear connection K. Then all natural 2-vector fields on TM are of the form*

$$\Lambda(\gamma,K)=\gamma_1(h(u))\,\Lambda(g,K)+\gamma_2(h(u))\,u^H\wedge u^V \tag{2.3}$$

where γ_1,γ_2 are smooth real functions defined on $\mathbb{R}$. In coordinates

$$\Lambda(\gamma,K)=\big(\gamma_1(h(u))\,g^{ij}+\gamma_2(h(u))\,\dot{x}^i\dot{x}^j\big)(\partial_i+K_i{}^m{}_k\dot{x}^k\dot{\partial}_m)\wedge\dot{\partial}_j\,. \tag{2.4}$$ □

3. Natural Poisson structures

In this Section we shall describe conditions for the nondegenerate $\Lambda(\gamma, K)$ to define a Poisson structure on TM, i.e. conditions for $[\Lambda(\gamma, K), \Lambda(\gamma, K)] = 0$.

LEMMA 3.1. *The 2-vector field $\Lambda(\gamma, K)$ is of maximal rank if and only if $\gamma_1(t) \neq 0$ and $\gamma_1(t) + 2t\gamma_2(t) \neq 0$ for any $t \in \mathbb{R}$.*

PROOF. It can be easily proved in coordinates. □

LEMMA 3.2. *We have*

$$\begin{aligned}
[\Lambda(\gamma, K), \Lambda(\gamma, K)] &= \gamma_1^2(h)\, [\Lambda(g, K), \Lambda(g, K)] \\
&+ \gamma_1(h)\dot\gamma_1(h)\, ((\nabla g)(u,u)^\sharp)^V \wedge \Lambda(g, K) + \gamma_1(h)\dot\gamma_2(h)\, ((\nabla g)(u,u)^\sharp)^V \wedge u^H \wedge u^V \\
&+ 2\big(\gamma_1(h)\gamma_2(h) - \gamma_1(h)\dot\gamma_1(h) - 2h\gamma_2(h)\dot\gamma_1(h)\big)\, u^H \wedge \Lambda(g, K) \\
&+ \gamma_2(h)\dot\gamma_1(h)\, (\nabla g)(u,u)(u)\, u^V \wedge \Lambda(g, K) + 2\gamma_1(h)\gamma_2(h)\, [\Lambda(g, K), u^H] \wedge u^V .
\end{aligned}$$

PROOF. It can be proved by using properties of the S-N bracket. □

THEOREM 3.3. *The nondegenerate 2-vector field $\Lambda(\gamma, K)$ defines a Poisson structure on TM if and only if the conditions* (A), (B) *and*

$$\text{(C)} \qquad \gamma_1(t)\gamma_2(t) - \gamma_1(t)\dot\gamma_1(t) - 2t\gamma_2(t)\dot\gamma_1(t) = 0$$

are satisfied for any $t \in \mathbb{R}$.

PROOF. It follows from Theorem 2.1, Lemmas 2.4, 2.6 and 3.2. □

REMARK 3.4. Let us note that the conditions of Theorem 3.3 for nondegenerate natural Poisson structures are equivalent with conditions for natural symplectic structures given in [**1**].

4. Natural Jacobi structures

In this Section we shall describe conditions for E and $\Lambda(\gamma, K)$ to define a Jacobi structure on TM, i.e. we have to find conditions for E and $\Lambda(\gamma, K)$ such that $[E, \Lambda] = 0$, $[\Lambda, \Lambda] = 2E \wedge \Lambda$.

LEMMA 4.1. *We have*

$$\begin{aligned}
[E, \Lambda(\gamma, K)] &= \alpha(h)\gamma_1(h)\, [u^H, \Lambda(g, K)] \\
&+ \Big(\frac{1}{2}\alpha(h)\dot\gamma_1(h)\, \nabla g(u,u)(u) + 2h\beta(h)\dot\gamma_1(h) - \beta(h)\gamma_1(h)\Big)\, \Lambda(g, K) \\
&+ \Big(\frac{1}{2}\alpha(h)\dot\gamma_2(h)\, \nabla g(u,u)(u) + 2h\beta(h)\dot\gamma_2(h) + \beta(h)\gamma_2(h) \\
&\quad - 2h\dot\beta(h)\gamma_2(h) - \frac{1}{2}\dot\alpha(h)\gamma_2(h)\, \nabla g(u,u)(u) - \dot\beta(h)\gamma_1(h)\Big)\, u^H \wedge u^V \\
&+ \frac{1}{2}\dot\alpha(h)\gamma_1(h)(\nabla g(u,u)^\sharp)^V \wedge u^H + \frac{1}{2}\dot\beta(h)\gamma_1(h)\, (\nabla g(u,u)^\sharp)^V \wedge u^V .
\end{aligned}$$

PROOF. It can be proved by using properties of the S-N bracket. □

LEMMA 4.2. *Let $\Lambda(\gamma, K)$ be of maximal rank. We have*

$$[E, \Lambda(\gamma, K)] = 0$$

if and only if one of the following groups of conditions is satisfied:

I. $\alpha \equiv 0$, $\beta \equiv 0$;

II. $\nabla g = 0$ *and for all* $t \in \mathbb{R}$ *the following equalities are satisfied*

$$\beta(\gamma_1 - 2t\dot\gamma_1) = 0\,, \quad \beta(\gamma_2 + 2t\dot\gamma_2) - \dot\beta(\gamma_1 + 2t\gamma_2) = 0\,.$$

PROOF. It is a consequence of Lemmas 2.5 and 4.1. □

LEMMA 4.3. *Let $\Lambda(\gamma, K)$ be of maximal rank. We have*

$$[\Lambda(\gamma, K), \Lambda(\gamma, K)] = 2E \wedge \Lambda(\gamma, K)$$

if and only if the conditions (A), (B) *and*

$$\alpha(t)\gamma_1(t) = \gamma_1(t)\gamma_2(t) - \gamma_1(t)\dot\gamma_1(t) - 2t\gamma_2(t)\dot\gamma_1(t)\,, \qquad \beta(t) = 0 \tag{4.1}$$

are satisfied for any $t \in \mathbb{R}$.

PROOF. We have

$$2E \wedge \Lambda(\gamma, K) = 2\alpha(h(u))\gamma_1(h(u))\, u^H \wedge \Lambda(g, K) + 2\beta(h(u))\gamma_1(h(u))\, u^V \wedge \Lambda(g, K)\,.$$

Comparing this with expression of $[\Lambda(\gamma, K), \Lambda(\gamma, K)]$ from Lemma 3.2 we get Lemma 4.3. □

THEOREM 4.4. *Let $\Lambda(\gamma, K)$ be of maximal rank and E be a non-zero natural vector field. E and $\Lambda(\gamma, K)$ define a Jacobi structure on TM if and only if K is the metric connection, $E = \alpha(h)\, u^H$, $\alpha \not\equiv 0$, where*

$$\alpha(t)\gamma_1(t) = \gamma_1(t)\gamma_2(t) - \gamma_1(t)\dot\gamma_1(t) - 2t\gamma_2(t)\dot\gamma_1(t) \tag{D}$$

is satisfied for any $t \in \mathbb{R}$.

PROOF. Theorem 4.4 follows from Lemmas 4.2 and 4.3. □

References

1. J. Janyška, *Remarks on symplectic and contact 2-forms in relativistic theories,* Bollettino U.M.I. (7) 9–B (1995), 587–616.
2. J. Janyška, *Natural symplectic structures on the tangent bundle of a space-time,* Proceedings of the Winter School Geometry and Topology (Srní, 1995), Supplemento ai Rendiconti del Circolo Matematico di Palermo, Serie II **43** (1996), pp. 153–162.
3. J. Janyška, *Natural vector fields and 2-vector fields on the tangent bundle of a pseudo-Riemannian manifold,* Archivum Mathematicum (Brno) **37** (2001), pp. 143–160.
4. I. Kolář, P. W. Michor and J. Slovák, Natural Operations in Differential Geometry, Springer–Verlag 1993.
5. D. Krupka and J. Janyška, Lectures on Differential Invariants, Folia Fac. Sci. Nat. Univ. Purkynianae Brunensis, Brno 1990.
6. P. Libermann and Ch. M. Marle, Symplectic Geometry and Analytical Mechanics, Reidel Publ., Dordrecht 1987.
7. A. Lichnerowicz, *Les variétés de Jacobi et leurs algèbres de Lie associées*, J. Math. Pures et Appl., **57** (1978), pp. 453–488.
8. A. Nijenhuis, *Natural bundles and their general properties*, Diff. Geom., in honour of K. Yano, Kinokuniya, Tokyo 1972, pp. 317–334.
9. I. Vaisman, Lectures on the Geometry of Poisson Manifolds, Birkhäuser, Verlag 1994.

DEPARTMENT OF MATHEMATICS, MASARYK UNIVERSITY
JANÁČKOVO NÁM 2A, 662 95 BRNO, CZECH REPUBLIC
E-mail address: janyska@math.muni.cz

Contemporary Mathematics
Volume **288**, 2001

On Alfred Gray's elliptical catenoid

Hugo Jiménez and Santiago López de Medrano

ABSTRACT. We give a parametrization, using Jacobi's elliptic functions, of Alfred Gray's Elliptical Catenoid and Elliptical Hellicoid that avoids some problems present in the original depiction of these surfaces.

1. Introduction. Björling's Problem

One of Alfred Gray's favorite tools was the solution to Björling's problem: *Given a planar analytic curve, find a minimal surface in* $\mathbb{R}^3$ *that contains it as a geodesic.* The Weierstrass representation provides an explicit solution: if $(x(t), y(t))$ is a parametrization of the curve, the parametrization

$$\Phi(z) = \Re(x(z), y(z), i\int_p^z \sqrt{x'(u)^2 + y'(u)^2}du) \tag{1.1}$$

gives the solution, where $x(z)$, $y(z)$ are extensions of $x(t)$, $y(t)$ to functions of a complex variable z. We will call it the Björling surface of the curve. When one takes, instead of the real part, the imaginary part of the expression, one gets another minimal surface, known as the conjugate surface of the first one.

It is well known how Alfred used this to produce many beautiful surfaces. But he could use it also as a theoretical tool: one day we asked Alfred what order of contact could a minimal surface have at a self-intersection point; without a second of thought he replied: "Any order of contact: take a planar curve with a self-contact of order n and solve Björling's problem". An instant theorem!

Formula (1) can be directly given to the computer. However, the fact that the integrand is in many cases multivalued poses some problems when we try to get the global picture. Take, for instance, the parabola $(2t, t^2)$. The formula gives the following parametrization of its Björling surface

$$\Phi(z) = \Re(2z, z^2, 2i\int_0^z \sqrt{1+u^2}du)$$

which the reader can immediately give to his favorite graphics package ... and get the wrong picture! (Figure 1a, where the thick line shows the parabola). The

2000 *Mathematics Subject Classification.* Primary 53A10, 53-04; Secondary 33E05.
Key words and phrases. minimal surfaces, geodesics, elliptic functions.
Work partially sponsored by grants 27969-E (Conacyt) and IN-119298 (Papiit-UNAM).

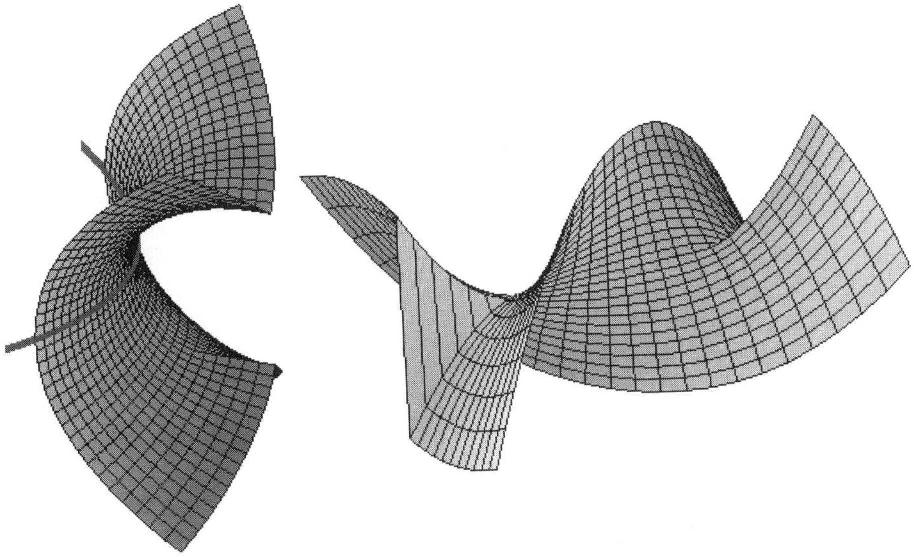

a *b*

FIGURE 1

problem has to do with the integrand that is multivalued and branches at the points i and $-i$. When we integrate from 0 to, say, $2i$ along different paths we can get different values, all with the same imaginary part. This produces a sharp edge, which is impossible in a minimal surface. The problem is more evident in the conjugate surface: we get a discontinuity which the computer fills with a planar face (Figure 1b).

The problem can be easily solved by a simple substitution $z = \sinh(w)$ which makes the function single valued and the surface becomes:

$$\Phi(w) = \Re(2\sinh(w), \sinh(w)^2, 2i(\cosh(w)\sinh(w) + w))$$

which gives the correct picture (Figure 2a). The surface continues beyond the sharp edges which become lines of self-intersection and then takes a turn, giving a periodic pattern. The surface turns out to be the well-known Catalan's surface that contains a cycloid perpendicular to the parabola which is another planar geodesic of the surface. Figure 2b shows the correct conjugate surface.

2. Björling's Duality

The abovc example shows an interesting relation between the parabola and the cycloid: they have the same Björling surface. Recall that a plane cuts a minimal surface along a geodesic if, and only if, it is a plane of symmetry of the surface. Now, the Björling surface of a curve that has a line of symmetry has two planes of symmetry: the original one containing the curve and the plane perpendicular to it that contains the line of symmetry. Therefore, the second plane intersects the surface in another geodesic with a line of symmetry. In formula (1) the original curve is obtained by restricting to real values of z and the orthogonal geodesic by

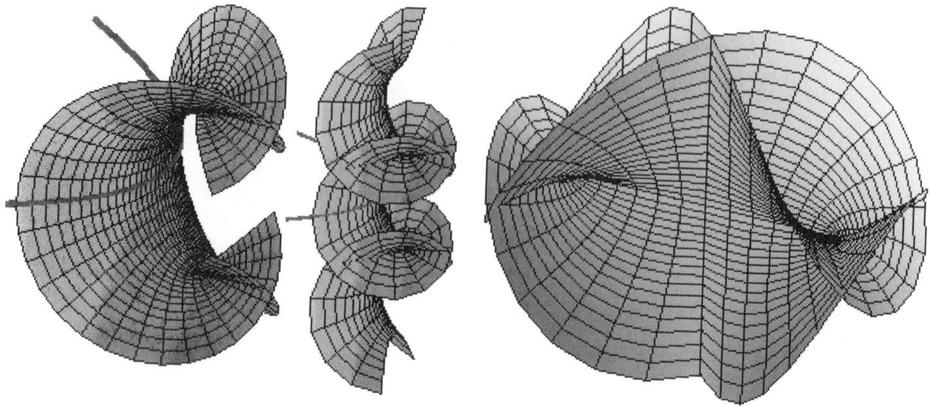

FIGURE 2. a) Two views of Catalan's surface. b) The conjugate surface.

restricting to purely imaginary values of z. The two curves are in a sense *dual* to each other. To be more precise: the duality is between objects consisting each of a curve and a point of intersection of the curve with a line of symmetry. We can call those pairs *Björling dual* to each other. For example, the circle and the catenary are Björling dual (the common surface is the catenoid) and so are the parabola and the cycloid with its line of symmetry that cuts it on one of the highest (smooth) points of the cycle.

3. The Elliptical Catenoid

Let us apply formula (1) to the ellipse with semi-axes a and b, $a > b$, and excentricity e. We obtain

$$\Phi(z) = \Re(b\cos(z), a\sin(z), i\ a \int_0^z \sqrt{1 - e^2\sin(u)^2}du)$$

and with it the computer produces Figure 3a. One observes that this surface has various *channels* where one of the principal curvatures is close to zero and the other one is very big. Again, this can not be a minimal surface! (One obtains a channel or an edge depending on the parity of the size of the grid used to plot the surface). The conjugate surface presents again a discontinuity which is filled by the computer with a flat face (Figure 3b).

Alfred's depiction in [G] of these surfaces, called by him *Elliptical Catenoid* and *Elliptical Hellicoid*, have the same problems. (Those two loves of his, the computer and Björling's problem, did not get together as smoothly as he thought). Both Maple (which we use) and Mathematica give the same picture. Maybe Alfred was aware of the problem since in this case he used an elliptic function instead of applying directly the formula. The analysis must, however, be carried some steps further:

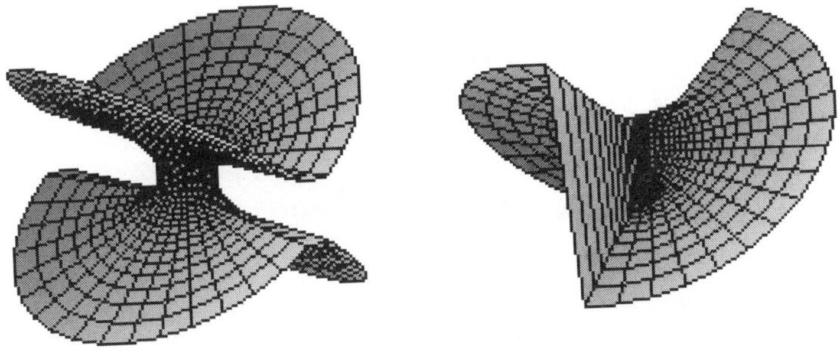

a *b*

FIGURE 3

We use Jacobi's elliptic functions (for details, see [L], [M]): Jacobi's elliptic sine $\phi = sn(v,k)$, being the inverse function of the integral

$$v = \int_0^\phi \frac{dt}{\sqrt{1-t^2}\sqrt{1-k^2t^2}},$$

Jacobi's elliptic cosine

$$cn(v,k) = \sqrt{1 - sn^2(v,k)}$$

and Jacobi's *delta* function

$$dn(v,k) = \sqrt{1 - k^2 sn^2(v,k)}.$$

Like the trigonometric functions (which we obtain as the particular case $k = 0$) they can be extended to complex functions, but the extensions are now meromorphic and doubly periodic instead of being holomorphic and periodic.

With the new parameter u related to z by $sn(u,e) = \sin(z)$ we obtain the parametrization:

$$\Psi(u) = \Re(b\ cn(u,e), a\ sn(u,e), i\ a \int_0^u dn^2(\sigma,e)\ d\sigma)$$

The integrand is still a meromorphic function, but its residues are all 0 ([L], p.241), so the integral is a well-defined function of u. This parametrization produces the correct picture of the Elliptical Catenoid. The ellipse has two lines of symmetry and, correspondingly, two Björling duals; one resembles a cycloid, the other one a catenary. Figure 4 shows the correct pictures of the Elliptical Catenoid and Elliptical Hellicoid:

The same method gives the Björling surface of the hyperbola (where now we take $k = 1/e < 1$):

$$\Psi(v) = \Re(b\ dn(v/k,k), i\ a\ k\ sn(v/k,k), a \int_0^v cn^2(\eta/k,k)\ d\eta).$$

The dual curve of the hyperbola again resembles a cycloid.

The analysis of the Björling surfaces of the conics will be developed fully in a forthcoming article.

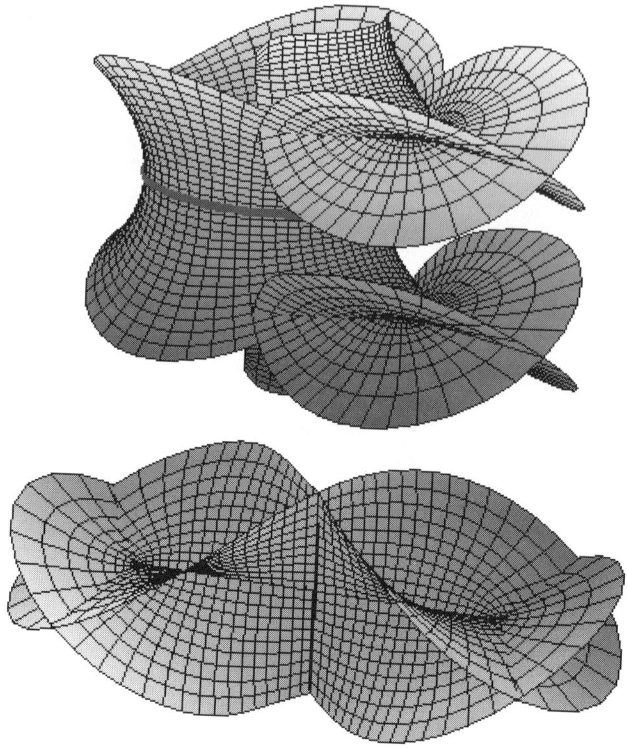

FIGURE 4. The Elliptical Catenoid and the Elliptical Hellicoid.

References

[G] Alfred Gray, *Modern Differential Geometry of Curves and Surfaces with Mathematica*, CRC Press, 2nd Edition, 1998.

[L] Lawden, D. *Elliptic Functions and Applications*, App.Math.Sci. **80**, Springer-Verlag 1989.

[M] Markushevich, A.I., *The Remarkable Sine Functions*, Elsevier, 1966.

FACULTAD DE CIENCIAS, UNAM, MÉXICO.
E-mail address: hugo@nisamox.fciencias.unam.mx

FACULTAD DE CIENCIAS E INSTITUTO DE MATEMÁTICAS, UNAM.
E-mail address: santiago@matem.unam.mx

Contemporary Mathematics
Volume **288**, 2001

Sub-Riemannian Geometry and Non-holonomic Mechanics

Jair Koiller, Paulo R. Rodrigues, and Paulo Pitanga

Abstract. Sub-Riemannian (SR) and Non-holonomic mechanics (NH) use same ingredients, a metric g on a configuration space Q and a distribution $E \subset TQ$, but lead to different geodesic equations. We outline some directions of research being pursued by our team (which includes our collaborators Richard Montgomery and Kurt Ehlers). Sub-Riemannian balls generated with ***Mathematica***^{TM} illustrate that it can be a very useful tool to understand pictorically the geometry of sub-Riemannian metrics. On a different tack, we explain a basic result in E. Cartan's paper (at the 1928 International Congress of Mathematicians) on non-holonomic mechanics. Here, computer algebra must play a role in further developments.

Alfred Gray was deeply involved in the interplay between Mechanics and Differential Geometry, and a pioneer in using ***Mathematica***^{TM} *as a tool to understand and teach them. We dedicate this paper to his memory.*

1. Sub-Riemannian geometry × Non-holonomic mechanics

Let (Q, g) be an n-dimensional Riemannian manifold. Assume that Q admits a distribution $\mathcal{E} : q \in Q \to E_q \subset T_qQ$ of constant rank $m < n$. We say that Q is *polarized* by $\mathcal{E}$ and that the Lie algebra of vector fields taken as sections of E over Q span the tangent space of Q at each point. The triple $(Q, g, \mathcal{E})$ defines two different mathematical structures. SR arises by considering the constrained variational principle, and NH by assuming D'Alembert's principle of virtual work. See the book edited by A. Bellaïche and J-J. Risler [**4**] and references therein for the former, and the book by Ju. I. Neimark and N.A. Fufaev [**18**] for the latter.

A (piecewise) smooth curve c: $[a, b] \to Q$ is *horizontal* or *polarized* if the tangent vectors are in E, i.e $\dot{c}(t) \in E_{c(t)} \subset TQ$, for almost all $t \in [a, b] \subset \mathbb{R}$. If Q is connected then any two points in Q can be joined by a smooth piecewise curve in Q where each smooth piece is a segment of an integral curve of one of the vector fields that span $\mathcal{E}$. A Hopf-Rinow result holds in SR but not in NH. In SR one defines the *length* of c by $\ell(c) = \int \sqrt{g(\dot{c}(t), \dot{c}(t))}$ and the *distance* of $x \neq y \in Q$ is $d(x, y) = \inf\,(\ell(c))$, taken over all horizontal curves connecting x to y. The curve that defines the distance of x to y is said a *shape minimizing geodesic*, or *sub-riemannian geodesic*.

In contradistinction, for NH-geodesics are not length extremizers. Cartan [**8**] showed that D'Alembert's principle can be casted in geometric terms with the

---

1991 *Mathematics Subject Classification.* MSC 2000: 37J60, 70F25, 53C17, 53C07.
Key words and phrases. Sub-Riemannian geometry, Non-holonomic mechanics.
The first author was supported in part by a fellowship from FAPERJ.

notion of a *non-holonomic connection*. Let a mechanical system with kinetic energy $T = \frac{1}{2}g(\dot q, \dot q)$ and external forces f (in contravariant form), be subjected to constraints defined by the distribution E. By D'Alembert's principle, the constraining force must be perpendicular to the constraint subspace, since it does not produce work. Unless the distribution is integrable, this gives different equations from those of the variational SR viewpoint (*vakonomic*, in Arnold's terminology). The non-holonomic dynamics is given by $D_{\dot c}\dot c = f^{\|}$. The right hand side is the ortogonal projection over E, and D is the *non-holonomic connection*, defined by $D_X Y = (\nabla_X Y)^{\|}$, $Y \in E$, where ∇ is the Levi-Civita connection of g.

Convention: lower case roman characters i, j, k run from 1 to m, greek characters α, β, γ, etc., run from $m+1$ to n, capital roman letters I, J, K, etc., run from 1 to n. Summation over repeated indices is assumed unless otherwise stated. Consider a local frame e_I on an open subset of Q and its dual coframe ω_J. Thus e_i span the constraint distribution E. In terms of Cartan's moving frames, the non-holonomic connection is defined by connection forms ω_{ij}, simply by abandoning those containing "greek indices" in $\nabla_X e_I = \omega_{IJ}(X)e_J$. The non-holonomic connection also appears in SR, as we now describe.

2. Sub-Riemannian: constrained Euler-Lagrange equations

E can be defined by a set of 1-forms ω^α, $1 \le \alpha \le n-m$, $\omega^\alpha_q(X) = 0$, $X \in E_q \subset T_qQ$, and we set

$$F_q = \text{span}\,\{e_\alpha(q) = g_\star(w^\alpha_q, \bullet)\},\ \ E_q = \text{span}\,\{e_i(q); 1 \le i \le m\},$$

so that $TQ = E \oplus F$. This spliting defines an *almost product structure* with corresponding bundle projectors (the interplay of projectors with covariant derivatives is one of the constant themes of our study group). Let $L = L(q(t), \dot q(t)) = \sqrt{g(\dot q(t), \dot q(t))}$ be the unconstrained Lagrangian and we consider the following constrained Lagrangian problem ([**20**])

$$\overline{L}(q(t), \dot q(t)) = L(q(t), \dot q(t)) + \lambda_\alpha\, g(\, w^\alpha, \dot q(t)\,)\ w^\alpha(\dot q(t))\,.$$

THEOREM 2.1. $\dot q(t) = u^i(t)\, e_i$ *satisfies*

$$\dot u^i(t) + \Gamma^i_{jk}\, u^j(t)\, u^k(t) - C^\alpha_{ij}\, u^j(t)\, \lambda_\alpha(t) = 0 \tag{2.1}$$

$$\dot\lambda_\alpha(t) - \Gamma^\alpha_{ij}\, u^i(t)\, u^j(t) - C^\beta_{\alpha a}\, u^i(t)\, \lambda_\beta(t) = 0 \tag{2.2}$$

$$\Gamma^i_{jk} = g(\nabla_{e_k} e_j, e_i) =: \omega_{ji}(e_k),\ \Gamma^\alpha_{ij} = g(\nabla_{e_j} e_i, e_\alpha) =: \omega_{i\alpha}(e_j) \tag{2.3}$$

$$C^\alpha_{ji} = g([e_j, e_i], e_\alpha),\ C^\beta_{i\alpha} = g([e_i, e_\alpha], e_\beta). \tag{2.4}$$

2.1. Example: the Heisenberg algebra (see [5], [1]). Let Q be the Euclidean space $\mathbb{R}^3$, $\Omega = (1/2)\,(dx \wedge dy - dy \wedge dx)$ and $\omega = dz - y\,dx + x\,dy$ the contact form with $\Omega = (1/2)\,d\omega$. Denote $e_1 = X_1 = \partial_x + y\partial_z$, $e_2 = X_2 = \partial_y - x\partial_z$ the horizontal vectors fields and $e_3 = Y_3 = \partial_z$ the vertical vector field. It is easy to verify that

$$[X_1, X_2] = 2\,Y_3,\ [X_1, Y_3] = 0,\ [X_2, Y_3] = 0\,.$$

This is just the Heisenberg algebra, a very well known non-holonomic Lie algebra, pervasive both in pure and in applied mathematics. One of us (PP) is developing a model for the Yang-Mills-Higgs heavy magnetic monopole using the Heisenberg algebra.

Then the equations for $\dot{q} = u_1 X_1 + u_1 X_2$ are

$$\dot{u}_1 = C^3_{12} u_2 \lambda_3 \, , \; \dot{u}_2 = C^3_{21} u_1 \lambda_3 \, , \; \dot{\lambda}_3 = C^3_{31} u_1 \lambda_3 + C^3_{32} u_2 \lambda_3, \tag{2.5}$$

A direct computation gives

$$C^3_{12} = -C^3_{21} = 2 \, , \; C^3_{31} = 0 \, , \; C^3_{32} = 0 \, .$$

so we have

$$\dot{x} = -u_2 \, , \; \dot{y} = u_1 \, , \; \dot{z} = x u_1 + y u_2, \; \dot{u}_1 = 2 u_2 \lambda_3 \, , \; \dot{u}_2 = -2 u_1 \lambda_3 \, , \; \dot{\lambda}_3 = 0 \, . \tag{2.6}$$

Thus we may write $u_1 = r \, cos \alpha(t)$, $u_2 = r \, sin \alpha(t)$, with $r = 1$, and it follows that

$$\dot{\alpha}(t) = \lambda, \text{ i.e, } \alpha(t) = \lambda t + \phi. \tag{2.7}$$

Integrating the system $\dot{x} = cos\alpha$; $\dot{y} = sin\alpha$; $\dot{z} = ycos\alpha - xsin\alpha$ one obtains the parametric equations of the geodesics, with $\lambda t \equiv \theta$.

$$x = a \left[\frac{\sin(\theta + \phi) - sin\phi}{\theta} \right], \tag{2.8}$$

$$y = a[\frac{cos(\phi) - \cos(\theta + \phi)}{\theta}], \tag{2.9}$$

$$z = a^2 (\frac{\theta + \sin \theta}{\theta^2}). \tag{2.10}$$

If we set $\Omega = dx \wedge dy$, $\Omega = -d\omega$, then $\omega = dz - x \, dy$. In such a case the last equation is replaced by (see [**20**])

$$z = a^2 \left[\frac{2\theta + sin(2\theta + 2\phi) - sin2\phi + 4cos\phi(-sin(\theta + \phi) + sin\phi)}{4\theta^2} \right],$$

since the horizontal space is now spanned by $X_1 = \partial_x, X_2 = \partial_y + x \, \partial_z$ and $Y_3 = \partial_z$. We feel that our choice gives a more symmetric expression for plotting figures, which we obtained using ***Mathematica***^{TM}.

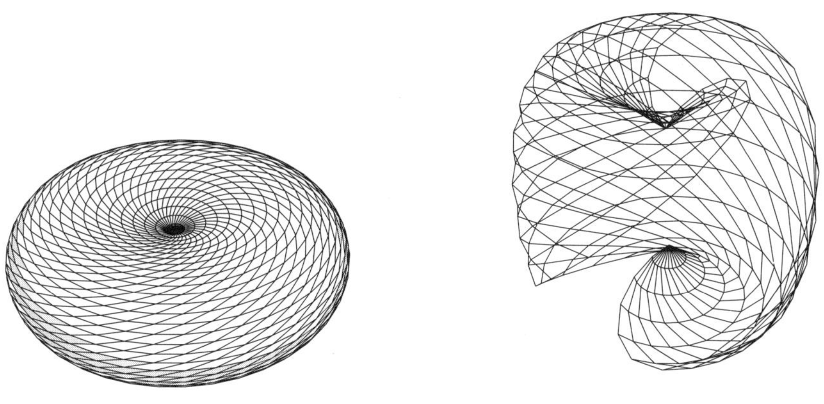

FIGURE 1. The "apple"(the geodesic sub-Riemannian ball) and the "apple" for the parameters $-2\pi \leq \theta \leq 2\pi, \; 0 \leq \phi \leq 1.35 \, \pi$.

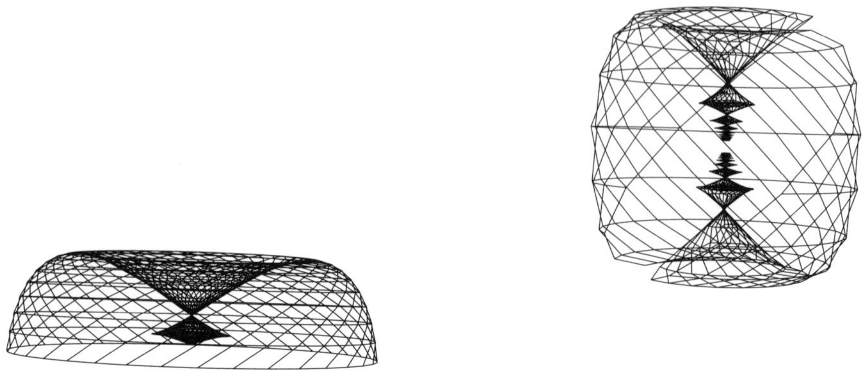

FIGURE 2. The interior of the "apple" for the parameters $0 \le \theta \le 4\pi$, $0 \le \phi \le 2\pi$ and $-20\pi \le \theta \le 22\pi$, $0 \le \phi \le 1.65\pi$.

3. Hodograph of the D'Alembert-Lagrange solution curves.

Concerning NH, we consider the following remark by Cartan ([**8**], §7): *"La trajectoire du système matériel, supposé soumis à des forces données de travail elémentaire* $\sum \overline{P}_i \overline{\omega}_i$*, se développe suivant la trajectoire d'un point matériel de masse 1 placé dans l'espace euclidien à m dimensions et soumis à la force de composan tes* $\overline{P}_i$*."* Thus Cartan proposes to transfer the dynamics to a fixed m-dimensional arena. To understand this statement in modern terms, firstly we describe *parallel transport*:

PROPOSITION 3.1. Given a curve $\gamma(t) \in \mathbb{R}^m \equiv E_{q_o}$, consider the nonautonomous system of ODEs in the frame manifold $\mathrm{Fr}(E)$:

$$X = (P\,\dot\gamma(t))_i\, e_i \;,\; \dot P = (-\omega_{ij}(X))P \;,\; c(0) = q_o,\; P(0) = \mathrm{id}\;. \tag{3.1}$$

It gives the developed curve $c(t)$ on Q and an attached parallel frame

$$(\overline{e}_1, \cdots, \overline{e}_m) = (e_1, \cdots, e_m)\, P\;. \tag{3.2}$$

For a line $\gamma(t) = t\, v$ passing through the origin in $\mathbb{R}^m$ we obtain the non-holonomic geodesic starting at q with velocity $\dot c(0) = v$.

Let γ be the "hodograph" of $c(t)$ to $E_{c(0)} \equiv \mathbb{R}^m$. This means that $c(t)$ in Q is recovered from γ via parallel transport. Fix a *constant* frame $\overline{f}_1, \cdots, \overline{f}_m$ on $\mathbb{R}^m$ so $\gamma(t) = \gamma_i(t)\overline{f}_i$. The moving frame is in a sense "the" correct reference frame for the dynamics. Let $\overline{e}_i$ be the corresponding parallel frame along $c(t)$ Decompose

$$F^{\|} = f_i \overline{e}_i\;. \tag{3.3}$$

THEOREM 3.2. (Cartan, [**8**], §7).

$$\ddot\gamma_i(t) = f_i(c(t))\;. \tag{3.4}$$

Equations (3.4) should be solved simultaneously with (3.1).

For details, see [**13**].

References

[1] A. Agrachev, *Methods of control theory in nonholonomic geometry*, Proc. ICM-94, Birkhouser, Zurich, 1995.

[2] L. Bates, R. Cushman, *What is a completely integrable nonholonomic dynamicalsystem?*, Rep. Math. Phys. **44**, 1/2, (1999) 29–35.

[3] L.Bates, J. Śniatycki, *Nonholonomic reduction*, Rep. Math. Phys. **32** (1992), 99–115.

[4] A. Belaïche, J.J. Risler, *Sub-Riemannian Geometry*, Birkhäuser, Basel, 1996.

[5] R. Brockett, *Control theory and singular Riemannian geometry*, in P. Hilton, G. Young, eds., New directions in Applied Mathematics, Springer-Verlag, 1981.

[6] F. Cantrijn, J. Cortés, M. de Léon, D. Martín de Diego, *On the geometry of generalized Chaplygin systems*, preprint.

[7] F. Cantrijn, M. de León, D. Martín de Diego, *On almost-Poisson structures in nonholonomic nechanics*, Nonlinearity **12**, (1999) 721–737.

[8] E. Cartan, *Sur la represéntation géométrique des systèmes matériels non holonomes*, Proc. Int. Congr. Math., vol. 4, Bologna, 1928, pp. 253–261.

[9] R. Cushman, D. Kemppainen, J. Śniatycki, L. Bates, *Geometry of nonholonomic constraints*, Rep. Math. Phys. **36**: (2/3), (1995), 275–286.

[10] J. A. Fax, R. M. Murray, *Optimal control of affine connection control systems: a variational approach*, IEEE CDC Proceedings, 2000.

[11] R. Gardner, *The Method of Equivalence and its Applications*, SIAM, 1989.

[12] J. Koiller, *Reduction of some classical non-holonomic systems with symmetry*, Arch. Rational Mech. Anal. **118**, (1992), 113–148.

[13] J. Koiller, P. Rodrigues, P. Pitanga, *Non-holonomic connections following Élie Cartan*, An. Acad. Bras. Cienc., **73**, (2001), 165–190.

[14] W.S. Koon, J.E. Marsden, *Poisson reduction of nonholonomic mechanical systems with symmetry*, Rep. Math. Phys. **42**: (1/2), (1998), 101–134

[15] P.S. Krishnaprasad, A. Bloch, J.E. Marsden, R.M. Murray, *Non-holonomic mechanics with symmetry, Arch. Rat. Mech. Anal.*, **13**:1 (1996), 21–99.

[16] M. León, D. Martín de Diego, *On the geometry of non-holonomic Lagrangian systems*, J. Math. Phys. **37**, (1996) 3389–3414.

[17] Z. Li, J.F. Canny, *Non-holonomic Motion Planning*, Kluwer, 1993.

[18] Ju.I. Neimark, N.A. Fufaev, *Dynamics of Nonholonomic systems*, AMS Translations of Mathematical Monographs **33**, 1972.

[19] A.J. van der Schaft, B.M. Maschke, *On the Hamiltonian formulation of nonholonomic mechanical systems*, Rep. Math. Phys., **34** (1994), 225–233.

[20] A.M. Vershik, V.Y. Gershkovich, *Nonholonomic Dynamical Systems, Geometry of Distributions and Variational Problems*, Dynamical Systems VII, vol, 16, 4-79, Encyclopaedia of Mathematical Sciences, V. I. Arnol'd & S. P. Novikov, Eds.

J.K. - Laboratório Nacional de Computação Científica, Av. Getulio Vargas 333, 25651-070 Petrópolis - Brazil

E-mail address: jair@lncc.br, jair@impa.br

P.R.R. - Departamento de Geometria, Instituto de Matemática,, Universidade Federal Fluminense, 24020-140 Niterói, RJ - Brazil

E-mail address: rodrigues@mat.uff.br, rodriguespr@ax.apc.org

P.P. - Instituto de Fisica, Universidade Federal do Rio de Janeiro, Caixa Postal 68528, Cidade Universitária, 21945-970, Rio de Janeiro, RJ - Brazil

E-mail address: pitanga@if.ufrj.br, pitanga@hotmail.com

Contemporary Mathematics
Volume **288**, 2001

Weil Algebras of Generalized Higher Order Velocities Bundles

Miroslav Kureš

ABSTRACT. The generalized higher order velocities bundles including nonholonomic and semiholonomic ones are described and their Weil algebras are studied. The homogeneity of such Weil algebras is proved.

1. The expression of Weil algebras

The theory of Weil bundles plays a very important role in differential geometry at present and we recommend the monograph of I. Kolář, P.W. Michor and J. Slovák, [**2**], for a self-contained introduction to this theory. We recall some known facts about Weil algebras. We take the algebra $\mathbb{R}[t^1, \dots, t^k]$ of all polynomials with k variables and its ideal $\mathfrak{i}$ satisfying $\langle t^1, \dots, t^k \rangle^{r+1} \subset \mathfrak{i} \subset \langle t^1, \dots, t^k \rangle$, where $\langle t^1, \dots, t^k \rangle$ is the ideal of all polynomials without an absolute term and $\langle t^1, \dots, t^k \rangle^{r+1}$ is its $(r+1)$–th power. Such an ideal $\mathfrak{i}$ is said to be a *Weil ideal* and the factor algebra $A = \mathbb{R}[t^1, \dots, t^k]/\mathfrak{i}$ is called a *Weil algebra.* If k is the minimal number of variables $t^1, \dots, t^k$ for the expressing of Weil algebra $A = \mathbb{R}[t^1, \dots, t^k]/\mathfrak{i}$, we call $t^1, \dots, t^k$ the *essential variables* (with respect to A). We have a canonical decomposition $A = \mathbb{R} \oplus \mathfrak{n}$, where $\mathfrak{n} = \langle t^1, \dots, t^k \rangle/\mathfrak{i}$ is the ideal of all nilpotent elements of A.

1.1. The Weil algebra $\tilde{\mathbb{D}}^r_k$. We study the *nonholonomic bundle of k–dimensional velocities of r–th order* $\tilde{T}^r_k M = \tilde{J}^r_0(\mathbb{R}^k, M)$. The functor $\tilde{T}^r_k$ is naturally equivalent to the functor $\underbrace{T^1_k \dots T^1_k}_{r\text{–times}}$. A delicate difference between them is described in [**1**] and we shall not distinguish these functors. Given some local coordinates x^i on M $i = 1, \dots, m = \dim M$ and t^j on $\mathbb{R}^k$, $j = 1, \dots, k$, the iterated differentiation of $x^i(t^1, \dots, t^k)$ determines the induced coordinates on $\tilde{T}^r_k M$ as $y^i_{j_1 \dots j_r}$, $j_1, \dots, j_r = 0, 1, \dots, k$, which are not symmetric in the subscripts.

Write $\tilde{\mathbb{D}}^r_k$ for the Weil algebra of $\tilde{T}^r_k$. The element of $\tilde{\mathbb{D}}^r_k = \tilde{T}^r_k \mathbb{R}$ has a form

$$a + a_{j_1 \dots j_r} t^{j_1 \dots j_r}. \tag{1}$$

2000 *Mathematics Subject Classification.* 58A32, 58A20.

Key words and phrases. Bundle of velocities, nonholonomic jet, Weil algebra.

The author was supported by GA CR, grants Nos. 201/99/0296, 201/99/P065. This contribution was presented at the International Congress on Differential Geometry in memory of Alfred Gray, Bilbao, 2000. The author's participation on this congress was also supported by the grant of the Scientific Committee ICDG.

This paper is in final form and no version of it will be submitted for publication elsewhere.

If we evaluate the product of two elements and if we compare coefficients, we obtain generating equations for $\tilde{\mathbb{D}}^r_k$ as

$$(2)\quad t^{j_1\dots j_r}t^{\bar{j}_1\dots\bar{j}_r} = \begin{cases} 0 & \text{if } j_q \neq 0 \text{ and } \bar{j}_q \neq 0 \text{ for any } q, 1 \leq q \leq r \\ t^{j_1+\bar{j}_1\dots j_r+\bar{j}_r} & \text{if either } j_q = 0 \text{ or } \bar{j}_q = 0 \text{ for all } q, 1 \leq q \leq r. \end{cases}$$

It implies that essential variables in the Weil algebra have only one non-zero j_q and the second power of them vanish. If we denote such variables by t^l_q $(1 \leq l \leq k)$, we have the expression of $\tilde{\mathbb{D}}^r_k$ as

$$(3)\qquad \mathbb{R}[t^1_1,\dots,t^k_r]/\langle\langle t^1_1,\dots,t^k_1\rangle^2,\dots,\langle t^1_r,\dots,t^k_r\rangle^2\rangle,$$

which is well-known.

1.2. Projections. For every s, $0 \leq s \leq r$, we denote by $\pi^s\colon \tilde{T}^s_k M \to M$ the canonical projection to the base. Further, we denote $\pi^s_b\colon= \pi^s_{\tilde{T}^b_k M}\colon \tilde{T}^s_k(\tilde{T}^b_k M) \to \tilde{T}^b_k M$ projection with $\tilde{T}^b_k M$ as the base space, ${}_a\pi^s\colon= \tilde{T}^a_k\pi^s\colon \tilde{T}^a_k(\tilde{T}^s_k M) \to \tilde{T}^a_k M$ induced projection originating by the posterior application of the functor $\tilde{T}^a_k$ and ${}_a\pi^s_b\colon= \tilde{T}^a\pi^s_{\tilde{T}^b M}$ the general case containing both previous cases. If a or b equal zero, we do not write them. In local coordinates, just the coordinates with s adjoining zero subscripts (the $(b+1)$–th subscript is at the beginning) are remained after the application of ${}_a\pi^s_b$. Projections $\tilde{T}^p M \to \tilde{T}^q M$ are of a type ${}_a\pi^s_b$ or they are a composition of projections of such types. The *composition* ${}_{a_1}\pi^{s_1}_{b_1} \circ {}_{a_2}\pi^{s_2}_{b_2}$ is defined only for $a_1 + b_1 + s_1 = a_2 + b_2$ and the following assertion holds.

LEMMA 1. *Let $s_1 > 0, s_2 > 0$ and $a_1 + b_1 + s_1 = a_2 + b_2$. Then*
1. *if $a_1 \leq a_2$ and $b_1 \leq b_2$, then ${}_{a_1}\pi^{s_1}_{b_1} \circ {}_{a_2}\pi^{s_2}_{b_2} = {}_{a_1}\pi^{s_1+s_2}_{b_1}$;*
2. *if $a_1 > a_2$ or $b_1 > b_2$, then do not exist such a and b, for which ${}_{a_1}\pi^{s_1}_{b_1} \circ {}_{a_2}\pi^{s_2}_{b_2} = {}_a\pi^{s_1+s_2}_b$.*

PROOF. 1. ${}_{a_1}\pi^{s_1}_{b_1} \circ {}_{a_2}\pi^{s_2}_{b_2} = {}_{a_1}(\pi^{s_1}_{b_1} \circ {}_{a_2-a_1}\pi^{s_2}_{b_2}) = {}_{a_1}((\pi^{s_1} \circ {}_{a_2-a_1}\pi^{s_2}_{b_2-b_1})_{b_1}) = {}_{a_1}((\pi^{s_1+s_2})_{b_1}) = {}_{a_1}\pi^{s_1+s_2}_{b_1}$.

2. We assume ${}_{a_1}\pi^{s_1}_{b_1} \circ {}_{a_2}\pi^{s_2}_{b_2} = {}_a\pi^{s_1+s_2}_b$. As $a_1+b_1+s_1 = a_2+b_2$, it is not possible $a_1 > a_2$ along with $b_1 > b_2$. Let us suppose $a_1 \leq a_2$, $b_1 > b_2$. Then

$${}_{a_1}\pi^{s_1}_{b_1} \circ {}_{a_2}\pi^{s_2}_{b_2} = {}_{a_1}(\pi^{s_1}_{b_1} \circ {}_{a_2-a_1}\pi^{s_2}_{b_2}) = {}_{a_1}((\pi^{s_1}_{b_1-b_2} \circ {}_{a_2-a_1}\pi^{s_2})_{b_2}).$$

But we assume ${}_{a_1}\pi^{s_1}_{b_1} \circ {}_{a_2}\pi^{s_2}_{b_2} = {}_a((\pi^{s_1+s_2})_b)$. The projections on right hand sides represent projections onto bundles of $s_1 + s_2$ lower order. Hence the application of the iterated tangent functor gives $a = a_1$ and $b = b_2$. As the target bundles must be the same, we have also $a + b = a_1 + b_1$. So, $b_1 = b_2$ and this is the contradiction. The case $a_1 > a_2$, $b_1 \leq b_2$ is analogous. □

1.3. The Weil algebra $\bar{\mathbb{D}}^r_k$. In general, we can obtain subbundles of $\tilde{T}^r_k M$ by equalizations of some projections. An element $Z \in \tilde{T}^r_k M$ is called the *semiholonomic k–dimensional velocity* of r–th order, if for all $q = 1,\dots,r$

$$\pi^1_{r-1}(Z) = {}_{q-1}\pi^r_{r-q}(Z)$$

is satisfied. We denote by $\bar{T}^r_k M$ the bundle of semiholonomic k–dimensional velocities of r–th order. In local coordinates, we identify the coordinates subscripts of which become equal if we delete zeros in each of them. Write $\bar{\mathbb{D}}^r_k$ for the Weil algebra of $\bar{T}^r_k$.

For $r = 2$, the identification means $t^{j0} = t^{0j}$ and we write simply t^j. We obtain generating equations (2) for $\bar{\mathbb{D}}_k^r$ as

$$(4)\qquad \begin{aligned} 2t^{j_1}t^{j_2} &= t^{j_1j_2} + t^{j_2j_1} \\ t^{j_1}t^{j_2j_3} &= 0 \\ t^{j_1j_2}t^{j_3j_4} &= 0. \end{aligned}$$

Then we have t^{j_1} and $t^{j_1j_2}$ (for $j_1 < j_2$) as essential variables and

$$(5)\qquad t^{j_1j_2} = \begin{cases} t^{j_1}t^{j_2} & \text{for } j_1 = j_2 \\ 2t^{j_1}t^{j_2} - t^{j_2j_1} & \text{for } j_1 > j_2. \end{cases}$$

Hence

$$(6)\qquad \bar{\mathbb{D}}_k^2 = \mathbb{R}[t^{j_1j_2}, t^{j_3}]/\langle\langle t^{j_1j_2}\rangle^2, t^{j_1j_2}t^{j_3}, \langle t^{j_3}\rangle^3\rangle \quad (\text{only } j_1 < j_2\ ).$$

In the general case, the generating equations are similar as (4) (see also the proof of Proposition 3), but it is not trivial to determine, which variables are essential. We remark that there is another general expression of Weil algebra $\bar{\mathbb{D}}_k^r$ in [**3**], in which so-called colored simplexes are used in indexes of variables.

1.4. Involutions. We denote by $\kappa\colon \tilde{T}_k^2 \to \tilde{T}_k^2$ the canonical involution. κ generalizes the classical canonical involution for $k = 1$. Further, we follow the idea of the notation of projections and denote by ${}_a\kappa_b := \tilde{T}_k^a \kappa_{\tilde{T}_k^b M}\colon \tilde{T}_k^a(\tilde{T}_k^2(\tilde{T}_k^b M)) \to \tilde{T}_k^a(\tilde{T}_k^2(\tilde{T}_k^b M))$ a number of induced involutions. In local coordinates, ${}_a\kappa_b$ interchanges coordinates with the following properties of subscripts:

1. the $(b+1)$–th subscript belonged to the first coordinate and the $(b+2)$–th subscript belonged to the second one are zeros
2. the $(b+2)$–th subscript belonged to the first coordinate equals the $(b+1)$–th subscript belonged to the second one
3. all other subscripts belonged to the first coordinate are the same as subscripts belonged to the second one

1.5. The Weil algebra $\overset{\omega}{\mathbb{D}}{}_k^r$. For every $\omega = 1, \ldots, r$, let us denote by $\overset{\omega}{T}{}_k^r M$ the subset of all $Z \in \tilde{T}_k^r M$ satisfying the *ω-condition*

$$ {}_{\omega-1}\pi_{r-\omega}^1(Z) = {}_{q-1}\pi_{r-q}^1(Z)$$

for all q, $q = \omega, \ldots, r$. $\overset{\omega}{T}{}_k^r M$ are Weil bundles and $\overset{1}{T}{}_k^r M = \bar{T}_k^r M$, $\overset{r}{T}{}_k^r M = \tilde{T}_k^r M$. Analogously, we can define the bundle denoted $\underset{\omega}{T}{}_k^r M$ by the *bottom ω-condition*

$$ {}_{\omega-1}\pi_{r-\omega}^1(Z) = {}_{q-1}\pi_{r-q}^1(Z)$$

for all q, $q = 1, \ldots, \omega$. However, then holds:

LEMMA 2. *For every $\omega = 1, \ldots, r$, we can construct $\underset{\omega}{T}{}_k^r M$ only by the ω-condition and involutions.*

PROOF. It is evident that $\underset{1}{T}{}_k^r M = \overset{r}{T}{}_k^r M = \tilde{T}_k^r M$ and $\underset{r}{T}{}_k^r M = \overset{1}{T}{}_k^r M = \bar{T}_k^r M$. In general, we show that the bundles $\overset{\omega}{T}{}_k^r M$ and $\underset{r-\omega+1}{T}{}_k^r M$ are isomorphic for all $\omega = 1, \ldots, r$. In local coordinates, the ω-condition reads as the identification of coordinates subscripts of which become equal if we delete zeros from the 1–st to the $(r-\omega+1)$–th subscripts in each of them. On the other hand, the bottom ω-condition reads as the identification of coordinates subscripts of which become equal if we delete zeros from $(r-\omega+1)$–th to the r–th subscripts in each of them and, especially, the bottom $(r-\omega+1)$-condition reads as the identification of coordinates subscripts of which become equal if we delete zeros from ω–th to the r–th subscripts in each of them. Therefore, roughly speaking, it is sufficient transfer subscripts by involutions. Particularly, we transfer the $(r-\omega+1)$–th subscript to the r–th one by $\kappa_{r-2}\circ\cdots\circ{}_{\omega-2}\kappa_{r-\omega}$, and so on, the 1–st subscript to the ω–th by ${}_{r-\omega}\kappa_{\omega-2}\circ\cdots\circ{}_{r-2}\kappa$. If we use the short denotation $\bigcirc_{i=1}^{n}\phi_i=\phi_1\circ\cdots\circ\phi_n$ for compositions, we obtain:

$$\bigcirc_{j=\omega}^{r}\bigcirc_{i=j-\omega+2}^{j}{}_{i-2}\kappa_{r-i}(\overset{\omega}{T}{}_k^r M)=\underset{r-\omega+1}{T_k^r}M$$

□

As we expect, the lemma confirms that $\overset{\omega}{T}{}_k^r M$ represent the suitable generalization of nonholonomic and semiholonomic velocities bundles. It includes even the iterations, see 1.1. For the evaluating of Weil algebra $\overset{\omega}{\mathbb{D}}{}_k^r$ of $\overset{\omega}{T}{}_k^r$, we have generating equations for it of the type of generalized (4) with respect to the first $r-\omega+1$ subscripts and the equations (2) with respect to the remaining subscripts, see details in the proof of Theorem.

2. The width and the homogeneity of Weil algebras

2.1. Motivation and definitions. The study of the width and the homogeneity of Weil algebras is important for the classification of natural operations on bundles of contact elements of type of the described Weil algebras, which can be viewed as a generalization of classical contact higher order contact elements introduced by C. Ehresmann. The newest results in this classification have obtained quite recently by W.M. Mikulski and the author.

The *width* $\mathrm{w}(A)$ of a Weil algebra A is defined as the dimension of the real vector space $\mathfrak{n}/\mathfrak{n}^2$, where $\mathfrak{n}$ is the nilpotent ideal of A. Let A be a Weil algebra with $\mathrm{w}(A)=k$. We say that A is *homogeneous* if there are homogeneous polynomials $P_1,\dots,P_l\colon\mathbb{R}^k\to\mathbb{R}$ such that $A=\mathbb{R}[t^1,\dots,t^k]/\langle P_1,\dots,P_l\rangle$.

We recall that for the Weil ideal $\mathfrak{i}$ $\langle t^1,\dots,t^k\rangle^{r+1}\subset\mathfrak{i}\subset\langle t^1,\dots,t^k\rangle$ is satisfied. The condition $\langle t^1,\dots,t^k\rangle^{r+1}\subset\mathfrak{i}$ secures a finite codimension of $\mathfrak{i}$. If even $\mathfrak{i}\subset\langle t^1,\dots,t^k\rangle^2$, then automatically $\mathrm{w}(A)=k$, see **[2]**. Trivially, $\mathrm{w}(\mathbb{D}_k^r)=k$ and $\mathbb{D}_k^r$ is homogeneous, $\mathbb{D}_k^r$ being the Weil algebras of classical velocities functor T_k^r.

2.2. The homogeneity of Weil algebras $\tilde{\mathbb{D}}_k^r$, $\bar{\mathbb{D}}_k^r$ and $\overset{\omega}{\mathbb{D}}{}_k^r$. We obtain the following results immediately:

PROPOSITION 1. *The Weil algebra $\tilde{\mathbb{D}}_k^r$ is homogeneous and* $\mathrm{w}(\tilde{\mathbb{D}}_k^r)=rk$.

PROOF. The assertion follows directly from (3). □

PROPOSITION 2. *The Weil algebra $\bar{\mathbb{D}}_k^2$ is homogeneous and* $\mathrm{w}(\bar{\mathbb{D}}_k^2)=\frac{k(k+1)}{2}$.

PROOF. The assertion follows directly from (6). □

Furthermore, we have the following result.

PROPOSITION 3. *The Weil algebra $\bar{\mathbb{D}}_k^r$ is homogeneous.*

PROOF. We have proved, that $\bar{\mathbb{D}}_k^2$ is homogeneous Weil algebra. By induction, we assume that homogeneity of $\bar{\mathbb{D}}_k^{r-1}$ is proved. The equations generating $\bar{\mathbb{D}}_k^r$ contain the equations generating $\bar{\mathbb{D}}_k^{r-1}$ and the added equations

$$\kappa t^{j_1\dots j_\mu} t^{j_{\mu+1}\dots j_r} = \sum_\sigma \kappa_\sigma t^{\sigma(j_1\dots j_r)} \tag{7}$$

$$t^{j_1\dots j_r} t^{\bar{j}_1\dots \bar{j}_\nu} = 0 \tag{8}$$

where $1 \le \mu \le r-1$, $1 \le \nu \le r$, σ denotes a certain permutation of superscripts, $\kappa_\sigma \in \mathbb{N}$ any coefficient and $\kappa = \sum_\sigma \kappa_\sigma$. On the base of equations (7), we set new essential variables with r non-zero superscripts. The rest of variables with r non-zero superscripts are linear combinations of these new essential variables and products $t^{j_1\dots j_\mu} t^{j_{\mu+1}\dots j_r}$. So, we have used the equations (7) for the described expression of variables and that is why they are not the generating equations from now on. Now, if an essential variable appears in any equation (8), we have the generating equation. But if the other (non-essential) variable with r non-zero superscripts appears in any equation (8), we do not obtain any new information, because such a variable is a linear combination of new essential variables and products $t^{j_1\dots j_\mu} t^{j_{\mu+1}\dots j_r}$ and that is why we have not the generating equation for $\bar{\mathbb{D}}_k^r$ in this case.

If we summarize, the new generating equations are only such equations (8), in which a new essential variable appears. It means the ideal $\mathfrak{i}$ is generated only by products of essential variables, i.e. by homogeneous polynomials. □

We believe that $\overset{\omega}{T}{}_k^r M$ can be viewed as one of basic examples of Weil bundles and our main result is:

THEOREM 1. *The Weil algebra $\overset{\omega}{\mathbb{D}}{}_k^r$ is homogeneous.*

PROOF. It is sufficient to assume $1 < \omega < r$ as we have proved Proposition 1 and Proposition 3.
The first step. The coordinates $t^{j_1\dots j_{r-\omega+1} j_{r-\omega+2}\dots j_r}$ are liable to (2) with respect to superscripts $j_{r-\omega+2}\dots j_r$ and, analogously, we make do only with coordinates denoted $t_q^{j_1\dots j_{r-\omega+1} l}$, $1 \le l \le k$, $r-\omega+2 \le q \le r$ and $\langle t_{r-\omega+2}^{-1}, \dots, t_{r-\omega+2}^{-k}\rangle^2, \dots, \langle t_r^{-1}, \dots, t_r^{-k}\rangle^2$ share the generating of the ideal $\mathfrak{i}$. ("—" means any $j_1 \dots j_{r-\omega+1}$.)
The second step. We delete zeros among $j_1 \dots j_{r-\omega+1}$ and by the expressing of essential variables as in the proof of Proposition 3 we complete the generators of $\mathfrak{i}$ only by homogeneous polynomials. □

REMARK 1. *We do not use the width $\mathrm{w}(\overset{\omega}{\mathbb{D}}{}_k^r)$ explicitly, evaluation of which is the open problem.*

References

1. Kolář, I., Bundle functors of the jet type, in: Differential Geometry and Applications (Proc. Sat. Confer. of ICM, 1998), Brno 1999, pp. 231–237
2. Kolář, I., Michor, P.W. and Slovák, J., Natural Operations in Differential Geometry, Springer Verlag 1993
3. Kureš, M., On the simplicial structure of some Weil bundles, Supplem. ai Rendic. del Circ. Matem. di Palermo 63, 2000, pp. 131–140

DEPARTMENT OF MATHEMATICS, BRNO UNIVERSITY OF TECHNOLOGY, TECHNICKÁ 2, 61669 BRNO, CZECH REPUBLIC
E-mail address: kures@mat.fme.vutbr.cz

Contemporary Mathematics
Volume **288**, 2001

The Variety of Lie Algebras from Certain Riemannian Viewpoint

Jorge Lauret

ABSTRACT. Each point of the variety of real Lie algebras is naturally identified with a left invariant riemannian metric on a Lie group. We report several interplays between invariant-theoretic aspects of this variety, like degenerations and closed orbits, and riemannian geometry, like Ricci soliton metrics, Einstein metrics and pinching curvature conditions.

1. Interpretation

Let $\mathcal{L}$ be the space of all n-dimensional real Lie algebras. If $\mathfrak{g} = \mathbb{R}^n$ we can identify $\mathcal{L}$ with the set of Lie brackets on $\mathfrak{g}$, which is an algebraic subset of $\Lambda^2\mathfrak{g}^*\otimes\mathfrak{g} = \mathbb{R}^{\frac{n^3-n^2}{2}}$ since the Jacobi identity is polynomial. We give a riemannian meaning to each point μ of the algebraic variety $\mathcal{L}$ in the following way: if $\langle\cdot,\cdot\rangle$ denotes a fixed inner product on $\mathfrak{g}$ then we identify each $\mu \in \mathcal{L}$ with $(G_\mu, \langle\cdot,\cdot\rangle)$, the corresponding simply connected Lie group G_μ with Lie algebra μ endowed with the left invariant riemannian metric determined by $\langle\cdot,\cdot\rangle$.

The aim of these notes is to report some interactions found between this riemannian interpretation of $\mathcal{L}$ and their algebraic aspects. The 'change of basis' action of $\mathsf{Gl}(n)$ on $\mathcal{L}$ given by $\varphi.\mu = \varphi\mu(\varphi^{-1}\cdot,\varphi^{-1}\cdot)$ has the following riemannian interpretation: for any $\varphi \in \mathsf{Gl}(n)$, the map φ^{-1} determines a riemannian isometry

$$(G_{\varphi.\mu}, \langle\cdot,\cdot\rangle) \longrightarrow (G_\mu, \varphi^{-1}.\langle\cdot,\cdot\rangle), \tag{1.1}$$

where $\varphi^{-1}.\langle\cdot,\cdot\rangle = \langle\varphi\cdot,\varphi\cdot\rangle$. In fact, $\varphi^{-1} : (\mathfrak{g}, \varphi.\mu) \longrightarrow (\mathfrak{g}, \mu)$ is a Lie algebra isomorphism and $\varphi^{-1} : (\mathfrak{g}, \langle\cdot,\cdot\rangle) \longrightarrow (\mathfrak{g}, \varphi^{-1}.\langle\cdot,\cdot\rangle)$ is an isometry of inner product spaces, and hence it follows from the left invariance of the metrics that φ^{-1} defines by exponentiation a riemannian isometry between $(G_{\varphi.\mu}, \langle\cdot,\cdot\rangle)$ and $(G_\mu, \varphi^{-1}.\langle\cdot,\cdot\rangle)$. We note that on the left of (1.1) the action of $\mathsf{Gl}(n)$ gives the orbit $\mathcal{O}(\mu) = \mathsf{Gl}(n).\mu$, and on the right this action gives the set of all inner products on $\mathfrak{g}$. We therefore can say that the orbit $\mathcal{O}(\mu)$ contains all the left invariant riemannian metrics on G_μ, and hence $\mathcal{L}$ can be viewed as the space of all simply connected Lie groups

2000 *Mathematics Subject Classification.* Primary: 22E60, 53C30; Secondary: 14L30.
Key words and phrases. Lie algebras, variety, degenerations, closed orbits, left invariant, riemannian metrics, Ricci soliton, Einstein solvmanifolds.
Partially supported by CONICET, FONCyT and SeCyT UNC (Argentina).

endowed with a left invariant riemannian metric. Recall that if φ belongs to the orthogonal group $\mathsf{O}(n) = \mathsf{O}(\mathfrak{g}, \langle\cdot,\cdot\rangle)$, then $\varphi.\mu$ and μ are isometric.

2. A riemannian functional on $\mathcal{L}$

The Ricci curvature operator of μ can be viewed as a symmetric transformation $\mathrm{Ric}_\mu : \mathfrak{g} \longrightarrow \mathfrak{g}$ defined by

$$\mathrm{Ric}_\mu = \mathrm{R}_\mu - \frac{1}{2}B_\mu - D_\mu, \tag{2.1}$$

where B_μ is the Killing form of μ in terms of $\langle\cdot,\cdot\rangle$, D_μ is the symmetric part of $\mathrm{ad}\, Z_\mu$, $Z_\mu \in \mathfrak{g}$ defined by $\langle Z_\mu, X\rangle = \mathrm{tr}(\mathrm{ad}_\mu X)$, $X \in \mathfrak{g}$ and R_μ is given by

$$\begin{aligned}\langle \mathrm{R}_\mu X, Y\rangle = & -\tfrac{1}{2}\sum_{ij}\langle\mu(X, X_i), X_j\rangle\langle\mu(Y, X_i), X_j\rangle \\ & +\tfrac{1}{4}\sum_{ij}\langle\mu(X_i, X_j), X\rangle\langle\mu(X_i, X_j), Y\rangle,\end{aligned} \tag{2.2}$$

for all $X, Y \in \mathfrak{g}$, where $\{X_1, ..., X_n\}$ is any orthonormal basis of $(\mathfrak{g}, \langle\cdot,\cdot\rangle)$ (see [**B**, 7.39]). Note that $Z_\mu = 0$ if and only if μ is unimodular and that if μ is nilpotent then $\mathrm{Ric}_\mu = \mathrm{R}_\mu$. It is clear that R_μ (and also Ric_μ) can be defined, at least formally, for every element $\mu \in \Lambda^2\mathfrak{g}^* \otimes \mathfrak{g}$.

There is a well known relationship between the small degree Chevalley Lie algebra cohomology (i.e. relative to the adjoint representation, see [**Ja**]), and the tangent geometry of the algebraic variety $\mathcal{L}$. For instance, the tangent space $T_\mu\mathcal{O}(\mu)$ coincides with the space of coboundaries since for any $A \in \mathfrak{gl}(\mathfrak{g})$ we have that $\frac{\mathrm{d}}{\mathrm{d}t}|_0 e^{tA}.\mu = -\delta_\mu(A)$, where $\delta_\mu : C^1 = \mathrm{End}(\mathfrak{n}) \longrightarrow C^2 = \Lambda^2\mathfrak{g}^* \otimes \mathfrak{g}$ is the cohomology coboundary operator of the Lie algebra μ defined by

$$\delta_\mu(A)(X, Y) = \mu(AX, Y) + \mu(X, AY) - A\mu(X, Y)$$

for all $A \in \mathfrak{gl}(\mathfrak{g})$, $X, Y \in \mathfrak{g}$. In the next lemma we show a first relationship between a 'riemannian' object, the curvature tensor R_μ defined via the Ricci curvature (see (2.2)), and the 'algebraic' concept of Chevalley cohomology.

LEMMA 2.1. [**L2**] *If $F : \Lambda^2\mathfrak{g}^* \otimes \mathfrak{g} \longrightarrow \mathbb{R}$ is defined by $F(\mu) = \mathrm{tr}\, \mathrm{R}_\mu^2$ then*

$$\mathrm{grad}(F)_\mu = -\delta_\mu(\mathrm{R}_\mu), \qquad \forall \quad \mu \in \Lambda^2\mathfrak{g}^* \otimes \mathfrak{g}.$$

We are interested in changes of F along the orbit $\mathcal{O}(\mu)$, which is identified via (1.1) with the space of all left invariant riemannian metrics on G_μ. Since F is a homogeneous polynomial, we shall restrict ourselves to the sphere S of $\Lambda^2\mathfrak{g}^* \otimes \mathfrak{g}$. Using Lemma 2.1, it is easy to prove the following characterization of the critical points of $F|_S$.

THEOREM 2.2. [**L4**] *Let F be the functional on $\Lambda^2\mathfrak{g}^* \otimes \mathfrak{g}$ given by $F(\mu) = \mathrm{tr}\, \mathrm{R}_\mu^2$ and let S denote the sphere of $\Lambda^2\mathfrak{g}^* \otimes \mathfrak{g}$. Then for $\mu \in \mathcal{L} \cap S$ the following conditions are equivalent:*

(i) $\mathrm{R}_\mu \in \mathbb{R}I \oplus \mathrm{Der}(\mu)$.
(ii) *μ is a critical point of $F|_S$.*
(iii) *μ is a critical point of $F|_{\mathcal{O}(\mu)\cap S}$.*

A $\mu \in \mathcal{L} \cap S$ satisfying any of the above condition will be called *critical.* The equivalence between (ii) and (iii) follows from the fact that $\operatorname{grad}(F)_\mu \in \mathrm{T}_\mu \mathcal{O}(\mu)$ for any μ. Thus it is easy to prove that the $\pm \operatorname{grad}(F)$ flows ψ_t^+ and ψ_t^- starting from a point $\mu_0 \in \mathcal{L}$ lies in the orbit $\mathcal{O}(\mu_0)$ for every $t \in [0, \infty)$. Henceforth, if ψ_t^+ or ψ_t^- converges, or just a subsequence converges, as it is the case when we restrict ourselves to the sphere S of $\Lambda^2\mathfrak{g}^* \otimes \mathfrak{g}$, then the flow will determine a degeneration of μ_0, that is, a Lie bracket $\mu \in \overline{\mathcal{O}(\mu_0)}$ (see §5). Thus the critical points of $F|_S$ also present some interest from the algebraic viewpoint.

In the study of Einstein solvmanifolds, Jens Heber [**H**] considers certain riemannian functionals on the space of left invariant metrics of a given solvable Lie group. Using some concavity properties of one of these functionals (see [**H**, 3.3]), we can give a proof of the following uniqueness result, which is in its turn inspired on the proof of the uniqueness up to isometry of a standard Einstein left invariant metric on a solvable Lie group (see [**H**, 5.1(1)]).

THEOREM 2.3. [**L4**] *A critical μ is unique in $\mathcal{O}(\mu) \cap S$ up to the action of $\mathrm{O}(n)$, that is, if μ is critical then $\lambda \in \mathcal{O}(\mu) \cap S$ is critical if and only if $\lambda = \varphi.\mu$ for some $\varphi \in \mathrm{O}(n)$.*

A critical point of $F|_{\mathcal{O}(\mu)\cap S}$ can be recognized as a distinguished left invariant metric on G_μ. The above theorem says that such a privileged metric, if there is one, it is unique up to isometry and scaling. We have some reasons to believe that a critical point of $F|_{\mathcal{O}(\mu)\cap S}$ is always a minimum of F restricted to $\mathcal{O}(\mu) \cap S$.

3. Ricci solitons

We will concentrate in the following two sections on the critical Lie brackets which are nilpotent. As we shall see, they have very nice geometric characterizations. If $\mu \in \mathcal{L}$ is nilpotent, then R_μ coincide with the Ricci operator, so $F(\mu)$ is precisely the square norm of the Ricci curvature tensor.

In 1982 R. Hamilton [**Ha1**] introduced the *Ricci flow* for a curve $g = g(t)$ of riemannian metrics on a differentiable manifold M defined by the equation

$$\frac{\partial g}{\partial t} = -2\,\mathrm{ric}_g, \tag{3.1}$$

where ric_g is the Ricci curvature tensor. The Ricci flow has proven to be a very useful tool to improve metrics in riemannian and Kähler geometry (we refer to the surveys [**CaCh**] and [**Ha3**] for further information). When M is compact, one often considers the *normalized Ricci flow*

$$\frac{\partial g}{\partial t} = -2\,\mathrm{ric}_g + \frac{2\,\mathrm{sc}(g)}{n} g, \tag{3.2}$$

where $\mathrm{sc}(g)$ is the average of the scalar curvature. Under this normalized flow the volume of the solution metric is constant in time, and equations (3.1) and (3.2) differ only by a change of scale in space by a function of t and a change of parametrization in time (see [**Ha1**]). Note that the fixed points of the normalized Ricci flow (3.2) are precisely the Einstein metrics. Close to being a fixed point, a *Ricci soliton* (or self-similar solution) is a special solution of the (normalized) Ricci flow moving along the equation by diffeomorphisms, that is, where the metric $g(t)$ is the pull-back of the initial metric $g(0)$ by a one-parameter group of diffeomorphisms.

More generally, a solution to the Ricci flow (3.1) which moves by a diffeomorphism and also shrinks or expands by a factor at the same time is called a *homothetic*

Ricci soliton (see [**Ha3**]). If φ_t is a one-parameter group of diffeomorphisms generated by some vector field $X \in \chi(M)$, then it is easy to see that $g(t) = c(t)\varphi_t^* g$ is a solution to the Ricci flow (3.1) for some factor $c(t)$ depending on time (i.e. g is a homothetic Ricci soliton) if and only if

$$\operatorname{ric}_g = cg + \mathrm{L}_X\, g \quad \text{for some } c \in \mathbb{R}, \tag{3.3}$$

where $\mathrm{L}_X\, g$ denotes the usual Lie derivative. Metrics satisfying (3.3) for some vector field X are called *quasi-Einstein* metrics, and they have been considered in physics by D. Friedan [**F**].

Searching for a distinguished left invariant metric on a given nilpotent Lie group, we found the following relationship.

THEOREM 3.1. [**L2**] *A nilpotent $\mu \in \mathcal{L} \cap S$ is critical if and only if μ is a homothetic Ricci soliton (or a quasi-Einstein metric).*

4. Einstein solvmanifolds.

A *solvmanifold* is a solvable Lie group S endowed with a left invariant riemannian metric, and S is called *standard* if $\mathfrak{a} = [\mathfrak{s}, \mathfrak{s}]^\perp$ is abelian, where $\mathfrak{s}$ is the Lie algebra of S. We observe that all known examples of noncompact homogeneous Einstein spaces are isometric to a standard Einstein solvmanifold.

THEOREM 4.1. [**L2**] *A nilpotent $\mu \in \mathcal{L}$ is critical if and only if μ admits a standard metric solvable extension ($\mathfrak{s} = \mathfrak{a} \oplus \mathfrak{g}, \langle\cdot,\cdot\rangle'$) (i.e. $[\cdot,\cdot]|_{\mathfrak{g}\times\mathfrak{g}} = \mu$ and $\langle\cdot,\cdot\rangle'|_{\mathfrak{g}\times\mathfrak{g}} = \langle\cdot,\cdot\rangle$) whose corresponding solvmanifold $(S, \langle\cdot,\cdot\rangle')$ is Einstein.*

In view of this characterization, we realize that a large class of nilpotent Lie algebras are critical.

EXAMPLE 4.2. (i) Since any irreducible symmetric space G/K of non-compact type is Einstein, we have that the Lie algebras of the N-groups of the corresponding Iwasawa decomposition $G = KAN$ are critical.

(ii) It is proved in [**H**] that there exist high dimensional families of standard Einstein solvmanifolds which are deformations of the N-groups above in the rank-one case. By Theorem 4.1, such families of deformations all correspond to nilpotent critical Lie brackets.

(iii) It is well known that any homogeneous space with non-positive sectional curvature is isometric to a solvmanifold $S = AN$. Moreover, certain distinguished horosphere of S can be identified with the nilpotent Lie group N endowed with the restricted left invariant metric (see [**E**]). Thus the Lie algebras of the horospheres of Einstein non-positive curved riemannian spaces are all critical.

(iv) Heisenberg-type Lie algebras are critical. In this case, the metric solvable extensions which are Einstein solvmanifolds are precisely the Damek-Ricci spaces (see [**BTV**]).

(v) Any nilpotent μ which is naturally reductive is critical. Indeed, such a Lie brackets are in one-to-one correspondence with representations of compact Lie groups (see [**L1**]) and it has been proved in [**EH**] that certain solvable extensions are Einstein.

However, if μ is characteristically nilpotent (c.f. $\mathrm{Der}(\mu)$ is nilpotent, [**Hak, T**]), then μ can never be critical, since D_μ is a symmetric nonzero derivation.

Einstein metrics are often considered as the nicest, or most privileged metrics on a given differentiable manifold (see for instance [**B**, Introduction]). One of the justifications is that they are precisely the critical points of the functional of total scalar curvature on the space of all metrics of volume one (see also [**B**, 4.21]).

Let $\mathcal{N}$ denote the subvariety of $\mathcal{L}$ which consists of the nilpotent Lie brackets. We fix an inner product vector space

$$(\mathfrak{s} = \mathbb{R}H \oplus \mathfrak{g}, \langle\cdot,\cdot\rangle'), \quad \text{with } \langle H, \mathfrak{g}\rangle' = 0,\ \langle H, H\rangle' = 1 \text{ and } \langle\cdot,\cdot\rangle'|_{\mathfrak{g}\times\mathfrak{g}} = \langle\cdot,\cdot\rangle.$$

For each $\mu \in \mathcal{N}$, one can prove that there exists a unique symmetric $D_\mu \in \mathrm{Der}(\mu)$ such that S_μ satisfies certain necessary condition to be Einstein. Here S_μ denotes the solvable Lie group with Lie algebra $(\mathfrak{s}, [\cdot,\cdot])$ defined by $[H, X] = D_\mu X$, $[X, Y] = \mu(X, Y)$ for all $X, Y \in \mathfrak{g}$, and it is endowed with the left invariant riemannian metric determined by $\langle\cdot,\cdot\rangle'$. Every $(n+1)$-dimensional rank-one Einstein solvmanifold can be modeled as S_μ for a suitable $\mu \in \mathcal{N}$. The following result, characterizes among the set $\mathcal{N} = \{S_\mu : \mu \in \mathcal{N}\}$ those which are Einstein.

THEOREM 4.3. [**L3**] *For an element $\mu \in \mathcal{N} \cap \mathsf{S} \subset \Lambda^2\mathfrak{g}^* \otimes \mathfrak{g}$, where S denotes the sphere of $\Lambda^2\mathfrak{g}^* \otimes \mathfrak{g}$, the following conditions are equivalent:*

(i) *μ is critical.*
(ii) *S_μ is an Einstein space.*
(iii) *μ is a critical point of the modified scalar curvature functional $\tilde{sc} : \Lambda^2\mathfrak{g}^* \otimes \mathfrak{g} \longrightarrow \mathbb{R}$, $\tilde{sc}(\mu) = \tilde{sc}(S_\mu)$, restricted to S (see [**H**]).*

The key to the proof of the above result is the somewhat mysterious relationship given in Lemma 2.1 between the Chevalley cohomology of μ and the Ricci tensor of the corresponding nilpotent Lie group N_μ endowed with the left invariant metric determined by $\langle\cdot,\cdot\rangle$.

Up to now, besides the irreducible symmetric spaces of non-compact type, in almost all the known examples of Einstein solvmanifolds, the nilradical $[\mathfrak{s}, \mathfrak{s}]$ is two-step nilpotent. As an application of the variational result given in Theorem 4.3, we have found several examples of rank-one Einstein solvmanifolds with nilradical of all possible steps of nilpotency.

THEOREM 4.4. [**L5**] *For any nilpotent Lie algebra having a codimension one abelian ideal, or having dimension ≤ 5, there exists a rank-one solvable extension which admit an Einstein left invariant metric.*

5. Degenerations

We say that μ *degenerates to* λ (denoted by $\mu \longrightarrow \lambda$) if $\lambda \in \overline{\mathcal{O}(\mu)}$, the closure of the $\mathsf{Gl}(n)$-orbit $\mathcal{O}(\mu)$ with respect to the usual (metric) topology of $\Lambda^2\mathfrak{g}^*\otimes\mathfrak{g}$. In other words, $\mu \longrightarrow \lambda$ if there exists a sequence $\{\varphi_n\} \subset \mathsf{Gl}(n)$ such that $\lim_{n\to\infty} \varphi_n.\mu = \lambda$. The degeneration $\mu \longrightarrow \lambda$ is nontrivial when λ lies in the boundary of $\mathcal{O}(\mu)$. In this case, it is easy to see that the entire orbit $\mathcal{O}(\lambda)$ lies in the boundary of $\mathcal{O}(\mu)$. Every degeneration will be assumed nontrivial. If $\mu \longrightarrow \lambda$ then we may say that in some sense, λ is 'more abelian' than μ (see [**BS**]).

The study of degenerations have attracted considerable attention in the last decades (see for instance [**GO, S, BS, SY**]). The idea of degeneration (or contraction) was really introduced by theoretical physicists in [**IW**]. For instance, the knowledge of all the Lie algebras which can be contracted into the Poincare

Lie algebra may present some interest from the viewpoint of general relativity and dynamical groups (see [**LN, Sa**]).

When μ degenerates to λ, it follows that $\dim \mathrm{Der}(\mu) < \dim \mathrm{Der}(\lambda)$, and hence the following concept of level of an element $\mu \in \mathcal{L}$ is well defined. The maximum length k of a sequence

$$\mu = \mu_0 \longrightarrow \mu_1 \longrightarrow \dots \longrightarrow \mu_{k-1} \longrightarrow \mu_k = 0$$

of (nontrivial) degenerations is called the *level* of μ. It will be usually denoted by $\mathrm{level}(\mu)$.

A problem particularly interesting from the physical point of view, is to find all the Lie algebras that degenerates to a given one (see [**Sa**]). We are interested in the Lie brackets μ with $\mathrm{level}(\mu) = 1$, that is, the only possible degeneration of μ is $\mu \longrightarrow 0$, and so they constitute the Lie algebras having the bigger sets of Lie algebras degenerating on it. Let μ_{heis}, μ_{hyp} denote the Lie brackets defined by

$$\mu_{heis}(X_1, X_2) = X_3, \qquad \mu_{hyp}(X_1, X_i) = X_i, \qquad i = 2, \dots, n,$$

and zero otherwise. Note that μ_{heis} is isomorphic to $\mathfrak{h}_3 \oplus \mathbb{R}^{n-3}$, where $\mathfrak{h}_3$ is the 3-dimensional Heisenberg Lie algebra and $\mathbb{R}^{n-3}$ is an abelian factor, and μ_{hyp} is isomorphic to the real hyperbolic Lie algebra. The proof of the following classification is based in techniques used by J. Milnor [**M**] in the study of curvature properties of riemannian left invariant metrics on Lie groups.

THEOREM 5.1. [**L4**] *Every Lie bracket which is not in $\mathcal{O}(\mu_{hyp})$ degenerates to μ_{heis}. Moreover, μ_{heis} and μ_{hyp} are the only n-dimensional Lie algebras of level* 1 *for $n \geq 3$.*

The results on the critical point behavior of the functional $F|_S$ given in §2, can be applied to prove the following interaction.

THEOREM 5.2. [**L4**] *A $\mu \in \mathcal{L} \cap S$ has* $\mathrm{level}(\mu) = 1$ *if and only if there is only one left invariant riemannian metric on G_μ up to isometry and scaling.*

We therefore obtain that $G_{\mu_{heis}}$ and $G_{\mu_{hyp}}$ are the only simply connected Lie groups having only one left invariant riemannian metric up to isometry and scaling.

6. Curvature pinched conditions

All the geometric quantities associated to a $\mu \in \mathcal{L}$, like sectional, Ricci and scalar riemannian curvatures, depend continuously on μ. Consequently, if $\mu \longrightarrow \lambda$ and there is a $\lambda_0 \in \mathcal{O}(\lambda)$ satisfying certain strict curvature inequality, then the orbit $\mathcal{O}(\mu)$ must contain an element satisfying the same inequality. In other words, if $\mu \longrightarrow \lambda$ and G_λ admits a left invariant riemannian metric satisfying a strict pinched curvature condition, then so G_μ. We then can use well known algebraic characterizations of Lie groups having metrics with special curvature properties (see for instance [**He, EH**]), to decide in the negative, whether a Lie algebra degenerates to another one. We have obtained using this method the following result.

THEOREM 6.1. [**L4**] *A Lie algebra $\mu \in \mathcal{L}$ degenerates to μ_{hyp} if and only if there exists $A \in \mathfrak{g}$ such that $\mathfrak{g} = \mathbb{R}A \oplus \mu(\mathfrak{g}, \mathfrak{g})$, $\mu(\mathfrak{g}, \mathfrak{g})$ is abelian and has a basis such that $\mathrm{ad}_\mu A|_{\mu(\mathfrak{g},\mathfrak{g})}$ is the direct sum of Jordan blocks $J_{r_1}, \dots, J_{r_k}$ with eigenvalue* 1*, with $r_1 \geq \dots \geq r_k$. Moreover, two of such a Lie algebras μ and μ' are isomorphic if and only if $r_i = r_i'$ for all $i = 1, \dots, k$.*

We also determined all the possible degenerations for 3-dimensional real Lie algebras by using the above theorem and some Ricci curvature estimations for left invariant metrics given in [**RoSl**].

7. Minimal vectors and closed orbits

A $\mathsf{Gl}(n)$-orbit on $\mathcal{L}$ can never be closed, unless it is $\{0\}$. Indeed, if $\varphi_t = t^{-1}I$ then $\varphi_t.\mu \to 0$ when $t \to 0$ and so $0 \in \overline{\mathcal{O}(\mu)}$ for every $\mu \in \mathcal{L}$. A natural question is then what happens if we do not allow the multiplication by a scalar, considering for instance $\mathsf{Sl}(n)$-orbits instead of $\mathsf{Gl}(n)$-orbits. Under what conditions on μ we would have that $0 \in \overline{\mathsf{Sl}(n).\mu}$?, when $\mathsf{Sl}(n).\mu$ will be closed?.

Applying some results on the minimal vectors of representations of real reductive Lie groups given in [**N, Sch, RS**] to the action of the reductive subgroup $\mathsf{Sl}(n) \subset \mathsf{Gl}(n)$ on $\mathcal{L} \subset \Lambda^2\mathfrak{g}^* \otimes \mathfrak{g}$, we found some relationships with the geometric interpretation of $\mathcal{L}$ considered in this paper.

THEOREM 7.1. [**L4**] *The orbit* $\mathsf{Sl}(n).\mu$ *is closed if and only if the corresponding Lie group* G_μ *admits a left invariant metric* $\langle\cdot,\cdot\rangle$ *such that its curvature operator* $\mathrm{R}_{\langle\cdot,\cdot\rangle}$ *is a multiple of the identity.*

As an immediate consequence, we obtain that if $\mu \in \mathcal{L}$ is nilpotent (and $\mu \neq 0$) then the orbit $\mathsf{Sl}(n).\mu$ is not closed, and moreover, $0 \in \overline{\mathsf{Sl}(n).\mu}$. Indeed, R_μ coincides with the Ricci operator when μ is nilpotent, and hence R_μ can never be a multiple of the identity since by a result due to J. Milnor [**M**], a non-abelian nilpotent Lie group can never admit an *Einstein* left invariant riemannian metric. However, the closure $\overline{\mathsf{Sl}(n).\mu}$ must contain a closed orbit (see [**RS**]), and so using that the set of nilpotent Lie brackets $\mathcal{N}$ is closed, we obtain that $0 \in \overline{\mathsf{Sl}(n).\mu}$ for any nilpotent $\mu \in \mathcal{L}$.

On the other hand, if $\mu \in \mathcal{L}$ is semisimple then it is easy to see that we can define, via the Killing form, a left invariant metric $\langle\cdot,\cdot\rangle_1$ on G_μ satisfying that $\mathrm{R}_{\langle\cdot,\cdot\rangle_1}$ is a multiple of the identity. By the characterization given above, we deduce that $\mathsf{Sl}(n).\mu$ is closed.

The riemannian characterization given in Proposition 7.1, allowed us to apply some ideas and techniques due to I. Dotti [**D**], used in the study of the Ricci curvature of unimodular Lie groups, in order to prove the following general result.

THEOREM 7.2. [**L4**] *Let* $\mu \in \mathcal{L}$, $\mu \neq 0$.

(i) *The orbit* $\mathsf{Sl}(n).\mu$ *is closed if and only if* μ *is semisimple.*
(ii) *If* μ *is not semisimple then* $0 \in \overline{\mathsf{Sl}(n).\mu}$.

References

[BTV] J. Berndt, F. Tricerri, L. Vanhecke, *Generalized Heisenberg groups and Damek-Ricci harmonic spaces*, Lect. Notes in Math. **1598** (1995), Springer-Verlag Berlin Heidelberg.

[B] A. Besse, *Einstein manifolds*, Ergeb. Math., vol. 10, (1987), Springer-Verlag, Berlin-Heidelberg.

[BS] D. Burde, C. Steinhoff, *Classification of orbit clousures of 4-dimensional complex Lie algebras*, J. of Algebra **214**(1999), 729–739.

[CaCh] H-D. Cao, B. Chow, *Recent developments on the Ricci flow*, Bull. A.M.S. **36**, no. 1 (1999), 59–74.

[D] I. Dotti, *Ricci curvature of left invariant metrics on solvable unimodular Lie Groups*, Math. Z. **180** (1982), 257–263.

[E] P. Eberlein, *Geometry of nonpositively curved manifolds*, Chicago Lectures in Mathematics, The University of Chicago Press., Chicago (1996).

[EH] P. Eberlien, J. Heber, *Quarter pinched homogeneous spaces of negative curvature*, Internat. J. Math. **7**(1996), 441–500.

[F] D. Friedan, *Nonlinear models in* $2+\epsilon$ *dimensions*, Annals of Physics **163** (1985), 318–419.

[GO] F. Grunewald, J. O'Halloran, *Deformations of Lie algebras*, J. of Algebra **162** (1993), 210–224.

[Hak] Y. Hakimjanov, *Variété des lois d'algebres de Lie nilpotentes*, Geom. Dedicata **40** (1991), 269–295.

[Ha1] R. Hamilton, *Three-manifolds with positive Ricci curvature*, J. Differential Geom. **17** (1982), 255–306.

[Ha3] R. Hamilton, *The formation of singularities in the Ricci flow*, Surveys in Differential Geometry II (1995), 7–136.

[H] J. Heber, *Noncompact homogeneous Einstein spaces*, Invent. math. **133** (1998), 279–352.

[He] E. Heintze, *On homogeneous manifolds of negative curvature*, Math. Ann. **211** (1974), 23–34.

[IW] E. Inonu, E.P. Wigner, Proc. Natl. Acad. Sci. U. S. **39**, 510 (1953) (see also **40**, 119 (1954)).

[Ja] N. Jacobson, *Lie algebras*, Interscience Publishers, New York (1962).

[L1] J. Lauret, *Homogeneous nilmanifolds attached to representations of compact Lie groups*, Manuscripta Math. **99** (1999), 287–309.

[L2] J. Lauret, *Ricci soliton homogeneous nilmanifolds*, Math. Annalen **319** (2001), 715–733.

[L3] J. Lauret, *Standard Einstein solvmanifolds as critical points*, to appear in Quart. J. Math.

[L4] J. Lauret, *Degenerations of Lie algebras and geometry of Lie groups*, preprint 2001.

[L5] J. Lauret, *Finding Einstein solvmanifolds by a variational method*, preprint 2001.

[LN] M. Levy-Nahas, *Deformation and contraction of Lie algebras*, J. Math. Phys. **8** (1967), 1211–1222.

[M] J. Milnor, *Curvature of Left-invariant Metrics on Lie Groups*, Adv. in Math. **21** (1976), 293–329.

[N] A. Neeman, *The topology of quotient varieties*, Ann. Math. **122** (1985), 419–459.

[RS] R.W. Richardson, P.J. Slodowy, *Minimum vectors for real reductive algebraic groups*, J. London Math. Soc. (2) **42** (1990), 409–429.

[RoSl] E.D. Rodionov, V.V. Slavskii, *Curvatures estimations of left invariant Riemannian metrics on three dimensional Lie groups*, Proc. of the 7th Int. Conf. Diff. Geom. and Appl., Brno, Czech. Republic, 1998, 111–126.

[Sa] E.J. Saletan, *Contraction of Lie groups*, J. Math. Phys. **2** (1961), 1–21.

[Sch] G. Schwarz, *The topology of algebraic quotients*, Progress in Mathematics **80** (1989), 135–152, Birkhauser 1989.

[S] C. Seeley, *Degenerations of 6-dimensional nilpotent Lie algebras over* $\mathbb{C}$, Commun. in Algebra **18** (1990), no. 10, 3493–3505.

[SY] C. Seeley, S. S.-T. Yau, *Variation of complex structures and variation of Lie algebras*, Invent. math. **99** (1990), 545–565.

[T] S. Togo, *On the derivation algebras of Lie algebras*, Can. J. Math., **13** (1961), 201–216.

Acknowledgement. I wish to thank the Departamento de Matemáticas of the Universidad del País Vasco, for giving me a fellowship to participate of this wonderful congress.

CIEM, FaMAF, Universidad Nacional de Córdoba, 5000 Córdoba, Argentina
E-mail address: lauret@mate.uncor.edu

Contemporary Mathematics
Volume **288**, 2001

How to Use MATHEMATICA™ to Find Cyclic Surfaces of Constant Curvature in Lorentz-Minkowski Space

Rafael López

ABSTRACT. The development in recent years of computer tools make it possible for geometers to approach old and new subjects with different methods than some decades ago. Recent advances in symbolic computation as well as in computer graphics open new perspectives in research in differential geometry. Using the software MATHEMATICA, we investigate a new family of maximal surfaces in Lorentz-Minkowski space. A maximal surface is a spacelike surface with zero mean curvature. The consequent visualisation of these surfaces goes hand-in-hand with the understanding of the geometry behind the computer construction. An advantage that MATHEMATICA offers is that runs on standard PC and Workstation, without the necessity to use big computers, which are only available for a small number of researchers. Gray's book [**2**] is a good example of the potentialities that MATHEMATICA offers in this sense.

1. Statement of results

In 1744 Euler successfully applied the calculus of variations to the problem of the determination of the minimum area surface bounded by two parallel coaxial circles. He proved that the catenoid is the only surface of revolution with the minimum area property. In the 1770s, Meusnier gave the geometric interpretation of this result: the catenoid is the only non-planar surface of revolution in Euclidean space with zero mean curvature H. In the 1860s Riemann found all minimal (non-rotational) surfaces foliated by circles in parallel planes [**11**]. Each one of such surfaces is an embedded surface invariant by a group of translations. This surface has planar ends with the exception of a discrete set of straight lines a discrete set of heights.

A cyclic surface M in Euclidean three-dimensional space $\mathbf{E}^3$ is a surface foliated by a smooth one-parameter family of circular arcs. This means that there is a one-parameter family of planes that intersect the surface M in circular arcs. Surfaces of revolution are the best known examples of cyclic surfaces. Enneper showed that in a cyclic minimal surface, the planes of the foliation must be parallel. Thus, the catenoid and Riemann minimal examples are the only cyclic surfaces in $\mathbf{E}^3$ with zero mean curvature.

In 1989 Nitsche considered non-zero constant mean curvature cyclic surfaces [**10**]. He proved that all cyclic surfaces with non-zero constant mean curvature

2000 *Mathematics Subject Classification.* Primary 53A10; Secondary 53C42.
Key words and phrases. cyclic surface, curvature, Lorentz-Minkowski space.
Financial support for this work came from a DGICYT Grant No. PB97-0785.

are surfaces of revolution, that is, Delaunay surfaces. Recently, the present author studied cyclic surfaces with constant Gaussian curvature ([**8**]). This situation is similar to what happens in the mean curvature case:

> *The only cyclic surfaces with non-zero constant Gaussian curvature K are the surfaces of revolution. If $K \equiv 0$, the surface is a cone or a cylinder over a circle.*

Let $\mathbf{L}^3$ denote the Lorentz-Minkowski 3-dimensional space with the metric $\langle x, y\rangle = x_1y_1 + x_2y_2 - x_3y_3$, for vectors $x = (x_1, x_2, x_3)$, $y = (y_1, y_2, y_3)$ in terms of natural coordinates. A circle in $\mathbf{L}^3$ is a (non-degenerate) planar curve having constant nonzero curvature. Though there is just one kind of circle in Euclidean ambient space, the nature of a circle in $\mathbf{L}^3$ varies with the causal character (spacelike, timelike or lightlike) of the plane containing the curve. After a rigid motion in $\mathbf{L}^3$, we assume that this plane is $x_3 = 0$, $x_1 = 0$ or $x_2 - x_3 = 0$, respectively. Then circles are Euclidean circles, hyperbolas and parabolas if the plane containing the circle is spacelike, timelike or lightlike, respectively.

As in Euclidean ambient space, a *cyclic surface* in $\mathbf{L}^3$ is a surface foliated by a smooth one-parameter family of circles. In this ambient space, the geometric behaviour of surfaces of revolution depends on the causal character of its axis of rotation. Denote by H and K the mean curvature and the Gaussian curvature of a non-degenerate surface in $\mathbf{L}^3$ respectively. A variety of results have been developed by the author in a sequence of papers (see [**3**], [**4**], [**5**],[**6**],[**7**],[**9**]). In this note, we can summarize these results as follows:

THEOREM 1. *All (spacelike or timelike) cyclic surfaces in $\mathbf{L}^3$ with either $H =$ constant $\neq 0$ or $K = \text{cte} \neq 0$ are surfaces of revolution.*

THEOREM 2. *Let M be a (spacelike or timelike) cyclic surface in $\mathbf{L}^3$ with either $H = 0$ or $K = 0$. Then the planes containing the circles must be parallel.*

In particular, we give a complete description of cyclic surfaces with zero Gaussian curvature and zero mean curvature. In the latter case, we obtain a new family of maximal surfaces foliated by circles and that play the same role as Riemann minimal examples of Euclidean space. Figure 2 shows one of such maximal surfaces of Riemann-type of $\mathbf{L}^3$.

2. What part does `MATHEMATICA` play in proofs?

2.1. `MATHEMATICA` and symbolic computations. We outline here the structure of the proof of Theorem 1. For simplicity, we discuss the case that the Gaussian curvature K is a non-zero constant. A simple topological argument allows us to assume that the foliating planes have the same causal character. Here we only consider the case that all these planes are spacelike. Assume the planes are not parallel. Denote by u the parameter of the family of these planes and let $\mathbf{N}(u)$ denote a unit vector field orthogonal to each u-plane. Then the derivative $\mathbf{N}'(u)$ does not vanish in some real interval. This assures that the curve Γ with $\mathbf{t} = \mathbf{N}$ as unit tangent vector field is not a straight-line and thus, there exists a Frenet frame $\{\mathbf{t}, \mathbf{n}, \mathbf{b}\}$ of Γ in each point. A suitable parametrization of our surface adapted in terms of this frame is given by:

$$\mathbf{X}(u, v) = \mathbf{c}(u) + r(u)(\cos v\ \mathbf{n}(u) + \sin v\ \mathbf{b}(u)).$$

It follows from the classical local theory that the Gaussian curvature of the surface M is given by

$$(1)\quad 0 = KW^2 - \det(\mathbf{X}_u, \mathbf{X}_v, \mathbf{X}_{uu}) \det(\mathbf{X}_u, \mathbf{X}_v, \mathbf{X}_{vv}) + \det(\mathbf{X}_u, \mathbf{X}_v, \mathbf{X}_{uv})^2,$$

where $W^2 = EG - F^2$ denotes the first fundamental form in M. Since M is a non-degenerate surface, $W \neq 0$ on M (if M is spacelike, $W > 0$ and if M is timelike then $W < 0$). Let $\mathbf{c}' = \alpha\mathbf{t} + \beta\mathbf{n} + \gamma\mathbf{b}$. After a homothety, we assume $K = -1$ or 1. Let us compute K using identity (1). This computation gives a 4-degree polynomial in $\cos nv, \sin nv$, where $0 \leq n \leq 4$. Thus each one of the seven coefficients in (1) vanishes in the whole u-interval.

Difficulties arise from the fact that the explicit computation of these coefficients is a long and hard task to do by hand. This is the place where `MATHEMATICA` helps the geometer. `MATHEMATICA` works with algebraic manipulations that would be almost impossible by hand (not to mention the possibility of human error). The computer simplifies all these computations making it unnecessary to do long checks by hand. Easy commands such as `Coefficient`, `Expand`, or `Det` solve these problems in a few lines. So, it is possible to use identity (1) to compute the Gaussian curvature. Once we obtain the seven coefficients, we impose that condition that they vanish in the u-interval. An easy argument provides a contradiction with the fact that Γ is not a straight line. Thus the foliating planes must be parallel.

Assume that the planes of the foliation are parallel. We suppose that the foliating planes are horizontal. Then M is locally described as

$$\mathbf{X}(u, v) = (a(u), b(u), u) + r(u)(\cos v, \sin v, 0),$$

where a, b and $r > 0$ are smooth functions on u. Again, let us compute K (recall $K = \pm 1$) according to (1) and the corresponding coefficients of the polynomial in $\cos nv$ and $\sin nv$. We conclude that the surface must be rotational.

2.2. `MATHEMATICA` and visualization of surfaces. Let us now consider a second application of `MATHEMATICA` in our research. It will consist in obtaining computer images of the Riemann-type surfaces of $\mathbf{L}^3$ that will appear below. The scheme is as follows. These surfaces are usually defined in terms of elliptic integrals. The special library of `MATHEMATICA` solves numerically the calculations, producing drawings for visualising these surfaces. Computer-generated images are not restricted to the role of visualizing these surfaces, but these images allow us a visual description of their geometrical properties, such as the periodicity, embeddness or the existence of planar ends. Having observed these properties, we then can proceed to construct to prove them mathematically.

We restrict the study to those surfaces that have properties similar to the Riemann minimal examples of $\mathbf{E}^3$. Let us consider a maximal surface foliated by circles in spacelike planes. It follows from Theorem 2 that the foliating planes are parallel. After a rigid motion in $\mathbf{L}^3$, the surface can be written as

$$\mathbf{X}(u, v) = (c(u), 0, u) + r(u)(\cos v, \sin v, 0),$$

where c and r are smooth functions on u. If $H = 0$, we obtain the following two differential equations:

$$(2)\quad -1 + c'^2 + r'^2 - rr'' = 0 \qquad 2r'c' - rc'' = 0.$$

The second differential equation has a first integral $c' = ar^2$, $a > 0$. After a change of variables, it follows that the parametrization of the surface takes the form (see

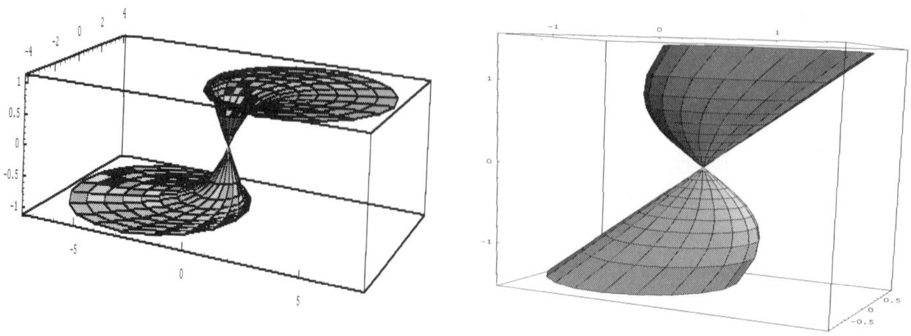

FIGURE 1. Surfaces for $(a, b) = (1, 1)$ and $(a, b) = (1, -1)$

[4] for more details)

$$\mathbf{X}(t, v) = \left(\int \frac{t^2}{\sqrt{a^2t^4 + 2bt^2 + 1}}dt + t\cos v, t\sin v, \int \frac{dt}{\sqrt{a^2t^4 + 2bt^2 + 1}}\right) \tag{3}$$

where a, b are real constants. Indeed, the surface is described in terms of elliptic integrals in the same that occurs in $\mathbf{E}^3$ with the Riemann minimal examples. There is an additional condition on (2), namely that the surface is spacelike. However, we carry out the study without any restriction on the causal character of the surface. So, we consider surfaces with zero mean curvature and with mixed structure (spacelike and timelike behaviour).

We are not so much interested in the image of a particular couple (a, b), but rather we would like to study the behaviour of the maximal surfaces and their qualitative properties. The values $(1, 1)$, and $(1, -1)$ for (a, b) illustrate the general situation. Figure 1 (a) is a spacelike surface, whereas Figure 1 (b) is timelike. In fact, the parametrizations are respectively

$$\mathbf{X}(t, v) = (t - \arctan t + t\cos v, t\sin v, \arctan t) \qquad \text{(spacelike surface)}$$

$$\mathbf{X}(t, v) = (t + \frac{1}{2}\log\left(\frac{1-t}{1+t}\right) + t\cos v, t\sin v, \frac{1}{2}\log\left(\frac{1-t}{1+t}\right)) \qquad \text{(timelike surface)}$$

The images produced by `MATHEMATICA` for different values of a and b are the objects that we use to understand them. Looking these pictures, we observe similar qualitative properties that we then prove mathematically. In general, the parametrizations given by (3) for such surfaces involve elliptic integrals, as occurs with the Riemann minimal examples in Euclidean 3-space (see [4]). Figure 1 (a) is an embedded surface with two (horizontal) planar ends and one singularity at the origin. By reflections, one can define a complete maximal surface invariant by a family of translations. Finally, a computer program working with visual images, as for example `CorelDRAW` v. 8.0, allows us to obtain an image of the whole surface. See Figure 2.

3. Conclusions and outlook

In the proofs of our results, the software `MATHEMATICA` plays an important role in a twofold sense. First, we need to realise an explicit computation of the curvature for a surface in $\mathbf{L}^3$. This requires a tedious symbolic calculation to do by hand and

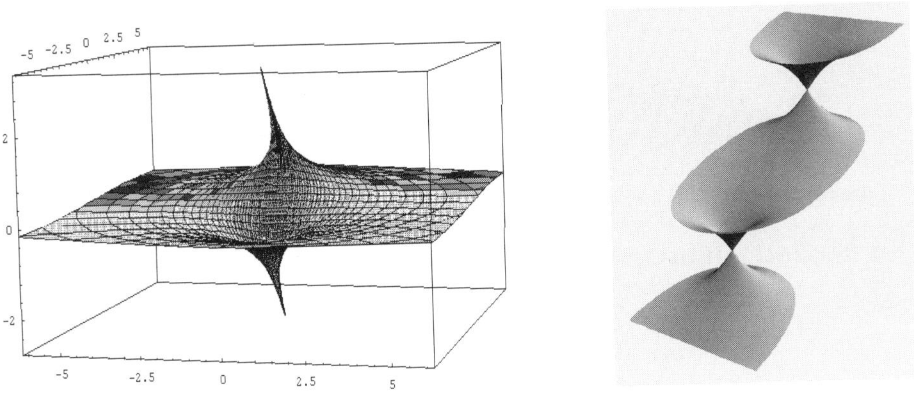

FIGURE 2. A piece of a Riemann-type surface and the whole surface

whose work is enormously simplified by MATHEMATICA. In this sense, MATHEMATICA plays the function that originally was assigned to computers: a way to save efforts.

On the other hand, MATHEMATICA offers a second aspect when we construct the images of the surfaces that we study. These images give us the clues to their qualitative properties, which we then establish mathematically. Now the computer plays a function that goes hand in hand with the process of understanding of our investigations. It is in processes of visualization of this sort where the recent advances in computers will enrich the development of geometry.

References

[1] A. Enneper, *Die cyklischen Flächen*, Z. Math. Phys. **14** (1869) 393–421.
[2] A. Gray, *Differential Geometry and Its Applications with Mathematica*, CRC Press, London, 1996.
[3] F. J. López, R. López, R. Souam, *Maximal surfaccs of Riemann type in Lorentz-Minkowski space* $\mathbf{L}^3$, Michigan Math. J. **47** (2000), 469-497.
[4] R. López, *Constant mean curvature hypersurfaces foliated by spheres*, Diff. Geom. Appl. **11** (1999), 245–256.
[5] R. López, *Constant mean curvature surfaces foliated by circles in Lorentz-Minkowski space*, Geom. Dedicata, **76** (1999), 81–95.
[6] R. López, *Timelike surfaces in Lorentz 3-space with constant mean curvature*, Tohoku Math. J. **52** (2000), 515–532.
[7] R. López, *Cyclic hypersurfaces of constant curvature*, to appear in Minimal Surfaces, Geometric Analysis and Symplectic Geometry, Advances Studies in Pure Mathematics Series.
[8] R. López, *Cyclic surfaces of constant Gauss curvature*, to appear in Houston J. Math. **27** (2001).
[9] R. López, *Surfaces of constant Gauss curvature in Lorentz-Minkowski 3-space*, to appear in Rocky Mount. J. Math.
[10] J.C.C. Nitsche, *Cyclic surfaces of constant mean curvature*, Nachr. Akad. Wiss. Göttingen Math. Phys, II **1** (1989) 1–5.
[11] B. Riemann, *Über die Flächen vom Kleinsten Inhalt bei Gegebener Begrenzung*, Abh. Königl. Ges. d. Wissensch. Göttingen, Mathema. Cl. **13** (1868) 329–333.

DEPARTAMENTO DE GEOMETRÍA Y TOPOLOGÍA, UNIVERSIDAD DE GRANADA, 18071 GRANADA. SPAIN

E-mail address: rcamino@ugr.es

Contemporary Mathematics
Volume **288**, 2001

The Fuglede-Ishihara and Baird-Eells Theorems for $p > 1$

E. Loubeau

ABSTRACT. We introduce the notion of p-harmonic morphisms ($p > 1$), generalising harmonic morphisms, and show that the Fuglede-Ishihara Characterisation remains valid. This allows us to consider the p-stress-energy tensor and prove that its main properties stand implying, in particular, a generalisation of the Baird-Eells Theorem.

1. Harmonic maps

Consider two Riemannian manifolds (M, g) and (N, h), the energy of a C^2-map $\phi : (M, g) \to (N, h)$ is:

$$E(\phi) = \frac{1}{2} \int_M |d\phi_x|^2 \, v_g.$$

When M is not compact, we integrate on a compact subset $K \subset M$.
A map is called *harmonic* if it is an extremum, in its homotopy class, of the energy functional E (or $E(K)$ for all compact subsets $K \subset M$).
The Euler-Lagrange equation corresponding to this functional is the *tension field*, a system of semi-linear second order elliptic partial differential equations:

$$\tau(\phi) = \text{trace}\nabla d\phi.$$

The first problem solved by Eells and Sampson was existence, more specifically within a given homotopy class. Using a heat equation method, they established:

THEOREM 1.1. **[E-S]** *Let (M, g) be a compact manifold and (N, h) a compact manifold with non-positive Riemannian sectional curvature. Then any continuous map from M to N has a harmonic representant, of minimal energy, in its homotopy class. Or, in other words, any map can be continuously deformed into a map of minimal energy, perforce harmonic.*

If ϕ is an immersion, then its mean curvature is, up to a constant, the trace of the second fundamental form $\nabla d\phi$ and therefore a Riemannian immersion is harmonic if and only if it is minimal ([**E-S**]).
More generally, define, for a real number $p > 1$, the p-energy of ϕ as:

2000 *Mathematics Subject Classification.* 58E20, 53C.
Key words and phrases. p-harmonic morphisms, p-harmonic maps, harmonic morphisms.

$$E_p(\phi) = \frac{1}{p} \int_M |d\phi_x|^p \, v_g.$$

A map is called a *p-harmonic map* if it is a critical point, in its homotopy class, of E_p.
The Euler-Lagrange equation of the p-energy functional is the p-tension field:

$$\tau_p(\phi) = |d\phi|^{p-2} \left[\tau(\phi) + \tfrac{p-2}{2} \, d\phi \, \mathrm{grad}(|d\phi|^2)\right],$$

where τ is the tension field for harmonic maps. The system τ_p is still elliptic, away from degenerate points (i.e. for which $d\phi \equiv 0$), but is no longer semi-linear, only quasi-linear. As for $p = 2$, a map is p-harmonic if and only if $\tau_p(\phi) = 0$.
An existence theorem emulating the Eells-Sampson Theorem was proved by Duzaar and Fuchs:

THEOREM 1.2. [**D-F**] *Let (M, g) be a compact manifold and (N, h) a compact manifold with non-positive Riemannian sectional curvature. Then, for $p \geq 2$, any continuous map from M to N has a p-harmonic representant, of minimal p-energy, in its homotopy class.*

In spite of many partial results on the regularity of p-harmonic maps, their main characteristic is that a continuous weakly p-harmonic map might not be $C^{1,1}$, even outside a set of measure zero ([**T**]), but for $p = 2$ they are smooth.

2. Harmonic morphisms

DEFINITION 2.1. Let $\phi : (M, g) \to (N, h)$ be a continuous mapping between Riemannian manifolds. Then ϕ is called a *harmonic morphism* if for any harmonic function $f : U \subset N \to \mathbb{R}$, its pull-back by ϕ, $f \circ \phi : \phi^{-1}(U) \subset M \to \mathbb{R}$ is harmonic as well.

DEFINITION 2.2. A C^1-map $\phi : (M^m, g) \to (N^n, h)$ is called *horizontally weakly conformal* if, at each point $x \in M$, either $d\phi_x \equiv 0$ or the linear map

$$d\phi_x\big|_{(\ker d\phi_x)^\perp} : (\ker d\phi_x)^\perp \to T_{\phi(x)}N$$

is a surjective and conformal map.
The conformal factor is usually called the *dilation* and denoted by $\lambda(x)$.

It is clear that if $m < n$, then a horizontally weakly conformal map is constant. Rather surprisingly, harmonic morphisms can be shown to be harmonic maps, this is the Fuglede-Ishihara Characterisation:

THEOREM 2.3. [**F78, I**] *A mapping between Riemannian manifolds is a harmonic morphism if and only if it is a horizontally weakly conformal harmonic map.*

This characterisation gives us another:

THEOREM 2.4. *A continuous map $\phi : (M, g) \to (N, h)$ is a harmonic morphism if and only if there exists a (unique and continuous) function λ such that:*

$$^M\Delta(f \circ \phi) = \lambda^2 (^N\Delta f) \circ \phi$$

for all $f \in C^2(U \subset N, \mathbb{R})$.

An idea originating from physics is that the stress at a point of an elastic body can be described by a 2-tensor, that the laws of conservation make it symmetric and the state of equilibrium translates into this tensor being divergence free. This point of view was applied by Baird and Eells to harmonic maps in [**B-E**].

Definition 2.5. Let $\phi : (M,g) \to (N,h)$ be a C^1-map between Riemannian manifolds. Its *stress-energy tensor* is defined to be:

$$S_\phi = e_\phi \, g - \phi^* \, h,$$

where $e_\phi = \frac{1}{2}|d\phi_x|^2$ is the energy density of ϕ.

A remarkable feature (easily proven with normal coordinates) is that:

$$\mathrm{div} S_\phi = - \langle d\phi, \tau(\phi) \rangle ,$$

so we recover the physical interpretation in term of equilibrium, i.e. if a map is harmonic then its stress-energy tensor is divergence free.
Conversely, if ϕ is a submersion (almost everywhere) and $\mathrm{div} S_\phi = 0$ then ϕ is harmonic.
This led to the Baird-Eells Theorem:

Theorem 2.6. [**B-E**] *Let $\phi : (M^m, g) \to (N^n, h)$ be a submersion which is a harmonic morphism. Then:*

1. *if $n = 2$, the fibres are minimal.*
2. *if $n \neq 2$, the following properties are equivalent:*
   (a) *the fibres are minimal submanifolds.*
   (b) *∇e_ϕ is a vertical field.*
   (c) *the horizontal distribution has mean curvature $\frac{1}{2}\frac{\nabla e_\phi}{e_\phi}$.*

This can be seen, especially for the case $n = 2$, as a converse of the property of Riemannian immersions.
The proof, as will be outlined further down, is a direct computation in an orthonormal frame field.

3. p-harmonic morphisms

Definition 3.1. A C^2-map $\phi : (M,g) \to (N,h)$ is called a *p-harmonic morphism* if it pulls back local p-harmonic functions on M onto p-harmonic functions on N.

As can be expected, the Fuglede-Ishihara Theorem can be extended to p-harmonic morphisms.

Theorem 3.2. [**Lo, B-L**] *A map is a p-harmonic morphism if and only if it is a horizontally weakly conformal p-harmonic map.*

This characterisation enables us to broaden some properties of harmonic morphisms.

Theorem 3.3. *A C^2-map $\phi : (M,g) \to (N,h)$ is a p-harmonic morphism if and only if there exists a (unique and continuous) function λ such that:*

$$^M\Delta_p(f \circ \phi) = \lambda^p(^N\Delta_p f) \circ \phi \tag{3.1}$$

for all $f \in C^2(U \subset N, \mathbb{R})$.

Since Equation (3.1) remains valid for maps, we have:

Theorem 3.4. *Let $\phi : (M,g) \to (N,h)$ be a p-harmonic morphism. Then ϕ pulls back p-harmonic* maps *onto p-harmonic* maps.

Besides, we can compare p-harmonic morphisms for different values of p.

THEOREM 3.5. *Let $p, q > 1, p \neq q$. A p-harmonic morphism ϕ is also a q-harmonic morphism if and only if it is horizontally homothetic, i.e. $d\phi(\nabla\lambda^2) = 0$.*

This property opens up the door to many examples of p-harmonic morphisms, by simply taking horizontally homothetic harmonic morphisms: Riemannian submersions with minimal fibres or the Hopf map $H : \mathbb{S}^3 \to \mathbb{S}^2$, $(z, w) \mapsto (2zw, |z|^2 - |w|^2)$ $(\lambda^2 = 4)$.

4. p-Stress-energy tensor

It is not difficult to figure out what the proper extension of the stress-energy tensor for harmonic maps has to be:
Let $\phi : (M, g) \to (N, h)$ be a C^1-map. Its *p-stress-energy tensor* is:

$$S_p = e_p\, g - |d\phi|^{p-2}\, \phi^* h,$$

where $e_p = \frac{1}{p}|d\phi|^p$.
This first appeared in [**B-G**] and [**T94**]. Further, one can show that:

$$\mathrm{div} S_p = -\left\langle d\phi, \tau_p(\phi)\right\rangle,$$

using either the method of [**B-G**] or normal coordinates. Here, we outline the latter. Let $q \in M$ and choose normal coordinates $(x_i)_{i=1,\ldots,m}$ around q and $(y^\alpha)_{\alpha=1,\ldots,n}$ around $\phi(q) \in N$.
Then the ith-component of $\mathrm{div}S_p$ is, at the point q:

$$\begin{aligned}(\mathrm{div}S_p)_i &= g^{kl}\frac{\partial}{\partial x^l}((S_p)_{ki})\\ &= g^{kl}g^{bc}h_{\alpha\beta}\phi^\alpha_{bl}\phi^\beta_c|d\phi|^{p-2}g_{ki} - (p-2)g^{kl}g^{bc}h_{\alpha\beta}\phi^\alpha_{bl}\phi^\beta_c|d\phi|^{p-4}(\phi^* h)_{ki}\\ &= - < d\phi, \tau_p(\phi) > .\end{aligned}$$

From this result we obtain a Baird-Eells Theorem for p-harmonic morphisms (compare with [**T94**, Proposition 7]):

THEOREM 4.1. *Let $\phi : (M^m, g) \to (N^n, h)$ be a submersive p-harmonic morphism. Then:*

1. *if $n = p$, the fibres are minimal.*
2. *if $n \neq p$, the following conditions are equivalent:*
   (a) *the fibres are minimal submanifolds.*
   (b) *∇e_p is a vertical field.*
   (c) *the horizontal distribution has mean curvature $\frac{1}{p}\frac{\nabla e_p}{e_p}$.*

PROOF. The proof of [**B-E**] can be followed verbatim:
If $m = n$, the fibres are points and ϕ is conformal $(m = n = 2)$ or homothetic, so we will assume that $m > n$.
Consider an adapted frame $(X_i, X_r)_{i=1,\ldots,n,r=n+1,\ldots,m}$, i.e. such that the X_i's are horizontal and the X_r's are vertical. Then

$$\begin{aligned}(\mathrm{div}S_p)(X_b) &= (\nabla_{X_a}S_p)(X_b, X_a)\\ &= \nabla_{X_b}e_p - |d\phi|^{p-2}\left[X_a(\phi^*h(X_b, X_a)) - \phi^*h(\nabla_{X_a}X_b, X_a)\right.\\ &\quad \left.-\phi^*h(X_b, \nabla_{X_a}X_a) - \nabla_{X_a}|d\phi|^{p-2}\left(\phi^*h(X_b, X_a)\right)\right].\end{aligned}$$

If we choose $X_b = X_j$ horizontal, then:

$$(\mathrm{div}S_p)(X_j) = (n-p)\nabla_{X_j}e_p + |d\phi|^p(m-n)\mathrm{MC}(X_j),$$

where $\mathrm{MC}(X_j)$ is the mean curvature of the fibre in the X_j-direction.
If $n = p$, we have

$$0 = (\mathrm{div}S_p)(X_j) = |d\phi|^p(m-n)\mathrm{MC}(X_j),$$

i.e. the fibres are minimal.
If $n \neq p$, the fibres are minimal if and only if $\nabla_{X_j} e_p = 0$, that is ,∇e_p is vertical.
On the other hand, if we take $X_b = X_r \in V$, then for ∇e_p orthogonal to X_r ($\nabla_{X_r} e_p = 0$), we have:

$$0 = (\mathrm{div}S_p)(X_r) = -\frac{|d\phi|^p}{n} g(X_r, \nabla_{X_i} X_i),$$

with a sum on the $X_i \in H$, meaning that the mean curvature of the horizontal distribution is zero in the X_r-direction.
If $\nabla_{X_r} e_p = \alpha$, then:

$$0 = (\mathrm{div}S_p)(X_r) = \alpha - \frac{|d\phi|^p}{n} g(X_r, \nabla_{X_i} X_i).$$

Therefore the mean curvature satisfies:

$$\frac{1}{n}\sum_i \nabla_{X_i} X_i = g(X_r, \frac{1}{n}\sum_i \nabla_{X_i} X_i) X_r = \frac{\alpha_r}{|d\phi|^p} X_r = \frac{\nabla e_p}{|d\phi|^p}.$$

□

References

[B-E] P. Baird and J. Eells, *A conservation law for harmonic maps.* In E. Looijenga, D. Siersma, and F. Takens, editors, Geometry Symposium. Utrecht 1980, volume 894 of Lecture Notes in Math., Springer, Berlin, Heidelberg, New York, 1981, pp. 1–25.

[B-G] P. Baird and S. Gudmundsson, *p-Harmonic maps and minimal submanifolds.* Math. Ann. **294** (1992), 611–624.

[B-L] J.M. Burel and E. Loubeau, *p-Harmonic morphisms: the $1 < p < 2$ case and a non-trivial example.* In Proceedings of the 9th MSJ-IRI conference: Integrable systems in differential geometry, to appear.

[D-F] F. Duzaar and M. Fuchs, *Existence and regularity of functions which minimize certain energies in homotopy classes of mappings.* Asymptotic Anal. **5** (1991), 129–144.

[E-S] J. Eells and J.H. Sampson, *Harmonic mappings of Riemannian manifolds.* Amer. J. Math. **86** (1964), 109–160.

[F78] B. Fuglede, *Harmonic morphisms between Riemannian manifolds.* Ann. Inst. Fourier (Grenoble) **28** (1978), 107–144.

[I] T. Ishihara, *A mapping of Riemannian manifolds which preserves harmonic functions.* J. Math. Kyoto Univ. **19** (1979), 215–229.

[Lo] E. Loubeau, *On p-harmonic morphisms.* Differential Geometry and its Applications **12** (2000), 219–229.

[T94] H. Takeuchi, *Some conformal properties of p-harmonic maps and regularity for sphere-valued p-harmonic maps.* J. Math. Soc. Japan **46** (1994), 217–234.

[T] P. Tolksdorf, *Regularity for a more general class of quasilinear elliptic equations.* J. Diff. Equ. **51** (1984), 126–150.

Université de Bretagne Occidentale, UFR Sciences et Techniques, Departement de Mathematiques, 6, avenue Victor Le Gorgeu, BP 809, 29285 Brest Cedex, France

E-mail address: loubeau@univ-brest.fr

Contemporary Mathematics
Volume **288**, 2001

Cohomology of Nilmanifolds and Gontcharova's Theorem

Dmitri V. Millionschikov

ABSTRACT. Let M_n be a nilmanifold with a corresponding nilpotent Lie algebra $\mathcal{V}_n = \{< e_1, \dots, e_n >, [e_i, e_j] = (j-i)e_{i+j}, i+j \leq n\}$. We study the cohomology $H^*(M_n, \mathbb{R})$: we prove that a stable Betti number $b_q(M_n)$, $n >> q$, is equal to the $(q+2)$-th Fibonacci number F_{q+2}.

1. Introduction.

Nilmanifolds are a fruitful source of examples of symplectic manifolds with no Kähler structure or non formal symplectic manifolds (see [**TO**], [**CFG**] for references). I. Babenko and I. Taimanov described in [**BaT1**], [**BaT2**] examples of non formal simply connected symplectic manifolds. They considered an interesting family of nilmanifolds M_n that we'll discuss below. I. Babenko and I. Taimanov used only the existence of a non-trivial triple Massey product in $H^*(M_n, \mathbb{R})$ in their proof and they didn't study the cohomology $H^*(M_n, \mathbb{R})$. The question of calculation of $H^*(M_n, \mathbb{R})$ appears to be very interesting and important.

2. Nilmanifolds and symplectic structures.

DEFINITION 2.1. A nilmanifold M is a compact homogeneous space of the form G/Γ, where G is a simply connected nilpotent Lie group and Γ is a lattice in G.

EXAMPLE 2.2. A n–dimensional torus $T^n = \mathbb{R}^n/\mathbb{Z}^n$.

EXAMPLE 2.3. The Heisenberg manifold $M_3 = \mathcal{H}_3/\Gamma_3$, where $\mathcal{H}_3$ is the group of all matrices of the form

$$\begin{pmatrix} 1 & x & z \\ 0 & 1 & y \\ 0 & 0 & 1 \end{pmatrix}, \quad x, y, z \in \mathbb{R},$$

and a lattice Γ_3 is a subgroup of matrices with $x, y, z \in \mathbb{Z}$.

Now we introduce the main example of this article.

2000 *Mathematics Subject Classification.* Primary 17B30, 17B56; Secondary 55P62, 55P20.

Key words and phrases. Nilpotent Lie algebras, cohomology, nilmanifolds.

Partially supported by the Russian Foundation for Fundamental Research, grant no. 99-01-00090.

EXAMPLE 2.4. Let $\mathcal{V}_n$ be a nilpotent Lie algebra with a basis $e_1, e_2, \ldots, e_n$ and a Lie bracket:

$$[e_i, e_j] = \begin{cases} (j-i)e_{i+j}, i+j \le n; \\ 0, i+j > n. \end{cases}$$

One can also define a group structure $*$ in the vector space $\mathcal{V}_n$ using the Campbell-Hausdorff formula:

$$x * y = x + y + \frac{1}{2}[x,y] + \frac{1}{12}([x,[x,y]] - [y,[y,x]]) + \ldots.$$

This nilpotent group $V_n = (\mathcal{V}_n, *)$ has a discrete subgroup (lattice) Γ_n and according to A.Maltsev's theorem ([**Mal**]) quotient spaces $M_n = V_n/\Gamma_n$ are compact manifolds (nilmanifolds).

REMARK 2.5. The manifolds M_n are Eilenberg-MacLane spaces $K(\Gamma_n, 1)$.

It was Victor Buchstaber who introduced these manifolds in [**Bu1**], [**Bu2**] for the study of the Landweber-Novikov algebra. V. Buchstaber proposed also a very interesting interpretation of V_n as a group of polynomial real line infinitesimal transformations

$$\{t \to P_{\mathbf{x}}(t) = t + x_1 t^2 + \cdots + x_n t^{n+1}\}$$

with a nilpotent group structure defined by

$$(P_{\mathbf{x}} * Q_{\mathbf{y}})(t) := Q_{\mathbf{y}}(P_{\mathbf{x}}(t)) \mod t^{n+2},$$

where we denote by $\mathbf{x} = (x_1, x_2, \ldots, x_n)$ a vector in the $\mathcal{V}_n$.

The cohomology $H^*(M_n, \mathbb{R})$ is isomorphic to the cohomology $H^*(\mathcal{V}_n)$ of the corresponding Lie algebra $\mathcal{V}_n$ (Nomizu theorem [**Nz**]). One of the most important Babenko-Taimanov's observations was

REMARK 2.6. ([**BaT1**]) M_{2k} is a symplectic manifold with a left-invariant symplectic form

$$\Omega_{2k+1} = (2k-1)\omega_1 \wedge \omega_{2k} + (2k-3)\omega_2 \wedge \omega_{2k-1} + \ldots + \omega_k \wedge \omega_{k+1},$$

where $\omega_1, \ldots, \omega_{2k}$ is a dual basis of left-invariant 1-forms to the $e_1, \ldots, e_{2k}$.

REMARK 2.7. A rational minimal model $\mathcal{M}_{M_n}$ for M_n has generators $x_1, \ldots, x_n$ of degree 1, with a differential

$$dx_1 = 0, dx_2 = 0, dx_k = \sum_{i<j, i+j=k} (j-i)x_i \wedge x_j, k = 3, \ldots n.$$

In fact I. Babenko, I. Taimanov in [**BaT1**], [**BaT2**] applied the symplectic blow-up procedure to the symplectic embedding $M_{2k} \to \mathbb{C}P^{2k+1}$ (as it was done by D. McDuff in [**McD**] where she described the first example of simply connected symplectic non Kähler manifold).

3. Cohomology of infinite dimensional Lie algebras and Gontcharova's theorem.

The problem of calculation of the cohomology of the $\mathcal{V}_n$ is closely related to Gontcharova's theorem in the infinite dimensional Lie algebras theory.

DEFINITION 3.1. L_k is an infinite dimensional Lie algebra of polynomial vector fields on $\mathbb{R}$ with a trivial k-jet at the origin. L_k has a basis

$$e_i = x^{i+1}\frac{d}{dx}, i = k, k+1, \ldots; \ [e_i, e_j] = (j-i)e_{i+j}.$$

So the algebra $\mathcal{V}_n$ is a quotient algebra $\mathcal{V}_n = L_1/L_{n+1}$.

The problem of calculation of $H^*(L_k)$ was solved by Lidiya Gontcharova in 1973. In [**G**] Gontcharova used the second grading p of a complex of continuous cochains $C^*(L_k)$ $(p(\omega_{i_1}\wedge\ldots\wedge\omega_{i_q}) = i_1+\cdots+i_q)$ for her computation of the bigraded cohomology $H^q_p(L_k)$.

THEOREM 3.2. *(L.Gontcharova,[**G**])*

$$1)\ \dim H^q_k(L_1) = \begin{cases} 1, & \text{if } k = \frac{3q^2\pm q}{2}, \\ 0, & \text{otherwise.} \end{cases}$$

Thus $\dim H^q(L_1) = 2$ *for all* $q \geq 1$.

$$2)\ \dim H^q(L_k) = \binom{q+k-1}{k-1} + \binom{q+k-2}{k-1}.$$

REMARK 3.3. One can consider a formula for Euler characteristic χ_N of the complex $C^*_N(L_1)$ of cochains of the second grading $p = N$ and then define a corresponding generating function:

$$\sum_N \chi_N t^N = \prod_{n=1}^{\infty}(1-t^n) = 1 - t - t^2 + t^5 + t^7 + \ldots + (-1)^q(t^{\frac{3q^2-q}{2}} + t^{\frac{3q^2+q}{2}}) + \ldots$$

This is the well-known Euler identity in combinatorics. From topological point of view this identity is an equality of two expressions of Euler characteristic χ_N (the right part of the formula is an alternated sum of Betti numbers, see [**Fu**] for details).

Numbers $\frac{3q^2\pm q}{2}$ are so called Euler pentagonal numbers. From their properties it follows that the algebra $H^*(L_1)$ has a trivial multiplication (see [**FFR**]).

THEOREM 3.4. *A stable Betti number* $b_q(M_n) = \dim(H^q(\mathcal{V}_n))$, $n >> q$, *is equal to the (q+2)-th Fibonacci number* F_{q+2} :

$$b_1(M_n) = \dim(H^1(\mathcal{V}_n)) = 2,\ \ b_2(M_n) = \dim(H^2(\mathcal{V}_n)) = 3,$$

$$b_3(M_n) = \dim(H^3(\mathcal{V}_n)) = 5, \ldots, b_q(M_n) = \dim(H^q(\mathcal{V}_n)) = F_{q+2}, n >> q.$$

PROOF. (Sketch of the proof). Let us consider the Serre-Hochschild spectral sequence of a pair of Lie algebras (L_1, L_{n+1}) (see [**HS**]) that converges to the $H^*(L_1)$.

In fact the first row of E_2 is our answer. We compute here $H^q(L_1/L_{n+1})$ only for $n >> q$.

LEMMA 3.5. $H^q(L_1) = E^{q,0}_\infty = E^{q,0}_3$. *In other words: Gontcharova's cocycles of small degrees* $q << n$ *are in the first row of* E_2 *and survive till* E_∞.

LEMMA 3.6. $H^q(L_1/L_{n+1}) = Imd_2H^{q-2}(L_1/L_{n+1}, H^1(L_{n+1})) \oplus H^q(L_1)$. *In other words: to get a basis of* $H^q(L_1/L_{n+1})$ *for large values* $n >> q$ *we have to add to two Gontcharova's cocycles of degree* q *a basis of* $Imd_2H^{q-2}(L_1/L_{n+1}, H^1(L_{n+1}))$.

EXAMPLE 3.7. Let us consider dimensions $q = 1, 2, 3$ and $n >> q$:

1) $H^1(\mathcal{V}_n)$ is generated by two classes $[\omega_1]$ and $[\omega_2]$ – Gontcharova's classes $g_1 = [\omega_1], g_2 = [\omega_2]$.

$$E_3 = E_4 = \cdots = E_\infty.$$

0	0	0	0	...
0	0	0	0	...
0	ω_1, ω_2	g_5, g_7	g_{12}, g_{15}	...

FIGURE 1. Gontcharova's cocycles and E_∞.

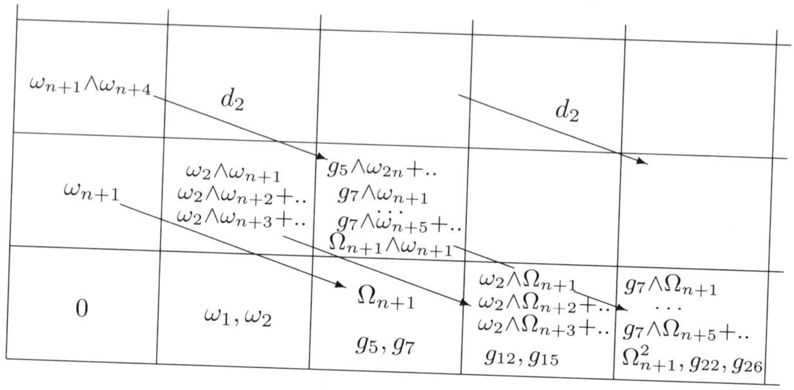

FIGURE 2. The term E_2 of the spectral sequence.

2) The basis of $H^2(\mathcal{V}_n)$ consists of two Gontcharova's classes $g_5 = [\omega_1\wedge\omega_4]$, $g_7 = [\omega_2\wedge\omega_5 + \frac{3}{2}\omega_1\wedge\omega_6]$ and the third generator is the "symplectic class" $[\Omega_{n+1}]$, where $\Omega_{n+1} = (n-1)\omega_1 \wedge \omega_n + (n-3)\omega_2 \wedge \omega_{n-1} + \dots$.

3) In $H^3(\mathcal{V}_n)$ to the classes $g_{12} = [\omega_1\wedge\omega_4\wedge\omega_7 + \dots], g_{15} = [\omega_2\wedge\omega_5\wedge\omega_8 + \dots]$ (degree 3 with second gradings p=12,15 respectively, see [**G**] for details) we add three new ones

$$[\omega_2]\wedge[\Omega_{n+1}], [\omega_2\wedge\Omega_{n+2} - n\omega_3\wedge\Omega_{n+1}],$$

$$[\omega_2\wedge\Omega_{n+3} - (n+1)\omega_3\wedge\Omega_{n+2} + \frac{n(n+1)}{2}\omega_4\wedge\Omega_{n+1}],$$

where $\Omega_{n+2}, \Omega_{n+3}$ are projections of "non-existing" $d\omega_{n+2}, d\omega_{n+3}$ to the $C^2(\mathcal{V}_n)$:

$$\Omega_{n+2} = (n-2)\omega_2\wedge\omega_n + (n-4)\omega_3\wedge\omega_{n-1} + \dots; \Omega_{n+3} = (n-3)\omega_3\wedge\omega_n + \dots.$$

REMARK 3.8. An elementary calculation shows that the cohomology algebra $H^*(\mathcal{V}_n)$ does not satisfy the hard Lefschetz condition:

$$[\omega_1]\wedge[\Omega_{n+1}] = 0, \ \omega_1\wedge\Omega_{n+1} = \frac{1}{n}d\Omega_{n+2}.$$

LEMMA 3.9.

$$H^q(\mathcal{V}_n) = H^q(L_1) \oplus Imd_2 = H^q(L_1) \oplus H^{q-2}(L_2) \oplus H^{q-4}(L_3) \oplus \dots.$$

The last proposition together with Gontcharova's theorem concludes the proof, namely, it gives us a representation of Fibonacci numbers by means of binomial coefficients:

$$F_{q+2} = \binom{q}{0} + \binom{q-1}{0} + \binom{q-1}{1} + \binom{q-2}{1} + \binom{q-2}{2} + \binom{q-3}{2} + \dots.$$

It is a finite sum that can be illustrated by means of the Pascal triangle:

$$\begin{array}{ccccc} 1 & & & & \\ 1 & 1 & & & \\ 1 & 2 & 1 & & \\ 1 & 3 & 3 & 1 & \\ 1 & 4 & 6 & 4 & 1 \end{array} \quad \Rightarrow \quad \begin{array}{ccccc} * & & & & \\ * & * & & & \\ * & 2 & 1 & & \\ 1 & 3 & * & * & \\ 1 & * & * & * & * \end{array}.$$

EXAMPLE 3.10. Let us consider a sum of the fourth and the third diagonals of the Pascal triangle

$$F_6 = \{1+3+1\} + \{1+2\} = 8,$$

or in terms of binomial coefficients

$$F_6 = F_{4+2} = \left\{\binom{4}{0} + \binom{3}{1} + \binom{2}{2}\right\} + \left\{\binom{3}{0} + \binom{2}{1}\right\} = 8.$$

□

References

[BaT1] I.K. Babenko and I.A. Taimanov. *On the existence of nonformal simply connected symplectic manifolds*, Russian Math. Surveys **53**:4 (1998), 1082-1083.

[BaT2] I.K. Babenko and I.A. Taimanov. *On nonformal simply connected symplectic manifolds*, SFB 2888, Preprint 358, 1998.

[Bu1] V.M. Buchstaber, A.V. Shokurov, *The Landweber–Novikov algebra and formal vector fields on the line*, Funct. Anal. and Appl. **12**:3 (1978), 1–11.

[Bu2] V.M. Buchstaber, *The groups of polynomial transformations of the line, nonformal symplectic manifolds, and the Landweber-Novikov algebra*, Russian Math. Surveys **54** (1999).

[CFG] L.A. Cordero, M. Fernandez, A. Gray, *Symplectic manifolds with no Kähler structure*, Topology **25** (1986), 375–380.

[DGMS] P. Deligne, P. Griffiths, J. Morgan, D. Sullivan, *Real homotopy theory of Kähler structure*, Invent. Math. **19** (1975), 245–274.

[FFR] B.L. Feigin, D.B. Fuchs, V.S. Retakh, *Massey operations in the cohomology of the infinite dimensional Lie algebra L_1*, Rokhlin Seminar, Lecture Notes in Math. **1346** (1988), 13–31.

[Fu] D.B. Fuchs, *Cohomology of the infinitedimensional Lie algebras*, Consultant Bureau, New York, 1987.

[G] L.V. Gontcharova, *Cohomology of Lie algebras of formal vector fields on the line*, Funct. Anal. and Appl. **7**:2 (1973), 6–14.

[HS] G.P. Hochschild, J.-P. Serre, *Cohomology of Lie algebras*, Ann. Math. **57**:3 (1953), 367–401.

[Mal] A. Malcev, *On a class of homogeneous spaces*, Izvestia Akad. Nauk SSSR Ser. Mat. **3** (1949), 9-32 (Russian); English translation: Amer. Math. Soc. Transl. (1) **9** (1962), 276–307.

[McD] D. McDuff, *Examples of symplectic simply-connected manifolds with no Kähler structure*, J. Diff. Geom. **20** (1984), 267–277.

[Mill] D.V. Millionschikov, *Cohomology of nilmanifolds and Gontcharova's theorem*, Russian Math. Surveys, to appear.

[Nz] K. Nomizu, *On the cohomology of homogeneous spaces of nilpotent Lie groups*, Ann. of Math. **59** (1954), 531–538.

[TO] A. Tralle, J. Oprea, *Symplectic manifolds with no Kähler structure*, Lecture Notes in Math. **1661** (1997).

DEPARTMENT OF MATHEMATICS AND MECHANICS, MOSCOW STATE UNIVERSITY, 119899 MOSCOW, RUSSIA

E-mail address: million@mech.math.msu.su

Contemporary Mathematics
Volume **288**, 2001

Asymptotically Holomorphic Embeddings of Contact Manifolds in Projective Spaces

Vicente Muñoz, Fran Presas, and Ignacio Sols

Abstract. Using asymptotically holomorphic techniques for contact manifolds, we construct embeddings of a $(2n-1)$-dimensional cooriented contact manifold (C,θ) into the projective space $\mathbb{C}P^{2n+1}$ which are weakly dominate.

In [**MPS**] the authors produce symplectic embeddings of a $2n$-symplectic manifold (M,ω) with $[\omega]/2\pi$ an integer cohomology class into the projective space $\mathbb{C}P^{2n+1}$, such that the pull-back of the Fubini-Study symplectic form is cohomologous to $k\omega$ for k large. The particularity of these newly constructed embeddings is that they are almost holomorphic in the following sense: let J be an ω-compatible almost complex structure on M and fix an hermitian line bundle $L \to M$ with $c_1(L) = [\omega]/2\pi$, and a connection on L with curvature $-i\omega$. Then we find sequences of sections $s_k = (s_k^0, s_k^1, \dots, s_k^{2n+1})$ of the bundles $\mathbb{C}^{2n+2} \otimes L^{\otimes k}$ such that

(1) s_k are asymptotically holomorphic, i.e., $|\nabla^p s_k| = O(1)$ and $|\nabla^{p-1}\bar{\partial} s_k| = O(k^{-1/2})$, for all $p \geq 1$.
(2) s_k are projectizable, i.e., there is an uniform $\gamma > 0$ (independent of k) such that $|s_k| \geq \gamma$ on the whole of M. Then we have defined the maps $\phi_k = \mathbb{P}(s_k) = [s_k^0, \dots, s_k^{2n+1}] : M \to \mathbb{C}P^{2n+1}$.
(3) s_k are generic, i.e., there is an uniform $\gamma > 0$ such that $|\bigwedge^n \partial\phi_k| \geq \gamma$ at all points of M. This makes the differential $d\phi_k$ injective (moreover it multiplies the length of tangent vectors by an amount bounded below uniformly in k). This guarantees that the tangent spaces $(\phi_k)_* T_x M \subset T_{\phi_k(x)}\mathbb{C}P^{2n+1}$ are within distance $O(k^{-1/2})$ of being complex.

The method for constructing these sequences of embeddings is an extension of that of Donaldson [**D1**] introduced to obtain asymptotically holomorphic sequences of sections s_k of $L^{\otimes k}$ which are transverse to zero in a controlled sense (i.e., there is some $\eta > 0$ such that $|\partial s_k| > \eta$ at the points where $|s_k| < \eta$). The sets of zeroes of s_k, $W_k = Z(s_k)$, are then nearly complex, in the sense that their tangent hyperplanes are within an angle $O(k^{-1/2})$ of being complex. In particular this gives a construction of symplectic codimension 2 submanifolds.

2000 *Mathematics Subject Classification.* Primary 53D35.
Key words and phrases. Contact manifold, asymptotically holomorphic, embedding.
Partially supported by The European Contract Human Potential Programme, Research Training Network HPRN-CT-2000-00101.

Our result follows the streamline of the works of Donaldson [**D2**] where asymptotically holomorphic sequences of sections (s_k^0, s_k^1) of $\mathbb{C}^2 \otimes L^{\otimes k}$ are obtained to get a map $M \to \mathbb{C}P^1$ which is a symplectic Lefschetz fibration, and of Auroux [**Au**] where asymptotically holomorphic sequences of sections (s_k^0, s_k^1, s_k^2) of $\mathbb{C}^3 \otimes L^{\otimes k}$ give ramified asymptotically holomorphic coverings $M \to \mathbb{C}P^2$ for a 4-manifold M. In these two cases there are singularities which are shown to be of generic holomorphic type. In our case, in contrast, the main task is the removal of all singularities of the maps from M into the projective space.

Borthwick and Uribe also construct asymptotically holomorphic embeddings into a projective space [**BU**, theorem 1.2], using a different kind of ideas related to geometric quantization. Their result is stronger since their embeddings are not only asymptotically holomorphic but also asymptotically isometric (although in this case the dimension of the projective space grows with k).

The asymptotically holomorphic theory has been extended to the case of contact manifolds by Ibort, Martínez and Presas [**IMP**]. Consider a cooriented contact manifold (C, θ) of dimension $2n-1$. Let L be a trivial (complex) line bundle with a connection with curvature form $-id\theta$. Now fix an almost complex structure J in the symplectic bundle $D = \ker\theta$ which is compatible with the symplectic form $\omega = d\theta$. Let g_D be the corresponding metric on D and define the metric $g = g_D + \theta \otimes \theta$ on C. The operator $\bar{\partial}_D$ is defined as the antiholomorphic part of ∇ restricted to $D \subset TC$. Using the metrics $g_k = kg$, we have the following definition

Definition 1. A sequence of embeddings $\phi_k : C \to \mathbb{C}P^{2n+1}$ is asymptotically holomorphic if it satisfies the following conditions:

(1) $d\phi_k$ is injective and $d\phi_k|_{D_x} : D_x \subset T_xC \to T_{\phi_k(x)}\mathbb{C}P^{2n+1}$ has a left inverse θ_k of norm less than γ^{-1} for some small uniform $\gamma > 0$, for all $x \in C$.
(2) $|(\phi_k)_*J - J_0|_{g_k} = O(k^{-1/2})$ on the subspace $(\phi_k)_*D_x$, for all $x \in C$.
(3) $|\nabla^p \phi_k|_{g_k} = O(1)$ and $|\nabla^{p-1}\bar{\partial}\phi_k| = O(k^{-1/2})$, for all $p \geq 1$.

The result that we want to mention in this note is the following:

Theorem 2. *There is a sequence of asymptotically holomorphic embeddings $\phi_k : C \to \mathbb{C}P^{2n+1}$ such that the pull-back of the Fubini-Study symplectic form restricted to D is cohomologically trivial.*

Denoting by H the hyperplane bundle of $\mathbb{C}P^{2n+1}$, we actually have that $\phi_k^* H = L^{\otimes k}$ (so $\phi_k^*[\omega_{FS}] = [k\omega]$). When the embeddings are asymptotically holomorphic then $\phi_k(C)$ are submanifolds of the projective space such that the distribution $(\phi_k)_* D \subset T\mathbb{C}P^{2n+1}$ is within $O(k^{-1/2})$ of being complex. The proof of theorem 2 follows the lines of the proof of [**MPS**, theorem 2.11] translated into the context of contact manifolds.

Definition 3. A sequence of sections s_k of $\mathbb{C}^N \otimes L^{\otimes k}$ is asymptotically contact-holomorphic if $|\nabla^p s_k| = O(1)$ and $|\nabla^p \bar{\partial}_D s_k| = O(k^{-1/2})$, for all $p \geq 0$.

Definition 4. An asymptotically contact-holomorphic sequence of sections s_k of $\mathbb{C}^{2n+2} \otimes L^{\otimes k}$ is γ-projectizable if $|s_k| \geq \gamma$ at all points of C and for all k large enough.

We start with any asymptotically holomorphic sequence of sections s_k and we perturb it as in [**MPS**, section 2.3] using the following result

PROPOSITION 5 (Proposition 4.4 in [**Pr**]). *Let $f\colon B\times[0,1]\to\mathbb{C}^m$ be a complex valued function, where B is the ball of radius 1 in $\mathbb{C}^n$. Let $0<\delta<1/2$ and $\sigma=\delta(\log(\delta^{-1}))^{-p}$, where p is a suitable fixed integer depending only on the dimensions n,m. Assume that $f_s(z)=f(z,s)$ satisfies the following bounds over $B\times[0,1]$,*

$$|f_s|\le 1,\ |\bar\partial f_s|\le\sigma,\ |\nabla\bar\partial f_s|\le\sigma.$$

Then there exists a smooth curve $w\colon[0,1]\to\mathbb{C}^m$ such that $|w|<\delta$ and the function $f_s-w(s)$ is σ-transverse to zero over the ball $B(0,1/2)$. Moreover, if $|\partial f_s/\partial s|<1$ and $|\partial\nabla f_s/\partial s|<1$, we can choose w such that $|d^iw/ds^i|<\Phi(\delta)$, $i=1,2$, $d^jw/ds^j(0)=0$ and $d^jw/ds^j(1)=0$ for all $j\in\mathbb{N}$, where Φ is a function depending only on the dimensions n,m.

Recall that f is σ-transverse to zero over U if for every $x\in U$ such that $|f(x)|<\sigma$ we have that df is surjective and it has a right inverse with norm less than σ^{-1}. When $m>n$ this condition reduces to $|f(x)|\ge\sigma$ for every $x\in U$. In our case we take an asymptotically holomorphic sequence of sections and perturb it using proposition 5 to get another sequence s_k which is γ-projectizable for some $\gamma>0$. This allows to construct a map $\phi_k=\mathbb{P}(s_k):C\to\mathbb{C}P^{2n+1}$ which we want to perturb into an immersion. Comparing with the result for symplectic manifolds [**MPS**, section 2.3] we see that we lose the isotopy result of the constructed sequences of sections.

DEFINITION 6. The sequence s_k is γ-generic if at every point of C, $d\phi_k$ is injective and $|\bigwedge^{n-1}(\partial_D\phi_k)|\ge\gamma$ (i.e., $d\phi_k|_D$ multiplies the length of vectors by an amount bounded below uniformly) for all k large enough.

In a way similar to that in the symplectic case [**MPS**, section 2.4] we may perturb a sequence of sections until we get a γ-generic one (for some small $\gamma>0$) in a process of $n-1$ steps. The main point again is to use the local transversality result in proposition 5. This gives an immersion $\phi_k:C\to\mathbb{C}P^{2n+1}$ for k large. Finally a very small perturbation in C^1-norm yields an embedding ϕ_k without destroying the conditions of projectizability and genericity previously obtained.

The globalization method described in [**IMP**] and used to prove theorem 2 produces, starting with any asymptotically holomorphic sequence of sections s_k of $\mathbb{C}^{2n+2}\otimes L^{\otimes k}$, another asymptotically holomorphic sequence $\sigma_k=s_k+\tau_k$ with $|\tau_k|_{C^0,g_k}<\epsilon$ (for any small $\epsilon>0$), such that $\mathbb{P}(\sigma_k)$ is an asymptotically holomorphic sequence of embeddings. We only control the C^0-norm of τ_k, and we may control the C^1-norm of τ_k in the directions of D (i.e., $|\nabla_D\tau_k|$), but not the C^1-norm of τ_k in the Reeb direction. Also we lose the isotopy result of the constructed sequences of sections.

From theorem 2 we obtain the following

COROLLARY 7. *For k large enough, there are embeddings $\phi_k:C\to\mathbb{C}P^{2n+1}$ which are weakly dominate, i.e., $(\phi_k^*\omega_{FS})|_D$ is non-degenerate.*

This follows directly from the asymptotic holomorphicity of the map (note that if the map is holomorphic the result is trivially true).

REMARK 8. If C admits a weakly dominate embedding into $\mathbb{C}P^{2n+1}$ then the bundle D acquires a symplectic structure given by the restriction of the Fubini-Study symplectic form. Therefore C is cooriented.

Can we perturb the sequence of embeddings to assure that the restriction of the Fubini-Study form to D is exactly $kd\theta$? There is no topological constraint for doing this as the restriction of the Fubini-Study form is cohomologous to $kd\theta$. (In [**Et**] there are examples when this does not happen for a codimension 1 contact manifold C in a symplectic manifold.) To apply Moser's trick in an open neighborhood of the contact manifold, one has to check that the fields defined to apply Moser's trick are complete and this is not clear.

However, if one is able to construct a stronger kind of embedding produced following the ideas of Borthwick and Uribe [**BU**] one would be able to get this result directly.

PROPOSITION 9. *Let $X \subset \mathbb{C}P^{2n+1}$ be a complex projective submanifold. Then there are asymptotically holomorphic embeddings $\phi_k : C \to \mathbb{C}P^{2n+1}$ such that $\phi_k(C) \bigcap X$ is a contact submanifold of $\phi_k(C)$ for k large.*

This is a generalization of the ideas contained in [**MPS**, section 3]. We start with an asymptotically holomorphic embedding $\phi_k = \mathbb{P}(s_k)$ as constructed in proposition 2 and we look for another embedding $\hat{\phi}_k = \mathbb{P}(\sigma_k)$ with $|s_k - \sigma_k|_{C^0,g_k} < \epsilon$ (for some $\epsilon > 0$) such that

$$\angle((\hat{\phi}_k)_* D_x, T_{\hat{\phi}_k(x)} X) \geq \eta, \text{ for every } x \in C,$$

for some uniform $\eta > 0$. This is achieved as in [**MPS**, Theorem 3.12] by using proposition 5. The embedded manifolds $\hat{\phi}_k(C) \subset \mathbb{C}P^{2n+1}$ then satisfy that $(\hat{\phi}_k)_* D_x \cap T_{\hat{\phi}_k(x)} X$ are within distance $O(k^{-1/2})$ of being complex. Hence $\hat{\phi}_k(C) \cap X$ are contact submanifolds of $\hat{\phi}_k(C)$.

This construction may provide new examples of contact submanifolds in the spirit of the symplectic examples produced in [**MPS**]. Also it may be extended to the case when C is not cooriented by taking the double (cooriented) cover $\hat{C}$, embedding it into $\mathbb{C}P^{2n+1}$ in a $\mathbb{Z}_2$-equivariant way (where $\mathbb{Z}_2$ acts in $\mathbb{C}P^{2n+1}$ by the involution $I[z_0, \ldots, z_{2n+1}] = [\bar{z}_0, \ldots, \bar{z}_{2n+1}]$), and intersecting with a real subvariety $X \subset \mathbb{C}P^{2n+1}$ (i.e., an algebraic subvariety invariant by I).

References

[Au] D. Auroux, *Symplectic 4-manifolds as branched coverings of $\mathbb{C}P^2$*. Invent. Math. **139** (2000), 551–602.

[BU] D. Borthwick and A. Uribe, *Nearly kählerian embeddings of symplectic manifolds*. Asian J. Math. **4** (2000), 599–620.

[D1] S. K. Donaldson, *Symplectic submanifolds and almost-complex geometry*. J. Diff. Geom. **44** (1996), 666–705.

[D2] ———, *Lefschetz pencils on symplectic manifolds*. J. Diff. Geom. **53** (1999), 205–236.

[Et] J. Etnyre, *Convexity in contact geometry*. Topology Appl. **88** (1998), 3–25.

[IMP] A. Ibort, D. Martínez and F. Presas, *On the construction of contact submanifolds with prescribed topology*. To appear in J. Diff. Geom.

[MPS] V. Muñoz, F. Presas and I. Sols, *Almost holomorphic embeddings in grassmannians with applications to singular symplectic submanifolds*. math.DG/0002212

[Pr] F. Presas, *Lefschetz type pencils on contact manifolds*. math.SG/0007034

Departamento de Matemáticas, Facultad de Ciencias, Universidad Autónoma de Madrid, 28049 Madrid, Spain
E-mail address: vicente.munoz@uam.es

Departamento de Álgebra, Facultad de Ciencias Matemáticas, Universidad Complutense de Madrid, 28040 Madrid, Spain
E-mail address: fpm@eucmos.sim.ucm.es

Departamento de Álgebra, Facultad de Ciencias Matemáticas, Universidad Complutense de Madrid, 28040 Madrid, Spain
E-mail address: sols@mat.ucm.es

Contemporary Mathematics
Volume **288**, 2001

Convergent star products on Fréchet linear Poisson algebras of Heisenberg type

Hideki Omori*, Yoshiaki Maeda†, Naoya Miyazaki, and Akira Yoshioka§

ABSTRACT. We discuss the convergence problem for the star product on the linear Poisson algebra associated to the Heisenberg Lie algebra of classes of entire functions with growth conditions. The critical growth exponents for extending the star product on these classes is determined. We also study collapsing phenomena for the convergence of the star product.

Introduction

The purpose of this paper is to give a concrete example of deformation quantization of a Poisson algebra with a convergent star product in the Fréchet category. In [**5**], using the Moyal product, we found classes of entire functions on the complex 2-plane which have convergent star products. Anomalous non-associativity phenomena were also studied. In this article, we study a new example of deformation quantization of Fréchet Poisson algebras associated with the Heisenberg Lie algebra.

We first recall the definition of deformation quantization of Fréchet Poisson algebras proposed in [8]. Let $\mathcal{F}$ be a commutative Fréchet algebra over $\mathbb{C}$. $\mathcal{F}$ is a Fréchet Poisson algebra if $\mathcal{F}$ has a continuous Poisson bracket operation $\{\,,\,\}: \mathcal{F}\times\mathcal{F} \longrightarrow \mathcal{F}$, that is, a skew-symmetric biderivation satisfying the Jacobi identity.

The deformation quantization initiated in [1] is a formal deformation of Poisson algebras, which entails the construction of an associative (non-commutative) product with a formal deformation parameter, called the star product, on a given Poisson algebra. In contrast, we propose a stricter notion of deformation quantization with a convergent star product.

Let us recall our defintion of a deformation quantization of a Fréchet Poisson algebra $(\mathcal{F}, \cdot, \{\,,\,\})$ [**5**]:

2000 *Mathematics Subject Classification.* Primary 53D55, 53D10; Secondary 46L65.
Key words and phrases. Deformation quantization, noncommutative geometry.
* Partially supported by Grant-in-Aid for Scientific Research (#09440041.), Ministry of Education , Science and Culture, Japan.
† Partially supported by Grant-in-Aid for Scientific Research (#12440022.), Ministry of Education , Science and Culture, Japan.
§ Partially supported by Grant-in-Aid for Scientific Research (#11640095.), Ministry of Education , Science and Culture, Japan.

DEFINITION 0.1. Let $\hbar \in \mathbb{R}$, and let $(\mathcal{F}, \cdot, \{,\})$ be a Fréchet Poisson algebra. $(\mathcal{F}, *_\hbar)$ is called a deformation quantization of $(\mathcal{F}, \cdot, \{,\})$ if it satisfies the following:

- For each $\hbar$, $(\mathcal{F}, *_\hbar)$ is an associative Fréchet algebra.
- $f *_\hbar g$ is continuous in $\hbar$, and $f *_\hbar g \to f \cdot g$ as $\hbar \to 0$, for every $f, g \in \mathcal{F}$.
- $\frac{1}{\hbar i}(f *_\hbar g - f \cdot g) \to \frac{1}{2}\{f, g\}$ as $\hbar \to 0$ for every $f, g \in \mathcal{F}$.

In treating noncommutative tori, Rieffel [8] proposed a similar definition in the C^*-algebraic framework. Natsume-Nest [2] and Natsume [3] also extended this idea to more general settings. We work here in the Fréchet category, as this provides fruitful examples and has the specific property of handling deformation quantizations with convergent star products.

We give here only statements of results without proofs. Details will appear in [7].

1. Deformation quantization of Fréchet Poisson algebras

1.1. Heisenberg Lie algebra. The 3-dimensional Heisenberg Lie algebra $\mathfrak{g}$ over $\mathbb{C}$ is the Lie algebra generated by x, y, z with the relation $[x, y] = z$. Viewing these generators as linear functions on the dual space $\mathfrak{g}^*$, we have a linear Poisson structure on $\mathfrak{g}^* = \mathbb{C}^3$ with the Poisson bracket $\{x, y\}_H = z$. The Fréchet Poisson algebras we discuss in this paper are as follows: consider the complex 3-space $\mathbb{C}^3$. Define a Poisson bracket on $\mathbb{C}^3$ by:

$$\{f, g\}_H = z(f(\overleftarrow{\partial_x} \cdot \overrightarrow{\partial_y} - \overleftarrow{\partial_y} \cdot \overrightarrow{\partial_x})g), \tag{1.1}$$

for functions $f{=}f(z, x, y)$ and $g{=}g(z, x, y)$, where $\overleftarrow{\partial_x} \cdot \overrightarrow{\partial_y} - \overleftarrow{\partial_y} \cdot \overrightarrow{\partial_x}$ stands for the bidifferential operator:

$$f(\overleftarrow{\partial_x} \cdot \overrightarrow{\partial_y} - \overleftarrow{\partial_y} \cdot \overrightarrow{\partial_x})g = \partial_x f \cdot \partial_y g - \partial_y f \cdot \partial_x g. \tag{1.2}$$

We have

$$\{z, x\}_H = 0, \quad \{z, y\}_H = 0, \quad \{x, y\}_H = z, \tag{1.3}$$

which gives a linear Poisson structure on $\mathbb{C}^3$ associated with the Heisenberg Lie algebra. We will study examples of deformation quantizations of the Poisson algebra given by (1.1).

1.2. Fréchet Poisson algebras of $\mathbb{C}^{n+1}$. We first introduce a system of semi-norms on the set of polynomials, and then obtain Fréchet algebras by taking completions. Let $\mathbb{C}^{n+1}$ be complex $(n+1)$-space with complex coordinates $(x_0, x_1, \cdots, x_n)$, and let $\mathcal{P}(\mathbb{C}^{n+1})$ be the set of all polynomials on $\mathbb{C}^{n+1}$. The usual pointwise multiplication $(f \cdot g)(x) = f(x) \cdot g(x)$ for polynomial functions f, g gives a commutative, associative structure on $\mathcal{P}(\mathbb{C}^{n+1})$.

To define a system of semi-norms, we use the following notation: $\tilde{p}$ and $\tilde{b}$ denote $(n+1)$-tuples $\tilde{p} = (p_0, p_1, \cdots, p_n)$ and $\tilde{b} = (b_0, b_1, \cdots, b_n)$ with $p_i > 0$ and $b_i > 0$ for $0 \le i \le n$. Forgetting p_0 and b_0, we let $\tilde{p}_*$, $\tilde{b}_*$ denote $(p_1, \cdots, p_n)$, $(b_1, \cdots, b_n)$, respectively.

DEFINITION 1.1. Let r_0 and N_0 be positive real numbers and non-negative integers, respectively. We define semi-norms $||\cdot||_{\tilde{p},\tilde{b}}$, $||\cdot||_{\tilde{p}_*,\tilde{b}_*,r_0}$ and $||\cdot||_{\tilde{p}_*,\tilde{b}_*,N_0}$ on

$\mathcal{P}(\mathbb{C}^{n+1})$ as follows:

$$||f||_{\tilde{p},\tilde{b}} = \sup_{(x_0,\cdots,x_n)\in\mathbb{C}^{n+1}} |f| \exp(-\sum_{i=0}^{n} b_i|x_i|^{p_i}), \tag{1.4}$$

$$||f||_{\tilde{p}_*,\tilde{b}_*,r_0} = \sup_{|x_0|\le r_0} \sup_{(x_1,\cdots,x_n)\in\mathbb{C}^{n}} |f| \exp(-\sum_{i=1}^{n} b_i|x_i|^{p_i}), \tag{1.5}$$

$$||f||_{\tilde{p}_*,\tilde{b}_*,N_0} = \sum_{k=0}^{N_0} ||f_k(x_1,\cdots,x_n)||_{\tilde{p}_*,\tilde{b}_*}, \tag{1.6}$$

where we expand $f = f(x_0, x_1, \cdots, x_n) = \sum_{k=0}^{\infty} f_k(x_1,\cdots,x_n)x_0^k$ as a power series in x_0.

We denote the completions of $\mathcal{P}(\mathbb{C}^{n+1})$ under the systems of seminorms $\{||\cdot||_{\tilde{p},\tilde{b}}\}_{\tilde{b}}$, $\{||\cdot||_{\tilde{p}_*,\tilde{b}_*,r_0}\}_{\tilde{b}_*,r_0}$, $\{||\cdot||_{\tilde{p}_*,\tilde{b}_*,N_0}\}_{\tilde{b}_*,N_0}$ by

(E.1): $\mathcal{E}_{\tilde{p}}(\mathbb{C}^{n+1})$, (E.2): $\mathcal{E}_{Hol,\tilde{p}_*}(\mathbb{C}^{n+1})$, (E.3): $\mathcal{E}_{\infty,\tilde{p}_*}(\mathbb{C}^{n+1})$,

respectively, and we let $\mathcal{E}_\Omega(\mathbb{C}^{n+1})$ denote any one of $\mathcal{E}_{\tilde{p}}(\mathbb{C}^{n+1})$, $\mathcal{E}_{Hol,\tilde{p}_*}(\mathbb{C}^{n+1})$ and $\mathcal{E}_{\infty,\tilde{p}_*}(\mathbb{C}^{n+1})$.

LEMMA 1.2. *$\mathcal{E}_\Omega(\mathbb{C}^{n+1})$ is a commutative, associative Fréchet algebra.*

We remark that $\mathcal{E}_{\tilde{p}}(\mathbb{C}^{n+1})$ and $\mathcal{E}_{Hol,\tilde{p}_*}(\mathbb{C}^{n+1})$ are subalgebras of the algebra of all entire functions $\mathcal{E}(\mathbb{C}^{n+1})$ on $\mathbb{C}^{n+1}$, and that $\mathcal{E}_{\infty,\tilde{p}_*}(\mathbb{C}^{n+1})$ is the space $\mathcal{E}_{\tilde{p}_*}(\mathbb{C}^n)[[x_0]]$ of all formal power series in x_0 with coefficients in $\mathcal{E}_{\tilde{p}_*}(\mathbb{C}^n)$ with the x_0-adic direct product topology.

In this paper, we are mainly concerned with the case of $n = 2$ and $p_1 = p_2$. In light of this, we set $p = (p_0, p_1, p_1)$ and $p_* = (p_1, p_1)$.

LEMMA 1.3. *Let $\mathcal{E}_\Omega(\mathbb{C}^3)$ be one of* (E.1)–(E.3). *Then $(\mathcal{E}_\Omega(\mathbb{C}^3), \cdot, \{\,,\,\}_H)$ is a Fréchet Poisson algebra.*

1.3. Deformation quantization of $(\mathcal{E}_\Omega(\mathbb{C}^3), \cdot, \{\,,\,\}_H)$. Let $(\mathcal{E}_\Omega(\mathbb{C}^3), \cdot, \{\,,\,\}_H)$ be a Fréchet Poisson algebra as in Lemma 1.3. We introduce a noncommutative product $*$ which gives a deformation quantization of $(\mathcal{E}_\Omega(\mathbb{C}^3), \cdot, \{\,,\,\}_H)$ as follows: for $\hbar = 1$ and $f, g \in \mathcal{P}(\mathbb{C}^3)$, set

$$f * g = \sum_{p=0}^{\infty} \frac{(iz)^p}{2^p p!} (f(\overleftarrow{\partial_x} \cdot \overrightarrow{\partial_y} - \overleftarrow{\partial_y} \cdot \overrightarrow{\partial_x})^p g). \tag{1.7}$$

This is also called the Moyal product formula.

As before for $f, g \in \mathcal{P}(\mathbb{C}^3)$, the product (1.7) is associative.

Moreover, if we replace z by $\hbar z$, then (1.7) gives a deformation quantization of $(C^\infty(\mathbb{C}^3), \cdot, \{,\}_H)$ (cf. [1], etc.). In this paper, we focus on the question of how (1.7) extends to the Fréchet Poisson algebra $(\mathcal{E}_\Omega(\mathbb{C}^3), \cdot, \{\,,\}_H)$. We concentrate on the case of the weight $p = (p_0, p_1, p_1)$, i.e. $p_1 = p_2$. Specifically, we state the following:

THEOREM 1.4. *Let $(\mathcal{E}_\Omega(\mathbb{C}^3), \cdot, \{,\}_H)$ be a Fréchet Poisson algebra given by Lemma 1.3. Assume that $p = (p_0, p_1, p_1)$. Then the following properties hold:*
(1) $(\mathcal{E}_\Omega(\mathbb{C}^3), *)$ is an associative Fréchet algebra if and only if Ω satisfies one of the following:

(A1) *For* $\Omega = p = (p_0, p_1, p_1)$, $0 < p_1 \leq \dfrac{2p_0}{p_0+1}$,
(A2) *For* $\Omega = (Hol, p_*)$, $p_* = (p_1, p_1)$, $0 < p_1 \leq 2$,
(A3) *For* $\Omega = (\infty, p_*)$, $p_* = (p_1, p_1)$, $0 < p_1$.

(2) *In either of* (A1)–(A3), $(\mathcal{E}_\Omega(\mathbb{C}^3), *)$ *has the following properties (where* $\mathcal{E}_{\Omega_*}(\mathbb{C}^2) = \mathcal{E}_{p_*}(\mathbb{C}^2)$*):*

(i) $[z, \mathcal{E}_\Omega(\mathbb{C}^3)] = 0$, *i.e.* z *is a central element.*
(ii) $[\mathcal{E}_\Omega(\mathbb{C}^3), \mathcal{E}_\Omega(\mathbb{C}^3)] \subset z*\mathcal{E}_\Omega(\mathbb{C}^3)$.
(iii) $\mathcal{E}_\Omega(\mathbb{C}^3) = \mathcal{E}_{\Omega_*}(\mathbb{C}^2) \oplus z*\mathcal{E}_\Omega(\mathbb{C}^3)$ *(topological direct sum).*
(iv) $z*$, *and* $*z$ *are continuous linear isomorphisms of* $\mathcal{E}_\Omega(\mathbb{C}^3)$ *onto* $z*\mathcal{E}_\Omega(\mathbb{C}^3)$.
(v) $a \to \bar{a}$ *gives an involutive anti-automorphism of* $(\mathcal{E}_\Omega(\mathbb{C}^3), *)$ *such that* $\bar{z} = z$, $\bar{u} = u$ *and* $\bar{v} = v$, *but* $\bar{i} = -i$.
(vi) $\bigcap_{m\geq 0} z^m * \mathcal{E}_\Omega(\mathbb{C}^3) = \{0\}$.

Properties (i)–(vi) imply that $(\mathcal{E}_\Omega(\mathbb{C}^3), *)$ is a regulated algebra in the sense of [**4**].

Note also that r eplacing z by $\hbar z$ in (1.7) gives a product $*_\hbar$.

Theorem 1.4 gives a typical deformation quantization of Fréchet Poisson algebras for $\mathcal{E}_\Omega(\mathbb{C}^3)$:

COROLLARY 1.5. *Assume* $(\mathcal{E}_p(\mathbb{C}^3), \cdot, \{,\}_H)$ *satisfies (A1) of Theorem 1.4. Then* $(\mathcal{E}_p(\mathbb{C}^3), *_\hbar)$ *is a deformation quantization of the Fréchet Poisson algebra* $(\mathcal{E}_p(\mathbb{C}^3), \cdot, \{,\}_H)$.

2. Non-associativity phenomena

The growth conditions in Theorem 1.4 are the best possible: if $p_1 > \frac{2p_0}{p_0+1}$ or $p > 2$ in the case $\Omega = (Hol, p)$, then the product $f * g$ diverges for some f, g. In this section, we exhibit some strange phenomena in the space $\mathcal{E}_{Hol,p}(\mathbb{C}^3)$ for $p > 2$.

We showed in [**5**] that the Moyal product makes sense as a convergent product on the space of entire functions on $\mathbb{C}^2$ of order < 2, but that associativity fails for entire functions of order ≥ 2. The techniques in [**5**] and [**6**] carry over to produce similar phenomena here.

We briefly explain this in the case (A2). Let $\mathcal{E}_{Hol,2+}(\mathbb{C}^3) \bigcap \mathcal{E}_{Hol,2+}(\mathbb{C}^3)$. We see that $e^{tuv} \notin \mathcal{E}_{Hol,2}(\mathbb{C}^3)$, but $e^{tuv} \in \mathcal{E}_{Hol,2+}(\mathbb{C}^3)$. Even in $\mathcal{E}_{Hol,2+}(\mathbb{C}^3)$, the product $*$ has the following properties:

(a) $f * g$ is defined if one of f, g is a polynomial,
(b) the associativity of $*$ holds if two of f, g, h are polynomials.

Properties (a), (b) hold even if the space $\mathcal{E}(\mathbb{C}^3)$ is replaced by $Hol(U)$, the space of holomorphic functions on any open submanifold U of $\mathbb{C}^3$.

If $U = \mathbb{C}_\times \times \mathbb{C}^2$, where $\mathbb{C}_\times = \mathbb{C} - \{0\}$, we see that z is invertible and

$$x^\bullet = \frac{1}{x}(1 - e^{-\frac{2i}{z}xy}), \quad x^\circ = \frac{1}{x}(1 - e^{\frac{2i}{z}xy}) \in Hol(U),$$

and it is easy to see by (1.7) with $\hbar = 1$ that

$$x^\bullet * x = 1 = x * x^\circ, \quad x * x^\bullet = 1 - 2e^{\frac{2i}{z}xy}, \quad x^\circ * x = 1 - 2e^{\frac{2i}{z}xy}. \tag{2.1}$$

This means that the coordinate function x has both a left inverse and a right inverse. Since $x^\bullet \neq x^\circ$, associativity fails:

$$(x^\bullet * x) * x^\circ \neq x^\bullet * (x * x^\circ). \tag{2.2}$$

If we restrict our system to the "submanifold" $z = a \neq 0$, this procedure produces a failure of associativity as in (2.2). For any fixed $a \neq 0$, set

$$\tilde{x}^\bullet = \frac{1}{x}\left(1 - e^{-\frac{2i}{a}xy}\right), \quad \tilde{x}^\circ = \frac{1}{x}\left(1 - e^{\frac{2i}{a}xy}\right) \in \mathcal{E}_{Hol,2+}(\mathbf{C}^3).$$

Then we have the failure of associativity,

$$(\tilde{x}^\bullet * x) * \tilde{x}^\circ \neq \tilde{x}^\bullet * (x * \tilde{x}^\circ),$$

because inserting $z = a$ gives the associativity failure in (2.2). Thus we see that $\mathcal{E}_{Hol,2+}(\mathbf{C}^3)$ is not an associative algebra, and in fact, we see that $\tilde{x}^\bullet * \tilde{x}^\circ$ diverges.

References

[1] F. Bayen, M. Flato, C. Fronsdal, A.Lichnerowicz and D.Sternheimer: *Deformation theory and quantization, I,II*, Ann. Phys. **111** (1977), 61-151.

[2] T. Natsume: *C^*-algebraic deformation quantization and the index theorem*, preprint.

[3] T. Natsume and R. Nest: *C^*-algebraic deformation quantization and the index theorem for families*, in preparation.

[4] H. Omori, Y. Maeda, N. Miyazaki and A. Yoshioka: *Noncommutative 3-sphere: A model of noncommutative contact algebras*, J. Math. Soc. Japan **50** (1998), 915-943

[5] H. Omori, Y. Maeda, N. Miyazaki and A. Yoshioka: *Convergence of the Moyal product defined on the Fréchet Poisson algebras of entire functions*, to appear.

[6] H. Omori, Y. Maeda, N. Miyazaki and A. Yoshioka: *Singular systems of exponential functions*, Mathematical Physics Studies, **23** (2000), 169-187.

[7] H. Omori, Y. Maeda, N. Miyazaki and A. Yoshioka: *Deformation quantization of Frëchet Poisson algebras of Heisenberg type*, to appear.

[8] M. Rieffel: *Deformation quantization for actions of $\mathbf{R}^n$*, Memoirs of A.M.S. **106**, 1993.

Department of Mathematics, Science University of Tokyo, Noda, Chiba, 278-8510, Japan

E-mail address: omori@ma.noda.sut.ac.jp

Department of Mathematics, Faculty of Science and Technology, Keio University, Hiyoshi, Yokohama, 223-8825, Japan

E-mail address: maeda@math.keio.ac.jp

Department of Mathematics, Faculty of Economics, Keio University, Hiyoshi, Yokohama, 223-8521, Japan

E-mail address: miyazaki@math.hc.keio.ac.jp

Department of Mathematics, Science University of Tokyo, Kagurazaka, Tokyo, 102-8601, Japan

E-mail address: yoshioka@rs.kagu.sut.ac.jp

Contemporary Mathematics
Volume **288**, 2001

Resolutions of 2-Polyhedra by Fake Surfaces and Embeddings into $\mathbb{R}^4$

Alexander Onischenko, Dušan Repovš and Arkady Skopenkov

ABSTRACT. We present several corollaries (concerning embeddings of 2-polyhedra into $\mathbb{R}^4$) of the following result due to the second and the third author: *for every 2-polyhedron P there exists a fake surface Q and an onto map $f : Q \to P$ with contractible preimages.* We show that the class of fake surface in this theorem cannot be replaced by a certain interesting smaller class of 2-polyhedra. We announce (with indication of proof) that *any non-locally flat PL sphere $S^2 \subset \mathbb{R}^4$ has a regular neighborhood, non-homeomorphic to $S^2 \times D^2$.*

A finite 2-polyhedron Q is called a *fake surface* if each of its points has a neighbourhood homeomorphic to one of the following: D^2, (triod)$\times I$ or the cone over the complete graph with four vertices (see the figure below). We shall refer to these points as points of type 1, 2 and 3, respectively. Soap films in $\mathbb{R}^3$ exhibit singularities precisely of type 2 and 3. This notion of a soap film from differential geometry has proven to be an important tool and object of investigation in algebraic and geometric topology.

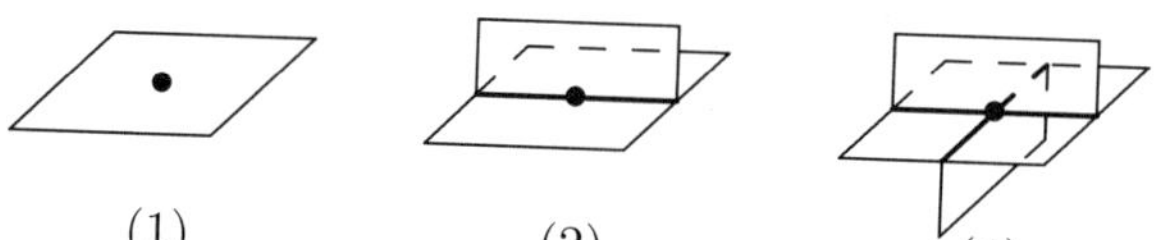

(1) (2) (3)

By Q' we shall denote the *intrinsic 1-skeleton* of a fake surface Q, i.e. the set of points of type 2 or 3. Obviously, Q' is a graph whose vertices have degree 1, 2 or 4. By Q'' we shall denote the finite set of points of Q, which have type 3. A fake surface Q is called a *special 2-polyhedron*, if both $Q - Q'$ and $Q' - Q''$ are *trivial*, i.e. they are a disjoint union of open 2- and 1- disks, respectively.

2000 *Mathematics Subject Classification.* Primary: 57M20, 57Q40; Secondary: 57N60, 57M05, 57M25, 57Q05, 57N40.

Key words and phrases. Cell-like resolution, fake surface, special polyhedron, regular neighborhood, codimension 2, fundamental group.

Onischenko was supported by the ISSEP Grant No. 98-s. Repovš was supported by the Ministry for Science and Technology of the Republic of Slovenia. Skopenkov was supported by the Russian Fundamental Research Grant No. 99-01-00009.

The concept of fake surfaces and special polyhedra as 'general position' 2-polyhedra is formalized by the following result:

Theorem 0. *a) Every 3-manifold has a spine which is a special polyhedron* [HAMS93; Theorem I.3.1.a].

b) If two 3-manifolds have the same special spine, then they are homeomorphic [Cas65].

c) Every finite 2-polyhedron is homotopy equivalent to a special polyhedron [Wri77; Proposition 1].

d) For every 2-polyhedron P there exists a fake surface Q and an onto map $f : Q \to P$ with contractible preimages (they are actually either points or arcs or 2-disks). Moreover, if P is dimensionally homogeneous and has a trivial manifold set, then we can make Q to be special [ReSk00; Theorem 1].

A polyhedron P in a 3-manifold M is called a *spine* of M, if M (or $M - \text{int}D^3$, if M is closed) is a regular neighborhood of P. The *manifold set* of a 2-polyhedron is the set of points having a neighborhood homeomorphic to D^2. A 2-polyhedron P is said to be *dimensionaly homogeneous* if every point of P has an arbitrarily small 2-dimensional neighbourhood. Theorem 0.c can be derived from [Wri77; Proposition 1] (or from Theorem 0.d) by applying the construction from [HAMS93; p.37].

The purpose of this paper is to announce several new results related to Theorem 0. For most of them we give a complete proof, for some we present only the main idea of the proof. Our first result was motivated (like Theorem 0.b) by the following problem [Hor85] [Rep88] [HAMS93; I.49].

Problem. *Find conditions on a compact polyhedron P and a PL manifold M under which regular neighborhoods in M of each two homotopic PL embeddings $P \to M$ are PL homeomorphic.*

This problem can be regarded as a PL version of the Massey problem on uniqueness of the normal bundle for distinct embeddings of a manifold [Mas59].

Theorem 1. *a) There exists a special 2-polyhedron Q and two PL embeddings $Q \to \mathbb{R}^4$ with non-homeomorphic regular neighborhoods.*

b) Any non-locally flat PL sphere $S^2 \subset \mathbb{R}^4$ has a regular neighborhood, non-homeomorphic to $S^2 \times D^2$.

Theorem 1.b is true even for all $k \geq 2$ and any PL embedding $S^k \subset S^{k+2}$ having an isolated singularity $S^{k-1} \subset S^{k+1}$ such that $\pi_1(S^{k+1} - S^{k-1}) \not\cong \mathbb{Z}$. Theorem 1 is sharp in the sense that for $m \neq 4$, every special 2-polyhedron (or even fake surface) Q and every two homotopic PL embeddings of Q into an m-manifold, their regular neighborhoods are homeomorphic [LiSi69] [RBS99] [ReSk]. Also, every PL sphere $S^n \subset \mathbb{R}^m$ has a PL standard regular neighborhood if either $m - n \neq 2$ or $S^n \subset \mathbb{R}^m$ is locally flat [Wal67] [Zee63].

Theorem 1.a is a corollary of the ambient version of Theorem 0.d – for details see below. We shall illustrate the idea of proof of Theorem 1.b by proving the same result for *some* sphere $S^2 \subset S^4$, which already gives a counterexample to the above PL analogue of the Massey problem (for the complete proof see [ReSk]).

Proof of Theorem 1.a. Let P be the Dunce hat, i.e. the quotient space of the 2-dimensional triangle ABC obtained by the identification $AB = AC = BC$ (in this direction). By [Zee63] [Hor85] there are two PL embeddings $f, g : P \to \mathbb{R}^4$ with distinct regular neighborhoods M_1 and M_2. Let Q be the special resolution of P given by Theorem 2.a below. Since Q is a spine of both M_1 and M_2, these two embeddings $Q \subset M_i \subset \mathbb{R}^4$ are as desired.

Proof of Theorem 1.b (for *some* sphere $S^2 \subset S^4$). Take the suspension of a non-trivial knot $S^1 \subset S^3$ such that $\pi_1(S^3 - S^1) \neq \mathbb{Z}$ (in fact, any non-trivial knot satisfies this assumption). For a polyhedron $P \subset M$ we denote by $R_M(P)$ the regular neighborhood of P in M. Consider the decomposition $S^4 = D^4_+ \cup D^4_-$ such that $D^4_+ \cap D^4_- = \partial D^4_\pm = S^3$. Let $B^2_\pm = D^4_\pm \cap S^2$ be the cone over S^1. Let $B^4_\pm = R_{D^4_\pm}(B^2_\pm)$ so that $B^4_\pm \cap \partial D^4_\pm = R_{S^3}(S^1)$.

Denote $C_\pm = \partial B^4_\pm - \mathrm{Int}(R_{\partial B^4_\pm}(S^1))$. Then $R_{S^4}(S^2) = B^4_+ \cup B^4_-$ and $\partial R_{S^4}(S^2) = C_+ \cup C_-$. Since $B^2_\pm$ is a cone it follows that $D^4_\pm \searrow B^2_\pm$, therefore $D^4_\pm$ is also a regular neighborhood of $B^2_\pm$ in $D^4_\pm$. Hence by the uniqueness of regular neighborhoods, $(\partial B^4_\pm, S^1) \cong (S^3, S^1)$ and so $\pi_1(C_\pm) \cong \pi_1(S^3 - S^1)$.

The composition $C_+ \to C_+ \cup C_- \stackrel{r}{\to} C_+$ of the inclusion and the 'symmetric' retraction is a homeomorphism. So the induced composition $\pi_1(S^3 - S^1) \to \pi_1(C_+ \cup C_-) \stackrel{r_*}{\to} \pi_1(S^3 - S^1)$ is an isomorphism, hence r_* is an epimorphism. Therefore $\pi_1(\partial R_{S^4}(S^2)) \not\cong \mathbb{Z}$ and we are done. □

Our second result (already used in the proof of Theorem 1.a) concerns the ambient version of Theorem 0.d.

Theorem 2. *a) For each $m \geq 4$ and every 2-polyhedron P there is an onto map $f : Q \to P$ with contractible preimages such that Q is a fake surface and any m-manifold M having P as the spine, also possesses a spine homeomorphic to Q.*

b) There is a 2-polyhedron P such that for each contractible resolution $f : Q \to P$ by a fake surface Q there is a 3-manifold M which has P as the spine, but possesses no spine homeomorphic to Q.

Theorem 2.a is proved analogously as Theorem 0.d, by using general position. Theorem 2.b follows by taking P to be a common spine of non-homeomorphic manifolds M_1 and M_2 (see examples in [Rep88] [HAMS93] [MPR89] [CLR97]). Since M_1 and M_2 are not homeomorphic, it follows by [RBS99] that they cannot have the same spine, which would be a fake surface.

Our third result is motivated by the well-known fact that every 2-manifold embeds into $\mathbb{R}^4$.

Theorem 3. *There exists a fake surface (even a special 2-polyhedron) Q which does not embed into $\mathbb{R}^4$.*

Proof. Let Q be a resolution, given by Theorem 0.d, of the 2-skeleton P of the standard 6-simplex. Suppose that Q embedded into $\mathbb{R}^4$. It is clear from the proof of Theorem 0.d that the non-trivial preimages of the resolution are those of the points of the 1-skeleton of P. Hence by contracting in $\mathbb{R}^4$ the preimages of the resolution we would obtain $\mathbb{R}^4$ (cf. [BDVW; the 1-LCC shrinking theorems for ANR's and the remark at the bottom of p.2], in which P is embedded. The latter is well-known to be impossible. Contradiction. □

Our fourth result shows that in Theorem 0.c-d the class of fake surfaces cannot be replaced by a certain smaller class of 2-polyhedra. Clearly, there exists a 2-polyhedron P which is not homotopy equivalent to any surface. Indeed, we can take any polyhedron P with $\pi_1(P) \cong \mathbb{Z}_3$, because $\mathbb{Z}_3$ is not the fundamental group of any surface.

Also, there exists a 2-polyhedron P which is not homotopy equivalent to any fake surface without points of type 3. Indeed, we can take as P any 2-dimensional spine of any homology 3-sphere, because [La00; Proposition 1.1] proved that the fundamental group of any fake surface without points of type 3 can never be a non-trivial perfect group. All homology groups used in this paper are assumed to have $\mathbb{Z}$-coefficients.

Theorem 4. *a) For each k there exists a connected 2-polyhedron P which is not homotopy equivalent to any fake surface with at most k vertices of type 3.*

b) If Q is a fake surface such that $H_1(Q) \cong 0$ (e.g. a fake surface spine of a homology 3-sphere), then $\operatorname{rk}\pi_1(Q) \le 20|Q''|$.

Proof. Take $P = K_1 \vee K_2 \vee \dots \vee K_{20k+1}$, where K_i is a 2-spine of a non-trivial homology 3-sphere. By [LySh77; Chapter 4, Corolary 1.9],

$$\pi_1(P) \cong \pi_1(K_1) * \pi_1(K_2) * \dots * \pi_1(K_{20k+1})$$

has at least $20k+1$ generators, so (a) follows from (b).

Our proof of (b) is a reduction to the case $|Q''| = 0$ proved in [La00; Proposition 1.1]. For a space X consisting of connected components $X_1, \dots, X_n$ we denote $\pi_1(X) := \oplus\pi_1(X_i)$. The following inequality holds for finite polyhedra X and Y (neither X nor Y nor $X \cap Y$ are assumed to be connected):

$$(*) \qquad \operatorname{rk}\pi_1(X \cup Y) \le \operatorname{rk}\pi_1(X) + \operatorname{rk}\pi_1(Y) + 2\operatorname{rk} H_0(X \cap Y).$$

The inequality is proved by adding to both X and Y a point and the union of arcs joining this point to each connected component of $X \cap Y$, and then applying the van Kampen theorem to the modified X, Y and (now connected) $X \cap Y$.

Let G be the subgraph of Q' obtained by deleting isolated circles. Then $Q' - G$ does not contain any points of type 3. Let N be a regular neighbourhood of G in Q. Then $\partial N = N \cap \operatorname{Cl}(Q - N)$ is homeomorphic to a disjoint union of n circles. Denote $v = |Q''|$. By (*), it suffices to prove that:

(1) $\operatorname{rk}\pi_1(N) = \operatorname{rk}\pi_1(G) \le 2v$;

(2) $n = \operatorname{rk} H_0(\partial N) \le 6v$; and

(3) $\operatorname{rk}\pi_1(\operatorname{Cl}(Q - N)) \le n$.

In order to prove (1), observe that all vertices of G have degree 4, hence G has $2v$ edges. Since G does not have any isolated circles, it follows that the number of connected components of G is at most v. Therefore $\operatorname{rk}\pi_1(G) \le 2v - v + v = 2v$.

In order to prove (2), observe that n does not exceed the number of pairs (e, s) where e is an edge of G and s is a boundary circle of ∂N, going close to e. Since for each of $2v$ edges e of G there are at most three circles s close to it, there are $6v$ such pairs, and (3) follows.

In order to prove (3), let $D = D_1 \sqcup \dots \sqcup D_n$, identify ∂D and ∂N and let $X = \operatorname{Cl}(Q - N) \cup D$. Since $Q \cup D$ is obtained from Q by attaching disks and $H_1(Q) = 0$, it follows that $H_1(Q \cup D) = 0$. Then the Mayer-Vietoris sequence for $Q \cup D = (N \cup D) \cup X$ with $(N \cup D) \cap X = D$ implies that $H_1(X) = 0$. Since the fake surface X does not contain any points of type 3, it follows by [La00; Proposition 1.1] that $\pi_1(X) = 0$. Hence $\operatorname{rk}\pi_1(\operatorname{Cl}(Q - N)) \le n$. $\square$

Note that the estimate in Theorem 4.b can be improved to $\operatorname{rk}\pi_1(Q) \le 8|Q''|$.

We conclude this paper by a conjecture on a higher-dimensional generalization of Theorem 0.d. Note that the class of 'resolving' polyhedra from our conjecture does not coincide with the class of higher-dimensional special polyhedra [Mat73]. Let Θ^k be the union of S^k with $k+1$ disks D^k attached to S^k along the main equator $S^{k-1} \subset S^k$.

Conjecture. *For every n-polyhedron P there is an onto map $f : Q \to P$ with contractible preimages and such that every point $x \in Q$ has a regular neighbourhood homeomorphic to the product $I^{n-k-1} \times \mathrm{Con}\,\Theta^k$.*

References

[BDVW] M. Bestvina, R. J. Daverman, G. Venema and J. Walsh, *A 4-dimensional 1-LCC shrinking theorem*, Topol. Appl. **110** (2001), 3–20.

[Cas65] B. G. Casler, *An embedding theorem for connected 3-manifolds with boundary*, Proc. Amer. Math. Soc. **16** (1965), 559–556.

[CLR97] A. Cavicchioli, W. B. R. Lickorish and D. Repovš, *On the equivalent spines problem*, Boll. Unione Mat. Ital. **A11** (1997), 775–788.

[HAMS93] C. Hog-Angeloni, W. Metzler and A. J. Sieradski, Eds., *Two-dimensional Homotopy and Combinatorial Group Theory*, London Math. Soc. Lect. Notes **197**, Cambridge Univ. Press, Cambridge, 1993.

[Hor85] K. Horvatič, *Regular neighborhoods of homeomorphic polyhedra*, Berichte D. Math.-Stat. Sektion im Forschungszentrum Graz **251** (1985), 1–16.

[La00] F. F. Lasheras, *Universal covers and 3-manifolds*, J. Pure Appl. Alg. **151** (2000), 163–172.

[LiSi69] W. B. R. Lickorish and L. C. Siebenmann, *Regular neighborhoods and the stable range*, Trans. Amer. Math. Soc. **139** (1969), 207–230.

[LySh77] R. C. Lyndon and P. E. Shupp, *Combinatorial Group Theory*, Springer-Verlag, Berlin - Heidelberg - New York, 1977.

[Mas59] W. S. Massey, *On the normal bundle of a sphere imbedded in Euclidean space*, Proc. Amer. Math. Soc. **10** (1959), 959–964.

[Mat73] S. V. Matveev, *Special skeletons of PL manifolds*, Mat. Sbornik **92** (1973), 287–293 (in Russian).

[MPR89] W. J. R. Mitchell, J. H. Przytycki and D. Repovš, *On spines of knot spaces*, Bull. Acad. Polon. Sci. **37:7–12** (1989), 563–565.

[Rep88] D. Repovš, *Regular neighborhoods of homotopically PL embedded compacta in 3-manifolds*, Proc. Conf. Topol. Trieste 1986, Suppl. Rend. Circ. Mat. Palermo **18** (1988), 415–422.

[RBS99] D. Repovš, N. Brodsky and A. B. Skopenkov, *A classification of 3-thickenings of 2-polyhedra*, Topol. Appl. **94** (1999), 307–314.

[ReSk00] D. Repovš and A. Skopenkov, *Cell-like resolutions of polyhedra by special ones*, Colloq. Math. **86** (2000), 231-237.

[ReSk] D. Repovš and A. Skopenkov, *PL embeddings of spheres in codimension 2 with nonstandard regular neighborhoods*, preprint, Univ. of Ljubljana, 1999.

[Wal67] C. T. C. Wall, *Locally flat PL submanifolds with codimension two*, Proc. Camb. Phil. Soc. **63** (1967), 5–8.

[Wri77] P. Wright, *Covering 2-dimensional polyhedra by 3-manifolds spines*, Topology **16** (1977), 435–439.

[Zee63] E. C. Zeeman, *Unknotting combinatorial balls*, Ann. of Math. (2) **78** (1963), 501–526.

[Zee64] E. C. Zeeman, *On the dunce hat*, Topology **2** (1964), 341–358.

D. Repovš, Institute for Mathematics, Physics and Mechanics, University of Ljubljana, P. O. Box 2964, 1001 Ljubljana, Slovenia.

E-mail address: dusan.repovs@fmf.uni-lj.si

A. Onischenko and A. Skopenkov, Department of Differential Geometry, Faculty of Mechanics and Mathematics, Moscow State University, Moscow, Russia 119899.

E-mail address: onisch@mccme.ru, skopenko@mccme.ru

Contemporary Mathematics
Volume **288**, 2001

Cayley 4-Frames and a Quaternion Kähler Reduction Related to Spin(7)

Liviu Ornea and Paolo Piccinni

ABSTRACT. The object of this note is CAYLEY, the Grassmannian of the oriented 4-planes in $\mathbb{R}^8$ that are closed under the three-fold cross product. We describe an action of $U(1) \times Sp(1)$ on the quaternionic projective space $\mathbb{H}P^7$, that allows to obtain a $\mathbb{Z}_2$-quotient of CAYLEY by quaternion Kähler reduction.

1. Introduction

The existence of only two exceptional cross products - in $\mathbb{R}^7$ and in $\mathbb{R}^8$, with two and three factors respectively - attracted the interest of Alfred Gray in the sixties, and this was one of his approaches to the study of holonomies G_2 and Spin(7) on Riemannian manifolds [**5**], [**10**]. About one decade later, the symmetric space structure of the Grassmannians of those planes in $\mathbb{R}^7$ or $\mathbb{R}^8$ that are closed under such cross products was recognized in the positive quaternion Kähler manifolds $\frac{G_2}{SO(4)}$ and $\frac{Spin(7)}{(Sp(1)\times Sp(1)\times Sp(1))/\mathbb{Z}_2}$ [**14**], [**11**]. The latter of these manifolds is in fact isometric to the Grassmannian $Gr_4(\mathbb{R}^7) = \frac{SO(7)}{SO(3)\times SO(4)}$ of oriented 4-planes in $\mathbb{R}^7$, but its rôle as "exceptional Grassmannian" of some distinguished 4-planes in $\mathbb{R}^8$ is so interesting to make it deserving of notations like CAY or CAYLEY in papers on calibrations [**6**], [**9**].

In this note we describe how CAYLEY can be obtained – up to a $\mathbb{Z}_2$-quotient – through a quaternion Kähler reduction of the projective space $\mathbb{H}P^7$, acted on by a group isomorphic to $U(1) \times Sp(1)$. This action is similar to that used by P. Kobak and A. Swann in $\mathbb{H}P^6$, obtaining a $\mathbb{Z}_3$-quotient of $\frac{G_2}{SO(4)}$ by quaternion Kähler reduction [**12**]. In the later note [**13**] a different action of the same group in $\mathbb{H}P^7$ is described, obtaining this time as a reduction a $\mathbb{Z}_2$-quotient of $Gr_4(\mathbb{R}^7)$. Our reduction is in fact equivalent to this latter, via the isometry CAYLEY $\cong Gr_4(\mathbb{R}^7)$, although our definition of the action is much closer to the point of view of the former article [**12**]. We observe also that, through the same isometry with this

---

2000 *Mathematics Subject Classification.* Primary 53C26; Secondary 53C27, 53C38.
Key words and phrases. Quaternion Kähler manifold, 3-Sasakian manifold, moment map, reduction, Spin(7), Cayley numbers.
The authors are members of EDGE, Research Training Network HRPN-CT-2000-00101, supported by The European Human Potential Programme.

real Grassmannian, one can obtain CAYLEY as a reduction of $\mathbb{H}P^6$ acted on by Sp(1), following one of the classical procedures described in [**1**] (and in this way no finite quotient is involved). However, the reduction we are going to describe here – without making use of the isometry CAYLEY $\cong \mathrm{Gr}_4(\mathbb{R}^7)$ – has the advantage of admitting two interesting generalizations.

One of them is a consequence of the possibility of introducing weights in the action of the U(1)-factor. This allows to obtain a family of 12-dimensional quaternion Kähler orbifolds, some of them admitting smooth 3-Sasakian manifolds over them. These smooth 15-dimensional manifolds are, together with similar 11-dimensional manifolds related to G_2, the first examples of 3-Sasakian manifolds which are neither homogeneous nor toric [**3**].

The other possibility of extending the present reduction is obtained by looking at two other quaternion Kähler Wolf spaces. Since CAYLEY can be regarded as the manifold of the hypercomplex 4-planes in $\mathbb{R}^8$ identified with the real vector space of Cayley numbers $\mathbb{C}\mathrm{a}$ (Proposition 3.1 below), higher dimensional analogues of it are the manifolds $\frac{\mathrm{Spin}(9)}{(\mathrm{Sp}(2)\times\mathrm{Sp}(1)\times\mathrm{Sp}(1))/\mathbb{Z}_2} \cong \mathrm{Gr}_4(\mathbb{R}^9)$ and the exceptional Wolf space $\frac{\mathrm{F}_4}{\mathrm{Sp}(3)\cdot\mathrm{Sp}(1)}$, geometrically the manifolds of the $\mathbb{H}P^1 \subset \mathbb{C}\mathrm{a}P^1$ and of the $\mathbb{H}P^2 \subset \mathbb{C}\mathrm{a}P^2$, respectively. Some issues related to this second generalization will be studied in a future work [**4**].

Acknowledgements. The first author acknowledges financial support by C.N.R. of Italy and by the Cultural Agreement between Università di Roma "La Sapienza" and University of Bucharest. Both authors thank Kris Galicki for helpful and stimulating conversations and Robert Bryant for a kind observation.

2. Preliminaries on Cayley numbers

Let $\mathbb{C}\mathrm{a}$ be the algebra of Cayley numbers, and let $\{1, \mathrm{i}, \mathrm{j}, \mathrm{k}, \mathrm{e}, \mathrm{f} = \mathrm{ie}, \mathrm{g} = \mathrm{je}, \mathrm{h} = \mathrm{ke}\}$ be its canonical basis over $\mathbb{R}$. The multiplication is given by

$$xy = (ac - \bar{d}b) + (b\bar{c} + da)\mathrm{e},$$

where $x = a + b\mathrm{e}, y = c + d\mathrm{e} \in \mathbb{C}\mathrm{a}$ are written through the identification $\mathbb{C}\mathrm{a} \cong \mathbb{H}^2$ with pairs of quaternions. The quaternionic conjugation (already used in the previous formula) induces a conjugation in $\mathbb{C}\mathrm{a}$: $\bar{x} = \bar{a} - b\mathrm{e}$, allowing to write the non-commutativity rule: $\overline{xy} = \bar{y}\,\bar{x}$.

The non-associativity of $\mathbb{C}\mathrm{a}$ gives rise to the associator $[x, y, z] = (xy)z - x(yz)$, alternating form that vanishes whenever two of its arguments are either equal or conjugate. Geometrically, the associator defines the class of *associative 3-planes* in $\mathbb{R}^7 \cong \mathrm{Im}\mathbb{C}\mathrm{a}$, defined in orthonormal bases by $[x, y, z] = 0$. They are characterized as the 3-planes of $\mathbb{R}^7$ closed with respect to the *two-fold cross product*:

$$x \times y = \mathrm{Im}(\bar{y}x),$$

that is so more generally defined for any $x, y \in \mathbb{R}^8 \cong \mathbb{C}\mathrm{a}$. Note that if x, y are orthogonal and in $\mathbb{R}^7 \cong \mathrm{Im}\mathbb{C}\mathrm{a}$, the cross product is simply xy.

The *three-fold cross product* in $\mathbb{C}\mathrm{a} \cong \mathbb{R}^8$ is defined by the formula

$$x \times y \times z = \frac{1}{2}(x(\bar{y}z) - z(\bar{y}x)),$$

reducing to $x(\bar{y}z)$ for x, y, z orthogonal.

The following properties hold whenever x, y are orthogonal and for any $w \in \mathbb{C}\mathrm{a}$:

$$x(\overline{y}w) = -y(\overline{x}w), \qquad (w\overline{y})x = -(w\overline{x})y. \tag{2.1}$$

Moreover, for any $x, y, z \in \mathbb{C}\mathrm{a}$:

$$(xy)(zx) = x(yz)x. \tag{2.2}$$

(A reference for these preliminaries is [**11**], Appendix IV).

3. The Stiefel manifold of Cayley 4-frames

A 4-plane ζ of $\mathbb{R}^8$ that is closed under the three-fold cross product is called a *Cayley 4-plane* and it is oriented by choices of bases $\{w = x \times y \times z, x, y, z\}$. The manifold of the Cayley 4-planes in $\mathbb{R}^8$ is CAYLEY $= \frac{\mathrm{Spin}(7)}{(\mathrm{Sp}(1)\times\mathrm{Sp}(1)\times\mathrm{Sp}(1)/\mathbb{Z}_2)}$, 12-dimensional quaternionic submanifold of the Grassmannian $\mathrm{Gr}_4(\mathbb{R}^8)$ of oriented 4-planes in $\mathbb{R}^8$ ([**14**], p. 262 or [**11**], p. 123).

Proposition IV,1.27 in [**11**] states that a 4-plane ζ in $\mathbb{R}^8$ is Cayley if and only if $-\zeta$ is closed under the complex structures defined by the 2-planes $\alpha \subset \zeta$. This fact can be reformulated as follows.

PROPOSITION 3.1. *A 4-plane ζ in $\mathbb{R}^8$ is Cayley if and only if any triple of mutually orthogonal 2-planes $\alpha, \beta, \gamma \subset \zeta$, all intersecting in a line, defines a hypercomplex structure on ζ.*

PROOF. For any ζ, $\dim(\zeta \cap \mathrm{Im}\mathbb{C}\mathrm{a})$ is either 3 or 4. Thus, if $\zeta \in$ CAYLEY we may select orthonormal imaginary octonions $x, y, z \in \zeta$ such that $\{x \times y \times z, x, y, z\}$ is an oriented basis of ζ. If $u = x \times (x \times y \times z) = y \times z$, $v = y \times (x \times y \times z) = z \times x$, $w = z \times (x \times y \times z) = x \times y$ we have $u, v, w \in S^6$, and their corresponding complex structures J_u, J_v, J_w are associated to the 2-planes $\alpha = \mathrm{span}\{x \times y \times z, x\}$, $\beta = \mathrm{span}\{x \times y \times z, y\}$, $\gamma = \mathrm{span}\{x \times y \times z, z\}$. Since $J_u y = -z$, $J_v x = z$, $J_w x = -y$, then (J_u, J_v, J_w) satisfy $J_v \circ J_u = -J_u \circ J_v = J_w$, i.e. it is a hypercomplex structure on ζ. The converse follows from the aforementioned characterization in [**11**], p. 119. □

Our construction of (u, v, w) out of $\zeta = \mathrm{span}\ \{x \times y \times z, x, y, z\}$ corresponds to the isometry $\sim$: CAYLEY $\to \mathrm{Gr}_3(\mathbb{R}^7)$ of [**6**], p. 11. The image under $\sim$ of the $\zeta \in$ CAYLEY can be interpreted as a *tricomplex section* of S^6, oriented orthonormal bases of the $\zeta^\sim$ being triples u, v, w of unit octonions non necessarily satisfying the hypercomplex relations. The non-associativity of $\mathbb{C}\mathrm{a}$ allows and ensures that such triples define a hypercomplex structure on ζ. An example is the Cayley 4-plane $\zeta = \mathrm{span}\ \{1 - \mathrm{h}, \mathrm{i} + \mathrm{g}, \mathrm{j} - \mathrm{f}, \mathrm{k} + \mathrm{e}\}$: our procedure gives the tricomplex triple $(u, v, w) = (\mathrm{i}, \mathrm{j}, \mathrm{e})$, whose associated $(J_\mathrm{i}, J_\mathrm{j}, J_\mathrm{e})$ is hypercomplex on ζ.

This discussion permits to describe the inverse of the isometry $\sim$, as follows.

COROLLARY 3.1. *Given a tricomplex section of S^6 with oriented orthonormal basis (u, v, w), there is a unique Cayley 4-plane ζ in $\mathbb{R}^8$ on which (J_u, J_v, J_w) is hypercomplex.*

DEFINITION 3.1. A *Cayley 4-frame* in $\mathbb{R}^8$ is an oriented orthonormal 4-frame in a Cayley 4-plane ζ, hence a frame $\{x, I_1 x, I_2 x, I_3 x\}$, where (I_1, I_2, I_3) is the hypercomplex structure of ζ.

By the action of $\mathrm{Spin}(7) \supset \mathrm{G}_2 \supset \mathrm{SU}(3)$ on the spheres $S^7 \supset S^6 \supset S^5$, the latter with isotropy $\mathrm{SU}(2) \cong \mathrm{Sp}(1)$, we have:

PROPOSITION 3.2. *The Stiefel manifold of Cayley 4-frames in* $\mathbb{R}^8$ *is the homogeneous space* $V = \frac{\mathrm{Spin}(7)}{\mathrm{Sp}(1)}$.

Observe finally that:

PROPOSITION 3.3. *An orthonormal frame* $\{f_1, f_2, f_3, f_4\}$ *in* $\mathbb{R}^8$ *is a Cayley 4-frame if and only if* $\overline{f}_2 f_1 = \overline{f}_3 f_4$.

4. CAYLEY and a reduction of $\mathbb{H}P^7$.

We now show how a $\mathbb{Z}_2$-quotient of CAYLEY can be obtained as quaternion Kähler reduction of $\mathbb{H}P^7$ by the action of $\mathrm{U}(1) \times \mathrm{Sp}(1)$. According to [**2**], we first reduce (by the same group) the 3-Sasakian manifold which stands over $\mathbb{H}P^7$, namely the sphere S^{31}. Then we interpret the quotient of S^{31} by $\mathrm{U}(1) \times \mathrm{Sp}(1)$ as the total space of an SO(3)-bundle over a quaternion Kähler orbifold which is the quotient of $\mathbb{H}P^7$ by the same group.

The 3-Sasakian sphere S^{31} is acted on by $\mathrm{U}(1) \times \mathrm{Sp}(1)$ as follows. The factor Sp(1) acts by right multiplication on $\vec{h} = (h_\alpha) \in S^{31} \subset \mathbb{H}^8$, and the moment map $\mu : S^{31} \to \mathbb{R}^9$ of the action reads:

$$\mu(\vec{h}) = (\sum_{\alpha=1}^{8} \overline{h}_\alpha \mathrm{i} h_\alpha, \quad \sum_{\alpha=1}^{8} \overline{h}_\alpha \mathrm{j} h_\alpha, \quad \sum_{\alpha=1}^{8} \overline{h}_\alpha \mathrm{k} h_\alpha).$$

By writing $\vec{h} = \vec{a} + \vec{b}\mathrm{i} + \vec{c}\mathrm{j} + \vec{d}\mathrm{k}$, it is easy to see that $\mu^{-1}(0)$ coincides with the Stiefel manifold of oriented (renormalized) orthonormal 4-frames in $\mathbb{R}^8$ [**2**].

We then act by the factor U(1), rotating pairs of coordinates. This is explicitly described by $\vec{h} \mapsto \mathrm{diag}\,(A(\theta), A(\theta), A(\theta), A(\theta)) \cdot \vec{h}$, where $A(\theta) = \left(\begin{smallmatrix} \cos\theta & -\sin\theta \\ \sin\theta & \cos\theta \end{smallmatrix}\right)$, $\theta \in \mathbb{R}$. The associated moment map $\nu : S^{31} \to \mathbb{R}^3$ is now:

$$\nu(\vec{h}) = \sum_{\beta=1}^{4} (\overline{h}_{2\beta-1} h_{2\beta} - \overline{h}_{2\beta} h_{2\beta-1}).$$

We are interested in the common zero set $\mathcal{N} = \mu^{-1}(0) \cap \nu^{-1}(0)$.

PROPOSITION 4.1. $\mathcal{N} = \mathrm{U}(1) \cdot V$, *where* $\cdot$ *is the action of* U(1) *on Cayley 4-frames.*

PROOF. The inclusion $V \subset \mathcal{N}$ can be checked either by direct computation, using Proposition 3.3, or by a standard choice of the frame, like $(1, \mathrm{i}, \mathrm{j}, \mathrm{k})$, and the observation that $\nu(\vec{h}) = \nu(\vec{a}+\vec{b}\mathrm{i}+\vec{c}\mathrm{j}+\vec{d}\mathrm{k}) = 0$ is invariant under right multiplication of $\vec{a}, \vec{b}, \vec{c}, \vec{d}$ by any $u \in S^6$, and hence by Spin(7) (cf. [**11**], p. 121). It follows also $\mathrm{U}(1) \cdot V \subset \mathcal{N}$, by the U(1)-equivariance of ν.

Conversely, to see that $\mathcal{N} \subset \mathrm{U}(1) \cdot V$, refer to a standard choice of three vectors to be substituted in the moment map equation $\nu(\vec{h}) = \nu(\vec{a} + \vec{b}\mathrm{i} + \vec{c}\mathrm{j} + \vec{d}\mathrm{k}) = 0$, assuming $f_2 = \vec{b} = \mathrm{j}$, $f_3 = \vec{c} = \mathrm{e}$, $f_4 = \vec{d} = \mathrm{g}$, (cf. the similar proof of the G_2-case in [**12**]). Then the equation $\nu(\vec{h}) = 0$ and the orthonormality of the frame give $\vec{f_1} = \vec{a} = \cos\theta + \sin\theta\mathrm{i}$. Then it is easy to check that the element $\mathrm{e}^{-\mathrm{i}\frac{\theta}{2}}$ of U(1) transforms $(\cos\theta + \sin\theta\mathrm{i}, \mathrm{j}, \mathrm{e}, \mathrm{g})$ into a Cayley 4-frame. □

Observe now that $\mathrm{U}(1) \cap \mathrm{Spin}(7) = \mathrm{U}(1) \cap \mathrm{SU}(4) = \mathbb{Z}_4$ with generator $\tau = \mathrm{e}^{\mathrm{i}\frac{\pi}{2}}$, and that under the action of τ on V, a Cayley 4-frame (f_1, f_2, f_3, f_4) is transformed into another frame of the same Cayley 4-plane if and only if (f_1, f_2, f_3, f_4) is complex unitary, i.e. an element of $\frac{\mathrm{SU}(4)}{\mathrm{Sp}(1)}$. Also, of course $\tau^2 = -1$, so that any Cayley 4-plane is fixed under it. This explains the following description of the orbits of the $\mathrm{U}(1) \times \mathrm{Sp}(1)$-action on $\mathcal{N}$: points in $\frac{\mathrm{SU}(4)}{\mathrm{Sp}(1)} \subset V$ generate orbits that are the fixed points of an induced action of $\mathbb{Z}_2$ on all the orbits of $\mathcal{N}$, and a 3-Sasakian orbifold $\mathbb{Z}_2 \backslash \frac{\mathrm{Spin}(7)}{\mathrm{Spin}(4)}$ is obtained as quotient. We state the corresponding quaternion Kähler reduction.

THEOREM 4.1. *The quaternion Kähler quotient of* $\mathbb{H}P^7$ *by the described action of* $\mathrm{U}(1) \times \mathrm{Sp}(1)$ *is an orbifold* $\mathbb{Z}_2\backslash$CAYLEY, *with a singular stratum isometric to the complex Grassmannian* $\frac{\mathrm{SU}(4)}{S(\mathrm{U}(2)\times \mathrm{U}(2))}$.

REMARK 4.2. By identifying any ζ with its orthogonal complement $\zeta^\perp$, one obtains a *smooth* $\mathbb{Z}_2$-quotient of CAYLEY. Since $\perp$ corresponds to the change of orientation on 4-planes in $\mathbb{R}^7$ ([**6**], p. 11), this smooth $\mathbb{Z}_2$-quotient of CAYLEY is the locally quaternion Kähler Grassmannian of *unoriented* 4-planes in $\mathbb{R}^7$.

The $\mathbb{Z}_2\backslash$ CAYLEY given by Theorem 4.1 is not smooth, its construction yielding the stratified space $\mathcal{M}_{\mathrm{reg}} \cup \mathrm{Gr}_2(\mathbb{C}^4)$. The singular stratum $\mathrm{Gr}_2(\mathbb{C}^4)$ corresponds, under the isometry CAYLEY $\cong \mathrm{Gr}_4(\mathbb{R}^7)$, to the standard $\mathrm{Gr}_4(\mathbb{R}^6) \subset \mathrm{Gr}_4(\mathbb{R}^7)$. Thus the orbifold $\mathbb{Z}_2\backslash$CAYLEY in Theorem 4.1 is isometric to the singular quotient $\mathrm{Gr}_4(\mathbb{R}^7)/\sigma_{\mathbb{R}^6}$ by the symmetry $\sigma_{\mathbb{R}^6}$ with respect to $\mathbb{R}^6 \subset \mathbb{R}^7$.

References

[1] Ch.P. Boyer, K. Galicki, 3-*Sasakian manifolds*, Essays on Einstein Manifolds, (C. LeBrun and M. Wang Eds), Surveys in Differential Geometry, vol. V, Int. Press (2000).

[2] Ch.P. Boyer, K. Galicki, B. Mann, *The Geometry and Topology of 3-Sasakian manifolds*, J. Reine Angew. Math., **455** (1994), 183-220.

[3] Ch. P. Boyer, K. Galicki, P. Piccinni, *3-Sasakian Geometry, nilpotent orbits and exceptional quotients*, prepubl. IHES, M/00/51 (2000), also math.DG/0007184.

[4] Ch.P. Boyer, K. Galicki, L. Ornea, P. Piccinni, *Geometry of exceptional quotients*, in preparation.

[5] R. Brown, A. Gray, *Vector cross products*, Comm. Math. Helv. **42** (1967), 222-236.

[6] R. Bryant, R. Harvey, *Submanifolds in hyperKähler Geometry*, J. Amer. Math. Soc. **1**(1989),1- 31.

[7] M. Fernandez, *A classification of Riemannian manifolds with structure group* Spin(7), Ann. Mat. Pura Appl. **143** (1986), 101-122.

[8] K. Galicki, H.B. Lawson, *Quaternionic reduction and quaternionic orbifolds*, Math. Ann. **282** (1988), 1-21.

[9] H. Gluck, D. Mackenzie, F. Morgan, *Volume-minimizing cycles in Grassmann manifolds*, Duke Math. J. **79** (1995), 335-404.

[10] A. Gray, *Vector cross products on manifolds*, Trans. Am. Math. Soc. **141** (1969), 465-504; **148** (1970), 625.

[11] R. Harvey, H.B. Lawson Jr, *Calibrated Geometries*, Acta Math. **148** (1982), 47-157.

[12] P.Z. Kobak, A. Swann, *Quaternionic Geometry of a nilpotent variety*, Math. Ann. **297** (1993),747-763.

[13] P.Z. Kobak, A. Swann, *Exceptional hyperKähler reductions*, Twistor Newsl. **44** (1998), 23-26.

[14] S. Marchiafava, *Alcune osservazioni riguardanti i gruppi di Lie* G_2 *e Spin*(7), *candidati a gruppi di olonomia*, Ann. Mat. Pura. Appl., **129** (1981), 247-264.

[15] L. Ornea, P. Piccinni, *On some moment maps and induced Hopf bundles in the quaternionic projective space*, Int. J. Math. **11** (2000), 925-942..

UNIVERSITY OF BUCHAREST, FACULTY OF MATHEMATICS, 14 ACADEMIEI STR., 70109 BUCHAREST, ROMANIA

E-mail address: lornea@imar.ro

DIPARTIMENTO DI MATEMATICA, UNIVERSITÀ DEGLI STUDI DI ROMA "LA SAPIENZA", PIAZZALE ALDO MORO 2, I-00185, ROMA, ITALY

E-mail address: piccinni@mat.uniroma1.it

Contemporary Mathematics
Volume **288**, 2001

Old and New Structures on Products of Spheres

Maurizio Parton

ABSTRACT. A classical theorem of Kervaire states that products of spheres are parallelizable if and only if at least one of the factors has odd dimension. In this note an explicit parallelization is given, and it is then used to describe G-structures on products of two spheres, for $G = \mathrm{U}(n)$, $\mathrm{Sp}(n)$, G_2, $\mathrm{Spin}(7)$, $\mathrm{Spin}(9)$. This approach gives an alternative description of the classical Calabi-Eckmann structures, and of some G_2, $\mathrm{Spin}(7)$, $\mathrm{Spin}(9)$-structures on $S^6 \times S^1$, $S^7 \times S^1$, $S^{15} \times S^1$ respectively. In other products of spheres some new G-structures are obtained.

1. Introduction

It is a classical result in Algebraic Topology that spheres S^n are parallelizable only in dimension $n = 1, 3, 7$. As for the products of two or more spheres M. Kervaire proved in the fifties the following:

THEOREM 1.1 ([**Ker56**]). *The manifold* $S^{n_1} \times \cdots \times S^{n_r}$, $r \geq 2$, *is parallelizable if and only if at least one of the* n_i *is odd.*

The paper [**Bru92**] is the only reference the author knows to provide explicit parallelizations $\mathcal{B}$ on some products of spheres, namely, whenever one of the factors is S^1, S^3, S^5, S^7.

The aim of the present note is to define significant G-structures on products of spheres of suitable dimension (section 3 and 4), and this is done by writing explicit parallelizations for all parallelizable products of spheres (section 2, in particular theorem 2.2).

The never-vanishing vector field on the odd-dimensional sphere is used to write an explicit isomorphism between $T(S^m \times S^n)$ and $S^m \times S^n \times \mathbb{R}^{m+n}$. Pulling back the standard basis of $\mathbb{R}^{m+n}$ one obtains explicit orthonormal parallelizations $\mathcal{P}$ on $S^m \times S^n$.

The G-structures considered here are: $G = \mathrm{U}((m+n)/2)$, if both the dimensions are odd (almost-Hermitian structures on $S^m \times S^n$); $G = \mathrm{Sp}((m+n)/4)$, if both

2000 *Mathematics Subject Classification.* Primary 53C10; Secondary 57R25.
Key words and phrases. Products of spheres, parallelization, G-structures, Calabi-Eckmann structures.

This note contains some results from my PhD thesis, that was partially supported by the national project *Proprietà Geometriche delle Varietà Reali e Complesse* of M.U.R.S.T., Italy. A special thanks is due to my advisor Paolo Piccinni for his constant help.

TABLE 1. Properties of structures. References are those where the structures were originally introduced.

spheres	G	frame	orbit	type
$S^{2n-1} \times S^1$	$\mathrm{U}(n)$	$\mathcal{B}$	$\mathrm{O}_{2n} I_{\mathcal{B}}$	Hopf Hermitian*
$S^{4n-1} \times S^1$	$\mathrm{Sp}(n)$	$\mathcal{B}$	$\mathrm{O}_{4n} H_{\mathcal{B}}$	Hopf hyperhermitian*
$S^6 \times S^1$	G_2	$\mathcal{B}$	$\mathrm{O}_7 \varphi_{\mathcal{B}}$	locally conformal parallel* [**Cab97**]
$S^7 \times S^1$	$\mathrm{Spin}(7)$	$\mathcal{B}$	$\mathrm{O}_8 \phi_{\mathcal{B}}$	locally conformal parallel* [**Cab95**]
$S^{15} \times S^1$	$\mathrm{Spin}(9)$	$\mathcal{B}$	$\mathrm{O}_{15} \Phi_{\mathcal{B}}$	locally conformal parallel* [**Fri99**]
$S^m \times S^n_{m+n\equiv_2 0}$	$\mathrm{U}(\frac{m+n}{2})$	$\mathcal{P}$	$\mathfrak{S}_{m+n} I_{\mathcal{P}}$	integrable if and only if...**
$S^m \times S^n_{m+n\equiv_4 0}$	$\mathrm{Sp}(\frac{m+n}{4})$	$\mathcal{P}$	$\mathfrak{S}_{m+n} H_{\mathcal{P}}$	non-integrable
$S^m \times S^n_{m+n=7}$	G_2	$\mathcal{P}$	$\mathfrak{S}_7 \varphi_{\mathcal{P}}$	general type
$S^m \times S^n_{m+n=8}$	$\mathrm{Spin}(7)$	$\mathcal{P}$	$\mathfrak{S}_8 \phi_{\mathcal{P}}$	general type
$S^m \times S^n_{m+n=15}$	$\mathrm{Spin}(9)$	$\mathcal{P}$	$\mathfrak{S}_{15} \Phi_{\mathcal{P}}$	?

*all structures in the orbit are isomorphic.

**see theorem 3.2. All integrable structures are isomorphic to Calabi-Eckmann.

the dimensions are odd and $m+n=0 \mod 4$ (almost-hyperhermitian structures on $S^m \times S^n$); $G = \mathrm{G}_2$, $\mathrm{Spin}(7)$, $\mathrm{Spin}(9)$ on the 7-dimensional, 8-dimensional, 16-dimensional products $S^m \times S^n$ respectively.

The results obtained are summarized in Table 1. It refers to the frames $\mathcal{B}$ and $\mathcal{P}$ already mentioned, to their associated G-structures $I_{\mathcal{B}}$, $H_{\mathcal{B}}$, ..., $\Phi_{\mathcal{P}}$ and to the orbits of these structures by the standard action on $\mathbb{R}^{m+n}$ of the orthogonal and the symmetric group O_{m+n}, $\mathfrak{S}_{m+n}$.

In some cases, the orbit contains classical structures, some of which were originally defined in a non-elementary way: this approach provides then a straightforward encoding of these classical structures (theorem 3.3 and remark 4.2). In the other cases, this construction provides new structures on products of spheres (theorem 4.3). The last part of this note is devoted to explain Table 1, that henceforth will be referred to as the *table*. More details can be found in [**Par00**].

2. Explicit parallelizations on $S^m \times S^n$

Let $n=1$. The vector fields $\{x_i \partial_{x_i}\}_{i=1,\dots,m+1}$ on $\mathbb{R}^{m+1} \setminus 0$ are Γ-equivariant, where Γ is the infinite cyclic group generated by multiplication by $e^{2\pi}$. Hence, they define a parallelization $\mathcal{B} = \{b_1, \dots, b_{m+1}\}$ on $S^m \times S^1 = (\mathbb{R}^{m+1} \setminus 0)/\Gamma$. Denoting by $T = -y_2 \partial_{y_1} + y_1 \partial_{y_2}$ the never-vanishing versor field on S^1, it is easy to see that

$$b_i = (\text{orthogonal projection of } \partial_{x_i} \text{ on } S^m) + x_i T, \qquad i = 1, \dots, m+1;$$

hence in particular $\mathcal{B}$ is an orthonormal frame.

Let now n be any odd integer. Denote by T the never-vanishing vector field on S^n given by the complex multiplication, and by t_j its coordinates:

$$T = \sum_{j=1}^{n+1} t_j \partial_{y_j} = -y_2 \partial_{y_1} + y_1 \partial_{y_2} + \cdots - y_{n+1} \partial_{y_n} + y_n \partial_{y_{n+1}}.$$

A key role here is played by the *meridian vector fields* (see [**Bru92**]):

$$M_i = \text{orthogonal projection of } \partial_{x_i} \text{ on } S^m \in \mathfrak{X}(S^m), \qquad i = 1, \dots, m+1,$$
$$N_j = \text{orthogonal projection of } \partial_{y_j} \text{ on } S^n \in \mathfrak{X}(S^n), \qquad j = 1, \dots, n+1,$$

and by the following:

LEMMA 2.1. *Let α, β be vector bundles, and let ε^k be the trivial rank k vector bundle. Then*

$$\alpha \times (\beta \oplus \varepsilon^k) \simeq (\alpha \oplus \varepsilon^k) \times \beta.$$

Split $T(S^n)$ in $\eta \oplus \langle T \rangle$, then use the lemma 2.1 to shift on the left the trivial summand. Since $T(S^m) \oplus \varepsilon^1$ is a trivial vector bundle, a rank 2 trivial summand can be shifted on the right: now remark that $\eta \oplus \varepsilon^2$ is trivial. This argument gives an isomorphism (actually an isometry) between $T(S^m \times S^n)$ and $\varepsilon^{m-1} \times \varepsilon^{n+1}$. Then pull back the sections $\{\partial_{x_1}, \dots, \partial_{x_{m-1}}, \partial_{y_1}, \dots, \partial_{y_{n+1}}\}$ of $\varepsilon^{m-1} \times \varepsilon^{n+1}$ to obtain the explicit parallelization $\mathcal{P}$ given by the following:

THEOREM 2.2. *The vector fields $\{p_1, \dots, p_{m+n}\} \in \mathfrak{X}(S^m \times S^n)$ given by*

$$p_i = M_i + x_i T, \qquad i = 1, \dots, m-1,$$
$$p_{m-1+j} = y_j M_m + t_j M_{m+1} + (t_j x_{m+1} + y_j x_m - t_j)T + N_j, \qquad j = 1, \dots, n+1.$$

define an orthonormal frame $\mathcal{P}$ of $S^m \times S^n$, for any odd n.

PROOF. Once observed that

$$M_i = \partial_{x_i} - x_i M, \qquad i = 1, \dots, m+1,$$
$$N_j = \partial_{y_j} - y_j N, \qquad j = 1, \dots, n+1,$$

where M and N denote the normal versor field of $S^m \subset \mathbb{R}^{m+1}$ and $S^n \subset \mathbb{R}^{n+1}$ respectively, the proof follows from a direct computation. □

3. Hermitian and hyperhermitian structures

The following statement is a consequence of the definition of $\mathcal{B}$. It gives lines 1 and 2 of the table.

THEOREM 3.1. *The almost-Hermitian structure $I_{\mathcal{B}}$ on $S^{2n-1} \times S^1$ coincide with the Hopf Hermitian structure given by the identification $S^{2n-1} \times S^1 = (\mathbb{C}^n \setminus 0)/\Gamma$. The same way, the almost-hyperhermitian structure $H_{\mathcal{B}}$ on $S^{4n-1} \times S^1$ coincide with the Hopf hyperhermitian structure given by the identification $S^{4n-1} \times S^1 = (\mathbb{H}^n \setminus 0)/\Gamma$.*

As for the frame $\mathcal{P}$, the following statement gives line 6 of the table.

THEOREM 3.2. *An almost-Hermitian structure I on $S^m \times S^n$ in the symmetric orbit $\mathfrak{S}_{m+n} I_{\mathcal{P}}$, m, n odd, is integrable if and only if*

$$I(p_{m-1+j}) = \pm p_{m+j}, \qquad j \text{ odd}, 1 \le j \le n+1, \tag{3.1}$$

where the sign is the same for all j.

PROOF. The "if" part relies on the following fact: there exists a covering of $S^m \times S^n$ by submanifolds whose tangent spaces span all $T(S^m \times S^n)$ and on these submanifolds I is integrable.

The "only if" part is given by a case by case computation here sketched. Suppose that the condition of the theorem is not satisfied. Then, taking $-I$ in case, there exists an odd $j \in \{1, \ldots, n+1\}$ such that one of the following conditions holds true:

(1) there exist $i, k \in \{1, \ldots, m-1\}$, $i \neq k$ such that
$$I(p_{m-1+j}) = p_i \qquad \text{and} \qquad I(p_{m+j}) = \pm p_k;$$

(2) there exist $i \in \{1, \ldots, m-1\}$, $k \in \{1, \ldots, n+1\}$, $k \neq j, j+1$ such that
$$I(p_{m-1+j}) = p_i \qquad \text{and} \qquad I(p_{m+j}) = \pm p_{m-1+k};$$

(3) there exist $i \in \{1, \ldots, n+1\}$, $k \in \{1, \ldots, m-1\}$, $i \neq j, j+1$ such that
$$I(p_{m-1+j}) = p_{m-1+i} \qquad \text{and} \qquad I(p_{m+j}) = \pm p_k;$$

(4) there exist $i, k \in \{1, \ldots, n+1\}$, $i, k \neq j, j+1$, $i \neq k$ such that
$$I(p_{m-1+j}) = p_{m-1+i} \qquad \text{and} \qquad I(p_{m+j}) = \pm p_{m-1+k}.$$

The torsion tensor can then be computed in each case and, in particular, one obtains:

(1) $\langle N(p_{m-1+j}, p_{m+j}), p_k \rangle = 2(\pm x_i(1 - y_j^2 - y_{j+1}^2) + x_k(1 - 2(y_j^2 + y_{j+1}^2))) \neq 0$
(2) $\langle N(p_{m-1+j}, p_{m+j}), p_i \rangle (y_j = y_{j+1} = y_k = 0, t_k = 1) = 2(x_i \mp x_{m+1}) \neq 0$
(3) $\langle N(p_{m-1+j}, p_{m+j}), p_k \rangle (y_j = y_{j+1} = y_i = 0, t_i = 1) = 2(x_k \pm x_{m+1}) \neq 0$
(4) $\langle N(p_{m-1+j}, p_{m+j}), p_{m-1+i} \rangle (y_j = y_{j+1} = t_i = x_m = 0, y_i = x_{m+1} = 1) = \mp 2t_k \neq 0$ □

Since an almost-hyperhermitian structure H cannot fix a 2-dimensional distribution, the previous theorem gives also line 7 of the table.

Let now S be the never-vanishing vector field given by complex multiplication on S^m. Denote by $I^{m,n}$ the Calabi-Eckmann Hermitian structure given by $I^{m,n}(T) = S$.

THEOREM 3.3. *The Calabi-Eckmann complex structure $I^{m,n}$ on $S^m \times S^n$ coincides with the almost-Hermitian structure $I_{\mathcal{P}}$, namely*
$$I^{m,n}(p_i) = p_{i+1}, \qquad \text{if } i \text{ is odd},$$
$$I^{m,n}(p_i) = -p_{i-1}, \qquad \text{if } i \text{ is even}.$$

PROOF. Using the Hopf fibrations of S^m and S^n, one is left to show that
$$I_{\mathcal{P}}|_{S^m \times S^1} = I^{m,n}|_{S^m \times S^1},$$
$$I_{\mathcal{P}}|_{S^1 \times S^n} = I^{m,n}|_{S^1 \times S^n},$$
and formulas $b_m = \sum_{j=1}^{n+1} y_j p_{m-1+j}$, $b_{m+1} = \sum_{j=1}^{n+1} t_j p_{m-1+j}$ gives $I_{\mathcal{P}}|_{S^m \times S^1} = I_{\mathcal{B}} = I^{m,1} = I^{m,n}|_{S^m \times S^1}$. The same way, using the frame $\mathcal{B}$ on $S^1 \times S^n$ given by $b_j = N_j - y_j S$, $j = 1, \ldots, n+1$, one gets $I_{\mathcal{P}}|_{S^1 \times S^n} = I_{\mathcal{B}} = I^{1,n} = I^{m,n}|_{S^1 \times S^n}$, and this completes the proof. □

The following corollary proves the second footnote in the table.

COROLLARY 3.4. *All integrable almost-Hermitian structures on $S^m \times S^n$ in the symmetric orbit $\mathfrak{S}_{m+n} I_{\mathcal{P}}$, m, n odd, are biholomorphic to the Calabi-Eckmann Hermitian structure $I^{m,n}$.*

PROOF. Given any permutation π of $\{1, \ldots, m+1\}$, the automorphism f_π of $S^m \times S^n$ given by $(x_1, \ldots, x_{m+1}, y) \mapsto (x_{\pi(1)}, \ldots, x_{\pi(m+1)}, y)$ satisfies

$$df_\pi(p_{\pi(i)}) = p_i, \qquad i = 1, \ldots, m-1.$$

If I is an integrable almost-Hermitian structure in $\mathfrak{S}_{m+n} I_{\mathcal{P}}$, then a permutation $\pi = \pi(I)$ can be defined such that f_π is a biholomorphism between I and $I_{\mathcal{P}}$. □

4. Special structures

THEOREM 4.1. *The* G_2*-structure* $\varphi_{\mathcal{B}}$ *on* $S^6 \times S^1$*, the* Spin(7)*-structure* $\phi_{\mathcal{B}}$ *on* $S^7 \times S^1$ *and the* Spin(9)*-structure* $\Phi_{\mathcal{B}}$ *on* $S^{15} \times S^1$ *are locally conformal parallel.*

PROOF. Denote by $\tilde{\mathcal{B}}$ the frame $\{x_i \partial_{x_i}\}$. The proof follows from the fact that the G_2-structure $\varphi_{\tilde{\mathcal{B}}}$ on $\mathbb{R}^7 \setminus 0$, the Spin(7)-structure $\phi_{\tilde{\mathcal{B}}}$ on $\mathbb{R}^8 \setminus 0$ and the Spin(9)-structure $\Phi_{\tilde{\mathcal{B}}}$ on $\mathbb{R}^{16} \setminus 0$ are globally conformal parallel. □

REMARK 4.2. The previous theorem partially proves lines 3, 4, 5 of the table. That $\varphi_{\mathcal{B}}$, $\phi_{\mathcal{B}}$ and $\Phi_{\mathcal{B}}$ actually coincide with the structures defined in [**Cab97**], [**Cab95**] and [**Fri99**] respectively is a more ticklish question, and it is here just stated.

Only lines 8 and 9 are then left over. Recall that a G-structure of general type, for $G = G_2$ or Spin(7), means a structure which does not belong to any proper subclass of $\mathcal{W}$ in the classifications given in [**FG82**] and [**Fer86**]. Using structure equations for $\mathcal{B}$ and $\mathcal{P}$, the problem of checking that a fixed structure belongs to a particular subclass can be translated in a linear algebra problem, and can be solved by a computer calculation. Since the symmetric group is finite, symmetric orbits can be tackled this way, and this is how the following theorem was proved:

THEOREM 4.3. *The* G_2*-structures in* $\mathfrak{S}_7 \varphi_{\mathcal{P}}$ *on* $S^4 \times S^3$ *and* $S^2 \times S^5$ *are all of general type. The* Spin(7)*-structures in* $\mathfrak{S}_8 \phi_{\mathcal{P}}$ *on* $S^7 \times S^1$, $S^5 \times S^3$, $S^3 \times S^5$ *and* $S^1 \times S^7$ *are all of general type.*

References

[Bru92] M. Bruni, *Sulla parallelizzazione esplicita dei prodotti di sfere*, Rend. di Mat., serie VII **12** (1992), 405–423.

[Cab95] F. M. Cabrera, *On Riemannian manifolds with Spin(7)-structure*, Publ. Math. Debrecen **46** (1995), 271–283.

[Cab97] F. M. Cabrera, *Orientable Hypersurfaces of Riemannian Manifolds with Spin(7)-Structure*, Acta Math. Hung. **76** (1997), no. 3, 235–247 (English).

[Fer86] M. Fernández, *A Classification of Riemannian Manifolds with Structure Group Spin(7)*, Ann. di Mat. Pura e Appl. **148** (1986), 101–122.

[FG82] M. Fernández and A. Gray, *Riemannian manifolds with structure group* G_2*.*, Ann. Mat. Pura Appl., IV. Ser. **132** (1982), 19–45 (English).

[Fri99] Th. Friedrich, *Weak Spin(9)-Structures on 16-dimensional Riemannian Manifolds*, 1–35 (English), electronic preprint: math.DG/9912112 15 Dec 1999.

[Ker56] M. Kervaire, *Courbure intégrale généralisée et homotopie*, Math. Ann. **131** (1956), 219–252 (French).

[Par00] M. Parton, *Hermitian and special structures on products of spheres*, tesi di dottorato, Pisa 2001.

DIPARTIMENTO DI MATEMATICA, UNIVERSITÀ DEGLI STUDI DI PISA, VIA F. BUONARROTI 2, 56127 PISA (ITALY).

E-mail address: parton@dm.unipi.it

Contemporary Mathematics
Volume **288**, 2001

1-homogeneous Sprays in Finsler Manifolds

L. Del Riego

ABSTRACT. The aim of this article is to show that in a Finsler manifold it is possible to define a 1-homogeneous spray. This spray is underlying in the usual studies, which generally consider as *geodesic spray* a quadratic spray; assuming this, the geodesics have **constant** Finslerian speed.

1. Homogeneous Sprays

We shall work with real, smooth, i.e. C^∞, finite dimensional manifolds M; let TM and TTM be the tangent and bitangent spaces, respectively. We recall that in local coordinates, if $(U,(x^i))$ is a chart for M, and $\pi : TM \to M$ is the tangent bundle, a local induced chart on TM is $(\pi^{-1}(U),(x^i,y^i(x^i)))$ where (y^i) are the components of a tangent vector in the natural basis $\frac{\partial}{\partial x_i}$. In the same way, we construct the local induced coordinates on TTM, $(x^i,y^i,X^i(x^i,y^i),Y^i(x^i,y^i))$; the functions in the second slot y^i are called fibre components and those in the third slot, the X^i, are called horizontal.

We have two bundle structures:

$$(TTM,\pi_{TM},TM) : (x^i,y^i,X^i,Y^i) \mapsto (x^i,y^i),$$
$$(TTM,\pi^T,TM) : (x^i,y^i,X^i,Y^i) \mapsto (x^i,X^i).$$

DEFINITION 1.1. A spray S over M is a C^∞-section of both bundles, i.e., $\pi_{TM} \cdot S = id_{TM}$ and $\pi^T \cdot S = id_{TM}$.

Their local expression is given by

$$S : TM \to TTM : (x^i,y^i,y^i,Y^i(x^i,y^i)) .$$

We will call Y^i the defining function for a spray S. Sprays **always** have the fibre component and the horizontal component equal.

These sections have been named 'semi-sprays' by some authors, and 'projective' sprays by others.

DEFINITION 1.2. A spray S is (completely) homogeneous of degree m if in some choice of local coordinates its functions $Y^i(x^i,y^i)$ are completely homogeneous of degree m with respect to the fiber components , i.e. if its defining function satisfies

$$Y^i(x^i,ay^i) = a^m Y^i(x^i,y^i) \quad \forall\, (x^i,y^i) \in TM, \forall\, a \in \Re, \quad \text{i.e.}$$

1991 *Mathematics Subject Classification.* Primary 53C15; Secondary 53C22, 53B40, 53D25.
Partially supported by CONACYT project 26594 E and FAI. UASLP.

$$y^j \frac{\partial Y^i}{\partial y^j} = mY^i.$$

In fact, such coordinates are not unique; any other coordinate system related by block-diagonal Jacobian matrices are compatible [**DRP2**]. As we shall see below, the integral curves of any spray will be invariant under affine changes of its parameters.

Before proceeding, we must note that the notion of homogeneity is used by several authors in a non-equivalent setting at least in the following senses.

1. $a > 0$ — This is the version usually referred to as *positive homogeneity*, commonly used in analysis (Euler's theorem).
2. $a \neq 0$ —This version is most familiar from projective geometry and homogeneous coordinates there. It is the one found in most mathematical reference books. We will call it simply *homogeneity*.
3. $a \in \Re$ —This version is *required* for ordinary quadratic sprays in order to obtain the correspondence with linear connections. We will call it *complete homogeneity*.
4. *Absolute homogeneity* will mean
$$Y(x, ay) = |a|Y(x, y), \quad a \in \Re.$$
5. The Liouville vector field
$$C : TM \to TTM : (x^i, y^i) \mapsto (x^i, y^i, 0, y^i)$$
is also used: S is a *homogeneous vector field* if
$$[C, S] = (m - 1)S.$$

The difference between homogeneity and complete homogeneity is minor; essentially, it is just the difference between working on $TM - 0$ and on TM. The difference between homogeneity and the others is much more significant: (4) makes the difference between 'forward' and 'backward' directions, whereas (5) requires that the horizontal part of a vector field S over TTM is homogeneous of degree $m - 1$ in the fiber component. Only quadratic sprays, i.e. sprays which are homogeneous of degree two in the sense of (3) are homogenenous vector fields in the sense of (5).

Because of this ambiguity, we will stick throughout this paper with *complete homogeneity*, i.e., Definition 1.2.

Before proceeding further, recall that quadratic sprays were introduced by Ambrose, Palais and Singer [**APS**], who showed that:

- the *set* of quadratic sprays S is in one to one correspondence with the *set* of symmetric linear connections Γ_S.
- The integral curves of a quadratic spray S are the geodesics of its symmetric linear connection Γ_S; these solution curves are invariant under affine coordinate changes of its parameter.

2. Sprays in Finsler Manifolds

Homogeneous sprays S of any degree of homogeneity are still projective sprays and their integral curves are still the geodesics of a manifold with connection. Finsler manifolds are defined by a function who is homogeneous of degree one. We claim that there is a complete 1-homogeneous spray defined in a general Finsler manifold, in the following way.

Recall that in Finsler geometry, if the manifold M has local coordinates $(U, (x^i))$, the tangent space $Mink(x)$ with x^i fixed and (y^i) arbitrary is called the *Minkowskian space* at this x^i [**As**]. The function

$$F : TM - \{0\} \to \Re,$$

smooth only off the zero section where it is continuous, is called the Finsler function. It has the following properties:

1. $\det\left(\frac{\partial^2 F^2}{\partial y^i \partial y^j}\right) \neq 0.$
2. $F(x^i, ky^i) = kF(x^i, y^i) \quad k \in \Re.$

and it defines the Finslerian metric tensor by:

$$g_{ij} = \frac{1}{2}\frac{\partial^2 F^2}{\partial y^i \partial y^j} = g_{ji}$$

The dependence of $g_{ij}(x^i, y^i)$ on y^i at a fixed x^i may be treated as a dependence on the coordinates y^i adopted in $Mink(x)$. The Finslerian metric tensor $g_{ij}(x, y)$ at a point x^i will play the role of the metric tensor in the tangent Minkowskian space $Mink(x)$; it defines the square of the length of an infinitesimal displacement dy^i in $Mink(x)$ as the usual rule, applied to $Mink(x)$, i.e.,

$$ds^2_{Mink} = g_{ij}(x, y)dy^i dy^j.$$

It is also possible to define $g^{ki}(x^i, y^i)$ as $g^{ki}(x^i, y^i)g_{ij}(x^i, y^i) = \delta^k_j$.

The Finslerian geodesics come from a variational problem; if $F > 0$, the geodesics afford the minimum of the functional; in the indefinite case one uses stationarity. They are the solutions of the following system of differential equations of order two, [**BCS**, page 116]:

$$\frac{d^2 c^i}{dt^2} + \gamma^i_{jk}(c, \dot{c})\dot{c}^j \dot{c}^k - \frac{d\,(\ln F(c, \dot{c}))}{dt}\dot{c}^i = 0, \quad \text{where} \tag{2.1}$$

$$\gamma^i_{jk}(x, y) = \frac{1}{2}g^{ik}(x, y)\left(\frac{\partial g_{jm}}{\partial x^k} + \frac{\partial g_{km}}{\partial x^j} - \frac{\partial g_{jk}}{\partial x^m}\right) = \gamma^i_{kj}(x, y)$$

are called the *Finslerian Christoffel symbols* of the second kind [**BCS**, page 33].

It is well known that on Finsler manifolds there are several non-linear connections defined, but also, remarkably, there is a quadratic spray [**BCS, A, Sh**], given by:

$$G^i_{jk}(x, y) = \frac{1}{2}\gamma^i_{jk}(x, y)\, y^j y^k,$$

which by [**APS**] would induce a linear connection.

Observe that this quadratic spray, at a point $(c(t), \dot{c}(t))$, is one of the components of the general geodesic Equation 2.1.

But a **natural** spray to be taken is not necessarily a quadratic one, unless you consider **constant finslerian speed** geodesics. But there is a 1-homogeneous spray which is defined in the general case, whose value at a special point appears as the last of the components in the Equation 2.1, and is defined as follows:

THEOREM 2.1. *Let (M, F) be a Finsler manifold. Then*

$$S(x, y) = \left(\frac{1}{F}\frac{\partial F}{\partial y^i}\right)(x, y)\; y^i$$

is a spray homogeneous of degree one over M.

The value of this 1-homogeneous spray is not always zero. For example, some Finsler manifolds studies (among others, Antonelli [**A**]), take the 1-homogeneous spray $S(x,y) = \lambda y^i,\ \ \lambda \in \Re$. In fact this author has succesfully applied KCC-theory to various Volterra-Hamilton systems in physical and biological applications, whose equation is of the form:

$$\frac{dN^2}{dt^2} = \lambda_i N^i - G^i_{jk}(x,N)N^j N^k + \rho^i,$$

where $\lambda_i N^i$ (the 1-homogeneous spray) represents growth rates, the quadratic term $G^i_{jk}(x,N)N^jN^k$ the density drop, and ρ^i the external environment. The Finsler function F in this cases generally depends only on the manifold variable x.

It has also been shown, [**BCS**, page 125], that even when the function F which defines the Finsler manifold depends both on x^i and on the fibre components y^i, the Finslerian geodesics can be parametrized so that they become autoparallel curves; then they are **constant Finslerian speed** geodesics, and they have the form:

$$\frac{d^2c^i}{dt^2} + G^i_{jk}|_{(c,T)}\, \dot{c}^j \dot{c}^k = 0,$$

where $T = \dot{c}^i \frac{\partial}{\partial x^i}$ (compare [**Sh**, page 4]). But it is not always 'useful' to have geodesics with constant Finslerian speed. For special Finsler spaces, such as a Randers space of Douglas type, a parametrization with constant Riemannian speed is much more telling.

This means that there is a 1-homogeneous spray S on M when there is absolutely no assumption about their speed, but that it has a Finslerian parametrization with respect to which it *can* appear as a quadratic spray. In a forthcoming paper we will explore the relationship between these 'two' sprays in special Finsler manifolds.

I want to thank P.E. Parker, Wichita State University, for his help, both with the English, but also for fruitful discussions. To the referee, for very useful comments resulting in the improvement of this paper, my thanks.

References

[A] P.L. Antonelli, R.S. Ingarden and M. Matsumoto, *The Theory of Sprays and Finsler Spaces with Applications in Physics and Biology*, Kluwer Academic Publishers, Dordecht, 1993.
[As] G.S. Asanov, *Finsler Geometry, Relativity and Gauge Theories*, Fundamental Theories of Physics, Kluwer Academic Publishers, 1985.
[APS] W. Ambrose, R. S. Palais and I. M. Singer, *Sprays*, Anais Acad. Brasil Ciênc., **32** (1960), 163–178.
[BCS] D. Bao. S.S. Chern and S. Shen, *An Introduction to Riemann-Finsler Geometry*, Graduate Texts in Mathematics 200, Springer-Verlag, 2000.
[BC] F. Brickell and R. S. Clark, *Differentiable Manifolds*, New York: Van Nostrand, 1970.
[DRP1] L. Del Riego and P. E. Parker, *Pseudoconvex and Disprisoning Homogeneous Sprays*, Geom. Dedicata **55** (1995) 211–220.
[DRP2] L. Del Riego and P. E. Parker, Nonlinear Connections and Sprays, in preparation.
[Sh] Zhongmin Shen, *Geometric Meanings of Curvatures in Finsler Geometry*, Winter School on Geometry and Physics at Srni in Czech Republic. (math.DG/0011136).

FACULTAD DE CIENCIAS
Current address: Universidad Autónoma de San Luis Potosí, 78290 San Luis Potosí, SLP, MEXICO
E-mail address: `lilia@galia.fc.uaslp.mx`

Contemporary Mathematics
Volume **288**, 2001

The Gysin Sequence for Riemannian Flows

José Ignacio Royo Prieto

ABSTRACT. Let $\mathcal{F}$ be a riemannian flow on a closed manifold M. The Gysin sequence relating the de Rham cohomology of M and the basic cohomology of $\mathcal{F}$ has been constructed for isometric flows in [**7**]. In this paper, we extend this result to riemannian flows. We also give a geometric characterization of the vanishing of the Euler class, similar to the one given in [**11**], and describe an example of a non-isometric riemannian flow with nonzero Euler class.

1. Introduction

Let M be a $(n+1)$-dimensional closed smooth manifold. A smooth action $\Psi : \mathbb{R} \times M \longrightarrow M$ without fixed points defines a 1-dimensional foliation $\mathcal{F}$. We are interested in the relation between the de Rham cohomology of M and the cohomology of the orbit space $M/\mathcal{F}$. Since this space may be somewhat pathological from the cohomological point of view, we shall use the *basic cohomology*, which has proven to be a rich and adapted invariant for the study of $M/\mathcal{F}$. When the action is periodic, we have indeed a circle action, and the mentioned relation is given by the classical *Gysin sequence*. In [**7**], the Gysin sequence has also been constructed for isometric flows, i.e., actions $\mathbb{R} \times (M, \mu) \longrightarrow (M, \mu)$ preserving a riemannian metric μ on M. We denote by $\Omega^*(M/\mathcal{F})$ the complex of *basic forms* and by $H^*(M/\mathcal{F})$ its cohomology. The Gysin sequence obtained is:

(1.1)
$$\cdots \to H^i(M/\mathcal{F}) \longrightarrow H^i(M) \longrightarrow H^{i-1}(M/\mathcal{F}) \xrightarrow{\wedge[d\chi]} H^{i+1}(M/\mathcal{F}) \to \cdots$$

where the connecting morphism is multiplication by the *Euler class* $[d\chi] \in H^2(M/\mathcal{F})$, being χ the characteristic form of the flow. The main tool used to get this is the quasi-isomorphism (i.e., isomorphism in cohomology):

$$(\Omega^*(M), d) \simeq (\Omega^*(M/\mathcal{F}) \oplus \Omega^{*-1}(M/\mathcal{F}), D), \tag{1.2}$$

where the differential D is given by $D(\alpha, \beta) = (d\alpha + d\chi \wedge \beta, -d\beta)$. In the proof, it is crucial that the closure of the 1-parameter group $\{\Psi_t\}_t$ in the group of diffeomorphisms of M is compact.

2000 *Mathematics Subject Classification.* Primary 58A12, 37C85; Secondary 57R30.
Key words and phrases. Basic cohomology, Riemannian flows.
The author wishes to thank Prof. M. Saralegi for his helpful discussions and suggestions. This research has been supported by Gobierno Vasco-Eusko Jaurlaritza and UPV 127.310 EA005/99.

In this paper we extend the scope of the Gysin sequence to the case of riemannian flows. A flow $\mathcal{F}$ on M is *riemannian* if there exists a riemannian metric μ which is *bundle-like*, i.e., if the orthogonal component of μ is invariant (note that in the isometric case, the entire metric must be invariant). There are two main differences with the isometric case. In one hand, $d\chi$ is basic for an isometric flow, but in the riemannian case, we have a decomposition $d\chi = \mathfrak{e} + \kappa \wedge \chi$, where $\mathfrak{e}$ and κ are, respectively, the *Euler form* and the *mean curvature form* of the flow. We can suppose these two forms to be basic for a suitable bundle-like metric μ (see [**5**]). On the other hand, the closure of $\{\Psi_t\}_t$ is no longer necessarily compact. So, we need a different approach, using strongly the local structure of the flow, to get a decomposition like (1.2). The quasi-isomorphism we get is:

$$(\Omega^*(M), d) \simeq (\Omega^*(M/\mathcal{F}) \oplus \Omega^{*-1}(M/\mathcal{F}), D), \tag{1.3}$$

with a new differential $D(\alpha, \beta) = (d\alpha + \mathfrak{e} \wedge \beta, -d\beta + \kappa \wedge \beta)$, where the new term $\kappa \wedge \beta$ appears, showing the difference with the isometric case. The Gysin sequence derived canonically is:

$$\cdots \to H^i(M/\mathcal{F}) \longrightarrow H^i(M) \longrightarrow H^{i-1}_{\kappa}(M/\mathcal{F}) \xrightarrow{\wedge[\mathfrak{e}]} H^{i+1}(M/\mathcal{F}) \to \cdots$$

where $H^*_{\kappa}(M/\mathcal{F})$ denotes the *twisted cohomology*, and the connecting morphism is, up to sign, multiplication by the *Euler class* $[\mathfrak{e}] \in H^2_{-\kappa}(M/\mathcal{F})$. We also give a geometric characterization of the vanishing of the Euler class similar to the one given in [**11**]. The vanishing of the Euler class is equivalent to the presence of a foliation transverse to the flow, with an additional condition, stated in Theorem 4.2. These results have been announced in [**9**]. Here, we also describe an example of a non-isometric riemannian flow with nonzero Euler class.

2. Riemannian flows

In this section, we review some well known facts about riemannian flows. Recall that a flow is a 1-dimensional oriented foliation. A flow $\mathcal{F}$ is *riemannian* if there exists a holonomy invariant riemannian metric μ. Such a metric is said to be *bundle-like*. We can choose a smooth nonsingular vector field X defining $\mathcal{F}$ such that $\mu(X, X) = 1$. We call the 1-form $\chi = i_X\mu$ the *characteristic form* of the flow, where i_X denotes the contraction by X. The *mean curvature form* of $\mathcal{F}$ is defined by $\kappa = L_X\chi$, where L_X is the Lie derivative respect to X. These two forms depend on the flow $\mathcal{F}$ and the metric μ. A form $\omega \in \Omega^*(M)$ is *basic* if for every vector field V tangent to $\mathcal{F}$, we have $i_V\omega = i_V d\omega = 0$. We denote the complex of basic forms as $\Omega^*(M/\mathcal{F})$, and its cohomology by $H^*(M/\mathcal{F})$.

It has been shown in [**5**] that we can choose a bundle-like metric μ such that κ is basic. In this conditions, κ is closed, and defines a class $[\kappa] \in H^1(M/\mathcal{F})$, which does not depend on the metric μ (see [**1**]), and that vanishes if and only if the flow is isometric. We shall call this invariant of $\mathcal{F}$ the *Álvarez class*. We have the decomposition:

$$d\chi = \mathfrak{e} + \chi \wedge \kappa. \tag{2.1}$$

The 2-form $\mathfrak{e}$ thus defined is called the *Euler form*, and satisfies $d\mathfrak{e} = -\kappa \wedge \mathfrak{e}$. We denote by $H^*_{\kappa}(M/\mathcal{F})$ the *twisted cohomology*, which is the dual of the basic cohomology (see [**8**]), i.e. $H^i(M/\mathcal{F}) \cong H^{n-i}_{\kappa}(M/\mathcal{F})$ for every $i \geq 0$. The twisted cohomology can be described as the cohomology of the basic complex endowed with the *κ-twisted differential* $d_\kappa\omega = d\omega - \kappa \wedge \omega$. We also define the *$(-\kappa)$-twisted differential* $d_{-\kappa}\omega = d\omega + \kappa \wedge \omega$.

The local structure of a riemannian flow has been described by Carrière in [**4**]. Let L be a leaf of the flow. Then, the closure of L is diffeomorphic to a torus $\mathbb{T}^{k+1}$, and there exists a neighbourhood U of $\overline{L}$ such that $\mathcal{F}$ restricted to U is conjugated to the flow obtained by suspension of a diffeomorphism (T, R) of $\mathbb{T}^k \times \mathbb{D}^{n-k}$, where T is an irrational traslation of $\mathbb{T}^k$ and R a rotation of $\mathbb{R}^{n-k}$. We shall refer to such a neighbourhood $U \cong \mathbb{T}^{k+1} \times \mathbb{D}^{n-k}$ as a *Carrière neighbourhood*. If we consider the metric induced in U by the flat metric of $\mathbb{T}^{k+1} \times \mathbb{D}^{n-k}$, both the mean curvature form and the Euler form vanish. Hence, the flow restricted to U is isometric.

3. Gysin sequence

The *short Gysin sequence* is the following sequence of differential complexes:

$$0 \longrightarrow \Omega^*(M/\mathcal{F}) \xrightarrow{\ \iota\ } \Omega^*(M) \xrightarrow{\ \rho\ } \Omega^*(M)/\Omega^*(M/\mathcal{F}) \longrightarrow 0, \tag{3.1}$$

where ρ is the proyection induced by the inclusion ι. The cohomology of the complex $\Omega^*(M)/\Omega^*(M/\mathcal{F})$ is called the *$\mathcal{F}$-relative de Rham cohomology*, and some of its properties have been studied in [**10**]. Our aim is to describe it in terms of basic forms. To achieve this, we use the following:

PROPOSITION 3.1. *In the above conditions, we have the quasi-isomorphism:*

$$\begin{array}{rcl} \Phi : (\Omega^*(M/\mathcal{F}) \oplus \Omega^{*-1}(M/\mathcal{F}), D) & \longrightarrow & (\Omega^*(M), d) \\ (\alpha, \beta) & \longmapsto & \alpha + \chi \wedge \beta \end{array} \tag{3.2}$$

where the differential D is defined by $D(\alpha, \beta) = (d\alpha + \mathfrak{e} \wedge \beta, -d\beta + \kappa \wedge \beta)$.

PROOF. We can consider the basis of the stratified manifold $M/\overline{\mathcal{F}}$ consisting of projections of Carrière neighbourhoods. Applying a Mayer-Vietoris argument (see [**3**], p.289) to $M/\overline{\mathcal{F}}$, it suffices to prove the result for a Carrière neighbourhood U, and this holds due to the isometric case. □

From 3.2 and 3.1, we obtain that the $\mathcal{F}$-relative cohomology and the twisted cohomology are quasi-isomorphic. This leads us to the *long Gysin sequence*:

THEOREM 3.2 (Gysin sequence). *Let $\mathcal{F}$ be a riemannian foliation on the closed manifold M, and choose a metric μ with basic mean curvature form. Then, we have the following long exact sequence:*

$$\cdots \to H^i(M/\mathcal{F}) \longrightarrow H^i(M) \longrightarrow H^{i-1}_\kappa(M/\mathcal{F}) \xrightarrow{\ \wedge[\mathfrak{e}]\ } H^{i+1}(M/\mathcal{F}) \to \cdots \tag{3.3}$$

where the connecting morphism is, up to sign, multiplication by the Euler class $[\mathfrak{e}] \in H^2_{-\kappa}(M/\mathcal{F})$.

In the case of an isometric flow, we can choose a metric such that the mean curvature form vanishes. So, (3.3) generalizes (1.1).

REMARK 3.3. The long Gysin sequence does not depend on the metric μ.

REMARK 3.4. From the quasi-isomorphism (3.2) and the duality result of [**8**], we obtain that the terms $E_2^{\cdot,1}$ and $E_2^{\cdot,0}$ in the spectral sequence of a riemannian flow are dual. This has been proven with more generality in [**2**] using analytic techniques, different from our geometric approach.

4. Vanishing of the Euler class

The Euler form $\mathfrak{e}$ vanishes if and only if the orthogonal complement of $\mathcal{F}$ is involutive. These equivalent conditions have been weakened in [**11**] for isometric flows. In that case, the Euler class $[d\chi] \in H^2(M/\mathcal{F})$ is an invariant of the foliation, and its vanishing is equivalent to the presence of a fibration transverse to $\mathcal{F}$. For a riemannian flow, we prove that the Euler class $[\mathfrak{e}] \in H^2_{-\kappa}(M/\mathcal{F})$ does not depend on the metric in the sense of the following:

PROPOSITION 4.1. *Let μ_1 and μ_2 be two bundle-like metrics with basic mean curvature forms κ_1 and κ_2. Consider the canonical (up to a multiplicative positive constant) isomorphism:*

$$\varphi : H^2_{-\kappa_2}(M/\mathcal{F}) \longrightarrow H^2_{-\kappa_1}(M/\mathcal{F}) \tag{4.1}$$

given by $\varphi([\omega]) = [e^f \omega]$, where $df = \kappa_1 - \kappa_2$. Then, $\varphi([\mathfrak{e}_2])$ and $[\mathfrak{e}_1]$ are proportional. In particular, the vanishing of the Euler class does not depend on μ, but just on $\mathcal{F}$.

PROOF. Proceeding as in Proposition 3.1, we get the *twisted Gysin sequence:*

$$\longrightarrow H^i_{-\kappa}(M/\mathcal{F}) \longrightarrow H^i_{-\kappa}(M) \longrightarrow H^{i-1}(M/\mathcal{F}) \xrightarrow{\wedge[\mathfrak{e}]} H^{i+1}_{-\kappa}(M/\mathcal{F}) \longrightarrow$$

for every basic κ. The isomorphism (4.1) induces a chain morphism between the twisted Gysin sequences for κ_1 and κ_2, yielding the result. □

Let $\mathcal{G}$ be a foliation transverse to a nonsingular vector field X defining $\mathcal{F}$. There is a unique 1-form ω determined by $\ker \omega = T\mathcal{G}$ and $i_X\omega = 1$. We mean by a *torsion* of $\mathcal{G}$ respect to X a 1-form τ such that $d\omega = \omega \wedge \tau$. Note that if a torsion is basic, for $\mathcal{F}$, then it must be $L_X\omega$.

THEOREM 4.2. *For a riemannian flow, the following conditions are equivalent:*

(1) *the Euler class vanishes;*
(2) *there exists a foliation $\mathcal{G}$ transverse to $\mathcal{F}$, with basic torsion.*

PROOF. Assume (1) and take $\gamma \in \Omega^1(M/\mathcal{F})$ such that $d_{-\kappa}\gamma = \mathfrak{e}$. Then, $\omega = \chi - \gamma$ is integrable, defining a foliation $\mathcal{G}$ that satisfies (2). For the converse, pick a bundle-like metric μ and construct $\nu = \omega \otimes \omega + \mu_{\mathcal{G}}$. Then, ν is bundle-like, κ_ν is basic, and $\mathfrak{e}_\nu = 0$. □

REMARK 4.3. The vanishing of the Euler class of $\mathcal{F}$ implies the vanishing of the Godbillon-Vey class of the transverse foliation $\mathcal{G}$ described in Theorem 4.2.

5. An example

The vanishing of the Euler class and the Álvarez class are independent. A trivial bundle and the Hopf fibration are immediate examples of isometric flows with zero and nonzero Euler class, respectively. In [**4**], Carrière exhibits a non-isometric riemannian flow whose Euler class is zero. In this section, we describe a riemannian flow where neither the Álvarez class or the Euler class vanish.

Consider the matrix $B = \begin{pmatrix} 2 & 1 \\ 1 & 1 \end{pmatrix} \in SL(2,\mathbb{Z})$, and its two real eigenvalues λ_1 and λ_2, which satisfy $0 < \lambda_1 < 1 < \lambda_2$ and $\lambda_1\lambda_2 = 1$. Denote by $\{(a_1, b_1), (a_2, b_2)\}$ an orthonormal basis of eigenvectors. The matrixes of $SL(4,\mathbb{Z})$ given by:

$$A = \begin{pmatrix} B & 0 \\ 0 & B \end{pmatrix} \quad \text{and} \quad I = \begin{pmatrix} Id_2 & Id_2 \\ 0 & Id_2 \end{pmatrix},$$

define automorphisms of the torus $\mathbb{T}^4$. We define M^6 as the suspension of the representation $\rho : \mathbb{Z}^2 \longrightarrow \mathrm{Diff}(\mathbb{T}^4)$ given by $\rho(k,l) = A^k \circ I^l$. More precisely, M^6 is the orbit space of the action $\Upsilon : \mathbb{Z}^2 \times (\mathbb{T}^4 \times \mathbb{R}^2) \longrightarrow (\mathbb{T}^4 \times \mathbb{R}^2)$, given by:

$$\Upsilon((k,l),[y_1,y_2,z_1,z_2],(t,x)) = (\rho(k,l)([y_1,y_2,z_1,z_2]),(x+k,t+l)).$$

Notice that there is a fibration $\pi : M^6 \longrightarrow \mathbb{T}^2$ with fiber $\mathbb{T}^4$. This manifold was introduced in another context in [**6**], where some of its properties are studied. A parallelization of M^6 is given by:

$$\text{(5.1)}\quad \begin{aligned} &X = \frac{\partial}{\partial t}, \quad T = \frac{\partial}{\partial x}, \quad Y_i = \lambda_i^t(a_i\frac{\partial}{\partial y_1} + b_i\frac{\partial}{\partial y_2}), \\ &Z_i = x\lambda_i^t(a_i\frac{\partial}{\partial y_1} + b_i\frac{\partial}{\partial y_2}) + \lambda_i^t(a_i\frac{\partial}{\partial z_1} + b_i\frac{\partial}{\partial z_2}). \end{aligned} \qquad (i=1,2)$$

Denote by $\{\alpha,\beta,\gamma_1,\gamma_2,\delta_1,\delta_2\}$ the dual basis of 1-forms. Let $\mathcal{F}$ be the flow defined by Y_1, and μ the metric such that (5.1) is an orthonormal parallelism. One can check that $L_{Y_1}\mu_Q = 0$, and so, μ is bundle-like. From (5.1), we obtain for $\mathcal{F}$:

$$\text{(5.2)}\qquad \chi = \gamma_1, \quad \kappa = (\log\lambda_1)\alpha, \quad \mathfrak{e} = -\beta\wedge\delta_1.$$

Notice that both κ and $\mathfrak{e}$ are basic.

PROPOSITION 5.1. *The Álvarez class of $\mathcal{F}$ is not zero.*

PROOF. Since $\iota^* : H^1(M^6/\mathcal{F}) \longrightarrow H^1(M^6)$ is injective, it suffices to prove that $[\alpha] \in H^1(M^6)$ is not zero. This holds because $[\alpha] = \pi^*([\zeta])$, where ζ is one of the generators of $H^1(\mathbb{T}^2)$. □

PROPOSITION 5.2. *The Euler class of $\mathcal{F}$ is not zero.*

PROOF. The cohomology of $(\Omega^*(M^6/\mathcal{F}), d_{-\kappa})$ can be computed using π-basic functions as coefficients. Suppose that there exists a form $\omega \in \Omega^1(M^6/\mathcal{F})$ such that $d_{-\kappa}\omega = \mathfrak{e}$. Using (5.2) we get the equations $X(f) = 1$ and $T(f) = 0$ on $\mathbb{T}^2$, being $f = i_{Z_1}\omega$. This is a contradiction. □

References

[1] J.A. Álvarez, *The basic component of the mean curvature form of riemannian foliations*, Annals of Global Analysis and Geometry **10** (1992), 179–194.

[2] ———, *Duality in the spectral sequence of riemannian foliations*, Amer. J. Math. **111** (1989), 905–925.

[3] G.E. Bredon, *Topology and Geometry*, Springer Verlag (1993).

[4] Y. Carrière, *Flots riemanniens*, Astérisque **116** (1984), 31–52

[5] D. Domínguez, *Finiteness and tautness for riemannian foliations*, Amer. J. Math. **120** (1998), 1237–1276.

[6] M. Fernández, M. de León, M. Saralegi, *A six dimensional compact symplectic solvmanifold without Kähler structures*, Osaka J. Math. **33** (1996), 19–35.

[7] F. Kamber, P. Tondeur, *Duality theorems for foliations*, Astérisque **116** (1984), 458–471.

[8] ———, *Foliations and Metrics*, Birkhäusser Progr. Math. **32** (1983), 103–152.

[9] J.I. Royo Prieto, *The Euler class for riemannian flows*, C.R.Acad.Sci.Paris **332** (2001), 45–50.

[10] K. Rummler, *Quelques notions simples en géometrie riemannienne et leurs applications aux feuilletages compacts*, Comment. Math. Helv. **54** (1979), 224–239.

[11] M. Saralegi, *The Euler class for isometric flows*, Pitman Research Notes in Math. **131** (1985), 220-227.

DEPARTAMENTO DE MATEMÁTICAS, UNIVERSIDAD DEL PAÍS VASCO-EUSKAL HERRIKO UNIBERTSITATEA, APARTADO 644, 48080 BILBAO, SPAIN

E-mail address: mtbroprj@lg.ehu.es

Contemporary Mathematics
Volume **288**, 2001

On Contact Groupoids and Legendre Bisections

Tomasz Rybicki

Abstract. By means of the convenient setting in global analysis it is observed that the group of global bisections of a Lie groupoid is a Lie group modeled on the space of sections of the associated algebroid. The Lie algebroid corresponding to a contact groupoid is identified with the algebroid of the Jacobi structure on the space of units. This enables to introduce a Lie group structure on the group of Legendre bisections of a contact groupoid. Some conclusions concerning the integrability of Poisson algebra are possible.

1. Introduction

Let $\Gamma = (\Gamma \rightrightarrows M, \alpha, \beta, m, i)$ be a Lie groupoid, where α (resp. β) is a source (resp. target) map, m is a multiplication, and i is an inversion. Recall that Γ is a smooth manifold, the space of units M is a Hausdorff submanifold, α, β are surmersions of Γ onto M, and m, i are smooth. The domain of m is $\Gamma_2 = \{(x,y) \in \Gamma : \alpha(x) = \beta(y)\}$.

It is well known [**2**] that for any Lie groupoid Γ the tangent space $T\Gamma = (T\Gamma \rightrightarrows TM, T\alpha, T\beta, \oplus, I)$ carries a structure of Lie groupoid, too. Here the multiplication $\oplus$ is given by

$$X \oplus Y = (\frac{\mathrm{d}}{\mathrm{d}\,t}(x(t).y(t))|_{t=0},$$

where $X = \frac{\mathrm{d}\,x}{\mathrm{d}\,t}|_{t=0}$, $Y = \frac{\mathrm{d}\,y}{\mathrm{d}\,t}|_{t=0}$, $\alpha(x(t)) = \beta(y(t))$, and the inversion $IX = \frac{\mathrm{d}\,x}{\mathrm{d}\,t}x(t)^{-1}|_{t=0}$ if $X = \frac{\mathrm{d}\,x}{\mathrm{d}\,t}|_{t=0}$.

Definition. ([**9**], [**5**]) A Lie groupoid Γ endowed with a contact form θ (i.e. $\theta \wedge \mathrm{d}\,\theta^n \neq 0$, where $\dim \Gamma = 2n+1$) is called a *contact groupoid* if there are smooth mappings $u, v : \Gamma \to \mathbb{R}_*$ (actually groupoid morphisms) such that the following hold:

(i) $\theta(X_x \oplus Y_y) = u(x)\theta(Y_y) + v(y)\theta(X_x)$;
(ii) the inversion i is a conformal θ-morphism.

2000 *Mathematics Subject Classification.* Primary 22E65; Secondary 53D10.

Key words and phrases. Regular Lie group, convenient settimg, Jacobi manifold, contact groupoid, Legendre bisection, Poisson algebra, integrability.

Supported by the AGH grant n.11.420.04

Basic facts concerning contact groupoids have been established by P. Libermann in [**11**], [**12**]. Contact groupoids are essentially related to Jacobi manifolds, cf. sect.2.

In this note we appeal to the convenient setting of the infinite dimensional Lie theory [**10**] due to A. Kriegl and P. Michor. The convenient setting is based on Boman's theorem which states that a mapping $f : \mathbb{R}^n \to \mathbb{R}$ is smooth whenever $f \circ c$ is smooth for any smooth curve $c : \mathbb{R} \to \mathbb{R}$. In [**10**] a great progress has been made in discription of the Lie group structure on $\mathrm{Diff}(M)$. Our aim is to extend [**10**] to more general context, namely for $\mathrm{Bis}(\Gamma)$, the group of all bisections of a Lie groupoid Γ. Recall that a submanifold $S \subset \Gamma$ belongs to $\mathrm{Bis}(\Gamma)$ iff both restrictions $\alpha|S, \beta|S$ are diffeomorphisms onto M.

THEOREM 1. *The group* $\mathrm{Bis}(\Gamma)$ *admits a regular Lie group structure modelled on the space of sections of the associated Lie algebroid* $\mathcal{A}(\Gamma)$.

We remind that a convenient Lie group G is called *regular* if for $\mathfrak{g} = T_e G$ there exists a bijective evolution map

$$\mathrm{evol}^r : C^\infty(\mathbb{R}, \mathfrak{g}) \to C^\infty((\mathbb{R}, 0), (G, e)),$$

and its evaluation at 1 is smooth. In the next section we indicate that Theorem 1 can be generalized to the group of Legendre bisections of a contact groupoid.

Finally we wish to notice that there are some "weak" settings of the infinite dimensional Lie theory (which usually do not correspond to each other). One example is the notion of diffeological groups due to J. M. Souriau [**20**] (see also [**4**]); another is the concept of generalized Lie groups of H. Omori [**14**]. The advantage of these settings is that Lie group structures are defined without using charts. However not all concepts of the "usual" infinite dimensional Lie theory could be possible in such frameworks.

In terms of diffeological Lie groups the group of Legendre bisections has been studied extensively by P. Dazord in [**5**]. We would like here to indicate that his results can be derived on the ground of usual Lie theory as well. For the case of symplectic groupoids, see [**18**].

2. The group of Legendre bisections

Recall that $S \subset M$, where (M, θ) is a contact manifold, is called a Legendre submanifold if it is a maximal integral submanifold of the symplectic distribution $\ker(\theta)$.

A *Jacobi structure* on a manifold M is a pair (Λ, E), where Λ is a 2-vector field on M, E is a vector field on M, and the equalities

$$[\Lambda, \Lambda] = 2E \wedge \Lambda, \quad [E, \Lambda] = 0$$

are satisfied, where $[.,.]$ is the Schouten-Nijenhuis bracket. Specifically transitive odd-dimensional Jacobi manifolds coincide with contact manifolds. It is well known that Jacobi strucures corresponds bijectively to Lie algebra brackets on $C^\infty(M)$ which are local, i.e.

$$\mathrm{supp}\{u, v\} \subset (\mathrm{supp}(u) \cap \mathrm{supp}(v)), \quad u, v \in C^\infty(M).$$

If $E = 0$ then the bracket $\{.,.\}$ becomes a derivation on each argument and defines a *Poisson structure* on M.

PROPOSITION. *Let $\Gamma = (\Gamma, \theta, u, v)$ be a contact groupoid over M. Then M is a Legendre submanifold. Furthermore, M is canonically equipped with a Jacobi structure (Λ, E) such that α (resp. β) is a conformal Jacobi u-morphism (resp. $(-v)$-morphism). We will denote Γ by $(\Gamma, \theta) \rightrightarrows (M, \Lambda, E)$.*

Given two contact manifolds (M_i, θ_i), $i = 1, 2$, one defines a contact form θ on $M_1 \times M_2 \times \mathbb{R}_*$ by $\theta = \pi_1^*\theta_1 - t\pi_2^*\theta_2$, where π_i is the canonical projection on M_i. Now in view of [**8**] $f : M_1 \to M_2$ is a contact ϕ-diffeomorphism, $\phi \in C^\infty(M_1)$, if and only

$$\operatorname{graph}(f) := \{(x, f(x), \phi(x))\,|x \in M_1\}$$

is a Legendre submanifold of $M_1 \times M_2 \times \mathbb{R}_*$.

EXAMPLE. If (M, θ) is a contact manifold then $\Gamma = M \times M \times \mathbb{R}_*$ with $\hat{\theta} = \pi_1^*\theta - t\pi_2^*\theta$ is the *coarse contact groupoid*. Here the multiplication and the inversion are given by

$$(x_1, y_1, a_1).(x_2, y_2, a_2) = (x_1, y_2, a_1a_2) \quad \text{iff} \quad y_1 = x_2,$$
$$(x, y, a)^{-1} = (y, x, a^{-1}).$$

So Legendre bisections generalize contactomorphisms.

For a contact groupoid (Γ, θ) the symbol $\operatorname{Bis}(\Gamma, \theta)$ will stand for the set of all Legendre bisections. Then there is the left representation in the unit space (M, Λ, E)

$$\phi^l : \operatorname{Bis}(\Gamma, \theta) \ni S \to \phi^l_S := \beta \circ \Phi^l_S|_M \in \operatorname{Diff}(M, \Lambda, E),$$

where $\Phi^l_S : x \to S.x$, and the analogous right one. One interesting problem is whether any "Jacobi flow" can be locally lifted to "Legendre bisection flow". For the Poisson case, see [**21**].

It is fundamental that to any Lie groupoid $\Gamma \rightrightarrows M$ is assigned the associated algebroid $\mathcal{A}(\Gamma)$, namely $\mathcal{A}(\Gamma) = (\ker T\beta, [[,]], T\alpha)$ where $[[,]]$ is a Lie algebra bracket on $\operatorname{Sect}(\ker T\beta)$ introduced by means of left invariant vector fields and of the identification $T\Gamma|_M/TM \simeq \ker T\beta$. In our case it takes a special form.

THEOREM 2. [**9**] *(i) If (M, Λ, E) is a Jacobi manifold the jet bundle $J^1(M, \mathbb{R}) \to M$ is endowed with an algebroid structure such that the jet mapping*

$$C^\infty(M) \to C^\infty(M \leftarrow J^1(M, \mathbb{R}))$$

is a Lie algebra monomorphism.

(ii) The algebroid associated with a contact groupoid $(\Gamma, \theta) \rightrightarrows (M, \Lambda, E)$ is canonically isomorphic to the algebroid associated with (M, Λ, E).

Observe that $J^1(M, \mathbb{R}) \simeq T^*M \times \mathbb{R}$ is endowed with the canonical contact structure $\theta_M - \mathrm{d}t$, where θ_M is the canonical 1-form on T^*M. The following Lychagin's lemma establishes the existence of canonical neighborhoods of a Legendre submanifold.

LEMMA. [**13**] *If S is a Legendre submanifold of a contact manifold (M, θ) then there exist an open neighborhood U of S, an open neighborhood V of the zero section in $T^*S \times \mathbb{R}$, and a diffeomorphism $\phi : U \to V$ such that $\phi|_S = \mathrm{id}_S$ and $\phi^*(\theta_S - \mathrm{d}t) = \theta$.*

The proof of our main result appeals to the convenient setting of the infinite dimensional Lie theory and, specifically, to Theorem 1, as well as to the above facts concerning contact groupoids. Here we set $\mathrm{Bis}(\Gamma,\theta)_c = (\phi^l)^{-1}(\mathrm{Diff}_c(M,\Lambda,E))$, the subgroup of all *compactly controlled* elements.

THEOREM 3. *Given a contact groupoid* $(\Gamma,\theta) \rightrightarrows (M,\Lambda,E)$, *the groups* $\mathrm{Bis}(\Gamma,\theta)$ *and* $\mathrm{Bis}(\Gamma,\theta)_c$ *are regular Lie groups, both with the Lie algebra* $C_c^\infty(M)$.

The chart at $e = M$ is built by means of the monomorphism and the isomorphism mentioned in Theorem 2.

REMARK. One can consider as well contact groupoids in the wider sense, that is given by a distribution of hyperplanes $\mathcal{H}$. A Lie groupoid Γ equipped with $\mathcal{H}$ is contact if

(i) $X, Y \in \mathcal{H} \Rightarrow X \oplus Y \in \mathcal{H}$,

(ii) i preserves $\mathcal{H}$.

Then analogous statements remain true but the space of units is now endowed with a structure of Jacobi bundle. The group $\mathrm{Bis}(\Gamma,\theta)$ still admits a regular Lie group structure.

3. Integrability of the Poisson algebra

The third theorem of Lie asserts that any finite dimensional Lie algebra is actually the Lie algebra of a Lie group. Since a famous paper [**6**] it is well known that, in general, this theorem is no longer true in the infinite dimensional case. However there are several generalizations of this theorem, cf.[**1**], [**3**], [**7**], [**15**], [**16**], [**17**]. Recently very general approach to the problem has been proposed in [**19**].

Let us recall that a Poisson manifold (M,Λ) is *prequantizable* if there exists a (global) symplectic groupoid $(\Gamma_0,\omega) \rightrightarrows (M,\Lambda)$ such that (Γ_0,ω) is prequatizable. If Γ is separated, the latter means that $[\omega] \in H^2(\Gamma_0,\mathbb{Z})$. Then there is a contact groupoid $(\Gamma,\theta) \rightrightarrows (M,\Lambda)$, called a natural contact groupoid over a prequantizable Poisson manifold (M,Λ), cf. [**21**]. In [**5**] P. Dazord gave conditions characterizing prequantizable Poisson manifols.

If (M,Λ) is prequantizable, the exact sequence of Lie algebras

$$0 \to \mathbb{R} \to C_c^\infty(M) \xrightarrow{\mathrm{d}} B\Omega_c^1(M) \to 0,$$

where $B\Omega_c^1(M)$ is the Lie algebra of exact 1-forms with compact support on M, can be integrated to the exact sequence of regular Lie groups

$$1 \to S^1 \to \mathrm{Bis}(\Gamma,\theta) \to \mathrm{Bis}^{exact}(\Gamma_0,\omega) \to 1,$$

where $\mathrm{Bis}^{exact}(\Gamma_0,\omega)$ is the group of all exact Lagrangian bisections of the symplectic groupoid (Γ_0,ω), cf. [**2**]. Therefore Theorem 3 can be reformulated as follows (see also [**5**])

THEOREM 4. *For any prequantizable Poisson manifold* (M,Λ) *its compactly supported Poisson algebra can be integrated to a regular Lie group* $\mathrm{Bis}(\Gamma,\theta)$, *where* $(\Gamma,\theta) \rightrightarrows (M,\Lambda)$ *is the natural contact groupoid over* (M,Λ).

Notice that a main result of [**1**] states that the Poisson algebra of any symplectic manifold is integrable, and in [**7**] this was extended for locally conformal symplectic structures.

References

1. A. Banyaga, P. Donato, *Some remarks on the integration of the Poisson algebra*, J. Geom. Phys. **19** (1996), 368-378.
2. A. Coste, P. Dazord, A. Weinstein, *Groupoides symplectiques*, Publ. Dpt. Mathématiques, Univ. C. Bernard - Lyon 1 **2/A** (1987), 1-62.
3. P. Dazord, *Groupoides symplectiques et troisième théorème de Lie "non-lineaire"*, Géométrie Symplectiques et Mécanique, Springer Lect. Notes in Math. **1416**, 39-74.
4. P. Dazord, *Lie groups and algebras in infinite dimension: A new approach*, Contemporary Math. **179** (1994), 17-44.
5. P. Dazord, *Sur l'intégration des algèbres de Lie locales et la préquantification*, Bull. Sci. math. **121** (1997), 423-462.
6. W. T. van Est, T. J. Korthagen, *Non enlargible Lie algebra*, Indag. Math. **26** (1964), 15-31.
7. S. Haller, T. Rybicki, *Integrability of the Poisson algebra on a locally conformal symplectic manifold*, Rend. Circ. Mat. Palermo, suppl. **63** (2000), 89-96.
8. R. Ibáñez, M. de León, J. C. Marrero, D. Martin de Diego, *Coisotropic and Legendre-Lagrangian submanifolds and conformal Jacobi morphisms*, J. Phys. A **30** (1997), 5427–5444.
9. Y. Kerbrat, Z. Souici-Benhammadi, *Variétés de Jacobi et groupoides de contact*, C.R. Acad. Sc. Paris **317** (1993), 81-86.
10. A. Kriegl, P. W. Michor, *The Convenient Setting of Global Analysis*, American Mathematical Society, 1997.
11. P. Libermann, *On contact and symplectic groupoids*, Proc. of the conference "Diff. Geom. and Appl.", Opava, 1992 (O. Kowalski and D. Krupka, eds), Silesian Univ. in Opava (1993), 29-45.
12. P. Libermann, *On contact groupoids and their symplectification*, Analysis and Geometry on Foliated Manifolds, Proc. Intern. Colloq. on Diff. Geom., Santiago de Compostela, 1994, World Sci., Singapore, 1995.
13. V. V. Lychagin, *On sufficient orbits of a group of contact diffeomorphisms*, Math. USSR Sb. **33** (1977), 223-242.
14. H. Omori, *Infinite-Dimensional Lie Groups*, Transl. of Math. Monographs A.M.S., 1997.
15. J. Pradines, *Troisième théorème de Lie sur les groupoides différentiables*, C.R. Acad. Sc. Paris **267** (1968), 21-23.
16. T. Ratiu, R. Schmid, *The differentiable structure of three remarkable diffeomorphism groups*, Math. Z. **177** (1981), 81-100.
17. T. Rybicki, *On foliated, Poisson and Hamiltonian diffeomorphisms*, Diff. Geom. Appl. (2001).
18. T. Rybicki, *On the group of Lagrangian bisections of a symplectic groupoid*, Banach Center Publ. (in press).
19. T. Rybicki, *An infinite dimensional version of the third Lie theorem*, Rend. Circ. Mat. Palermo, suppl. (in press).
20. J. M. Souriau, *Groupes différentiels de physique mathématique*, South Rhone seminar on geometry, II, Travaux en Cours, Hermann, Paris (1984), 73-119.
21. A. Weinstein, P. Xu, *Extensions of symplectic groupoids and quantization*, J. Reine Angew. Math. **417** (1991), 159-189.

Dpt. of Appl. Mathematics at AGH, al. Mickiewicza 30, 30-059 Kraków, Poland
E-mail address: tomasz@uci.agh.edu.pl

Contemporary Mathematics
Volume **288**, 2001

Affine Maximal Tori Intersecting a Fixed One

Marcos Salvai

Abstract. H. Gluck and F. Warner characterized the oriented great circle fibrations of the three-sphere. In a previous paper we generalized partially their result, obtaining, for any compact connected semisimple Lie group G, infinite dimensional spaces of concrete examples of fibrations of G by Weyl-oriented affine maximal tori. In this note, we study the space of such tori intersecting a fixed one, with the hope that it could help characterize those fibrations.

Let G be a compact connected semisimple Lie group. A maximal torus in G is a maximal abelian Lie subgroup of G. A subset S of G is an affine maximal torus if there exist $g, h \in G$ such that gSh^{-1} is a maximal torus of G. Equivalently, it is a maximal connected totally geodesic flat submanifold of G, provided that the group is endowed with a bi-invariant Riemannian metric.

H. Gluck and F. Warner [**1**] characterized the oriented great circle fibrations of the three-sphere. In [**3**] we generalized partially their result, obtaining, for any compact connected semisimple Lie group G, infinite dimensional spaces of concrete examples of fibrations of G by Weyl-oriented affine maximal tori (in our setting, the convenient generalization of an oriented great circle is a Weyl-oriented affine maximal torus, see below).

A tangent vector to G is said to be regular if it is tangent to a unique affine maximal torus. Let $\mathcal{R}$ denote the set of regular tangent vectors. An affine Weyl chamber is a connected component of $\mathcal{R} \cap T_pS$, where S is an affine maximal torus and $p \in S$. Given an affine Weyl chamber C, there exists a unique affine maximal torus S such that C is contained in TS.

Fix a maximal torus T. A Weyl-oriented (briefly, W-oriented) affine maximal torus is a pair $(gT, hT) \in (G/T) \times (G/T)$. In [**3**], it was actually defined independently of the choice of T, as a pair (S, ρ), where S is an affine maximal torus of G and ρ is a continuous section of affine Weyl chambers tangent to S. For some fixed Weyl chamber $C_0 \subset T_eT$, (gT, hT) corresponds to the torus $S = gTh^{-1}$ equipped with the affine Weyl chamber section $\rho\left(guh^{-1}\right) = dL_{gu}dR_{h^{-1}}C_o$ $(u \in T)$.

Let $\mathcal{T} \cong (G/T) \times (G/T)$ denote the set of all W-oriented affine maximal tori of G. By convention, the intersection of two W-oriented tori in $\mathcal{T}$ is the intersection

---

2000 *Mathematics Subject Classification.* Primary 22E15.

Key words and phrases. compact Lie group, affine maximal torus .

Partially supported by CONICOR, CONICET, CIEM (CONICET) and SECYT (UNC).

of the underlying tori. In [**3**], we found sufficient conditions for a subset $\mathcal{F}$ of $\mathcal{T}$ to be the space of fibers of a fibration of G by W-oriented affine maximal tori. Obviously, two different tori in $\mathcal{F}$ must have empty intersection. In this note, we study the space of such tori intersecting a fixed one, with the hope that it could help characterize those fibrations.

Let $T_o = (T, T) \in \mathcal{T}$ and $\mathcal{I} = \{S \in \mathcal{T} \mid S \cap T_o \neq \emptyset\}$. Let $\mathfrak{g}, \mathfrak{t}$ be the Lie algebras of G, T, respectively, and let $\mathfrak{m}$ be the orthogonal complement of $\mathfrak{t}$ in $\mathfrak{g}$ (with respect to the opposite of the Killing form of $\mathfrak{g}$), which may be identified in a natural way with $T_T(G/T)$. Let $\mathcal{J} = \{(g, ug) \in G \times G \mid g \in G, u \in T\}$ and $P : G \times G \to \mathcal{T}$, $P(g, h) = (gT, hT)$ be the canonical projection.

THEOREM 1. (a) *$\mathcal{J}$ is a submanifold of $G \times G$ which projects to $\mathcal{I}$ under P.*

(b) *T_o is a non-manifold point of $\mathcal{I}$ and*

$$T_{T_o}\mathcal{I} = \{(X, \mathrm{Ad}\,(u)\,X) \mid X \in \mathfrak{m}, u \in T\},$$

which is a full cone in $\mathfrak{m} \times \mathfrak{m}$.

(c) *The set $\{(gT, ugT) \in \mathcal{T} \mid \mathfrak{t} \cap \mathrm{Ad}\,(g)\,\mathfrak{t} = \{0\}\}$ is a submanifold of $\mathcal{T}$ which is open and dense in $\mathcal{I}$.*

PROOF. a) $\mathcal{J}$ is a submanifold of $G \times G$ since the function $F : T \times G \to G \times G$ defined by $F(u, g) = (g, ug)$ is a bijection onto $\mathcal{J}$ and

$$dF_{(u,g)}(Z, X) = (X, dL_u X + dR_g Z) = 0$$

only if $X = Z = 0$ (L_k, R_k denote left and right multiplication by k, respectively). Moreover, if $T \cap gTh^{-1} \neq \emptyset$, there exist $u, v \in T$ such that $u^{-1} = gvh^{-1}$, hence $hT = ugT$. Conversely, $u^{-1} \in gT\,(ug)^{-1}$. Therefore, $\mathcal{I} = P(\mathcal{J})$.

b) $T_{T_o}\mathcal{I}$ is the set of velocities of smooth curves γ in $\mathcal{I}$ with $\gamma(0) = T_o$. Clearly, $t \mapsto (\exp(tX)\,T, u\exp(tX)\,T)$ is a curve in $\mathcal{I}$ with velocity $(X, \mathrm{Ad}\,(u)\,X)$, for all $X \in \mathfrak{m}$, $u \in T$. Conversely, let c be a curve in G/T with $c(0) = T$ and u a curve in T. Taking the horizontal lift of c through e, we may write $c(t) = g(t)\,T$ with $g(0) = e$ and $\dot{g}(0) = X \in \mathfrak{m}$. Hence,

$$\begin{aligned} \left.\frac{d}{dt}\right|_0 (g(t)\,T, u(t)\,g(t)\,T) &= \left(X, d\pi\left(dL_{u(0)}X + dR_{g(0)}\dot{u}(0)\right)\right) \\ &= (X, \mathrm{Ad}\,(u(0))\,X). \end{aligned}$$

Next, we show that $T_{T_o}\mathcal{I}$ is full in $T_{T_o}\mathcal{T} \cong \mathfrak{m} \times \mathfrak{m}$. Suppose that there exist $X_o, Y_o \in \mathfrak{m}$ such that $\langle (X, \mathrm{Ad}\,(u)\,X), (X_o, Y_o)\rangle = 0$ for all $u \in T$, $X \in \mathfrak{m}$, in particular for $X = X_o$. We have $Y_o = -X_o$, since the diagonal $\Delta(\mathfrak{m})$ is included in $T_{T_o}\mathcal{T}$. Hence, $\langle \mathrm{Ad}\,(u)\,X_o, X_o\rangle = \|X_o\|^2$ for all $u \in T$. Thus, $\mathrm{Ad}\,(u)\,X_o = X_o$ for all $u \in T$, since $\mathrm{Ad}\,(u)$ is an orthogonal operator of $\mathfrak{m}$. This implies that $X_o = 0$. Finally, observe that $(X, 0) \notin T_{T_o}\mathcal{T}$ for all $0 \neq X \in \mathfrak{m}$, hence $T_{T_o}\mathcal{I}$ is not a vector subspace of $T_{T_o}\mathcal{T}$.

c) Clearly, $\mathrm{Ker}\, dP_{(g,ug)} = \{(dL_g Z_1, dL_{ug} Z_2) \mid Z_1, Z_2 \in \mathfrak{t}\}$ and $T_{(g,ug)}\mathcal{J}$ is the image of $dF_{(u,g)}$, which consists of the elements

$$(dL_g V, dL_u dL_g V + dR_g dL_u Z) = \left(dL_g V, dL_{ug}\left(V + \mathrm{Ad}\,\left(g^{-1}\right) Z\right)\right),$$

with $V \in \mathfrak{g}$, $Z \in \mathfrak{t}$. Hence,

$$\begin{aligned} &\left(T_{(g,ug)}\mathcal{J}\right) \cap \left(\mathrm{Ker}\, dP_{(g,ug)}\right) = \\ &= \left\{(dL_g Z, dL_{ug}(Z + Z')) \mid Z \in \mathfrak{t}, Z' \in \mathfrak{t} \cap \mathrm{Ad}\,\left(g^{-1}\right)\mathfrak{t}\right\}, \end{aligned}$$

whose dimension is

$$\dim \mathfrak{t} + \dim (\mathfrak{t} \cap \mathrm{Ad}\,(g)\,\mathfrak{t})\,. \tag{1}$$

Now, $G_o = \{g \in G \mid \mathfrak{t} \cap \mathrm{Ad}\,(g)\,\mathfrak{t} = \{0\}\}$ is an open dense subset of G. Indeed, let $\Phi = \{\alpha_1, \ldots, \alpha_k\}$ be a basis of the root system associated with $\mathfrak{g}$, $\mathfrak{t}$ [**2**] and denote $\mathfrak{t}^j = \cap_{i\neq j} \mathrm{Ker}\,\alpha_i$. The complement of G_o is the union of

$$G^j = \left\{g \in G \mid \mathrm{Ad}\,(g)\,\mathfrak{t} \cap \mathfrak{t} \supset \mathfrak{t}^j\right\},$$

$j = 1, \ldots, k$. Each G^j is contained in the union of the stabilizers of $\cap_{i\in J} \mathrm{Ker}\,\alpha_i$, with $J \subset \Phi - \{\alpha_j\}$. Therefore, $\mathcal{J}_o = F\,(T \times G_o)$ is open and dense in $\mathcal{J}$ and (c) follows, since by (1), dP has maximal rank on $\mathcal{J}_o$. □

Consider on G/T any fixed G-invariant metric d and let

$$\mathcal{D} = \{(x, y) \in (G/T) \times (G/T) \mid d\,(x, T) = d\,(y, T)\}\,.$$

We showed (and used) in [**3**] that $\mathcal{I} \subset \mathcal{D}$. In contrast with [**1**], where it is proved that equality holds for $G = S^3$, we have:

PROPOSITION 2. *$\mathcal{I} = \mathcal{D}$ if and only if $G = S^3$.*

PROOF. For $G = S^3$, Gluck and Warner proved in [**1**] that $\mathcal{I} = \mathcal{D}$. Let U be a normal ball in G/T centered at T and denote $V = U \times U - \{T_o\}$, which is an open subset of $(G/T) \times (G/T) \cong \mathcal{T}$. Next, we show by means of the Implicit Function Theorem that $\mathcal{D} \cap V$ is a hypersurface of $\mathcal{T}$. It is a level set of the function $F : V \to \mathbf{R}$ defined by $F\,(x, y) = d\,(x, T) - d\,(y, T)$, which is smooth and satisfies $dF_{(x,y)} \neq 0$ for all $(x, y) \in V$. Indeed, we may suppose without loss of generality that $x \neq T$. If γ is the geodesic in U satisfying $\gamma\,(0) = T$ and $\gamma\,(1) = x$, we have that

$$\begin{aligned} dF_{(x,y)}\,(\gamma'\,(1)\,, 0) &= \left.\frac{d}{dt}\right|_{t=1} d\,(\gamma\,(t)\,, T) - d\,(y, T) \\ &= \left.\frac{d}{dt}\right|_{t=1} t\,\|\gamma'\,(0)\| = \|\gamma'\,(0)\| \neq 0. \end{aligned}$$

On the other hand, T acts on G/T on the left by isometries and T is a fixed point of this action. We know by Theorem 1 (a) that $\mathcal{I} = \{(x, ux) \mid x \in G/T, u \in T\}$. Let $H : (G/T) \times T \to (G/T) \times (G/T)$ be defined by $H\,(x, u) = (x, ux)$. H is smooth and its image is $\mathcal{I}$, which is included in $\mathcal{D}$, since $d(x, T) = d(ux, uT) = d(ux, T)$ by the G-invariance of d. Now, if G is not S^3, straightforward arguments using the root system associated with $\mathfrak{g}$ and $\mathfrak{t}$ yield that $\dim\,(G/T) \times T = \dim G < 2 \dim\,(G/T) - 1 = \dim \mathcal{D} \cap V$. Therefore, in this case, $\mathcal{I}$ is strictly contained in $\mathcal{D}$. □

References

[1] H. Gluck, F. Warner, *Great circle fibrations of the three-sphere,* Duke Math. J. **50** (1983), 107–132.

[2] S. Helgason, *Differential geometry, Lie groups, and symmetric spaces,* Academic Press, Boston, etc, 1978.

[3] M. Salvai, *Affine maximal torus fibrations of a compact Lie group*, preprint.

FAMAF, CIUDAD UNIVERSITARIA, 5000 CÓRDOBA, ARGENTINA

E-mail address: salvai@mate.uncor.edu

Contemporary Mathematics
Volume **288**, 2001

On the Asymptotic Series of the Heat Content

Alessandro Savo

Abstract. We show how to express the coefficients of the asymptotic series of the heat content, as computed in our previous paper [**8**] by distance function methods, in terms of the curvatures of the manifold.

Let Ω be a compact Riemannian manifold with smooth boundary $\partial\Omega$, let $\phi(x)$ be a smooth function on Ω, and let $\phi(t,x)$ denote the solution of the heat equation on Ω with initial conditions $\phi(x)$ and Dirichlet boundary conditions, so that $\phi(t,x)$ satisfies the problem

$$\begin{cases} (\Delta + \frac{\partial}{\partial t})\phi(t,x) = 0 \\ \phi(0,x) = \phi(x) \quad \text{for all} \quad x \in \Omega \\ \phi(t,y) = 0 \quad \text{for all} \quad t > 0,\, y \in \partial\Omega. \end{cases} \tag{1}$$

The *heat content* function (corresponding to the initial data ϕ) is the integral

$$H_\phi(t) = \int_\Omega \phi(t,x)\, dx, \tag{2}$$

where dx is the Riemannian measure. The heat content admits, as $t \to 0$, an asymptotic expansion of type:

$$H_\phi(t) \sim \int_\Omega \phi - \sum_{k=1}^{\infty} \beta_k(\phi) t^{k/2}. \tag{3}$$

The calculation of the coefficients $\beta_k(\phi)$ in (3) has attracted the interest of several mathematicians, especially when the initial data is unital: $\phi \equiv 1$. For euclidean domains, $\beta_1(1)$ was computed in [**1**] and $\beta_2(1)$ was computed in [**5**]. For general manifolds, the coefficients $\beta_k(\phi)$, for $k \le 4$, were computed in [**2**] and [**3**], and $\beta_5(1)$ is computed in [**4**] (see also [**7**, Theorem 0.3]). In all these works, the coefficients are written in terms of the initial data ϕ and its derivatives, the curvature tensor and its covariant derivatives, the second fundamental form of $\partial\Omega$ and its (tangential) covariant derivatives.

2000 *Mathematics Subject Classification.* 58J35, 58J37.
Key words and phrases. Heat equation, heat content, asymptotic expansions.
Work partially supported by GNSAGA and MURST, Italy

We gave, in [**8**], an algorithm for the calculation of all $\beta_k(\phi)$'s. The algorithm is proved using distance function methods. However, in that paper, the coefficients are expressed in terms of a certain algebra of operators (recalled below in Theorem 2). The scope of this note is to describe an effective procedure which converts the coefficients $\beta_k(\phi)$, as found in [**8**], into a more geometric form involving the curvatures of Ω. We sketch the essential steps of the procedure in the next section, and we write the first six coefficients of (3), when $\phi \equiv 1$, in the desired form: see Theorem 1 below.

Let us set some notation. Let ν be the inward unit normal to $\partial\Omega$, and let S denote the shape operator of $\partial\Omega$: it acts on the (tangential) vector field $\bar{X}$ on $\partial\Omega$ by the rule $S(\bar{X}) = -\nabla_{\bar{X}}\nu$, where ∇ is the Levi-Civita connection of Ω. By R_ν we denote the endomorphism of $T\Omega$ which takes the vector field X to $R_\nu(X) \overset{def}{=} R(\nu, X)\nu$, where R is the curvature tensor of Ω (so R_ν is the tensor $R(\nu,\cdot,\nu,\cdot)$ and its trace is the Ricci tensor at ν). Let $\bar{\nabla}$ denote the Levi-Civita connection on $\partial\Omega$, and let $\bar{\Delta}$ be the tangential Laplacian, that is, the Laplacian on $\partial\Omega$. Set for brevity $\eta = \mathrm{tr} S = (\dim(\Omega) - 1)$ times the mean curvature function of $\partial\Omega$. All traces are taken with respect to an orthonormal frame in $\partial\Omega$. Then:

THEOREM 1. *The coefficients $\beta_1(1), \dots, \beta_6(1)$ are:*

$$\beta_1(1) = \frac{2}{\sqrt{\pi}}\mathrm{Vol}(\partial\Omega);$$

$$\beta_2(1) = -\frac{1}{2}\int_{\partial\Omega}\eta;$$

$$\beta_3(1) = -\frac{1}{6\sqrt{\pi}}\int_{\partial\Omega}\left[2\,\mathrm{tr}(R_\nu + S^2) - \eta^2\right];$$

$$\beta_4(1) = \frac{1}{16}\int_{\partial\Omega}\left[\eta\,\mathrm{tr}(R_\nu + S^2) - \mathrm{tr}(\nabla_\nu R_\nu + 2S\circ R_\nu + 2S^3)\right];$$

$$\begin{aligned}\beta_5(1) = \frac{1}{240\sqrt{\pi}}\int_{\partial\Omega}&\Big[-8\,\mathrm{tr}(\nabla_\nu^2 R_\nu + 2R_\nu^2 + 8S^2\circ R_\nu + 2S\circ\nabla_\nu R_\nu + 6S^4)\\ &+ 8\eta\,\mathrm{tr}(\nabla_\nu R_\nu + 2S\circ R_\nu + 2S^3) + 4\eta^2\,\mathrm{tr}(R_\nu + S^2)\\ &+4[\mathrm{tr}(R_\nu + S^2)]^2 - \eta^4 + 8\|\bar{\nabla}\eta\|^2\Big];\end{aligned}$$

$$\begin{aligned}\beta_6(1) = -\frac{1}{192}\int_{\partial\Omega}&\Big[\mathrm{tr}(\nabla_\nu^3 R_\nu + 6R_\nu\circ\nabla_\nu R_\nu + 16S\circ R_\nu^2 + 40S^3\circ R_\nu)\\ &+ \mathrm{tr}(2S\circ\nabla_\nu^2 R_\nu + 10S^2\circ\nabla_\nu R_\nu + 24S^5) - \eta\,\mathrm{tr}(\nabla_\nu^2 R_\nu + 2R_\nu^2)\\ &- \eta\,\mathrm{tr}(8S^2\circ R_\nu + 2S\circ\nabla_\nu R_\nu + 6S^4) - \eta^2\,\mathrm{tr}(\nabla_\nu R_\nu + 2S\circ R_\nu)\\ &- \eta^2\,\mathrm{tr}(2S^3) - \mathrm{tr}(R_\nu + S^2)\mathrm{tr}(\nabla_\nu R_\nu + 2S\circ R_\nu + 2S^3)\\ &+ \eta^3\,\mathrm{tr}(R_\nu + S^2) - 2\eta[\mathrm{tr}(R_\nu + S^2)]^2 - 6\langle\bar{\nabla}\eta, S(\bar{\nabla}\eta)\rangle\\ &-5\,\mathrm{tr}(R_\nu + S^2)\,\bar{\Delta}\eta + \eta^2\bar{\Delta}\eta\Big].\end{aligned}$$

Various generalizations of the definition of the heat content (2) have been considered, most notably when the Laplacian on functions is replaced by an operator of Laplace type acting on a vector bundle on Ω: see [**2**],[**3**] and the survey paper [**7**], which also provides an extensive bibliography. When the boundary is not smooth, we refer to [6] for the heat content asymptotics of a planar polygonal region, and to [9] for the first three terms of the heat content asymptotics of a convex polyhedral body of arbitrary dimension.

Let us now go back to the smooth case, and recall the main facts from [8]. Let ρ be the distance function from $\partial\Omega$. Fix a tubular neighborhood U of $\partial\Omega$ where ρ is smooth (which exists because $\partial\Omega$ is smooth), and define the operator N acting on $\psi \in C^\infty(U)$ by the formula:

$$N\psi = 2\langle\nabla\psi, \nabla\rho\rangle - \psi\Delta\rho. \tag{4}$$

Let $\mathcal{A}$ be the (non-commutative) $\mathbf{R}$–algebra of operators acting on $C^\infty(U)$ and generated by N and the Laplacian Δ. As N has order one and Δ has order two, this algebra is endowed with a natural grading $\mathcal{A} = \oplus_{n=0}^\infty \mathcal{A}_n$, where $\mathcal{A}_n$ is spanned by the homogeneous monomials (words) of order n in N and Δ. The action of this algebra on the initial data $\phi \in C^\infty(U)$ determines the heat content asymptotics, as the main result of [8] shows:

THEOREM 2.
1. *For all $k \geq 1$, there exists a differential operator $D_k \in \mathcal{A}_{k-1}$, such that*

$$\beta_k(\phi) = \int_{\partial\Omega} D_k\phi.$$

2. *For all $k \geq 1$, D_k can be computed explicitly in terms of $D_1, \ldots, D_{k-1}$.*

See Theorem 2.1 in [8] for the relevant recursive formulas, and for the explicit calculation of $D_1, \ldots, D_8$. For concreteness, we write below the first six coefficients:

TABLE 1.

$$\beta_1(\phi) = \frac{2}{\sqrt{\pi}} \int_{\partial\Omega} \phi;$$

$$\beta_2(\phi) = \frac{1}{2} \int_{\partial\Omega} N\phi;$$

$$\beta_3(\phi) = \frac{1}{6\sqrt{\pi}} \int_{\partial\Omega} (N^2 - 4\Delta)\phi;$$

$$\beta_4(\phi) = -\frac{1}{16} \int_{\partial\Omega} (\Delta N + 3N\Delta)\phi;$$

$$\beta_5(\phi) = -\frac{1}{240\sqrt{\pi}} \int_{\partial\Omega} (N^4 + 16N^2\Delta + 8N\Delta N - 48\Delta^2)\phi;$$

$$\beta_6(\phi) = \frac{1}{768} \int_{\partial\Omega} (\Delta N^3 - N^3\Delta + N\Delta N^2 - N^2\Delta N + 40N\Delta^2 + 8\Delta^2 N + 16\Delta N\Delta)\phi.$$

A look at Table 1 (note that $\beta_6(1)$ contains only four terms there) and Theorem 1 shows how Theorem 2 provides a rather synthetic presentation of the heat content invariants. As $\mathcal{A}_n = \Delta \cdot \mathcal{A}_{n-2} \oplus N \cdot \mathcal{A}_{n-1}$, we see in fact that $\dim\mathcal{A}_n = \dim\mathcal{A}_{n-2} + \dim\mathcal{A}_{n-1}$; hence the number of terms in $\beta_k(\phi)$, as written in Theorem 2, grows at most like the sequence of the Fibonacci numbers. Finally, the role played by the boundary operator N and the algebra $\mathcal{A}$ gives a new point of view when dealing with the heat content asymptotics.

The conversion to curvature invariants

We will consider the foliation of a small tubular neighborhood U of $\partial\Omega$ by the level sets of the distance function $\rho = \text{dist}(\cdot, \partial\Omega)$. The vector field $\nu = \nabla\rho$ is smooth on U, has unit length, and is everywhere normal to the leaves. A vector field X is *tangential* on U iff $\langle X, \nu\rangle = 0$ on U (so X is everywhere tangent to the leaves). We denote tangential fields by an overbar: $\bar{X}, \bar{Y}$ etc. The shape operator of $\partial\Omega$ extends to a field of endomorphisms, acting on tangential fields, by setting $S(\bar{X}) = -\nabla_{\bar{X}}\nu$: at $p \in \rho^{-1}(r)$ it is clearly the shape operator of the hypersurface $\rho^{-1}(r)$. We again denote its trace by η, which is then a smooth function on U. If $\bar{\delta}$ denotes the tangential divergence, one has by definition $\bar{\Delta} = \bar{\delta}\circ\bar{\nabla}$, and one proves the formula:

(5) $$\Delta\phi = \bar{\Delta}\phi + \Delta_{rad}\phi$$

where: $\Delta_{rad}\phi = -\dfrac{\partial^2\phi}{\partial\nu^2} + \eta\dfrac{\partial\phi}{\partial\nu}$. Note that, in particular, $\Delta\rho = \text{tr}S = \eta$ everywhere on U. Note that, in this notation, $N\phi = 2\dfrac{\partial\phi}{\partial\nu} - \eta\phi$ on U (see (4)).

Given the splitting (5), we can in turn split the operator D_k of Theorem 2 as a sum $D_{k,rad} + D_{k,tan}$ where $D_{k,rad}$ contains only N and Δ_{rad}, and where $D_{k,tan}$ contains at least one $\bar{\Delta}$ in all of its monomials. We then need to express $D_{k,rad}\phi$ and $D_{k,tan}\phi$ in the desired geometric form, that is, the function ϕ and its covariant derivatives, the curvature tensor and its covariant derivatives, the second fundamental form (or the shape operator) and its tangential covariant derivatives.

Now observe that $D_{k,rad}\phi$ contains only radial (that is, normal) derivatives of ϕ and η. To express this functions by the curvature invariants, it is enough to do so for the normal derivatives $\dfrac{\partial^k\eta}{\partial\nu^k}$. This is achieved by using a well-known formula for the covariant derivative of S. When acting on tangential fields, one has, on U:

(6) $$\nabla_\nu S = R_\nu + S^2.$$

As the covariant derivative ∇_ν commutes with taking traces, one has then $\dfrac{\partial\eta}{\partial\nu} = \text{tr}(\nabla_\nu S)$; hence, by (6), each $\dfrac{\partial^k\eta}{\partial\nu^k}$ will be the trace of an endomorphism which is a sum of compositions of covariant derivatives of R_ν, in the normal direction, with powers of the shape operator S. For example:

$$\frac{\partial^2\eta}{\partial\nu^2} = \text{tr}(\nabla_\nu R_\nu + 2S\circ R_\nu + 2S^3).$$

This takes care of the radial part of $D_k\phi$. For its tangential part, we need to express the k−th normal derivative of the tangential Laplacian in the desired form. For this, we use the following:

COMMUTATION RULES.
1. *For all tangential fields $\bar{X}$ on U one has:*

$$\frac{\partial}{\partial\nu}(\bar{\delta}\bar{X}) = \bar{\delta}(\nabla_\nu\bar{X}) + \langle\bar{X}, \bar{\nabla}\eta\rangle + \bar{\delta}(S(\bar{X})).$$

2. *For all $\phi \in C^\infty(U)$ one has:* $\nabla_\nu \bar\nabla\phi = \bar\nabla(\frac{\partial\phi}{\partial\nu}) + S(\bar\nabla\phi)$, *hence:*

$$\frac{\partial}{\partial\nu}\bar\Delta\phi = \bar\Delta(\frac{\partial\phi}{\partial\nu}) + \langle\bar\nabla\phi, \bar\nabla\eta\rangle + 2\bar\delta(S(\bar\nabla\phi)).$$

Repeated application of the commutation rules above, together with (6), will yield the desired geometric form also for $D_{k,tan}\phi$.

References

[1] van den Berg, M., Davies, E.B., *Heat flow out of regions in* $\mathbf{R}^m$, Math. Z. **202** (1989), 463-482.

[2] van den Berg, M., Desjardins, S., Gilkey, P., *Functorality and heat content asymptotics for operators of Laplace type*, Topological Methods in Nonlinear Analysis **2** (1993), 147-162.

[3] van den Berg, M., Gilkey, P.B., *Heat content asymptotics for a Riemannian manifold with boundary*, J. Funct. Anal. **120** (1994), 48-71.

[4] van den Berg, M., Gilkey, P.B., *The heat equation with inhomogeneous Dirichlet boundary conditions*, Comm. Anal. Geom. **7, n. 2** (1999), 279-294.

[5] van den Berg, M., Le Gall, P.B., *Mean curvature and the heat equation*, Math.Z. **215** (1994), 437-464.

[6] van den Berg, M., Srisatkunarajah, S., *Heat flow and brownian motion for a region in* $\mathbf{R}^2$ *with a polygonal boundary*, Prob. Theory Rel. Fields **86** (1990), 41-52.

[7] Gilkey, P., *Heat content asymptotics*, Contemporary Mathematics **242** (1999), 125-133.

[8] Savo, A., *Uniform estimates and the whole asymptotic series of the heat content on manifolds*, Geometriae Dedicata **73** (1998), 181-214.

[9] Savo, A., *A mean-value lemma and applications*, Bulletin de la Soc. Math. de France (to appear).

Dipartimento di Metodi e Modelli Matematici, Università di Roma, La Sapienza, Via Antonio Scarpa 16, 00161 Roma

E-mail address: savo@dmmm.uniroma1.it

Contemporary Mathematics
Volume **288**, 2001

Visualization of Flat Slant Surfaces in $\mathbb{C}^2$

Yoshihiko Tazawa

Abstract. This article is a summary of the presentation at the Congress on Differential Geometry in Memory of Alfred Gray. The first half of the presentation is the introduction of a book [4], and the second half is about the visualization of flat slant surfaces in $\mathbb{C}^2$ given in [6]. This is one of the sequential talks ([2], [3], [5], etc) about curves and surfaces by a computer algebra, and the author would like to emphasize the advantage of using computers in differential geometry, in a slightly different way from Professor Gray's book [1]. The features of the talk are numerical calculation and computer graphics, animations in particular. The full commands for the calculation and graphics are downloadable at `http://math.kn.dendai.ac.jp/tazawa/bilbao/`.

1. Theory of curves and surfaces

The book [4] is a textbook on classical differential geometry of curves and surfaces based on *Mathematica*. The use of computers enables us to extend the range of examples to randomly generated objects, to do experimental approach, and to visualize geometrical objects. An appendix CD-ROM is attached that contains all commands and graphics. The outline of the book is stated on the web page.

2. Flat slant surfaces in $\mathbb{C}^2$

Flat slant surfaces, namely flat surfaces with constant Wirtinger angle, in $\mathbb{C}^2$ are classified into the five types by B.-y. Chen in [6]. Here we visualize them through animations by projecting to $\mathbb{R}^3$ after rotating in $\mathbb{C}^2 = \mathbb{R}^4$. We use numerical integration to figure out the shapes of these surfaces.

Let $\theta \in (0, \frac{\pi}{2}]$. The surfaces $\mathbf{x}(x, y)$ defined in the following are all flat and θ-slant in $\mathbb{C}^2$. For the details about the concrete construction of these surfaces, see the web page.

Type (I). A unit speed curve $z(s)$ in the unit hypersphere $S^3 \subset \mathbb{C}^2$ is called a *θ-Legendre curve*, if $\langle iz(s), z'(s)\rangle = \cos\theta$. For any θ-Legendre curve $z(s)$ and any function $\beta(u)$, define

$$\mathbf{x}(x, y) = z(y)x + \int_0^y \beta(u)z'(u)du.$$

This is a kind of tangent developable surface.

2000 *Mathematics Subject Classification.* Primary 53B25; Secondary 53B35.
Key words and phrases. curves and surfaces in Euclidean space, slant immersion, flat surface.

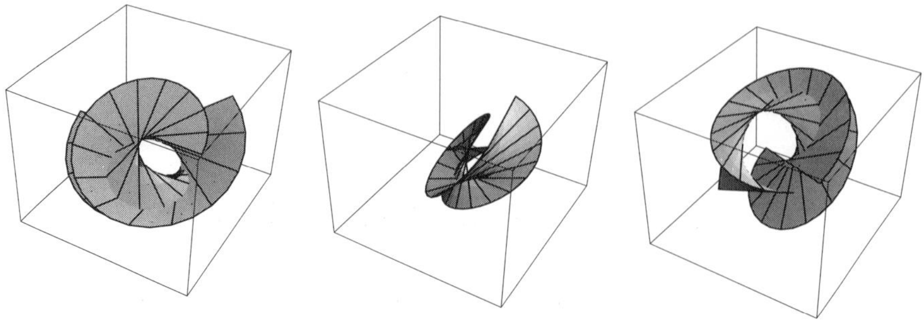

FIGURE 1. A flat slant surface of type (I)

Type (II). For any given nonzero function $\varphi(y)$ around 0, define

$$\mathbf{x}(x,y) = \left(x + i\,\cos\theta \int_0^y \frac{dt}{\varphi(t)},\ \sin\theta \int_0^y \frac{e^{i\,(\csc\theta)y}}{\varphi(t)} dt\right).$$

This is a kind of cylindrical surface.

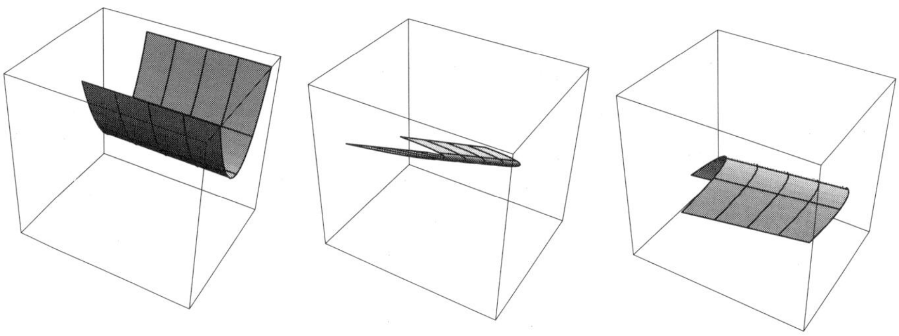

FIGURE 2. A flat slant surface of type (II)

Type (III). For two functions $\mu(x) \neq 0$ and $k(y)$ defined around 0 such that $\psi = k(y) + \cot\theta \int_0^x \frac{dx}{\mu} \neq 0$, define

$$\mathbf{x}(x,y) = \left(i \int_0^y k(y) e^{i\,y\csc\theta} dy + e^{i\,y\csc\theta} \cos\theta \int_0^x \frac{dx}{\mu(x)},\ \sin\theta \int_0^x \frac{e^{i\,x\csc\theta}}{\mu(x)} dx\right).$$

Type (IV). For two functions $u(x)$ and $v(y)$ defined around 0 with $v(0) \neq 0$ and $v'(y) \neq 0$, define

$$\mathbf{x}(x,y) = \left(\int_0^x u(x) e^{i\,x\csc\theta} dx - i\,v(y) e^{i\,x\csc\theta} \sin\theta + i\,v(0)\sin\theta,\right.$$

$$\left.\sin\theta\tan\theta \int_0^y v'(y) e^{i\,y\csc\theta} dy\right).$$

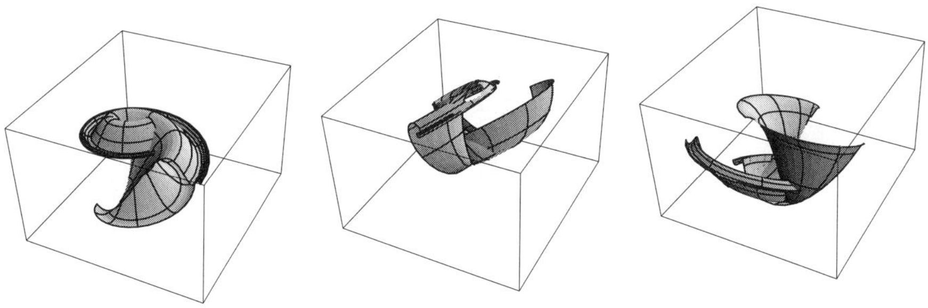

FIGURE 3. A flat slant surface of type (III)

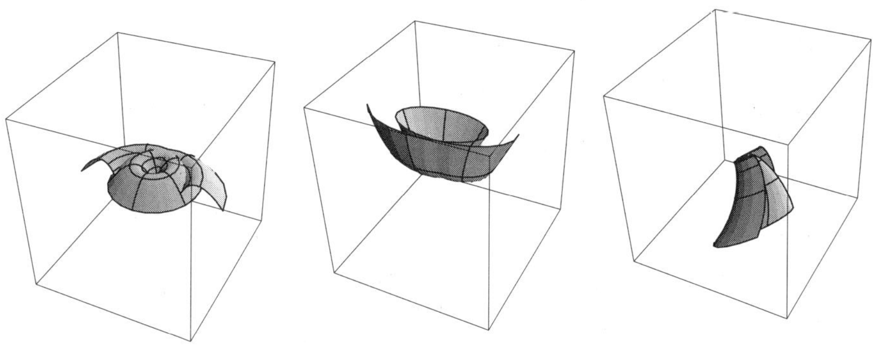

FIGURE 4. A flat slant surface of type (IV)

Type (V). Let $f(u)$ be a nonzero function defined around 0 such that $f \neq \cot\theta$, and $\rho = \rho(x, y)$ be a solution of the wave equation:

$$\rho_{xy} - \frac{f'(x-y)}{f(x-y)}\,\rho_y + \{(\cot\theta)f(x-y) - f^2(x-y)\}\rho = 0$$

with $\rho_y \neq 0$, and K be a $\mathbb{C}^2$-valued solution of the ordinary differential equation:

$$K''(u) + (i\csc\theta - \frac{f'(u)}{f(u) - \cot\theta})K'(u) + f(u)(f(u) - \cot\theta)K(u) = 0,$$

satisfying $|K|^2 = 1$ and $|K'|^2 = (f - \cot\theta)^2$. Define

$$\mathbf{x}(x, y) = \int_0^y \left\{ \frac{\rho_y K'(x-y)}{f(\cot\theta - f)} e^{i\,x\csc\theta} - \int_0^x (\rho K(x-y))_y\, e^{i\,x\csc\theta} dx \right\} dy + \int_0^x \rho K(x-y) e^{i\,x\csc\theta} dx, \quad f = f(x-y).$$

In this article, each animation is indicated by three scenes. After entering the commands that can be obtained through web, the reader can watch the animations on a monitor.

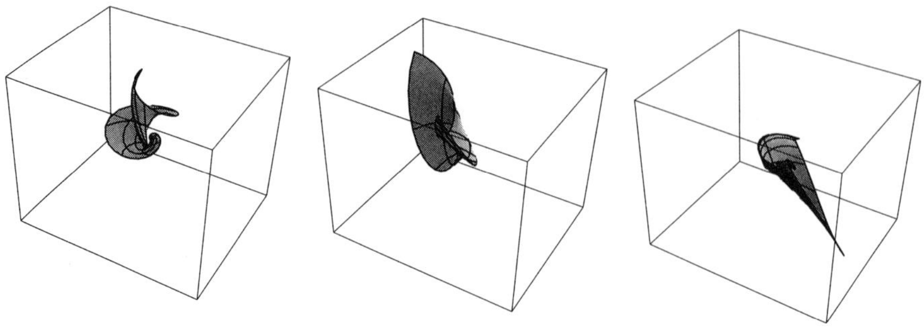

FIGURE 5. A flat slant surface of type (V)

References

1. A. Gray. *Modern differential geometry of curves and surfaces with Mathematica* , Second Edition, CRC Press, 1998.
2. Y. Tazawa. Construction and visualization of slant immersions, *Geometry and Topology of Submanifolds*, VII, World Scientific, pp257-260, 1995.
3. Y. Tazawa. Curves and surfaces in the three dimensional sphere in the space of quaternions, *Innovation In Mathematics, Proceeding of the Second International Mathematica Symposium*, pp459-466, 1997.
4. Y. Tazawa. *Theory of Curves and Surfaces, An Introduction to Classical Differential Geometry by Mathematica*, Pearson Education, 1999 (in Japanese).
5. Y. Tazawa. Computers and differential geometry, to appear in *Atti del Congresso Internazionale di Geometria Differenziale in onore di Pasquale Calapso.*
6. B. Y. Chen. Flat slant surfaces in $\mathbb{C}^2$, preprint.

DEPARTMENT OF INFORMATION ENVIRONMENT INTEGRATION AND DESIGN, SCHOOL OF INFORMATION ENVIRONMENT, TOKYO DENKI UNIVERSITY, 2-1200 MUZAI-GAKUENDAI, INZAI, CHIBA PREFECTURE, 270-1382, JAPAN

E-mail address: tazawa@cck.dendai.ac.jp

Contemporary Mathematics
Volume **288**, 2001

On Solvable Lie Groups without Lattices

Aleksy Tralle

Dedicated to the memory of Alfred Gray

ABSTRACT. In this article we show that solvable Lie group G^{bg} constructed in the paper of Benson and Gordon [**BG**] has no lattices. This yields a negative answer to a question posed by several authors [**BG, FLS, TO**] in the context of symplectic geometry. The result is of conceptual interest, because it is obtained by a delicate application of the Félix-Thomas theorem on models of non-nilpotent fibrations.

1. Introduction and preliminaries

This paper is devoted to a question posed by several authors and related to solvmanifolds. This class of manifolds was studied and used in many geometric contexts by Alfred Gray and his co-authors [**AFGM, CFG, FG, FGG**]. These works have had an essential influence on the direction of research presented also in this article.

Solvmanifolds, i.e. factor-spaces of solvable Lie groups G over discrete co-compact subgroups (lattices) Γ constitute a large and important subclass of the class of aspherical manifolds. This subclass is interesting not only on its own right, but very often serves as a source of testing examples in many areas of geometry (see, e.g. [**FGG, FLS, Ha, IRTU1, IRTU2, O, RT, TO**]).

Formally, the present paper contains one explicit example of a completely solvable Lie group without lattices, answering (in a negative way) to a question posed in [**BG, FG, FLS, TO**]. In greater detail, in [**BG**] the following completely solvable Lie algebra was constructed:

$$\mathfrak{g}^{bg} = \{S, T, X_1, Y_1, Z_1, X_2, Y_2, Z_2\},$$

$$[X_1, Y_1] = Z_1, \ [X_2, Y_2] = Z_2,$$

$$[S, X_1] = X_1, [S, X_2] = -X_2, \ [S, Y_1] = -2Y_1, \ [S, Y_2] = 2Y_2,$$

2000 *Mathematics Subject Classification.* 53C15, 55P62.

Key words and phrases. Solvable Lie group, lattice, K.S. model of fibration.

This work was supported by the Polish Committee for the Scientific Research (KBN), research project no. 2P03A 027.

$$[S, Z_1] = -Z_1, \ [S, Z_2] = Z_2.$$

Here $\mathfrak{g}^{bg}$ is defined by the Lie brackets of the base elements, and all the brackets which are not written, are assumed to be zero.

It was noted in [**BG**] that this algebra is unimodular (i.e. $\mathrm{Tr}\, ad\, X = 0$, for all $X \in \mathfrak{g}^{bg}$). Hence, the corresponding simply connected Lie group G^{bg} satisfies the necessary condition for the existence of a lattice. However, the authors pointed out that it is not known if there is a lattice in G^{bg} or not. If G^{bg} admitted a lattice, one would obtain an example of (aspherical) closed symplectic manifold which is formal but non-Lefschetz (this observation was made by Fernández, de León and Saralegui [**FLS**], cf. also [**AFGM**]). It is not known if such *aspherical* closed symplectic manifolds exist. The importance of constructing them was discussed in [**IRTU1**]. In general, relations between formality and the hard Lefschetz property are not yet clear (cf. [**AFGM, FLS, IRTU1, M**]). For example, there is a conjecture in [**M**] that any closed symplectic manifold satisfying the hard Lefschetz property must be formal. Our result indicates the difficulties of constructing examples of that kind in the class of solvmanifolds.

Although we give an answer to a question posed in [**BG, FG, TO**], the paper contains a result also of conceptual interest. The reason is that it is obtained by a delicate application of rational homotopy theory. In greater detail, if the given Lie group G^{bg} contained a lattice Γ, we could associate with G^{bg}/Γ the *Mostow fibration*

$$N/\Gamma \cap N \to G^{bg}/\Gamma \to G^{bg}/\Gamma N$$

and calculate the minimal model of the total space G^{bg}/Γ in two different ways. The first of these uses the Chevalley-Eilenberg complex $(\Lambda\mathfrak{g}^*, \delta)$, and the second one – the Félix-Thomas refinement of the method of KS-extensions [**FT**]. Since these two methods yield different results, we obtain a contradiction.

In this paper we work with *completely solvable* Lie groups G. It means that the corresponding Lie algebra $\mathfrak{g}$ satisfies the property that any adjoint linear operator $ad\, S : \mathfrak{g} \to \mathfrak{g}$, $S \in \mathfrak{g}$ has only real eigenvalues.

To get the main theorem of this article, we use some techniques from rational homotopy theory [**FHT, H, L,TO**]. We use freely the notion of the *minimal model* of a space X and denote it $\mathcal{M}(X)$. The minimal model of a differential graded algebra, say $(\mathcal{A}, d)$ is denoted as $\mathcal{M}(\mathcal{A})$. Also, the notion of the *K.S. extension* [**H**] is used without explanations. We just mention that K.S. extensions yield models of fibrations. Note that all differential graded algebras are considered over $\mathbb{R}$. The model of arbitrary fibration is given by the *Félix-Thomas theorem* [**FT**].

Theorem (Félix-Thomas). *Let $F \to E \to B$ be a fibration and let U denote the largest $\pi_1(B)$-submodule of $H^*(F)$ on which $\pi_1(B)$ acts nilpotently. Assume that the following conditions are satisfied:*

(i) *$H^*(F)$ is a vector space of finite type;*

(ii) *B is a nilpotent space (i.e. $\pi_1(B)$ is a nilpotent group which acts nilpotently on $\pi_i(B)$ for all $i \geq 1$).*

Then, in the K.S. model of the fibration

$$\begin{array}{ccccc} A(B) & \longrightarrow & A(E) & \longrightarrow & A(F) \\ \uparrow & & \uparrow & & \uparrow{\scriptstyle\psi} \\ (\Lambda V, d_V) & \longrightarrow & (\Lambda(V \oplus Y), D) & \longrightarrow & (\Lambda Y, d_Y) \end{array}$$

the differential graded algebra homomorphism $\psi : (\Lambda Y, d_Y) \to A(F)$ *induces an isomorphism*

$$\psi^* : H^*(\Lambda Y, d_Y) \cong U.$$

In the sequel, we will need the fact, that for solvmanifolds G/Γ of completely solvable Lie groups the standard Chevalley–Eilenberg complex $(\Lambda\mathfrak{g}^*, \delta)$ for the Lie algebra cohomology $H^*(\mathfrak{g})$ is also a free model for the manifold G/Γ. In particular, $\mathcal{M}(G/\Gamma) \cong \mathcal{M}(\Lambda\mathfrak{g}^*, \delta)$ (see [**TO**] for details).

2. Result

THEOREM. *The solvable Lie group* G^{bg} *has no lattices.*

OUTLINE OF PROOF.

For any solvable Lie group G, we denote by N its nilradical, i.e. the (unique) maximal connected nilpotent Lie subgroup. It is not difficult to check that G^{bg} is completely solvable and that it is splittable into a semidirect product of a one-dimensional subgroup $A \cong \mathbb{R}$ and the nilradical N: $G^{bg} = A \cdot N$, where $N = \mathbb{R} \times N_3 \times N_3$, with N_3 being the 3-dimensional Heisenberg group. One can also prove that any *completely solvable* $G = A \cdot N$ such that $\dim A = 1$ admitting a lattice Γ, admits also a splitting $A_1 \cdot N$ such that $\Gamma = \mathbb{Z}^{A_1} \cdot (\Gamma \cap N)$, i.e., the splitting of G, which is compatible with the splitting of the lattice. Here $A_1 \cong \mathbb{R}$, $\mathbb{Z}^{A_1} \cong \mathbb{Z}$ is a lattice in $A_1 \subset G$.

Assume that G^{bg} admits a lattice. Then, one can consider the Mostow fibration

$$N/\Gamma \cap N \to G^{bg}/\Gamma \to G^{bg}/N\Gamma = S^1$$

and apply to it the Félix-Thomas theorem. To write the corresponding model, we need, first, the $\pi_1(G^{bg}/N\Gamma = S^1) = \mathbb{Z}$-action on the cohomology $H^*(N/\Gamma \cap N) = H^*(\mathfrak{n})$. Since we can consider the compatible splitting of G^{bg}, it is possible to calculate the above action using Theorem 2.4 in [**OT**]. Consider the homomorphism $\varphi : A_1 \to \mathrm{Aut}(N)$ given by the rule $\varphi(a_1)(n) = a_1 \cdot n \cdot a_1^{-1}$. From Theorem 2.4 in [**OT**] we get the following.

Let G/Γ *be a compact completely solvable solvmanifold such that* $\dim G/N = 1$. *Then, for the corresponding Mostow fibration,* $\pi_1(S^1)$*-action on* $H^*(N/\Gamma_N)$ *is given by:*

1. *taking the compatible splitting* $A_1 \cdot N$,
2. *restricting* $\bar{\varphi} = d\varphi : A_1 \to \mathrm{Aut}\,(\mathfrak{n})$ *to* $\psi : \mathbb{Z}^{A_1} \to \mathrm{Aut}\,(\mathfrak{n})$,
3. *taking the dual automorphism* $\psi^t : \mathbb{Z}^{A_1} \to \mathrm{Aut}(\mathfrak{n}^*)$,
4. *extending it to an automorphism of the exterior algebra* $\Lambda\psi^t : \mathbb{Z}^{A_1} \to \mathrm{Aut}(\Lambda\mathfrak{n}^*)$ *as differential graded algebras,*
5. *and taking the induced automorphisms on cohomology* $(\Lambda\psi^t)^* : \mathbb{Z}^{A_1} \to \mathrm{Aut}(H^*(\Lambda\mathfrak{n}^*, \delta))$.

Let $G^{bg} = A \cdot N$ be the "initial" splitting of G^{bg}. Consider the corresponding homomorphism $\theta : A \to \mathrm{Aut}(N), \theta(a)(n) = a \cdot n \cdot a^{-1}$. On can check that the $\pi_1(S^1)$-action on $H^*(N/\Gamma_N)$ is expressed through initial homomorphism θ as follows: $(\Lambda\psi^t)^*(a_1) = (\Lambda\bar{\theta}^t)^*(a) \circ (\Lambda\alpha^t)$, where the homomorphism ψ is a restriction of φ onto $\mathbb{Z}^{A_1}$, and $\alpha \in \mathrm{Aut}(\mathfrak{n})$ is a unipotent automorphism of the Lie algebra $\mathfrak{n}$.

Here $a_1 \in \mathbb{Z}^{A_1} \subset A_1$ is a generator of $\mathbb{Z}^{A_1} \cong \mathbb{Z}$ and $a \in A$ is an element uniquely determined by a_1.

Having the $\pi_1(S^1)$-action described as above, one can calculate the maximal nilpotent submodule U in the Félix-Thomas theorem explicitly. The result is given as follows.

$$U = \langle b, u_1, u_2, u_3, u_4, v_1, v_2, v_3 \rangle$$

is a graded commutative algebra generated by elements $u_i, i = 1,2,3,4$ of degree 2 and elements $v_j, j = 1,2,3$ of degree 4 and one element b of degree 1. The generators of even degrees satisfy the following relations:

$$\begin{aligned} u_i^2 = u_1u_3 = u_2u_3 = u_3u_4 = 0,\ i = 1, ..., 4;\\ u_1v_1 = u_1v_2 = u_2v_1 = u_3v_2 = \\ = u_3v_3 = u_4v_1 = u_4v_2 = u_4v_3 = 0;\\ v_sv_t = 0 \text{ for all } s,t. \end{aligned}$$

It is not possible to calculate $(\Lambda Y, d)$ in the model given in the Félix-Thomas theorem explicitly, using the only known condition $H^*(\Lambda Y, d) \cong U$. However, one can use the fact that $(\Lambda Y, d)$ is a perturbation of the Halperin-Stasheff bigraded model of U (see [**HS**] for the details of the construction). This information, together with a method of calculating of the minimal model of a free differential algebra [**L,TO**] yields the following result

$$\mathcal{A}(G/\Gamma) = (\Lambda(a,b) \otimes \Lambda \bar{Y}, D), \bar{Y} = \oplus_{ij} \bar{Y}_i^j,\ \bar{Y}_0^4 = \langle v_1, v_2, v_3 \rangle.$$

On the other hand, if G contained Γ, the corresponding solvmanifold would be completely solvable, and as a consequence, it would have the same minimal model as $(\Lambda\mathfrak{g}^*, \delta)$. The latter minimal model was calculated by Fernández, de León and Saralegui [**FLS**]. It was shown in this article that

$$\mathcal{M}(\Lambda\mathfrak{g}^*, \delta) \cong (\Lambda(a,b) \otimes \Lambda Z, d),\ Z = \oplus_i^j Z_i^j, Z_0^4 = \{e\}.$$

Compairing the degrees of generators we see that $\mathcal{M}(G/\Gamma) \ncong \mathcal{M}(\Lambda\mathfrak{g}^*, \delta)$. Hence we have obtained a contradiction caused by the assumption that G^{ab} admitted a lattice.

REMARK. Vladimir Gorbatsevich informed me, that there is yet another way to prove this theorem, using direct calculations with the cohomology of lattices and spectral sequences, as well as a description of all $\mathbb{Q}$-forms on $\mathfrak{n}_3 \oplus \mathfrak{n}_3$ and a proof that any automorphism of any such rational form cannot be conjugate to the automorphism determining the given solvable Lie algebra.

In this paper we give an outline of proof which enables an expert to fill in all the steps. The details will appear elsewhere and they are available in [**T**].

References

[AFGM] L.C. de Andrés, M. Fernández, A. Gray and J.J. Mencia, *Some six-dimensional compact symplectic and complex solvmanifolds*, Rendiconti di Matematica, Serie VII **12** (1992), 59-67.

[BG] C. Benson and C. Gordon, *Kähler structures on solvmanifolds for completely solvable Lie groups*, Proc. Amer. Math. Soc. **108** (1990), 971-990.

[CFG] L. Cordero, M. Fernández and A. Gray, *Lie groups with no left invariant complex structures*, Portugaliae Mathematica **47** (1990).

[FG] M. Fernández, A. Gray, *Compact symplectic four dimensional solvmanifolds not admitting complex structure*, Geom. Dedicata **34** (1990), 295-299.

[FGG] M. Fernández, M. Gotay and A. Gray, *Four-dimensional compact parallelizable symplectic and complex manifolds*, Proc. Amer. Math. Soc. **103** (1988), 1209-1212.

[FHT] Y. Félix, S. Halperin and J.-C. Thomas, *Rational Homotopy Theory*, Springer, Graduate Texts Math., 205, Berlin, 2000.

[FLS] M. Fernández, M. de León and M. Saralegui, *A six dimensional compact symplectic solvmanifold without Kähler structures*, Osaka J. Math. **33** (1996), 19-35.

[FT] Y. Félix and J.-C. Thomas, *Le Tor differentielle d'une fibration non-nilpotente*, J. Pure and Appl. Alg. **38** (1985), 217-233.

[H] S. Halperin, *Lectures on Minimal Models*, vol. 9-10, Mémoire Soc. Math. France, 1982.

[Ha] K. Hasegawa, *A class of compact Kählerian solvmanifolds and a general conjecture*, Geom.Dedicata **78** (1999), 253-258.

[HS] S. Halperin and J. Stasheff, *Obstructions to homotopy equivalences*, Advances Math. **32** (1979), 233-279.

[IRTU1] R. Ibáñez, Y. Rudyak, A. Tralle and L. Ugarte, *On certain geometric and homotopy properties of closed symplectic manifolds*, to appear in Topology and its Appl., available as math. SG/0002071 from xxx.lanl.gov.

[IRTU2] R. Ibáñez, Y. Rudyak, A. Tralle and L. Ugarte, *On symplectically harmonic forms on 6-dimensional nilmanifolds*, Comment. Math. Helv. **76** (2001), 89-109.

[L] D. Lehmann, *Theorie homotopique des formes differentielles (d'aprés D. Sullivan)*, Asterisque, Soc. Math. France **45** (1977).

[M] S. Merkulov, *Formality of canonical symplectic complexes and Frobenius manifolds*, Internat. Math. Res. Notices **14** (1998), 727-733.

[O] J. Oprea, *Homotopy theory of circle actions on symplectic manifolds*, Banach Center Publ. **45** (1998), 63-86.

[OT] J. Oprea and A. Tralle, *Koszul-Sullivan models and the cohomology of certain solvmanifolds*, Annals Global Anal. Geom. **15** (1997), 347-360.

[RT] Yu. Rudyak and A. Tralle, *On symplectic manifolds with aspherical symplectic form*, Topol. Methods Nonlinear Anal. **14** (1999), 353-362.

[T] A. Tralle, *A note on solvable Lie groups without lattices and the Félix-Thomas models of fibrations*, Preprint, available as math.DG/0009105 from xxx.lanl.gov.

[TO] A. Tralle and J. Oprea, *Symplectic manifolds with no Kähler structure*, Lect. Notes Math. 1661, Springer, Berlin, 1997.

University of Warmia and Mazury in Olsztyn, Żołnierska 14A, 10-561 Olsztyn, Poland

E-mail address: tralle@matman.uwm.edu.pl

Contemporary Mathematics
Volume **288**, 2001

Where does Geometry go?
A research and education perspective

Jean-Pierre BOURGUIGNON, Eugenio CALABI, James EELLS,
Oscar GARCIA-PRADA, Misha GROMOV.

ABSTRACT. After brief statements by panel members, the topic of the round table has been discussed with the audience. The main points that emerged from the exchange are: the evolution of geometry in the past century and years to come, the extent to which geometers are doing enough efforts to keep informed of geometric works done by other scientists, the impact of computer tools on the future of geometry, the future ability of mathematicians, and geometers in particular, to attract good students.

Jean-Pierre Bourguignon (JPB): It was the organisers' idea to close the conference with a round table. They wanted to take advantage of the proclamation of the year 2000 World Mathematical Year to look back and also to give us an apportunity to try and project ourselves into the future.

For a mathematician it is very tempting to draw a perspective comparing the year 2000 to the year 1900, which is remembered through David Hilbert's lecture at the International Congress of Mathematicians held that year in Paris, in which he proposed his famous list of 23 problems. If one now looks back at Hilbert's list, it is a fact that geometry as a whole is not that much present. Indeed many more problems are devoted to algebra. It is interesting to contrast this with the fact that the 19th century has been extremely fruitful as far as geometry is concerned. Recall, for example, that in the early part of that century the fundamental barrier of having a unique model for geometry had been broken. This major event in the history of mathematics was followed up by many others. We can quote a few: the groundbreaking work of Bernhard Riemann establishing Riemannian geometry as a comprehensive subject in which curvature plays a central role; the Erlangen programme of Felix Klein, identifying a geometry with the set of properties invariant under a group action; and something perhaps less well known, the systematic use of geometric models to describe natural phenomenon by Hermann von Helmholtz.

1991 *Mathematics Subject classification.* Primary 53; Secondary 51, 52, 97.
The participants to the round table thank Marisa Fernandez and her co-organizers for the great efforts they put into many details to set up the conference which resulted in a wonderfully congenial atmosphere.

The century closed on a great variety of works, from which I would like to single out the work of Sophus Lie on continuous transformation groups.

At that time, that is 1900, nobody could really foresee that in the 20th century geometry would undergo a major evolution and play a central role in the development of mathematics. This is indeed the impression that comes out of Henri Poincaré's attempt to propose some avenues for the future of mathematics in the 20th century in his opening address to the Paris Congress. Nevertheless, it is a fact that this century has really been a great century for geometry, in particular in its relations with physics, to start with nothing less than the big boost given to these relations by Albert Einstein's theory of general relativity.

So, here we are today, trying to look into the future, and asked to point to some directions in which, according to our opinion, geometry will develop.

We know how daring this exercise is. Nevertheless, I personally find such an endeavour very appropriate at the end of a conference devoted to the memory of Alfred Gray. Alfred definitely contributed to many areas of geometry. At the end of his life he was actively engaged in convincing colleagues that one should really take advantage of the new tools provided by computers to make good propaganda for geometry in order to help form new insights through images.

The format proposed is to have the panelists make brief statements. The most important part of the morning should be what I hope will be a very lively and open discussion with people in the audience.

Eugenio Calabi (EC): Over the years, I came to the conclusion that one can only speak very subjectively about what constitutes geometry. I see it as a field in which the primary source of intuition is ultimately tied with our sensory perceptions of the world. Of course, as we reach more abstract areas, one has to interpret what sensory experience means. And I've tried to make that as visible as possible and to communicate this view. I must say that communication can be difficult in that respect, something that made me very much concerned with the process of education in this area... although I've never been known to be an effective lecturer at the undergraduate level.

The idea of communicating what is our source of intuition has been changed by the big impact the advent of computers had - note that it is only twenty years ago that I saw the first one in operation. And that affects only the last one fifth of the century. In the future it will have an impact that is now difficult to evaluate in full. For me, geometry is as essential as other areas of mathematics such as analysis, algebra, combinatorics, statistics, probability. It is an applied field because being very largely problem-solution oriented. We are bugged by different problems that we try to solve, and we try to master the tools that seem to bring us closer to the solutions or give us more insight on them until, being very optimistic, we really make a breakthrough. But I've gone through periods when I felt I had to study algebra and analysis, functional analysis and so on.

Aside from that, I would like to know how my colleagues sitting here feel about the same problem: what constitutes geometry? It's certainly very different from what I thought of as geometry when I was a high-school, or even a college, student.

Oscar Garcia-Prada (OGP): The point I want to emphasise is the interplay between geometry and physics, and in particular how physics is a source of very challenging and deep problems in geometry.

Here are a few instances of this interplay tailored to my taste and to my knowledge and limitations. They are coming most of the time from quantum physics, and in particular from the understanding of the geometry of gauge theories:

- the influence of general relativity on differential geometry;
- the influence of classical mechanics on symplectic geometry;
- the appearance of instantons in Yang-Mills' theories and their use by Simon Donaldson to actually define invariants of 4-manifolds, using the moduli spaces of instantons, as structuring objects.

In many of the examples, I see a path going from physics to differential equations and partial differential equations, continuing to differential geometry and algebraic geometry, and finally arriving at topology. In fact this connection stresses the deep relations existing between differential geometry and algebraic geometry. In particular in the example I just mentioned, the moduli of instantons has a reincarnation in algebraic geometry in the moduli space of algebraic bundles. And this link was the tool used by Donaldson to compute his invariants.

There are many instances where topological field theories have been used to define knot invariants such as Jones-Witten theory, leading to substantial applications to low dimensional geometry and topology. There are also important applications from physics to higher dimensional geometry such as the contribution of Yang Mills' theories to the study of Kähler groups via the theory of Higgs bundles initiated by Hitchin and Simpson, leading to the solution of problems that people like Gromov and others had studied.

The list is very long. Coming back to relativity theory I want to mention the development of twistor theory by Penrose some thirty years ago. For him this was a way of understanding space-time in a deeper way, trying to solve Einstein's equations by converting them into complex geometric terms, in a more fundamental space that he calls twistor space. This dictionary has become an important tool in the study of differential, in particular Riemannian, geometry, for the study of subfields such as quaternionic manifolds, harmonic maps and many other things. This lead to the explicit solution of instanton equations via the beautiful construction of Atiyah-Drinfeld-Hitchin-Manin.

These are also instances in which somehow an important aspect of the interaction is a general idea of symmetry, such as the recent and challenging idea of supersymmetry and the understanding of the geometry underpinning it. Many beautiful theories are coming from supersymmetric theories, like e.g. Seiberg-Witten theory and a deeper understanding of Morse theory. In the same list one finds the systematic study of Calabi-Yau manifolds by physicists and their discovery of a new type of symmetry, mirror symmetry, with a new emphasis on interesting objects such as special Lagrangian submanifolds.

The point I want to make is that today a working geometer should always be looking for new ideas coming from physics. Since we are supposed to discuss here also about education and how the teaching of geometry should be done in universities and schools, I would like to point out that physics must be part of the curriculum of the undergraduate degree in mathematics. And this is almost never the case in Spain.

James Eells (JE): When I got this invitation from Marisa, or I should say order from Marisa, to participate in this round table, of course I accepted. It's hard to refuse Marisa, as you know, but I certainly entered with a good deal of trepidation.

My real, serious question is: is this worthwhile at all? For instance, I found that I could take the list of topics and suggestions that Jean-Pierre [Bourguignon] had put down and, systematically going through it, I could replace "geometry" by any one of many other terms of significance. I mentioned at breakfast this morning that one could replace it by "beauty" as a concept and get just as interesting and in my view just as meaningful a formulation of a problem.

I was a little bit nervous about this round table because the one thing that I would find quite unfortunate and inappropriate would be that we should legislate too much, in our own minds at least, on what geometry is and where it's going, and what should be included, let's say at the research level. The basic viewpoint that I personally have is that whatever geometry is, in my study about whatever subject I deal with, it is very important to think geometrically. I can't do it otherwise. And certainly when I think of analysis, as I remarked to Jean-Pierre [Bourguignon], I certainly think in geometric terms. I had the very good fortune of having three very great influences in my formative years in mathematics. Coming from English Literature I went to Harvard and studied with Hassler H. Whitney. We never spoke a word of mathematics, I might say, in the forty years of our friendship, but of much else. And then, surprising for many, Jean Leray, who treated me like a son. I don't know how many people can say that, but I'm very proud to be able to say it from the heart. Then Victor Hurewicz. If any of you ever knew Victor Hurewicz you knew a man of just unbounded love for all human aspects of life. These mathematicians certainly taught me above all not to make sharp delimitations of what is geometry or what is algebra or what is what. If you look at the works of any of those three splendid teachers of mine, you'll see how much they influenced me in being absolutely unrestricted in their spheres of interest.

The most important single point about which I had concerns here was that this limitation should not be formulated in any way that is morally binding or even predicated in our thinking. And with that in mind I pass the word to Misha Gromov, who is certainly one of the most universal amongst us, and I dare say he will have very clearly defined views.

Misha Gromov (MG): I would like to continue what I was saying in my lecture *"Separating inside and outside views of mathematics"*. You have just heard the inside view. So let me propose an outside one. It will be a specific outside view on geometry and geometers given by one of the most powerful geometric minds of our time whom I met. You may not know him, his name is Manfred Eigen. He studied physics and invented a new type of spectrography. This is a remarkable technique, based on geometry. And then he worked on chemical evolution. But in some sense the kind of geometry he does, geometers don't do it and don't know it. When I first met him, I didn't quite understand. I asked him how he understood geometry so well, and then I realised, after somebody explained to me later, his major work in physics is based on how the geometry of high dimensional space gets projected to lower dimensions. He doesn't just analyse them. He found specific formulas and technology for how to measure relaxation times for interactions between molecules, and also the association/disociation of water.

His next contribution was to the theory of chemical evolution. He actually made some specific things with his hands, and gave a very elegant model of evolution specifically applicable for viruses and bacteria. It very well applies to the evolution of geometry and geometers. The model is a kind of random cloud of points moving in a landscape following a gradient exhibiting small perturbations. This determines some critical values, and depending on these a variety of different behaviours. And when a parameter reaches a specific value, ones goes to a deep well and stay there for a very long time. One may become extinct if there is no fluctuation there. If one gets out, successful development will occur, depending on a particular range of parameters, provided there are not too spread out in order that species preserve their identity, and they rarely make big jumps in order not to stagnate.

When I think of geometers, I see a major point in the jumps. We should give more attention to them. In my view, we are stagnating. We don't make jumps. We sit in a very narrow niche and we don't even know about great geometric developments happening next door.

This is my criticism of present day geometers. My suggestion is that we must make efforts to allow big jumps, and not idealise what we do and say how great we are. This is really why I feel we are in a still very primitive stage. We don't know the basic facts about space. Since Riemann we haven't progressed at all, and even spaces studied by other people like physicists, e.g., the huge evolution spaces studied by Eigen, we are not even aware of them. Really we are in very poor shape. That's my statement.

JPB: Thank you, Misha, for these provocative thoughts.

We should really start the discussion now. I would just add for myself two comments. The first one is that it's extremely hard to identify what we need to do and to make predictions. It's very important that we explore many predictions so that we don't have only a one-way street ahead of us.

Nevertheless, we still make decisions, e.g., we chose topics of courses and we design curricula. This is in fact our everyday responsibility as teachers. We actually train the next generation in a certain way, and this is not neutral. Of course training is not the end of the story because, as you know, the best students are the ones who, after some time, completely forget the details of the training and just keep its spirit. That was my first point.

The second point concerns boundaries. Certainly one of the worst things that could happen for the development of our discipline would be to dissect mathematics in a way that finally creates very strong separations between its subfields. In relation with this question, at the end of the 20th century, within geometry one witnessed a blurring of the boundaries between the various types of geometry studied today. One has for example realised that complex geometry is not that far away from real geometry, that symplectic geometry is not that far away from Riemannian geometry in some ways. Another idea is exemplified by the study of moduli spaces of structures. It's indeed important, when studying a specific structure, to look at neighbouring ones, something that physicists have been using consistently over the years with great success. Mathematicians have been slow in getting interested in this type of questions, and were more or less forced into it, because they were asked very specific questions about, say, dimensions of moduli spaces or more specific information about them. But it took time to them to really take it as an interesting object in itself. In some sense my second comment echoes what Misha [Gromov] said. It concerns the extreme emphasis that has been put for a long time on continuous structures and, correlatively, the great neglect in which in some countries, France is one of them, the discrete side of things and combinatorics in various disguises have been kept, completely separated from geometry. The result is that they remain largely ignored by many colleagues, and often considered as weak mathematics. This is one of the walls that, I think, has to be put down. We need to help our students form a better global idea of the whole field.

Let us now listen to questions, comments, statements, everything from the audience.

Luc Lemaire *(Université libre de Bruxelles)*: I have a specific question. Non commutative geometry has not been mentioned by anybody so far, and although it's difficult to make predictions about the future, do people in the panel feel that non commutative geometry is an important subject that will develop. Should we ask our students to make the rather heavy investment that studying this subject represents.

JPB: Two comments on this. The first one is that, as Misha [Gromov] said, since Riemann we haven't had any real new insight into the structure of space. It's true that non commutative geometry forces people who want to use it to change their approaches and their feelings about space.

The second comment is that non commutative geometry is a structure in which the continuum and the discrete are unified, precisely one of the features that I put forth in my preliminary point. When developing non commutative geometry, Alain Connes, and also others, were definitely inspired by physics. In the last three or four years, it has been amazing to see a number of physicists, coming from different viewpoints, who got interested in formulating their problem in the non commutative world. This applies to Ed Witten, originally skeptical about non commutative geometry, who realised that results he recently achieved could be obtained following a non commutative path.

Misha [Gromov] was asking us to look for such enlargements. In my opinion non commutative geometry is definitely a direction, but there should be steps in other directions too. I don't believe it would be a good thing to think only of a one dimensional space. One has to be ready to go in several different dimensions.

EC: This reminds me still of the other question about intergenerational communication - the form of teaching, I mean. Each generation grows up and is educated with one view of mathematics and experiences of the evolution of it, and then has to try and communicate to the next generation how one's own feelings are. I remember myself when I was in graduate school. I would be starting a course on which I thought I knew something to begin with, and when a slightly different viewpoint was being presented in the lecture room my private reaction to that was one of cold hostility. My first impression was always that the lecturer was trying to put something over on us. Almost invariably I was wrong. It was just that I had to modify my way of thinking, my way of looking at the same subject. The small amount of technical things I would learn was really over-ridden by these changes.

My next thought is that, human beings having roughly the same capacity for learning from one generation to another, we have to learn as teachers what is important and decide what is best understandable if not forgotten in our own teaching, even though we may retain some love for it. I still love the idea of geometrical constructions with ruler and compass, even though nobody ever teaches them nowadays. I would like to hear people express their views on what areas of geometry not only should be emphasised, but what areas can or may be omitted or at least areas we could take the risk of not communicating in detail.

Antonio Ros *(Granada University)* **:** I'd like to introduce an idea that is closer to the area I am interested in at the present date, which is geometry of surfaces in 3-space. This area is somewhat special inside the world of differential geometry because we have at our universities an undergraduate course called something like "classical differential geometry of curves and surfaces". In many countries and many institutions we exist as professional geometers because this course belongs to the curriculum, and if tomorrow this course disappears, then our universities and our government may not want to have differential geometers any more. In my opinion our future is linked to the future of this particular course.

On the other hand almost everywhere in the world a new geometry is emerging very quickly. I mean the kind of computational and discrete geometry that puts more emphasis on applications. For instance, every day one can hear that *"video games are a direct consequence of computational geometry"*, and this sort of things. Presidents of universities may start to wonder about the type of geometry a student must learn. And here is my question: what is the attitude we must take in order to preserve classical differential geometry and in order to convince people that this geometry is also a useful one? Basically, must we or even can we do something about this question?

JE: Had I spoken three minutes longer - and I won't speak three minutes now - I would have remarked about a concrete example or two, amongst many available at this meeting, to illustrate my point about the basic unity in mathematics as someone with a geometric taste. I would have chosen two spectacularly fine examples: One by Antonio Ros, who gave a lecture which had both a geometric spirit and an analytical one; And the other one by Eugenio Calabi (may I say "of course"), whose presentation was both analytical and geometrical, and peculiarly "Calabian". I say this with a good deal of affection.

But it does seem to me that a basic unity is at work, e.g. when one considers the basic influence geometry has when one does analysis, or probability theory, or algebra, or when one works in the direction of Jean Leray. This inter-relationship is vastly more important than course assignations. This is a point I wanted to make, and that Antonio [Ros] himself made very, very clear to me in his lecture. I told him so afterwards, so excuse the embarrassment of repeating it now, but I thought this was worth it.

Santiago Lopez de Medrano *(University Autonoma de Mexico)***:** We must observe that we are reaching the end of the 20th century with a good balance between - there are different ways of saying this - geometry and analysis, pictures and theorems. But we must remember that this hasn't been the case all the time. Let me recall that Lagrange was very proud of his book on celestial mechanics, that didn't have a single picture in it. Of course, out of the work of Lagrange came a beautiful new geometry that should be called Lagrangian, but is now called symplectic.

In the middle of this century there was a big attack on geometry from which we are still suffering in schools. The emphasis was then put on anything that had to do with logic and rigour, and pictures and geometry and the thinking around geometric figures was practically banned. In research mathematics the stress was put more on algebraic and logical aspects than on geometric ones. So it is very good that we have arrived at a point where, in my opinion, there is a good balance, and of course Alfred Gray was one of the proponents of improving the balance between these aspects. But, as Professor Ros mentioned, there are new challenges, there are new dangers, maybe we are now being pushed to a situation where we will have beautiful pictures but no theorems. And in particular in teaching, there is a very strong emphasis on computers and calculators in school, and of course it is not necessarily the best aspects of calculators and computers that are taught. So we have to be very careful about this. We still have to fight for geometry. Everyone of us that had something to do with setting up programmes in high schools and universities has witnessed that geometry is always under attack.

JPB: Let me just try to pick up what you've just said. It's true that in some cases there are attempts to really wipe out geometry from programmes, but doesn't that mean also that we are just off the main stream and that we, geometers, have not made the necessary efforts to really incorporate into our teaching, or even into the tools that we are using fluently, some new tools that some people feel are really relevant for many things, including modelling. It is quite clear that doing modelling today without at some stage using a computer is just inappropriate. You know that the computer is just doing what you ask it to do. But for the moment, for example, the fact that the level of development of adapted software able to deal comfortably with images is not so satisfactory may also be the proof of our limited engagement. So there is always some balance between the pressure from outside, which most of the time is over-simplifying the situation, and that coming from inside. Incorporating computer tools into teaching is something which can in the end introduce into the minds of students the idea that after all you don't need to think geometrically since a computer is doing that for you. But, if you ease too much the understanding of geometric concepts, you may actually achieve the exact opposite.

So are there some people who have personal experience about this kind of thing or strong personal opinions? In my opinion, in the future it's something we will not be allowed to neglect. We will even have to make sustained efforts in this direction.

Karsten Grove *(University of Maryland)*: I would like to return to the beginning of the discussion of the panel, because I believe it is actually destructive to leave it where it was. From a personal point of view I consider it is very important in research to be always critical about where you are, what you do, but also to enjoy it. And the same thing goes for talking about geometry and mathematics in general. We heard from the panel what a great century this has been for geometry on the one hand, and then we hear from Misha [Gromov] what a terrible shape geometry is in. It would be too bad to leave it like that. I would like to have a much more balanced view. We should recognise and be happy about what we all know about geometry. But at the same time we should be critical as Misha is and try to do better.

On the other hand, it is always good to try and have some ideas about how we would do better. And I would like to hear more about the positive side of these views, namely not to say *"we're in terrible shape"* but *"let's see what we can really do"*. That would be very important.

Jair Koiller *(Laboratorio Nacional de Computacao Cientifica, Brazil)*: I'd like to go on with the line of questioning of Professor Grove, actually. Just to provoke Professor Gromov, because his talk was so inspiring about the views of in and out math, I think he's hiding something, because I keep hearing about the necessity of new ideas of space. We heard this twice today and once in his talk. Is it possible that he can give some hints of what the structure of space is that he has in mind, learning from his context with biologists? So that we will have a more optimistic view.

MG: You know there are two separate issues here. One is purely mathematical - what type of spaces you may envisage or expect to be there if you don't know much about them - and the other is what sorts of geometric problems are being considered not only in biology but also in everyday physics that we are not aware of. There is remarkable geometry going on, as I mentioned, in the work of Eigen. Recently there was also some very remarkable work about how one can apply geometry - and exactly the kind of geometry we are doing here - to kinetic equations. They were looked at from a geometric angle, namely via some projection on the space of solutions to the Boltzmann equation. It's very simple and clear but, when you write it down, you have huge, complicated equations, which happen to be pertinent. They describe correctly the kinetics of chemical reactions, and people doing that can improve the efficiency of these reactions and can control them better, giving some 20% better input. We don't know this geometry. If we were more involved in that, we would not complain that universities don't take us seriously. We could come to them and say *"you see, we can improve your chemistry"*, something we just do not do. I learned this recently and feel really ashamed that I never knew it before. One point for us generally is we have to learn for our research before we start thinking about how we teach our students. In my opinion we do not know enough geometry to say what geometry is anyway. These are just two instances that I have learned - I'm certain there are billions of them in science - of remarkable geometric structures, which are extremely useful and new, and we don't know about them. We have not so much to teach but to learn. That's one part. I have been doing this for a few years and found more and more things others use in elementary courses that we still don't know about. This is unpleasant to discover if you have lived all your life as a geometer.

Secondly, the geometry of space requires a more technical discussion. I can just make some suggestions on the situations where you expect the emergence of new spaces. And one of them is class field theory. There are indications of the existence of a very remarkable space in the little glimpses we have on the other side of the mirror. We need to find the proper lens. Definitely this must be part of a much bigger body of knowledge containing a mixture of symmetries. Objects of the same nature seems to appear in pure differential geometry, e.g. in Hodge theory. This determines much of the remarkable structure we observe, and we don't know what it is.

JPB: If you will allow me one word on this. Instead of claiming we already know what space is about, another approach is to look where challenges are, because very often mathematics has been changing because mathematicians were facing big challenges. Just to give an example, I was really struck by the lecture given on data analysis by David Leigh Donoho, professor of statistics at Stanford University, at the American Mathematical Society meeting *"Mathematical Challenges of the 21st century"* last August. He presented very challenging problems which in my opinion require developing some kind of a geometric vision. In the area of data analysis, because of the availability of new computer tools one can present the data more and more using pictures, and the question is then: can one have geometric structures emerge from these pictures? It is likely that an approach completely different from the one we have been developing so far will be required. Requests for such a treatment will be coming from many different fields, e.g. seismology, astronomy, molecular biology, in particular genomics, in fact from all areas which can generate huge amounts of data. There is a need to organise these data, in order to deal with them as a global object with an internal geometry that needs to be identified.

This will require on the one hand we be not too conservative and do not claim too easily we already have all the tools, and on the other hand we be not so destructive that we claim everything we learned has to be forgotten. For me such a mix is needed, but a good way to have our minds pressed to go deeper is definitely having to face some big challenges.

Joseph Wolf *(University of California at Berkeley)***:** One of the things that I wanted to comment on is the use of geometry and special functions. A lot of special function theory can be put into the context of looking at a manifold of some sort, for instance a sphere, and looking at eigenfunctions of the Laplacian. Up to now, people have mostly looked at the spectrum of the Laplacian. They haven't actually looked at the eigenfunctions, except in very specific situations, for example spherical harmonics.

Now, in those specific situations often there is good classical theory which is extremely boring unless you are going to use it for something. But then it turns out to be related to something more geometric, some group action or something of that sort. Very often you have an interesting geometric situation which you discover by computing some explicit examples such as Riemannian symmetric spaces, and then turn out to be - with the right viewpoint - something more general in Riemannian geometry, for example product decomposition or de Rham theory.

So, what I suggest is that one looks to the possibility that you have a surface or a Riemannian manifold of some sort which is described in a very geometric way, and then you could hope to look for the Laplacian and try to find the eigenfunctions using the computer in some way. This is an area which simply has not been explored except, as I say, in a few special cases where there are auxiliary tools. Computer technology should provide more and more exciting ways of penetrating this open domain as it improves in the future.

EC: I just want to bring back this idea that we do have a mixed view of what geometry really is. On the one hand, we have the viewpoint that geometry evolves from axioms to theorems. On the other hand, there is the definition at the opposite end that geometry is a part of mathematics which draws its intuition ultimately from our perceptions of physical experiences with the real world. Now, you can be extreme in one way or the other. From the standpoint of the perception from the physical world, that empirical view of geometry could be regarded as science fiction, if you push it to extremes. Real numbers which were initially conceived as ratios of measurement of quantities became mathematically efficient only when you apply the Jonathan Swift principle that everything can be re-scaled up and down indefinitely often, which is in fact completely contrary to our current conception of the real world. So this ambiguous origin, namely being on the one hand very rigidly tied up with logical reasoning and rooted imaginatively in our perceptions of the world, is in part the charm of geometry, but more generally of mathematics.

MG: For me, the position of logic versus intuition is purely emotional. You cannot justify it in any way, and moreover this may be a negative phenomenon because we try to use words with connotations and conceptions that actually don't exist.

EC: I will agree with you that logic is a process of instinctive acceptance, according to rules that most of us agree upon. Otherwise, if everybody challenges every argument that's being presented ...

MG: No but there is a particular specific issue here. We believe there is a foundation of logic, however we fail to prove it, and the proof would be making an artificial proof system.

Joseph Wolf: One has to admit that there is some basic level on which we understand each other, or at least understand the words that other people are using. So you cannot say that our notion of logic is completely wrong. Perhaps it leaves something to be desired.

MG: Indeed it is not completely wrong, but it's not what helps us understand how mathematics work. For example in communications we get the vibration of air and the theory of sound. There again is a basic ingredient which probably makes sense of everything we know. First it is the absence - manifested by the absence of a computer program of this nature. People believe that logic responds to thinking, they try to come up with such a programme and fail. So when you're correcting it, it's like religious belief. But if you want to make a scientific theory, completely forget it and go from a different angle.

Simon Gindikin *(Rutgers University)*: For me, we have to face a very painful problem, about which our friends from Spain and Mexico already talked. What we can observe now is that mathematics has lower and lower priority in the world and also that less and less young people want to become mathematicians. This is at least the case in the USA where I am now. I have worked for 40 years in mathematics, and sometimes I think this was the best time. It may be that the best possibility for mathematics is in totalitarian countries, as it was in the Soviet Union, where the best young people went into mathematics. I believe we have a problem there. For example in the USA at my university nobody would think to teach physics as part of the education of professional mathematicians, and only 5 students every year take the differential geometry course.

There is one point we did not discuss here, namely the relationship between pure and applied mathematics. I understand perfectly that to sell mathematics and obtain some grants we need to insist that mathematics has a lot of applications. But let us be honest with ourselves: how did the whole thing start? In Plato's academy future kings studied geometry as first subject, not military arts or economics. But after they developed their minds, they could do many things. We now observe that, when good mathematicians after their PhD's cannot find a reasonable job, they go to Wall Street and apply not exactly what they studied during their PhD or graduate education but use the mind that was the result of their mathematical education. This helps even in such a strange environment as Wall Street.

So, in my opinion, it is very important to keep the traditional role of a mathematical education, not as direct and applied information, but as part of the general culture, for the sake of developing reasoning ability. And I must say I am very pessimistic at least in my environment in New Jersey.

JPB: I'll challenge what you just said. My starting point is that, in recent years, the call for mathematical knowledge in a very broad sense came from many different areas of industry and services with an unprecedented strength. Usually, when a mathematician is employed by a company, his training, in particular if he has gone as far as obtaining a PhD, is used more comprehensively than through his or her academic speciality. But at the same time the knowledge needed to deal with very different types of problems that one meets outside academia is extremely broad. Note that this also corresponds to the present situation in mathematics where ones has to borrow knowledge from many different sub-areas. So I don't feel appropriate to state that, if one is employed as a mathematician somewhere, one has to forget one's mathematics. It has to be used in a different way, but certainly for more than just the training to reasoning. One is actually forced to use various sorts of mathematics, some of which simple, some more advanced, and sometimes even some waiting to be developed.

This new situation has to do with resources presently at the top of the shop list for industry: more and more formal data, more and more things of a virtual nature, more and more complex systems... Mathematics is precisely the science which has been developed to deal with such data. So, in a sense, in terms of interactions with society in general and more specifically concerning jobs to be offered to students, I would be more worried about the future of my discipline if I was a physicist.

It is true though that we, as mathematicians, tend to see the choice made by some of our very best students to take jobs outside academia as a loss for the whole community. My viewpoint on this is that we have to be less definitive about it, because, if we have people in these positions who have a great respect for mathematics because they enjoyed doing it, they will be our friends, or even our ambassadors. They will help us develop mathematics possibly in different ways. One of the best ways in which mathematics has developed over the centuries has been by coping with challenges coming from many different areas. It is true that sciences, and natural sciences in particular, have been one of the absolute main sources of challenges for mathematicians, but there are many others. In his lecture, Misha [Gromov] has been stressing challenges coming from biology, but there will also be many challenges coming from the services, human activities that didn't exist at that level and with this extension in the past. As you probably know, the development of probability theory came originally from problems connected to census, insurance, and demography.

Therefore, I would be less pessimistic than Simon Gindikin. Again we, as teachers, must take a broader view on what training is to be offered to our students and what relations we can keep with them. Of course when they go and work in companies, we don't see them as often as if they were in research institutes, but we have to find mechanisms by which we still are in contact with them, so that they can come back to us when they see challenges they feel are deep enough to be addressed in themselves. I am sure many such challenges will be brought to us. I say this on the basis of my personal experience with students at L'École Polytechnique. Some of them having left completely the scientific world and working as engineers come up to me with really challenging questions. In fact they have still sufficient trust in mathematics to call me up and pose questions to me as a mathematician. This is one of the reasons why I am less pessimistic than Simon, but I may be wrong.

MG: I absolutely agree with what Jean-Pierre [Bourguignon] said but the main point is that it's up to us to do this, and to broaden our perspectives. And we will have to work hard to achieve this. In particular we will have to learn things we presently don't know.

Luc Lemaire: I also agree with the point of view just expressed about the way to attract people to mathematics. If we present the situation in the way we see it, we will be tempted to say that the good guys remain at the universities and somehow are paid to teach and given free time to do research. This leads to a very small mathematical family not really connected to the outside world, and in particular it makes us regret when mathematicians leave our small world. The risk is that we shall dry out somehow, and also we won't be able to attract young people. On the other hand, if more and more often good mathematicians go outside academia and obtain positions in which they can use their mathematics in very different ways, in ways visibly useful for society, a proof being that they are given good jobs, then you can expect to have more people starting to study mathematics. And so, even if we seem to lose some students who go away, we can win a much larger number of them because more will come and study mathematics.

Concerning the question raised of the disappearance of geometry in high-school programmes, the problem is not limited to geometry. In fact, in many countries, governments tend to press down the level of difficulty of high-school studies, as a form of demagogy to please parents. Children have to be able to pass and obtain good marks in exams without really working or thinking hard. And so everything that is too difficult is sort of pushed away from the programmes. And of course geometry is an example of something difficult because one has to think when practicing it. You don't have a crank that you just turn to obtain the answer. Some types of algebraic manipulations and computations can be easy, not all of them of course, and they are easier to teach. When students have made ten of them, they can make the eleventh one, and so everybody is happy because they can pass the exams. For me, the fight to maintain geometry in curricula should really be the fight to maintain opportunities to think and face some difficulties to solve problems. Whether they are problems in geometry or in probability or in combinatorics is really not so important for me. It is really the teaching of minimally difficult things which is presently put in jeopardy in high schools.

EC: We've been touching on our role in dealing, as Misha [Gromov] put it earlier, with mathematics and on the need for mathematicians to talk to the outside world. I have often been asked *"What good is what you're doing as a mathematician?"*, and I've tried to work out the most convincing answers. We live in a complicated world where nobody does one thing completely. I mean there is a tradesman, a builder and a carpenter. They need tools. Some people are attracted to working with their hands and doing carpentry and building, and others prefer being toolmakers. There are people who love making money out of oil, but they also need prospectors and wildcatters to find the oil. In some sense, we are at the wildcatter end of the industrial world. But we have to sell the fact that even an oil magnate would not get very far unless he did some prospecting for unexpected sources of oil. We are the wildcatters in the sense that we have to look for the new sources of information for science.

Santiago Lopez de Medrano: The point of Professor Gindikin was not that it was bad that some mathematicians leave universities and go and do different things. For me the important point is that something is happening in teaching the way we do that stops students from going into mathematics, into that real mathematics that will be useful in universities and outside.

The Chairman asked about experiences with teaching and computers. Let me go back to Alfred Gray, who, as you will all agree, was enthusiastic about everything. But one thing he wasn't enthusiastic about, when he visited us, was going back home to teach. No, he wasn't enthusiastic about that. He wasn't teaching differential geometry of curves and surfaces. He was teaching some low-level course with calculators, and he said *"I have to go back and teach students how to push buttons"*. That's exactly what he said.

Given the programmes, given the structure of the curriculum and the current general philosophy of education, many mathematicians will have to do teach some very mechanical aspects of mathematics. In that way we are not creating the geometers who will be very good for research and for applications and all that. This is the point.

JPB: You may be right in stating or formulating the question this way but I don't believe that this state of affairs is completely out of our hands. It depends a little bit on how we project ourselves and whether we make the personal investment into having some knowledge of what is around us, and how we can approach physics and other disciplines from that point of view. We need to be probably more open than we are in general. From that point of view the key issue can be what sort of documents are available to really get more knowledge about other things, but also how can we create events in which we really get some inside view on how mathematics is used elsewhere. If one really wants colleagues from the mathematical community absorb information on how mathematics gets used here or there, one really needs more than personal contacts, documents in particular. Documents of this nature are rarely available. As a community we probably have to create these opportunities, both in reading, in writing and in hearing and meeting the right people.

Joseph Wolf: When you come to a problem in applying mathematics or where you're asked questions about possible mathematical tools in some industrial or biophysical setting, usually the problem is very difficult to understand. The hard part is often not the mathematical tool but to understand well enough what the problem isto pick out the appropriate tool. Some of our students who do not find jobs in academia go into industry or analogous settings. We need to have them come back to us and talk to us. This is extremely important, because it helps build a bridge through which one can have the sufficient background in some non academic setting so that one can understand questions and perhaps have the possibility of contributing something to such discussions.

Helaman Ferguson *(Washington)*: I am a sculptor and in my sculpture I celebrate mathematics. So I would like to return to the title of this discussion: Where does geometry go? We are living in a golden age of mathematics. I don't think anyone would contest that. Not only is it a golden age because of the theorems and achievements but also because of the problems that we have in mathematics, and the ones that we are confronted with coming from outside. As a group of humans on the planet, we are communicating much better and much more clearly than we ever have at any time in the past. As mathematicians, we get to view problems in other disciplines that we would not ordinarily be able to do. This brings me back to the question of where does geometry go? In my work as a mathematical sculptor working in stone primarily, I endeavour to celebrate mathematics, celebrate the theorems and also celebrate the problems.

I have a patron who has acquired 200 acres of property in Maryland, and he has invited me to develop it as a mathematical theme sculpture park. And it is my particular goal to put in good settings in various places of this property, mathematical theorems - multi-tone granite theorems. And also multi-tone granite problems as well.

Geometry can go on this 200 acres somewhere, and so I am inviting you to suggest what you think are the most beautiful theorems and what are the most beautiful problems in mathematics and geometry specifically and communicate those to me by e-mail, or some of you may have some immediate thoughts right now. But I am inviting your participation in this mathematical garden of delights.

Joseph Wolf: Let me say something optimistic, although perhaps it doesn't fit in too well with some things that other people have said. I think that one of the things that led to the flowering of geometry was the Russian launch of the Sputnik. Indeed, as a result, it was no longer so bizarre to study mathematics, it wasn't so weird to learn something about science, and in the USA at least many more students in secondary schools and universities were exposed to much more science than they would have been otherwise. Suddenly, it became respectable to study, to learn some science, even to go to a library on a weekend. Of course I realise this is quite different in other countries, and in particular was very different in Russia, where the intellectual tradition was rather different.

But now what has happened is that respectability has shifted more towards applications, more towards things that you deal with directly with computers. I don't really think that in the long run the study of geometry is going downhill. It is perhaps an oscillation which will rebound, perhaps in view of applicability, in view of some technological developments that are difficult to foresee at the moment.

Luc Lemaire: If I can also make an optimistic statement, when I think of what I have simply observed in the last 30 years in geometry, I really feel I have seen some fantastic things. I could for example take the work of Donaldson on 4-dimensional manifolds, since he's not here. It is fantastic and completely unexpected. It was certainly not planned in advance, and it is not a panel which said 20 years before *"we should study such and such equations of physics because it will be relevant to understand the geometry of 4-dimensional manifolds"*. That is of course part of the excitement, that there are all those unexpected discoveries.

About the future I am very impatient to see what surprises will come, and what new results will come, and that cannot be foreseen at all now. If they could be foreseen, they would not be that interesting!

JPB: The topic we tried to focus on was definitely where does geometry go, for the good reason that, a priori, we are the experts who should be able to say something about this question. At least we should be able to come up with some vision of this. What we heard shows that we are definitely all conscious of the fact that it's very important to keep geometry open, and not to restrict it to things we have been doing. Each one of us has his or her own little garden that we like to cultivate, and having these little gardens is certainly extremely useful, because we need to have fun when we do mathematics, when we do our research. But we should know that our little garden lies on a much bigger planet, and we need to get to know the other side of the planet.

The other point made concerns the need for this openness in relation with the training of students. Eugene [Calabi] said it very well: at some point you really need to communicate to the students what is really most important, and what you really consider the heart of the matter, which is why you like things and the things you do. At some stage this has to be carried across to the students, and this is not so easy. Very often students don't immediately realise what is the weight going with this message. Sometimes they even get frustrated or irritated by the faces that each of us puts on this side or the other side of the questions we discuss. But quite often students, after some time, realise there is a message in what we say, which finally gets carried across.

Concerning the development with computers or other new sides of society, different views have been expressed, and in my opinion it is good that it remains this way: some people are really fearing that the technology will taking over so much that finally the heart of the matter, that Simon Gindikin called old fashioned geometry, will be lost in the battle. But even if you take technology in a strict sense, as Professor Lopez de Medrano remarked, to teach how to push buttons require knowing which button there is to push, and this already requires some thinking.

If one moves the thinking from one place to the other, it is still fine, but with the deluge of pictures, under which we will be living, we as geometers will not be able to be out of that. We will have to look at these pictures and be able to say something about them. What lies ahead of us is more challenges and, as I tried to suggest, many other new disciplines are going to challenge us as geometers.

There are definitely some disciplines which for the moment have not been so much inspiring or so much challenging for geometers. One of them is certainly medicine, but I'm pretty sure with all the imaging which goes into it, very soon there will be really very big challenges coming from there which will require the interest of a large number of mathematicians. At some stage we will have to use a number of things we know about pictures, illusions with pictures and so on, because for medical doctors this will be absolutely crucial in their practical work.

Altogether, I feel that we have to keep this openness, and also a sense of responsibility according to which, if some bad things happen, we accept a share of them. Therefore, if there are not just bad devils trying to kill geometry from outside, it is not true either that we are defending an old field a bit out of breath and trying to survive. In my view the correct picture is rather different: we have to look outward, identify new challenges, really face them and at the same time prepare our students to look forward to more work and more joy while doing it. This last point is extremely important for me because this is one face of passion.

IHÉS, 35, route de Chartres, F-91440 Bures-sur-Yvette, (France)
E-mail address: jpb@ihes.fr

Selected Titles in This Series

(Continued from the front of this publication)

263 **Yan Guo, Editor,** Nonlinear wave equations, 2000

262 **Paul Igodt, Herbert Abels, Yves Félix, and Fritz Grunewald, Editors,** Crystallographic groups and their generalizations, 2000

261 **Gregory Budzban, Philip Feinsilver, and Arun Mukherjea, Editors,** Probability on algebraic structures, 2000

260 **Salvador Pérez-Esteva and Carlos Villegas-Blas, Editors,** First summer school in analysis and mathematical physics: Quantization, the Segal-Bargmann transform and semiclassical analysis, 2000

259 **D. V. Huynh, S. K. Jain, and S. R. López-Permouth, Editors,** Algebra and its applications, 2000

258 **Karsten Grove, Ib Henning Madsen, and Erik Kjær Pedersen, Editors,** Geometry and topology: Aarhus, 2000

257 **Peter A. Cholak, Steffen Lempp, Manuel Lerman, and Richard A. Shore, Editors,** Computability theory and its applications: Current trends and open problems, 2000

256 **Irwin Kra and Bernard Maskit, Editors,** In the tradition of Ahlfors and Bers: Proceedings of the first Ahlfors-Bers colloquium, 2000

255 **Jerry Bona, Katarzyna Saxton, and Ralph Saxton, Editors,** Nonlinear PDE's, dynamics and continuum physics, 2000

254 **Mourad E. H. Ismail and Dennis W. Stanton, Editors,** q-series from a contemporary perspective, 2000

253 **Charles N. Delzell and James J. Madden, Editors,** Real algebraic geometry and ordered structures, 2000

252 **Nathaniel Dean, Cassandra M. McZeal, and Pamela J. Williams, Editors,** African Americans in Mathematics II, 1999

251 **Eric L. Grinberg, Shiferaw Berhanu, Marvin I. Knopp, Gerardo A. Mendoza, and Eric Todd Quinto, Editors,** Analysis, geometry, number theory: The Mathematics of Leon Ehrenpreis, 2000

250 **Robert H. Gilman, Editor,** Groups, languages and geometry, 1999

249 **Myung-Hwan Kim, John S. Hsia, Yoshiyuki Kitaoka, and Rainer Schulze-Pillot, Editors,** Integral quadratic forms and lattices, 1999

248 **Naihuan Jing and Kailash C. Misra, Editors,** Recent developments in quantum affine algebras and related topics, 1999

247 **Lawrence Wasson Baggett and David Royal Larson, Editors,** The functional and harmonic analysis of wavelets and frames, 1999

246 **Marcy Barge and Krystyna Kuperberg, Editors,** Geometry and topology in dynamics, 1999

245 **Michael D. Fried, Editor,** Applications of curves over finite fields, 1999

244 **Leovigildo Alonso Tarrío, Ana Jeremías López, and Joseph Lipman,** Studies in duality on noetherian formal schemes and non-noetherian ordinary schemes, 1999

243 **Tsit Yuan Lam and Andy R. Magid, Editors,** Algebra, K-theory, groups, and education, 1999

242 **Bernhelm Booss-Bavnbek and Krzysztof Wojciechowski, Editors,** Geometric aspects of partial differential equations, 1999

241 **Piotr Pragacz, Michał Szurek, and Jarosław Wiśniewski, Editors,** Algebraic geometry: Hirzebruch 70, 1999